223834

This book is to be returned on or before
the last date stamped below.

1 7 MAY 2007

D1576667

LIBRARY RESOURCE CENTRE
CANOLFAN ADNODDAU LLYFRGELL

THE ENCYCLOPAEDIA OF
MAMMALS:2

CANOLFAN ADNODDAU LLYFRGELL

THE ENCYCLOPAEDIA OF
MAMMALS:2

Edited by Dr David Macdonald

Guild Publishing · London

COLEG LLA... ...LD COLLEGE
LIBRAR... ...CE CENTRE
CANOLFA... ...LLY FRG...

NF   11/97

599 MAC   043874

Project Editor: Graham Bateman
Editors: Peter Forbes, Bill MacKeith, Robert Peberdy
Art Editor: Jerry Burman
Picture Research: Linda Proud, Alison Renney
Production: Clive Sparling
Design: Chris Munday

AN EQUINOX BOOK

This edition published 1985
by Book Club Associates
by arrangement with George Allen & Unwin

Planned and produced by:
Equinox (Oxford) Ltd
Littlegate House
St Ebbe's Street
Oxford OX1 1SQ

Copyright © Equinox (Oxford) Ltd. 1984

Color and line artwork panels © Priscilla Barrett,
1984

All rights reserved. No part of this publication may be
reproduced, stored in a retrieval system, or
transmitted, in any form or by any means, electronic,
mechanical, photocopying, recording, or otherwise,
without the permission of the Publishers.

Origination by Alpha Reprographics Ltd, Harefield,
Middx; Excel Litho Ltd, Slough, Bucks; Fotographics,
Hong Kong.

Filmset by Keyspools Ltd, Golborne, Lancs, England.

Printed in Spain by Heraclio Fournier S.A. Vitoria.

CAERLLANDRILLO COLLEGE
LIBRARY RESOURCE CENTRE
CANOLFAN ADNODDAU LLYFRGELL

Advisory Editors

Dr Diana Bell
University of East Anglia
England

Dr Gordon Corbet
British Museum (Natural History)
England

Dr Chris Dickman
University of Oxford
England

Dr Jean Dorst
Muséum National d'Histoire Naturelle
Paris
France

Dr G. Dubost
Muséum National d'Histoire Naturelle
Brunoy
France

Dr John F. Eisenberg
The Florida State Museum
USA

Professor Valerius Geist
The University of Calgary
Canada

Dr Peter J. Jarman
The University of New England
Australia

Dr Sandro Lovari
Universita degli Studi di Parma
Italy

Dr Pat Morris
Royal Holloway College
England

Dr Martyn G. Murray
University of Cambridge
England

Professor Dr Jochen Niethammer
Zoologisches Institut der Universität
Bonn
West Germany

Dr Eleanor Russell
CSIRO, Helena Valley
Australia

Professor Dr Uwe Schmidt
Zoologisches Institut der Universität
Bonn
West Germany

Dr Bob E. Stebbings
Monks Wood Experimental Station
England

Artwork panels

Priscilla Barrett

with additional panels by:
Michael R. Long
Denys Ovenden
Dick Twinney
Graham Allen

Left: Honey possum (Oxford Scientific Films); half-title Red squirrel (Nature—Agence Photographique)
pii-iii: White rhinoceroses (Nature—Agence Photographique).

CONTRIBUTORS

RJvA Rudi J. van Aarde
University of Pretoria
Pretoria
South Africa

GA Greta Ågren
University of Stockholm
Stockholm
Sweden

MLA M. L. Augee
University of New South Wales
Kensington, NSW
Australia

RFWB Richard F. W. Barnes BSc PhD
Karisoke Research Centre
Ruhengeri, Rwanda
East Africa

CJB Christopher J. Barnard BSc DPhil
University of Nottingham
England

CB Claude Baudoin
Université de Franche-Comté
Besançon
France

GEB Gary E. Belovsky
University of Michigan
Ann Arbor, Michigan
USA

RJB R. J. Berry PhD
University College
London
England

IRB Ian R. Bishop
British Museum (Natural History)
London
England

JWB Jack W. Bradbury PhD
University of California
San Diego, California
USA

TMB Thomas M. Butynski
Kibale Forest Project
Fort Portal
Uganda

HC Hernan Castellanos
Caracas
Venezuela

JAC Joe A. Chapman PhD
Utah State University
Logan, Utah
USA

RAC Rosemary A. Cockerill PhD
Cambridge
England

GBC Gordon B. Corbet PhD
British Museum (Natural History)
London
England

DPC David P. Cowan PhD
Ministry of Agriculture, Fisheries
and Food
Guildford
England

DHMC David H. M. Cumming DPhil
Department of National Parks
and Wildlife Management
Causeway
Harare
Zimbabwe

MJD Michael J. Delany MSc DSc
University of Bradford
Bradford
England

CRD Christopher R. Dickman PhD
University of Oxford
England

ACD Adrian C. Dubock PhD
Plant Protection Division
Imperial Chemical Industries PLC
Haslemere
England

GD G. Dubost PhD
Muséum National d'Histoire
Naturelle
Brunoy
France

JFE John F. Eisenberg PhD
Florida State University
University of Florida
Gainesville, Florida
USA

JE James Evans
US Fish and Wildlife Service
Olympia, Washington
USA

JF Julie Feaver PhD
University of Cambridge
England

TEF Theodore E. Fleming PhD
University of Miami
Coral Gables, Florida
USA

HF Hans Frädrich PhD
Zoologischer Garten
Berlin
West Germany

WLF William L. Franklin MS PhD
Iowa State University
Ames, Iowa
USA

PJG Peter J. Garson BSc DPhil
University of Newcastle upon Tyne
Newcastle upon Tyne
England

VG Valerius Geist PhD
University of Calgary
Calgary
Canada

WG Wilma George DPhil
University of Oxford
England

GG Greg Gordon
Queensland National Parks
and Wildlife Service
Brisbane, Queensland
Australia

MLG Martyn L. Gorman PhD
University of Aberdeen
Scotland

TRG Tom R. Grant
University of New South Wales
Kensington, NSW
Australia

SJGH Stephen J. G. Hall MA
University of Cambridge
England

EH Emilio Herrera BSc
University of Oxford
England

HEH Harry E. Hodgdon
School of Forestry Resources
North Carolina State University
Raleigh, North Carolina
USA

HNH Hendrik N. Hoeck PhD
Universität Konstanz
Konstanz
West Germany

DJH Donna J. Howell
Southern Methodist University
Dallas, Texas
USA

UWH U. William Huck
Princeton University
Princeton, New Jersey
USA

CJ Christine Janis MA PhD
Brown University
Rhode Island
USA

PJJ Peter J. Jarman BA PhD
University of New England
Armidale, NSW
Australia

JUMJ Jennifer U. M. Jarvis
University of Cape Town
Rondebosch
South Africa

TKa Takeo Kawamichi
University of Osaka
Osaka
Japan

LBK Lloyd B. Keith PhD
University of Wisconsin-Madison
Madison, Wisconsin
USA

REK Robert E. Kenward DPhil
Institute of Terrestrial Ecology
Abbott's Ripton
England

JK Jonathan Kingdon PhD
University of Oxford
England

DWK David W. Kitchen PhD
Humboldt State University
Arcata, California
USA

KK Karl Kranz
Smithsonian Institution
Washington DC
USA

CJK Charles J. Krebs PhD
Division of Wildlife and
Rangelands Research
CSIRO
Lyneham, A.C.T.
Australia

MAK Margaret A. Kuyper BSc MSc
Sidmouth, Devon
England

TEL Thomas E. Lacher, Jr BSc PhD
Huxley College of Environmental
Studies
Bellingham, Washington
USA

RAL Richard A. Lancia
School of Forestry Resources
North Carolina State University
Raleigh, North Carolina
USA

RML Richard M. Laws PhD FRS
British Antarctic Survey
Cambridge
England

AKL A. K. Lee
Monash University
Clayton, Victoria
Australia

WL Walter Leuthold PhD
Zurich
Switzerland

DFL Dale F. Lott
University of California
USA

SL Sandro Lovari PhD
Istituto di Zoologia
Parma
Italy

GFM Gary F. McCracken
University of Tennessee
Knoxville, Tennessee
USA

DWM David W. Macdonald MA DPhil
University of Oxford
England

JMcI John McIlroy PhD
Division of Wildlife and
Rangelands Research
CSIRO
Lyneham, A.C.T.
Australia

KMack Kathy Mackinnon MA DPhil
Bogor
Indonesia

RM Roger Martin
Monash University
Clayton, Victoria
Australia

MGM Martyn G. Murray PhD
University of Cambridge
England

PN Paul Newton BSc
University of Oxford
England

MEN Martin E. Nicoll BSc PhD
Smithsonian Institution
Washington DC
USA

MAO'C Margaret A. O'Connell PhD
National Zoological Park
Smithsonian Institution
Washington DC
USA

BPO'R Brian P. O'Regan BA MPhil
University of the Witwatersrand
Johannesburg
South Africa

NO-S Norman Owen-Smith PhD
University of the Witwatersrand
Johannesburg
South Africa

JLP James L. Patton PhD
University of California
Berkeley, California
USA

RAP Robin A. Pellew PhD
University of Cambridge
England

WEP   William E. Poole
Division of Wildlife and
Rangelands Research
CSIRO, Lyneham, A.C.T.
Australia

RJP   Rory J. Putman BA DPhil
University of Southampton
Southampton
England

KR   Katherine Ralls PhD
Smithsonian Institution
Washington DC
USA

GBR   Galen B. Rathbun BA PhD
US Fish and Wildlife Service
Gainesville, Florida
USA

DR   Dan Rubinstein PhD
University of Princeton
Princeton, New Jersey
USA

Wild asses (P. Barrett—see p484).

EMR   Eleanor M. Russell
Division of Wildlife and
Rangelands Research
CSIRO
Midland, Perth
Western Australia

MJR   Michael J. Ryan
Milwaukee Public Museum
Milwaukee, Wisconsin
USA

ES   Eberhard Schneider
University of Göttingen
Göttingen
West Germany

PWS   Paul W. Sherman PhD
Cornell University
Ithaca, New York
USA

AS   Andrew Smith PhD
University of New England
Armidale, NSW
Australia

ATS   Andrew T. Smith BA PhD
Arizona State University
Tempe, Arizona
USA

MSP   Mark Stanley Price DPhil
Office of the Adviser for
Conservation of the Environment
Oman

RES   Robert E. Stebbings PhD
Monks Wood Experimental Station
England

DMS   D. Michael Stoddart BSc PhD
King's College
London
England

ABT   Andrew B. Taber BA
University of Oxford
England

DCT   Dennis C. Turner ScD
Universität Zürich
Zurich
Switzerland

MDT   Merlin D. Tuttle PhD
Milwaukee Public Museum
Milwaukee, Wisconsin
USA

RU   Rod Underwood MSc MIBiol BSc
University of Cambridge
England

JOW   John O. Whitaker, Jr PhD
Indiana State University
Terre Haute, Indiana
USA

PW   Peter Wirtz PhD
Institut für Biologie
Freiburg
West Germany

CAW   Charles A. Woods
Florida State Museum
University of Florida
Gainesville, Florida
USA

AW   Andrew Wroot
Royal Holloway College
University of London
Egham, Surrey
England

CONTENTS

Star-nosed mole (Denys Ovenden—see p767).

Left: Jackrabbit (Bruce Coleman).

PREFACE

To say that *The Encyclopaedia of Mammals* (volumes 1 and 2) covers all known members of the class Mammalia is an accurate but arid summary of these books. The newest discoveries of modern biology weave a thread among this array of 4,000 or so species, giving insight into the lives of animals so intricate in their adaptations that the reality renders our wildest fables dull.

In this second volume we shall include three main groups of mammal—the large herbivores (such as horses, deer, rhinos, hippos, elephants), small herbivores (such as squirrels, voles, mice and rabbits) and terrestrial insect-eaters (such as shrews, moles, anteaters), as well as marsupials and bats. The first volume covered the carnivores, sea mammals and primates.

Throughout the animal kingdom shape and form reflect function: size, color, tooth and claw are all sculptured by evolution, fitting each species to a particular lifestyle. Few mammals, however, flaunt their evolutionary adornments so flagrantly as do many of the large herbivores (ungulates), for it is among this group that conspicuous antlers and fearsome tusks are commonplace. The functions of bony crowns of the warthog, spreading palmate fingers above the moose's head or violent stilettos on the duiker's forehead doubtless include armament against both predator and rival, and the intricate elegance of their design is a forceful reminder of the power of natural selection.

The ungulates include five orders only distantly united by common descent. One order (the Tubulidentata) contains only one species—the bizarre aardvark which has evolved to eat just ants and termites. Two of the other orders (Proboscidea, Hydracoidea) include creatures as diverse as towering elephants and their closest surviving relative the diminutive hyraxes. Both of the two large ungulate orders, the odd- and the even-toed ungulates (Perissodactyla, Artiodactyla), include familiar members (horses and cows, respectively). They also include the terrestrial giants—rhinoceroses and hippopotamuses—and the variety of deer, antelopes, horses and pigs is as great in shape and form as it is in behavior. One thing unites these ungulates: that is their adaptation to herbivory—eating plants.

In a world cloaked in greenery these vegetarians might be thought to have opted for an easy lifestyle. On the contrary, although their "prey" may be neither fleet of foot nor have acute "senses," herbivores have to join battle with a devilish armory of spikes, thorns, poisons and unexpected collaborations, all deployed via life histories which are intricately adapted to outmaneuver the plant's predators. What is more, the cellulose walls which encase plant cells are unassailable to the normal mammalian digestive system, thereby reducing the mightiest elephant to dependence on an alliance with minuscule bacteria in its gut, which have the capacity to digest these resilient tissues. The armory of adaptation and counter-adaptation arrayed between large herbivores and their food has been further refined by the continual pressure on most of them to dodge attacks from their own predators. The result is an overwhelming diversity of form and function.

In the second section two orders of small herbivores are covered (rodents, rabbits and hares) and the elephant-shrews. Their small size makes them difficult to study, but their lifestyles are nevertheless intriguing.

Inconspicuous scuttlings of mice and the pestilential image of

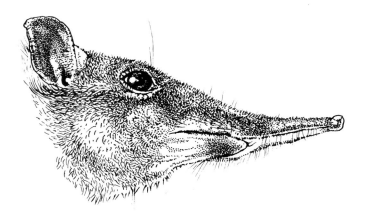

rats that characterize these creatures in the public's mind belie the extraordinary interest of the rodent order. This group of mammals embraces some 1,700 species. Rodents are captivating for at least two reasons. First, they are outstandingly successful. Not only does one form or another occupy almost every habitable cranny of the earth's terrestrial environments, but many do so with conspicuous success while every man's hand is turned against them. Their adaptability, and particularly their ability to breed with haste and profligacy, is stunning. Second, rodents are captivating because so many of them surprise us by not conforming at all to our stereotypes—some burrow like moles, others climb, while others hunt fish in streams; in South America some rodents are the size of sheep, while others are, at a glance, indistinguishable from small antelopes. Their societies, as these pages illustrate, are no less varied than their bodies and give us an insight into a community in microcosm that reverberates with life beneath the grassy swathes of the countryside.

The second major group of small herbivores share most of the characteristics of rodents—global distribution, small size, incredible breeding capacity and pestilential propensities. These are the lagomorphs—the rabbits and hares. Although most conform to the "bunny" stereotype, one group—the pikas—look more like hamsters, and are to be found mostly living on rocky outcrops in North America and Asia.

Finally in this section, we deal with the elephant-shrews (order Macroscelidia), which are a group of uncertain evolutionary affinities. Confined to Africa, they have characteristics of several groups—snouts like elephants, the diet of shrews and the body form of a small antelope with a rat-like tail. Their intriguing lifestyles live up to this unusual image.

The final section covers a series of orders (insectivores, anteaters, bats, monotremes and marsupials). Herein lie accounts of mammals which fly, burrow, run and swim, which can navigate, hunt and communicate with ultrasounds and echoes quite beyond human perception and stretching our comprehension; some are nearly naked, some blind, some are poisonous; others lay eggs and one, the recently discovered Kitti's hog-nosed bat, weighs just 1.5g (0.05oz) and is the smallest mammal in the world—a record held until 1974 by a shrew. That most of the species introduced in this section may be unfamiliar makes them all the more intriguing.

The Insectivora (shrews, hedgehogs, moles) include small predatory members who not only bear "primitive" resemblance to the very earliest mammals, but which, with their tiny size and racing metabolisms, push modern mammalian body processes to the very limit.

Over one quarter of mammal species are bats. Whenever the climate allows their activity they have filled the aerial niche. They have done so with such success that within this vast assemblage are to be found plant and fruit eaters, insectivores, carnivores—even blood eaters—and those that are opportunists. For a group so large, knowledge of their lifestyles is slim, but within these pages will be found insights into how they are adapted to life on the wing and how their societies work.

A taste for ants and termites, lack of teeth and possession of a long, sticky tongue are the hallmarks of two other orders—the Edentata (anteaters, armadillos) and Pholidota (pangolins). These two groups have evolved to occupy the "anteating" niche in South America and Africa, respectively, and this again emphasizes how evolution molds animals with different origins into similar form.

To many people, marsupials are no more than kangaroos and wallabies. However, this group actually includes over 260 species, in all manner of shapes and sizes, and represents one of the major surviving subdivisions of the mammals. Among the marsupials there are species living in almost all of the niches which placental mammals occupy elsewhere. The two types, placentals and marsupials, represent a worldwide, evolutionary experiment in the making of mammals: the fact that the results are often strikingly similar animals with quite distinct origins but adapted in the same way to cope with the same niche is compelling evidence for the processes of evolution. Today, it seems that the third such "experiment" in mammalhood,

which produced the egg-laying monotremes (platypus, echidnas) is imperiled, since only three species survive.

In planning this encyclopedia our aim was precise and rather ambitious. Recently, zoologists have had increasing success in discovering facts about the behavior and ecology of wild mammals, and in developing new and intriguing ideas to explain their discoveries. Yet so scattered is the information that much of it fails to reach even professional zoologists, far less percolate through to a general readership. Worse still, there is the idea that the public will, or can, only accept watered-down science, diluted to triviality and carefully sieved free of challenge. Our aim, then, has been to gather the newest information and ideas and to present them lucidly, entertainingly, but uncompromisingly. If we have succeeded, these pages will refresh the professional as much as they enthrall the schoolchild. The best way to convey the excitement of discovery is through the discoverers, and for this reason the book is written by the researchers themselves. Aside from any other qualities, this encyclopedia must be unique in its class in amassing so much first-hand experience and international expertise.

There is no single "correct" classification of the animal kingdom, and so we have had to select from the views of different taxonomists. In general we follow Corbet and Hill (see Bibliography) for the arrangement of families and orders. However, within family or species entries authors have been encouraged to follow the latest views on "their" groups.

This encyclopedia is structured at a number of levels. Firstly, for each order or group of orders there will be a general essay which highlights common features and main variations of the biology (particularly body plan), ecology, and behavior of the

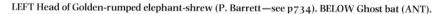

LEFT Head of Golden-rumped elephant-shrew (P. Barrett—see p734). BELOW Ghost bat (ANT).

group concerned and its evolution. For some orders—the bats, elephant-shrews and monotremes—these order essays will be a complete review of the group, but for the other groups it acts as a pivot for the next level of entries which it introduces. Thus the hoofed mammals are divided into two orders—the odd-toed and even-toed ungulates—and the rodents into three suborders—the squirrel-like, mouse-like and cavy-like rodents. These essays cover various themes and highlight the particular interest of given groups, but each invariably includes a distribution map, summary of species or species groupings, description of skull, dentition and unusual skeletal features of representative species and color artwork or photographs that enhance the text.

The bulk of this encyclopedia is devoted to families of species. The text on these pages covers details of physical features, distribution, evolutionary history, diet and feeding behavior, social dynamics and spatial organization, classification, conservation and relationships with man.

An information panel precedes the textual discussion of each species or group. This provides easy reference to the main features of distribution, habitat, dimensions, coat and duration of gestation and life span, and includes a map of natural distribution (not introductions to other areas by man), and a scale drawing comparing the size of the species with that of a whole, half or one quarter six-foot (1.8m) human or a 12in (30cm) human foot. Where there are silhouettes of two animals, they are the largest and smallest representatives of the group. Where the panel covers a large group of species, the species listed as examples are those referred to in the accompanying text. For large groups, the detailed descriptions of species are provided in a separate Table of Species. Unless otherwise stated, dimensions given are for both males and females. Where there is a difference in size between sexes, the scale drawings show males.

Some orders contain the vast majority of mammalian species, ie rodents, insectivores, bats and marsupials. For most of these large groups information on individual species is restricted or does not exist. However, to provide comprehensive coverage the scientific name, common name (where relevant) and distribution of all members of these groups are listed in the Appendix.

Every so often a really remarkable study of a species emerges. Some of these studies are so distinctive that they have been allocated two whole pages so that the authors may develop their stories. The topics of these "special features" give insight into evolutionary processes at work throughout all mammals, and span social organization, foraging behavior, breeding biology and conservation. Similar themes are also developed in smaller "box features" alongside most of the major texts.

Most people will never have seen many of the animals mentioned in these pages, but that does not necessarily mean they have to remain remote or unreal in our thoughts. Such detachment vanishes as these animals are brought vividly to life not only by many photographs, but especially by the color and line artwork. These illustrations are the fruits of great labor: they are accurate in minute detail and more importantly they are dynamic—each animal is engaged in an activity or shown in a posture that enhances points made in the text. Furthermore, the species have been chosen as representatives of their group. Unless otherwise stated, all animals depicted on a single

panel are to scale: actual dimensions will be found in the text. Simpler line drawings illustrate particular aspects of behavior, or anatomical distinctions between otherwise similar species. Similarly, we have sought photographs that are more than mere zoo portraits, and which illustrate the animal in its habitat, emphasizing typical behavior; the photographs are accompanied by captions which complement the main text.

The reader may marvel at the stories of adaptation, the process of natural selection and the beauty of the creatures found herein, but he should also be fearful for them. Again and again authors return to the need to conserve species threatened with extinction and by mismanagement. About 150 of the species and subspecies described in these pages are listed in the Appendices I to III of the Convention on International Trade in Endangered Species of Wild Flora and Fauna (CITES). The *Red Data Book* of the International Union for the Conservation of Nature and Natural Resources (IUCN) lists about 200 of these species as at risk. In this book the following symbols are used to show the status accorded to species by the IUCN at the time of going to press: Ⓔ = Endangered—in danger of extinction unless causal factors are modified (these may include habitat destruction and direct exploitation by man). Ⓥ = Vulnerable— likely to become endangered in the near future. Ⓡ = Rare, but neither endangered nor vulnerable at present. Ⓘ = Indeterminate—insufficient information available, but known to be in one of the above categories. ? = Suspected but not definitely known to fall into one of the above categories. We are indebted to Jane Thornback of the IUCN Monitoring Centre, Cambridge, England, for giving us the very latest information on status. The symbol ⊡ indicates entire species, genera or families, in addition to those listed in the *Red Data Book*, that are listed by CITES. Some species and subspecies that have ⒺⓍ or probably have ⒺⓍ? become extinct in the past 100 years are also indicated.

The success of books like this rests upon the integrated efforts of many people with diverse skills. I want to thank all of the authors for their enthusiastic cooperation, Priscilla Barrett and other artists for their painstaking attention to detail in the artwork, and all the other people, too numerous to list, who have helped to shape these volumes. I also thank my wife, Jenny, for her tolerance of the avalanches of manuscripts which, for two years, have cascaded from every surface of our home. In particular I am grateful to Dr Graham Bateman, Senior Natural History Editor at Equinox, for his good-humored navigation as we have paddled our editorial boat hither and thither, fearful of being smashed on the rocks of gauche sensationalism or becalmed in the stagnant pools of obsessive detail. Hopefully, the reader will judge that we have charted a course where the current is invigorating and the waters clear.

David W. Macdonald
DEPARTMENT OF ZOOLOGY
OXFORD

LEFT Gemsbok (A. Bannister).

LARGE HERBIVORES

PRIMITIVE UNGULATES

ORDERS: PROBOSCIDEA, HYRACOIDEA, TUBULIDENTATA†

Three orders: 6 genera; 14 species.

Elephants

Order: Proboscidea.
Two species: **African elephant** (*Loxodonta africana*), **Asian elephant** (*Elephas maximus*).

†A further order, the aquatic Sirenia, is considered to belong with the primitive ungulates, but is discussed only generally here.

Hyraxes

Order: Hyracoidea.
Eleven species in 3 genera.
includes **Johnston's hyrax** (*Procavia johnstoni*).
Bruce's yellow-spotted hyrax (*Heterohyrax brucei*).

Aardvark

Order: Tubulidentata.
One species: **aardvark** (*Orycteropus afer*).

THE relationship between the huge elephant and the tiny hyrax is not immediately obvious. Yet these creatures are primitive ungulates, and they, together with the aardvark, are thought to be more closely related to each other than to the other mammals. The aquatic sea cows of the order Sirenia (dugong and manatees) also share the same lineage as evidenced by the affinities between their ancestors.

The earliest ungulates, the Condylarthra, appeared in the early Paleocene (about 65 million years ago). They were the ancestors of the modern ungulates, the Perissodactyla (odd-toed ungulates) and Artiodactyla (even-toed ungulates). The Tubulidentata (now represented only by the aardvark) diverged early on during the Paleocene from the Condylarthra, and specialized in feeding on termites and ants.

Despite its resemblance to the other anteaters, the aardvark is not related to them at all. Its dentition is unique in that the front teeth are lacking and the peg-like molars and premolars are formed from columns of dentine. Fossil evidence suggests that members of this order spread from Africa to Europe and Asia during the late Miocene (about 8 million years ago), but three of the four genera recognized are now extinct.

Another Paleocene offshoot from the Condylarthra gave rise to the Paenungulata (the primitive ungulates or sub-ungulates) in Africa, during the continent's isolation. By the early Eocene (about 54 million years ago), the Paenungulata had separated into three distinct orders, the Hyracoidea (hyraxes), Sirenia (sea cows), and Proboscidea (trunked mammals, today represented only by the elephants).

The Hyracoidea proliferated about 40 million years ago, but spread no further than Africa and the eastern Mediterranean. Some of them became as large as tapirs, but today only the smaller forms remain. The decline of the Hyracoidea during the Miocene, 25 million years ago, coincided with the radiation of the Artiodactyla, against whom it is likely that they were unable to compete.

The Proboscidea were a very successful order that went through a period of rapid radiation in Africa and then spread across the globe except for Australia and Antarctica. The most obvious feature of the Proboscidea is of course their large size (see p454). Associated with this are their flattened soles, elongated limb bones, and the modifications to the head and associated structures. These include the evolution of the trunk and tusks, the elongated jaw and the specialized dentition, and the enlarged skull and shortened neck. In the middle Pleistocene the family Elephantidae enjoyed a period of very rapid evolution and radiation. Today only two species of elephant remain.

The three orders of the Paenungulata appear very different, but they share a few common anatomical and morphological features. None of them has a clavicle, and the primitive claws are more like nails than hooves. They all have four toes in the

▶ **The last outposts** of the hyraxes are the rocky outcrops or kopjes of East Africa. These are Johnston's hyrax and Bruce's yellow-spotted hyrax.

▼ **Evolution of the elephant.** Beginning with the small tapir-like *Mŏenitherium* (1) in the early Oligocene (38 million years ago), proboscideans became a large, widespread group in the Pleistocene (2 million years ago). *Trilophodon* (2) was one of a family of long-jawed gamphotheres found in Eurasia, Africa, and North America from the Miocene to the Pleistocene (26–2 million years ago). *Platybelodon* (3) was a "shovel-tusked" gamphothere found in the late Miocene and Pliocene (about 12–7 million years ago) of Asia and North America. The Imperial mammoth (*Mammithera imperater*) (4), the largest ever proboscidean, flourished in the Pleistocene of Eurasia, Africa and North America. Unlike the earlier forms, it had high-crowned teeth like those of the modern African elephant (*Loxodonta africana*) (5).

forelegs and three toes in the hindlegs (elephants vary from this pattern); digits with short flattened nails (in hyraxes, the inner digit of the hindfoot bears a long curved claw); 20–22 ribs; similar anatomy of both placenta and womb; females have two teats between the forelegs (hyraxes have an additional two or four on the belly); the testes remain in the body cavity close to the kidneys. The orders are also related biochemically.

They all show developments of the grinding teeth and incisors, and they lack the other front teeth. The elephant's incisors have become its characteristic tusks, the dugong has a pair of upper incisor tusks, while the manatee has no front teeth at all. The hyrax has an enlarged upper incisor. All have transverse ridges upon their grinding teeth. The dugong has no premolars but large cusped molars. Elephants and manatees have unique dentition: in both species, the teeth form at the back of the jaw and are then pushed forward along the jaw. As they move forward, they are worn down with use. The elephant has six teeth in each jaw, only one being fully operational at a time, while the manatee has over 20, with about half a dozen in use at a time. Even more remarkable were the teeth of the Steller's sea cows: there were none at all. Instead it had horny grinding plates.

The Sirenia seem to have more in common with the Proboscidea than with the Hyracoidea. The Sirenia and Proboscidea may have diverged in the late Eocene from an ancestor whose fossils appear in the Fayum swamps of Egypt.   RFWB

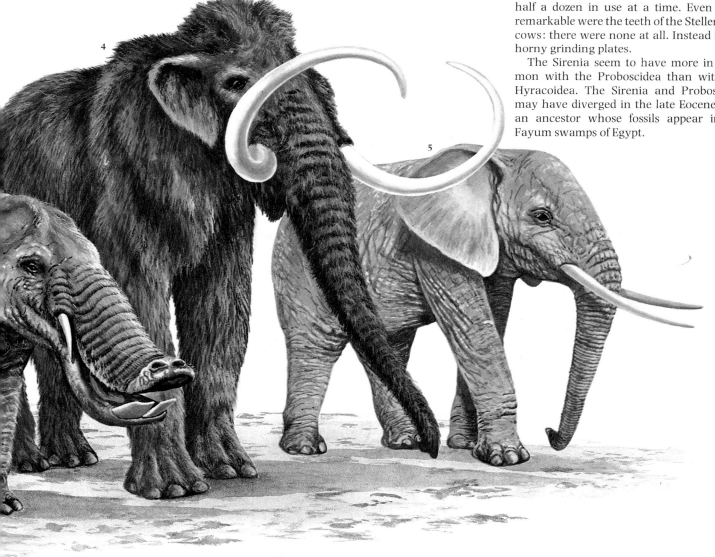

ELEPHANTS

Order: Proboscidea
Family: Elephantidae.
Two species in 2 genera.
Distribution: Africa south of the Sahara;
Indian subcontinent, Indochina, Malaysia,
Indonesia, S China.

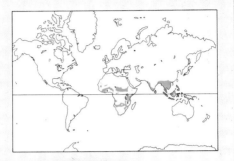

African elephant ☑ ⊡

Loxodonta africana
Distribution: Africa south of the Sahara.

Habitat: savanna grassland; forest.

Size: male head-body length 6–7.5m
(20–24.5ft), height 3.3m (10.8ft), weight up to
6,000kg (13,200lb); female head-body length
0.6m (2ft) shorter, weight 3,000kg (6,600lb).

Skin: sparsely endowed with hair; gray-black
when young, becoming pinkish white with age.

Gestation: 22 months.

Longevity: 60 years (more than 80 years in
captivity).
Subspecies: 2. **Savanna** or **Bush elephant**
(*L.a. africana*); E, C and S Africa. **Forest
elephant** (*L.a. cyclotis*); C and W Africa. The so-
called Cape elephant is regarded by some
authorities as a full subspecies. Once threatened
with extinction, it is now mainly restricted to
the Addo National Park in South Africa, where
it is rigorously protected and numbers are
increasing.

Asian elephant ☑ ⊡

Elephas maximus
Distribution: Indian subcontinent, Indochina,
Malaysia, Indonesia, S China.

Habitat: forest.

Size: head-body length 5.5–6.4m (18–21ft),
height 2.5–3m (8.2–9.8ft); weight up to
5,000kg (11,000lb).

Skin, gestation, and longevity: as for African
elephant.

Subspecies: 4. **Indian elephant** (*E.m.
bengalensis*), **Ceylon elephant** (*E.m. maximus*),
Sumatran elephant (*E.m. sumatrana*),
Malaysian elephant (*E.m. hirsutus*).

☑ Endangered.　☑ Vulnerable.　⊡ CITES listed.

ELEPHANTS have always been regarded
with awe and fascination, mainly because of their great size—they are the largest
living land mammals—and because of their
trunk and formidable tusks; but also
because of their longevity, their ability to
learn and remember and their adaptability
as working animals. For millennia, their
great strength has been exploited in agriculture and warfare and even today, notably in
the Indian subcontinent, they are still important economically and as cultural symbols. But a continuing demand for elephant
tusks, still the main source of commercial
ivory, has been largely responsible for a
drastic decline in elephant populations over
the past hundred years.

In the recent past the Asian elephant
ranged from Mesopotamia (now Iraq) in the
west, throughout Asia south of the
Himalayas to northern China. Today there
are fewer than 50,000 wild Asian elephants
remaining in isolated refuges in hilly or
mountainous parts of the Indian subcontinent, Sri-Lanka, Indochina, Malaysia, Indonesia and southern China.

Of all elephants, the Savanna elephant is
the best understood—simply because it is
easier to study behavior in the open grasslands of eastern Africa than in the denser
forest habitats of the Forest and Asian
elephants.

Body size, the most conspicuous feature of
elephants, continues to increase throughout
life, so that the biggest elephant in a group is
also likely to be the oldest. The largest and
heaviest elephants alive today are African
Savanna bulls. The largest known specimen, killed in Angola in 1955 and now on
display in the Smithsonian Institute, Washington DC, weighed 10,000kg (22,050lb)
and measured 4m (13.1ft) at the shoulder.
There have been several reports in the past
century of so-called Pygmy elephants, with
an adult shoulder height of less than 2m
(6.6ft). It has been suggested that these
represent a separate species or subspecies,
but the current view is that they are merely
abnormally small individuals which occasionally appear at random in herds of
normal-sized individuals.

The characteristic form of the skull, jaws,
teeth, tusks, ears and digestive system of
elephants are all part of the adaptive complex associated with the evolution of large
body size (see p454). The skull, jaws and
teeth form a specialized system for crushing
coarse plant material. The skull is disproportionately large compared with the size of the
brain and has evolved to support the trunk
and heavy dentition. It is, however, rela-

tively light due to the presence in the upper
cranium of interlinked air-cells and cavities.
Asian elephants have two characteristic
dome-shaped protuberances above the eyes.

The tusks are elongated upper incisor
teeth. They first appear at the age of about 2
years and they grow throughout life so that
by the age of 60 a bull's tusks may average
60kg (132lb) each and a cow's 9kg (20lb)
each. In very old individuals they have been
known to reach 130kg (287lb) and attain a
length of 3.5m (7.7ft). Such massive "tus-
kers" have always been prime targets for
ivory and big game hunters, with the result
that few such specimens remain in the wild.
In general, the tusks of Asian bull elephants
are smaller than in their African counterparts, and among bull Ceylon elephants
they are formed in only 10 percent of
individuals. In the Forest elephant they are
thinner, more downward pointing and composed of even harder ivory than in the
Savanna subspecies. Elephant ivory is a
unique mixture of dentine, cartilaginous
material and calcium salts, and a transverse
section through a tusk shows a regular
diamond pattern, not seen in the tusks of
any other mammal. The tusks are mainly
used in feeding, for such purposes as prising
off the bark of trees or digging for roots, and
in social encounters, as an instrument of

▲ **The majesty** of the African elephant, seen here in Amboseli National Park, Kenya, with Mount Kilimanjaro towering behind it. The huge ears, which provide such a distinctive frontal appearance, function as radiators for the animal's bulky body, losing heat from their vast surfaces.

◄ **African and Asian elephants** compared. The Asian elephant (BOTTOM) is smaller than the African (TOP), has a convex back and much smaller ears. The trunk has two lips in the African elephant and one in the Asian. The small tusks of female Asian elephants are not visible beyond the lips.

display, or they may be used as a weapon.

The upper lip and the nose have become elongated and muscularized in elephants to form a trunk. Unlike other herbivores, the elephant cannot reach the ground with its mouth, because its neck is too short. The option of evolving an elongated neck was not open to the early Proboscidea because of the weight of their heavy cranial and jaw structures. The trunk enables elephants to feed from the ground. It is also used for feeding from trees and shrubs, for breaking off branches and picking leaves, shoots and fruits. Though powerful enough to lift whole trees, the trunk, with the nostrils at its tip, is

also an acutely sensitive organ of smell and touch. Smell plays an important part in social contacts within a herd and in the detection of external threats. As to touch, the trunk's prehensile finger-like lips, endowed with fine sensory hairs, can pick up very small objects.

Further uses of the trunk include drinking, greeting, caressing and threatening, squirting water and throwing dust over its owner, and the forming and amplifying of vocalizations. Elephants drink by sucking water into their trunks then squirting it into their mouths; they also squirt water over their backs to cool themselves. At times of water shortage they sometimes spray themselves with the regurgitated water contents of their stomach. The trunk can also serve as a snorkel, enabling an individual to breathe if submerged, perhaps during a river crossing.

The elephant's large ears perform the same function as a car's radiator: they prevent overheating, always a danger in a large compact body with its relatively low rate of heat loss. They are well supplied with blood and can be fanned to increase the cooling flow of air over them. By evolving ears which substantially increase the body surface area, and so the rate of cooling, elephants have overcome one of the most important limitations to the evolution of large body size (see box). The ears are largest in the African Savanna elephant, which probably reflects the more open habitat in which this species lives. They are also more triangular than in the Forest elephant. Elephants have a keen sense of hearing and communicate extensively by means of vocalizations, particularly the forest-dwelling forms.

The massive body is supported by pillar-like legs with thick, heavy bones. The bone structure of the foot is intermediate between

that of man (plantigrade, where the heel rests on the ground) and that of the horse (digitigrade, where the heel is raised off the ground). The phalanges (fingers and toes) are embedded in a soft cushion of white elastic fibers enclosed within a fatty matrix. This enables the elephant to steal silently through the bush. The large surface area of the sole spreads the weight of their huge

▲ ▶ **Versatile trunk.** Really an extended, muscularized upper lip, the elephant's trunk is a sensitive all-purpose organ, giving it greater skill in handling food and other objects than any other ungulate. ABOVE An elephant having a mud-bath. The skin is sensitive, and dust- and mud-baths help to keep it free from parasites. ABOVE RIGHT An elephant drinking by squirting water, previously sucked into its trunk, into its mouth. BELOW RIGHT The trunk also allows them to feed on a wide range of food, from grasses to the leaves of trees.

▷ **Wind-sniffers.** OVERLEAF The trunk is an extremely sensitive sensory organ. These elephants are obviously picking up something interesting on the breeze.

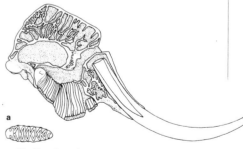

a

▲ **Skull and teeth.** The elephant's skull is massive, comprising 12–25 percent of its body weight. It would be even heavier if it were not for an extensive network of air-cells and cavities. The dental formula is I1/0, C0/0, P3/3, M3/3. The single upper incisor grows into the tusk and the molars (**a**) fall out at the front when worn down, being replaced from behind. Only one tooth on each side, above and below, is in use at any one time.

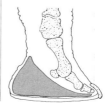

◀ **The elephant's foot** is broad and the digits are embedded in a fatty matrix (green). The huge weight of the animal is spread so well that it hardly leaves any track marks.

Evolution of Large Body Size

The elephant family were once highly successful, and during their peak they spread to all parts of the globe except Australia, New Zealand and Antarctica. Until the Pleistocene (about 2 million years ago), modern elephants occupied a range of habitats from desert to montane forest throughout Africa and southern Asia. This success was related to their most outstanding feature: the evolution of large body size.

In order to understand this, it is necessary to look at the early large herbivore community.

Different herbivore species in the same habitat avoid competing with each other for food by eating different plant groups (eg grasses, herbs, shrubs or trees), different species, or different parts (eg stem, leaf, fruit or flower) of the same species. The first large mammal herbivores in Africa were perissodactyls (eg the horses) which arose in the Eocene (54–38 million years ago) and dominated the large herbivore community until the coming of the ruminants (artiodactyls, eg the antelopes) in the early Oligocene (37 million years ago). It is likely that while each perissodactyl species ate a wide range of plants, taking the coarser parts, each ruminant species ate a narrower range, taking the softer parts.

The Proboscidea arose in the late Oligocene, followed by the Elephantidae in the Lower Miocene, so the ancestors of the elephants

arrived at a time when the large herbivore community had long been dominated by the perissodactyls and when the highly successful ruminants were continuing to evolve and to colonize new ecological niches. As non-ruminants, the Elephantidae were able to feed on plant foods which were too coarse for ruminants to live on. But this brought them into competition with the other non-ruminants there, the perissodactyls.

For a given digestive system, differences in metabolic rate enable a large animal to feed on less nutritious plant parts than can a smaller animal. Thus there was a strong selective pressure for the Elephantidae to increase their body size and so reduce the competition with perissodactyls. The most nutritious plant parts, such as leaf shoots and fruits, are produced only at certain seasons and even then may be sparse and widely scattered, but coarse plants are more abundantly distributed both in space and time. The elephants' evolutionary strategy thus enabled them to feed on plant parts which were not only abundant but available all the year round. In particular, it enabled them to feed on the woody parts of trees and shrubs. Thus they were able to tap a resource which other mammalian herbivores could neither reach nor digest. At the same time, they were able to eat rich plant parts (eg fruits) whenever these were available. This enabled elephants to thrive in a wide range of habitats.

and shrubs—twigs, branches and bark. They will also eat large quantities of flowers and fruits when these are available and they will dig for roots, especially after the first rains of the season. Asian elephants eat a similar range of foods, one of the most important being bamboo. Because of their large body size and rapid rate of throughput, elephants need large amounts of food: an adult requires about 150kg (330lb) of food a day. However, half the food leaves the body undigested.

Elephants cannot go for long without water, but because they are large, they can travel long distances each day between their water supplies and favorable feeding areas. They require 70–90 liters (19–24gal) of fluid each day, and at times of drought will dig holes in dry riverbeds with their trunks and tusks to find water. They also require shade, and during the hot season they spend the middle part of the day resting under trees to avoid overheating.

Elephants require a large home range in order to find enough food, water and shade in all seasons of the year. In one study, the home ranges of African elephants in a woodland and bushland habitat were found to average 750sq km (290sq mi) in an area of abundant food and water and 1,600sq km (617sq mi) in a more arid area. They respond very rapidly to sudden rainfall, often traveling long distances, up to 30km (19mi) to reach a spot where an isolated shower has fallen, in order to exploit the lush growth of grass which soon follows.

In both Asian and African forests, elephants often follow the same paths when moving from one place to another; over several generations this results in wide so-called "elephant roads" that can be found cutting through even the densest jungles.

Elephants are herd animals and display complex social behavior. Early hunters and naturalists spoke of a "herd bull" or "sire bull" which acted as a permanent leader and defender of the elephant herd. But several recent studies in East Africa have shown that bull and cow African elephants tend to live apart. Female elephants live in family units which are groups of closely related adult cows and their immature offspring. The adults are either sisters or mothers and daughters. A typical family unit may consist of two or three sisters and their offspring, or of one old cow and one or two adult daughters and their offspring. When the female offspring reach maturity, they stay with the family unit and start to breed. As the family unit grows in size, a subgroup of young adult cows gradually

bulk over such a wide area that on firm ground they leave hardly any track marks. Forest and Asian elephants usually have five toes on the forefoot and four toes on the hindfoot. The Savanna elephant typically has only four toes on the forefoot and three on the hindfoot.

Elephants walk at about 4–6km/h (2.5–3.7mph), but have been known to maintain double this speed for several hours. A charging or fleeing elephant can reach 40km/h (25mph), which means that over short distances they can easily outstrip a human sprinter.

The skin is 2–4cm (0.8–1.6in) thick and sparsely endowed with hair. Despite the thickness of the skin, it is highly sensitive and requires frequent bathing, massaging and powdering with dust to keep it free from parasites and diseases.

Elephants have a non-ruminant digestive system similar to that of horses. Microbial fermentation takes place in the cecum, which is an enlarged sac at the junction of the small and large intestines.

In the wet season, African Savanna elephants eat mainly grasses. They also eat small amounts of leaves from a wide range of trees and shrubs. After the rains have ended and the grasses have withered and died, they turn to feeding on the woody parts of trees

separates to form their own family unit. As a result, family units in the same area are often related.

The family unit is led by the oldest female, the matriarch. The social bonds between the members of the family are very strong. Cooperative behavior, particularly in the protection and guidance of the young, is frequently shown. When a hazard is detected (for example human scent) the group bunches with calves in the center and the matriarch facing the direction of the threat. If the matriarch decides to retreat, the unit runs in a very tight bunch. If she decides to confront the threat, the herd closely observe the outcome—which is normally the retreat of the threat. If one member is shot or wounded the rest of the group frequently comes to the aid of the stricken member even in the face of considerable danger—a behavior pattern which clearly plays into the hands of hunters.

If the matriarch is older than 50 years she may be reproductively inactive, for elephants, like humans, may survive after they are physically too old to reproduce. In other words, there is a menopause. This is a consequence of the elephant's longevity, which is in turn related to the evolution of large body size. Survival over a long lifespan both requires and facilitates the acquisition of considerable experience. By continuing to guide her family unit long after she is too old to breed, the matriarch can enhance the survival of her offspring by providing them with the benefits of her accumulated experience: her knowledge of their home range, of seasonal water sources and ephemeral food supplies, and of sources of danger and ways of avoiding them.

In contrast to their sisters, when the young male elephants reach puberty they leave the natal group. Adult bulls tend to live alone or in small temporary bull groups which are constantly changing in numbers and composition. Social bonds between bulls are weak and there seems to be little cooperative behavior.

Elephants, like other animals, communicate through sight, sound and smell. The most common vocalization is a growl emanating from the larynx (this is what hunters used to call the "tummy rumble"), a sound that carries for up to 1km (0.6m). It may be used as a warning, or to maintain contact with other elephants. When feeding in dense bush, the members of a group monitor each other's positions by low growls. They vocalize less frequently when the bush is more open and the group members can see one another. The trunk is used as a resonating chamber to amplify bellows or screams so as to convey a variety of emotions. The characteristic loud trumpeting of elephants is mainly used when they are excited, surprised, about to attack, or when an individual is widely separated from its herd.

Visual messages are conveyed by changes in posture and the position of the tail, head, ears and trunk. The ears and trunk evolved primarily for other purposes (as described earlier) but have a secondary value in communication. Elephants often touch each other using the trunk. Touch is especially important in mother-infant relations. The mother continually touches and guides her infant with her trunk. When elephants meet they often greet each other by touching the other's mouth with the tip of the trunk.

Most elephant populations show an annual reproductive cycle which corresponds to the seasonal availability of food and water. During the dry season, the population suffers a period of nutritional stress and cows cease to ovulate. When the rains break and the food supply improves, a period of one or two months of good feeding is needed to raise the females' body fat above the level necessary for ovulation. Thus females are in heat during the second half of the rainy season and the first months of the dry season.

Bulls of the Asian elephant have long been known to exhibit a condition during rutting called "musth"—a period of high male hormone (testosterone) levels, aggressive behavior, pronounced secretions from the temporal gland, and an increase in sexual activity. Musth usually lasts two or three months and tends to coincide with periods of high rainfall. Recently, African elephants have been found to show the same phenomenon. Another notable feature of bull elephants is that their testes remain within the body cavity throughout life and do not descend into a scrotum at puberty.

During the mating season, each female may be in heat for a few days only so the distribution of sexually receptive cows is constantly changing in space and time. Bulls travel long distances each day in order to monitor the changing reproductive status of cows in their home range. The bull who can travel longest and farthest during the mating season will find the greatest number of receptive females. The ability to travel fast at a low unit-energy expenditure is a further advantage of a large body. Having found a receptive cow, a bull will have to compete with other bulls for mating opportunities. Usually it is the largest bull who succeeds in

copulating. It is this competition between males for females that has conferred evolutionary advantage on the size difference between males and females.

Following the long gestation, there is a long period of juvenile dependency. The infant suckles (with its mouth, not its trunk) from the paired breasts between the mother's forelegs, for three or four years. Sexual maturity is reached at about 10 years of age, but it may be delayed for several years during drought or periods of high population density. Once she starts to breed, a cow may produce an infant every three or four years, although this period may also be extended when times are bad. The period of greatest female fecundity is between 25 and 45 years of age.

The long gestation period means that the infant is born nearly two years later, in the wet season, when conditions are optimal for its survival. In particular, abundant green food ensures that the mother will lactate successfully during the early months. During a birth, other cows may collect around the newborn elephant and so-called "midwives" may assist at the birth by removing the fetal membrane. Others may help the infant to its feet, and this marks the start of a joint family responsibility for the young of a group.

At birth, the African elephant weighs about 120kg (265lb) and the Asian about 100kg (220lb). The young elephant grows

▲ **Fighting among elephants** is usually restricted to play ritual, but on rare occasions it can be serious and even result in the death of one of the combatants if a tusk penetrates a vital organ. Fighting comprises a series of charges and head-to-head shoving matches which often involve wrestling with tusk and trunk.

Until about 10,000 years ago, man the hunter no doubt managed to kill some elephants as a source of meat, but it was not until modern man—the farmer—appeared that the first conflicts occurred (raiding elephants can destroy a crop overnight).

The first record of elephants being used as beasts of burden is found 5,500 years ago in the Indus Valley. Their natural character-istics of longevity, immense strength and placidity, and their ability to learn and remember, have been exploited up to the present day. Asian elephants have been kept in captivity continuously since that time, yet have never been completely domesticated. Although they are bred to a limited extent in captivity, there has always been a big de-mand for elephants captured from the wild, since an elephant has to be 10 years old before training can start and is not really useful until it is 20 years old. Also, bulls are less suited than cows to the captive way of life, which further reduces the chances of captive breeding. African elephants have had a more erratic record as beasts of burden. The most famous where those used by the Carthaginian leader Hannibal in the wars against the Romans, but recently

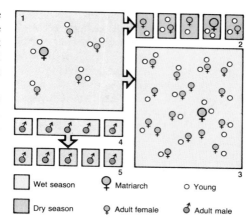

| Wet season | ♀ Matriarch | ○ Young |
| Dry season | ♀ Adult female | ♂ Adult male |

▲ ▼ **Group size in elephants.** (1) A typical family unit comprises closely related cows (with one dominant cow, the matriarch), and their offspring. (2) When food is scarce, the family groups tend to split up to forage. (3) In the wet season, family units may merge to form groups of 50 or more. Bulls leave the family group at puberty to join small, loose bull herds (4) or live alone (5). BELOW Mature bull elephants spend much of their time alone, moving between the female family units, searching for receptive cows.

rapidly, reaching a weight of 1,000kg (2,200lb) by the time it is 6 years old. The rate of growth decreases after about 15 years but growth continues throughout life. In addition, males experience a post-pubertal growth spurt between 20 and 30 years of age, which accounts for the size difference between them and females. Although the potential life span is about 60 years, half the wild elephants born die by the age of 15 years and only about one-fifth survive to reach 30 years of age.

The occurrence of elephant graveyards has been widely reported throughout the ages. However, there is no scientific evidence to support the theory that eleph-ants migrate to specific sites to die. Large collections of bones can be accounted for in three ways: they may be the site of a mass slaughter of a herd by ivory traders or poachers or simply collections of bones washed to one site from a far wider area by flood waters or river systems; finally, they may be the site of the only water hole at a time of drought—elephants may have gath-ered there only to find food in such short supply, possibly due to bush fires, that they have starved to death.

Man's relationship with the elephants is beset by contradictions. On the one hand, elephants are valuable beasts of burden which need to be conserved; on the other hand, man is bringing them near to exter-mination in his thirst for land and ivory.

▲ **Ceremonial elephants.** Asian elephants still feature in ceremonies on the Indian subcontinent, as in this festival in Sri Lanka.

▼ **Working elephant.** Although declining in importance, the Asian elephant still has an economic role to play, particularly in the timber forests of Southeast Asia and the Indian subcontinent, where rough terrain and a lack of roads make it impossible to use tractors and lorries.

efforts have been made to use them in Zaire.

Elephants have always had another, more elevated, role in human affairs. Viewed with wonder and surrounded by mystique throughout its history, the elephant has found its way into the culture-myths and religions of all regions in which it has existed and even in modern times it features prominently in local religious symbolism and ceremonial.

In industrialized societies, elephants are held in such high regard as a popular spectacle, that for several centuries no circus or zoo has been without them. However, in the past, such establishments have depended almost entirely on the importation of wild stock and, as this stock diminishes, more attention will need to be paid to the breeding of elephants in captivity.

Although young elephants are often killed by lions, hyenas or tigers, the elephant's most dangerous enemy is man. In the 7th century BC, hunting for ivory caused the extinction of elephants in western Asia. In India, elephant numbers have steadily declined during the last millennium, as a result of sport hunting, ivory hunting and the spread of agriculture and pastoralism. In Africa, the Arab ivory trade, which started in the 17th century, caused a rapid decline of elephants in West Africa.

The colonial era, with the opening up of previously inaccessible areas and the introduction of modern technology, especially high-powered rifles, accelerated the decline of elephants. In Asia, this happened in the second half of the 19th century. In Africa, the destruction of elephants was highest between 1900 and 1910. Today, continuing deforestation and the spread of roads, farms and towns into former elephant habitats threaten both species by restricting their range, cutting off seasonal migration routes and bringing elephants into more frequent conflict with man. The human populations of the African countries south of the Sahara are doubling every 25–30 years, so intensifying the demand for land and the pressure on elephant habitats.

The worldwide economic recession of the early 1970s encouraged investors to switch to ivory as a wealth store. The sudden increase in the world price of ivory stimulated a wave of illegal hunting for ivory in Africa which is now causing a dramatic decline in elephant numbers. For instance, the elephant population of Kenya fell from 167,000 in 1973 to 60,000 in 1980. A recent wide-ranging survey estimates that 1.3 million elephants survived in Africa in 1980, but the fear is that this may be a considerable overestimate, bearing in mind the alarming rate at which elephants are still being killed (50,000–150,000 each year). In Asia, scarcely 50,000 elephants remain in the wild.

As their former habitats are destroyed and their numbers are cut by the greed for ivory, the only hope for the future conservation of elephants lies in the national parks. Ironically, in Africa, the existence of some national parks set up specifically to protect elephants is threatened by the elephants themselves. When a national park is created, mortality is reduced because poaching is controlled and access to water is guaranteed. The higher survival rates, especially among juveniles, causes an increase in elephant numbers within the park. At the same time, continuing human harassment outside drives elephants into the sanctuary of the park. Elephants have density-dependent mechanisms which regulate the population size. At high population densities there is an increase in the age of puberty, an increase in the interval between births, an earlier age of menopause and an increase in infant mortality. But under artificial conditions these mechanisms act too slowly. At

▲ Destructive elephants. In the dry season, elephants feed on trees, stripping and eating bark, demolishing whole trees to reach the leaves and twigs. In some parks, the elephant population is large enough to cause a rapid loss of the tree population and a conversion of the original habitat from woodland or bushland to grassland. The vegetation changes are probably accompanied by changes in the insect, bird and mammal populations, and possibly by changes in the soils and water table. Often the trends are exacerbated by fires which may be started by poachers or which may sweep in from outside the park. Fires prevent regeneration of woody species, kill trees already damaged by elephants, and destroy dry season browse, so that elephants are forced to concentrate on unburned areas, so causing even greater pressure on the vegetation.

high elephant densities, more trees and shrubs are killed by their feeding than can be replaced by natural regeneration. This results in a conversion of the habitat from woodland or bushland to grassland.

Fears that national parks would be irreversibly damaged by elephants caused a fierce controversy in the 1960s. Some conservationists argued that the elephant increase was due primarily to human activities and therefore humans should cull elephants to redress the ecological balance. They argued that while killing elephants is repugnant, especially in a national park, culling is necessary to prevent the loss of plants and animals that the park is supposed to conserve. Other conservationists argued that all animal and plant populations fluctuate naturally and that the elephant increase was just one phase of a natural cycle: soon numbers would go down, trees and shrubs would regenerate, and so there was no justification for culling. But in the last few years another factor has emerged: a lengthy drought has brought many elephants close to starvation, and in 1983 in Tsavo National Park, Kenya, some culling of dying

elephants took place, leaving healthier herds.

African elephants could, in theory, be farmed, but the main drawback is their long reproductive cycle. However, in some overpopulated reserves, culled animals have already become a managed source of income—the ivory is sold at auction, the meat is dried and sold locally, the fat is turned into cooking oil, and the skin tanned and used for leather goods.

Since the late 1960s, the continent's political upheavals have also contributed to the elephant problem: the various conflicts and civil wars in Eritrea, Somalia, Sudan, Chad, Uganda, Zimbabwe, Mozambique, and Angola made semi-automatic and automatic weapons widely available throughout eastern and central Africa. These weapons are now falling into the hands of commercial poachers, making them the final arbiters of wildlife management in Africa. National park rangers, who are nearly always poorly armed and badly equipped, cannot contend with this new development. For the elephant, faced with modern weapons and the continuing loss of its habitat, the future is grim indeed.   RFWB

HYRAXES

Order: Hyracoidea
Family: Procaviidae.
Eleven species in 3 genera.
Distribution: Africa and the Middle East.

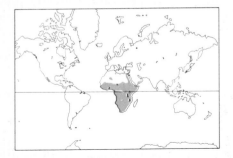

Rock hyraxes or dassies
Genus *Procavia*
Species: **Abyssinian hyrax** (*P. habessinica*),
Cape hyrax (*P. capensis*), **Johnston's hyrax**
(*P. johnstoni*), **Kaokoveld hyrax** (*P. welwitschii*),
Western hyrax (*P. ruficeps*).
Distribution: SW and NE Africa, Sinai to
Lebanon and SE Arabian Peninsula. Habitat:
rock boulders in vegetation zones from arid to
the alpine zone of Mt. Kenya (3,200–4,200m;
10,500–13,800ft); active in daytime. Size:
head-body length 44–54cm (17–21in); weight
1.8–5.4kg (4–12lb). Coat: Light to dark brown;
dorsal spot dark brown in Cape hyrax,
yellowish-orange in Western hyrax. Gestation:
210–240 days. Longevity: 9–12 years.

Bush hyraxes
Genus *Heterohyrax*
Species: **Ahaggar hyrax** (*H. antinae*), **Bruce's
yellow-spotted hyrax** (*H. brucei*), **Matadi
hyrax** (*H. chapini*).
Distribution: SW, SE to NE Africa. Habitat: rock
boulders and outcrops in different vegetation
zones in E Africa, sometimes in hollow trees;
active in daytime. Size: head-body length
32–47cm (12.5–18.5in); weight 1.3–2.4kg
(2.9–5.3lb). Coat: light gray, underparts white,
dorsal spot yellow. Gestation: about 230 days.
Longevity: 10–12 years.

Tree hyraxes
Genus *Dendrohyrax*
Species: **Eastern tree hyrax** (*D. validus*),
Southern tree hyrax (*D. arboreus*), **Western
tree hyrax** (*D. dorsalis*).
Distribution: SE and E Africa (Southern tree
hyrax), W and C Africa (Western tree hyrax),
Kilimanjaro, Meru, Usambara, Zanzibar,
Pemba and Kenya coast (Eastern tree hyrax).
Habitat: evergreen forests up to about 3,650m
(12,000ft) (Southern tree hyrax among rock
boulders in Ruwenzori). Size: head-body length
32–60cm (12.5–24in); weight 1.7–4.5kg
(3.7–9.9lb). Longevity: greater than 10 years.
Coat: long, soft and dark brown; dorsal spot
light to dark yellow, from 20–40mm long
(Eastern tree hyrax) to 40–75mm (Western tree
hyrax). Gestation: 220–240 days.

"THE high mountains are for the wild
goats; the rocks are a refuge for the
conies"—so runs the biblical characterization of the hyrax (Psalms 104:18). In
Phoenician and Hebrew, hyraxes are
known as *shaphan*, meaning "the hidden
one." Some 3,000 years ago, Phoenician
seamen explored the Mediterranean, sailing
westward from their homeland on the coast
of Syria. They found land where they saw
many animals which they thought were
hyraxes, and so they called the place
"Ishaphan"—Island of the Hyrax. The Romans later modified the name to Hispania.
But the animals were really rabbits, not
hyraxes, and so the name Spain derives from
a faulty observation!

The odd appearance of the hyrax has
caused even further confusion. Their superficial similarity to rodents led Storr, in 1780,
mistakenly to link them with guinea pigs of
the genus *Cavia*, and he thus gave them the
family name of Procaviidae or "before the
guinea pigs." Later, the mistake was discovered and the group was given the equally
misleading name of hyrax, which means
"shrew mouse."

Hyraxes are small and solidly built, with a
short rudimentary stump for a tail. Males
and females are approximately the same
size. The feet have rubbery pads containing
numerous sweat glands, and are ill-
equipped for digging. While the animal is
running, the feet sweat, which greatly enhances its climbing ability. Species living in
arid and warm zones have short fur, while
tree hyraxes and the species in alpine areas
have thick, soft fur. Hyraxes have long
tactile hairs at intervals all over their bodies,
probably for orientation in dark fissures and
holes. They have a dorsal gland, surrounded
by a light-colored circle of hairs which
stiffen when the animal is excited. This circle
is most conspicuous in the Western tree
hyrax and least so in the Cape hyrax.

Early this century, fossil evidence showed
that the hyraxes share many common features of primitive ungulates, especially
elephants, and the related sirenians, and, as
more recent research has indicated, also
with the aardvark. Fossil beds in the Fayum,
Egypt, show that 40 million years ago
hyraxes were the most important medium-
sized grazing and browsing ungulates during that time. Then there were at least six
genera, ranging in size from that of contemporary hyraxes to that of a tapir. During the
Miocene (about 25 million years ago), at the
time of the first radiation of the bovids,
hyrax populations were greatly reduced,
surviving only among rocks and trees—

habitats that were not invaded by bovids.

Contemporary hyraxes retain primitive
features, notably an inefficient feeding
mechanism, which involves cropping with
the molars instead of the incisors used by
modern hoofed mammals, poor regulation
of body temperature and short feet.

In the Pliocene (7–2 million years ago),
hyraxes were both widespread and diverse:
they radiated from southern Europe to
China, and one fossil form, *Pliohyrax
graecus*, was probably aquatic. Yet today
they are confined to Africa and the Middle
East.

The rock hyraxes have the widest geographical and altitudinal distribution, while
the bush hyraxes are largely confined to the
eastern parts of Africa. Both are dependent
on the presence of suitable refuges in rocky
outcrops (kopjes) and cliffs. As their name

▲ **Hyrax shows its teeth.** Hyraxes, like this Johnston's hyrax, can fight vigorously if cornered, biting savagely with their incisors. Curiously, these teeth are not much used in cropping, the molars being used instead, a relatively inefficient cropping method.

suggests, the tree hyraxes are found in arboreal habitats of Africa, but in the alpine areas of the Ruwenzori Mountains they are also rock dwellers. The Eastern tree hyrax might be the earliest type of forest-living tree hyrax, being a member of the primitive fauna and flora of the islands of Zanzibar and Pemba.

Hyraxes feed on a wide variety of plants. Rock hyraxes feed mainly on grass, which is a relatively coarse material, and therefore have hypsodont dentition (high crowns with relatively short roots), whereas the browsing bush hyraxes and tree hyraxes consume softer food and have a brachydont dentition (short crowns with relatively long roots).

Hyraxes do not ruminate. Their gut is complex, with three separate areas of microbial digestion, and their ability to digest fiber

efficiently is similar to that of ruminants. Their efficient kidneys allow them to exist on minimal moisture intake. In addition, they have a high capacity for concentrating urea and electrolytes, and excrete large amounts of undissolved calcium carbonate. As hyraxes have the habit of always urinating in the same place, crystallized calcium carbonate forms deposits which whiten the cliffs. These crystals were used as medicine by several South African tribes and by Europeans.

Hyraxes have a poor ability to regulate body temperature, and a low metabolic rate. Body temperature is maintained mainly by gregarious huddling, long periods of inactivity, basking and relatively short periods of activity. In summary, their physiology allows them to exist in very dry areas with food of poor quality, but they are dependent on shelter which provides them with an environment of relatively constant temperature and humidity.

Different species of hyraxes can co-exist in the same habitat (see box).

Groups of Bruce's yellow-spotted hyrax and Johnston's hyrax live on rock outcrops in the Serengeti National Park, Tanzania. Together, these two species are the most important resident herbivores of the kopjes. Their numbers depend upon the area of the kopje. The population density ranges for bush hyraxes from 20–53 animals per hectare and for the rock hyraxes from 5–40 animals per hectare. The group size varies between 5 and 34 for the former and for the latter between 2 and 26. Taking the two species together, this is comparable to the density of wildebeest in the long grass plains surrounding the kopjes. Among the different groups, the adult sex ratio is skewed in favor of females (1.5–3.2:1 for Bruce's yellow-spotted hyrax and 1.5–2:1 for Johnston's hyrax), while the sex ratio of newborns is 1:1.

The social organization varies in relation to living space. On kopjes smaller than 4,000sq m (43,000sq ft), both rock and bush hyraxes live in cohesive and stable family groups, consisting of 3–7 related adult females, one adult territorial male, dispersing males and the juveniles of both sexes. Larger kopjes may support several family groups, each occupying a traditional range. The territorial male repels all intruding males from an area largely encompassing the females' core area (average for bush hyraxes, 2,100sq m; 22,600sq ft; 27 animals, and for rock hyraxes 4,250sq m; 45,750sq ft; 4 animals).

The females' home ranges are not defended and may overlap. Rarely, an adult female will join a group, and such females are eventually incorporated into the female group. Females become receptive about once a year, and a peak in births seems to coincide with rainfall. Within a family group, the pregnant females all give birth within a period of about three weeks. The number of young per female bush hyrax

Kopje Cohabitants

The dense vegetation of the Serengeti kopjes supports two species of hyraxes—the gray-brown Bruce's yellow-spotted hyrax and the larger, dark brown Johnston's hyrax—living together in harmony. Whenever two or more closely related species live together permanently in a confined habitat, at least some of their basic needs like food and space resources must differ, otherwise one species will eventually exclude the other.

When bush and rock hyraxes occur together they live in close contact. In the early mornings they huddle together, after spending the night in the same holes. They use the same urinating and defecating places. Newborns are greeted and sniffed intensively by members of both species. The juveniles associate and form a nursery group; they play together with no apparent hindrance as play elements in both species are similar. Most of

their vocalizations are also similar, such as sounds used in threat, fear, alertness and contact situations. Such a close association has never been recorded between any other two mammal species except primates. However, bush and rock hyraxes do differ in key behavior patterns. Firstly, they do not interbreed, because their mating behavior is different and they have also different sex organ anatomy: the penis of the bush hyraxes being long and complex, with a thin appendage at the end, arising within a cup-like glans penis, and that of the rock hyrax being short and simple. Secondly, the male territorial call, which might function as a "keep out" sign is also different, and, finally, the bush hyrax browses on leaves while the rock hyrax feeds mainly on grass. The latter is probably the main factor that allows both species to live together.

varies between 1 and 3 (mean 1.6) and in rock hyraxes between 1 and 4 (mean 2.4). The young are fully developed at birth, and suckling young of both species assume a strict teat order. Weaning occurs at 1–5 months and both sexes reach sexual maturity at about 16–17 months of age. Upon sexual maturity, females usually join the adult female group, while males disperse before they reach 30 months. Adult females live significantly longer than adult males.

▼ **Mating in hyraxes** is brief and vigorous. The penis anatomy varies between the three genera. In rock hyraxes it is short, simply built and elliptical in cross section; in tree hyraxes it is similarly built and slightly curved; in bush hyraxes, shown here, it is long and complex: on the end of the penis, and arising within a cup-like glans penis is a short, thin appendage, which has the penis opening. (1) The male presses the penis against the vagina. (2) Violent copulation, in which the male leaves the ground. (3) The female moves forward causing the male to withdraw.

There are four classes of mature male: territorial, peripheral, early and late dispersers. Territorial males are the most dominant. Their aggressive behavior towards other adult males escalates in the mating season, when the weight of their testes increases twenty-fold. These males monopolize receptive females and show a preference for copulating with females over 28 months of age. A territorial male monopolizes "his" female group year round, and repels other males from sleeping holes, basking places and feeding areas. Males can fight to the death, although this is probably quite rare. While his group members feed, a territorial male will often stand guard on a high rock and be the first to call in case of danger. The males utter the territorial call all year round.

On small kopjes, peripheral males are those which are unable to settle, but which on large kopjes can occupy areas on the periphery of the territorial males' territories. They live solitarily, and the highest ranking among them takes over a female group when a territorial male disappears. These males show no seasonality in aggression but call only in the mating season. Most of their mating attempts and copulations are with females younger than 28 months.

The majority of juvenile males—the early dispersers—leave their birth sites at 16–24 months old, soon after reaching sexual maturity. The late dispersers leave a year later, but before they are 30 months old. Before leaving their birth sites, both early and late dispersers have ranges which overlap with their mothers' home ranges. They disperse in the mating season to become peripheral males. Almost no threat, submissive and fleeing behavior has been observed between territorial males and late dispersers.

Individuals of rock and bush hyraxes were observed to disperse over a distance of at least 2km (1.2mi). However, the further a dispersing animal has to travel across the open grass plains, where there is little cover and few hiding places, the greater are its chances of dying, either through predation or as a result of its inability to cope with temperature stress.

The most important predator of hyraxes is the Verreaux eagle, which feeds almost exclusively on them. Other predators are the Martial and Tawny eagles, leopards, lions, jackals, Spotted hyena and several snake species. External parasites such as ticks, lice, mites and fleas, and internal parasites such as nematodes and cestodes also probably play an important role in hyrax mortality.

In Kenya and Ethiopia it was found that rock and tree hyraxes might be an important reservoir for the parasitic disease leishmaniasis.

The Eastern tree hyrax is heavily hunted for its fur, in the forest belt around Mt. Kilimanjaro; 48 animals yield one rug. Because the forests are disappearing at an alarming rate in Africa, the tree hyraxes are probably the most endangered of all hyraxes. HNH

▼ **Bush hyrax in a tree.** A young Bruce's yellow-spotted hyrax climbing. Despite their name, bush hyraxes generally inhabit the same rocky outcrops as rock hyraxes. They do, however, sometimes inhabit hollow trees.

AARDVARK

Order: Tubulidentata
Family: Orycteropodidae.
Orycteropus afer
Aardvark, antbear or **Earth pig**
Distribution: Africa south of the Sahara, excluding deserts.

Habitat: mainly open woodland, scrub and grassland; rarer in rain forest; avoids rocky hills.

Size: head-body length 105–130cm (41–51in); tail length 45–63cm (18–25in); weight 40–65kg (88–143lb). Sexes same size.
Skin: pale yellowish gray with head and tail off-white or buffy white (the gray to reddish brown color often seen results from staining by soil, which occurs while the animal is burrowing). Females tend to be lighter in color.

Gestation: 7 months.

Longevity: up to 10 years in captivity.

Subspecies: 18 have been listed but most may be invalid. There is insufficient knowledge about the animal for firm conclusions to be drawn.

▶ **Nocturnal pursuits of the aardvark.** Once it has found a termite mound, an aardvark takes up a sitting position and inserts its mouth and nose, creating a V-shaped furrow. Although it has not been directly observed, termites and ants are presumably taken in by the long sticky tongue. Feeding bouts, lasting 20 seconds to seven minutes, are interrupted by short bursts of active digging. Digging may continue until the whole body enters the excavation. Termite mounds are, however, not totally destroyed during a single visit and an aardvark will feed on a single mound on consecutive nights.

Few people have had the fortune of a close encounter with one of Africa's most bizarre and specialized mammals: the aardvark. This nocturnal, secretive termite- and ant-eating mammal is the only living member of the order Tubulidentata. Thanks to its elusiveness, it is one of the least known of all living mammals.

Superficially, aardvark resemble pigs, in possessing a tubular snout and long ears. Their pale, yellowish body is arched and covered with coarse hair which is short on the face and on the tapering tail but long on its powerful limbs. These are short, with four digits on the front feet and five on the back. The claws are long and spoon-shaped, with sharp edges. The elongated head terminates in a long, flexible snout and a blunt, pig-like muzzle. A dense mat of hair surrounds the nostrils and acts as a dust filter during burrowing. The wall between the nasal slits is equiped with a series of thick fleshy processes which probably have sensory capabilities. Aardvark have no incisor or canine teeth and their continuously growing, open-rooted cheek teeth consist of two upper and two lower premolars and three upper and three lower molars in each jaw half. The cheek teeth differ from those of other mammals in that the dentine is not surrounded by enamel but by cementum.

The aardvark has a sticky tongue—round, thin and long—and well-developed salivary glands. Its stomach has a muscular pyloric area, which functions like a gizzard, grinding up the food. Aardvark therefore do not need to chew their food. Both males and females have anal scent glands which emit a strong-smelling yellowish secretion.

Aardvark feed predominantly on ants and termites, with ants dominating the diet during the dry season and termites during the wet. When foraging, the aardvark keeps its snout close to the ground, and its tubular ears pointed forwards. This indicates that smell and hearing play an important part in locating food. Aardvark follow a zigzag course when seeking food, and the same route may be used on consecutive nights. While foraging, they frequently pause to explore their immediate surroundings, by rapidly digging a V-shaped furrow with the forefeet and by sniffing it intensively.

Little information is available on reproduction, but young are probably born just before or during the rainy season, when termites become more available. Only one young, with a weight of approximately 2kg (4lb), is born at a time. It will accompany its mother when two weeks old and start digging its own burrows at about six months, but may stay with the mother until the onset of the next mating season.

Aardvark are almost exclusively nocturnal and solitary. Two individuals tracked by radio in the Transvaal (South Africa) were more active during the first part of the night (20.00–24.00 hours). They foraged on both dark and bright moonlit nights but took shelter in one of several burrow systems within their home range during spells of adverse weather or when disturbed. They foraged over distances varying from 2–5km (1.2–3mi) per night. Other studies suggest that aardvark may range as far as 15km (9mi) during a 10-hour foraging period—even as far as 30km (19mi) a night.

The only evidence of aardvark is normally their burrows, of which there are three types: the burrows made when looking for food; larger temporary sites which may be used for refuge and which occur throughout the home range; and permanent refuge sites where young are born. The latter often have more than one entrance and are frequently modified through digging, extend deeply into the ground and comprise an extensive burrow system up to 13m (43ft) long. An aardvark can excavate a burrow very quickly and, depending on the soil type, can dig itself in within 5–20 minutes. Droppings are deposited in shallow digs throughout their range and covered with soil.

Aardvark share their habitat with a variety of other termite- and ant-eating animals, such as hyenas, jackals, vultures, storks, geese, pangolin, Bat-eared fox and aardwolf, but all of these also take other prey, which reduces competition with the aardvark. In being a specialized feeder, aardvark are extremely vulnerable to habitat changes. While intensive crop farming over vast areas may reduce aardvark density, increased cattle herding, which through trampling creates better conditions for the termites, may increase their numbers. In general, however, until more is known about the behavior and ecology of the aardvark, little progress can be made in formulating management policies.

Apart from aardvark flesh, which is said to taste like that of pork, various parts of the aardvark's body are prized. Its teeth are worn on necklaces by the Margbetu, Ayanda and Logo tribes of Zaire to prevent illness and as a good-luck charm. Its bristly hair is sometimes reduced to powder and, when added to local beer, regarded as a potent poison. It is also believed that the harvest will be increased when aardvark claws are put into baskets used to collect flying termites for food.   RJvA

THE HOOFED MAMMALS

**ORDERS: PERISSODACTYLA
AND ARTIODACTYLA**
Thirteen families: 82 genera: 203 species.

Odd-toed Ungulates
Order: Perissodactyla—perissodactyls

Asses, Horses and Zebras
Family: Equidae—equids
Seven species in 1 genus.
Includes **African ass** (*Equus africanus*),
Domestic horse (*E. caballus*), **Grevy's zebra**
(*E. grevyi*), **Plains zebra** (*E. burchelli*).

Tapirs
Family: Tapiridae
Four species in 1 genus.
Includes **Brazilian tapir** (*Tapirus terrestris*).

Rhinoceroses
Family: Rhinocerotidae
Five species in 4 genera.
Includes **White rhino** (*Ceratotherium simum*).

Even-toed Ungulates
Order: Artiodactyla—artiodactyls

Suborder: Tylopoda—tylopods
Camels
Family: Camelidae—camelids
Six species in 3 genera.
Includes **Bactrian camel** (*Camelus bactrianus*),
guanaco (*Lama guanicoë*).

"UNGULATE" is a general term given to all those groups of mammals which have substituted hooves for claws during their evolution. This character appears to follow from a commitment to a terrestrial lifestyle, with rapid locomotion and a herbivorous diet. Ungulates are relatively large animals, none less than 1kg (2.2lb) in body weight, and they comprise the majority of terrestrial mammals over 50 kg (110lb). Living ungulates belong to two different orders which diverged from a common hoofed ancestor some 60 million years ago. Despite the superficial similarities between horses and cows, rhinos and hippos, tapirs and pigs, the former of each pair belongs to the Perissodactyla (odd-toed ungulates), and the latter to the Artiodactyla (even-toed ungulates), and the similarities between them have largely come about due to convergent evolution.

The Ungulate Body Plan
Despite the variety of bodily shapes and adornments, there is an underlying common theme to the two lineages of modern ungulates. From a one-kilogram chevrotain to a three-tonne hippopotamus, and from the ponderous rhinoceros to the graceful horse, ungulates generally have long-muzzled heads, held horizontally on the neck, barrel-shaped bodies carried on forelegs and hindlegs of roughly equal length, and small tails. Their skin is quite thick and tends to carry a coat of hairs (which may be air-filled for insulation) rather than soft fur. Compared to the primitive mammalian limb pattern, in which the foot has five digits, all of which are placed on the ground in locomotion, all ungulates have thickened, hard-edged keratinous hooves. There is a reduction in the number of toes, and a lengthening of the metapodials (the long bones in the fleshy parts of human hands and feet), with the resultant lifting of the foot, so that the animal is in effect balancing on the tips of its toes. The evolutionary climax of this type of limb—termed unguligrade—can be seen in the long slender limbs of horses and antelopes, where it bestows both speed and endurance.

As the names imply, odd- and even-toed ungulates differ in the type and degree of modification, and the number of toes. Even-toed ungulates (artiodactyls) have four or two weight-bearing toes on each foot, the weight-bearing axis of the limb passing between the third and fourth digits. All modern artiodactyls have lost the first digit. Odd-toed ungulates (perissodactyls) have a single toe or three toes together bearing the weight of the animal, with the axis of the limb passing through the middle (or single). All modern species have lost the first and fifth digits in the hindfoot, and the first digit in the forefoot. The metapodials are unfused and relatively short.

The earliest horses were three-toed ungulates, but since the Oligocene (35 million years ago) horses have borne their weight on the single third toe, with ligaments rather than a fleshy pad for support. In the fossil three-toed horses the second and fourth digits were much reduced, although they bore fully formed hooves and would have contacted the ground to provide additional support in extreme extension of the front foot in locomotion, such as in galloping and jumping; the metapodials were greatly elongated (although not fused together) to form a long slender limb. All living species of equids have reduced these side toes to proximal splint bones, and bear their entire weight at all times on an enlarged single hoof.

Senses
Ungulates have good but not exceptional hearing, small ears which can be rotated to detect the direction of a sound, an apparently very good sense of smell, and excellent eyesight. The eyes function well by day and

▼ **Built for running,** these gemsbok in Etosha National Park epitomize the grace of the antelopes, the most successful of the hoofed mammals.

Suborder: Suina—suoids

Pigs
Family: Suidae
Nine species in 5 genera.
Includes **Wild boar** (*Sus scrofa*).

Peccaries
Family: Tayassuidae
Three species in 2 genera.
Inludes **Collared peccary** (*Tayassu tajacu*).

Hippopotamuses
Family: Hippopotamidae
Two species in 2 genera.
Includes **hippopotamus** (*Hippopotamus amphibius*).

Suborder: Ruminantia—ruminants

Chevrotains
Family: Tragulidae—tragulids
Four species in 2 genera.
Includes **Water chevrotain** (*Hyemoschus aquaticus*).

Musk deer
Family: Moschidae—moschids
Three species in 1 genus.
Includes **Musk deer** (*Moschus moschiferus*).

Deer
Family: Cervidae—cervids
Thirty-four species in 14 genera.
Includes **Red deer** (*Cervus elaphus*),
Reindeer (*Rangifer tarandus*).

Giraffe
Family: Giraffidae—giraffids
Two species in 2 genera.
Includes **giraffe** (*Giraffa camelopardalis*).

Cattle, antelope, sheep, goats
Family: Bovidae—bovids
One hundred and twenty-four species in 45 genera.
Includes **Pronghorn** (*Antilocapra americana*),
eland (*Taurotragus oryx*), **impala** (*Aepyceros melampus*), **kob** (*Kobus kob*), **oryx** (*Oryx gazella*),
Thomson's gazelle (*Gazella thomsonii*).

night, and give a fair degree of binocular vision, especially in open-country species, allowing the animals to judge distance and speed accurately. Their communication depends mainly upon sight and sound, with some use of scent marks, in open-country species; forest-dwelling artiodactyls are more dependent upon scent for social signalling. Perissodactyls lack the diversity of scent glands found on the feet and faces of many artiodactyls. They rely instead more on auditory communication, with frequent vocalizations and the production of a large variety of sounds, and they produce a much greater variety of facial expressions than do artiodacyls.

Food and Feeding
The evolution of the ungulate limb illustrates adaptation to a mobile open-country existence. Evolution of their teeth, skulls and digestive anatomy parallels changes in their locomotion. All ungulates are terrestrial herbivores, feeding on leaves, flowers, fruits or seeds of trees, herbs and grasses (although pigs and peccaries are characteristically more omnivorous, and may include roots, tubers and animals in the diet). With rare exceptions, all ungulates stand on the ground to feed; even the aquatic hippopotamus feeds on land. They cannot use forelimbs to manipulate food, as do primates and squirrels; nor do they fell trees to reach foliage, as do elephants or beavers. Their food has to be taken directly from the plant, or off the ground if it has fallen, with the lips, teeth and tongue, and these are appropriately modified.

Even though ungulates are herbivorous, not all plant food is of similar nutritive value. Plants tend to be abundant in carbohydrates such as sugars and starches, which are easily digested sources of energy. However, they are low in fat, and frequently low in protein, which is the source of the building blocks of amino acids essential for growth and repair of body tissues. The absence of abundant fat does not seem to constitute a critical problem for ungulates, and many have lost the gall bladder, which in other mammals is the source of bile salts that emulsify and break down fats.

But obtaining sufficient protein in the diet is a critical matter for all herbivores. This is particularly true for the smaller ungulates, which have relatively greater requirements and higher turnover of nutrients, and small antelopes have very occasionally been observed to catch and kill birds as an additional source of protein. Pigs will also consume carrion.

The most abundant source of vegetable

protein is in seeds, but these are small and widely dispersed. The more easily available sources of vegetation, such as leaves, are composed primarily of carbohydrates, especially when they are mature. The carbohydrates in vegetation are available in two forms: in the soluble cell contents, and in the fibrous cell wall casing of cellulose, which is indigestible to most mammals.

Ungulate dentition is adapted to grinding so as to mechanically disrupt the cell wall to release the digestible contents. The back of an ungulate's mouth functions like a mill, with large flat square molars which reduce plant matter to fine particles. In conjunction with this, the jaw musculature and the configuration of the jaw joint are modified so that the lower jaw can be moved across the upper with a sweeping transverse grinding motion, in contrast to the more up-and-down motion in other mammals that simply cuts and pulps the food. The high crowned (hypsodont) cheek teeth are made to last a lifetime of continual abrasion. In ungulates there is typically a gap between the milling molars and the plucking incisors. Whether or not canine teeth are retained depends upon their use as weapons; they appear to have no feeding function in ungulates,

although in ruminants (members of the artiodactyl suborder Ruminantia) the lower canines are retained and modified to form part of the lower incisor row.

Artiodactyls such as pigs and peccaries, which select only non-fibrous vegetation such as fruit and roots, do not digest the cellulose content of vegetation, and have a digestive system which resembles that of other mammals. However, other ungulates have a more fibrous diet so must be able to digest the large quantities of cellulose that they must ingest along with the more easily digestible parts of the plant. To achieve this the ingested food is fermented by bacteria somewhere along the digestive tract, transforming it into products which can then be absorbed and utilized.

There is a critical difference between the complex digestive systems of perissodactyls and those suborders of artiodactyls which eat fibrous vegetation (Ruminantia and Tylopoda). The "ruminant" artiodactyls have their fermentation chamber containing microorganisms situated within a complex multi-chambered stomach. The ruminant itself can digest both the continually multiplying microorganisms that overspill into the rest of the digestive tract and the products of

THE UNGULATE BODY PLAN

▶ **Teeth.** Primitive herbivorous mammals have molars with separate cusps (bunodont), designed to pulp and crush relatively soft food. This type is seen in pigs (**a**). Fibrous vegetation is tough and ungulates have developed modifications of the bunodont pattern. In perissodactyls, such as the rhinoceros (**b**), shearing edges (lophs) have formed by a coalescing of the cusps to form two crosswise lophs and one lengthwise (lophodont). In horses (**c**) the lophs are very complex and folded (hypsodont). In ruminant artiodactyls, such as the ox (**d**), the cusps take on a crescent shape (selenodont).

▼ **Jaws.** The different modes of feeding of perissodactyls and ruminant artiodactyls are reflected in the size of the jaw and musculature. Non-ruminant grazers, like the horse (**a**), have to consume large quantities of tough fibrous food and the lower jaw is very deep and the masseter muscle, primarily used in closing the jaws, is very large. Ruminants, like the giraffe (**b**), spend much of their time chewing the already half-digested cud and the lower law and masseter muscle are much less pronounced.

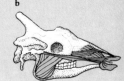

▲ **Mode of feeding.** Most perissodactyls (with the exception of some rhinos) retain both sets of incisors and use the upper lip extensively in feeding, like the horse (**a**). Ruminant artiodactyls, like the giraffe (**b**), have lost the upper incisors and make extensive use of a prehensile tongue rather than the upper lip. The resulting differences in facial musculature mean that perissodactyls have a much greater variety of facial expressions, used to communicate with each other, than do artiodactyls.

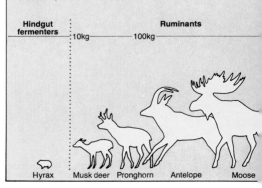

Hindgut fermenters	Ruminants
10kg	100kg

Hyrax   Musk deer   Pronghorn   Antelope   Moose

Hoofed Mammals' Feet

In the hoofed mammals the primitive mammalian foot (**a**) has been modified in various ways. The toes are reduced in number, and the long bones (metapodials) much extended. The foot is held lifted with only the tips of the toes on the ground (unguligrade). The joint surfaces are restricted so that the limbs cannot be rotated or moved in or out of the body to any great extent—the

prime movement is thus fore and aft, which facilitates fast running at the expense of climbing and digging. In the generalized ungulate feet (**b–e**), one or two digits are lost, the metapodials somewhat elongated, the tarsal bones are more ordered: (**b**) tapir, (**c**) pig, (**d**) peccary, (**e**) chevrotain. Rhinos (**f**) and hippos (**g**) have feet specialized for weight bearing (graviportal), with short digits and a spreading foot in which the side toes touch the ground when standing. In camels (**h**) the metapodials are long and fused for most of their length into a single bone. The most drastic modifications occur in the hoofed mammals adapted to fast running (cursorial). In the horse (**i**) the metapodials are totally fused and the digits are reduced to one (digits 2 and 4 are retained as vestigial splint bones). Deer (**j**) retain the side toes, and the metapodials are only partly fused. In the pronghorn (**k**) the fused metapodials are long.

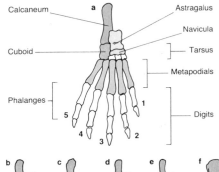

Calcaneum — a
Cuboid
Phalanges
5   4   3   2   1
Astragalus
Navicula
Tarsus
Metapodials
Digits

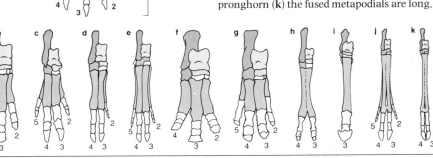

▼ **Digestive systems.** Ungulates have evolved two very different systems for dealing with the relatively indigestible cellulose in their highly fibrous food: hindgut fermentation (**a**) and rumination (**b**). In the hindgut fermenters (perissodactyls) food is completely digested in the stomach, and passes to the large intestine and cecum, where microorganisms ferment the ingested cellulose. Ruminants have a more complex digestive system and retain food in the gut for much longer. Food passes initially to the first stomach chamber (rumen) where it is fermented by microorganisms, and is regurgitated to be chewed and mixed with saliva. It then passes back to the second chamber (reticulum), bypassing the rumen. Bacteria spill over with the food and accompany it through the third (omasum) and fourth (abomasum) stomachs. Digestion is completed in the abomasum and nutrients are absorbed in the small intestine. Some additional fermentation and absorption occur in the cecum.

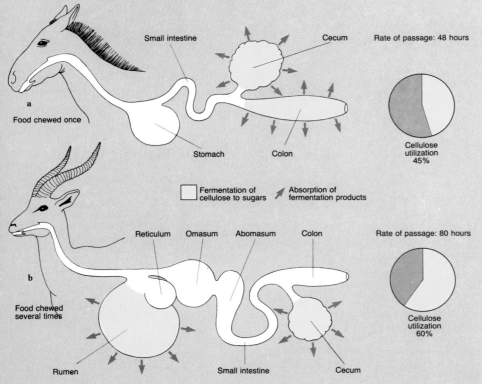

their fermentation of cellulose. The complex stomach allows food to be differentiated, digested food passing through a sieve-like structure to the posterior "true" stomach and the intestines, whereas undigested food is set aside to ferment, and is regurgitated to the mouth to be chewed again. This act is called rumination, familiar in domestic ruminants as "chewing the cud." This fermentation system is very efficient, making maximum use of the available cellulose, but is limited in that the food is retained for a very long time (up to four days), so that there is a lower limit to the amount of food that can be eaten and processed in a given period by a ruminant than a non-ruminant.

In contrast, in perissodactyls the hindgut areas of the cecum and colon (large intestine) are the site of fermentation. Although the processes of fermentation are biochemically identical to the ruminant process, the digestion of cellulose is less efficient, as the food is only retained for about half the length of time. However, this does mean that a greater quantity of food can be consumed per day, as the turnover rate is greater.

When comparing these ruminant and hindgut systems of fermentation, it is apparent that the latter is less efficient in the utilization of young, short herbage, which is high in protein and can yield all the requisite nutrients in small quantities. These can be utilized more efficiently by a ruminant. However, hindgut fermenters are at an advantage where food is of limited quality and high in fiber, thus necessitating a high intake to obtain sufficient protein, provided that this food is not also limited in quantity. Ruminants, on the other hand, are at an advantage in environments where food is of limited quantity, but where the quality is relatively high, for example desert inhabitants such as the oryx and camel, or arctic tundra inhabitants such as reindeer or Musk ox. In habitats such as the tropical savannas of Africa, where both types of animal co-exist, there is a partitioning of cropping, zebras (perissodactyls) for example eating the poorer quality old foliage at the top of the grass stand, and gazelles and wildebeest (artiodactyls) eating the higher quality new foliage uncovered by the zebras.

Absorption of the products of protein digestion is comparable in ruminants and hindgut fermenters, but ruminants can additionally recycle urea, a nitrogen-rich waste product that is normally excreted in the urine, using it to feed the microorganisms which they later digest. An important result of this difference is that perissodactyls

▼ **Consequences of body size.** Ruminants must pay a price for their highly efficient digestive system: it takes a long time for food to pass through their gut. In large herbivores, this price outweighs the advantages of efficient digestion. This is because large animals require absolutely more food per day than small ones, and they are therefore forced to accept low-quality food which can be gathered in large quantity. It takes so long to process low-quality food in a ruminant system that a greater net intake of nutrients is achieved with a simple gut and fast throughput. Hippos are the largest ruminants and appear to be an exception; however, they do not chew the cud and they have a fairly fast passage through the digestive system. Small animals require more food *per kilogram of body-weight* than large ones. Very small herbivores can select a high quality diet which can be digested easily without time-consuming rumination. This is why no ruminants are less that 5kg (11lb) in weight.

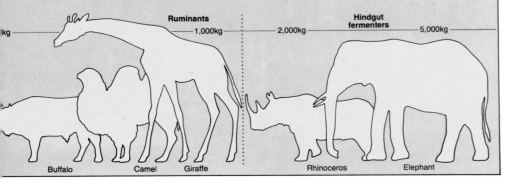

such as asses which inhabit desert areas, need to drink daily to produce sufficient water to balance the urea in the urine, whereas ruminants, such as the oryx and camel, need to drink only occasionally.

Perissodactyls can make maximum use of fruit in their diet, as its essential nutrients, particularly sugars, are absorbed before the region of fermentation is reached. Ruminants ferment fruit in the fore-stomach and hence lose much of its nutritional value. The only ruminants that gain benefit from eating fruits are very small, such as chevrotains and duikers, which eat little fibrous food and have a small rumen.

There are more species of ruminants than there are of the hindgut-fermenting perissodactyls. The reason for this seems to be that the ability of ruminants to recycle nitrogen and to digest the protein-rich bacteria frees them from having to obtain all their essential amino acids from their diet, which means that they can afford to specialize on a narrower range of plant species in the environment. This in turn means that ruminants can subdivide the available niches in a habitat in a finer fashion.

Many living ungulates, such as tapirs, giraffes and many species of deer, are browsers, eating the leaves of dicotyledonous plants such as trees and shrubs. This was probably the original diet of all perissodactyls and ruminant artiodactyls.

But in the Miocene (about 20 million years ago) the monocotyledonous grasses emerged as abundant land plants, which comprise an excellent and abundant source of nutritious, although cellulose-rich, vegetation. The main difference between grasses and dicotyledonous plants as a dietary source lies in the structure and distribution patterns of the plants.

Grasses grow in great expanses, but their nutritive leaves are at the base of the plant, and valuable feeding time can be wasted in searching for them through a stand of fibrous stems of low nutritional value. Leaves on trees and shrubs are on the perimeter of the plant and are more easily accessible, but the actual plants themselves are spread further apart and must be located. Moreover, a bitten grass leaf can grow from the base to replace itself rapidly, but a tree or shrub grows from the apex, and cannot rapidly regenerate. As a consequence, trees and shrubs tend to arm themselves with thorns and unpalatable chemicals in the leaves to discourage destruction, which may make them hard to feed on. In contrast, grasses do not seem to "mind" being eaten so much, and are not similarly

defended. However, grass is only a really good nutritional source for certain parts of the year, in the growing season, whereas trees and shrubs provide a more predictable source of nutrients that are available the whole year round.

Most ungulate species have opted for becoming either predominantly browsers or predominantly grazers, and the most successful and abundant have been the grazers, the bovids and the horses, which radiated with the Miocene grasslands, evolving the mouth parts and digestion to cope with the high cellulose and high silica content of grasses, and the agility to avoid predators which were attendant risks of feeding in the exposed conditions where grasses grow.

▲ **Eocene ungulates.** During the Eocene (54–38 million years ago), the first hoofed mammals, which had evolved in the preceding Mesozoic and Paleocene, rapidly evolved to fill a wide variety of environmental niches. The perissodactyls (odd-toed ungulates) and the artiodactyls (even-toed ungulates), appeared simultaneously in Europe and North America. Odd-toed ungulates predominated in these early forms. (1) *Uintathere*, a large grotesque herbivore. (2) *Dolichorhinus*, a titanothere from the late Eocene. (3) *Eohippus*, the "Dawn horse" from the lower Eocene. (4) *Hyrachyus*, a small "running" rhinoceros. (5) *Amynodentopsis*, a semi-aquatic rhinoceros. (6) *Phenacodus*, a primitive hoofed herbivore. (7) *Meniscotherium*, a small browsing herbivore.

▼ **The evolution of hoofed mammals.**

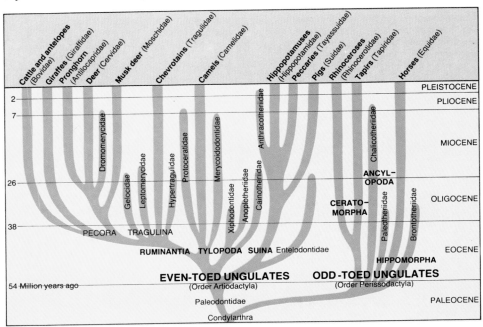

Evolution

Perissodactyls and artiodactyls first radiated in the early Eocene in the Northern Hemisphere 54 million years ago, at a time when Africa was isolated from Eurasia, and North America was isolated from South America, although a limited amount of interchange of animals was possible between the northern continents. The climate was warmer than today, and less differentiated latitudinally, with tropical forest reaching northwards to within the Arctic circle.

Eocene perissodactyls were a wide range of body sizes, ranging from about 5kg (11lb) to about 1,000kg (2,200lb), and were the first ungulates to adopt a diet of relatively fibrous vegetation (although all of them were browsers).

In North America the small Eocene equids and tapirs were selective browsers; with tapirs being more exclusively leaf-eaters, but equids handling a greater amount of fiber. Rhinos and chalicotheres were medium-sized browsers. The brontotheres were large animals, earlier forms being smaller and hornless, but later ones attaining the size of a White rhino and developing Y-shaped bony horns on the nose. Their teeth suggest a mixed diet of leaves and fruit. In Europe the brontotheres were absent, and their fruit specialist niche was probably taken by the endemic equid-related paleotheres, some of which sported a tapir-like proboscis. The only tapirs in the European faunas were the large and aberrant lophiodonts, which mimicked the rhinos of North America and Asia, and the European lineage of equids (which were all extinct by the end of the Eocene) diverged to fill the niches taken by both equids and tapirs in North America. In contrast, equids and paleotheres were largely absent from Asia, where the tapirs experienced the peak of their diversity during the Eocene, producing rabbit-sized and gazelle-like forms as well as more normal varieties.

In contrast to this blossoming of perissodactyl diversity, Eocene artiodactyls were all small (under 5kg), and were omnivorous or fruit-eating. This primary omnivorous tendency is retained today in the pig-like artiodactyls. However, there was never an equivalent radiation of omnivorous perissodactyls.

At the end of the Eocene (38 million years ago), the global climate changed dramatically, resulting in a seasonal climate in the northern continents. The ungulates were forced to adapt their foraging habits to seasonal growth in vegetation, Herbivores, such as primates, dependent on year-round availability of fruit disappeared from the northern continents at this time, migrating with the retreating tropical forests to the southern continents. As there was no direct land route between northern and southern continents at this time, larger and less agile animals such as ungulates apparently failed to do so. In the absence of a year-round supply of non-fibrous growing parts of the plants a considerable increase in body size occurred amongst the artiodactyls, with one lineage (the suborder Suina) remaining as omnivores and the others (ruminants) becoming specialized herbivores. With this initital increase in body size, it was possible for the ruminant artiodactyls to evolve their characteristic foregut site of fermentation to accompany this more fibrous diet as the disadvantages of being a ruminant (ie inability to avoid fermenting all food with a consequent loss of nutrients) was outweighed by the greater capacity to take in food and lower energy demands per unit of volume of the larger animal. A rumen may have been additionally advantageous in enabling them to conserve protein by recycling urea, and also to detoxify plant secondary compounds.

The evolutionary radiation of the ruminant artiodactyls had little effect on the equids, who had always taken a more fibrous diet than other perissodactyls, and who now continued to specialize on diets too coarse for ruminants to handle. However, the tapir radiation was badly affected. Although the tapirs managed to survive the Tertiary, their real adaptive response to the

▼ **Oligocene ungulates.** The Oligocene (38–26 million years ago) was a period of major change in the world's climate and fauna. The climate became cooler, with an ice cap forming at the South Pole and lowering the sea levels all over the world. The dense lowland forests gave way to more open woodland. In this changing environment many archaic mammals became extinct while the ancestors of modern forms made their first appearance. The predominance of odd-toed ungulates began to wane, with the even-toed forms developing in size and diversity. Nevertheless, some of the odd-toed ungulates also grew in size, the horses, the brontotheres and the giant hornless rhinoceroses of Asia, such as *Indricotherium*, being typical. Among the even-toed forms were the first ruminants, creatures such as the small oreodont *Merycoidodon*. (**1**) *Indricotherium*, a giant hornless rhinoceros. (**2**) *Mesohippus*, a three-toed horse. (**3**) *Cainotherium*, a hare-like small ruminant. (**4**) *Hyracodon*, a three-toed rhino. (**5**) *Merycoidodon*, an oreodont. (**6**) *Poebrotherium*, an early camelid. (**7**) *Brontops*, a giant titanothere.

environmental changes of the late Eocene was to give rise to the rhinos, whose larger body size, and hence lower food requirements per unit body weight, rendered them immune from direct ruminant competition, and made them better able to cope with seasonally variable vegetation.

The subtropical woodlands of the Northern Hemisphere during the Oligocene (38–26 million years ago) were good habitat for the omnivorous and larger browsing ungulates. This time represented the peak of rhino diversity in the Northern Hemisphere. Besides the ''true'' rhino lineage that survives today, there were large hippo-like amynodont rhinos, small pony-sized hyracodont rhinos, and small tapir-like true rhinos called diceratheres, none of which survived past the early Miocene. Omnivorous suoid entelodonts and anthracotheres were common across the Northern Hemisphere, and in North America hyrax-like tylopod oreodonts were the dominant ungulates.

In contrast to the Oligocene radiation of rhinos, suoids and oreodonts, the equids and ruminant artiodactyls were all of small body size and limited diversity. In North America, the equids were all three-toed browsers with low-crowned teeth, although there was a moderate diversity of small camelids that already showed the long legs and neck typical of open habitat animals. In addition, small traguloids (chevrotains) were abund-

ant, and horns were developed in the camel-related protoceratids.

The early Miocene (25 million years ago) saw a gradual climatic change in the northern latitudes, with the subtropical woodland replaced by a more open savanna type of habitat. The entelodont suoids, and the amynodont and hyracodont rhinos became extinct, the oreodonts greatly reduced in numbers and diversity, and the anthracotheres confined to the tropical regions of Africa and southern Asia. The North American continent suffered early extinction of rhinos and chalicotheres during the Miocene, probably because it was still isolated from South America, and provided no tropical refuge for animals during periods of climatic fluctuations. However, Eurasia was in contact with Africa by this time, many Eurasian ungulate families made their first appearance in Africa, and the Old World suffered less dramatic reduction in the numbers of large browsing ungulates during the Tertiary.

The Miocene climatic changes encouraged the spread of the grasslands, and with them the explosive radiation of the equids and ruminant artiodactyls. In the Old World the horned pecoran families (ancestors of giraffes, deer, musk deer, bovids), larger than the earlier hornless forms, appeared and diversified. In North America, the camelids also increased in size and

▲ **Miocene ungulates.** All the modern groups of hoofed mammals evolved during the Miocene (26–7 million years ago) and the following epoch, with even-toed ungulates now outnumbering odd-toed. (**1**) *Moropus*, a claw-bearing chalicothere. (**2**) *Synthetoceras*, a grotesque four-horned protoceratid ruminant. (**3**) *Alticamelus*, a giraffe-like camelid. (**4**) *Cranioceras*, a deer-like protoceratid ruminant. (**5**) *Hypohippus*, a woodland browsing horse. (**6**) *Teloceros*, a hippo-like amphibious rhinoceros. (**7**) *Blastomeryx*, a primitive deer-like ruminant. (**8**) *Dinohyus*, a giant pig-like entelodont. (**9**) *Dicrocerus*, a primitive cervoid or deer.

▶ **Pliocene ungulates.** The Pliocene (7–2 million years ago) saw the emergence of the first great grass plains, dominated by large herds of antelopes, mastodonts (primitive ungulates) and three-toed horses. The even-toed ungulates continued their dramatic development, especially of deer, giraffids and antelopes, included in which were a large number of primitive bovids. (**1**) *Paleotragus*, an okapi-like giraffid. (**2**) *Helladotherium*, a short-necked giraffid. (**3**) *Paleoreas*, an early antelope. (**4**) *Dicerorhinus*, a small long-legged rhinoceros. (**5**) *Hipparion*, a three-toed horse. (**6**) *Pliohippus*, the first one-toed horse. (**7**) *Alticornis*, a typical early pronghorn. (**8**) *Duiker*, a small antelope.

diversified into a number of lineages, including forms paralleling African ungulates like gazelles, giraffes and eland, and the newly immigrant deer-like artiodactyls (cervoids) diversified into forms paralleling the smaller grazing and browsing African antelopes. While bovids are the most diverse and geographically abundant of the pecoran families today, in the global woodland savanna habitat of the Miocene the cervoids predominated. Pigs and giraffes were also more diverse during the Miocene than they are today. The giraffes, which first appeared in the early Miocene of Africa, initially consisted of two distinct lineages: the long-necked, high browsing giraffes, surviving in reduced diversity today and especially—in the Pliocene of southern Eurasia—a lineage of low-level browsing giraffes, the sivatheres, with shorter necks, higher crowned teeth, and moose-like palmate horns which became extinct about a million years ago.

The first hypsodont grazing equids appeared at the close of the early Miocene in North America, and the late Miocene represented the peak of equid diversity, when there were at least six genera sharing the same habitats in North America, including a large persistently browsing form (*Hypohippus*) and a gazelle-sized three-toed grazing form (*Calippus*). It seems likely that the North American savannas were more arid than the African ones, resulting in a type of tough vegetation more suitable for sustaining equids than ruminant artiodactyls. In contrast to the Old World ruminants, none of the medium- to large-sized North American camelids or cervoids were grazers, and the Miocene radiation of the grazing equids in North America can be seen as equivalent to the more recent radiation of grazing bovids in the African savannas. The true Old World savanna fauna began to emerge in the late Miocene with the invasion of the three-toed grazing equid *Hipparion* and diversification of the large grazing bovids.

The diversity of the endemic North American ungulates declined in the Plio-Pleistocene, with the savanna habitat giving way to prairie. However, species of the larger camelids and one-toed grazing equids retained a moderate diversity and abundance, both families invading South America when these continents became connected in the early Pleistocene. About 10,000 years ago an unknown catastrophe wiped out the endemic North American ungulates, leaving the pronghorn as the only survivor of the late Tertiary radiations, and leaving us with a reduced impression of the suc-

cess of equids and camelids in diversifying.

In contrast, the Old World retained a diversity of tropical and subtropical habitats throughout the Plio-Pleistocene, and it was during this time that the bovids experienced their explosive radiation. These subtropical savanna faunas, containing not only bovids but a diversity of giraffids and the equid *Hipparion*, were widespread in Africa and Southern Eurasia during the Pliocene, but in the Pleistocene Ice Ages the geographical range and diversity of the giraffids decreased, whereas those of the more temperate habitat cervids (true deer) expanded with giant forms such as the Irish "elk" *Megaceros* and the giant moose *Cervalces*. Cervids and bovids migrated into North America in the Pleistocene, but cervids alone reached South America. Both equids and rhinos were abundant as temperate and cold-adapted forms in the Pleistocene of Eurasia and included the massive woolly rhino *Coelodonta* and the rhino *Elasmotherium* which had a long single horn and continuously growing cheek-teeth. However, the ranges of Old World perissodactyls were greatly reduced, possibly due to the influence of emerging human populations.

Ecology and Behavior

The social organization of different ungulate species is a consequence of the body size of the animal, its diet, and the structure of the habitat in which it lives. Such factors determine the relative advantages and disadvantages of group feeding, taking into account, for example, predator avoidance and food availability.

Small species are vulnerable to many more predators than are large ungulates, and often cannot avoid predation by flight or self-defense. They seek to avoid detection, and most are cryptically colored. They eat buds and berries which are scattered and scarce items of high protein content and best sought alone, preferably over familiar ground. A herd of such animals would be too scattered in its foraging to result in any advantage for predator detection or avoidance, and many such animals all feeding in one area would soon deplete all the available resources. Many small artiodactyls thus forage singly or in pairs throughout a so-called resource-defended territory. They live in tropical forest or woodland where such food is readily available, and cover is ample to hide from predators. In perissodactyls, a similar social system is seen in tapirs and the browsing rhinos (see pp496–497). For all animals of this type, contacts and conflicts between members of the same species are few, they tend to be solitary or monogamous in their reproductive behavior and show little difference in appearance or body size between the sexes.

Larger species of perissodactyls and artiodactyls can tolerate food of a lower protein content, and are better able to avoid predators. Consequently they can forage in open habitats, where their food is more abundant than is that of a small forest browser. Open-country ungulates can thus benefit from membership of a herd, where cohesion and communication between a group of animals can aid in early predator detection, and where any one individual is less likely to be the victim of an attack, and, because of the greater availability of food, they are less likely to interfere with each other's feeding behavior.

▲ **Pleistocene ungulates.** The Pleistocene, which began about 2 million years ago, was the period of the ice ages in the Northern Hemisphere. There was a tendency towards gigantism in many mammalian forms. (**1**) *Coelodonta*, the Wooly rhinoceros. (**2**) *Elasmotherium*, a six-foot-horned giant rhinoceros. (**3**) *Sivatherium*, a great antlered giraffid. (**4**) *Eucladoceros*, a large "bush-antlered" deer. (**5**) *Megaceros*, the giant Irish deer. (**6**) Aurochs. the ancestral bovid of modern European cattle. (**7**) *Capra ibex*, an early form of the modern ibex.

▼ **Prehistoric frieze.** Cave paintings at Lascaux, France, dating from about 18,000BC.

5

6

7

Nearly all large ruminant artiodactyls are found in herds, as are the grazing perissodactyls. The exceptions among the artiodactyls are specialist forest browsers, such as the okapi and the moose (see pp 532–533), which seek refuges—water, dense forest—to escape predators.

Among the larger artiodactyls, two distinct types of herds are apparent: fixed or temporary membership. In the former, closed-membership herd, the members are usually closely related, and show group defense against predators. Within this type of mixed sex herd, the males establish a dominance hierarchy to determine mating rights for the females when they come into heat. Although the females in such herds usually resemble the males in having horns, male dominance depends on size, and in such species the males are usually much larger than the females, and may continue to grow for most of their lives. Such species are mainly large-bodied open-habitat bovids, such as bison (see pp 554–555), buffalo and Musk oxen.

In artiodactyl species with open-membership herds, there are some species in which mating rites are tied to territorial possession, while in others it is tied strictly to rank. These territories are often mating areas that can support at most only one male, rather than a resource area for a harem. Thus females will enter and leave the male's territory, and the most successful males are those who hold the most attractive "property." Males must challenge other males for the rights to a territory, and rarely do they hold it for long, as its defense is exhausting. Such species tend to have the most marked difference in appearance and body size between the sexes, and the males have elaborate horns or antlers to engage in continual territorial combat. Most medium- to large-sized deer and antelope belong to this category. Some species, such as the migratory wildebeest, have this type of social organization, yet appear to lack clearly marked territories which are fixed in locality. In others, such as the kob, a "lek" system is seen, in which receptive females visit the holders of conventionalized, highly contested, close-packed mini-territories for mating.

In contrast to the ruminant artiodactyls, the typical perissodactyl social system (as exemplified by equids) is one where a single male consorts with a group of females, constituting a fixed membership harem with strong interpersonal bonds. Fights between males occur more rarely than in artiodactyls, as the male horse neither defends a fixed territory, nor does he have to continually contend with other adult males within the herd. Males without harems (usually younger males) form roving bachelor herds of more variable composition. A modified version of this social system is seen in the Wild ass, Grevy's zebra and the White rhino (see pp 496–497), where the males additionally defend a territorial area. However, no perissodactyl exhibits the pronounced sexual differences typical of ruminant artiodactyls.

Exclusive mating territories maintained by males are typical of medium-sized ruminants of tropical and subtropical woodland and savanna, but not of comparable perissodactyls in similar habitats. Probably a ruminant needs a smaller area than would a similar perissodactyl, because of its ability to survive on a smaller quantity of food per day. However, the situation is reversed in low productivity open-grassland when a perissodactyl would be able to maintain a smaller home range by dint of eating everything available, whereas the ruminant would have to forage further afield in order to locate food of suitable quality. Thus in the arid open grasslands, equids and the White rhino are territorial, whereas bison and buffalo maintain a non-territorial dominance hierarchy system.

Most ungulate females mature relatively quickly for their size, have long gestation periods, and bear one, large, well-developed juvenile at a birth. Pigs are an exception, in producing several piglets in a litter. All artiodactyl juveniles can see, hear, call out,

and stand to suckle soon after being born. All females, except pigs, make no nest but usually seek seclusion shortly before giving birth. Where cover is available many ungulates hide the young, the mother leaving it while she feeds. The calf hides immobile until her return. In contrast some of the largest, open-country, herd-forming species, for example wildebeest (artiodactyl) and horses (perissodactyl), which have evolved the ability to get to their feet and start moving remarkably soon after birth; some can run at near adult speed within half an hour of birth.

Despite relatively slow reproduction, artiodactyls are not very long-lived for their size; a 10-year-old impala, or a 15-year-old Red deer, would be old animals. Perissodactyls are longer-lived; up to 35 years in equids and 45 years in rhinos. As the most abundant large mammals, ungulates form the staple diet of most of the great terrestrial carnivores. For those which do not fall to predators, starvation is an annual threat since populations of many of them, as dominant herbivores in their communities, live close to the minimum yearly carrying capacity of the plant community. Although they have evolved a considerable capacity for storing energy in their body tissues, to be used when food is low, extra burdens of energy expenditure, such as suckling a calf or defending a territory, will weaken some individuals to the point where they are vulnerable to disease, predators or final starvation. At the last, starvation is inevitable when their abrasive diet finally reduces their grinding molars to eroded stumps which are useless for feeding.

Ungulates and Man
The genus *Homo* emerged near the peak of the bovid radiation and has shared the Pleistocene and recent fortunes of that family. Humans are partly responsible for the dwindling numbers of ungulate species from the Pleistocene. The dominant position of ungulates in most communities of large herbivores, their use of habitats, size, social organization and ecology, even their antlers and horns, have all exposed them to damaging interactions with humans. Yet those same characteristics have been the salvation of some.

The human ancestors whose remains have been found in eastern Africa lived in a community of ungulates similar to that of the same area today, but containing a number of now-extinct "giant" forms. Such large ungulates were at minimal individual risk from most predators. However,

although neither strong nor fast, humans excelled in three traits: perceptive and inventive intelligence; coordinated and mobile group hunting; and the use of artificial weapons, especially projectiles. These made humans the most generalized large predators in the community, able to hunt all prey in a wide variety of circumstances. From the glacial and interglacial epochs in Eurasia comes a wealth of evidence from societies culturally dependent upon the hunting of ungulates. Not only are there bones of the animals they killed, but the hunters also left their own record in the form of art: carvings, clay models and, most dramatic of all, vast cave-murals depicting many of the animals.

Small wonder, then, that the hunting of ungulates was the symbol of early heroism, and sport and trophy hunting are still widely accepted as justifiable ways of

▲ **A panorama of pastoralism.** Somali nomads with their docile and well-ordered cattle, sheep and camels at a water-hole in northeastern Kenya epitomize man's exploitation of hoofed mammals. These vast herds of domesticated stock have largely displaced wild hoofed mammals and profoundly affected the environment across much of North Africa, the Near East, Mediterranean Europe, and northern Asia.

▶ **Boar hunting.** A Persian bas-relief showing the Emperor Khusraun II (591–628AD) killing a Wild boar.

at other centers in its wide range in Asia at other dates. Cattle were domesticated by 6500BC from wild cattle or aurochs in Europe and the Near East, although other centers of domestication from rather different stock may have lead to the cattle of India and East Asia. These major domestications occurred at about the same time as domestication of wheat, barley and the dog, and preceded by 2,000–4,000 years the domestication of donkeys, horses, elephants and camels. In South America, domestic breeds of llamas appeared between 4000 and 2500BC, some 2,000 years after the domestication of maize. The temporarily settled conditions of a primitive, crop-growing society would have been ideal for the domestication of captured ungulates; indeed settlement rather than nomadic hunting would have made domestication necessary to ensure meat.

Horses and donkeys were the last of the common livestock animals to be domesticated, and they have been the least affected by human manipulation and artificial selection.

The first domestic horses appeared at about 4000BC, when they may have initially been used for food, but it was not until about 2000BC that the widespread use of the horse as a means of transport came about, which caused a revolution in the human mobility race and the development of modern techniques of warfare.

Some domesticated stock remained confined to the areas in which they were domesticated (yak, or dromedaries for example). But cattle, sheep and goats, with horses, donkeys and camels, supported human groups which formed distinct, nomadic pastoral cultures, depending almost entirely on their animals, not on crops. These people lived typically on milk from their stock, supplemented occasionally by meat or rarely by blood taken without killing the beast. They spread throughout the savanna, steppe and semidesert lands of Eurasia and Africa. Their cumulative effect has been to alter grossly the environment of Mediterranean Europe, the Near East, northern Asia, and much of Africa, by the combined grazing and browsing pressure of this spectrum of stock.

These changes to the landscape affected not only humans dependent on the stock, but also the wild herbivores of the area. Pastoralism everywhere has diminished wild ungulates to the point where communities remain only where physical barriers or the risk of disease have kept out humans and their stock.

enhancing a person's status and prestige.

The crucial event in the relationship between ungulates and man was domestication—some 15 species of ungulates have been domesticated. This has certainly occurred independently with different species—domestication of South American camelids is quite separate from any Old World domestication, for example. The features which occur most commonly amongst domesticated ungulates are a tendency, for the females at least, to occur in closed-membership herds, and for males to be non-territorial. These characteristics lend themselves to herding and tending by humans. In the Old World, sheep and goats were domesticated by 7500BC (perhaps much earlier), from mouflon and Wild goat respectively. The pig was domesticated from Wild boar by 7000BC in the Middle East, but was probably domesticated independently

CJ/PJJ

ODD-TOED UNGULATES

Order: Perissodactyla
Sixteen species in 6 genera and 3 families.
Distribution: Africa, Asia, S and C America.

Habitat: diverse, ranging from desert and grassland to tropical forest.

Size: head-body length from 180cm (71in) in the Mountain tapir to 370–400cm (145–160in) in the White rhino. Weight from 225kg (495lb) in the Brazilian tapir to 2,300kg (5,070lb) in the White rhino.

Asses, Horses and Zebras (family Equidae)
Seven species of the genus *Equus*: **African ass** (*E. africanus*), **Asiatic ass** (*E. hemionus*), **Domestic horse** (*E. caballus*), **Grevy's zebra** (*E. grevyi*), **Mountain zebra** (*E. zebra*), **Plains zebra** (*E. burchelli*), **Przewalski's horse** (*E. przewalskii*).

Tapirs (family Tapiridae)
Four species of the genus *Tapirus*: **Baird's tapir** (*T. bairdi*), **Brazilian tapir** (*T. terrestris*), **Malayan tapir** (*T. indicus*), **Mountain tapir** (*T. pinchaque*).

Rhinos (family Rhinocerotidae)
Five species in 4 genera: **Black rhinoceros** (*Diceros bicornis*), **Indian rhinoceros** (*Rhinoceros unicornis*), **Javan rhinoceros** (*Rhinoceros sondaicus*), **Sumatran rhinoceros** (*Dicerorhinus sumatrensis*), **White rhinoceros** (*Ceratotherium simum*).

THE odd-toed ungulates, the Perissodactyla, are a small order of mammals today, with the horses, and the closely related asses and zebras, being the only well-known and widely spread members of the group. The familiarity of the horse is, of course, largely a consequence of its domestication and use by humans, first as a means of transport, warfare and agricultural labor, and today predominantly for recreation and sport. Populations of wild equids are limited in their abundance and geographical distribution.

The other members of the order are animals that one does not usually associate with the graceful open-country horse: the ponderous and endangered rhinos of Africa and Asia, and the rare and elusive tapirs of the tropical forest of Malaysia and South America. But details of the anatomy of these animals, and overall similarities in their behavior and physiology, can be shown to unite them in a single order.

Today perissodactyls apparently run a poor second to the even-toed ungulates, artiodactyls, in terms of numbers of species, geographical distribution, variety of form and ecological diversity. But our present-day viewpoint belies the success of the odd-toed ungulates over geological time. They were the dominant ungulate order during the early Tertiary (54–25 million years ago), and their subsequent decline is more likely to have been due to climatic factors than to direct competition with artiodactyls.

Perissodactyls first appeared in the late Paleocene (58 million years ago) in North America, and by the early Eocene (54 million years ago) five out of the six known families were evident. Living perissodactyl families are the Tapiridae (tapirs), the Rhinocerotidae (rhinos) and the Equidae (horses, asses and zebras). Tapirs and rhinos are closely related, with rhinos representing an offshoot of the tapir family in the late Eocene. These two families are frequently grouped together as the suborder Ceratomorpha, distinguishing them from the equids in the suborder Hippomorpha. Rhinos are characteristically heavy bodied and show adaptations of the limbs for weight bearing (graviportal). In contrast, equids showed modifications for running from the start of their evolution, with a progressive tendency to lengthen the limbs and reduce the lateral digits.

Living tapirs are medium-sized animals, principally inhabiting tropical and subtropical woodland in Malaysia, and Central and South America, where they feed on a mixture of browse material and fruit. Tapirs have only inhabited South America since the Pleistocene (2 million years ago), and until the Pleistocene Ice Ages were also found in Europe and North America. Living rhinos are all large, although pony- and tapir-sized rhinos were common in the early Tertiary. They are found today in Africa and Southeast Asia, and occupy a variety of habitats ranging from forest to grassland in tropical and subtropical areas.

Their diet varies from browse, through a mixture of grass and browse, to grass exclusively, depending on the species. Rhinos first appeared in Africa at the start of the Miocene (25 million years ago), were found in North America until the end of the Miocene (7 million years ago), and were common throughout Eurasia until the end of the Pleistocene (10,000 years ago).

Living equids are medium sized, and they are all specialist grazers, inhabiting grassland ranging from woodland savanna to arid prairie in temperate and tropical regions. They were widely distributed in all continents except Australasia until the end of the Pleistocene, but today are found only in Africa and parts of Asia, although feral populations of horses and asses flourish in Europe, Australia, western North America

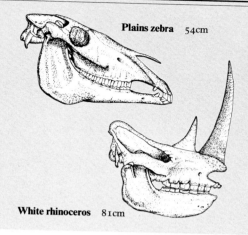

Plains zebra 54cm

White rhinoceros 81cm

Brazilian tapir 41cm

Skulls of Odd-toed Ungulates

Odd-toed ungulates may have low or high crowned cheek-teeth, but they all tend to have molarized premolars, with the molar cusps coalescing to form cutting ridges. Tapirs retain the complete mammalian dental formula, and have a fairly generalized skull shape. Rhinos are characterized by an especially deep and long occipital region (rear of the skull), associated with the large mass of neck musculature needed to hold up their massive head, and by their characteristically projecting and wrinkled nasal bones, on which the keratinous horns are attached. The incisors in rhinos are reduced to two upper and one lower, and tend to be lost entirely in the grazing species. The skulls of equids like the Plains zebra show a great elongation of the face, the posterior position of the orbits (to allow room for the exceedingly high-crowned cheek-teeth), the complete post-orbital bar, and the exceedingly deep and massive lower jaw.

▼ **Division of the spoils** among late Eocene perissodactyls in North America. The surviving perissodactyl species are rarely found in the same habitats but 40 million years ago, they, and species that are now extinct, were the dominant hoofed mammals in North America. TOP LEFT The extinct brontotheres took high-level browse, including about 50 per cent fruit. TOP RIGHT The extinct chalicotheres took high-level browse of moderately good quality, plus some fruit. MIDDLE Tapirs took low-medium-level browse, of good quality, preferably green shoots. BOTTOM RIGHT Equids took moderately high-fiber low-level browse, supplementing protein intake with buds, berries, some fruit etc. BOTTOM LEFT Rhinos took moderately high-fiber browse.

and South America. The center of equid evolution was North America, although an offshoot to the subsequently successful North American lineage was present in Eurasia in the Eocene. True equids (genus *Equus*) did not apear in the Old World until the Pleistocene (2 million years ago), but previous invasions of Eurasia were made by a three-toed browsing horse in the early Miocene, and of Eurasia and Africa by a three-toed grazing horse in the late Miocene, which survived into the Pleistocene and was found for a time alongside true equids. Equids became extinct in North America only about 10,000 years ago.

In addition to the three living families of perissodactyls, three of the original families are now extinct. Grouped with the equids in the suborder Hippomorpha are the rhino-like brontotheres, which lived in North America and Asia in the Eocene and early Oligocene, and tapir-like paleotheres, which lived during the same period in Europe. The family Chalicotheriidae occupied its own suborder Ancylopodia, as chalicotheres were sufficiently different from other per-issodactyls to warrant separate classifiction. They were largish animals, ranging from the size of a horse to the size of a giraffe, and had secondarily substituted claws for hooves on their feet. Chalicotheres were initially found in both North America and Eurasia; they first appeared in Africa along with rhinos in the Miocene, and became extinct in North America in the middle Miocene, but persisted into the Pleistocene in Asia and Africa. However, speculation abounds about their possible survival into more recent times. Chalicothere-like animals, with a horse-like head and bear-like feet, appear on plaques in Siberian tombs dating from the 5th century BC, and it has been suggested that the mysterious Kenyan "Nandi bear" could be a surviving chalicothere.

CJ

HORSES, ASSES AND ZEBRAS

Family: Equidae
Seven species in 1 genus.
Order: Perissodactyla.
Distribution: E Africa, Near East to Mongolia.

Habitat: from lush grasslands and savanna to sandy and stony deserts.

Size: ranges from head-body length 200cm (79in), tail length 42cm (16.5in) and weight 275kg (605lb) in the African ass to head-body length 275cm (108in), tail length 49cm (19in), and weight 405kg (885lb) in the Grevy's zebra.
Gestation: about 11½ months, but 12½ in Grevy's zebra.
Longevity: about 10–25 years (to 35 in captivity).

▶ **Fighting Plains zebras** at Amboseli National Park, Kenya. Such contests between males are common when females are ready to mate.

▼ **Plains zebras drinking** at a water hole at Umfolozi National Park, South Africa.

EVER since horses were first domesticated in Asia, five thousand years ago, they have served mankind as a beast of burden and as a means of transport; they have helped him till the soil and wage war; they have provided him with recreation and even companionship; and with mane waving and hooves thundering, they have served as a symbol of power, grace and freedom. But the domesticated horse is just one member of the once diverse family Equidae. Only six other species survive, some precariously.

All equids are medium-sized herbivores with long heads and necks and slender legs that bear the body's weight only on the middle digit of each hoofed foot. They possess both upper and lower incisors that clip vegetation and a battery of high-crowned, ridged, cheek-teeth that are used for grinding. The ears are moderately long and erect, but can be moved to localize sounds and to send visual signals. A mane covers the neck, but only in the Domestic horse does it fall to the side. On the other species it stands erect. All equids have long tails, which are covered with long flowing hair in horses, and short hair only at the tip of asses and zebras.

The species differ somewhat in size, and males are generally 10 percent larger than females. But the most striking feature that distinguishes species is coat color, the zebra's stripes being the most dramatic example (see pp486–487). The coats of horses and asses are more uniform in color: dun in Przewalski's horse, from tan to gray in asses.

Equids' eyes are set far back in the skull, giving a wide field of view. Their only blind spot lies directly behind the head, and they even have binocular vision in front. They probably can see color and although their daylight vision is most acute their night vision ranks with that of dogs and owls. They can detect subtle differences in food quality. Males use the flehmen or lip-curl response to assess the sexual state of females, and the vomeronasal or Jacobson's organ which is used in this is well developed. Equids can also detect sounds at great distance and by rotating their ears can locate the source without changing body orientation.

Moods are often indicated visually by changes in ear, mouth and tail positions. Smell assists individuals in keeping track of the movements of neighbors, since urine and feces bear social odors, but social contact is effected primarily by sounds. In horses and Plains zebra, mothers whinney when separated from their foals and nicker to warn them of danger. Males often nicker to declare their interest in a female and they squeal to warn competitors that further escalated combat is imminent. In asses and Grevy's zebra, males often bray when fighting, or calling to each other over long distances.

The Plains zebra occupies the lushest environments—the grasslands and drier savannas of East Africa from Kenya to the Cape. As its name implies, the Mountain zebra is restricted to two mountainous regions of southwest Africa where vegetation is abundant. The remaining species live in more arid environments with sparsely distributed vegetation. Przewalski's horse inhabits the semi-arid deserts of Mongolia, while the Asian wild ass inhabits the most arid deserts of central Asia and the Near East. The African wild ass, the least horse-like of the equids, roams the rocky deserts of North Africa. Generally, the ranges of the species do not overlap, the only exception being Grevy's and Plains zebra, which coexist in the semi-dry thorny scrubland of northern Kenya. Only the Domestic horse is found worldwide, and it has spawned feral populations in North America on the western plains and on east coast barrier islands, and in the mountains of western Australia.

The earliest of the horse-like ancestors, *Hyracotherium*, appeared in the Eocene, about 54 million years ago; it was a small dog-sized mammal which browsed on low shrubs of the forest floor, and had low-crowned teeth without the complex enamel ridges of modern equids. It had already lost two hind toes on its hindfeet and one on its forefeet, but the feet were still covered with soft pads. When grasses appeared in the Miocene, equids began to radiate. Continuously growing teeth with high crowns, complex grinding ridges and cement-filled interstices evolved with the opening out of the habitat, and the need to run from predators and to travel long distances in search of food and water led to many changes in body shape. Overall body size increased, which reduced relative nutritional demands. By the early Pleistocene (2 million years ago), the one-toed equids had spawned the genus *Equus*, which rapidly spread all over the world.

As environments changed, populations became isolated, giving rise to most of the living species. The first to split off from the equid stem after becoming single-toed was the Grevy's zebra, which, despite its stripes, is only distantly related to the other two zebra species. The only species that probably did not originate via geographic isolation is

the ancestor of the Domestic horse. It is thought to be directly descended from some mutant Przewalski's horses.

All equids are perissodactyls that forage primarily on fibrous foods. Although horses and zebras feed primarily on grasses and sedges, they will consume bark, leaves, buds, fruits and roots, which are common fare for the asses. Equids employ a hindgut fermentation system, in which plant cell walls are only incompletely digested but processing is rapid. As long as they are able to ingest large quantities of food, they can achieve extraction rates equal to those of ruminants. Because forage quality does not affect the process, equids can sustain themselves in more marginal habitats and on diets of lower quality than can ruminants. Equids do spend most of the day and night foraging. Even when vegetation is growing rapidly, equids forage for about 60 percent of the day (80 percent when conditions worsen). Although equids can survive on low quality diets, they prefer high quality, low fiber food.

Equids are highly social mammals that exhibit two basic patterns of social organization. In one, typified by the two horse species, as well as the Plains and Mountain zebra, adults live in groups of permanent membership, consisting of a male, a few females remain in the same harem throughout their adult lives. Each harem has a home range, which overlaps with those of neighbors. Home range size varies, depending on

---

Abbreviations: HBL = head-body length. TL = tail length. wt = weight. Approximate nonmetric equivalents: 2.5cm = 1in; 1kg = 2.2lb.
E Endangered. * CITES listed.
V Vulnerable.

Subgenus *Equus*
N and S America, Mongolia, Australia. General body form variable, in part due to domestication. Long tails with hairs reaching to the middle of leg. Usually solid coat colors.

Przewalski's horse E
Equus przewalskii
Przewalski's, Asiatic or Wild horse.

Mongolia near Altai mountains. Open plains and semidesert. HBL 210cm; TL 90cm; wt 350kg. Coat: dun on sides and back, becoming yellowish white on belly; dark brown erect mane with legs somewhat grayish on inside; thick-headed, short-legged and stocky; regarded as true wild horse.

Domestic horse
Equus caballus
Domestic or Feral horse.

N and S America and Australia. Open and mountainous temperate grasslands, occasionally semideserts. HBL 200cm; TL 90cm; wt 350–700kg. Coat: sandy to darkish brown; mane falls to side of neck. Dozens of varieties. Feral forms thick-headed and stocky, domestic breeds slender-headed and graceful-limbed.

Subgenus *Asinus*
N Africa, Near East, W and C Asia. Sparsely covered highland and lowland deserts. Horse-like, or even stockier, with long pointed ears, tufted tails, uneven mane and small feet.

African ass E
Equus africanus

Sudan, Ethiopia and Somalia. Rocky desert. HBL 200cm; TL 42cm; wt 275kg. Coat: grayish with white belly and dark stripe along back. Nubian subspecies with shoulder cross; Somali subspecies with leg bands. Smallest equid with narrowest feet. Subspecies: 3

Asiatic ass V *
Equus hemionus

Syria, Iran, N India, and Tibet. Highland and lowland deserts. HBL 210cm; TL 49cm; wt 290kg. Coat: summer, reddish brown, becoming lighter brown in winter; belly is white and has prominent dorsal stripe. Most horse-like of asses with broad round hoofs and larger than African species. Subspecies: 4.

Subgenus *Hippotigris*
E and S Africa. Resemble striped horses.

Plains zebra
Equus burchelli
Plains or Common zebra.

E Africa. Grasslands and savanna. HBL 230cm; TL 52cm; wt 235kg. Coat: sleek with broad vertical black and white stripes on body, becoming horizontal on haunches. Always fat looking, but short-legged and dumpy. Subspecies: 3.

Mountain zebra V
Equus zebra

SW Africa. Mountain grasslands. HBL 215cm; TL 50cm; wt 260kg. Coat: sleeker than Plains zebra with narrower stripes and a white belly. Thinner and sleeker than Plains zebra with narrower hoofs and dewlap under neck. Subspecies: 2.

Subgenus *Dolichohippus.*

Grevy's zebra E *
Equus grevyi
Grevy's or Imperial zebra.

Ethiopia, Somalia and N Kenya. Subdesert steppe and arid bushed grassland. HBL 275cm; TL 49cm; wt 405kg. Coat: narrow vertical black and white stripes on body, curving upwards on haunches; belly is white and mane prominent and erect. Mule-like in appearance with long narrow head and prominent broad ears.

the quality of the habitat, and varies in Plains zebra from 80–250sq km (31–97sq mi) in the Ngorongoro crater to over 350sq km (135sq mi) during the rainy season in the Serengeti. Since the habitat deteriorates during the dry season, the Serengeti zebra harems congregate and migrate *en masse* about 100km (62mi) to different habitats, where home ranges are often as large as 600sq km (230sq mi).

The second social system, typified by the asses and Grevy's zebra, involves more ephemeral adult associations, rarely lasting longer than a few months. Temporary aggregations of one or both sexes are common, but most adult males live alone within large territories. For Grevy's zebra these vary in size from 2–10 sq km (0.8–3.9sq mi), but for the asses they can be as large as 15sq km (5.8sq mi). Within territorial boundaries, which are marked with large piles of dung, owners obtain exclusive mating access to receptive females that wander through them. In both systems, surplus males live together in bachelor groups.

Social systems involving temporary groupings and solitary territorial males occur in drier habitats and in areas where resources are distributed in a more patchy fashion. Small, widely scattered patches of low quality vegetation preclude the formation of long-lasting associations by intensifying competition among females. And without female groups to defend, males must defend large areas containing the resources females require if they are to obtain a disproportionate share of the matings. Only larger, more evenly distributed resources allow females to feed in permanent groups, thus enabling harems to form.

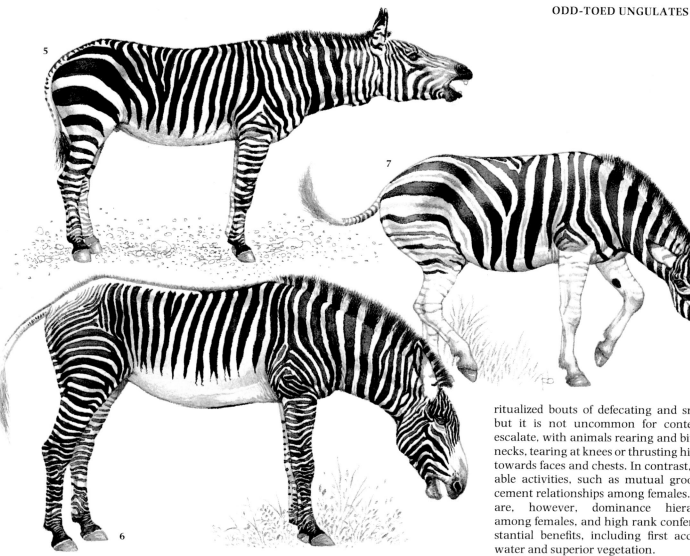

▲ Representative species of horses, asses and zebras. (1) Przewalski's wild horse (*Equus przewalskii*), the ancestor of all domestic horses, showing the stallion's bite threat. (2) A female African ass (*Equus africanus*), showing the kick-threat, with its ears held back. (3) A male onager, a subspecies of the Asiatic ass (*Equus hemionus*), adding to a dung pile, as a territorial mark. (4) The kiang, the largest subspecies of Asiatic ass, showing the flehmen reaction after smelling a female's urine. (5) A young male Mountain zebra (*Equus zebra*) showing a submissive face to an adult male. Note the dewlap and the grid-iron rump pattern. (6) A female Grevy's zebra (*Equus grevyi*) in heat and showing the receptive stance, with hindlegs slightly splayed and tail raised to one side. (7) A male Plains zebra (*Equus burchelli*) driving mares in a characteristic low-head posture, with ears held back.

Even for equids that maintain long-lasting associations, their groups are different from those of most mammals. Typically, daughters remain with their mothers, creating groups composed of close kin, but in equids groups are composed of non-relatives, since both sexes leave their natal area. Females emigrate when they become sexually mature at about two years old, and neighboring harem or bachelor males attempt to steal them. Males disperse by the fourth year to form bachelor associations. Only after a number of years in such associations are males able to defend territories, steal young females, or displace established harem males.

Mares usually bear only one foal. Only in the Grevy's zebra does gestation last slightly more than one year; since females come on heat 7–10 days after bearing a foal, mating and birth occur during the same season, coinciding with renewed growth of the vegetation. Reproductive competition among males for receptive females is keen. This begins with pushing contests, or ritualized bouts of defecating and sniffing, but it is not uncommon for contests to escalate, with animals rearing and biting at necks, tearing at knees or thrusting hindlegs towards faces and chests. In contrast, amicable activities, such as mutual grooming, cement relationships among females. There are, however, dominance hierarchies among females, and high rank confers substantial benefits, including first access to water and superior vegetation.

Young are up and about within an hour of birth. Within a few weeks they begin grazing, but are generally not weaned for 8–13 months. Females can breed annually, but most miss a year because of the strains of rearing foals.

Despite the proliferation of domestic horses, their wild relatives are in a precarious situation. No Przewalski's horses have been seen in their natural habitat since 1968, and only about 200 individuals, scattered among the world's zoos, separate the species from extinction. Many feral populations of Domestic horses which roam freely are treated as vermin. As for zebras, only the Plains zebra is plentiful and occupies much of its former range. Populations of Mountain zebra are small and are protected in national parks, but those of Grevy's zebra have been drastically reduced, since their beautiful coats fetch high prices. It would be tragic if these wonderful creatures were to go the way of the South African quagga, a yellowish-brown zebra with stripes only on the head, neck and forebody, which was exterminated in the 1880s.   DIR

The Zebra's Stripes

An aid to group cohesion?

The term zebra has no taxonomic meaning. It describes three equines that have stripes. On fossil and anatomical evidence, the Grevy's and Mountain zebras are as distantly related to each other as they are to horses and asses. However, the ancestor of all equines was probably striped.

What do zebras have in common apart from their stripes? They live in Africa and they are exceptionally social: groupings range from two to several hundreds; they disperse or aggregate in response to the vagaries of pasture, water and climate. Like all horses, they are nomadic grazers of coarse grasses. They are active, noisy and alert. They never attempt to conceal themselves or to "freeze" in response to predators and they prefer to rest, grouped, in exposed localities where they have the advantage of a good view at the cost of being conspicuous themselves. The widespread theory that their stripes are camouflage is therefore contradicted by the zebra's behavior.

Opinion is divided as to whether the stripes are a visual or chemical device. Proponents of the latter think that stripes may help the animals to regulate their body temperature, yet zebras live under many different climates, and the stripe patterns show no variation with climate. A suggestion that stripes may have evolved to deter harmful flies is based on the observation that signals of both chemical and optical origin influence flies (yet insects are no hazard over most of the zebras' very wide range of habitats).

The "visual" theorists try to imagine the effect of stripes on pests and predators. One ingenuous explanation has it that charging lions are unable to single out an individual because it merges with others in the herd; another suggests that the lion is dazzled or miscalculates its imaginary last leap. These theories founder on the observable confidence with which lions kill zebras and on the fact that in those places for which there are records zebras are killed broadly in proportion to their relative abundance.

An alternative approach has been to consider the intrinsic optical properties of stripes in relation to the physiology of the optical centers of the brain, rather than look for resemblances to anything else. Explanations of the stripes' function and the selective pressures that maintain them may be better sought in the behavior of those animals that are most exposed to seeing the stripes—not a passing predator, but the zebras themselves.

Research on vertebrate vision suggests that several kinds of primary nerve cells in the visual system are excited by crisp black and white stripes, notably the detectors of tonal contrasts, spatial frequencies, linear orientation, "edges" and the "flicker" effect of moving edges. So zebras within a herd cannot escape the visual stimulation of stripes, and there is evidence that they actively seek it. Zebras walk towards, stop and stand near one another with great consistency and show few signs of discrimination between individuals or sexes. They presumably register the totality that signals a particular fellow zebra, but their responses to artificially striped panels suggest that their attraction to stripes is more or less automatic.

The mechanism may originate in a transfer from relationships in infancy based mainly on touch (tactile) to looser, visually based associations in nibbling and grooming one another, but this is a one-to-one affair and tends to be restricted to mother-young sibling relationships.

As a foal grows older, its contacts become more numerous and casual. It is at this stage that a clear separation takes place between the tactile and visual components of the young zebra's behavior; it begins to make empty grooming gestures. Prompted by the approach of other zebras, this ritualized grimace is an exaggerated greeting that helps neutralize aggression. At a lower intensity, passive head-nodding and nib-

▲ **A black zebra?** The markings of zebras are unique to each individual and some startling patterns occur, as in this Chapman's zebra, a subspecies of the Plains zebra, with the upper parts almost solid black.

► **Reversed stripes.** The normal black-on-white pattern is reversed in this Plains zebra.

▼ **Visual dazzle.** A large herd of zebras presents a visually disturbing appearance to human eyes, rather like the effects deliberately created in op-art paintings, but these patterns seem to be restful to zebras.

bling at nothing are a common response to being surrounded by stripes.

It may be the association of visual stimulation with the security of mother and family that makes any other zebra attractive. This has an obvious utility whenever there are dense aggregations on temporary pastures, resting grounds or at water holes.

The link with grooming is important for explaining how stripes may have evolved and why they are found in equines but are generally absent in the even-toed ungulates, where grooming is not the main social adhesive. Many animals have visual "markers" to direct companions to particular parts of the body. In horses and zebras the preferred area for grooming is the mane and withers. Extreme bending at the base of the neck causes skin wrinkles, so it is possible that the evolutionary origins of stripes lie in enhancement of this natural characteristic. Once the optical mechanism was established on this small target area, its effectiveness would have been enhanced by spreading over the entire animal. Significantly, the three contemporary species are most alike in the flat panel area of neck and shoulder. All horses are follow-my-leader travelers and have rump patterns which are characteristic for each species.

For three very different equines to maintain crisp, evenly spaced black and white stripes there must be very strong selective pressure in favor of stripes. If animals with defective patterns consistently fail to achieve positive social relationships, their overall fitness is likely to suffer, and this could be the explanation for abnormal patterns being so rare (for example only seven black, white, dappled or marbled zebras out of one sample of several thousand animals, skins and photographs). When a black zebra was seen, in the Rukwa Valley, Tanzania, it tended towards a peripheral position whenever its group was buzzed by an aircraft or approached by a vehicle. It is possible that "visual grooming" masks the latent animosities of zebras, and a resurgence of intolerance may be a part of the selective forces that eliminate unstriped zebras.

But what of the zebras that have lost their stripes? Crisp black and white stripes can only be maintained in sleek tropical coats. Only high densities in relatively rich habitats would warrant such a conspicuous socializing mechanism. Stripes become redundant in low density desert-dwelling asses and they become inoperable in very cold climates with annual molts into shaggy winter coats; hence the breakdown of striping in horses and the Cape quagga.   JK

TAPIRS

Family: Tapiridae
Four species of the genus *Tapirus*.
Order: Perissodactyla.
Distribution: S and C America; SE Asia.

Size: head-body length
180–250cm (71–98in); tail
5–10cm (2–4in); shoulder
height 75–120cm (29–47in);
weight 225–300kg
(500–66olb).

Brazilian tapir ⊡

Tapirus terrestris

Brazilian or South American tapir.
Distribution: S America from Colombia and
Venezuela south to Paraguay and Brazil.
Habitat: wooded or grassy with permanent
water supply.
Coat: dark brown to reddish above, paler
below, short and bristly; low narrow mane.
Gestation: 390–400 days.
Longevity: 30 years.

Mountain tapir ⋁

Tapirus pinchaque

Mountain Woolly, or Andean tapir.
Distribution: Andes mountains in Colombia,
Equador, Peru and possibly W Venezuela up to
altitudes of 4,500m (14,750ft).
Habitat: mountain forests to above treeline.
Coat: reddish brown and thicker than other
species; white chin and ear fringes.
Gestation and longevity: as for Brazilian tapir.

Baird's tapir ⋁

Tapirus bairdi

Distribution: Mexico through C America to
Colombia and Equador west of Andes.
Habitat: swampy or hill forests.
Gestation and longevity: as for Brazilian tapir.
Coat: reddish brown and sparse; short thick
mane, white ear fringes.

Malayan tapir ⋿

Tapirus indicus

Malayan or Asian tapir.
Distribution: Burma and Thailand south to
Malaya and Sumatra.
Habitat: dense primary rain forests.
Size: head-body length 220–250cm (88–98in);
tail 5–10cm (2–4in); shoulder height
90–105cm (34–42in); weight 250–300kg
(550–66olb). Coat: middle part of body white,
fore and hind parts black.
Gestation: 390–395 days.
Longevity: 30 years.

⋁ Vulnerable. ⋿ Endangered. ⊡ CITES listed.

TAPIRS are among the most primitive large mammals in the world. There were members of the modern genus *Tapirus* roaming the Northern Hemisphere 20 million years ago and their descendants have changed little. Their scattered relict distribution is often cited as evidence for the existence of the supercontinent of Gondwanaland, on the assumption that tapirs reached their present homes overland before the continents drifted apart.

Their curious appearance has led people to liken tapirs to pigs and elephants, but in fact their closest relatives are the horses and the rhinoceros. All tapirs have a stout body, slightly higher at the rump than the shoulder, and the limbs are short and sturdy. This compact streamlined shape is ideal for pushing through the dense undergrowth of the forest floor, and similar body forms have evolved independently in two other quite unrelated South American species, the peccary (see pp 504–505) and capybara, which share the same habitat.

The neck of tapirs is short and the head extends into a short, fleshy trunk derived from the nose and upper lip, with the nostrils at the tip. This small proboscis helps them to sniff their way through the jungle and is a sensitive "finger" used to pull leaves and shoots within reach of the mouth. The ears protrude and are often tipped with white, and their hearing is good, although not as acute as their sense of smell. Vision is less important for these nocturnal animals and the eyes are small and lie deep in the socket, well protected from thorns. Baird's tapir and the Brazilian tapir both have short, bristly manes extending along the back of the neck, protecting the most vulnerable part of the body from the deadly bites of the main predator, the jaguar. The dramatic coloration of the Malayan tapir—middle part of the body white, fore and hind parts black—is also a protection against predators: the patches break up and obscure the body outline so that nocturnal predators fail to detect their prey. Tapir skin is tough and covered with sparse hairs; only the Mountain tapir has a thick coat, which protects it from the cold.

Tapir tracks, often the only evidence one sees of the animal's presence, are characteristically three-toed, although the forefeet have four toes and the hindfeet three. The fourth toe, which is slightly smaller than the others, and placed to one side, higher up on the foot, is functional only on soft ground. All toes have hooves and there is also a callous pad on the foot which supports some of the weight.

Tapirs are forest dwellers, active mainly at night when they roam into forest clearings and along river banks to feed. They are both browsers and grazers, feeding on grasses, aquatic vegetation, leaves, buds, soft twigs and fruits of low-growing shrubs, but they prefer to browse on green shoots. Animals follow a zig-zag course while feeding, moving continuously and taking only a few leaves from any one plant. In Mexico and South America, tapirs sometimes cause damage to young maize and other grain crops and in Malaya they are reputed to raid young rubber plantations.

Apart from mothers with young, tapirs are usually solitary. They range over wide areas. They are excellent swimmers and spend much time in water, feeding, cooling off, or ridding themselves of skin parasites. If alarmed, they may seek refuge in water and can stay submerged for several minutes; the Malayan tapir is reputed to walk on the river bottom like a hippopotamus. When the animals bathe, there is increased activity in the digestive tract and, like the hippo, they

4

▲ The species of tapirs. (1) Mountain tapir (*Tapirus pinchaque*). (2) Brazilian tapir (*Tapirus terrestris*). (3) Malayan tapir (*Tapirus indicus*). (4) Bairds's tapir (*Tapirus bairdii*) with young.

usually defecate in the water at the river's edge.

Tapirs are also good climbers, scrambling up river banks and steep mountain sides with great agility. They often follow the same routes and may wear paths to standing water. They mark their territories and daily routes with urine, as do rhinos. Tapirs walk with their nose close to the ground, probably to recognize their whereabouts and to detect the scent of other tapirs and predators. If threatened, they crash off into the bush or defend themselves by biting. Their main predators are big cats—the jaguar in the New World and the leopard and tiger in Asia. Bears sometimes prey on Mountain tapirs and caymans will attack young animals in the water.

Breeding occurs throughout the year, and females appear to be sexually receptive every two months or so. Mating is preceded by a noisy courtship, with the participants giving high-pitched squeals. Male and female stand head to tail, sniffing their partner's sexual parts and moving round in a circle at increasing speed. They nip each other's feet, ears and flanks (as do horses and zebras) and prod their mate's belly with their trunks.

Just before they give birth, pregnant mothers seek a secure lair in which to bear the single young (twins are rare). A female in her prime can produce an infant every 18 months. Whatever the species, all newborn tapirs have reddish-brown coats dappled with white spots and stripes, excellent camouflage in the mottled light and shade of the jungle undergrowth. At two months old, the pattern begins to fade, and by six months the youngster has assumed the adult coat. Youngsters stay with the mother till they are well grown, but at 6–8 months they may begin to travel independently; they are not sexually mature for another two to three years.

Although tapirs have survived for millions of years, their future is by no means secure. They are hunted extensively for food, sport and for their thick skins, which provide good quality leather, much prized for whips and bridles. By far the greatest threat is habitat destruction caused by logging, clearing land for agriculture or manmade developments such as the recent flooding of huge areas of forest for a new dam on the Paraguay/Brazil border. The tapirs' best hope for survival is in forest reserves like Taman Negara in Malaya, but even here there are pressures: part of Taman Negara is threatened with flooding as part of a hydroelectric project.   KMacK

RHINOCEROSES

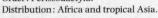

Family: Rhinocerotidae ·
Five species in 4 genera.
Order: Perissodactyla.
Distribution: Africa and tropical Asia.

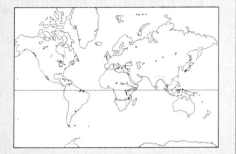

Size: head-body length from 250–315cm
(98–124in) in the Sumatran rhino to
370–400cm (146–158in) in the White rhino.
Weight from 800kg (1,765lb) in the Sumatran
rhino to 2,300kg in the White rhino.

· CITES listed.

To many people, rhinoceroses with their massive size, bare skin and grotesque appearance are reminiscent of the reptilian dinosaurs which were the dominant large animals of the world between 250 million and 100 million years ago. Though rhinos are certainly not reptiles, it is true that they are relicts from the past. Rhinos of various forms were far more abundant and diverse during the Tertiary era (40–2 million years ago), but in Europe the Woolly rhinoceros survived until the last Ice Age (about 15,000 years ago).

While the extinct rhinos varied in their possession of horns and in the arrangement of these horns, they were generally large. Together with the elephants and hippopotamuses, they represent a life-form which was much more abundant and diverse in the past: that of the giant plant-feeding animals, or "megaherbivores". Of the surviving species, two are on the brink of extinction, while the other three are becoming increasingly threatened.

The name "rhinoceros" derives from the distinctive horns on the snout. Unlike those of cattle, sheep and antelopes, rhino horns have no bony core: they consist merely of an aggregation of keratin fibers perched on a roughened area on the skull. Both African species and the Sumatran rhinoceros have two horns in tandem, with the front one generally the largest, while the Indian and Javan rhinos have only a single horn on the end of the nose. Rhinos have short stout limbs to support their massive weight. The three toes on each foot give their tracks a characteristic "ace-of-clubs" appearance. The Indian rhino has an armor-plated look, produced by the prominent folds on its skin and its lumpy surface. The White rhino has a prominent hump on the back of its neck, containing the ligament supporting the weight of its massive head. In both the White rhino and the Indian rhino, adult males are notably larger than females, while in the other rhino species both sexes are of similar size. The Black rhino has a prehensile upper lip for grasping the branch ends of woody plants, while the White rhino has a lengthened skull and broad lips for grazing the short grasses that it favors. In color, the two species are not notably different, and the popular names most probably arose from the local soil color tinting the first specimens seen.

Rhinos have poor vision, and are unable to detect a motionless person at a distance of more than 30m (100ft). The eyes are placed on either side of the head, so that to see straight in front the animals peer first with one eye, then with the other. Their hearing is good, their tubular ears swiveling to pick up the quietest sounds. However, it is their sense of smell upon which they mostly rely for knowledge of their surroundings: the volume of the olfactory passages in the snout exceeds that of the brain! When undisturbed, rhinos can sometimes be noisy animals: a variety of snorts, puffing sounds, roars, squeals, shrieks and honks have been described for various species.

A further peculiarity of rhinos is that, as in elephants, the testes do not descend into a scrotum. The penis, when retracted, points backwards so that the urine is directed to the rear by both sexes. Females possess two teats located between the hindlegs.

The five surviving species of rhinos fall into three distinct subfamilies which are only distantly related to one another. The Sumatran rhino is the only surviving member of the Dicerorhinae, which also included the extinct Woolly rhino and other Eurasian species. The Sumatran rhino itself is little different from forms which existed 40 million years ago. The Asian one-horned rhinos (Rhinocerinae) have an evolutionary

▲ **Rhino in repose.** The massive horns of this placid Black rhino in Amboseli National Park, Kenya, pose no threat to the oxpecker perched on its upper jaw. Oxpeckers remove parasites from the rhinos' skin and also take insects stirred up by their dust baths.

◄ **Charging Black rhino.** Black rhinos are more aggressive than White rhinos.

► **An Indian rhino** in characteristic habitat: tall swampy grassland.

history extending back to Oligocene deposits in India; the Javan rhino is more primitive than the Indian rhino, having changed little over the past 10 million years. The African two-horned rhinos evolved independently in Africa. The White rhino is an offshoot of the same stock as the Black rhino, having diverged during the course of the Pliocene (about 3 million years ago).

All rhinoceroses are herbivores dependent on plant foliage, and they need a large daily intake of food to support their great bulk. Because of their large size and hindgut fermentation, they can tolerate relatively high contents of fiber in their diet, but they prefer more nutritious leafy material when available. Both African species of rhino have lost their front teeth entirely; although Asian species retain incisor teeth and the Sumatran rhino canines too, these are modified for fighting rather than for food gathering. The broad lips of White rhinos give them a large area of bite, enabling them to obtain an adequate rate of intake from the short grass areas that they favor for much of the year. Black rhinos use their prehensile upper lips to increase the amount of food they gather per bite from woody plants. Indian rhinos use a prehensile upper lip to gather tall grasses and shrubs, but can fold the tip away when feeding on short grasses; woody browse comprises about 20 percent of their diet during the winter period. Both the Javan and Sumatran rhinos are entirely browsers, often breaking down saplings to feed on leaves and shoot ends. They also include certain fruits in their diet, as also do African Black rhinos and, to a lesser extent, Indian rhinos.

All rhinos are basically dependent upon water, drinking almost daily at small pools or rivers when these are readily available. But under arid conditions, both African species can survive for periods of 4–5 days between waterhole visits. Rhinos are also dependent on waterholes for wallowing. Indian rhinos in particular spend long periods lying in water, while the African species more commonly roll over to acquire a mud coat. While the water may provide some cooling, the mud coating probably serves mainly to give protection against biting flies (despite the thick hide, blood vessels lie just under the thin outer layer).

For large, long-lived mammals like rhinos, life-history processes tend to be protracted. Female White rhinos and Indian rhinos undergo their first sexual cycles at about 5 years of age and bear their first calves at 6–8 years. In the smaller Black rhino, females breed about a year younger

than these ages. A single birth is the rule. Intervals between successive offspring can be as short as 22 months, but more usually vary between 2 and 4 years in natural populations of these three species. The babies are relatively small at birth, weighing only about 4 percent of the mother's weight—about 65kg (143lb) in the case of the White rhino and Indian rhino and 40kg (88lb) in the Black rhino. Females seek seclusion from other rhinos around the time of the birth. White rhino calves can follow the mother about three days after birth. Indian rhino mothers sometimes move away for up to 800m (2,600ft), leaving calves lying alone. Calves of both Indian and White rhinos tend to run in front of the mother, while those of Black rhinos usually run behind. White rhino mothers stand protectively over their offspring should danger threaten.

Males first become sexually potent at about 7–8 years of age in the wild; but they are prevented from breeding by social factors until they can claim their first territories or dominant status at an age of about 10 years.

Births may take place in any month of the year. In the African rhinos, conceptions tend to peak during the rains so that a birth peak occurs from the end of the rainy season through to the middle of the dry season.

Rhino are basically solitary, except for the association between a mother and her most recent offspring, which usually ends shortly before the birth of the next offspring. In White rhinos, and to a lesser extent in Indian rhinos, immature animals pair up or occasionally form larger groups. The White rhino is the most sociable of the five species, and females lacking calves sometimes join up, while such females also accept the company of one or more immature animals. In this way, persistent groups numbering up to seven individuals may be formed. Larger temporary aggregations may be found around resting areas or favored feeding areas. Adult males of all species remain solitary, apart from temporary associations with females in heat (see pp496–497).

In White rhinos and Indian rhinos, females move over home ranges covering 9–15sq km (3.5–5.8sq mi), with temporary extensions when food and water supplies run out. The home ranges of Black rhino females vary from about 3sq km (1.2sq mi) in forest patches to nearly 90sq km (35sq mi) in arid regions. Female home ranges in all species overlap extensively and there is no indication of territoriality among females. White rhino females commonly

1

► **Species of rhinoceros.** (1) Indian rhinoceros (*Rhinoceros unicornis*). (2) Sumatran rhinoceros (*Dicerorhinus sumatrensis*). (3) White rhinoceros (*Ceratotherium simum*). (4) Javan rhinoceros (*Rhinoceros sondaiacus*). (5) Black rhinoceros (*Diceros bicornis*).

3

Abbreviations: HBL = head-body length. HT = height. TL = tail length. AH = anterior horn. PH = posterior horn. wt = weight.
Approximate nonmetric equivalents: 2.5cm = 1in; 1kg = 2.2lb. E Endangered. V Vulnerable.

Black rhinoceros V

Diceros bicornis
Black or Hooked-lipped rhinoceros,

Africa from the Cape to Somalia. From montane rain forest to arid scrublands; browser; more nocturnal than diurnal. HBL 286–305cm; HT 143–160cm; TL 60cm; AH 42–135cm; PH 20–50cm; wt 950–1,300kg. Coat: gray to brownish gray (varying with soil color); hairless. Gestation: 15 months. Longevity: 40 years.

White rhinoceros

Ceratotherium simum
White or Square-lipped rhinoceros.

S and NE Africa. Drier savannas; grazer; both diurnal and nocturnal. Male HBL 370–400cm;

HT 170–186cm; TL 70cm; AH 40–120cm; PH 16–40cm; wt up to 2,300kg. Female HBL 340–365cm; HT 160–177cm; AH 50–166cm; PH 16–40cm; wt up to 1,700kg. Coat: neutral gray, varying with soil color; almost hairless. Gestation: 16 months. Longevity: 45 years.

Indian rhinoceros E

Rhinoceros unicornis
Indian or Greater one-horned rhinoceros.

Floodplain grasslands; mainly a grazer; diurnal and nocturnal. Male HBL 368–380cm; HT 170–186cm; TL 70–80cm; horn 45cm; wt 2,200kg. Female HBL 310–340cm; HT 148–173cm; TL and horn as for males; wt 1,600kg. Coat: gray; hairless. Gestation: 16 months. Longevity: 45 years.

Javan rhinoceros E

Rhinoceros sondaicus
Javan or Lesser one-horned rhinoceros.

Southeast Asia. Lowland rain forests; browser; diurnal and nocturnal. HT up to 170cm; wt up to 1,400kg. Coat: gray, hairless.

Sumatran rhinoceros E

Dicerorhinus sumatrensis
Sumatran or Asian two-hroned rhinoceros.

Southeast Asia. Montane rain forests; browser; diurnal and nocturnal. HBL 250–315cm; HT up to 138cm; AH up to 38cm; wt up to 800kg. Coat: gray, sparsely covered with long hair. Gestation: 7–8 months. Longevity: 32 years.

engage in friendly nose-to-nose meetings, but Indian rhino females generally respond aggressively to any close approach. However, subadults of both species approach adult females, calves and other immature animals for nose-to-nose meetings and sometimes playful wrestling matches.

Males of all species sometimes fight viciously, inflicting gaping wounds. Both African species fight by jabbing one another with upward blows of their front horns. In contrast, the Asian species attack by jabbing open-mouthed with their lower incisor tusks, or, in the case of the Sumatran rhino, with the lower canines.

Black rhinos have a reputation for unprovoked aggression, but very often their charges are merely blind rushes designed to get rid of the intruder. However, if a human or a vehicle should fail to get out of their way, they can inflict much damage with their horns. Indian rhinos also frequently respond with aggressive rushes when disturbed, and may occasionally attack the elephants used as observation platforms in some of the sanctuaries where they occur. However, rhinos invariably come off second best in any fight with an elephant.

In contrast, the White rhino is mild and inoffensive by nature, and despite its large size is easily frightened off. Very often, a group of White rhinos will stand in a defensive formation with their rumps pressed together, facing outwards in different directions. While this formation may be successful against carnivores such as lions and hyenas, it is useless against a human armed with a gun.

Rhinos have been under threat from man for a long time. The three Asian species suffered a great reduction in numbers and considerable contraction of their ranges during the last century because of the local demand for their products. Following the advent of guns in Africa, the southern White rhino was reduced to the brink of extinction before the end of the 19th century. There is little local use in Africa, and products were generally exported. The Black rhino was exterminated in the Cape soon after the arrival of white settlers, but elsewhere in Africa remained widespread and fairly abundant until recently. However, with escalating trade between African countries and Asia, Black rhino numbers declined precipitously in East, Central and West Africa during the 1970s.

The reason behind the recent declines is the rapid increase in the value of rhino horn. While ground rhino horn is used as an aphrodisiac in parts of North India, its main

use in China and neighboring countries of the Far East is as a fever reducing agent. It is also used for headaches, heart and liver trouble, and for skin diseases. Many other rhino products, including the hooves, blood and urine, have reputed medicinal value in the East (but not in Africa). Chemically, rhino horn is composed of keratin, the same protein which forms the basis of hooves, fingernails and the outer horny covering of cattle and antelope horns, and there is no pharmacological basis for these uses; whatever success is achieved is probably pyschological.

However, it is the use of rhino horns to

▲ **Rhino companions.** Rhinos are rarely without a few oxpeckers (here, on the nose of the right-hand rhino) and Cattle egrets in attendance.

▶ **Rhino horns** at Tsavo National Park, Kenya. The rhinos' most distinctive feature could also prove to be their downfall, as demand continues for their use as aphrodisiacs, as other medicinal agents and as dagger handles.

about 15,000 animals. The Sumatran rhino is now restricted to perhaps 150 individuals scattered throughout Sumatra, Malaya, Thailand and Burma. The Javan rhino, formerly widely distributed from India and China southwards through Indonesia, is now confined to a remnant of 50 in the Udjong Kulon Reserve in western Java. The Indian rhino is restricted to a few reserves in Assam, west Bengal, and Nepal, with a total population of about 1,500.

The situation of the White rhino is very different. The White rhino had a strange distribution, occurring in southern Africa south of the Zambezi River, and then again in northeastern Africa west of the Nile. The northern race has suffered severely from poaching in recent years, and has declined to perhaps a few hundred individuals in Zaire and the Sudan. The southern population was almost exterminated during the last century, but effective protection after 1920 resulted in a steady increase in the sole surviving population in the Umfolozi Game Reserve. By the mid 1960s their numbers had risen from perhaps 200 to nearly 2,000. As a result, the famous "Operation Rhino" was initiated by the Natal Parks Board to capture White rhinos alive for restocking other parts of their former range. This proved so successful that the species could be removed from the endangered list. By 1982, 1,200 White rhinos still remained in the Hluhluwe-Umfolozi Reserve in South Africa, and an equal number in other conservation areas in southern Africa. In fact, the main threat is posed by habitat deterioration due to the high densities attained by the species in the Umfolozi Game Reserve. To help provide outlets for the surplus animals which still need to be removed annually, some old males are sold to safari operators to be shot later by licensed hunters.

This contrasting situation is the source of an embarrassing conflict in conservation circles. While international cooperation is being sought to stop illegal hunting of rhinos through much of Africa, White rhinos, once the most endangered species, can be hunted legally in South Africa for legitimate reasons. While conservationists debate the most effective action, the situation is rapidly becoming desperate for rhinos in most regions of their occurrence. These hulking but simple-minded creatures are ill-adapted to cope with modern man armed with sophisticated weapons, and unless illegal hunting is controlled, rhinos may no longer be around by the turn of the century.

make handles for the "jambia" daggers traditionally worn by men in North Yemen as a sign of status that is mainly responsible for the recent rise in prices. Between 1969 and 1977 horns representing the deaths of nearly 8,000 rhinos were imported into North Yemen alone. The increase in the demand for rhino horn can be attributed to the fivefold increase in per capita income in Yemen as a result of oil wealth in the region.

The bulk of the pressure from the trade in rhino horns falls on the Black rhino in Africa and the Sumatran rhino in Asia. The Black rhino is still the most abundant and widespread species, but has been reduced to

NO-S

Horn to Horn
Territoriality, dominance and breeding in rhinos

Two male White rhinos approach each other to stare silently, horn to horn, then back away to wipe their horns on the ground. This ritual confrontation is repeated many times for perhaps up to an hour, before the males move apart to return to the hearts of their domains. For the point at which this ceremony takes place is the common boundary between their respective territories.

Territory holders also exhibit specialized techniques of defecation and urination, which may serve to scent mark the territories. The droppings are deposited at fixed dungheaps or middens, and are scattered by backwardly directed kicking movements. Especially large dungheaps, with prominent hollows developed by the kicking action, are located in border regions. The urine is ejected in a powerful aerosol spray, and urination is commonly preceded by wiping the horn on the ground then scraping over the site with the legs. Territory holders spray-urinate particularly frequently while patrolling boundary regions.

In White rhinos, access by males to receptive females is controlled by the strict territorial system. Prime breeding males occupy mutually exclusive areas covering 80–260ha (200–650 acres). These males form consort attachments to any females coming into heat that they encounter, and endeavor to confine such females within the territory for 1–2 weeks, until the latter are ready for mating. However, if the female should happen to cross into a neighboring territory, the male does not follow and the next-door male joins her.

Within territories, one or more subordinate males may be resident. Subordinate males do not spray their urine or scatter their dung, and they do not consort with females. When confronted by the territory holder, a subordinate male stands defensively uttering loud roars and shrieks. Females use similar roars to warn off males that approach too near. Generally, confrontations are brief, but if the subordinate male is an intruder from another territory a more prolonged and tense confrontation ensues, which may develop into a fight.

A defeated territory holder ceases spray-urination and dung scattering and takes on the status of subordinate male. Territory holders outside their own territories, on their way to and from water, also do not spray-urinate until they regain their own territories. If a territory holder is confronted by another male on a distant territory, he adopts the submissive stance and roars of a subordinate male. However, on a neighboring territory he maintains a dominant posture, but backs away steadily towards his own territory.

These behavior patterns signify a relationship whereby each territory holder is supremely dominant within the spatial confines of his own territory. This dominance gives him the opportunity to court and mate with any receptive female encountered there without interference from other males. Males nearing maturity, and deposed territory holders, choose to settle within a particular territory where the owner eventually becomes habituated to the presence of the additional male, providing that he displays subordinacy whenever challenged. While thus temporarily foregoing mating opportunities, subordinate males may gain strength to enable them at a later stage to challenge successfully for the status of territory holder in a nearby territory.

In a relatively high density population in the Hluhluwe Reserve, Black rhino breeding males occupy mutually exclusive home

▲ **Rhino confrontation.** A dominant White rhino confronts two subordinate rhinos on his territory. Subordinate males are tolerated by the dominant male provided they behave in a suitably submissive fashion.

◄ **Mating in rhinos** TOP can be a prolonged business, with several hours of foreplay, and copulations often lasting for one hour.

◄ **Rhino ritual.** In their confrontations, rhinos repeat the same gestures over and over before one concedes: (1) horns forced against each other; (2) wiping the horns on the ground. (3) A dominant male proclaims his mastery by spray-urinating, while the subordinate male retreats. Only dominant males spray-urinate.

areas which are shared by non-breeding males. These areas cover 4sq km (1.5sq mi), and meetings between neighboring males are rarely witnessed. In other Black rhino populations, the home ranges of males overlap and no clear evidence for territoriality has been found. Some males emit their urine in the form of a backwardly directed spray, but both males and females scatter their dung. When a female is in heat, several males sometimes displace one another in succession, before one succeeds in mating. Horn jousting matches between male and female sometimes occur during courtship.

In Indian rhinos, males can be classified as "strong" or "weak," but rather than being discrete categories there seems to be a continuum between them. Strong males urinate in a powerful backwards jet, associate frequently with females, and only they copulate. Such males move over home ranges covering up to 6sq km (2.3sq mi), but these overlap with those of other strong males, and are also shared by weak males. However, neighboring strong males rarely fight one another, while strange males entering from elsewhere are viciously attacked. Fights between male and female, and prolonged and noisy chases covering distances of several kilometers, are features of courtship.

These differences in social system can be related to differences in the density and distribution of food resources. As short grass grazers, White rhinos build up local densities in excess of 5 animals per sq km (12.5 per sq mi), while the location of favorable feeding areas at particular seasons is relatively predictable. Indian rhinos achieve local densities nearly as high, but because they are dependent upon flood plain habitats the location of favorable feeding areas changes in an unpredictable way. This does not favor spatial localization by males. In addition, the vegetation is dense, so that males may be screened from sensory contact with one another even when quite close by. For browsers the density of accessible food is much lower than is the case for grazers, and Black rhinos rarely exceed local densities of 1 per sq km (2.5 per sq mi). As a result, individuals occupy fairly large ranges and seldom come into contact, so that there is less pressure for males to avoid potentially risky contacts with other powerful males. Thus the control of mating rights seems to be more fluid, with stronger males claiming females from weaker ones when they come into contact. It is possible that in low density populations of White rhinos the territorial system would be far less strongly expressed.

NO-S

EVEN-TOED UNGULATES

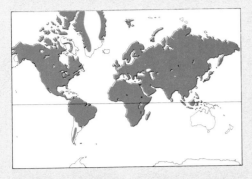

Order: Artiodactyla
One hundred and eighty-seven species in 76 genera and 10 families.
Distribution: worldwide, except Australia and Antarctica.

Habitat: very diverse.

Size: from head-body length 44–48cm (17–19in) in the Lesser mouse deer to 3.8–4.7m (12–15ft) in the giraffe. Weight from 1.7–2.6kg (4–5.7lb) in the Lesser mouse deer to 1,600–3,200kg (3,525–7,055lb) in the hippopotamus.

Pigs (family Suidae) – suids
Nine species in 5 genera.

Peccaries (family Tayassuidae) – tayassuids
Three species in 2 genera.

Hippopotamuses (family Hippopotamidae)
Two species in 2 genera.

Camels (family Camelidae) – camelids
Six species in 3 genera.

Chevrotains (family Tragulidae) – tragulids
Four species in 2 genera.

Musk deer (family Moschidae) – moschids
Three species in 1 genus.

Deer (family Cervidae) – cervids
Thirty-four species in 14 genera.

Giraffe (family Giraffidae)
Two species in 2 genera.

Bovids (family Bovidae)

Pronghorn (subfamily Antilocaprinae)
One species in 1 genus.

Wild cattle (subfamily Bovinae)
Twenty-three species in 8 genera.

Duikers (subfamily Cephalophinae)
Seventeen species in 2 genera.

Grazing antelope (subfamily Hippotraginae)
Twenty-four species in 11 genera.

THE even-toed ungulates, the Artiodactyla, are the most spectacular and diverse array of large, land-dwelling mammals alive today. Living in all habitats from rain forest to desert, from marshes to mountain crags, and found on all continents except Australasia and Antarctica, they dominated the mammal communities of the savannas where man's ancestors first arose. Indeed, they helped to mold the environment to which humans were adapted, and humans now control the environment in which dwindling communities of wild artiodactyls still survive. Early humans hunted them, and may have caused the extinction of some species. After the ice ages artiodactyls provided most of the large domesticated animals upon whose products and labor agricultural civilizations have depended.

Artiodactyls first appeared in the early Eocene (54 million years ago) in North America and Eurasia. They were at this time represented by small, short-legged animals, whose simple cusped teeth suggest a primate-like diet of soft herbage. By the late Eocene, the first members of the later, more advanced families were distinguishable, and by the Oligocene (35 million years ago) the three present-day suborders (Suina, Tylopoda, Ruminantia) were established. Piglike artiodactyls first appeared in Eurasia and Africa in the Oligocene, and ruminant artiodactyls made their first appearance in Africa in the early Miocene (25 million years ago).

The order Artiodactyla comprises two rather different types of animals, linked by similarities of anatomical structure. The suoids (suborder Suina), including pigs and their relatives, are primarily omnivorous animals, retaining low-crowned cheek teeth with simple cusps, large tusk-like canines, short limbs, and four toes on the feet, although some lineages may be more "progressive" in the modification of these features. In contrast, ruminant artiodactyls, comprising the suborders Tylopoda (camels) and Ruminantia (deer, giraffes and bovids) are specialized herbivores that have evolved a multi-chambered stomach and adopted the habit of chewing the cud in order to digest fibrous herbage. They have cheek teeth with ridges rather than cusps, that may be high crowned, and show a progressive tendency to elongate the limbs and to reduce the number of functional toes to two from the five found in primitive mammals.

Living suoids include the families Suidae (pigs), Tayassuidae (peccaries) and Hippopotamidae (hippos). Pigs are an exclusively Old World family. Peccaries are now confined to the New World, but were found in the Old World during the Tertiary. Members of both families are similar in having short legs, large heads and rooting snout, and are found in woodland, bushland and savannas in temperate and tropical habitats. In general, peccaries take a greater proportion of browse in their diet than the more omnivorous pigs, but some pigs, such as the African warthog, are specialist grazers. Hippos are today exclusively African tropical animals. The Pygmy hippo is a browser in the tropical forests, and the hippo is a semi-aquatic grazer. Hippos first appeared in the late Miocene (about 10 million years ago) in Africa, and radiated out into Eurasia in the

► **Artiodactyl horn types.** The extinct *Protoceratid* (1) had unbranched post-orbital horns and a curious forked nasal horn. *Dromomerycid* (2), another extinct species, had two unbranched supra-orbital horns and a single horn at the back of the skull. The Musk deer (3) is a primitive living artiodactyl with no horns and enlarged canines. The giraffe (4) has simple unbranched post-orbital horns in both male and female, covered by skin. The Roe deer (5) has branched post-orbital antlers (horn-like organs) which are shed yearly. Except in the reindeer only male deer have antlers. The pronghorn (6) has simple, unbranched, supra-orbital antlers with a keratinous cover that is shed annually. The female's antlers are smaller. Bovids, like the Common eland (7), have unbranched, post-orbital, keratin-covered horns that may be coiled or spiraled. Females in some species have horns.

Bone · Deciduous bone · Keratin · Deciduous keratin

Gazelles and Dwarf antelopes (subfamily Antilopinae)
Thirty species in 12 genera.

Goat antelopes (subfamily Caprinae)
Twenty-six species in 13 genera.

Skulls of Even-toed Ungulates

The skulls of pigs, like the babirusa, are characterized by a deep occiput (rear of the skull), an incomplete post-orbital bar, a short space between the front and back teeth, and the retention of large upper and lower canines. The teeth are usually low-crowned (except in some specialized grazers such as the warthog), simple cusped, and there is little tendency to molarize the premolars. The upper incisors are never entirely lost; the upper canines are characteristically curved upwards to form tusks in pigs, but point downwards in peccaries and hippos.

The skulls of ruminants, such as the Wild water buffalo and the Roe deer, are characterized by a shorter occiput, a complete post-orbital bar, a long gap between the front and back teeth and the tendency to reduce or completely lose the upper incisors. The molars may be low or high crowned, with the cusps coalesced into cutting ridges, and the premolars may be molarized. Ruminants resemble equids in the long face and posterior position of the orbits, but never develop the deep and massive lower jaw that characterizes horses.

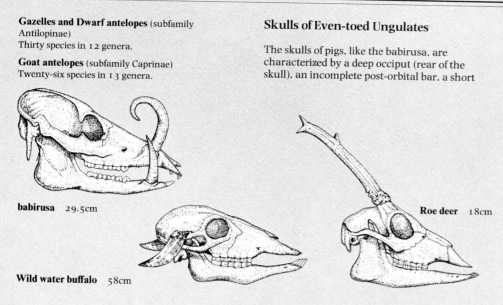

babirusa 29.5cm

Wild water buffalo 58cm

Roe deer 18cm

Plio-Pleistocene (about 2 million years ago).

The only living tylopods are the camels and llamas (family Camelidae), which are low-level feeders that take a mixture of herbs and fresh grass. Camels are found today in arid steppes and deserts in Asia and North Africa, and llamas in high altitude plains in South America, but until the Pleistocene (2 million years ago), camelids were an exclusively North American group. They first appeared in the late Eocene (40 million years ago), and only became extinct in North America 10,000 years ago.

The suborder Ruminantia can be further subdivided into the traguloids and the pecorans. Living traguloids are the chevrotains or mouse deer (family Tragulidae), which are very small animals—1–2kg (2.2–4.4lb)—found today in the Old World tropical forests; they eat mainly soft browse which requires little fermentation. They lack horns, but possess saber-like upper canines. Traguloids first appeared in the late Eocene, and were widespread in North America as well as Eurasia in the early Tertiary. Chevrotains first appeared in the Miocene, and have always been restricted to the Old World.

The pecorans comprise those ruminants which possess bony horns, and the five living families are distinguished primarily by their horn type. The first pecorans were the hornless gelocids of the Eurasian Oligocene, and members of the horned families did not begin to appear until the Miocene epoch.

Of the pecorans, giraffes (family Giraffidae) are high-level browsers, found today in African tropical forest (okapi) or savanna (giraffe), but they were also found in southern Eurasia during the Plio-Pleistocene. Deer (family Cervidae) are primarily small- to medium-sized browsers, with shorter legs and necks than giraffes, and are found today in temperate and tropical woodland in Eurasia, and North and South America. The Musk deer (family Moschidae), a small browser lacking horns and retaining large upper canines, is found today only in high-altitude forests in Asia, but in the Miocene there was a moderate radiation of moschids in North America, Europe and Africa.

The antelopes and cattle (family Bovidae) are today the most successful and diverse pecoran family, with a wide diversity of body sizes, forms and feeding adaptations. Most notably, they are the only family to have evolved medium- to large-sized open-habitat specialist grazers, such as the buffalo and the wildebeest. Bovids have been a primarily Old World group, reaching their peak of diversity in Africa, and they did not invade the New World until the Pleistocene. The pronghorn (subfamily Antilocaprinae) is also a surviving remnant of a once large radiation. Antilocaprids have always been exclusively North American, and the pronghorn is a medium-sized, long-legged, open-habitat, low-level browser on the western prairies. Here the pronghorn is regarded as a bovid, but some authorities consider it should be placed in its own family, suggesting a closer relationship with deer.

EXTINCT FORMS

LIVING FORMS

CJ/PJJ

WILD PIGS AND BOARS

Family: Suidae
Nine species in 5 genera.
Order: Artiodactyla.
Distribution: Europe, Asia, E Indies, Africa;
introduced into N and S America, Australia,
Tasmania, New Guinea and New Zealand.

Size: head-body length from 58–66cm
(23–26in) in the Pygmy hog to 130–210cm
(51–83in) in the Giant forest hog. Weight from
6–9kg (13–20lb) in the Pygmy hog to
130–275kg (285–605lb) in the Giant forest
hog.

▷ **Wild boar in a watery habitat.** Pigs are
the most omnivorous of the hoofed mammals
and like to root for food in moist soil.

▶ **The strange face** of the Giant forest hog,
somewhat like that of a gorilla or a bat, but
with tusks added!

▼ **Warthogs drinking.** Pigs usually forage in
family groups like this.

WHAT wild pigs lack in grace and beauty
they make up for in strength, adaptability and intelligence. They are admirably
adapted to range the forests, thickets, woodlands and grasslands which they haunt in
small bands, to duel for position and mates,
to fend off predators and to enjoy a catholic
diet. They are the most generalized of the
living even-toed hoofed mammals (artiodactyls) and have a simple stomach, four toes
on each foot and, in three of the genera, a
full set of teeth.

The living wild pigs are medium-sized
artiodactyls characterized by a large head,
short neck and powerful, but agile, body
with a coarse bristly coat. The eyes are small
and the expressive ears fairly long. The
prominent snout carries a distinctive set of
tusks (lower canine teeth) and ends in a
mobile, disk-like nose pierced by the nostrils.
The structure of the snout, tusks and facial
warts is intimately linked to diet, mode of
feeding and fighting style. The tassled tail
effectively swats flies and signals mood.

Key features of wild pigs are the well-
developed, upturned canines, molars with
rounded cusps, and a prenasal bone, which
supports the nose. Pigs walk on the third
and fourth digits of each foot, while the
smaller second and fifth digits are usually
clear of the ground. In warthogs, the first
and second molars regress and disappear as
the third molar elongates to fill the tooth
row; in all pigs males are larger than
females, with more pronounced tusks and
warts. In the genera *Sus* and *Potamochoerus*
the coat is striped at birth.

The senses of smell and hearing are well
developed. Pigs are highly vocal, and family
groups communicate incessantly by
squeaks, chirrups and grunts. A loud grunt
may herald alarm, while rhythmic grunts
characterize the courtship chant which, in
warthogs, sounds like the exhaust of a two-
stroke engine.

Pigs usually forage in family parties. The
Wild boar and bushpig feed on a wide range
of plant species and parts (fungi, ferns,
grasses, leaves, roots, bulbs and fruits) but
also take insect larvae, small vertebrates
(frogs and mice) and earthworms. They root
for much of their food in litter and moist
earth. The babirusa, the Giant forest hog
and the warthog are more specialized herbivores. The babirusa feeds largely on fruits
and grass; the Giant forest hog grazes
predominantly in evergreen pastures and
forest glades, seldom digging with its snout;

and warthogs feed almost entirely on grasses, plucking the growing tips with their unique incisors or with their lips. They take grass seeds at the end of the rains and during the dry season use the tough upper edge of the nose to scoop grass rhizomes out of the dry, sun-baked, savanna soils. Contrary to many popular accounts, warthogs seldom if ever use their tusks to dig out food.

Sexual maturity is attained at about 18 months, although males may only gain access to sows on reaching physical maturity at about 4 years old. Pigs may breed throughout the year in the moist tropics but in temperate and tropical areas of marked seasonality, mating occurs in the fall and sows farrow the following spring. Both the proportion of sows farrowing and the litter size (one or two in babirusa and up to 12 in the genus *Sus*) are influenced by prevailing environmental conditions. The young are born in a grass nest constructed by the mother or in a hole underground (warthog) and weigh between 500 and 900g (18–32oz). The piglets remain in the nest for about 10 days before following their mother. Each piglet has its own teat. Weaning occurs at about three months but young pigs remain with their mother in a closely knit family group until she is ready to farrow again. After farrowing in isolation, young sows may rejoin her. In this way, larger matriarchal herds or sounders are formed which may include several generations.

Abbreviations: HBL = head-body length. TL = tail length. wt = weight. Approximate nonmetric equivalents: 2.5cm = 1in; 1kg = 2.2lb.
[E] Endangered.   [V] Vulnerable.   [*] CITES listed.

Pygmy hog [E] [*]
Sus salvanius
Himalayan foothills of Assam. Tall savanna grassland. Active daytime and twilight. Omnivorous.
HBL 58–66cm; TL 3cm; wt 6–10kg.
Coat: blackish-brown bristles on gray-brown skin; no facial warts.
Gestation: about 100 days.
Mammae: 3 pairs.
Longevity: 10–12 years.

Wild boar
Sus scrofa
Wild boar or Eurasian wild pig.
Europe, N Africa, Asia, Sumatra, Japan, Taiwan. Introduced into N America. Feral domestic pigs in Australia, New Zealand and N and S America. Broad-leaved woodlands and steppe. Active daytime and twilight.
HBL 90–180cm; TL 30–40cm; wt 50–200kg.
Coat: brownish-gray bristles; short dense winter coat; no facial warts.
Gestation: 115 days.
Mammae: 6 pairs.
Longevity: 15–20 years.

Javan warty pig
Sus verrucosus
Java, Madura and Bawean. Forest, lowland grasslands and swamps, probably forest. HBL 90–160cm; TL ?; wt up to 185kg.
Coat: red or yellow hair with black tips; marked facial warts.
Gestation and longevity: probably as for Wild boar.

Bearded pig
Sus barbatus
Malaya, Sumatra and Borneo. Tropical forest, secondary forest and mangroves. Active daytime.
Coat: dark brown-gray with distinctive white beard on cheeks; marked facial warts.
Size, gestation and longevity: probably as for Wild boar.

Celebes wild pig
Sus celebensis
Sulawesi. Lowland and upland tropical forest. Smaller than Bearded pig. Well-developed facial warts.

Babirusa [V]
Babyrousa babyrussa
Sulawesi, Togian, Sulu and Buru Islands. Tropical forest. Active daytime. HBL 85–105cm; TL 27–32cm; wt up to 90kg.
Coat: sparse, short, white or gray bristles; no facial warts.
Gestation: 125–150 days.
Mammae: 1 pair.
Longevity: up to 24 years.

Bushpig
Potamochoerus porcus
Bushpig or Red river hog.
Subsaharan Africa and Madagascar. Forest, moist savanna woodlands and grasslands. Active day and night. HBL 100–150cm; TL 30–40cm; wt 50–120kg.
Coat: brick red to gray bristles with white, often patterned, face and mane; facial warts in male.
Gestation: 127 days.
Mammae: 3 pairs.
Longevity: 10–15 years.

Giant forest hog
Hylochoerus meinertzhageni
Central African Congo basin, parts of West and East Africa. Tropical forest and intermediate zone between forest and grassland. Active daytime.
HBL 130–210cm; TL 30–45cm; wt 130–275kg.
Coat: brown and black bristles; facial warts.
Gestation: 149–154 days.
Mammae: 3 pairs.
Longevity: not known.

Warthog
Phacochoerus aethiopicus
Subsaharan Africa. Savanna woodland and grasslands. Active daytime. HBL 110–135cm; TL about 40cm; wt 50–110kg.
Coat: black or white bristles on gray skin; pronounced facial warts.
Gestation: 170–175 days.
Mammae: 2 pairs.
Longevity: 12–15 years.

◄ **The warthog's tusks.** Although the upper tusks are more impressive, it is the smaller but sharper lower tusks that are the warthog's principal weapons.

▼ **Fighting styles of pigs.** The distinctive patterns of weapons and armor displayed by each species reflects the mode of combat. (1) In the Giant forest hog, contact is frontal and the top of the head is toughened. (2) In the warthog, contact is also frontal and the facial warts protect against the incurving tusks. (3) In the bushpig, snouts are crossed, sword-like, and are similarly protected by warts. (4) Wild boars slash at each other's shoulders, which are protected by thickened skin and matted hair.

In courtship, a chanting boar nudges the sow's flanks, sniffs her genital region, indulges in lateral displays and repeatedly attempts to rest his chin on her rump—a stimulus which causes a fully receptive sow to stand. The bushpig, warthog, and pigs of the genus *Sus*, and possibly other pigs, produce lip-gland pheromones which may be dispersed while chanting; *Sus* also produces a salivary foam. Mating may last 10 minutes and the spiral penis fits into a grooved cervix in which a plug forms after copulation.

The main social units are solitary boars, bachelor groups and matriarchal sounders comprising one or more adult sows with their young of various ages. Fragmentation of larger mother-daughter groups leads to kinship units or clans comprising a number of related sounders, with overlapping home ranges, which share feeding grounds, water holes, wallows, resting sites and sleeping dens. Wild pigs appear to be non-territorial, with home ranges of about 25ha (61 acres) in Pygmy hogs, 1–4sq km (0.4–1.5sq mi) in warthogs and 10–20sq km (3.8–7.7sq mi) in Wild boars. Home ranges are marked by lip glands, pre-orbital glands (warthog and Giant forest hog) or foot glands (bushpig). The mating system appears to be a roving dominance hierarchy among the males within a clan area. In northeast Borneo, the Bearded pig is noted for the large migrations it undertakes.

Habitat destruction threatens the Pygmy hog, and the Javan warty pig popupulation is critically low; they require both habitat protection and vigorous captive breeding programs. If the Javan warty pig is to be safe in the wild it requires translocation to protected islands which exclude the Wild boar as a competitor. The position of the African wild pigs and the widely distributed Wild boar is presently satisfactory.

Man has reared pigs for meat in Europe and Asia since neolithic times, and well-developed pig cultures still exist in parts of Indonesia. The Wild boar has been domesticated separately in Europe, India, China and Malaya. A domesticated form of the Celebes wild pig occurs in the Philippines, Sulawesi and neighboring islands.

The African wild pigs are symptomless carriers of African swine fever (ASF), a virus disease transmitted by the tampan, a soft-bodied tick, and lethal to domestic swine. ASF may acount for the absence of feral swine south of the Sahara. This has led to eradication campaigns against the bushpig, warthog and Giant forest hog, where the disease has threatened pig farming. African wild pigs also support blood sucking tsetse flies, the carriers of trypanosomes responsible for sleeping sickness in man and ngana in domestic livestock. The Savanna tsetse fly is particularly partial to warthog, and elimination of warthogs and bushpigs has played an important part in controlling ngana, which covers an area of Africa equivalent to the USA.

Whereas the warthog is readily exterminated, the bushpig has taken advantage of agricultural development in many parts of Africa and is notorious for its nocturnal raids on cultivated crops. In America, Australia and New Zealand, feral pigs cause damage to lambs, cultivated crops and local plant communities. DHMC

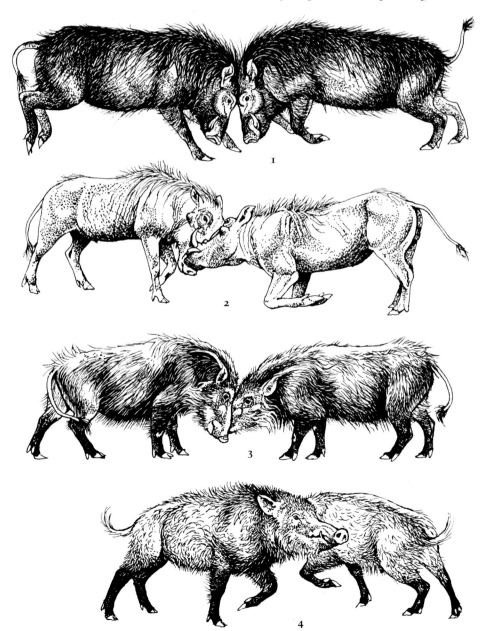

1

2

3

4

PECCARIES

Family: Tayassuidae
Three species in 2 genera.
Order: Artiodactyla.
Distribution: SW USA to N Argentina.

Collared peccary ▣
Tayassu tajacu
Collared peccary, báquiro, chacharo, javelina or javalí.
Distribution: SW USA to N Argentina.
Habitat: low and high dry tropical forest, wet
tropical forest, chaparral and oak-grasslands;
active mainly during day.
Size: head-body length 80–98cm; shoulder
height 30–40cm (12–16in); tail length
3.5–4.5cm (1.4–1.8in); weight 17–25kg
(38–55lb). Coat: grizzled gray with dark
grayish back; limbs blackish; whitish band
diagonally from middle back to chest; juveniles
russet colored and collar diffuse. Gestation: 142
days. Longevity: 8–10 years (21 years in
captivity).

White-lipped peccary
Tayassu pecari
White-lipped peccary, pecari, careto, cachete blanco,
cochino salvaje.
Distribution: SE Veracruz state, Mexico to
N Argentina.
Habitat: high dry tropical forest, wet and rain
tropical forest; active day and night.
Size: head-body length 100–120cm (39–48in);
shoulder height 50–60cm (20–24in); tail
length 4–5cm (1.6–2in); weight 25–40kg
(55–88lb). Coat: grayish black to dark brown,
with whitish bristles on lips, chin, throat and
rump; juveniles uniform gray. Gestation: 158
days. Longevity: unknown.

Chacoan peccary ▣
Catagonus wagneri
Chacoan or chaco peccary, pagua or tagua.
Distribution: Gran Chaco (N Argentina,
SE Bolivia, W Paraguay).
Habitat: thorny forest with isolated islands of
palm and grasses, active during day.
Size: head-body length 96–117cm (38–45in);
shoulder height 52–69cm (21–25in); tail
length 3–10cm (1.2–4in); weight 31–43kg
(68–95lb). Coat: grizzled gray-brown, black
and white, with black dorsal stripe and white
collar; limbs blackish; juveniles tan and black
with diffuse tan collar. Gestation: not known.
Longevity: not known.

▣ Vulnerable. ▣ CITES listed.

SOUTH American peasants claim "that the
jaguar only kills peccaries that remain
apart from the group." This saying has been
verified, since it is now known that individual peccaries will confront a predator, at
considerable risk to themselves, while other
members of their group escape. Such altruistic anti-predator behavior is just one of the
unusual features of the life-style of peccaries,
gregarious inhabitants of the tropical and
subtropical forests of the Americas. One of
the species—the Chacoan peccary—was discovered by scientists as recently as 1975.

Peccaries are medium-sized pig-like
animals with long, slender legs. The three
species differ in color as well as in size: the
Collared peccary has a white collar, the
White-lipped peccary white lips, and the
Chacoan peccary both the white collar and
lips. Compared to the two species of *Tayassu*,
the Chacoan peccary is slightly larger and
has a bigger skull, its dental crowns are
more developed, its eyes are located further
to the rear of the head and its snout is
longer—differences which justify placement
in the separate genus *Catagonus*. Males and
females are the same size in all species. Coat
color is different between juveniles and
adults in each species.

Both peccary genera evolved from the
North American genus *Platygonus* from the
upper Pliocene (about 5–2 million years
ago). *Platygonus* colonized South America
with the formation of the South American
land bridge. *Catagonus* is more closely related to *Platygonus* than is *Tayassu*.

Peccaries are omnivorous, but feed preferentially on roots, seeds and fruits. Occasionally, they also eat insects and other invertebrates. Cacti figure largely in the diets of
Chacoan and Collared peccaries. They have
three compartments in their stomachs in
which it is thought microbial flora digest
cellulose, as occurs in ruminants. Their
tusks are used to cut roots, which they
consume avidly. Their jaw muscles are well
developed and are strong enough for peccaries to crush tough seeds. The jaws of the
White-lipped peccary are especially powerful. Jaw movement is up-and-down rather
than gyratory, as in other hoofed mammals.
Peccaries are known to lick and eat soil,
possibly as a source of mineral nutrients.

Female Collared peccaries are sexually
mature at 33–34 weeks, males at 46–47
weeks. Copulation lasts just a few seconds
and is not preceded by any courtship displays. Females may mate with several
males, with adult males establishing a
hierarchy to prevent subordinates mating.

Unlike pigs, which are mostly solitary,

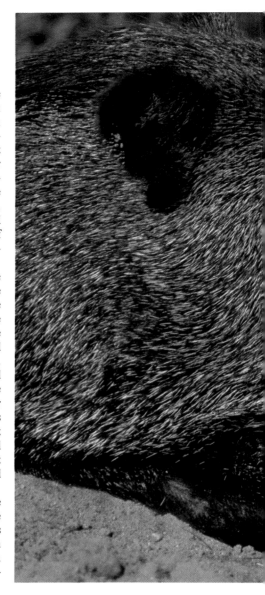

peccaries live in herds of 14–50 individuals
in the Collared peccary, of 100 or more in
the White-lipped peccary and of 2–10 in the
Chacoan peccary. Each herd is subdivided
into small family groups, which in the
Collared peccary comprise 13–14, including males, females, females with young, and
juveniles; females may outnumber males by
3:1. During the dry season, when food is not
abundant, these groups often separate to

▲ **The canines of peccaries,** visible here in this Collared peccary, are short, sharp, and conventional compared to the exotic tusks of the related pigs. They have 38 teeth but the first premolars are reduced. Their coat is a distinctive feature, comprising tough, thick hairs 2–22cm (0.8–8.6in) long, the longest on the mane and rump. A conspicuous gland, up to 7.5 × 1cm (3 × 0.4in) in size, is present on the rump.

◄ **A young Collared peccary,** at one month old. Young remain dependent on their mothers for about 24 weeks. Lactation lasts 6–8 weeks, although the young eat solid food within 3–4 weeks. Both parents and other group members help to care for the young. In the face of threatened danger, parents and non-parents alike will shelter the young between their rear legs, and the young are allowed unrestricted access to high-quality or limited food.

search for food. Even when a herd is resting, groups tend to disperse to different parts of the site. Such groups are enduring and are particularly cohesive when juveniles are present.

Peccaries occupy stable territories, which are strongly defended from intruders. Territories of the Collared peccary vary in size from 30–280ha (74–690 acres) depending on vegetation type and quantity and distribution of food resources. Secretions from the rump gland are used by adults and subadults to mark tree trunks and rocks within the territory. The core area, which is preferentially used, is also marked by dung piles, one of which is used throughout the year and others only seasonally. These dung piles are important parts of the forest habitat since they contain large numbers of undigested seeds which readily germinate to rejuvenate the forest and disperse the trees.

Cohesion within groups is reinforced by mutual marking by a gland below the eyes—individuals stand side-by-side, turning their faces to each other and rubbing

each other's glands. Such marking is believed to aid recognition between all members of the group. In the Collared peccary, boisterous play and mutual grooming and scratching with snouts also often occur at the start of a day, before feeding.

Groups of Collared and Chacoan peccaries apparently have no leader, but when moving from one feeding site to another they follow whichever individual happens to be at the front of the line, usually one of the females in the group, followed by other adult females, young and juveniles.

Peccaries are vocal animals, and six types of vocalization can be distinguished in the Collared peccary. A cough-like grouping call by an adult male recalls dispersed individuals to the group. A repeated dry short "woof" is the alarm call and a "laughing" call is used in aggressive encounters between individuals. A clear nasal sound is produced when peccaries eat. Infants that have strayed indicate their distress by emitting a shrill clucking call which prompts the mother to seek them out. A repeated rasping sound, produced by chattering the teeth, is emitted by animals that are angry or annoyed. During squabbles, the neck and back bristles are often raised.

The main predators of peccaries are mountain lions and jaguars. Two forms of anti-predator behavior have been observed. In the Collared peccary, if the predator gets very close before detection, a herd or group disperses in all directions, to confuse the assailant, while emitting the alarm call. If young are present and the habitat is dense, one individual, usually a subadult of either sex, may approach the predator, in an act of apparent altruism, while others escape as the predator's attention is diverted. The outcome for the vigilant animal may be fatal. When resting, some males generally remain vigilant around the periphery of groups, and these are replaced by rested males during the period.

Clearance of forest for crops and pastures is reducing peccary habitat, and management plans are needed to conserve peccary populations. Indians and peasants hunt peccaries in the tropics for their meat; their gregarious nature and wide distribution make them readily accessible. The Chacoan peccary is the most susceptible to human disturbance as its distribution is the most restricted. Collared peccaries are often considered pests because they eat and destroy plantations of yucca, corn, watermelons and legumes, to the point where local campaigns have been mounted to exterminate them.   HGC

HIPPOPOTAMUSES

Family Hippopotamidae
Two species in 2 genera.
Order: Artiodactyla.
Distribution: Africa.

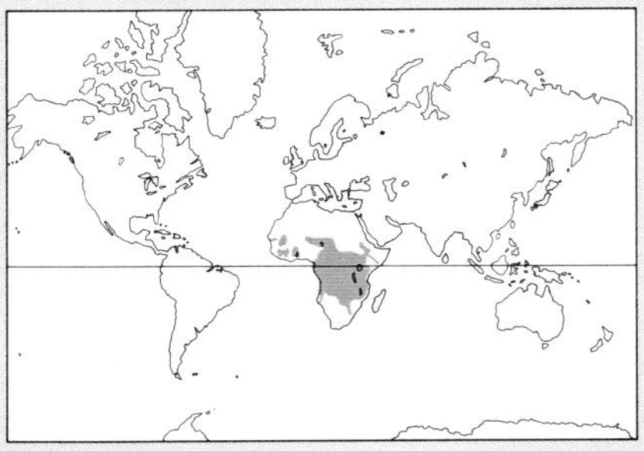

Pigmy hippopotamus [V] [*]
Choeropsis liberiensis
Distribution: Liberia and Ivory Coast; a few also in Sierra Leone and Guinea.

Habitat: lowland forests and swamps.

Size: head-body length 1.5–1.75m (4.9–5.7ft); height 75–100cm (30–39in); weight 180–275kg (397–605lb).

Skin: slaty greenish black, shading to creamy gray on the lower part of the body.

Gestation: 190–210 days.

Longevity: about 35 years (42 recorded in captivity).

Hippopotamus [*]
Hippopotamus amphibius
Distribution: W, C, E and S Africa.

Habitat: short grasslands (night); wallows, rivers, lakes (day).

Size: head-body length 3.3–3.45m (10.8–11.3ft); height 1.4m (4.6ft); weight male 1,600–3,200kg (3,525–7,055lb), female 1,400kg (3,086lb).

Skin: upper part of body gray-brown to blue-black; lower part pinkish; albinos are bright pink.

Gestation: about 240 days.

Longevity: about 45 years (49 recorded in captivity).

[V] Vulnerable. [*] CITES listed.

THE hippos are an example of the unusual phenomenon of a species pair, ie two closely related species that have adapted to different habitats (other examples of species pairs are Forest and Savanna elephants and buffalos). The smaller hippo occupies forest, the larger grassland. The hippo is also unusual for the way its daytime and nighttime activities, and feeding and breeding, take place in very different habitats: daytime activities in water, nighttime ones on land. This division of life zones is associated with a unique skin structure which causes a high rate of water loss when the animal is in the air and therefore makes it necessary for the hippo to live in water during the day.

The body of the hippo is barrel-shaped with short, stumpy legs. Its large head is adapted for an aquatic life, with eyes, ears and nostrils placed on the top of the head. It can submerge for up to five minutes. Its lips are up to 50cm (20in) broad and the jaws can open to 150 degrees wide. Its most characteristic vocalization is a series of staccato grunts.

The Pigmy hippopotamus resembles a young hippopotamus in size and anatomy, with proportionately longer legs and neck, smaller head and less prominent eyes which are placed to the side rather than on top of the head. It is less aquatic than the hippo-

► **Pigmy hippos,** showing the sleek skin which, like that of the hippopotamus, is covered in pores which exude a sticky fluid.

potamus; the feet of the Pigmy hippo are less webbed and the toes more free.

The skin of the hippo is smooth, with a thick dermis and an epidermis with unusually thin, smooth and compact surface layer—a unique adaptation to aquatic life. Because of the very thin cornified layer, the skin seems to act as a wick, allowing the transfer of water, and the rate of water loss through the epidermis in dry air is several times greater than in other mammals, losing about 12mg from every 5sq cm every 10 minutes—about 3–5 times the rate in man under similar conditions. The Pigmy hippo has a similar rate. This necessitates resort to a very humid or aquatic habitat during the day, otherwise the animal would rapidly dehydrate. Another characteristic of the skin is the absence of sebaceous glands or true temperature-regulating sweat glands. Instead there are glands below the skin which secrete a pink fluid which dries to form a lacquer on the surface. This gave rise in the past to the idea that the hippo "sweats blood." The "sweat" is very viscous, and highly alkaline. The red coloration is probably due to skin pigments similar to those which cause tanning in man. The pigment is not penetrated by ultraviolet radiation and

protects the skin from sunburn. It may also have a protective function against infection, because even large wounds are always clean and free from pus, despite the filthy water of wallows and lakes.

The fossil record shows that there were once several species of hippos. Only two have survived, of which the Pigmy hippo is not only the smaller but more primitive. It probably bears the closer resemblance to the ancestral hippo, which, like the Pigmy, lived in forests. A speculative view of the evolution of the hippos envisages their origin in a small forest-living form similar to the Pigmy hippo. In the course of time, forests were replaced by grasslands and the ancestral hippo eventually moved out of the forest to occupy gallery forest or thickets along the rivers and lakes by day, grazing in open grasslands by night. This gave access to the abundant food supply of the tropical savanna grassland and was accompanied by development of much larger size, but the animal's unusual physiology required terrestrial feeding at night and resort to water during the day. There was a behavioral adaptation to the grassland environment rather than a physiological one and the consequent penalty is a limited geographical

▲ **Looking every inch an aquatic mammal,** a bull hippo stalks a river bed. The hippo is forced to spend the daytime in water because the skin loses water at a very high rate in air.

▼ **Hippos "blood."** The red exudate from the skin glands of hippos led to an early belief that they "sweated blood." The secretion is a protection against the sun and probably also against infections.

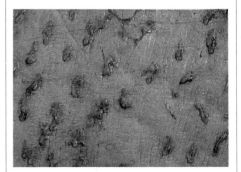

▷ **Nothing like mud!** OVERLEAF By using muddy wallows up to 10km (6.2mi) from water, hippos can extend their grazing range further inland.

range, restricted to the vicinity of water.

The hippos feed almost exclusively on terrestrial vegetation. Very little is known of the food of the Pigmy hippo, but it probably feeds on roots, grasses, shoots and fruits found on the forest floor and in swamps. The hippopotamus feeds almost exclusively on a number of species of short grasses, which are cropped by a plucking motion of the broad lips. This means that shallow-rooted grasses are selectively removed and at high population densities overgrazing and soil erosion ensues (see box).

Feeding is compressed into five or six hours of the night; the rest of the time is spent in the water. Food is digested in ruminant-like manner in compartmented stomachs, but the process seems inefficient. However, daily food requirement only averages 40kg (88lb) a night, equal to a dry weight of about $1-1\frac{1}{2}$ percent of body weight, compared with $2\frac{1}{2}$ percent for other hoofed mammals. This is probably because the way of life minimizes energy expenditure: the active feeding period is short and for the remainder of the 24 hours the hippo is resting or more or less inactive, in a supportive, stable, warm medium. Even energy requirements for muscle tone are reduced: the hippo most often opens its mouth to yawn!

Mating is correlated with dry seasons when the population is most concentrated. Both species prefer to mate in water, the female partly or wholly submerged, lifting her head from time to time to breathe. The proportion of females pregnant in one season can vary between only 6 percent in dry years and 37 percent in wetter periods. Sexual maturity is attained at an average of 7 years in the male (range 4–11 years) and 9 years in the female (range 7–15 years). Social maturity in the male is reached at about 20 years. In captivity, sexual maturity is attained in the Pigmy hippo at about 4–5 years.

The birth season occurs during the rains, resulting in a single peak of births in southern Africa and a double peak in East Africa. To give birth, a female leaves her group and a single young is born on land or in shallow water, weighing about 42kg (93lb). Occasionally, it is born underwater and has to paddle to the surface to take its first breath. Twins occur in less than 1 percent of pregnancies. The mother is very protective, and when she and her offspring rejoin the herd 10–14 days after birth the young may lie across the mother's back in deeper water. Suckling occurs on land, and in or under water, and the duration of lactation is about

8 months. Natural mortality in the first year is about 45 percent, reducing to 15 percent in the second year, then 4 percent per annum to about 30 years, subsequently increasing.

The average group size is 10–15 in hippos, the maximum 150 or so, according to locality and population density, but only 1–3 in the Pigmy hippo. In the latter, the triads are usually male, female and calf; their social behavior is almost unknown. The hippo, however, is either solitary (usually male individuals) or found in nursery groups and in bachelor groups. The solitary males are often territorial; other males maintain territories containing the nursery groups of females and young. In a stable lake environment some bulls maintain territories for at least eight years; in a less stable river environment most territories change ownership after a few months, but some are maintained for at least four years. Hippo densities in lake and river are respectively 7 and 33 per 100m (330ft) of shore. The territories are 250–500m (550–1,000ft) and 50–100m (110–220ft) in length respectively. Territorial bulls have exclusive mating rights in their territories but tolerate other bulls if they behave submissively. Non-territorial bulls do not breed. The frequency of fighting is reduced by an array of ritualized threat and intimidation displays, but when fights do occur they may last up to one and a half hours and are potentially lethal. The razor-sharp lower canines are up to 50cm (20in) long; the canine teeth have a combined weight of up to 1.1kg (2.4lb) in females and 2.1kg (4.6lb) in males. Defense is by the well-developed dermal shield thickest along the sides. Adult males can receive severe wounds and are often seen heavily scarred, but the skin possesses a remarkable ability to heal. Displays include mouth opening, dung scattering, forward rushes and dives, rearing up and splashing down, sudden emergence, throwing water (using the mouth as a bucket), blowing water through the nostrils, and vocalizations, especially a series of loud staccato grunts. Submissive behavior is a lowering of the head and body. Male territorial boundary meetings are ritualized: they stop, stare, present their rear ends and defecate while shaking the short, vertically flattened tail to spread the dung, before walking back to the center of their territories.

In the evenings, the aquatic groups break up and the animals go ashore, either singly or as females with their calves. They walk along regular trails marked by dung piles that serve as scent markers for orientation at

Conservation and Management

The highest density hippo populations in the world are found in East and Central Africa in the lakes and rivers of the western Rift Valley, particularly around Lakes Edward and George. Maximum grazing densities recorded in the Queen Elizabeth National Park were formerly over 31 per sq km (81 per sq mi), equivalent to 31,000kg per sq km (177,000lb per sq mi) of hippo alone, not counting other grazing and browsing hoofed mammals. They create an energy and nutrient sink from land to water by removing grass and defecating in the lakes. This promotes fish production by fertilizing the water.

Overgrazing within 3km (1.8mi) of the water's edge led to erosion gulleys and reduction of scrub ground cover. The Uganda National Parks Trustees initiated a hippo management culling program in 1957, the first such management for any species in Africa. Subsequently, from 1962–1966, an experimental management scheme was operated, with the aim of maintaining a range of different hippo grazing densities in a number of management areas from zero to 23 per sq km (60 per sq mi) while climate, habitat changes and populations of other herbivores were monitored to establish the influence of different hippo densities. About 1,000 hippos were shot annually over 5 years, and the meat was sold to local people. In one of the experimental areas there were 90 hippos in 4.4sq km (7sq mi) in 1957, and it was bare ground except for scattered thickets. By 1963, after the near elimination of hippos, the grass had returned and erosion was halted, while average numbers of other grazers increased from 40 to 179 and the total hoofed mammal biomass increased by 7.7 percent; this trend continued until 1967, when management culling ceased and reversion towards the previous situation began. Subsequently, the

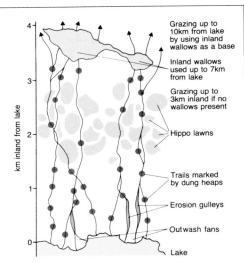

Grazing up to 10km from lake by using inland wallows as a base

Inland wallows used up to 7km from lake

Grazing up to 3km inland if no wallows present

Hippo lawns

Trails marked by dung heaps

Erosion gulleys

Outwash fans

Lake

breakdown of law and order in Uganda interrupted research and led to serious poaching.

As the hippo population was reduced, the mean age of sexual maturity declined from 12 years to under 10 years, and the proportion of calves increased from 6 percent to 14 percent. The influence of climatic change is difficult to establish in the short term, but for the rainfall levels of the 1960s the predicted grazing density for hippo in this area of Uganda was about 8 per sq km (21 per sq mi) if soil erosion was to be controlled, and if vegetation cover and variety, and herbivore numbers and diversity were to be maintained.

This work indicated the optimal rate of culling in one locality. However, the optimum level of hippo grazing density varies according to soil type, vegetation composition and climate; it needs to be established by management research for any particular area. Management culling remains a controversial issue, but almost all current knowledge about hippo populations is the result of such schemes.

▲ **Lunging hippos** contest a river territory. Hippo fights can be lethal, and dominant males can maintain territories for up to eight years.

◀ **Young hippos** ride piggy-back on their mothers in deeper water. Birth sometimes takes place in the water and some suckling is carried out there too.

night and have no territorial function. The paths have been measured to 2.8km (1.7mi) long and branch at the inland end, leading to the hippo lawns, which are discrete short-grass grazing areas. The animals share a common grazing range, where they feed; they then return to the water along the tracks during the night or early morning. Observations of identifiable individuals indicate that the composition of aquatic groups can remain fairly constant over a period of months, although they are very unstable compared with, for example, elephant family groups. The Pigmy hippo has similarly marked trails on the thinly vegetated forest floor or tunnel-like paths through denser vegetation.

Both species occupy wallows, which may be very large in the case of the hippo-potamus; smaller wallows are occupied pre-dominantly by males. This allows the grazing range to be extended inland up to 10km (6.2mi) or more. High hippo densities, in some areas 31 per sq km, have led to overgrazing and management culling (see box).

Both species play host to several animals. An association with a cyprinid fish, *Labio velifer*, has been described; it is sometimes temporarily attached and may graze on algae or other deposits on the hippo skin. Terrapins and young crocodiles bask on the backs of hippos, Hammer-headed storks and cattle egrets use them as a perch for fishing, and other birds, including oxpeckers, are associated. A parasitic fluke, *Oculotrema hippopotami*, is uniquely found attached to 90 percent of hippos' eyes; on average 8, but up to 41 have been found on one hippo.

RML

CAMELS AND LLAMAS

Family: Camelidae
Six species in 3 genera (including 3 domesticated species).
Order: Artiodactyla.
Distribution: SW Asia, N Africa, Mongolia, Andes.

Size: ranges from shoulder height 86–96cm (34–38in); weight 45–55kg (99lb–121lb) in the vicuna to height at hump 190–230cm (75–91in), weight 450–650kg (1,000–1,450lb) in the dromedary and Bactrian camel.

▼ **Well-wrapped up** in its winter coat, the Bactrian camel has an ungainly, shaggy appearance. In spring, when the winter coat is shed, they look even more tattered.

MEMBERS of the camel family are among the principal large mammalian herbivores of arid habitats, and they make a crucial contribution to man's existence and survival in desert environments. The domesticated one-humped dromedary of southwestern Asia and North Africa, and the two-humped Bactrian camel, which is still found wild in the Mongolian steppes, are well-known, but there are also four species in the New World: the "lamoids" or "cameloids." Of these, the vicuna and guanaco are wild and the llama and alpaca are domesticated.

The camel differs from other hoofed mammals in that the body load rests not on the hoofs but on the sole-pads, and only the front ends of the hoofs touch the ground. The South American cameloids are adapted to mountainous terrain, and the pads of their toes, which are not as wide as those of camels, are movable to assist them on rocky trails and gravel slopes. The split upper lip, the long curved neck and the lack of tensor skin between thigh and body, so that the legs look very long, are characteristic of camels. They move at an ambling pace. Camels are unique among mammals in having elliptical blood corpuscles. They all have an isolated upper incisor, which in the males of the lamoids is hooked and sharp-edged, like the tusk-like canines found in both jaws. Camels have horny callosities on the chest and leg joints.

The camelids first appeared in the late Eocene (among the earliest of the even-toed hoofed mammals). The camel family originated and evolved during 40–45 million years in North America, with key dispersals to South America and Asia occurring only 2–3 million years ago.

The relationships between the South American lamoids are confused. Fertile offspring are produced by all possible pure and hybrid matches between the four and they all have the same chromosome number (74), but there are major differences between wild vicuna and guanaco in the growth of their incisor teeth, and in their behavior. The long-held view that both the llama and alpaca are descended from the wild guanaco has recently been challenged by the suggestion that the fine-wooled alpaca is the product of cross-breeding between the vicuna and llama (or guanaco). Based upon wool length and body size, there are two widely recognized breeds each for the llama and alpaca. The *suri* alpaca is well known for both long-straight and wavy wool fibers. Today no llamas or alpacas live independently of man in their native Andean homeland.

The center of lamoid domestication may have been the Lake Titicaca pastoral region or perhaps the Junin Plateau, 100km (62mi) to the northwest, where South American camelid domestication has been dated at 4,000–5,000 years ago.

One school believes that the dromedary was first domesticated in central or southern Arabia before 2000BC, others as early as 4000BC, while some place it between the 13th and 12th centuries BC. From there, domesticated camels spread to Egypt and North Africa, and later to East Africa and India.

The Bactrian camel was domesticated independently of the dromedary, probably before 2500BC at one or more centers in the plateaux of northern Iran and southwestern Turkestan. From here they spread east to Iraq, India and China.

The South American cameloids are seasonal breeders and mate lying down on their chests—copulation lasting 10–20 minutes. They are induced ovulators, give birth to only a single offspring in a standing position and neither lick the newborn nor eat the afterbirth. The newborn is mobile and following its mother within 15–30 minutes. The female comes into heat again 24 hours after birth, but usually does not mate for two weeks. Large-bodied llamas that have access to high levels of nutrition may breed before one-year-old, but most female lamoids first breed as two-year-olds.

There is no information on the social organization of the domesticated llama and alpaca, because non-breeding males are castrated. However, circumstantial evidence suggests a territorial system, with

▲ **The proud, aloof face** of the dromedary. Actually, the facial characteristics which in human terms imply arrogance and disdain are meaningless in camelid terms. A camel with ears pressed flat to the body is far more likely to be hostile than one apparently sneering. Dromedary camels in Africa are generally maintained in a semi-wild state in which they are free to forage alone but dependent on man for water. During the rutting season they are guarded by their owners. Unguarded camels form stable groups composed of up to 30 animals. Free-grazing camels in North Africa form bachelor herds and mixed herds of perhaps one male plus 10–15 females and their young. Unlike the South American lamoids, male camels possess an occipital or poll gland immediately behind the head, which plays a role in the camel's mating behavior. The male camel also occasionally inflates his oral skin bladder ("dulaa") as a part of his rutting display towards other males.

males maintaining a harem of breeding females (polygynous).

The vicuna is strictly a grazer in alpine *puna* grassland habitats between 3,700–4,800m (12,000–15,700ft). They are sedentary, non-migratory with year-round defended feeding and sleeping territories. Populations are divided into male groups and family groups. The territories are defended by the territorial male and are ocupied by the male, several females and their offspring of less than one-year-old. The territory is in two parts—a feeding territory where the group spend most of the day, and a sleeping territory located in higher terrain; the corridor between the two is not defended. The feeding territory—average size 18.4ha (45 acres)—is a predictable source of food; is the place where mating and birth often occur; and is a socially stable site with the advantages of group living, where females can raise their offspring. Sleeping territories average 2.6ha (6.4 acres).

The guanaco, which is twice as large as the vicuna, is both a grazer and browser, may occur in either sedentary or migratory populations, is much more flexible in its habitat requirements, in that it occupies desert grasslands, savannas, shrublands, occasionally forest, and ranges in elevation between 0–4,250m (0–13,900ft). Unlike the vicuna, it does not need to drink. Its territorial system is similar to that of the llama and alpaca, but the number of animals using the feeding territory, unlike the vicuna, is not related to forage production. Vicuna and guanaco young are forcefully expelled from the family group by the adult male, as old juveniles in the vicuna, and as yearlings in the guanaco.

Camels mate throughout the year, but a peak of births coincides with the season of plant growth. Copulation again takes place lying down, and they may be induced ovulators. Litter size is one. Camels eat a wide variety of plants over expansive home ranges. They eat thorns, dry vegetation, and saltbush that other mammals avoid. Camels can endure long periods without water (up to 10 months if not working), and so can graze far from oases. When they do drink, they can consume 136 liters (30 gallons) within a short time. They conserve water by producing dry feces and little urine, and allow their day-time body temperature to rise by as much as 6–8°C (11–14.5°C) during hot weather to diminish the need for evaporative cooling by sweating, although sweat glands occur over most of their body for use when necessary. Their nostrils, which can be closed to keep out blowing sand, their nostril cavities, which reduce water loss by moistening inhaled air and cooling exhaled air, and the localized storage of energy-rich fats in the hump, help them to survive long periods without food. A thick fur and underwool provide warmth during cold desert nights and some daytime insulation against the heat.

Today there are approximately 21.5 million camelids. In South America there are an estimated 7.7 million lamoids, with 53 percent in Peru, 37 percent in Bolivia, 8 percent in Argentina, and 2 percent in Chile. The domestic llamas and alpacas (91 percent of the total) are far more numerous than the wild guanacos and vicunas (9 percent). Llamas (3.7 million) are slightly more abundant than alpacas (3.3 million), and guanacos (575,000) are much more common than vicunas (85,000). Most alpacas (91 percent) and vicunas (72 percent) are in Peru, the majority of South American's llamas (70 percent) are in Bolivia, and nearly all the guanacos (96 percent) are found in Argentina. In general, numbers of alpaca and vicuna are increasing due to the value of their wool, whereas llamas are declining as they are replaced by trucks and rail transport. The guanaco is decreasing, due to hunting for its wool and skin, and competition with livestock.

Ninety percent of the world's 14 million camels are dromedaries, with 63 percent of all camels in Africa. Sudan (2.8 million), Somalia (2.0 million), India (1.2 million),

and Ethiopia (0.9 million) have the largest populations of camels, with Somalia having the highest density (3.14 per sq km) by far. Numbers of camels have drastically declined over recent decades in some countries (eg Turkey, Iran and Syria), due in part to the forced settlement of nomads, but the worldwide camel population has remained relatively stable. A large but uncounted population of feral camels occupies the arid

▲ **The curiously craning gait** of vicuna, the smallest of the South American cameloids. Reduced to dangerous population levels in the 1960s, they are now recovering under protection.

▶ **Guanaco on the pampas.** Millions of these elegant camelids once roamed the South American plains, but their numbers are much reduced and they are still hunted in Argentina.

Abbreviations: SH = shoulder height.   WT = weight.   Approximate nonmetric equivalents 2.5cm = 1in; 1kg = 2.2lb.

[D] Domesticated.   [V] Vulnerable.   [•] CITES listed.

Llama [D]
Lama glama

Andes of C Peru, W Bolivia, NE Chile, NW Argentina. Alpine grassland and shrubland (2,300–4,000m. HBL 120–225cm (47–88in); SH 109–119cm; wt 130–155kg. Coat: uniform or multicolored white, brown, gray to black. Gestation: 348–368 days. Two breeds: *chaku, ccara*.

Alpaca [D]
Lama pacos

Andes of C Peru to W Bolivia. Alpine grassland, meadows and marshes at 4,400–4,800m. HBL 120–225cm (47–88in); SH 94–104cm;

wt 55–65kg. Coat: uniform or multicolored white, brown, gray to black; hair longer than llama. Gestation: 342–345 days. Two breeds: *huacaya, suri*

Guanaco [•]
Lama guanicöe

Andean foothills of Peru, Chile, Argentina, and Patagonia. Desert grassland, savanna, shrubland and occasional forest at 0–4,250m. SH 110–115cm; wt 100–120kg (220–265lb). Coat: uniform cinnamon brown with white undersides; head gray to black. Gestation: 345–360 days. Four questionable subspecies.

Vicuna [V]
Vicugna vicugna

High Andes of C Peru, W Bolivia, NE Chile, NW Argentina. Alpine *puna* grassland at 3,700–4,800m (12,100–15,750ft). SH 86–96cm; wt 45–55kg. Coat: uniform rich cinnamon with or without long white chest bib; undersides white. Gestation: 330–350 days. Two subspecies: Peruvian, Argentinean.

Dromedary [D]
Camelus dromedarius
Dromedary, Arabian camel or One-humped camel.

SW Asia and N Africa. Deserts. Feral

in Australia. Height at hump 190–230cm, wt 450–650kg. Coat color variable from white to medium brown, sometimes skewbald; short but longer on crown, neck, throat, rump and tail tip. Gestation: 390–410 days.

Bactrian camel [V]
Camelus bactrianus
Bactrian or Two-humped camel.

Mongolia. Steppe grassland. Height at hump 190–230cm; wt 450–650kg. Coat: uniform light to dark brown; short in summer with thin manes on chin, shoulders, hindlegs and humps; winter coat longer, thicker and darker in color. Gestation: 390–410 days.

and train transport. Llamas can carry loads of 25–60kg (55–132lb) for 15–30km (9–18mi) a day across rugged terrain. Llamas have been introduced into a number of countries in small numbers for recreational backpacking, wool production, and novelty pets. In the Andes, the woolly alpaca is replacing the llama as the most important domestic lamoid.

There were millions of guanacos in South America when the Spaniards arrived, perhaps as many as 35–50 million on the Patagonian pampas alone. Today, the guanaco is still the most widely distributed lamoid, but in greatly reduced numbers compared to historical times. The guanaco is protected in Chile and Peru, but not in Argentina, where tens of thousands of adult and juvenile (*chulengos*) guanaco pelts are exported annually.

From a population level of millions in the 1500s, to 400,000 in the early 1950s, then to less than 15,000 in the late 1960s, the vicuna was placed on the rare and endangered list by the International Union for the Conservation of Nature (IUCN) in 1969. Today the vicuna is fully protected and populations are increasing. World vicuna population size has recovered to over 80,000 and its status was changed by IUCN from endangered to vulnerable in 1981.

Dromedary camels are important beasts of burden throughout their range, but they are especially valued for their meat, wool, and milk—6 liters (1.3 gallons) per day for 9–18 months. Man in return provides water for the camels. The milk produced by camels is often the main source of nourishment for the desert nomads of the western Sahara. Nomads use camels as work animals primarily for moving camp and carrying water. They walk 30km (19mi) a day at a leisurely pace to allow for feeding and resting, and carry up to 100kg (220lb) per animal. Camels play a special role in social customs and rituals. They are required as gifts for marriage and reparation for injury and murder. White camels are favored most. Camels have been used to transport military material and personnel during conquests and wars of North Africa, beginning with Napoleon's campaign there in 1798.

The future of the camel in the Saharan countries depends upon the fate of the nomads. If these countries settle their nomads, the vast stretches of arid lands will be of little use. If the nomads are encouraged to continue their traditional ways, the desert will continue to serve as an important resource for millions of people and camels.

inland regions of the mainland of Australia.

The once vast range of the Bactrian camel has contracted severely in recent times, although some remain in Afghanistan, Iran, Turkey, and the USSR. Most of the camels in Mongolia and China are Bactrians, and there are probably less than 1,000 wild Bactrians in the trans-Altai Gobi Desert. These animals have slender build, short brown hair, short ears, small conical humps, small feet, no chest callosities, and no leg callosities.

The camelids' energy as pack animals and their products of meat, wool, hair, milk and fuel have been indispensable to man's successful habitation and survival in earth's temperate and high altitude deserts. The Inca Empire's culture and economy revolved around the llama, which provided the primary means of transportation and carrying goods. Domestic cameloid numbers declined drastically during the century following the Spanish invasion of the Central Andes. Uncontrolled shooting, disruption of the Inca social system, and the introduction of sheep were responsible for the crash.

Today, the docile llama is still actively used in the Andes as an important pack animal, though it is being replaced by trucks

WLF

CHEVROTAINS

Family: Tragulidae
Four species in 2 genera.
Order: Artiodactyla.
Distribution: tropical rain forests of Africa,
India and SE Asia.

Water chevrotain ▣
Hyemoschus aquaticus
C and W Africa. Tropical rain forest; nocturnal.
Size: head-body length 70–80cm (28–32in);
shoulder height 32–40cm (13–16in); tail
length 10–14cm (4–5.5in); weight 8–13kg
(18–29lb).
Coat: blackish red-brown, with lighter spots,
arranged in rows and continuous side stripes;
throat and chest with white herringbone
pattern. Gestation: 6–9 months. Longevity:
10–14 years.

Spotted mouse deer
Tragulus meminna
Spotted mouse deer or Indian spotted chevrotain.
Sri Lanka and India. Tropical rain forest,
particularly rocky; nocturnal. Size: head-body
length 50–58cm (20–23in); shoulder height
25–30cm (10–12in); tail length 3cm (1.2in);
weight about 3kg (6.6lb). Coat: reddish brown
with lighter spots and stripes; more or less
similar to Water chevrotain. Gestation:
probably 5 months.

Lesser mouse deer
Tragulus javanicus
Lesser mouse deer or Lesser Malay chevrotain.
SE Asia. Tropical rain forest and mangroves;
nocturnal. Size: head-body length 44–48cm
(17–19in); shoulder height 20cm (8in); tail
length 6.5–8cm (2.5–3.2in); weight 1.7–2.6kg.
(4–5.7lb). Coat: more or less uniform red with
characteristic herringbone pattern over the fore
part of the body. Gestation: probably 5 months.

Larger mouse deer
Tragulus napu
Larger mouse deer or Greater Malay chevrotain.
SE Asia excluding Java. Tropical rain forest;
mainly nocturnal. Size: head-body length
50–60cm (20–24in); shoulder height
30–35cm (12–14in); tail length 7–8cm
(2.8–3.2in); weight 4–6kg (9–13lb). Coat:
more or less uniform red. Gestation: 5 months.

▣ CITES listed.

THE smallest species no bigger than a
rabbit, virtually unchanged in 30 million years of evolution and intermediate in
form between pigs and deer—these are the
main claims to distinction of chevrotains,
diminutive inhabitants of Old World tropical
forests. During the Oligocene and Miocene
(38–7 million years ago), chevrotains had a
worldwide distribution, but today the four
species are restricted to the jungles of Africa
and Southeast Asia.

Chevrotains are among the smallest of
ruminants—hence the collective name
mouse deer—with the Lesser mouse deer the
smallest at a mere 2kg (4.4lb). All species
have a cumbersome build, with short, thin
legs and limited agility. Males are generally
smaller than females: in the Water chevrotain, 20 percent less in weight. The coat is
usually a shade of brown or reddish brown,
variously striped or spotted depending on
the species. They are shy, secretive creatures, mostly active at night. They inhabit
prime tropical forest, the three Asiatic
species showing a preference for rocky
habitats. The Water chevrotain is a good
swimmer. Their diet mainly comprises fallen
fruit, with some foliage.

Chevrotains are ruminants, ie their gut is
modifed for the fermentation of their food,
but they exhibit many non-ruminant characters, and should be regarded as the most
primitive ruminants, providing a living link
between non-ruminants and ruminants.
They have a four-chambered stomach,
although the third chamber is poorly developed. Other anatomical features chevrotains share with the ruminants are: a lack
of upper incisor teeth; incisor-like lower
canines adjoining a full set of upper incisors;
and only three premolar teeth. Behavioral
similarities with ruminants include: a typical (although extreme) solitary life-style for
a forest ruminant; single young; and inges-

tion of the placenta by females after birth.
More interestingly, chevrotains share a
number of (primitive) characters with non-
ruminants. These include: no horns or
antlers; continually growing projecting upper canines in males (peg-like in females);
premolars with sharp crowns; and all four
toes fully developed. Behavioral patterns
shared with pigs include: long copulation
and simple sexual behavior; lack of visual
displays, so communication is limited to
smells and cries; lack of specialized scent
glands below the eyes or between the toes;
and a habit of lying down rump first, then
retracting their forelimbs beneath them. Of
the four species, the Water chevrotain is the
most pig-like (ie primitive) since its forelimbs
lack a cannon bone and the skin on the
rump forms a tough shield which protects
the animal against canine teeth.

The reproductive biology of chevrotains is
poorly known. In the Water chevrotain and
Larger mouse deer, breeding occurs
throughout the year, with only one young
per litter; weaning occurs at 3 months in
both species with sexual maturity achieved
at 10 months and 4–5 months, respectively.
In the former species, young are known to
be hidden in undergrowth soon after birth,
but maternal care is limited to nursing. In
the Larger and Spotted mouse deer, mating
occurs within two days of birth. In the
Water chevrotain, the only mating "display" is a cry by the male akin to the *chant de
cour* of pigs which brings the courted female
to a standstill for copulation, which may last
2–5 minutes. Water chevrotains are solitary
except during the mating season, and
mostly show no interest, aggressive or
otherwise, in other members of their species.
Communication is by calls and scent.

Water chevrotains possess anal and preputial glands, and this species marks its
home range with urine and feces, the latter

▲ **The Lesser mouse deer,** the smallest of the chevrotains.

◀ **The Spotted mouse deer** ABOVE, a small primitive artiodactyl intermediate in form between pigs and deer.

impregnated with anal gland secretions. All species have a chin gland which is rudimentary in the Water chevrotain and Spotted mouse deer. In males of the other species it produces copious secretions which they use to mark a mate's or male antagonists's back during encounters, In the Water chevrotain, the heaviest and oldest animals are dominant. Owing to their solitary life, however, there is no established hierarchy. Fighting between males is reduced to a short rush, each antagonist biting his opponent all over his body with his sharp canines.

Spacing behavior is only known for Water chevrotains. They mainly inhabit terrestrial habitats, but will retreat to water when danger threatens. Home ranges are 23–28ha (60–70 acres) in males and 13–14ha (32–35 acres) for females, and they always border a watercourse at some point. Home ranges of adult females do not overlap; neither do those of males, although they do overlap those of females. There is no evidence of territorial defense. Population density varies from 7.7–28 per sq km (20–72 per sq mi) for the Water chevrotain in Gabon and 0.58 per sq km (1.5 per sq mi) for the Spotted mouse deer in Sri Lanka.

Their small size makes chevrotains easy and important prey for various predators, such as big snakes, crocodiles, eagles and forest-inhabiting cats. As with many other tropical forest species, the survival of these four living fossils depends on conservation of their habitat and restriction of hunting.

GD

MUSK DEER

Family: Moschidae ⊡
Three species of the genus *Moschus*.
Order: Artiodactyla.
Distribution: Asia.

Habitat: mountain forests; active in twilight.

Size: head-body length 80–100cm
(31–39in); tail 4–6cm
(1.5–2.5in); height 50–70cm
(20–28in); weight 7–17kg
(15.5–37.5lb).

Musk deer

Moschus moschiferus, M. berezovskii,
M. chrysogaster†
Distribution: Siberia, E Asia, Himalayas.
Coat: grayish-brown to golden, speckled;
striped at the under part of the neck.

Gestation: 150–180 days.

Longevity: 13 years (in captivity).

†As there is very little information on the
distribution, biology and morphology of the
three species, which until recently have been
considered a single one with many subspecies,
the above data refer to the genus as a whole.

⊡ CITES listed.

► **The "milk-step."** Suckling in Musk deer
is unusual in that the young touches the
mother's hindleg with a raised foreleg. This
gesture is similar to that seen during the
courting behavior of other hoofed mammals,
such as the gerenuk. The female usually gives
birth to one young, rarely twins. The newborn
calves are very small in comparison with their
mother —600–700g (20–25oz). During the
first weeks of life they are motionless and
inconspicuous among rocks or in dense
undergrowth. From time to time, the mother
visits the young and lets it suckle; after a month
it leaves the resting place and accompanies the
mother.

MUSK, a substance produced by a gland in
the male Musk deer, is an important
and expensive ingredient of the best perfumes. In 1972 an ounce of musk was more
valuable in Nepal than an ounce of gold!
Despite its economic importance and its
wide distribution, little is known of the
biology of the Musk deer. This is especially
regrettable, since in many regions it is
declining due to hunting or deforestation.

Musk deer are stockily built animals with
small heads. The hindlegs are about 5cm
(2in) longer than the forelegs, indicating a
tendency to move by leaping. The hooves,
including the lateral toes, are long and
slender. Most of the hair is very coarse.
Neither sex possesses antlers but the male
has long, saber-like upper canine teeth
which project well below the lips; in females
the canines are small and not visible. In this
respect, Musk deer resemble other primitive
deer like the Water-deer. They give a good
impression of the appearance of the earliest
predecessors of antler-bearing deer. Their
most notable, and unique, feature is the
musk bag or pod which develops when the
male has reached sexual maturity. This sac,
situated between the genitals and the umbilicus, is about the size of a clenched fist. Its
glands produce a secretion whose precise
function is unknown but which is probably
a signal for the females. In contrast to the
adults, newborn Musk deer have a distinctly
striped and spotted coat. Musk deer are shy
and furtive animals with a keen sense of
hearing.

The classification of the Musk deer has
long been confused. Formerly, only one
species was recognized, but recently the
existence of three different species has been
proven. The limits of their distribution are
unknown, so the map gives only an idea of
the distribution of the genus as a whole.

Musk deer, being primarily forest
animals, are never found in desert regions or
in areas with a dense human population.
Musk deer prefer dense vegetation, especially hills with rocky outcrops in coniferous,
mixed or deciduous forests. In the morning
and evening hours they leave their resting
places, situated among rocks or fallen trees,
in search of food. Their diet consists of
leaves, flowers, young shoots and grasses,
also of twigs, mosses and lichens, especially
in winter. In captivity, they readily accept
lettuce, carrots, potatoes, apples, rolled oats,
hay and alfalfa.

Musk deer are solitary for most of the
year. Outside the rutting season, more than
two or three are seldom seen together; such
groups usually consist of a female and her
young. During the rutting season, mainly
November/December in northern latitudes,
males run restlessly, pursuing females,

chasing them to the point of exhaustion, when they seek shelter. During this period, males fight for the females, and their teeth can inflict deep, and sometimes deadly, wounds on their opponent's neck and back. During the weeks of courtship, the males eat little, they are very excited and cover large distances. After the rutting period they return to their original territories.

Musk deer are strongly territorial. Within their home range they regularly use well-established trails which connect feeding places, hiding places and the latrines where their droppings are deposited. By continuous use, these spots may reach considerable size. Both sexes cover their droppings immediately after defecation by scratching the soil with their forelegs. Males scent mark tree trunks, twigs and stones within their territories by rubbing their tail gland against them, producing oily patches. When disturbed, they dart off in enormous bounds. Thanks to the strongly developed lateral toes, Musk deer climb well on precipitious crags, rocks and even trees. They also run readily on snow or heavy soil. Apart from man, they are subject to predation by lynx, wolf, marten, fox and probably some birds of prey.

From time immemorial, Musk deer have been hunted by man for musk, one of the few mammal substances used in perfumes. Musk is not only used in Europe as a base for exquisite perfumes but also in the Far East for a great variety of medicinal applications (sore throats, chills, fever and rheumatism). In a single year, Japan, for instance, imported 5,000kg (11,000lb) of musk which, at an ounce per pod, represents the astonishing total of 176,000 male Musk deer. As the animals are taken mainly with automatic snares set up on their trails, the total loss, including muskless females and young, must have been even higher. Because of the relentless persecution, Musk deer have already become rare in parts of their range.

It is, however, not necessary to kill the males in order to obtain the musk. Therefore, in 1958, the Chinese started a breeding program in the province of Sichuan. Between 1958 and 1965, 1,000 Musk deer were caught alive. In spite of heavy losses at the beginning, mainly during the transportation and acclimatization period, the Chinese have succeeded in breeding the species in vast enclosures. Although juvenile mortality is still high and longevity relatively short (the same applies to zoos, where Musk deer are seldom on exhibit), some satisfactory results have been achieved. When the sexual activity of the male is at its peak, the animals are caught by hand. The musk is then removed from the pod by a spoon inserted into the aperture of the sac, a procedure which takes only minutes; then the animal is released. The musk, a jelly-like, oily substance with a strong odor and reddish-brown color is subsequently dried, when it becomes a powdery mass which gradually turns black.

If the techniques of breeding and handling Musk deer continue to improve, their farming may help to reduce the pressure on wild populations and may make a valuable contribution to the conservation of this interesting little deer.   HF

▼ **The long, protruding canines** of the male Musk deer give it a lugubrious expression. The Musk deer is thought to be similar to the ancestors of the antler-bearing deer.

DEER

Family: Cervidae
Thirty-six species in 16 genera.
Order: Artiodactyla.
Distribution: N and S America, Eurasia,
NW Africa (introduced to Australasia).

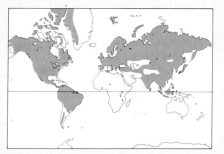

Habitat: mainly forest and woodland, but also
arctic tundra, grassland and mountain regions.

Size: shoulder height and weight from 38cm
(15in) and 8kg (17.5lb) in Southern pudu to
230cm (90in) and 800kg (1750lb) in the
moose.
Coat: mostly shades of gray, brown, red and
yellow; some adults and many young spotted.

Gestation: from 24 weeks in Water-deer to 40
weeks in Père David's deer.

Species include: **Water-deer** (*Hydropotes
inermis*); **muntjacs** (5 species of genus
Muntiacus); **Tufted deer** (*Dama dama*); **chital**
(*Axis axis*); **Swamp deer** (*Cervus duvauceli*); **Red
deer** (*Cervus elaphus*); **wapiti** (*Cervus
canadensis*); **Sika deer** (*Cervus nippon*); **Père
David's deer** (*Elaphurus davidiensis*); **Mule deer**
(*Odocoileus hemionus*); **White-tailed deer**
(*Odocoileus virginianus*); **Roe deer** (*Capreolus
capreolus*); **moose** (*Alces alces*); **reindeer**
(*Rangifer tarandus*); **Pampas deer** (*Ozotoceros
bezoarticus*); **huemuls** (2 species of genus
Hippocamelus); **Southern pudu** (*Pudu pudu*);
Northern pudu (*Pudu mephistophiles*); **Red
brocket** (*Mazama americana*); **Brown brocket**
(*Mazama gouazoubira*).

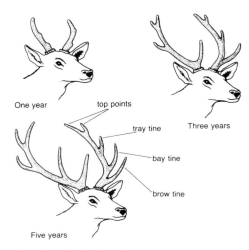

THE commonest view of a wild deer is of a rapidly disappearing rump, but even that is rare because deer are always on guard against predators. Yet cave paintings dating from 14,000 years ago and more modern images, such as The Monarch of the Glen, Bambi and Santa Claus' reindeer, show the interest which deer hold for man. Firstly deer were a source of food and then sport, to the extent that the Norman kings of England planted forests in the South of England to ensure a supply of deer for the chase. More recently deer have been regarded as creatures of interest in their own right and knowledge of their lives and behavior has broadened.

Deer are similar in appearance to other ruminants, particularly antelopes, with graceful, elongated bodies, slender legs and necks, short tails and angular heads. They have large, round eyes which are placed well to the side of the head, and triangular or ovoid ears which are set high on the head. The size of deer ranges from the moose to the Southern pudu which weighs only about 1 percent of the moose's 800kg (1,750lb).

The antlers of adult male deer distinguish them from other ruminants. These are bony structures like horns, but unlike horns they are shed and regrown each year. In form they range from single spikes with no branches, as in pudu, to the complex, branched structures of Red deer, Fallow deer and reindeer (or caribou). These last two species have palmate antlers where the angle between some of the branches is filled in, giving a flat "palm" with the points of the branches resembling fingers from the palm. Only one species—the Water-deer—lacks antlers, but the males have tusks, as do muntjacs and Tufted deer which have only small simple antlers.

Females do not have antlers except in the reindeer. Many explanations have been suggested for this; perhaps the most likely is that they help females compete for food in the large, mixed sex herds, unusual for deer, in late winter when the reindeer dig craters in the snow to reach the vegetation below.

Why do stags make such a huge investment in antler growth each year? Antlers are used as weapons in fights for females and are often damaged. In Red deer major damage to the antlers can reduce rutting success that season, and it would have the same effect in future years if antlers were not regrown each year. The effect of antlers on dominance is particularly clear at casting time in Red deer when the older, dominant stags who cast first can no longer displace younger ones with antlers.

The color of the coat varies through many shades of gray, brown, red and yellow. Usually the underparts are lighter than the back and flanks, and there is often a light colored area around the anus—the rump patch—which may be fringed with dark hair. The rump patch is often made conspicuous by alarmed deer, who raise their tails and rump hair as they flee. The young of many species have a pattern of light colored spots on a darker ground which improves their camouflage; this pattern is retained by adults in some species, eg Sika

◀ **Antler development in Red deer.** Deer are unique among ruminants in their annual cycle of antler growth and shedding. The process is under the control of the sexual and growth hormones. Fur-covered skin (velvet) carries blood to the growing antlers. In the fall the velvet dries up and the deer thrash their antlers against vegetation to remove it. In the winter the antlers are shed. Young deer take several years to develop a full set of antlers, with bifurcated points and projecting fines.

◀ **A forest of antlers.** This herd of Red deer stags on a deer farm seem almost part of the vegetation. Herds of stags have been kept for their velvet, an expensive ingredient of oriental medicines, and sometimes for their meat.

▼ **Gory antlers** of a reindeer bull which has just shed its velvet.

and Fallow deer. The coat is molted twice a year, in spring and fall. Deer are more or less completely covered in hair, but all species except the reindeer have a small naked patch on the muzzle.

In common with other ruminants, deer bite off their food between the lower incisors and a callous pad on the upper gum. The dental formula of adult deer is I0/3, C0/1, or C1/1, P3/3, M3/3. The presence of upper canines depends on the species, and in male Water-deer, muntjacs and Tufted deer they develop into tusks which may be used for fighting. The premolars and molars grind food during rumination (chewing the cud).

Deer use hearing, smell and sight to detect danger from predators. When feeding, the first warning a deer has of a predator is probably through sound or smell, as sight is restricted by the vegetation. Deer, however, raise their heads and scan their environment during feeding which probably increases their rate of predator detection. When a deer is alarmed it will raise its head high, stare hard at the source of alarm and rotate its ears forwards towards it. Although deer are quick to see and respond to movement, they may fail to detect, for example, a human stalker if he keeps quite still until the deer begins to feed again. If they hear or see movement deer rapidly flee.

Deer also use smell to gain information about members of their own species. A rutting Red deer stag detects which hinds in his harem are in heat by sniffing them or the earth where they have lain or urinated. Most species have interdigital (or foot) glands which probably leave scent trails, and all have facial (or pre-orbital) glands which produce strong-smelling secretions.

Vocal communication among females and young deer is more or less limited to bleats by which mothers and offspring locate each other and alarm barks which are given when they are disturbed. However, a wide range of vocalizations, including the whistling scream of Sika deer, barks of muntjac and the bellowing roar of Red deer, is produced by males competing for females in the breeding season.

Deer are typically woodland animals, not found far from cover in open country; the largest number of species occur in subtropical woodland. However, deer occur in almost all habitats from arctic tundra (reindeer) to open grassland in Argentina (Pampas deer) and deep forest (pudu). Deer are indigenous to Europe, Asia, North, Central and South America and Northern Africa. However, they have been introduced into Australia, New Zealand and New Guinea. There have also been translocations of species within the natural range of deer, for example England has feral muntjac originating from India, and there are feral Red deer and Fallow deer originating from Europe in South America.

Most species of deer can be distinguished on the basis of size, coat color or other external characteristics such as tail length and ear size. Some species are extremely easily recognizable—nobody could mistake a huge, ungainly deer with humped shoulders, a belled throat and palmate antlers for anything but a moose. In other cases the differences between species are more subtle, for example the Red and Brown brockets differ only in a slight variation in the shade of coat color. Size and form of the antlers are also distinctive: for example the dichotomously branched antler of an adult male Mule deer is quite different from that of the adult male White-tailed deer where tines branch off a main beam. But not all members of a species are adult males with antlers, and there may be confusion between the antlers of immature males of Mule and White-tailed deer; here a minor feature of the antler—the length of the basal snag (a small point)—differs between the species. However, this still leaves the problem of identification of deer without antlers, males after they have cast their antlers and females who never have them. In these circumstances tail length and the length of the metatarsal gland are used to distinguish Mule deer and White-tailed deer.

In some cases where two species are very similar but occupy distinct ranges, it is more convenient to separate them; for example, the two species of huemul that live in widely different parts of the Andes. In other instances no such geographical separation exists between very similar species. For example, Red deer are found from western Europe to central Asia and wapiti from central Asia to North America. Both show considerable variation between subspecies and localities but are basically similar to each other. On this basis some authorities prefer to classify them as the same species. However, since not only the coloration but also the vocalization of rutting males are different, they are treated here as separate species.

Some authorities regard the three species of musk deer (genus *Moschus*) as members of this family. However, others place them in a separate family, as here (see pp518–519).

Deer generally feed by grazing on grass or browsing on the shoots, twigs, leaves, flowers and fruit of herbs, shrubs and trees. However, chital and Swamp deer in India feed almost exclusively on grass (but see box) while the Indian muntjac prefers herbs, leaves, fruit, mushrooms and bark. Reindeer

and caribou scrape away snow in winter to get at lichens, which form a major part of their diet.

Within a species the type and amount of food eaten is determined by the individual's nutritional requirements and by food availability. Large animals such as deer have higher maintenance requirements than herbivores such as rodents, so in the many species where males are larger than females, males might be expected to need more food. However, deer need food for reproduction as well as for maintenance and comparisons of male and female nutritional requirements are complicated by differences in the extent and timing of the costs of reproduction. Consider Red deer: the highest requirements for energy and protein by hinds occur in spring and early summer during the last third of pregnancy and early lactation; the peak for stags is slightly later when they grow antlers and neck muscle in the approach to the rut. In a study of wild Scottish Red deer, stags spent more time in summer feeding than hinds without calves (10.4 and 9.8 hours per day), reflecting the high cost of milk production. In winter, stags increased the amount of time spent feeding to 12.88 hours per day; hinds with calves still fed for 11.08 hours per day even though most of the calves were weaned, probably because it took longer to ingest the same amount of food as the available vegetation declined.

Food availability is low in winter for many northern deer and they use more energy than they gain, thus losing weight. That reduced food intake is due to a reduction in appetite as well as in food availability was shown in captive Red and White-tailed deer which reduced intake in winter even when given unlimited food. Presumably in the natural state this helps prevent them from wasting more energy searching for low quality food than the food would provide.

Another response to low food availability is to switch to a different diet. Thus the

▲ **The noble features** of the Rusa deer, an Indonesian species that has been introduced to Australia, New Zealand, Fiji and New Guinea.

▼ **Representative species of deer.**
(1) Southern pudu (*Pudu pudu*). (2) Pampas deer (*Ozotoceros bezoarticus*). (3) Marsh deer (*Blastocerus dichotomous*). (4) Peruvian huemul (*Hippocamelus antisensis*). (5) Red brocket (*Mazama americana*). (6) White-tailed deer (*Odocoileus virginianus*). (7) Reindeer (*Rangifer tarandus*).

Chital and Langur Monkeys—An Unusual Association

A herd of large, spotted deer forage below a tree in an Indian jungle. Amongst them, and above, a troupe of gray monkeys feeds, a female of which slaps a deer buck that had approached her head-on. With a lunge, the monkey momentarily grabs the deer's antler and the deer backs away. The deer is a chital, and the monkey a Common langur, both common species of moist and dry deciduous forest on the Indian subcontinent.

In Central India, throughout the year, excepting the monsoon season of July–October, chital can be found in attendance below and around a langur troupe for as much as 8 hours a day, and up to 17 approaches a day by chital herds to a single monkey troupe have been observed. The chital browse upon foliage dropped from the trees by feeding langurs, particularly the leaf blades which the langurs reject, having eaten the stalks. A troupe of 20 langurs drops an estimated 1.5 tonnes of foliage each year, of which 0.8 tonnes is suitable forage for chital. This gleaned waste is particularly valuable for chital between November and June when grass is sparse. During the monsoon season, when other food is more plentiful, langur/chital associations decline dramatically.

Chital initiate the association and probably locate feeding langurs, usually from 50–75m (165–245ft) away, either by their movement, scent or perhaps even by the smell of broken foliage. Once located, chital move directly and quickly to take up stations below the monkeys.

At first sight, it would appear that the relationship is one-sided, with the chital gaining all the benefit. However, chital and langurs respond to each others' alarm calls, so early detection of predators such as tigers and leopards probably benefits both. Also, the sensory abilities of the two species are complementary, langurs having acute vision and chital a good sense of smell.

This association, with the benefit of concentrated and abundant food supply, allows chital to congregate in denser herds than is normal for this opportunistic herbivore. Chital grazing in the loose herds that are typical over most of their range are rarely overtly aggressive towards each other, but where they are closely packed below langurs, large bucks in particular often threaten does and young. In consequence, such bucks probably eat more of the langur-created waste and are thus at a competitive advantage compared to other bucks without access to these gleanings. Some langurs come to the ground to glean items dropped by their companions: thus an adult male langur might discard small fruit that a mother and young would descend to glean. On the ground, the males may react aggressively towards chital. Chital also occasionally feed on the waste dropped by fruit-eating birds and on wind-felled foliage, so they appear to be pre-adapted to associating with langurs due to their opportunistic feeding behavior. The association is probably mutually beneficial, but asymmetrical in that the chital benefit considerably more than the langurs. **PN**

composition of the diets of temperate deer often shows marked seasonal changes. Fallow deer for example, are principally grazers, with grass accounting for over 60 percent of their diet (estimated from stomach contents) in summer and never less than 20 percent. Yet in the fall when the availability of grass declines and that of fruits such as acorns and beech mast suddenly increases, the proportion of fruit in the diet goes up from zero to about 40 percent. Later in the winter browse from bramble, ivy and holly becomes more important as the fruit in turn is exhausted. In the spring, as new grass grows, browse declines in importance again even though browse plants are still available (see pp 530–531). Moose are also known to vary their diet throughout the year (see pp 532–533).

Natural selection operates on differences in the number of offspring individuals leave over their lifetimes. What causes such differences in deer? The answers to this question differ for each sex.

When a female deer reaches puberty, which may be as early as her first year in White-tailed deer or as late as her fourth in poorly nourished Red deer, she will conceive and bear offspring once a year at most. The number of offspring is usually only one or two per year, although some species such as the Water-deer commonly produce triplets. After gestation, lasting between 25 weeks (Water-deer) and 40 weeks (Père David's deer), the female gives birth, often leaving her normal range and becoming solitary to do so. The calf remains concealed for the first few weeks, emerging only when the mother visits to suckle it, but soon it joins the mother in her normal range. Lactation lasts several months, for example 7 months in the Red deer. Thus female deer are occupied with one or another aspect of reproduction for most of the year.

In contrast, males contribute nothing to the rearing of their offspring. Most temperate species are seasonal breeders and for a few weeks in the fall males compete for access to receptive females, either defending harems against other males (eg Red deer—see box), defending individual females whom they leave after copulation (eg White-tailed deer), or defending territories which overlap female ranges (eg muntjac). Although in many species yearling males are fertile, they do not usually succeed in gaining females until they reach adult size (eg at 5–6 years old in Red deer). Outside the breeding season males grow antlers, feed and build up reserves for the rut. Even in species where there is no fixed breeding season, for example chital, the amount of time males invest directly in reproduction is insignificant compared to that of females. Because of this difference the factors affect-

► **Male and female Roe deer.** Deer are strongly sexually dimorphic, with usually only the males developing antlers (the reindeer is an exception).

▼ **Species of deer** CONTINUED (**8**) moose (*Alces alces*). (**9**) Roe deer (*Capreolus capreolus*). (**10**) Chital (*Axis axis*). (**11**) Reeve's muntjac (*Muntiacus reevesi*). (**12**) Sika deer (*Cervus nippon*). (**13**) Tufted deer (*Elaphodus cephalophus*). (**14**) Père David's deer (*Elaphurus davidiensis*). (**15**) Water-deer (*Hydropotes inermis*).

8   9   10   11

ing lifetime reproductive success are different in males and females. In males, reproductive success depends on the number of receptive females to which a male has access. Females are highly unlikely to lack access to males and their reproductive success depends on their ability to rear offspring and the factors that affect that ability. A study of wild Red deer on the Isle of Rhum has attempted to tease out some of these factors. Estimates of lifetime reproductive success of stags ranged from 0–24 calves with an average of 4.2. Lifetime reproductive success depends on lifespan and the success a stag has within breeding seasons. Success within a season was related to harem size, the duration of harem holding and the timing of holding relative to the conception peak; the principal determinant of these aspects of rutting success was the ability to beat other males in fights. This in turn depended on body size and condition, and indeed lifetime reproductive success of stags was correlated with antler weight, which is an indicator of body size. As a rule, large stags leave more offspring than small ones.

The total number of offspring reared by females varied from 0–13 with an average of 4.5. The lifetime reproductive success of a hind is affected by her lifespan, the number of years in which she fails to conceive (usually because of poor condition) and the mortality among her calves. In fact fecundity was less important than calf mortality,

12   13   14   15

which was in turn related to birth date, late calves being more likely to die in their first winter. Conception date and therefore birth date were related to the mother's condition during the rut. The size and condition of calves was also related to the mother's condition, but not to her skeletal size. Thus mothers with best access to the food needed to maintain good body condition successfully rear the most calves.

Because body size is important to male reproductive success and early differences persist into adulthood, natural selection favors rapid growth in males. It also favors mothers who invest heavily in their sons' early growth since they will leave more grandchildren if their sons are large. There is evidence from the longer gestation length of male calves (236 days rather than 234 for females) and the higher proportion of mothers of male calves who are in such poor condition that they fail to conceive in the following rut that mothers do make a heavier early investment in sons than in daughters.

Skeletal size and early growth have less effect on hind reproductive success than access to food during reproduction. The importance of access to food may explain why hinds remain in their natal areas when adult and why mothers allow them to share their resources (see below). Such tolerance of adult daughters may balance the heavier early investment in sons.

The social organization of different species is probably related to their differing food supply. The food of browsing deer such as Roe deer or the brockets occurs in small, discrete patches, so close proximity of other animals is likely to interfere with feeding; these species live singly or in small groups. Deer which graze in more open habitats live in larger groups (eg wapiti, Sika deer); feeding interference is not as important for grazers and they may gain some protection from predators by their companions' vigilance. In most deer, males and females live apart outside the breeding season and the social organization of the two sexes may be very different, as a consideration of Red deer will illustrate.

Males live in bachelor groups whose membership varies from day to day but in which there are consistent, linear dominance hierarchies related to body size. Threats often displace the victim from a particular feeding site and are more frequent in winter when food is scarce than in summer (1.8 versus 1.0 threats per stag per hour). High ranking stags are thus able to monopolize good food patches. Stags do not

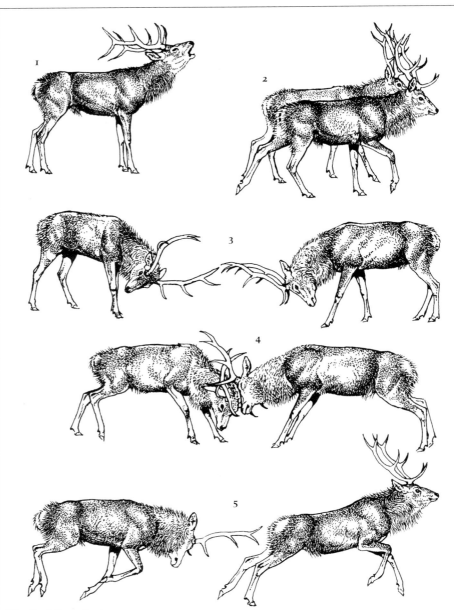

The Red Deer Rut

Rutting Red deer stags fight vigorously for the possession of harems. Usually a challenging stag approaches a harem holder and the two roar at each other for several minutes (1). Then the contest progresses to a parallel walk (2), the stags moving tensely up and down until one of them turns and faces his opponent, lowering his antlers (3). The other stag turns quickly, they lock antlers and push strenuously against each other, each trying to twist its opponent round and to gain the advantage of slope (4). When one succeeds in pushing the other rapidly backwards the loser disengages and runs off (5). Many contests do not reach the fight stage, but end after the roaring contest or the parallel walk with the challenging stag withdrawing.

A mature stag will fight about five times over the period of the rut, risking considerable danger and cost. In a study on the Isle of Rhum, Scotland, about 23 percent of mature stags were injured during the rut each year, up to 6 percent permanently. Even if he won the fight the harem holder often lost hinds to the young stags who lurk on the edges of harems waiting for the opportunity to steal them. With such costs it is not surprising that stags are reluctant to fight. Unequal contests between stags of widely differing fighting success were less common than well matched ones, suggesting that stags were able to assess their chances of winning.

Playback experiments showed that stags attempted to outdo their rivals in roaring frequency, and the roaring rates of individual stags were correlated with fighting success.

▲ **Wapiti in the snow.** In the Rocky Mountains of the USA and Canada, wapitis have been forced by farming into more inhospitable habitats. Many can die during harsh winters.

appear to distinguish relatives from non-relatives.

In contrast, hinds associate with relatives and threaten them less than non-relatives, and this is probably why threats are less frequent among hinds than among stags (0.37 threats per hind per hour in winter, 0.15 in summer). This sex difference in the treatment of relatives may occur because of kin selection, since a mother might reduce the number of grandchildren she leaves by denying her daughter access to food.

The two main factors which threaten deer populations are restriction and modification of habitat by human activity, and hunting, either as pests or for food, skins, antlers, etc. Nine species and several subspecies of deer are regarded as endangered.

Almost all species of deer have been hunted for meat, and many for their antlers as trophies. In parts of North America yearling wapiti males may be responsible for most of the following years' calves, as so many of the older males are shot for their antlers. The popularity of hunting has led to the introduction of deer to new ranges by man, and about eight species now exist in the wild in Australia and New Zealand, where deer are not indigenous.

Domestication of deer has until recently been restricted to the herding of reindeer by the Lapps for meat, hides and milk, and to the maintenance of deer (usually Red and Fallow deer) in parks for ornamental purposes. In the last two decades, however, there has been a major growth of Red deer farming in New Zealand, Australia, and Britain, chiefly for meat but also for the velvet covering the growing antlers which is believed by some, particularly in the Orient, to have aphrodisiac qualities.

Deer often come into conflict with man because they eat his crops, and they may also be pests of commercial forestry. Careful management is needed to ensure that the line between sensible pest control and persecution is not crossed, particularly in more vulnerable species.   RAC

Abbreviations: sh = shoulder height. wt = complete body weight. cwt = clean body weight, ie after removal of entrails. al = length of antler beam. pt = number of points on both antlers. ? = unknown. Dimensions are for males except where otherwise stated: females are usually smaller than males. Only males have antlers except in reindeer. Approximate nonmetric equivalents: 2.5cm = 1in; 1kg = 2.2lb.
E Endangered. V Vulnerable. R Rare. I Indeterminate. * = cites listed.

Subfamily Hydropotinae

Water-deer
Hydropotes inermis
Water-deer or Chinese water-deer

China, Korea. Swamps, reedbeds, grasslands. sh 52cm male, 48cm female; wt 11–14kg male, 8–11kg female. Antlers lacking in both sexes. Upper canines present and in males form tusks 7cm long. Coat: reddish brown in summer, dull brown in winter; young dull brown with light spots each side of midline. Gestation: 176 days. Subspecies: 2.

Subfamily Muntiacinae

Indian muntjac
Muntiacus muntjac

India, Sri Lanka, Tibet, SW China, Burma, Thailand, Vietnam, Malaya, Sumatra, Java, Borneo (introduced in England). Woodland and forest with good undergrowth. sh 57cm male, 49cm female; wt 18kg. Antlers usually single spikes; antler pedicels extend down face; al 17cm. pt 2–4. Upper canines present and in males form tusks 2.5cm long. Coat: dark chestnut on back toning to almost white on belly, darker in winter than in summer. Gestation: about 180 days. Subspecies: 15.

Reeves's muntjac
Muntiacus reevesi
Reeves's or Chinese muntjac or Barking deer.

E China, Formosa (introduced in England). Habitat, antlers and canines as Indian muntjac. sh 41cm; cwt 11kg. Coat: chestnut with darker limbs and black stripe along nape: rufous patch between pedicels on forehead; chin, throat and underside of tail white. Gestation: 210 days. Subspecies: 2.

Hairy-fronted muntjac I
Muntiacus crinifrons
Hairy-fronted or Black muntjac.

E China. Habitat, antlers and canines as Indian muntjac. sh 61cm; wt? al 6.5cm, with small projection from inner side of base. Coat: blackish brown with lighter head and neck. Tail longer than in other muntjacs.

Fea's muntjac E
Muntiacus feae

Thailand, Tenasserim. Similar to Indian muntjac in all respects but coat darker brown with yellow hairs up to center of the pedicels.

Roosevelt's muntjac
Muntiacus rooseveltorum

N Vietnam. Intermediate in size between Indian and Reeves's muntjac. Coat: redder than Reeves's muntjac. In other respects same as Indian muntjac.

Tufted deer
Elaphodus cephalophus

S, SE and C China, NE Burma. Forest up to 4,570m. sh 63cm; wt? Antlers and pedicels smaller than in muntjacs; antlers unbranched and often completely hidden by tufts of hair which grow from forehead. Upper canines present and in male form tusks 2.5cm long. Coat: deep chocolate brown, with gray neck and head; underparts, tips of ears and underside of tail white. Young have spots on back. Gestation: about 180 days. Subspecies: 3.

Subfamily Cervinae

Fallow deer *
Dama dama

Europe, Asia Minor, Iran (introduced to Australia, New Zealand). Woodland and woodland edge, scrub. sh 91cm male, 78cm female; wt 63–103kg male, 29–54kg female. Antlers branched and palmate: al 71cm. Coat: typically fawn with white spots on back and flanks in summer, grayish brown without spots in winter; rump patch white edged with black; black line down back and tail; belly white; color highly variable and melanistic (black), white and intermediate forms common. Young spotted in some color varieties. Gestation: 229–240 days. Subspecies: 2.

Chital
Axis axis
Chital, Axis deer or Spotted deer.

India, Sri Lanka (introduced to Australia). Forest edge, woodland. sh 91cm; wt 86kg; al 76cm; pt 6; beam of antler curves back and out in lyre shape. Coat: rufous fawn with white spots on back; no seasonal change in color. Young spotted. Gestation: 210–225 days. Subspecies: 2.

Hog deer
Axis porcinus

N India, Sri Lanka, Burma, Thailand, Vietnam (introduced to Australia). Grassland, paddyfields. sh 66–74cm; wt 36–45kg; al 39cm; pt 6. Coat:

yellowish brown with darker underparts, the metatarsal tufts lighter colored than the rest of the leg; young spotted. Build low and heavy with short face and legs. Gestation: about 180 days. Subspecies: 2.

Kuhl's deer R *
Axis kuhlii
Kuhl's or Bawean deer.

Bawean Island. sh 68cm; wt? al ? pt 6; pedicles long. Coat: uniform brown; tail long and bushy; young unspotted.

Calamian deer V
Axis calamianensis

Calamian Islands. Similar to Hog deer except face and ears shorter, and presence of white patch from lower jaw to the throat and white moustache mark.

Thorold's deer I
Cervus albirostris

Tibet. sh 122cm; wt? Antlers branched; al ?; pt 10. Coat: brown with creamy belly, white nose, lips, chin, throat, and white patch near ears; winter coat lighter in color than summer. Ears are narrow and lance shaped. Hooves are high, short and wide like those of cattle. Reversal of hair direction on withers gives the appearance of a hump.

Swamp deer E *
Cervus duvauceli
Swamp deer or barasingha.

N and C India, S Nepal. Swamps, grassy plains. sh 119–124cm; wt 172–182kg. al 89cm; pt 10–15. Coat: brown with yellowish underparts; in the hot season stags become reddish on the back. Young spotted. Subspecies: 2.

Red deer *
Cervus elaphus
Red deer, maral, hangul, shou, Bactrian deer or Yarkand deer.

Scandinavia, Europe, N Africa, Asia Minor, Tibet, Kashmir, Turkestan, Afghanistan (introduced to Australia and New Zealand). Woodland and woodland edge; open moorland in Scotland. Size and weight vary with subspecies and locality, eg: sh 122–127 in *C.e. hanglu*, cwt 76kg in *C.e. corsicanus*, cwt 272–299kg in *C.e. hippelaphus*: females are smaller than males. Antlers branched. varying with subspecies and locality, eg: al 123cm; pt 20 in *C.e. hippelaphus*; al 89cm; pt 12 in *C.e. scoticus*. Coat: red, liver or gray brown in summer, becoming darker and

grayer in winter; rump patch yellowish often with a dark caudal stripe; young spotted. Gestation: 230–240 days. Subspecies: 12.

Wapiti
Cervus canadensis
Wapiti or elk (elk in America only).

WN America, Tien Shan Mts to Manchuria and Mongolia, Kansu, China (introduced to New Zealand). Grassland, forest edge; in mountainous regions there is an upward summer movement. sh 130–152cm; wt 240–454kg; females smaller than males. Antlers branched; al 100cm; pt 12. Coat: in summer, light bay with darker head and legs; in winter, darker with gray underparts; rump patch light colored; young spotted. Gestation: 249–262 days. Subspecies: 13.

Eld's deer E *
Cervus eldi
Eld's deer or thamin.

Manipur, Thailand, Vietnam, Hainan Island, Burma, Tenasserim. Low, flat marshy country. sh 114cm; wt? Antlers: the long brow tine and the beam form a continuous bow shaped curve with forks at the end of the beam; al 99cm (does not include brow tine). Coat: in summer, red with pale brown underparts; in winter, dark brown with whitish underparts; white on chin, around eyes and margins of ears; females lighter colored than males. Young spotted. Walks on the undersides of its hardened pasterns as well as its hooves, a modification to marshy ground. Subspecies: 3.

Sika deer E
Cervus nippon
Sika or Japanese deer.

Japan, Vietnam, Formosa, Manchuria, Korea, N and SE China (introduced to New Zealand). Forest. sh 65–109cm; wt 48kg. al 28–81cm; pt 6–8 depending on race; top points tend to be palmate. Coat: in summer, chestnut-yellowish brown with white spots on sides and a white caudal disc edged with black; in winter, gray brown with spots less obvious; young spotted. Gestation: 217 days. Subspecies: 13.

Rusa deer
Cervus timorensis
Rusa or Timor deer.

Indonesian Archipelago (introduced to Australia, New Zealand, Fiji, New Guinea). Parkland, grassland, woods and forest. sh 98–110cm male;

86–98cm female; wt ? Antlers branched; AL 111cm; pt 6. Coat: brown with lighter underparts and under tail; young unspotted. Subspecies: 6.

Sambar
Cervus unicolor

From Philippines through Indonesia, S China, Burma to India and Sri Lanka (introduced to Australia and New Zealand). Woodland; avoids open scrub and heaviest forest. SH 61–142cm; wt 227–272kg. Antlers branched with long brow tine at acute angle to beam and forward pointing terminal forks; AL 60–100cm; pt 6. Coat: dark brown, with lighter yellow brown under chin, inside limbs, between buttocks and under tail; females and young lighter colored than males; young unspotted. Gestation: 240 days. Subspecies: 16.

Père David's deer
Elaphurus davidiensis

Formerly China, never known outside parks and zoos. SH 120cm; cwt 135kg. Antlers branched but with tines pointing backwards. AL 80cm. Coat: bright red with dark dorsal stripe in summer, iron gray in winter. Tail long. Hooves very wide. Young spotted. Gestation: 250 days.

Subfamily Odocoilinae

Mule deer [E]
Odocoileus hemionus
Mule or Black-tailed deer.

WN America, C America. Grassland, woodland. SH 102cm; wt 120kg. Antlers branch dichotomously, ie in arrangement of even forks not as tines from a main beam; pt 8–10. Coat: rusty red in summer, brownish gray in winter; face and throat whitish with black bar round chin and black patch on forehead; belly, inside of legs and rump patch white; tail white with black at tip or, in Black-tailed subspecies, with black extending up outer surface. Young spotted. Gestation: 203 days. Subspecies: 11.

White-tailed deer [E] [*]
Odocoileus virginianus

N and C America, northern parts of S America extending to Peru and Brazil (introduced to New Zealand and Scandinavia). In North America conifer swamps in winter, woodland edge spring-fall. In South America valleys near water in dry season, higher altitudes in rainy season.

SH 81–102cm; wt 18–136kg. Antlers branch from main beam, ie not dichotomously; beams curve forwards and inwards; basal snag longer than in Black-tailed deer; pt 7–8, fewer in southern races. Coat: reddish brown in summer, gray brown in winter; throat and inside ears with whitish patches; belly, inner thighs and underside of tail white; tail raised when fleeing showing white underside; young spotted. Metatarsal gland on hock shorter (2.5cm) than on Black-tailed deer (7.5–12.7cm). Gestation: 204 days. Subspecies: 38.

Roe deer
Capreolus capreolus

Europe, Asia Minor, Siberia, N Asia, Manchuria, China, Korea. Forest, forest edge, woodland, moorland. SH 64–89cm; wt 17–23kg. Antlers branched; pt 6. Coat: in summer, foxy red, with gray face, white chin and a black band from the angle of the mouth to the nostrils; in winter, grayish fawn with white rump patch and white patches on throat and gullet; no visible tail, but in winter females grow prominent anal tufts which are sometimes mistaken for tails; young spotted. After the rut in July–August implantation is delayed until December and kids are born in April–June. Gestation: 294 days (including 150 days pre-implantation). Subspecies: 3.

Moose
Alces alces
Moose or elk (elk in Europe only).

N Europe, E Siberia, Mongolia, Manchuria, Alaska, Canada, Wyoming, NE USA (introduced to New Zealand). SH 168–230cm; wt 400–800kg (females are about 25 percent smaller than males). Antlers large, branched and palmate; pt 18–20. Shoulders humped, muzzle pendulous, and from the throat hangs a growth of skin and hair—the "bell"—commonly up to 50cm long. Coat: blackish brown with lighter brown underparts, darker in summer than in winter; naked patch on the muzzle is extremely small; young unspotted. Gestation: 240–250 days. Subspecies: 6.

Reindeer
Rangifer tarandus
Reindeer or caribou.

Scandinavia, Spitzbergen, European Russia from Karelia to Sakhalin Island, Alaska, Canada, Greenland and adjacent islands (introduced to South Georgia in the South Atlantic).

Many of the reindeer in Europe and Scandinavia are domestic. Woodland or forest edge, all year for some races, but other races migrate to arctic tundra for summer. SH 107–127cm male, 94–114cm female; wt 91–272kg male. Antlers present in both sexes, smaller and with fewer points in the female; branched with a tendency for palmate top points; the brow points are palmate in the males; males shed antlers November–April, females May–June; AL up to 147cm in males; pt up to 44 in males. Coat: brown in summer, gray in winter, with white on rump, tail and above each hoof; neck paler and chest and legs darker; males have white manes in the rut; only deer with no naked patch on muzzle, which is fur-covered; young unspotted. Gestation: 210–240 days. Subspecies: 9.

Marsh deer [V] [*]
Blastocerus dichotomous

C Brazil to N Argentina. Marshes, floodplains, savannas. SH ?; wt ?; similar in size to small Red deer. AL 60cm; the brow tine forks and there is a terminal fork on the beam which gives pt 8. Coat: rufous in summer, duller brown in winter; lower legs dark; black band on the muzzle; white, woolly hair inside the ears; young unspotted.

Pampas deer [E]
Ozotoceros bezoarticus

Brazil, Argentina, Paraguay, Bolivia. Open, grassy plains. SH 69cm; wt ?; AL ?; pt 6. Coat: yellowish brown with white on the underparts and inside the ear; upper surface of the tail dark; hair forms whorls on the base of neck and in the center of back; young spotted. Subspecies: 3.

Chilean huemul [E] [*]
Hippocamelus bisulcus
Chilean huemul or Chilean guemal.

Chile, Argentina, High Andes. SH 91cm; wt ? Antlers branched; AL 28cm; pt up to 4. Coat: dark brown in summer, paler in winter; lower jaw, inside ears and lower part of tail white; young unspotted. Ears large and mule-like.

Peruvian huemul [V] [*]
Hippocamelus antisensis
Peruvian huemul or Peruvian guemal.

Peru, Ecuador, Bolivia, N Argentina, High Andes. wt slightly smaller than Chilean huemul. Antlers similar to Chilean huemul but bifurcation closer to coronet. Coat: similar to but paler

than Chilean huemul. Young unspotted.

Red brocket [*]
Mazama americana

C and S America from Mexico to Argentina. Dense mountain thickets. SH 71cm; wt 20kg. Antlers simple spikes; AL 10–13cm. Coat: red brown with grayish neck and white underparts; tail brown above, white below. Young spotted. Gestation: 220 days. Subspecies: 14.

Brown brocket
Mazama gouazoubira

C and S America from Mexico to Argentina. Mountain thickets, but more open than Red brocket. SH 35–61cm; wt 17kg. Antlers simple spikes; AL 10–13cm. Coat: similar to Red brocket but duller brown. Young spotted. Gestation: 206 days. Subspecies: 10.

Little red brocket
Mazama rufina

N Venezuela, Ecuador, SE Brazil. Forest thickets. SH 35cm; wt ? Antlers simple spikes; AL 7cm. Coat: dark chestnut with darker head and legs. Pre-orbital glands extremely large. Young spotted. Subspecies: 2.

Dwarf brocket
Mazama chunyi

N Bolivia and Peru. Andes. SH 35cm; wt ? Antlers simple spikes; AL ? Coat: cinnamon-rufous brown with buff throat, chest and inner legs, and white underside to tail; whorl on nape, supra-orbital streak and circumorbital band lacking; smaller and darker than Brown brocket. Tail shorter than in other brockets. Young spotted.

Southern pudu [E] [*]
Pudu pudu

Lower Andes of Chile and Argentina. Deep forest. SH 35–38cm; wt 6–8kg. Antlers simple spikes; AL 7–10cm. Coat: rufous or dark brown with paler sides, legs and feet; young spotted. Gestation: 210 days.

Northern pudu [I] [*]
Pudu mephistophiles

Lower Andes of Ecuador, Peru, Colombia. Deep forest. Slightly larger than Southern pudu. Antlers simple spikes; AL ? Coat: reddish brown with almost black head and feet. Young spotted. Subspecies: 2.

RAC

Harvest or Harm

Relationship between Fallow deer and man

Fallow deer are among the most familiar of the European deer, perhaps because their elegance has led to their establishment in park herds. Their special appeal to man has dramatically influenced their distribution: they were widespread throughout Europe some 100,000 years ago, but they probably became extinct in the last glaciation, except for a few small refuges in southern Europe. From these relict populations, the spread of Fallow deer to the rest of Europe, their introduction to the British Isles, North and South America, Africa, Australasia and other parts of Eurasia must all have been assisted by man.

In the wild state, Fallow deer are characteristic of mature woodlands; although they will colonize plantations (provided these contain some open areas), they prefer deciduous forests with established understory. These forests need not be particularly large, since they are used mainly for shelter. Fallow deer are primarily grazers, feeding in larger woodlands, on grassy rides or on ground vegetation between the trees, but often leaving the trees to graze on agricultural or other open land.

During the summer, the deer graze out in the open at dawn and dusk. During fall and winter, when grazing is less nutritious, they spend more time within woodland, feeding increasingly on woody browse and coming together in dense concentrations to exploit local areas of abundant mast (beechnuts). Fall is also the breeding season or rut, when bucks establish display grounds in traditional areas within the woodland and call to attract females.

Fallow deer are commonly regarded as a herding species, since they may be seen in concentrations of 40 or more, but in practice social organization is rather complex. The sexes remain separate for much of the year; adult males may travel together in bachelor groups, whereas females and young form separate herds, often in different areas to males. If a woodland is small, it may support only males or only females for much of the year.

Males enter females' areas to establish rutting stands early in fall and then adult groups of mixed sex may be observed through to early winter. Strictly speaking, these large groups are not herds but are usually chance aggregations where numbers of individuals or small groups temporarily coincide at favored feeding grounds. It now appears that the basic social unit is in fact the individual or the mother and fawn. Since the deer occupy overlapping home ranges, temporary associations of 7–14 animals may occur at certain times of year, but these are not of long duration and their composition in terms of individual animals is not constant.

The smaller social groups usually maintain their identity within the large aggregations at feeding grounds. Home ranges are usually comparatively small—20–50ha (50–123 acres) in females and 40–70ha (100–172acres) in males within the New Forest in Hampshire, England. Home ranges are probably larger in coniferous than in deciduous woodland, and when the deer form associations these too are larger in coniferous areas. The persistence of "herds" of mixed sex after the rut also appears to change with habitat-type.

With their preference for feeding on ground vegetation, Fallow deer in sufficient density can have a significant effect upon their environment. In two experimental areas of mature beech woodland in Hampshire, one occupied by Fallow deer at a density of approximately one per hectare (one per 2.5 acres), the other maintained free of all large herbivores, the grazed area is characterized by an almost total lack of herb-layer, absence of understory shrubs and complete lack of any regenerating tree

► **A tree trunk shredded** by the antlers of Fallow deer in a wood in Worcestershire, England. The damage is caused by the bucks thrashing their antlers in the trees in aggressive display or in clearing the remnants of the protective skin (velvet) from newly grown antlers.

◄ **Fallow deer in bracken.** BELOW during the summer the deer move out of the woodlands at dawn and dusk to graze, returning to the shelter of the woods to ruminate during the middle of the night and in the daylight hours.

▼ **Fallow deer in a glade.** Grassy rides and glades in woodland are especially favored by Fallow deer. Clearings provide good grazing and the surrounding trees protection.

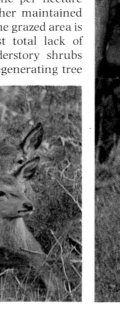

seedlings. This effect is extreme, since the density of deer is unusually high; nonetheless, the impact of the deer can be observed in other ways and in other environments. Where Fallow deer are established in small woodlands in agricultural areas, their habit of moving out into the surrounding farmland to feed may cause significant damage. Within woodland, male deer may inflict considerable damage on individual trees by thrashing them with their antlers, in aggressive display or in cleaning the remnants of the protective skin (velvet) from newly-grown antlers. In certain areas, therefore, where populations are well established, Fallow deer may be a pest both of agricultural and forest crops. As a result, most wild populations are subject to some degree of management for control. Man thus affects not only worldwide geographical distribution of this species, but local abundance too.

More recently, managers have recognized in their deer populations a valuable renewable resource. Instead of management being directed purely towards control, more populations are being harvested for profit. Similarly, the possibility of enclosing populations of Fallow deer and farming them under more intensive conditions as a new agricultural species is being explored. Fallow deer-farms have been established in Germany, Australia, New Zealand and the United Kingdom.　　　　　　　RP

Moose Foraging

How a moose chooses food

During the summer at Isle Royale, a wilderness in Lake Superior, on the USA–Canada border, moose move daily in a regular manner between aquatic and terrestrial (open-forest) feeding sites; deeper forest areas are used for resting. Their regularity in feeding is reminiscent of the regularity of a person's grocery shopping or a manager's operation of a factory. Both Adam Smith, the founder of the theory of capitalism, and Charles Darwin, the founder of the theory of evolution, noticed this similarity between animal feeding and human economic endeavors. These human endeavors follow a process of decision-making, mathematical or intuitive, ensuring that the groceries provide the best nutrition and/or pleasure and that the factory provides the best profit. Whether moose foraging is truly like such human decision-making—do moose choose the best diet?—was studied at Isle Royale, a 550sq km (212sq mi) site best known for its wolves but also the home of a large moose population.

In the operation of a factory, a manager's actions are limited by available capital, raw materials and labor force, provided the technology is available to convert these quantities into products. A moose in its feeding, likewise, is constrained by daily feeding time, digestive capacity, sodium requirements and energy needs.

A moose's daily feeding time is limited by loss and gain of body heat with the environment; on an average summer day moose can feed on land at air temperatures of $10-15°C$ ($50-59°F$) and in ponds from $15-21°C$ ($59-70°F$); otherwise it is either too hot or too cold to feed. This time restraint is inflexible but the moose makes the best use of the time available by its ability to crop different plant foods. Cropping rates are set by each food's abundance and by the moose's ability to collect each.

A moose is a ruminant, so its ability to digest food is limited by the amount of time the food remains in the rumen, one of four stomach chambers. The rumen's capacity (volume times emptying rate) imposes another constraint on food intake, and the rumen is filled to differing degrees by each plant food, as each differs in bulkiness.

Moose, like all mammals, require sodium for maintenance, growth and reproduction; at Isle Royale past glaciation has removed soil sodium, and oceanic salt spray does not add any. Sodium is certainly in short supply at Isle Royale in the sense that in order to satisfy its sodium needs a moose would have to eat more terrestrial vegetation each day than the rumen could hold. However, aqua-

tic vegetation, such as pondweed, Water lily, horsetails and bladderworts, growing in the beaver ponds, contains a higher sodium concentration. Moose eat aquatics to obtain the sodium, but eating these plants poses problems because they are available to the moose for only 120 days a year when conditions are warm enough for them to feed in the beaver ponds, and because aquatic plants are so bulky that they fill a moose's rumen five times faster than other vegetation. A moose must exceed its personal sodium needs to reproduce. Moose energy requirements for maintenance and growth must also be exceeded to allow reproduction; this is achieved by the ingestion of different plant foods, such as leaves of ash, maple and birch, each with a different energy content.

Given resource limitations, a factory manager seeks to produce a mix of products which maximizes profits. Animals, given their limitations, do not have such a clear goal, but have two possibilities. One goal is to maximize the intake of a nutritional component, generally energy. Maximum energy intake, given feeding limitations, provides a moose with the greatest energy for reproduction in summer and for winter survival, when food is in short supply. The other potential goal is to spend the least possible time in acquiring food. This enables

▲ **A moose browsing on land.** Moose take both leaves from trees and shrubs and also plants from the forest floor. Each day a moose consumes over 4kg (8.8lb) dry weight of terrestrial vegetation or the equivalent of over 20,000 leaves.

► **Aquatic foraging in moose** at Yellowstone National Park, USA. In a summer day, a moose will consume 870g (1.9lb) dry weight of sodium-rich aquatic plants, or over 1,100 plants.

► **Consumer choice.** Moose must meet their daily energy and sodium requirements, given a limited number of hours in which to feed and a limiting rate at which food can be digested. If the moose's goal is to achieve the maximum intake of energy, it will spend just long enough feeding on aquatics to satisfy its sodium requirement, and spend the rest of the time feeding on terrestrial vegetation, up to its digestive capacity. If the moose wishes to spend the minimum time in feeding (so that it can devote time to reproductive behavior), it will take the minimum of terrestrial vegetation and fill up with easily obtained, bulky aquatic vegetation. Observations of moose foraging show that they take the first course.

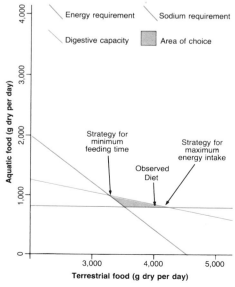

a moose to have more time for non-feeding activities, such as caring for young, and reduces exposure to hostile environmental factors, such as predators or excessive heat loss etc.

Over three years, measurements of moose physiology and behavior, and the summer environment, were made in a 3sq km (1.2sq mi) area at Isle Royale. A mathematical model of moose feeding was constructed with these data to predict what a moose should eat to satisfy either of the two potential goals. The model is identical to models used in industry to predict output!

Comparing a moose's observed diet with the model's predictions indicates that moose maximize energy intake. The diet contains enough aquatic vegetation to supply sodium demands, including sodium storage for times when aquatic plants are unavailable,

while terrestrial plants are sought for energy. Therefore, moose appear to make foraging decisions in the same way that decision-making occurs in human economic endeavors.

Why should a moose maximize energy intake? Evolution should have produced a moose diet which copes best, given limitations, with environmental factors controlling reproduction and survival. Because of mooses' large size, predation by wolves probably does not limit their survival and reproduction, with only young, old or unhealthy animals vulnerable to wolves. If such predation were important, moose might have adopted a stategy that minimized feeding time. Rather, the observed energy-maximized diet indicates that food limits moose survival and reproduction.

GEB

GIRAFFE AND OKAPI

Family: Giraffidae
Two species in 2 genera.
Order: Artiodactyla.
Distribution: Africa.

Giraffe
Giraffa camelopardalis
Distribution: Africa south of the Sahara.
Habitat: open woodland and wooded grassland.
Size: head-body length (male) 3.8–4.7m
(12–15ft); tail length (excluding tassel) (male)
80–100cm (31–39in); height to horn tips
(male) 4.7–5.3m (15–17ft), female 3.9–4.5m
(13–15ft); weight (male) 800–1,930kg
(1,765–4,255lb), (female) 550–1,180kg
(1,213–2,601lb).
Coat: patches of variable size and color (usually
orange-brown, russset or almost black)
separated by a network of cream-buff lines.
Gestation: 453–464 days.
Longevity: 25 years (to 28 in captivity).
Subspecies: **West African giraffe** (*G.c. peralta*),
Kordofan giraffe (*G.c. antiquorum*), **Nubian
giraffe** (*G.c. camelopardalis*), **Reticulated giraffe**
(*G.c. reticulata*), **Rothschild giraffe** (*G.c.
rothschildi*), **Masai giraffe** (*G.c. tippelskirchi*),
Thornicroft giraffe (*G.c. thornicrofti*), **Angolan
giraffe** (*G.c. angolensis*), **South African giraffe**
(*G.c. giraffa*).

Okapi
Okapia johnstoni
Distribution: N and NE Zaire.
Habitat: dense rain forest; prefers dense
undergrowth in secondary forest. Often
recorded in forest beside streams and rivers.
Size: head-body length 1.9–2m (6–7ft); tail
length (excluding tassel) 30–42cm (12–17in);
height 1.5–1.7m (5–6ft); weight 210–250kg
(465–550lb).
Coat: velvety dark chestnut, almost purplish
black, with conspicuous transverse black and
white stripes on hindquarters, lower rump, and
upper legs; the female is redder in color.
Gestation: 427–457 days (in captivity).
Longevity: 15 or more years (in captivity).

THE giraffe is often referred to as "the animal built by a committee," an assemblage of left-over parts, put together after the divine creator had run dry of ideas. This description belittles an animal that is superbly adapted for feeding on the high foliage beyond the reach of other hoofed mammals, and which is one of the more widespread and successful herbivores of the African savanna.

The curious anatomy of the giraffe is accentuated by the short length of the body in relation to the pronounced length of the neck. This foreshortening is exaggerated by the height of the legs, the forelegs being longer than the hind, so that in profile the animal slopes continuously from its horn tips to its tail. The neck has the usual seven vertebrae of most mammals, although each is greatly elongated. The thoracic vertebrae, especially numbers four and five, have large forward-facing dorsal spines which form the conspicuous shoulder hump, and serve as anchors for the attachment of the large muscles that support the head and neck. The tail has a meter-long terminal tuft of black hairs, producing an effective whisk against the tsetse fly. The soup-plate-sized hooves are used as offensive weapons, usually in the defense of the calves, for the powerful kick from the front feet can kill a lion. The hide is thick for defense.

A unique feature of giraffes is the progressive laying down of bone material around the skull, particularly of bulls, which in late life may be so covered with bony lumps and concretions that the original profile is obliterated. These bony growths are typically sited above each eye socket, centrally in the forehead region, and occasionally at the back of the skull, producing unusual "three-horned" and "five-horned" giraffes. An adult male skull (15kg; 33lb) may weigh three times as much as a female skull (4.5kg; 10lb) which has few bony growths.

The giraffe is one of the few ruminants born with horns. At birth, the cartilaginous cores lie flat upon the skull, but adopt the typical upright stance during the first week of life. Both sexes have horns which are unusual in being covered with skin, with a terminal tuft of black hairs. The horns start to become bony as soon as they are upright. The horns of the males are thicker and heavier, often fusing at their base and losing the terminal hair tuft. The horns and the head are used in fights between males.

Other peculiar anatomical features include the canine teeth, which are splayed out into two or three lobes to comb the leaves off shoots, and the long black tongue, which can be extended up to 46cm (18in) and is used to gather food into the mouth. To compensate for the sudden increase in blood pressure when the head is lowered, the giraffe has very elastic blood vessels and valves in the venous system of the neck.

Coat pattern and color are very variable, ranging from the large regular chestnut patches separated by a network of narrow white lines in the Reticulated giraffe, to the very irregular, star-shaped or jagged patches of the Masai giraffe, which may vary

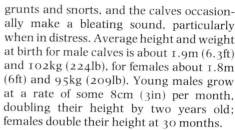

▷ **On the run.** OVERLEAF Giraffes do not form stable herds but they do band together in loose groups for protection against predators. Giraffes are extremely mobile and their long-range visual acuity is excellent, enabling groups several miles apart to be in communication.

◀ ▼ **Browsing styles.** The sex of a distant browsing giraffe can be reliably determined by stance, the female LEFT bending over the vegetation and the male feeding at full stretch BELOW. This reduces competition between the sexes.

in color from pale orange to almost black. Horn growth and coat pattern are dubious characteristics by which to classify giraffes, although they have been frequently used as such. Each giraffe has its own unique pattern of coat markings, like human fingerprints, which enables it to be distinguished from all other giraffes. The pattern remains constant from birth to death, although the color may change, so the animal can be recognized throughout its life.

The best developed sense of the giraffe is its excellent sight, although both hearing and the sense of smell are acute. Though usually silent, they can produce a variety of grunts and snorts, and the calves occasionally make a bleating sound, particularly when in distress. Average height and weight at birth for male calves is about 1.9m (6.3ft) and 102kg (224lb), for females about 1.8m (6ft) and 95kg (209lb). Young males grow at a rate of some 8cm (3in) per month, doubling their height by two years old; females double their height at 30 months.

The current distribution of giraffes is confined to the open woodlands and wooded grasslands of Africa south of the Sahara. They are typically associated with acacia (*Acacia*), myrrh (*Commiphora*), *Combretum* and open myrobalan (*Terminalia*) woodlands. Their distribution in the miombo (*Brachystegia*) woodlands of central and southern Africa is largely confined to the drainage lines where *Acacia* and *Commiphora* are common. They are absent from the central African rain forest.

Giraffes are exclusively browsers, although in the absence of browse in captivity, they will graze, often when lying down. They are highly selective feeders, the bulk of their diet comprising the leaves and shoots of trees and shrubs, supplemented by climbers, vines and some herbs. Flowers, seed pods and fruits are also eaten in season. Such material is rich in nutrients and does not show the same marked decline in quality as do the grasses during the drier months of the year. By shifting their feeding in the dry season into the woodlands beside streams and rivers, which continue to produce new leaf and shoot material throughout the year, giraffes are able to exploit a high-quality diet, and thus to breed, at all times of the year. They can also escape the seasonal limitation on breeding imposed upon most grazers by nutritional stress.

How the food is gathered depends upon the plant's defenses. With the acacias, protected by an array of long spiky and short recurved thorns, individual leaf whorls and shoot tips are bitten off, the tongue and sensitive lips being used to gather the material into the mouth. The roof of the mouth is heavily grooved, and this, together with the copious production of viscous saliva, enables the giraffe to compress and swallow the thorny food. With the thornless broad-leaf shrubs, individual shoots are pulled through the mouth, the leaves being combed off by the lobed canine teeth.

Giraffe cows feed for some 55 percent of each 24 hours, bulls for about 43 percent. Daytime feeding occurs mainly in the three-hour periods after dawn and dusk, with an increase in ruminating, or chewing the cud, during the hot midday period. Nocturnal behavior depends upon the amount of moonlight: more activity, including feeding, occurs during bright nights. Adult bulls each day consume some 19kg (42lb) dry weight, 66kg (145lb) fresh weight, cows some 16kg (35lb) dry, 58kg (128lb) fresh. Expressed as a proportion of body-weight, cows (2.1 percent per day) have a greater relative rate of food intake than bulls (1.6 percent). Although they can survive without water for relatively long periods, giraffes make regular visits to water, often trekking over long distances. Giraffes can rest standing, but they often lie down with their legs folded beneath them. The neck is held vertical except during short periods of sleep, usually of about five minutes duration, when the head is rested on the rump.

A characteristic feature of giraffe feeding is the vertical separation between bulls and cows. The typical feeding stance of an adult male is with the head and neck at full vertical stretch, often with the tongue extended, to reach the shoots on the underside of a mature tree canopy, up to 5.8m (19ft) above ground level. The typical stance of an adult female is with the neck curled over, feeding at body- or knee-height upon regenerating trees and low shrubs. So characteristic is this sexual disparity that a giraffe can be reliably sexed from its feeding behavior alone at distances too great for other characteristics to be seen.

Locomotion is peculiar in that the "pacing" walk of the giraffe involves swinging the two legs on the same side of the body forward at almost the same time. The left foreleg leaves the ground soon after the left hindleg has begun its swing forward. When galloping, the hindlegs are brought forward almost together and placed outside the front. Maximum galloping speed is 50–60km/h (31–37mph). Calves, having less inertia, sprint ahead of the adults.

Female giraffes conceive for the first time in their fifth year (48–60 months old), although sexual maturity may be delayed for a year or more by poor nutritional conditions. With a gestation period of 15 months, a mean interval between births of some 20 months, and a maximum longevity of 25 years, a cow may produce up to 12 calves in her lifetime, although about half that number is probably the average. Twin fetuses have been recorded at autopsy, but the live birth of twins is extremely rare. Males become sexually mature at about 42 months, but do not have the opportunity for successful mating until fully grown, at 8 or more years old. Sexual libido declines in very old males. Although early in courtship, subordinate bulls may consort with a receptive female, a succession of bulls of progressively increasing dominance status will temporarily consort with the cow. By the time she will stand to be mated, some 24 hours

Traditional Calving Grounds

Giraffes do not drop their calves just anywhere in the bush, for births are concentrated in discreet calving grounds. The home range of an adult female may include several such calving grounds, yet their total area probably accounts for less than 10 percent of the animal's range. From a sample of 89 calves born in the Serengeti Natonal Park, only five were located outside the calving areas.

Calving grounds show no constant feature, some comprising open grassy hilltops, others dense riverside thickets of young trees. Anecdotal evidence suggests that an area continues to be used for births for a relatively long time (at least 15 years), despite major habitat changes. A calving ground in the Serengeti has changed from open mature acacia parkland to dense regeneration thicket as a result of elephant activity, yet it is still used by giraffes as a birth area.

A probable explanation is that the concentration of births facilitates the formation of calf groups of creches, especially for calves born outside the main birth peaks. A detailed familiarity with the calving area may also have some survival value against predators at this time of extreme vulnerability.

A female giraffe is faithful to a specific calving ground. From a sample of 25 cows in the Serengeti, who were recorded as giving birth to second offspring, all the second births occurred in the same calving ground as the first. One particular female, with a home range of some 95sq km (37sq mi), dropped her second calf within 15m (50ft) of the birth-site of her previous calf born 18 months earlier. Surprisingly, the predators do not appear to have appreciated the significance of these calving grounds: if they did, dispersed births would become a more effective antipredator stategy.

Evidence from long-term radio-tracking suggest that female home ranges may not be static but drift. This drift may result in the calving ground patronized by a particular female moving out of her central core area into her buffer zone, and even out of her home range altogether. This explains the curious behavior of some heavily pregnant cows, particularly old females, who suddenly desert their usual haunts to give birth in a distant calving ground, despite the presence of a choice of well-used calving areas within their normal home range.

into heat, the most dominant bull of the locality is courting the female. A relatively small number of top bulls do all the mating.

Calves are born in isolation. The use by the cows of traditional calving grounds facilitates the formation of calf groups, which the newborn joins at about 1–2 weeks old. It is not unusual to see groups of very young calves, some still with the umbilical stump, apparently abandoned by their mothers in the middle of the day. In fact, the collective vigilance of these groups is very acute, and the predators are largely inactive in the heat of the day. The cows benefit by visiting distant feeding areas without having to spend time on the care of their offspring, resulting in good lactation.

Female giraffes are excellent mothers, and will vigorously defend their offspring against predators, kicking out with their feet. Lions are the principal predator, although newborns up to about three months old may also be killed by hyenas, leopards and even African wild dogs. In the Serengeti, the first-year calf mortality is about 58 percent. Some 50 percent of calves die in their first six months of life, with mortality greatest during the first month (22 percent). Mortality in the second and third years drops to about 8 percent and about 3 percent per annum in adults. Yet despite this apparently high rate of calf mortality, the giraffe population of the Serengeti is increasing at some 5–6 percent per annum. Male calves are weaned at about 15 months, female calves a couple of months later. No differences in the mortality of male and female calves have been observed.

Giraffes form scattered herds, the compositions of which are continually changing. In over 800 consecutive daily observations of an adult female in the Serengeti, the herd composition remained unchanged over the 24-hour period on only two occasions. The individual is thus the social unit in giraffe society, but the giraffe being a gregarious animal, individuals band together into loose groups for protection against predators. These loose associations are an adaptation to the mobility and extreme long-range visual acuity of giraffes, for groups several miles apart may be in visual communication.

Home ranges in giraffes are large, about 120sq km (46sq mi) for adult cows, but smaller in mature bulls, and larger in young males, who are great wanderers. Not all the range is equally used, for about 75 percent of sightings are confined to the central 30 percent of the area. This central core probably supplies all the requirements for day-to-day survival, but is surrounded by a familiar area which buffers the individual from the unknown world. Behavioral differences within the two areas are conspicuous, with more feeding in the central core and more alert vigilance and greater mobility in the buffer zone.

In areas where their home ranges overlap, individuals may associate, forming loose groupings. However, associations between certain individuals occur with a much higher frequency than between others, implying that giraffes have friends. This in turn suggests that they can recognize each other. Observations of their social behavior, particularly the investigation of unknown animals in their buffer zones, supports this suggestion. The exceptionally high frequency of association between pairs of adult cows may represent mother-daughter relationships.

► **Mating giraffes.** Males spend much of their time patrolling, looking for females in heat. A dominant male will mate with all such females he encounters.

► **A giraffe calf** BELOW at 5–7 days old. After weaning, young females remain within the home ranges of their mothers. Young males band together into all-male groups, and disperse from the home range where they were born in their third or fourth year.

▼ **Necking giraffes.** "Necking" is ritualized fighting indulged in mainly by young bulls to determine dominance. The necks are slowly intertwined, pushing from one side to the other, rather like a bout of arm-wrestling in humans.

Bulls are non-territorial, and amicably coexist together within overlapping home ranges. The reason for this harmony is the dominance hierarchy, the predominant feature of the bull's society. Each individual knows its relative status in the hierarchy, which minimizes aggression. Status is reinforced at encounters between two bulls by threat postures—standing tall to impress the rival. Cows also have a dominance hierarchy, although its manifestation is more subtle, usually being confined to the displacement of subordinate females at attractive feeding sites.

Young bulls have developed an elaborate ritual, called necking, to determine dominance. Necking is ritualized fight in which the two participants intertwine their necks in a slow-motion ballet. Its purpose is to assess the relative status of the two individuals. Occasionally, the intensity of the encounter may increase and the two combatants, standing shoulder to shoulder, exchange gentle blows upon each other's flanks with the horn tips. Such bouts are interspersed with much pushing, but usually degenerate into necking, which typically terminates with one animal enforcing his dominance by mounting the other.

Serious fighting, when sledgehammer blows are exchanged using the side of the head, is rare, being confined to occasions when the dominance hierarchy breaks down. This usually occurs with the intrusion of an unknown nomadic male that has no position in the local hierarchy. To protect the brain when fighting, the entire roof of the skull is pocketed with extensive cranial sinuses.

Juvenile females, after weaning, remain within the home ranges of their mothers. Juvenile males band together into all-male groups, and disperse from their natal areas in their third or fourth year. This is the age of most necking encounters, when the hierarchy is established. Top bulls can exert their dominance only over the relatively small core area of their home ranges, for their status declines as they move out into their buffer zones. There thus exists some latent territoriality, in that the dominance status has some geographical connotation. Maybe ancestral giraffes were territorial.

Within its core area, a dominant bull will mate with most of the cows that come into heat, provided that he can find them. Bulls therefore spend much of their time patrolling their core areas, reinforcing their dominance over other bulls, and seeking out females coming into heat. The bull's life-strategy is to feed for the minimum time necessary to attain the nutritional requirements for reproduction, leaving the maximum time for dominance assertion and finding cows. On the other hand, the female strategy is to feed for as long as possible so as to maximize her nutritional intake, to ensure year-round breeding and maximum reproductive success.

The constant searching for females results in the top bulls leading somewhat solitary lives. On sighting a group of giraffes, a mature bull will join the heard, testing each female in turn by sampling her urine with the flehmen, or lip-curl, response. If none of the females is in heat, the bull will soon leave the group to continue his solitary searching elsewhere. If a female is in reproductive condition, the bull will consort with the cow, displacing the subordinate bull already courting her.

Giraffes are still relatively common in East and South Africa, although their distribution in West Africa has been fragmented by poaching. In their stongholds, such as the Serengeti, they may reach densities of 1.5–2 animals per sq km (3.9–5.2 per sq mi), making an important contribution to the total animal biomass of the savanna. At such densities, their browsing can exert a significant brake upon the development of tree regeneration. In areas where elephants are destroying mature trees, this retarding of the growth of the young replacements can present a serious management problem. The problem is particularly acute in areas of high fire frequency, where heavy browsing prolongs the period that the regeneration is vulnerable to fire. A combination of burning and browsing can jeopardize the conservation of the woodlands, especially in areas where the existing mature canopy is being eliminated by elephants. But because giraffes do not actually kill trees, and merely act as a catalyst exacerbating the impact of other agents of mortality, the most effective solution to the problem is to remove the real tree killers, especially fire.

Giraffe meat has for long been sought by subsistence hunters. Certain tribes, such as the Baggara of southern Kordofan in Sudan, the Missiriés of Chad, and the Boran of southern Ethiopia, have traditionally hunted giraffe from horseback, the quarry receiving a special reverence and being of particular social significance to these tribes. But this mystique has not prevented local exterminations of giraffe by over-hunting. The tourist trade in giraffe-hair bracelets has also encouraged poaching, only the tail being removed from the carcasses.

Giraffes have important potential as a source of animal protein for human consumption. Because they feed upon a component of the vegetation that is unused by existing domestic animals, except possibly the camel, they do not compete with traditional livestock. Cattle ranchers and pastoralists have therefore been tolerant of giraffes, although they can damage stock fences. Giraffes rapidly become used to the presence of man and, as a free-range resource, can be easily cropped. At high densities, it has been estimated that giraffes could provide up to one-third of the meat requirements of a pastoral family, by exploiting the high browse which is currently neglected. This would reduce the dependence upon the sheep and goats that are responsible for much of the overgrazing and erosion problems in Africa today.   RP

▶ **The elusive okapi,** which looks somewhat like a cross between a giraffe and a zebra. Okapis are secretive, solitary animals, relying upon their acute senses of hearing and smell to evade predators; unlike giraffes, their sight is relatively poor. Okapis are generally described as being nocturnal but in undisturbed areas they may be active in the day. They are pure browsers, feeding upon the leaves of young shoots of forest trees. Seeds and fruit have also been recorded in their diet.

▼ **To drink,** giraffes have to splay their forelegs very widely and then bend the knees.

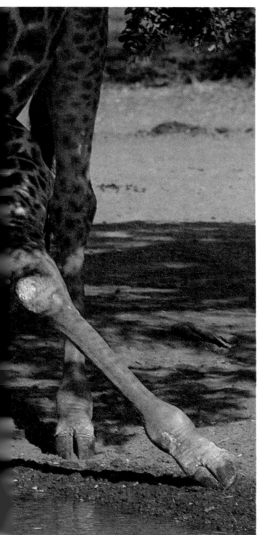

open savanna in the giraffe, forest in the okapi. Early reports suggested that okapis, especially the males, were nomadic. However, their regular use of tracks linking favorite feeding areas suggests a more sedentary way of life. Unlike giraffes, okapis have glands on their feet, and they have also been reported marking bushes with urine. These factors, together with their solitary existence, imply a social system based on male territoriality.

In captivity, heat lasts for up to a month, during which time the male consorts closely with the female. Females advertise their condition by urine marking and by calling, although at other times okapis are silent. The exceptionally long period of heat may be to ensure sufficient time for the male to locate the female and to overcome their solitary inclinations before mating. The initial approach and contact phases of okapi courtship are marked by more female aggression and male dominance display than in giraffes, presumbably to overcome the defensive reactions of typically solitary individuals. Male courtship behavior is more antelope-like, including flehmen (lip-curl), display of the white throat patch, leg kicking and head tossing. Male encounters, particularly in the presence of a female in heat, are often aggressive, the ritualized neck fight being reinforced by periodic charging and butting with the horns.

Calves are born with non-adult proportions, having a small head, a short neck and thick long legs. The conspicuous mane is much reduced in the adults. Calves remain hidden for the first few weeks. Vocalization is important to maintain contact between mother and offspring. In captivity, calves are weaned by six months. The horns of males develop between the first and third years. Leopards are the principal predator of calves. Like giraffes, females defend their offspring by kicking with their feet. The strong contrast of the zebra-like stripes of the back legs and flanks are probably important for calf imprinting and as a follow-me signal.

The okapi has been protected by government decree since 1933. Although the Zaire authorities have made real efforts to enforce this protection, it is difficult to prevent hunting in remote forest areas, particularly when the meat is so prized. Subsistence hunting by pygmies is unlikely to endanger the status of the species. This is not the case, however, with intensive commercial poaching, which now threatens all the larger forest mammals, including the okapi, in the more accessible areas.   RAP

The **okapi** remains one of the major zoological mysteries of Africa, so little is known about its behavior in the wild. The species was only officially "discovered" in 1901 by Sir Harry Johnston, the British explorer, whose interest had been aroused by the persistent rumors of a horse-like animal, living in the forests of the Belgian Congo, that was hunted by the pygmies. Okapi is the name that the pygmies have given to this creature.

Its present distribution is confined to the rain forest of northern Zaire, between the Oubangi and the Uele rivers in the west and north, and up to the Uganda border and the Semliki river in the East. Like most forest-dwelling large mammals, it is very localized, but in areas of suitable habitat is relatively common, particularly in the Ituri and Buta regions. Local population densities of 1–2.5 per sq km have been suggested. Suitable habitat comprises dense secondary forest where the young trees and shrubs increase food availability. Riverside woodland, and particularly clearings and glades where the penetration of light encourages the production of low browse, is favored. Closed high-canopy forest with little ground vegetation is not suitable habitat.

Characteristic giraffe-like features include skin-covered horns, lobed canine teeth, and an extendable long black tongue which is used to gather food into the mouth. Only males are horned although females sometimes have a variable pair of horn sheaths.

Differences in behavior and ecology between the two giraffid species arise from differences in the habitat in which they live:

PRONGHORN

Antilocapra americana
Pronghorn, prongbuck, antelope, berrendos
(Mexican), Tah-ah-Nah (Taos), Te-na (Piute),
Cha-o (Klamath).
Subfamily: Antilocaprinae.
Family: Bovidae.
Order: Artiodactyla.
Distribution: W USA and Canada, parts of
Mexico.

Habitat: open grass and brushlands, rarely
open coniferous forests; active in daytime to
twilight.

Size: head-to-tail length 141cm
(55in); shoulder height 87cm
(34in); weight 47–70kg
(103–154lb); horn length
(female) 4.2cm (1.5in); (male)
43cm (17in). Females fractionally
smaller than males.

Coat: upper body tan; belly, inner limbs,
rectangular area between shoulders and hip,
shield and crescent on throat, and rump white.
Face mask and subauricular gland on mature
males black.

Gestation: 252 days.

Longevity: 9–10 years (to 12 years in
captivity).

Subspecies: *A.a. americana*, *A.a. peninsularis* [E]
A.a. mexicana, *A.a. sonoriensis* [E].

[E] Endangered.

THE pronghorn is renowned in folklore and in fact for its speed, endurance and curiosity. Speeds of up to 86km/h (55mph) have been recorded and 70km/h (45mph) can be maintained for over 6.4km (4mi). They will approach and inspect moving objects, even predators, from long distances. Early settlers used this "curiosity" to attract pronghorns within gunshot range by flagging (tying handkerchiefs to poles and waving them in the air). Flagging is now illegal, but contributed to the pronghorn's decline from 40–50 million in 1850 to only 13,000 by 1920.

Pronghorns have chunky bodies with long, slim legs. Their long, pointed hooves are cloven and cushioned to take the shock of a stride that may reach 8m (27ft) at a full run. Both sexes have black horns that are shed and regrown annually. The buck's horns have enlarged forward-pointing prongs below backward-pointing hooks. The doe's horns are rarely long enough to develop prongs or hooks. Pronghorn eyes are unusually large and are set out from the skull, allowing a 360° field of vision. Long, black eyelashes act as sun-visors. The beautifully marked tan and white body and neck are unique among North American large mammals. These markings are related to the pronghorn's use of open prairie and brushlands and are important in communication. Bucks have a black face mask and two black patches beneath the ear. These black markings are used in courtship and dominance displays. Scent is an important, but little understood aspect of pronghorn communication. Males have nine skin glands (2 beneath the ears, 2 rump glands, 1 above the tail, 4 between the toes) and does six (4 between the toes, 2 rump). Rump glands produce alarm odors, and scent from the glands beneath the ears is used to mark territories and during courtship.

The pronghorn is the only living species of the subfamily Antilocaprinae which in the late Pleistocene (about 1.5 million years ago) had 5 genera and 11 species. The subfamily is North American in origin and distribution. Current subspecies differ slightly in size, color, and structure, with darker, larger forms in the north and paler, smaller forms in the south.

The pronghorn diet varies with the local flora, but on a plant-type basis they eat: forbs (herbs other than grass), shrubs, grasses, and other plants (cacti, domestic crops etc). Forbs are eaten from spring to late fall and are critical to good fawn production. Shrubs are eaten all year, but are most important in winter. Grasses are used in spring and other food types vary locally in importance. Succulence is an important criterion in deciding food choice, reproduction, timing of migrations, and placement and quality of territories. Pronghorn teeth are adapted to selective grazing and they grow continually in response to wear and provide an even surface for grinding rough vegetation.

Both sexes mature at 16 months, but does occasionally conceive at 5 months. Only dominant males breed and this usually delays a male's first breeding until 5 years of age or older. Twins are common on good range and females produce 4–7 eggs. Up to four fertilized eggs may implant and when this occurs the tip of the embryo in the upper chamber of the two-horned (compartmented) womb grows and pierces the fetal membranes of the embryo lower in the womb, eventually killing its sibling while still in the womb.

The rut occurs from late August till early October, with exact timing depending on latitude; it only lasts 2–3 weeks in any given area. Fawns are born in late May to early June and on good range weigh from 3.4–3.9kg (7.5–8.6lb) at birth. At two days old a fawn can run faster than a horse, but it lacks the stamina needed to keep up with a herd in flight; therefore it hides in vegetation until 21–26 days old. Fawns interact with their mothers for only 20–25 minutes each day; even after an older fawn joins a nursery herd this pattern continues. Does nurse, groom, distract predators from their fawns, and lead their fawns to food and water. Nursing and grooming continue for about 4–5 months, but development of aggressive

▲ **Pronghorns in open prairie.** There were perhaps 35 million pronghorns on the plains of North America before the arrival of settlers.

◄ **Lying out.** Newborn pronghorns hide until they are able to move with the herd.

▼ **A male pronghorn,** showing the curiously hooked horns. They are shed after the rut and the males join the female herds.

behavior may cause weaning of males 2–3 weeks earlier than females. Adult bucks show no direct paternal care.

Pronghorns and Roe deer are the only two territorial species of ungulates found in northern latitudes. Males defend and mark territories from early March to the end of rut even though mating only occurs in the fall. Territories may range in size from 0.23–4.34sq km (0.09–1.68sq mi). Doe-fawn herd home ranges usually include several territories and may range in size from 6.35–10.50sq km (2.45–4.05sq mi). Bachelor herd home ranges are located adjacent to territorial areas and are inferior to territorial areas in forage quality. Bachelor home ranges may vary in size from 5.1–12.9sq km (2–5sq mi).

Over 95 percent of all mating is on territories by territory holders. North American rangelands have relatively continuous forage distributions, but within these ranges there are pockets of better forage that a male can defend from rivals. These more productive areas occur in depressions where soils are more moist and richer in nutrients. There is a strong correlation between a male's dominance position established as a bachelor and the quality of the territory he acquires. Does mate most often with males on the best territories and may return to mate on the same territory throughout their lifetime. Therefore, 30–40 percent of all mating may occur on only the best one or two territories in a given area. A top male on a high quality territory might hold it up for 4–5 years before being replaced, and might sire from 15–30 percent of all fawns for a given year. In a given age-class of bucks, as many as 50 percent may never reproduce.

Early conservation efforts have brought pronghorns back from near extinction to a population of over 450,000. While this is low compared to their former abundance, it is a significant recovery and provides excellent recreational opportunities for hunters, photographers and nature watchers. However, oil exploration and strip mining for coal now pose threats to their habitat.

DWK

WILD CATTLE AND SPIRAL-HORNED ANTELOPES

Subfamily: Bovinae
Twenty three species in 8 genera.
Family: Bovidae.
Order: Artiodactyla.
Distribution: Africa, India, N America to Europe, Asia, Philippines and Indonesia.

Habitat: ranges from alpine tundra to tropical forest, sometimes near water.

Size: head-body length from 100cm (39in) in the Four-horned antelope to 240–340cm (95–134in) in the Cape buffalo; weight from 20kg (44lb) in the Four-horned antelope to 1,200kg (2,640lb) in the Wild water buffalo.

Four-horned antelopes
Tribe: Boselaphini.
Two species in 2 genera: **Four-horned antelope** (*Tetracerus quadricornis*); **nilgai** (*Boselaphus tragocamelus*);

Wild cattle
Tribe: Bovini.
Twelve species in 4 genera.
Species include: **African buffalo** (*Syncerus caffer*); **American bison** (*Bison bison*); **banteng** (*Bos javanicus*); **cattle** (*Bos primigentus*); **European bison** (*Bison bonasus*); **gaur** (*Bos gaurus*); **yak** (*Bos mutus*).

Spiral-horned antelopes
Tribe: Strepsicerotini.
Nine species in 2 genera.
Species include: **bongo** (*Tragelaphus euryceros*); **bushbuck** (*Tragelaphus scriptus*); **Common eland** (*Taurotragus oryx*); **Giant eland** (*Taurotragus derbianus*); **Greater kudu** (*Tragelaphus strepsiceros*); **Mountain nyala** (*Tragelaphus buxtoni*); **sitatunga** (*Tragelaphus spekei*).

► ▲ **The four-horned antelope** RIGHT and the nilgai ABOVE are the two remaining species of the primitive tribe Boselaphini from which modern cattle arose.

WHEN we think of cattle, our minds turn to peaceful herds of dairy cows or to placid beasts of burden. These are reasonable impressions, since there are 1.2 billion Domestic cattle in the world and 130 million water buffalo. In contrast, the kouprey, a very rare forest ox of Cambodia, probably numbers only a few dozen, if indeed it is not already extinct.

Cattle (tribe Bovini) evolved from animals resembling the present-day Four-horned antelope and the nilgai (tribe Boselaphini). The Bovini achieved great diversity in the warm Eurasian plains of the Pliocene (5–3 million years ago) and, in the form of the yak and bisons, they evolved into cold-tolerant species which could cope with the rapid climatic changes of the Pleistocene. Only the bison made the passage via Siberia and Alaska to the New World, spreading as far south as El Salvador.

The challenge of the Pleistocene was also met by the highly successful aurochs, ancestor of our cattle, which spread out from Asia, and after the last Ice Age occupied a vast range from the Atlantic to the Pacific, and from the northern tundras to India and North Africa. A distinct form of aurochs inhabited India, Asia and Africa. As agriculture spread, numbers declined and the last one died in 1627.

The banteng, the gaur, the yak and the Water buffalo have also been domesticated, and with domestication came experimentation. Hybrids between cattle and yak are economically important in Nepal and China. Attempts have been made in North America to develop a breed (called the cattalo) based on a bison-cattle cross. However, the male progeny of these hybrids are infertile.

In many places, domesticated animals have been allowed to revert to the wild state. In the Northern Territory of Australia, perhaps as many as 250,000 water buffalo roam free, descendents of animals left behind when military settlements were abandoned in the 19th century. The nilgai, although never domesticated, has been successfully introduced to Texas, where it lives as a wild animal. In Britain, many stately homes and castles had parks with herds of more-or-less wild cattle. The most ancient such herd is the Chillingham herd in Northumberland, England.

All species of wild cattle have a keen sense of smell which they use to detect enemies, communicate information to each other and find food—while grazing, they constantly sniff the pasture. Sight and hearing are generally good though not particularly

▲ **Yaks in harsh terrain** near Khumjung, Tibet. Cattle thrive in such extremes, maintaining their fermentation system in the rumen at 40°C (104°F).

▷ **The Viking helmet look** OVERLEAF of the African buffalo, seen here among Red oat grass and attended by oxpeckers.

Domestic cattle; their rumen contents ferment at 40°C (104°F). This "central heating system" means that some breeds do not have to generate extra heat (by shivering, eating more etc) even at −18°C (0.4°F), in still dry air. The environment of the yak or the bison can be very rigorous, and perhaps this is one reason why the embellishments of the bull bison and bull yak are hairy and heat-conserving (ie the beard and the long fringes on the body) rather than fleshy and heat-radiating (ie the dewlaps and dorsal ridges of the Asian wild cattle).

The Boselaphini have what is thought to be a primitive form of social behavior. The Four-horned antelope lives a solitary life but forms small groups during the rut, probably marking out territories with its facial and foot glands. Nilgai cows and calves tend to stay in herds. The bulls are solitary except during the rut, when they establish territories and gather breeding herds of up to 10 cows, fighting other bulls with their horns. Both species advertise their presence by defecating in particular places.

The Bovini, in contrast, have developed the herd life-style. Little is known of the anoas and the tamarau. In all the other species, females spend their lives in groups of stable composition organized on the basis of personal recognition and which consist of cows, their offspring and perhaps further generations. Young bulls generally leave at about 3 years. These groups are small (ie up to 10) in the gaur, banteng and African buffalo (Forest subspecies). The American and European bison usually live in larger groups (10–20). In all these species, temporary aggregations may form particularly in the breeding season. The African buffalo (Cape subspecies) lives in very large herds. In the open plains and wooded grasslands of the Serengeti, they average 350. Feral Water buffalo in Australia also live in an open habitat; their herds average 100–300, made up of family groups of up to 30 closely related animals. Bulls live alone or in small bachelor herds numbering up to 10–15 which join the female groups during the breeding season or, in the case of the Cape buffalo, when the rains start. In species which inhabit relatively dense cover (Wild water buffalo, banteng, gaur, European bison and forest-dwelling populations of the American bison), individual mature bulls have been seen in cow groups all year.

Males need to be able to assess each other's fighting potential and so avoid dangerous fights between adversaries of comparable status. Their facial musculature does not permit a wide range of expressions,

acute. Domestic cattle have partial color vision—they cannot distinguish red. The sense of taste and of touch on the lips are important in food selection, and the more succulent plant parts are always preferred.

When not feeding, the way that the Bovini spend their time is often determined by the need to avoid heat and insects. Although all species sweat, they frequently have to seek shade. Water buffalo are particularly sensitive to heat, and depend on wallowing; in many places, they are only active at night. Sleep has been studied in Domestic cattle—bouts typically last 2–8 minutes and total no more than one hour in 24.

Cold weather presents few problems to

so posture and movement are used instead. The few contests which escalate to actual combat may result in serious injuries. The horns of bovids are very effective anti-predator defenses. A herd of Cape buffalo is very likely to attack a lion; this is such an effective defensive unit that blind, lame or even three-legged individuals may continue to thrive within it. Tigers, however, frequently kill fully grown gaur, and solitary African buffalo bulls often fall victim to lions. Evidently the habit of associating in herds is a vital part of their defense strategy.

The alarm call of the Bovini is an explosive snort, quickly followed by the alarm posture in which the head is held high, facing the danger, with the body tensed. When running away, gaur have the unusual habit of thumping the ground with their forelegs in unison. The Cape buffalo's alarm call brings the whole herd to the defense of the frightened animal.

Animals living in cohesive herds have to coordinate their activities: among Cape buffalo the whole herd may change activity within a few minutes, for example from grazing to lying down, and the American bison's habit of stampeding is notorious.

This harmonization does not extend to breeding. The Boselaphini have long breeding seasons and produce litters of young—the nilgai produces one or, more commonly, two young, and the Four-horned antelope between one and three. Twins are rare in the Bovini, about one in 40 of calvings in Domestic cattle. These are unwelcome because if one twin is a male and the other a female the latter will usually be sterile; this is because in cattle most twins share a common blood supply and hormones and cells can be exchanged.

The length of the rut may depend to a great extent on the availability of food; year-round sufficiency of food leads to year-round breeding in Domestic cattle. Bison rut in the summer, and births coincide with the spring flush of grass.

A bull detects that a cow is in heat by sniffing her urine and genitals—while doing this, he displays the flehmen, or lip-curl, reaction which, with the pumping action of the tongue, forces scent to the Jacobson's organ above the palate. In all the Bovini the cow becomes attractive to the bull several hours before she ceases to run away whenever he tries to mount her. In commercial herds of cattle, receptive cows mount other cows—in the Chillingham wild cattle this is hardly ever seen because the bulls are efficient at detecting receptive cows; then they do their best to stop the cow from moving away or associating with other animals. The rutting behavior and fighting of the bull Bovini leads to a brief mating bond with a cow (see pp554–555). If he is not displaced by another bull, he will in due course mount her and inseminate her.

When about to give birth, Domestic cows and the Chillingham cows tend to move away from the herd. African buffalo usually calve within the herd and the cow stays and defends her calf fiercely if the herd should then move on. The gaur is perhaps less efficient at protection—tigers frequently find and kill calves. American bison cows sometimes leave the herd to give birth (lying down to calve); they, too, defend their calves. The Chillingham cows calve standing up; the cow eats the afterbirth. She licks the newborn calf to prevent infestation by flies and to stimulate it to defecate. The calf soon attempts to stand and reach the udder. When the calf has fed, the cow usually rejoins the herd and returns at intervals to feed the calf. After about four days, Chillingham calves, instead of lying down after being fed, follow their mother back to the herd. The other cows show great interest in the new arrival and the mother sometimes chases them away. The bulls are less interested.

In the African buffalo, fertility seems to stay constant as population density increases; in other species, cows may not come into heat when conditions are bad (this is noticeable in the banteng in Java and in the Chillingham wild cattle) but this is a response to the physical rather than to the social environment.

Some of the Bovini were once exceedingly numerous: 40–60 million bison roamed North America. In 1898 the explorer Wellby wrote of Tibet that "on one green hill there were I believe more yak visible than hill." By contrast, some species were probably always rather rare. The gaur, banteng and kouprey need grassy glades in forests, which were probably uncommon before man started praticing shifting agriculture. And crop-raiding aurochsen may have been the first domesticated cattle.

Paradoxically, the success of the Bovini is the cause of the main threat to the survival of some of the species. Any habitat that can support wild cattle can support Domestic cattle, and man's herds carry diseases that can wipe out wild populations. In countries hungry for meat and trophies, big wild hoofed mammals are a great temptation. For these reasons, the survival of most of the wild Bovini can only be assured in properly protected reserves. Indeed, the European bison died out in the wild in 1919 and the present thriving populations in Poland and the Caucasus (USSR) were built up from zoo stocks, then released back into the wild. This may be the only hope for the kouprey. There is also a need to conserve the rare and vanishing breeds of cattle. These, together with the wild species, comprise genetic banks which may well be a valuable source of new varieties and hybrids.   SJGH

The **spiral-horned antelopes** are an African offshoot from the boselaphine lineage but differ from the boselaphines and most African antelopes in apparently lacking territoriality. In this they resemble the Bovini, despite having a very different feeding adaptation, so that comparisons of this tribe with cattle and with other antelopes are likely to give new insights into the evolution of mammalian social organizations. The possibility of domesticating one species, the Common eland, may provide an alternative to cattle in marginal habitats.

More slenderly built than the Bovini and with long, narrow faces, the elegant spiral-horned antelopes are further distinguished by long horns, corkscrewing backwards in the plane of the face. All species are to some extent sexually dimorphic, males tending to be heavier and darker than females and either being the only sex with horns or else having the heavier, more robust horns. They have sharp senses: their greatly enlarged ears suggest that sound is particularly important, but they vocalize rarely. Calls include a few mother/infant calls, an alarm bark and some courtship calls, which often resemble maternal calls.

Even the gregarious, open-country elands are shy; smaller spiral-horned antelopes are mainly active either during the night or at dawn and dusk, and are rarely seen even when at high densities. All species have a white underside to the tail and several have short bushy tails which they turn up sharply when in flight, so showing a conspicuous white "flag."

Between them, the nine species of spiral-horned antelopes have exploited many of the major African habitats south of the Sahara. They are often found either in the zones between vegetation-types or in seasonally variable habitats, where their opportunistic feeding may give them an advantage over more specialized feeders. Their food preferences and their lack of the explosive speed in flight from predators found in many open-habitat species seem to have limited the smaller spiral-horned antelopes to woodlands and other closed

habitats. Even so, the bushbuck, normally found in dense woodland, can use local cover, such as that given by thickets associated with termites' nests in otherwise open grasslands, to exploit a wide range of habitats. The sitatunga lives in marshes, swamps and reed-beds, and the nyala in and around riverside forest and similar habitats in southeastern Africa. The Mountain nyala, only recognized as a species at the beginning of this century, has a limited distribution but is apparently well adapted to its stronghold in the heaths and forests of the Ethiopian highlands. The large, gregarious eland species and, to a certain extent, the kudus, exploit more open woodlands and even grasslands, so following the general trend for large antelope species to be found in more open, arid habitats than their smaller relations. Some nomadic Common eland herds penetrate into extremely dry thorn scrub areas of the Namib and Kalahari deserts, taking advantage of seasonal browse flushes and using water pans as refuges. However, the bongo (the largest known forest antelope) is found in various dense East, central and West African woodlands, a notable and so far unexplained exception to the large species/open habitat rule.

▲ **Wallowing water buffalo** in Sri Lanka. Water buffalo wallow in mud to form a protective cake on their skin which may shield the body from solar radiation as well as providing a protection against biting insects.

▶ **More graceful** than cattle, the Common eland has a long, narrow face and elegantly fluted spiral horns.

▶ **Horn-tangling in Common elands.** Male elands assess each other's status by means of ritual confrontations (horn-tangling). (1) As a preliminary they thresh their horns through aromatic shrubs, collecting gummy saps; (2) they then drive them into the mud, often where another eland has urinated; (3) the horn-tangling encounter itself involves pushing with the horns gently engaged. The smelly mud and sap on the horns advertises prowess at horn-tangling.

All spiral-horned antelopes feed opportunistically off a wide range of food types and species. Their ability to pick out scanty, high quality foods from much poorer surrounding vegetation has led to their being described as "foliage gleaners." They may take fruits, seed pods, flowers, leaves (both monocotyledon and dicotyledon species), bark and tubers. Horned species are known to break down otherwise out-of-reach branches and to dig up tubers.

As in most other antelope groups, the smallest, the bushbuck, probably has the highest quality diet and takes the least grass. The much larger eland species can subsist on much poorer forage but are still much more selective than are similarly sized cattle species, and all body structures related to food-processing reflect this. Spiral-horned antelopes have small, low crowned teeth, their muzzles are narrow and long, and their digestive systems are not particularly specialized for processing protein-poor, high-fiber food. Cattle species, on the other hand, are mainly grazers and so have to cope with highly fibrous food: they have much more massive teeth, broader muzzles and more elaborate stomachs.

Little is known of the social organization of some spiral-horned antelopes but only one species, the bushbuck, shows a weak form of the territoriality typically found in the boselaphines. Males generally seem to rely on established dominance hierarchies to determine access to females when in large herds, and on dominance displays when strange males meet, in the more solitary species. This convergence with cattle in social organization seems to have had anatomical consequences: the specialized facial and foot glands found in the territorial boselaphines are missing, and body size is emphasized by contrasting markings, by crests and by dewlaps as in the Bovini.

The spiral-horned antelopes tend to have smaller group sizes than do comparable cattle. This may follow from differences in feeding behavior: spiral-horned antelopes often move quite rapidly and erratically through dense woodland while searching for high quality food. If they were in the large herds typical of the larger Bovini, not only would group members have great trouble coordinating their activities in poor visibility, but also there would be great competition as many animals tried to glean scarce, high quality food items in the same area.

Females reach sexual maturity at about 18 months (bushbuck) to 3 years (elands). Males in most species continue to grow beyond sexual maturity and it is usually about this age that they start to become noticeably distinct from the females in coloring and build, although their horns will have been growing and differentiating since birth. Males probably breed successfully from 1–3 years after their female peers: they are capable of breeding much earlier, but social factors probably prevent this.

All spiral-horned antelopes almost invariably bear single calves. A given cow probably bears only one calf per year, although the calving rate for bushbuck, which have a gestation period of about five months and may be found in habitats which are only weakly seasonal, may approach two per year. Calving dates vary both with species and with habitat: populations in strongly seasonal habitats (desert fringes, extreme highlands, southern Africa generally) tend to have more sharply defined breeding seasons than those in less variable habitats (riverside forests, equatorial woodlands). In southern Africa, Greater kudu calves are often dropped in the middle or towards the end of the rainy season, while eland cows in the same area will give birth just before the rains begin, a difference of up to six months. Calves wean at from 3–6 months.

Spiral-horned antelopes, particularly the smaller ones, are eaten by a wide range of terrestrial predators. Even the elands, the largest of all antelopes, may be taken by lions or hyenas. The smaller species probably avoid predation mainly by conceal-ment—the sitatunga may actually sub-merge itself for some time, with just its nostrils above the water, while the larger species will usually flee. Eland cows are known to attack predators near their calves; other spiral-horned antelopes may also do this, but the main defense against predation for newborn calves is to lie concealed in dense undergrowth or tall grass away from the herd for up to a month, being visited by their dams several times a day to nurse. All species have a gruff alarm bark.

When calves join the herd after lying out, they often form a distinct nursery group and have very little to do with adults except for their mothers. Female calves tend to stay with their mothers more than males do, so that most female groups of the smaller species probably consist of related animals (from 2–3 in the bushbuck, up to 10 or 15 in the Greater kudu). Males usually split off from their original group, perhaps as soon as they are weaned, but in any event before they are sexually mature, and form groups of from 2–30 (depending on area and species) males of mixed ages. Older males are more solitary than young ones. The elands differ from this slightly in that small cow groups aggregate during the conception peak into herds of up to several hundreds; many males associate with these herds of females.

▲ **Eland herd.** Unusually for spiral-horned antelopes, elands form herds of up to several hundreds in the breeding season.

◄ **The Greater kudu** shows extreme differences between the sexes, with the females, as here, lacking horns and being considerably smaller than the males.

▼ **The Giant eland** is more of a woodland species than the Common eland, although seen here in wooded savanna.

Regardless of the season, female groups are often accompanied by one or more males, but this is often only a loose association and a given group of females may be accompanied by several different males over a matter of days. During the conception peak, however, a given male may, if not driven off, escort a female for several days. Calves are therefore probably fathered by the most dominant bull around when cows come into heat.

Most spiral-horned antelopes, particularly the elands and kudus, are highly mobile and probably have large home ranges compared with similarly sized antelopes. Most species are resident in a given area throughout the year, the ranges of adjacent groups overlapping considerably. The elands travel large distances through the year and some populations must be regarded as migratory, although they do not necessarily form large, conspicuous herds in their yearly moves from, for example, highland plateau grasslands to lowland savannas. Other species show a more subtle shift in habitat use: Greater kudus may move up and down hillsides and vary their use of areas around the hills to take advantage of seasonal changes in the vegetation that they consume.

All the species of spiral-horned antelope are fairly secure, at least in the short term, even the Mountain nyala, once thought to be highly endangered. Various populations within species are, however, less safe. In particular, the low-density, highly mobile elands are vulnerable to any agricultural development and to hunting. The refuges provided by nature and forestry reserves, and by some game ranching efforts, are particularly important, since the spiral-horned antelopes have largely disappeared from areas cleared for agriculture. The precarious state of the Western race of the Giant eland, which was probably originally a scattered series of small local populations, should warn against allowing other species to be whittled back to a few small strongholds.

Several species are known to have suffered badly from rinderpest. The various ranching experiments with elands and kudus also suggest that the spiral-horned antelopes are particularly vulnerable to heavy tick infestations and to resulting tick-borne diseases. Bushbuck and kudus may act as reservoirs of sleeping sickness in tsetse infested areas.

Perhaps because of their combination of size and elegance, the larger spiral-horned antelopes, particularly the elands, have a special place in African tradition. Rock paintings in the Kalahari include sketches of Common eland apparently in some form of domestic relationship with the local bushmen. In more recent times, herds of Common eland have been kept under various regimes as dairy animals (Askanya Nova, USSR), draft animals (Natal, South Africa), and as either a direct substitute for more conventional livestock in marginal habitats or else, as with many other species of hoofed mammals, as part of a game-cropping scheme. The success of such efforts seems to depend very much on local conditions and the individuals involved: large-scale plans for land use, now being prepared in several African countries, should provide more and better opportunities for such development. Experiments in domestication are doubly important since similar efforts may have already saved the blesbok and the Black wildebeest from extinction.

RU

THE 23 SPECIES OF CATTLE AND SPIRAL-HORNED ANTELOPES

Abbreviations: HBL = head-body length. TL = tail length. HL = horn length. SH = shoulder height.
HT = overall height. WT = weight. Approximate nonmetric equivalents: 2.5cm = 1in; 1kg = 2.2lb.
E Endangered. V Vulnerable. * CITES listed. D Domesticated.

Tribe Boselaphini

Both species in peninsular India. Relicts of the stock from which the Bovini arose. They have primitive skeletal, dental and behavioral characteristics; females not horned.

Genus *Tetracerus*

Four-horned antelope *

Tetracerus quadricornis
Four-horned antelope or chousingha.

Wooded, hilly country near water. HBL 100cm; TL 13cm; HT 60cm; HL (male) front pair 2–4cm, rear 8–10cm; WT 20kg. Coat: male dull red-brown above, white below, dark stripe down front of each leg; old males yellowish; females brownish bay. Facial and foot glands.

Genus *Boselaphus*

Nilgai

Boselaphus tragocamelus
Nilgai or Blue bull.

Thinly wooded country. Rather horse-like. HBL 200cm; TL 45–50cm; HT 120–150cm; HL (male) 15–20cm; WT male 240kg; female 120kg. Coat: coarse iron-gray; white markings on fetlocks, cheeks, ears; tuft of stiff black hairs on throat; young bulls and cows tawny.

Tribe Bovini
(Wild Cattle)

N America to Mexico; Africa, Europe, Asia, Philippines and Indonesia. Introduced to Australia and New Zealand. Stout bodies, low, wide skulls and smooth or keeled horns in both sexes, splaying out sideways from skull. No facial or foot glands. Sexual maturation and frequency of calving depend to a degree on health and feeding, but all those which have been studied can achieve maturity at 2 years and produce a calf a year for most of their life. Males may also be able to fertilize at 2 years but are usually prevented by social factors. Longevity around 15–20 years; this, and most of the body dimensions quoted may seldom if ever be achieved in most wild populations now.

Genus *Bubalus*
(Asiatic buffalos)
Stout, dark-colored bodies, hair generally sparse. Horn cores triangular in cross section. No boss of horn in either sex; no dewlaps or hump.

Wild water buffalo V *

Bubalus arnee (bubalis)
Wild water buffalo, carabao or arni.

Very widespread (Asia, S America, Europe, N Africa) as domestic form, of which there are two groups of breeds, the River and the Swamp buffaloes. Latter is feral in N Australia. The wild race survives as remnants in reserves or in remote places in India, Assam, Nepal, Burma, Indochina and Malaysia. Near (and in) large rivers in grass jungles and marshes. HBL 240–280cm, TL 60–85cm, HT 160–190cm, WT male 1,200kg, female 800kg. Coat: slaty black, legs dirty white up to hocks and knees; white chevron below jaw. Flexible fetlock joints make it nimble in the mud; horns heavy, backswept. Gestation 310–330 days.

Lowland anoa E

Bubalus depressicornis
Confined to swampy lowland forests of N Sulawesi (Celebes). A miniature Water buffalo—almost deer-like. Biology almost unknown. HBL about 180cm; TL about 40cm; HT about 86cm; HL male 30cm, female 25cm. Coat: black, sparse hair and white stockings; young woolly, brown. Horns flat and wrinkled.

Mountain anoa E

Bubalus quarlesi
Forest at high altitude in Sulawesi. Adults look like juvenile Lowland anoa. Biology almost unknown. HBL about 150cm; TL about 24cm; HT about 70cm; HL 15–20cm. Coat: legs same color as body which is brown-black; woolliness persisting into adulthood. Horns smooth and circular.

Tamarau E

Bubalus mindorensis
Tamarau or tamaraw.
Found only in forest on the island of Mindoro (Philippines). Biology unknown; said to be nocturnal and ferocious. HT 100cm; HL about 35–50cm. Coat: dark brown-grayish black. Horns stout. Gestation 276–315 days.

Genus *Bos*
(True cattle)
Dark short-haired coat (excepting yak and some Domestic cattle). Horn cores circular in cross section. No boss of horn in either sex; dewlaps and humps well developed in some species.

Banteng V

Bos javanicus
Banteng, tsaine, tembadau.

Isolated populations in Indochina and on the islands of Borneo, Java and Bali (domesticated as the Bali cattle). Feral in N Australia. Quite thick forest with glades. Very like Domestic cow in general proportions. HBL (mainland race) 190–225cm; TL 65–70cm; HT 160cm; WT 600–800kg. Coat: adult bulls dark chestnut; young bulls and cows reddish brown; all have white band around muzzle, white patch over eyelids, white stockings and white rump patch; males have horny bald patch of skin between horns, and a dorsal ridge and dewlap. Gestation about 285 days.

Gaur V

Bos gaurus
Gaur, Indian bison or seladang.

Scattered herds in peninsular India, a few surviving in Burma, W Malaysia and elsewhere in Indochina. Upland tropical forest with glades. HBL 250–300cm; TL 70–100cm; HT 170–200cm; HL (male) up to 80cm; WT male 940kg, female 700kg. Coat: adult bulls shiny black with white stockings and gray boss between the horns; young bulls and cows dark brown, also with white stockings. Huge head, deep massive body and sturdy limbs. Hump, formed by long extensions of the vertebrae, small dewlap below the chin and a large one draped between the forelegs; the horns sweep sideways and upwards. Gestation: 9 months.

Yak E

Bos mutus (grunniens)
Scattered localities on alpine tundra and ice deserts of the Tibetan Plateau at altitudes of 4,000–6,000m. Biology almost unknown, though domesticated yak have been studied. A 1938 report records a bull as 203cm high at the shoulder, with horns of 80cm and weighing 821kg; a cow as 156cm high, with horns of 51cm and weighing 306kg. Coat: shaggy fringes of coarse hair about body with dense undercoat of soft hair; blackish brown with white around the muzzle. Massively built with drooping head, high humped shoulders, straight back and short sturdy limbs. Gestation: 258 days.

Cattle D

Bos primigenius (taurus)
By convention, present-day cattle are given the same scientific name as that of their ancestor, the aurochs or urus which died out in 1627. All cattle are completely interfertile but the skull and blood proteins of the humped Zebu cattle are different from those of the humpless breeds; perhaps the domestications were of two races of the aurochs. The hump of the Zebu, unlike those of the other species mentioned in this table, is composed of muscle and fat and is not simply the result of long processes on the vertebrae. Domestic cattle are feral in many places, the most notable being Chillingham Park, N England. Cattle exist in many sizes and forms. There are long-horned and polled (hornless) breeds; generally shoulder height is 180–200cm and weight 450–900kg. Gestation: around 283 days.

Kouprey E

Bos sauveli
Kouprey or Cambodian forest ox.

A few living in forest glades and wooded savannas of Indochina. Discovered in 1937. HBL 210–220cm; TL 100–110cm; HT 170–190cm; WT 700–900kg. Coat: old bulls black or very dark brown; may have grayish patches on body; cows and young bulls gray, underparts lighter and chest and forelegs darker; both sexes have white stockings. Bulls have very long (over 40cm) dewlap hanging from neck; dorsal ridge not well developed; horns in female are lyre shaped; in males horns curve forwards and round, then up; horn tips are frayed.

Genus *Synceros*
(African buffalos)
Africa south of the Sahara. Bulky, dark. Horn cores triangular in cross section. Males have heavy boss of horn on top of head; this, with the elaborate submissive display differentiate the genus from *Bubalus*.

African buffalo

Synceros caffer
Generally considered as two subspecies: the Cape buffalo (*S.c. caffer*), and the Forest buffalo (*S.c. nanus*), with intermediate forms. The former lives in savannas and woodlands, the latter in forests nearer the Equator. The Cape buffalo has HBL 240–340cm; TL 75–110cm; HT 135–170cm; HL 50–150cm; WT male 680kg, female 480kg. Coat: brownish black. The Forest buffalo has HBL 220cm; TL 70cm; HT 100–120cm; WT male 270–320kg, female 265kg. Coat: reddish brown. Gestation: 340 days.

Genus *Bison*

(American and European bisons)
Reddish brown–dark brown coat;
long shaggy hairs on neck and head,
shoulders and forelegs, and a beard.
Short and broad skull, dorsal hump
formed by processes of the vertebrae.
Smooth horns, circular in cross
section, about 45cm. These species
are completely interfertile.

American bison ⋆

Bison bison
American bison or buffalo.

Widely considered to be two
subspecies, the Plains bison (*B.b.
bison*) and the rather larger and
darker Wood bison (*B.b. athabascae*),
which lives further north. N America.
Grassland, aspen parkland and
coniferous forests. Associated now
with prairies but inhabited forests as
well before its virtual extermination
last century. Now mainly in parks
and refuges. HBL 380cm; TL 90cm; ht
male 195cm; wt male 818kg, female
545kg. Gestation: 270–300 days.

European bison

Bison bonasus
European bison or wisent.

Became extinct in the wild in 1919
(Russian-Polish border) but was reestablished in Bialowieza Primeval
Forest and later in the Caucasus and
elsewhere in the USSR. Mixed woods
with undergrowth and open spaces.
HBL 290cm; TL 80cm; ht
180–195cm; wt male 800kg.
Gestation: 254–272 days.

Tribe Strepciserotini

(Spiral-horned antelopes)
Restricted to Africa. Medium to large
body size, more slenderly built than
the Bovini, with long necks and deep
bodies. Adult males larger than
females; sexes also differ in markings
and horn structure. Horns in males
only in most species. No distinct facial
or foot glands.

Genus *Tragelaphus*

(Kudus and nalas)

Sitatunga

Tragelaphus spekei

Swamps, reedbeds and marshes of the
Victoria, Congo and Zambezi-
Okavango river systems. Male HBL
150–170cm; TL 20–25cm; HL
45–90cm; wt 80–125kg. Female HBL
135–155cm; TL 20–25cm; wt
50–60kg. Coat: shaggy and slightly
oily; female lighter and redder in color
than the yellowish to dark gray-
brown males; spots and up to ten
white stripes on the body; dorsal crest
runs the length of the body and is
erectile; white patches on throat,
white spots on cheeks. Long splayed
hooves distribute weight, allowing the
animal to walk through mud without
sinking into the ground.

Nyala

Tragelaphus angasi

Riverside thicket and dense
vegetation in SE Africa. Male HBL
210cm; TL 43cm; SH 112cm; HL
65cm; wt 107kg. Female HBL 179cm;
TL 36cm; SH 97 cm; wt 62kg. Coat:
shaggy, dark gray-brown,
particularly along the underside of
the body and throat; usually several
poorly marked white vertical stripes;
long, conspicuous erectile crest,
brown on the neck and white along
the back; legs orange; white chevron
between the eyes; horns lyre-shaped
with a single complete turn, black
with whitish tips; females much
redder with clearly marked white
stripes, short coats, a less obvious
chevron, and generally resemble
bushbuck females.

Bushbuck

Tragelaphus scriptus

Locally throughout Africa south of
the Sahara, except for the arid
southwestern and northwestern
regions, in a wide range of habitats
whose common feature is dense
cover. Male HBL 115–145cm; TL
20–24cm; HL 25–57cm; wt 30–75kg.
Female HBL 110–130cm; TL
20–24cm; wt 24–42kg. Coat: short,
varying from bright chestnut to dark
brown, with white transverse and
vertical body stripes being either
clearly marked, broken, or reduced to
a few spots on the haunches; black
band from between the eyes to the
muzzle; white spot on cheek, two
white patches on the throat; adult
males are darker than females and
young, especially on the forequarters,
and the erectile crest is more
prominent.

Mountain nyala

Tragelaphus buxtoni

Highland forest and heathland of the
Arusi and Bale Mountains in Ethiopia.
HBL 190–250cm; HL up to 80cm.
Coat: shaggy, grayish brown; about
four ill-marked vertical white stripes;
white chevron between the eyes; two
white spots on the cheeks; two white
patches on the neck; short white
mane continued as a brown and
white crest. Horns one to one-and-a-
half fairly open turns. In many ways
resembles the Greater kudu more
closely than the nyala.

Lesser kudu

Tragelaphus imberbis

Thicket vegetation in Ethiopia,
Uganda, Sudan, Somalia, Kenya and
N and C Tanzania. HBL 160–175cm;
TL 26–30cm; HL 60–90cm; wt (male)
90–110kg, female 55–70kg. Coat:
sleek and short haired, brownish-gray
with 11–15 clearly marked vertical
white stripes; head darker with
incomplete white chevron between
the eyes; two white patches on the
neck; male's dorsal crest extends
forwards into a short mane; tail
bushy; small but clear spots on
cheeks; reddish tinge to legs; female
slightly more reddish than male.
Horns 2–3 open spirals.

Greater kudu

Tragelaphus strepsiceros

Woodland, especially in hilly, broken
ground, in E, C and S Africa. Male HBL
190–250cm; TL 37–48cm; HL
100–180cm; wt 190–315kg. Female
HBL 190–220cm; TL 37–48cm; HL
120–215kg. Coat: short, blue-gray to
reddish brown; 6–10 vertical white
body stripes; white chevron between
eyes; up to three white cheek spots;
dorsal crest extended by mane along
whole body; fringe of hairs from chin
to base of neck in males; females and
young redder than males.

Bongo

Tragelaphus euryceros

Discontinuously distributed in
lowland forest in E, C and W Africa;
found outside this habitat in S Sudan,
in small populations in montane or
highland forest in Kenya, and in the
Congo. Male HBL 220–235cm; TL
24–26cm; HL 60–100cm; wt
240–405kg. Female HBL 220–235cm;
TL 24–26cm; HL 60–100cm;
wt 210–253kg. Coat: bright chestnut
red, much darker in adult males; dark

muzzle, white chevron between eyes,
about 2 white cheek spots; lower neck
and undersides darker, whitish
crescent collar at base of neck; black
and white spinal crest and many
narrow but clear white vertical stripes
on the body; contrasting black and
white markings on the legs. Horns
present in both sexes, heavy and
smooth with an open spiral of one to
one-and-a-half turns. Tail long and
tufted at tip.

Genus *Taurotragus*

(Elands)

Common eland

Taurotragus oryx
Common or Cape eland.

Nomadic grassland and open
woodland; may have at least visited
all but the most arid of these habitats
in E, S and C Africa in the past; now
found only in game reserves and
ranches in some areas. Male HBL
250–340cm; TL 54–75cm; SH
135–178cm, HL 60–102cm; wt
400–950kg. Female HBL 200–280cm;
TL 54–75cm; SH 125–150cm, HL
60–140cm; wt 390–595kg. Coat:
light tan, darkening to gray in old
males; a few light stripes on the
forequarters (not found in adults of
the southern populations); black and
white leg markings; black tuft on end
of tail; black stripe along back,
merging into short mane; adult males
develop a tuft of frizzy hair on their
foreheads. Ears have much smaller,
more horse-like pinnae than other
spiral-horned antelopes.

Giant eland

Taurotragus derbianus

More of a woodland species than the
Common eland and found in small
populations in W and C Africa, with
larger and more secure populations in
E Africa, particularly Sudan. Male HBL
290cm; HL 150–175cm; TL
55–78cm; HL 80–110cm;
wt 450–900kg. Female HBL 220cm; ht
150cm; TL 55–78cm; HL 80–125cm;
wt 440kg. Coat: reddish brown,
becoming slate gray in adult males;
12–15 body stripes, white chevron
between eyes, white cheek spots,
black stripe along back merging into a
short mane, black collar around neck,
with contrasting white patches on
either side. The collar emphasizes the
dewlap, which begins under the chin
and finishes above the base of the
neck. Horns more strongly keeled
than in the Common eland, but more
slender and longer, even in the males.

SJGH/RU

Bison Breeding

Mating system of American bison

An all-out fight between American bison bulls is one of the great dramas in nature. They slam their heads together, their hooves churning up dust in enveloping clouds. Clumps of hair two or three times the size of a man's fist are sheared from their heads by their horns grinding against each other, and tossed into the air. They circle, trying to exploit the agility conferred by their small hindquarters, to drive a horn into the opponent's ribcage or flank. If one succeeds, the other may die of the wound.

Fighting is as costly as it is spectacular: it costs time and energy, and the combatants risk injury, even death. In general, costly behavior is rare. Fewer than 15 percent of bull bison disputes are settled by fights. All the rest are settled by signals of threat and yielding, of several degrees of intensity and sometimes surprising subtlety. A bull may threaten by standing broadside to his opponent, by bellowing, by rolling in dust (sometimes he urinates in the dust before rolling in it) and by approaching his opponent head on. Sometimes two bulls stand close together, heads to the side, and "not threaten," swinging their heads up and down in matched movements. To yield, a bull may simply withdraw, at other times he may duck his head and turn it away from his opponent. Sometimes he grazes rather ceremoniously.

An observer knowing these signals can recognize which bull of a pair is dominant. Dominance relations between bison bulls often change after only a few days, but they are nevertheless important. During the two weeks of each year when 90 percent of the breeding takes place, the dominant bull of any pair can displace the subordinant from a receptive cow. The more other bulls a particular bull dominates, the more cows to which he has priority. As a result, the one-third of the bulls that are the most dominant mate with about two-thirds of the cows each year. A few bulls serve as many as 10 cows in a season.

This variation in male reproductive success helps to explain the great differences in the bodies of bulls and cows. The bulls give no care to their young. They are specialized for breeding, and in bison society that means specialized for threatening and fighting. In 1871, Darwin pointed out that selection for such breeding characteristics is a special case of natural selection, which he named "sexual selection." The bull bison is an excellent example of a fighting specialist: male reproductive success is determined by competition for females; female reproductive success varies much less, and is deter-

mined by competition for food, not mates.

Bison mate in temporary one-to-one relationships called tending. The tending bull stands beside a particular cow, threatening bulls that approach. He blocks the cow's path if she starts to move away, and he tries to mount her. This relationship may last for only a few minutes before the bull is displaced by a more dominant bull, or leaves voluntarily to seek a more receptive cow. Few cows or bulls show any preference for particular partners.

Each sex follows its own strategy to maximize reproductive success. The cow has no more than one calf a year. So her strategy emphasizes the quality of the bull with whom she mates. Since more dominant males are likely to have more dominant sons, the female behaves so as to maximize the chance of being mated by a more dominant bull. For several hours

▲ **Classic confrontation.** Bison bulls locked in combat.

▶ **Stages of combat.** Bison bulls use a repertoire of threats in establishing dominance relations, and conflict rarely escalates to all-out fighting. In the broadside threat (1) the animals stand broadside to each other, either facing the same way or in opposite directions. They often bellow and try to present the most impressive profile, arching their backs and lifting their bellies. In the nod-threat (2) the animals stand facing each other and swing their heads from side-to-side in unison. A bull may signal submission during the threat interactions by various means: by turning the head away, by retreating, by lowering the head to graze etc. If no submissive signal is given fighting may begin (3).

istically and unselectively. After mating, a bull usually stays with the cow for half an hour or so, guarding her against the approach of other bulls, and making sure that it is his sperm that fertilizes her. The few cows that copulate more than once a year copulate the second time within an hour of the first, usually with the same bull. Therefore, while bison behave promiscuously in the sense that any cow may mate with any bull, the genetic outcome of their mating pattern is like that of a polygynous mating system, in which a dominant male prevents access by other males to the females in his harem.

Knowledge of their mating system is essential for the development of a proper conservation program for bison. In their natural state, bison lived in large, interacting populations. Genes were exchanged over great distances among millions of individuals. There was little chance of the inbreeding, or gene loss through random failure of some genes to be transmitted (genetic drift), that can occur in small, isolated populations. But today bison, like more and more wild animals, live in reserves, where their populations are often small and isolated. Inbreeding and gene loss are now quite possible, and both are harmful. Whether or not either will occur depends largely on the effective size of the breeding population, which in turn depends more on the breeding system than on the number of males and females of breeding age in the population. For example, if a population of ten males and ten females breeds monogamously, the effective breeding population is 20. But if they breed polygynously, with one dominant male mating with all the females, the effective breeding population is only 3.7. Knowledge of the social system, including the breeding system, of bison will help to determine how large populations must be to survive genetically in the long term. In this way it will make a major contribution to the long-term conservation of this no longer plentiful, but still magnificent, wild animal.   DFL

before she copulates, the female breaks away from tending bulls and runs through the herd, drawing the attention of all the bulls present, and inciting competition between them.

In contrast, bulls may sire several calves each year, so their strategy emphasizes quantity rather than quality. They move quickly from cow to cow, mating opportun-

2

3

DUIKERS

Subfamily: Cephalophinae
Seventeen species in 2 genera.
Family: Bovidae.
Order: Artiodactyla.
Distribution: Africa south of Sahara.

Habitat: mainly dense forest thickets.

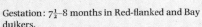

Size: ranges from head-body length
55–72cm (22–29in), tail length
7–12.5cm (3–5in) and weight
4–6kg (8.8–13.2lb) in the Blue
duiker to head-body length
115–145cm (45–57in), tail length
11–18cm (4.3–7in) and weight
45–80kg (100–176lb) in the
Yellow-backed duiker.

Gestation: 7½–8 months in Red-flanked and Bay
duikers.
Longevity: unknown in the wild; 10–15 years
in captivity.

Forest duikers
Sixteen species in genus *Cephalophus*. Species
include: **Blue duiker** (*C. monticola*); **Yellow-
backed duiker** (*C. sylvicultor*); **Bay duiker**
(*C. dorsalis*); **Maxwell's duiker** (*C. maxwelli*);
Zebra duiker (*C. zebra*); **Jentink's duiker**
(*C. jentinki*).

Savanna or bush duikers
One species: Common duiker (*Sylvicapra
grimmia*).

ALMOST anywhere in Africa, south of the Sahara desert, the observant traveler may catch a glimpse of one of many small to medium-sized antelopes disappearing into dense forest or thickets. These are the "duikers," an Afrikaans word meaning divers, so named for their habit of diving into cover when disturbed. The short forelegs, longer hindlegs, and arched body shape, enable duikers to slip easily through dense vegetation.

The sexes are similar in appearance, although females are often up to 4 percent longer than males. Both usually possess short, conical horns, but these are occasionally lacking in females. The coat is often reddish, but in some species it is blue-gray, black or striped. Duikers have the largest brains relative to body size of all antelopes.

Most duikers are forest duikers of the genus *Cephalophus*. The name *Cephalophus* refers to the crest of long hair between the horns. The genus *Sylvicapra* includes only one species, the Common duiker, which is the only species typically found in the savanna and open bush country.

Duikers are uncommon as fossils, but the living species retain many features found in the fossil remains of early bovids and are sometimes considered to be the most primitive living African antelopes.

Little is known about the natural history of most duiker species in the wild due to the impenetrability of their natural habitat and their secretive nature. The Yellow-backed duiker may be active during both night and day but some species, such as the Blue duiker, are active only during the day, while others, such as the Bay duiker, are active only at night. Duikers are primarily browsers and require high quality food because of their relatively small size. They feed on leaves, fruits, shoots, buds, seeds and bark. Surprisingly, they sometimes stalk and capture small birds and rodents and occasionally also eat insects and even carrion.

Duikers are not gregarious and are usually seen alone or in pairs. In the wild the Blue duiker is truly monogamous; pairs seem to mate for life and reside in small—2–4ha (5–10acre)—stable territories which are actively defended against other members of the same species by both male and female. In captivity, male Maxwell's duikers are most aggressive to unfamiliar males while females are more aggressive towards unfamiliar females. Observations on captive animals suggest that several other duiker species may also be monogamous.

In monogamous social systems, which are comparatively rare in mammals, males sometimes provide a great deal of care for the young, and the young remain in the social group for extended periods during which they often help to care for their younger siblings (obligate monogamy). However, in other monogamous species, males leave most responsibility for the care of the young to the female and young disperse from the social group before the birth of their younger siblings (facultative monogamy). In species exhibiting obligate monogamy, such as marmosets and tamarins, mates remain near one another

▼ **Representative species of duiker.** (1) Male Common duiker (*Sylvicapra grimmia*) "duiking" (fleeing). (2) Yellow-backed duiker (*Cephalophus sylvicultor*), showing the erect yellow rump patch. (3) Male Zebra duiker (*Cephalophus zebra*) marking a sapling with the preorbital gland. (4) Jentink's duiker (*Cephalophus jentinki*) suckling a calf. (5) Male Maxwell's duiker (*Cephalophus maxwelli*) about to rub his preorbital gland on a rival before fighting. (6) Red-flanked duiker (*Cephalophus rufilatus*) eating a bird.

and usually feed and rest together. Blue duikers (along with species in other groups, such as dikdiks and elephant shrews) exhibit facultative monogamy, in which mates do not remain near each other and often feed and rest at different times.

Duikers have only a single calf. Calves of all species spend little time with their mothers during the first few weeks of life, remaining well-hidden in vegetation. Young Blue duikers reach sexual maturity at about one year of age and leave their parents during the second year of life to attempt to find their own mates and territories. Larger species probably reach sexual maturity at an older age. Calves of some species, such as the Zebra duiker and Maxwell's duikers, resemble their parents in color while those of other species have distinctive juvenile coats. Young Jentink's and Bay duikers are very similar in appearance, with a uniform dark brown coat unlike that of the adults of either species. The yellow rump patch of the Yellow-backed duiker does not begin to appear until about one month of age and is not fully developed until about 10 months of age.

Duikers possess large scent glands beneath each eye. The structure of these glands differs from that of the preorbital glands found in other antelopes. The secretion, which may be clear or bluish in some species, is extruded through a series of pores instead of a single large opening. In captivity, many duiker species rub these glands on fences, trees and other objects in their enclosures. This has usually been interpreted as territorial marking. This behavior is very frequent in some species: in captivity, male Maxwell's duikers may mark as often as six times during a 10-minute interval.

Maxwell's duikers also press these glands on the glands of other individuals, first on one side and then on the other. This behavior is called mutual marking. Very forceful mutual marking has been observed in captivity between males as a prelude to fighting. A much more gentle form of mutual marking is often seen between males and females.

The rarest species is Jentink's duiker, which is considered to be endangered. It was not described until 1892 and even today few specimens exist in museums and zoos. The main threat to this species, as well as to other duiker species, is subsistence hunting. Duiker meat is well-liked by humans and duikers are easily dazzled at night with lights, making them comparatively easy to shoot or capture. Some species are also captured by driving them into nets.   KR/KK

THE 17 SPECIES OF DUIKERS

Abbreviations: HBL = head-body length. TL = tail length. wt = weight.
Approximate nonmetric equivalents: 2.5cm = 1in; 1kg = 2.2lb.
E Endangered. * CITES listed.

Genus *Cephalophus*
Forest Duikers

Horns usually in both sexes and in same plane as forehead. Ears short and rounded.

Maxwell's duiker
Cephalophus maxwelli

Nigeria west to Gambia and Senegal. Lowland forest and adjacent gallery forest. HBL 55–90cm; TL 8–10cm; wt 8–9kg. Coat: from gray-brown to blue-gray, with much variation. Horns sometimes absent in females.

Blue duiker *
Cephalophus monticola

Nigeria to Gabon, east to Kenya and south from this latitude to S Africa. Lowland forest. HBL 55–72cm; TL 7–12.5cm; wt 4–6kg. Coat: color variable: similar to Maxwell's duiker but may be bluer.

Black-fronted duiker
Cephalophus nigrifrons

Cameroun to Angola and east through Zaire to Kenya. Lowland, gallery, montane and marshy forests. HBL 85–107cm; TL 10–15cm; wt 13–16kg. Coat: reddish to dark brown, with reddish brown to black stripe from nose to horns.

Red-flanked duiker
Cephalophus rufilatus

Senegal to Cameroun east to Sudan and Uganda. Gallery forests and forest edges. HBL 60–70cm; TL 7–10cm; wt 9–12kg. Coat: reddish yellow to reddish brown with dark dorsal stripe from nose to tail. Horns regularly lacking in female.

Jentink's duiker E
Cephalophus jentinki

Liberia and W Ivory Coast. Lowland forest only. HBL about 135cm; TL about 15cm; wt up to 70kg. Coat: head and neck black, shoulders white, back and rump a grizzled "salt and pepper." A very rare species.

Yellow-backed duiker
Cephalophus sylvicultor

Guinea-Bissau east to Sudan and Uganda, south to Angola and Zambia. Wide variety of forest types and open bush. HBL 115–145cm; TL 11–18cm; wt 45–80kg. Coat: blackish brown except for yellow rump patch.

Abbott's duiker
Cephalophus spadix

Tanzania. High montane forest. HBL 100–120cm; TL 8–12cm; wt up to 60kg. Coat: dark chestnut brown to black.

Zebra duiker
Cephalophus zebra

Sierra Leone, Liberia and Ivory Coast. Lowland forest. HBL 85–90cm; TL about 15cm; wt 9–16kg. Coat: reddish brown with 12–15 black transverse stripes.

Black duiker
Cephalophus niger

Guinea east to Nigeria. Lowland forests. HBL 80–90cm; TL 12–14cm; wt 15–20kg. Coat: brownish black to black.

Ader's duiker
Cephalophus adersi

Zanzibar, coastal Kenya and Tanzania. HBL 66–72cm; TL 9–12cm; wt 6.5–12kg. Coat: tawny-red with white band on rump.

Red forest duiker
Cephalophus natalensis

Somalia south to Zimbabwe and Mozambique. Lowland, montane forest. HBL 70–100cm; TL 9–14cm; wt 10.5–12kg. Coat: orange-red to dark brown.

Peter's duiker
Cephalophus callipygus

Cameroun and Gabon east through Central African Republic and Zaire. Lowland forest. HBL 80–115cm; TL 10–16.5cm; wt 15–24kg. Coat: light to dark reddish brown.

Cephalophus weynsi

Zaire, Uganda, Rwanda and W Kenya. A very poorly known species which may be a distinct species or a form of either Peter's duiker or the Red forest duiker.

Bay duiker
Cephalophus dorsalis

Guinea-Bissau east to Zaire and south to Angola. Lowland forest. HBL 70–100cm; TL 8–15cm; wt 19–25kg. Coat: brownish yellow to brownish red; dorsal black stripe from nose to tail.

White-bellied duiker
Cephalophus leucogaster

Cameroun south and east into Zaire. Lowland forest. HBL 90–100cm; TL 12–15cm; wt 12–15kg. Coat: light to dark reddish brown with white chin, throat, and belly.

Ogilby's duiker
Cephalophus ogilbyi

Sierra Leone East to Cameroun and Gabon. Lowland forests. HBL 85–115cm; TL 12–15cm; wt 14–20kg. Coat: reddish orange with white stockings; dorsal black stripe from shoulder to tail.

Genus *Sylvicapra*
Savanna or Bush duikers

Horns usually in males only and directed upward. Ears longer and more pointed than in *Cephalophus*.

Common duiker
Sylvicapra grimmia

Subsaharan Africa except Zaire. Savanna and open bush. HBL 80–115cm; TL 10–22cm; wt 10–18kg. Coat: sandy-tan.

KR/KK

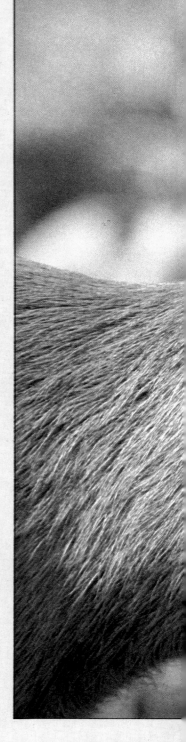

► **A Common duiker** in the Winterberg Mountains, Namibia.

GRAZING ANTELOPES

Habitat: dry and wet grasslands up to 5,000m (16,400ft).

Size: shoulder height from 65–76cm (26–30in) in the Mountain reedbuck to 126–145cm (49–57in) in the Roan antelope; weight from 23kg (51lb) in the Gray rhebok to 280kg (620lb) in the Roan antelope.

Subfamily: Hippotraginae
Twenty-four species in 11 genera.
Family: Bovidae.
Order: Artiodactyla.
Distribution: Africa, Arabia.

Reedbuck, waterbuck and rhebok
Tribe: Reduncini.
Nine species in 3 genera.
Species include: **Bohor reedbuck** (*Redunca redunca*), **Gray rhebok** (*Pelea capreolus*), **kob** (*Kobus kob*), **lechwe** (*Kobus leche*), **Mountain reedbuck** (*Redunca fulvorufula*), **Nile lechwe** (*Kobus megaceros*), **puku** (*Kobus vardoni*), **waterbuck** (*Kobus ellipsiprymnus*).

Gnus, hartebeest and impala
Tribe: Alcephalini.
Eight species in 5 genera.
Species include: **bontebok** (*Damaliscus dorcas*), **Brindled gnu** (*Connochaetes taurinus*), **Hartebeest** (*Alcephalus busephalus*), **hirola** (*Beatragus hunteri*), **impala** (*Aepyceros melampus*), **topi** (*Damaliscus lunatus*), **White-tailed gnu** (*Connochaetes gnou*).

Horse-like antelope
Tribe: Hippotragini.
Seven species in 3 genera.
Species include: **addax** (*Addax nasomaculatus*), **Arabian oryx** (*Oryx leucoryx*), **bluebuck** (*Hippotragus leucophaeus*), **gemsbok** (*Oryx gazella*), **Roan antelope** (*Hippotragus equinus*), **Sable antelope** (*Hippotragus niger*), **Scimitar oryx** (*Oryx dammah*).

▷ **Apparently marooned,** these waterbuck are browsing on a patch of aquatic vegetation. They are never far from water, which they need in quantity to accompany their high-protein diet.

▶ **The Bohor reedbuck,** an antelope of the northern savannas of Senegal, east to Sudan and south across Tanzania.

THE grasses of Africa lie in an unbroken prairie which stretches from the cold subdesert steppes of the Cape Province of South Africa, through the deciduous woodlands and open grasslands of the southern savannas, crosses the equator in East Africa, and spreads out across the northern savannas, reaching up to the Sahara desert. Even in the heart of the desert, the soils can throw up a rich growth of grass following the passage of a rain storm. Interspersed with the prairie grasses are the papyrus-filled lakes and swamps, marshes, water meadows, reed-beds, and the grass-bound flats and floodplains of Africa's great river systems, notably the Nile, Niger, Zaire and Zambesi. Rising above these wetlands, the Adamawa highlands of Cameroon, the highland massifs of Ethiopia and Kenya, and the Drakensberg Mountains of South Africa and Lesotho are clad in montane grasslands. Africa's diverse grasslands are home to the grazing antelopes (subfamily Hippotraginae), a group which have colonized every habitat from the inundated swamps of the Nile Sudd to the barren centers of the Namib and Sahara Deserts and the exposed mountain pastures up to 5,000m (16,400ft) on Mt. Kilimanjaro.

The wetlands and, rather unexpectedly, the montane grasslands are inhabited by the reedbucks, kobs and waterbuck, all from the tribe Reduncini. The smaller reedbucks have retained many primitive features, and the Mountain reedbuck particularly may be taken as an approximate model, in general appearance and behavior, of the ancestors of this group. Indeed, the existence of three highland populations, each separated by 2,000km (1,240mi) from its nearest neighbor, suggests that a single dominant ancestral stock was once widespread in all types of grassland, subsequently relinquishing the lowlands to more advanced forms. Adapted to a poor quality, fibrous diet, the diminutive Mountain reedbuck is particularly sedentary. The females and young either live within the territories of single resident males or range over a small number of neighboring territories, normally in groups of 2–6. The average size of territory varies from 10–15ha (25–37acres) in one Kenyan population of moderately high density (11 animals per sq km; 28 per sq mi). These territories are not marked by glandular secretions, dung piles or scrapes; the principal advertisement of presence is the animals' whistle. The population structure of the Gray rhebok, another small montane antelope, is very similar.

The Southern reedbuck and Bohor reedbuck are lowland species which occur in the southern and northern savannas respectively. Seldom far from water, they are typical inhabitants of floodplains and inundated grasslands. They are particularly active at night, emerging from dense cover to feed on open lawns, to the accompaniment of much whistling and bouncing. In farming areas, young cereal crops are a

A High Density Waterbuck Population

At Lake Nakuru National Park, Kenya, the density of waterbuck reaches up to 100 animals per sq km (250 per sq mi) in some areas. The park average of 30 animals per sq km (75 per sq mi) is so much higher than the more typical 1–2 animals per sq km (2.5–5 per sq mi) that there are grounds for suspecting the Nakuru waterbuck to differ in social behavior from other populations.

Male waterbuck usually occupy territories larger than 100ha (250 acres), but at Lake Nakuru the severity of competition probably explains why territories are smaller than elsewhere—10–40 hectares (25–100 acres)—and average duration of territory ownership is shorter (about 1.5 years). At any one moment, only a very small proportion (about 7 percent) of the adult males hold a territory, and only 20 percent of the males surviving to prime age ever become owners of a territory.

Fifty-three percent of the territories at Lake Nakuru contain one or more "satellite males," adult males subordinate to the territory holder and tolerated by him. Satellite males have access to the whole territory, and some have access to two adjacent territories. While tolerating his satellite(s) in the territory, a territory owner will threaten and chase out of the territory other adult males. Adult males which are neither territory holders nor satellites (over 80 percent of the adult male population) unite with young and juvenile males into bachelor herds, which rarely enter the territories of the dominant males.

Adult and young males attempting to enter a territory are often confronted and repelled by the satellite male instead of by the territory owner. Satellites apparently share in the defense of the territory and a territory holder having a satellite might thus save energy and decrease the risk of being wounded in a fight.

When a receptive female is in the territory, usually only the territory holder copulates with her. Occasionally, however, a satellite male manages to copulate with a receptive female while the territory holder is not close by. Waterbuck territories are situated along rivers and lakeshores where the grass is noticeably greener, so by being inside a territory satellite males gain access to better resources than bachelor males.

The biggest advantage for the satellite male is probably his high chance of being a territory owner himself. In 5 of 12 observed cases of change of territory ownership, a satellite became the new territory owner either on the territory it had already occupied as a satellite or on a territory adjacent to it. It can be calculated that the average probability of gaining possession of a territory is about 12 times greater for a satellite male than for a bachelor.   PW

favorite item of the diet. The primitive territorial system of the Mountain reedbuck is also found in these species. At low density, the territory of a male is shared by one resident female but home ranges were found to overlap in a high-density Zululand population, with 16.6 reedbuck per sq km (33 per sq mi). Even here, however, groups of more than three reedbuck were rare, showing that the socialization process is still strongly inhibited. Reedbuck do concentrate in much larger numbers on good pastures or when caught in open country after a fire has destroyed their normal cover of tall riverside grasses and reeds. These associations are unstable and social bonds are weak. Exceptionally high local densities of up to 110 Bohor reedbuck per sq km (285 per sq mi) occur at the end of the dry season along the upper tributaries of the Nile. In this area, groups of up to four females with attendant young are centered on the tiny territories of individual males.

Living near to rivers and lakes, but inhabiting the adjacent savanna and woodland, are the large, shaggy and slightly ungainly waterbuck. The French name for these animals is *Cobe onctueux* or "greasy kob," a reference to the oily musky secretion that covers the hairs of the coat and which is

detectable at 500m (about 1,600ft) in light airs. In captivity, waterbuck require a diet with much more protein than that of other bovids, for which they require a high water intake. In the wild, this constraint accounts for their localization close to permanent water. Their diet is normally made up from short and medium grasses, reeds and rushes, but it is supplemented by browsing and wading for aquatic vegetation in the dry season. Some riverside vegetation is usually present in the waterbuck's domain. By accepting a fairly mixed diet from this rich habitat, individuals can be fairly sedentary. Female home range size varies from 0.3sq km (0.12sq mi) in a Kenyan population of density 37–54 does per sq km, to 6 sq km (2.3sq mi) in a Ugandan population of density 4 does per sq km. The home range is typically shared by 5–8 does, although these animals do not move about as a close-knit

group; it overlaps with several of the smaller male territories, which average 13–240ha (32–593 acres) in size, depending on population density.

Waterbuck are moderately long-lived and males may hold territories from the age of 6–10 years. In East Africa, young adults of 5–6 years employ a variety of strategies to gain their own breeding territory within the permanently established network (see box).

Inhabiting the gently rolling hills and low-lying flats close to permanent water, the kob grazes on shorter savanna grass than the waterbuck. Frequently occurring at high density, the female kob moves in bands of 30–50 animals and for much of the year remains fairly sedentary. In the rainy season, herds of over 1,000 animals congregate locally, keeping the grasses short and in good growing condition for continuous cropping. In dry months, patches of green

▲ **Gestures of grazing antelopes.** (1) Southern reedbuck (*Redunca arundinum*) in the "proud posture." (2) Defassa waterbuck (*Kobus ellipsiprymnus defassa*) showing the dominance display. (3) Gray rhebok (*Pelea capreolus*) in the alert posture. (4) Uganda kob (*Kobus kob thomasi*) in the head-high approach to a female during the mating season. (5) Roan antelope (*Hippotragus equinus*) in the submissive posture. (6) Sable antelope (*Hippotragus niger*) presenting horns, a male dominance display. (7) Addax (*Addax nasomaculatus*) performing the flehmen test after sampling a female's urine. (8) Gemsbok (*Oryx gazella*) showing the ritual foreleg kick during courtship. (9) Coke's hartebeest (*Alcephalus busephalus cokii*) showing the submissive posture of a yearling. (10) A territorial male impala (*Aepyceros melampus*) roaring during the rutting season. (11) A male bontebok (*Damaliscus dorcas dorcas*) initiating butting by dropping to his knees. (12) Topi (*Damaliscus lunatus*) in the head-up approach to a female. (13) Brindled gnu (*Connochaetes taurinus*) in the ears-down courtship approach.

grass may draw kob from long distances, resulting again in the formation of large assemblages.

Adapted to the floodplains and seasonally inundated swamps, lechwe are the most specialized of the reduncine tribe. Their diet consists principally of grasses, with a distinct preference for leaf over stem. On the Kafue flats of Zambia the herds graze along the floodline in the dry season, typically in a depth of 5–20cm (2–8in) water, but a few animals venture so deep in search of food that water covers their backs. Aggregations of several hundreds (formerly thousands) occur, and mass movements have been observed in response to heavy rains and rising flood waters. In structure, lechwe populations resemble those of the related puku and kob. Close associations between individual females have not been observed and even the calf's bond with its mother is loose. Like the Uganda kob, males of Kafue lechwe may be found in conventional territorial breeding grounds (TGs). TGs of lechwe are only temporarily manned for a few weeks during the rut. Usually they hold 50–100 males within a circular area of approximately 500m (1,640ft) diameter. Large mixed herds of lechwe have been observed adjacent to these grounds.

The close adaptation of lechwe and the related Nile lechwe to their semi-aquatic habitat makes them particularly vulnerable to organized hunting. Traditional lechwe drives or *chilas* which took place along the Kafue River in the 1950s accounted for up to 3,000 animals per *chila*, while uncontrolled hunting of the Black lechwe in northern Zambia has reduced the population to a level barely one-tenth the estimated carrying capacity. Today, dams and drainage schemes are potentially just as devastating as overhunting. The Kafue Gorge hydro-electric scheme, which was completed in 1978, involved the building of dams at either end of the flats most favored by the lechwe. The total population of 94,000 lechwe subsequently dropped to about one half.

The fertile grasslands and woodlands of the moist southern and northern savannas are the residence of large herds of gnus, hartebeests and impala (tribe Alcelaphini). This tribe illustrates well the relationship between a population's ecology (distribution of available food, water, cover and other components of the habitat) and its social structure and mating system. Although its long face and sloping back mark the topi unmistakably as a true alcelaphine, it has comparatively generalized features. Widely distributed across the moist grasslands of Africa, it specializes on the green grass of valley bottoms and intermediate vegetation zones; sedentary populations occur within woodland, particularly where strips of woodland and grassland intersect. Depending on the population density, single bulls defend territories of 25–400ha (60–1,000 acres) which contain small groups of 10–20 females. The resident females are hierarchical and threaten, chase and even fight intruding females. Records of known individuals in the same territory for over three and a half years imply a remarkably stable group structure. In many parts of Africa, the seasonal availability of green grass is affected over large areas by annual droughts or floods. In those parts, the movement of topi between pastures is much more extensive, but follows, nonetheless, a predictable cyclic pattern. In the Akagera National Park of Rwanda, herds of up to 2,000 topi sweep over pastures on the broad valley bottoms. As the mating season draws near, males aggregate, up to 100 at a time, on traditional grounds similar to those of the Uganda kob. Each male defends a small territory or stamping ground which may be reduced to 25m (80ft) diameter in the densest clusters. The topi herds of the Ruwenzori National Park, Uganda, which number 3,000–4,000 animals, exploit a well-

defined area of 80sq km (30sq mi) of open grasslands. The animals apparently behave as a single population with no internal structure, and all individuals move freely over the entire plain. Since the herds have no regular directional movement, the bulls are not found on TGs. Instead, they stay with the main aggregations, and constantly engage in efforts to herd together "wards" of females and chase out other males. A typical ward is 80–100m (265–330ft) in diameter and attached to a specific piece of ground. In the middle of an aggregation, a ward contains 30–80 females, but as the herd moves on the wards slowly empty.

Though less selective feeders of medium and long savanna grassland, hartebeest are particularly fond of the edges of woods, scrub and grassland. Although basically sedentary, the female hartebeest in Nairobi National Park, Kenya, are locally quite active, moving in small groups within individual home ranges of 3.7–5.5sq km (1.4–2.1sq mi). At an overall density of 22 hartebeest per sq km, these ranges are too large for male defense. Bulls therefore defend small territories of average size 0.31sq km (0.12sq mi), which are not particularly associated with any one female group. In

▲ **Water-chase.** Lechwe running through a floodplain in Botswana. Lechwe are the most aquatic of the grazing antelopes.

▶ **With horns interlocking,** two male Uganda kobs dispute dominance. No discrete social units or stable associations have been recorded in either male or female kobs; however, it appears that herds are oriented about traditional breeding grounds (abbreviated "TGs"). In western Uganda, TGs contain 10–20 males on closely packed central territories of 15–35m (50–115ft) diameter, surrounded by a similar number of slightly larger and more widely spaced peripheral territories. Females visit TGs for mating all the year round, usually moving directly towards the central territories, triggering off a chain reaction of clashes and ritualized displays among the males. The whole TG is usually 200–400m (650–1,310ft) in diameter and situated on smooth, slightly raised ground which is well trampled and grazed short. Some TGs have been observed on the same spot for over 15 years and traced back a further 30 years through local inhabitants.

▶ **Migratory wildebeest crossing a river.** OVERLEAF There are still vast herds of the Brindled gnu or Blue wildebeest from northern South Africa to Kenya. Such a herd on the move is a dramatic sight. Wildebeest often perish in large numbers at river crossings.

fact, the average female home range includes over 20–30 male territories.

The open grasslands and woodlands of the southern savannas are also exploited by the bizarre-looking wildebeest or gnu, which are particularly common in areas where pruning, by fire and other herbivores, has maintained a short sward of grass. As with the topi, wildebeest populations may be either sedentary or nomadic, depending on the local distribution of rainfall and green grass. Again corresponding to the topi, discrete small herds of female wildebeest occur in sedentary populations, together with a permanent territorial network of bulls and segregated bachelor groups. However, wildebeest herds are not known to be closed to outsiders, nor do the bulls associate exclusively with a single female group.

No social structure has been detected among the vast nomadic assemblages of the Brindled gnu. Both sexes are present, and during the rut males establish temporary territories whenever the aggregations come to rest. These small territories are firmly attached to a fixed piece of ground but are seldom held for more than 10 hours.

The diet of cattle is broadly similar to that of the Alcelaphini, which consequently are viewed by livestock owners as competitors for dry-season forage. Over the past few decades, the total range of the alcelaphines has severely contracted with the increase in numbers of livestock, and several populations have been nearly exterminated. Particularly vulnerable is the hirola, a species confined to a small region of dry savanna on either side of the border between Kenya and Somalia. Its total Kenyan population declined from about 10,000 in 1973 to 2,385 in 1978. Concomitantly, the number of cattle sharing the same range increased from 200,000 head to 454,414 head. The vulnerability of the alcelaphines to loss of habitat is brought out by the recent history of the bontebok. This race also has a restricted coastal range, in this case along the southwestern Cape. Indiscriminate hunting and the enclosure of the best land for farms by the early settlers had already seriously reduced the numbers by 1830. In an attempt to avoid the inevitable path to extinction, the first Bontebok National Park was established in 1931 and stocked with the pathetic total of 17 animals. This move and the creation of a second Bontebok National Park in 1961 removed the subspecies from danger; at the end of 1969 the total number in the whole Cape Province was about 800.

The dominant antelope of the less fertile woodlands of central and southern Africa is the impala. Golden tan herds of 100 or more animals moving through the park-like woods provide an unforgettable sight. Impala are selective but opportunistic feeders, accepting a broad range of dietary items including grasses, browse leaves, flowers, fruits and seeds. In the Sengwa Research Area of northwestern Zimbabwe, their diet changed from 94 percent grass in the wet season to 69 percent herbs and woody browse in the dry season. Throughout their range, impala prefer zones between different vegetation types and are particularly abundant along evergreen riverside strips in the dry season. Here, a wide variety of plant species is available and the impala can meet their annual forage requirements in home ranges of 0.5–4.5sq km (0.2–1.7sq mi). Female group size varies according to the season, averaging between 7 and 33 impala in the Sengwa area, but despite this wide variation, the population has a distinctive clan structure. Male society is much looser and essentially independent from that of the females, except with regard to the social organization in mating.

The reproductive cycle of impala is closely linked to the annual pattern of rainfall. In the equatorial region, births occur in all months, but are concentrated around two peaks associated with two rainy seasons. In contrast, a single well-defined peak of births, lasting for two to three weeks, is observed in southern Africa, which has a single wet season. The difference in timing of breeding has far-reaching effects. In both regions, impala males are territorial and, in the manner of hartebeest, they defend a smaller area than that used by an individual female. In equatorial regions, however, territory size is up to five times larger and male tenure of territory is ten times longer, leading to a prolonged displacement of bachelor groups from the best pastures. Dominance interactions are less likely to lead to fights in the equatorial zone, and territories change ownership less frequently. The short intense rut of the southern populations is heralded by displays of contagious chasing and roaring among males, not dissimilar to (although less ritualized than) oryx tournaments. At its peak, up to 180 roars per hour have been logged during the impala rut. Males mobilize all their fat reserves for this impressive breeding effort, but although fights are not uncommon, serious injury is surprisingly rare. Six to seven months after the rutting peak, single impala lambs are born. It is now the mother's turn to mobilize her fat reserves to meet the demands of lactation.

Half the newborn impala are taken by predators in the first few weeks of life, and this high toll is a clue to the close synchronization of breeding of the mothers. Preliminary evidence suggests that the timing of conceptions in impala and wildebeest can be synchronized with the phase of the moon. By ensuring that her offspring is born at the same time as those of others, a mother takes advantage of the temporary excess which satiates the local predator community. Other social factors probably contribute to the advantage of synchronized breeding.

The dry regions of Africa are the province of the Hippotragini, a tribe of horse-like antelopes. The driest country of all is inhabited by the addax, a large white antelope with magnificent spiraling horns and an elegant chestnut wig. Surviving in waterless areas of the Sahara, particularly in dune regions, the addax is well adapted to heat, coarse foods and the absence of water. They are reputed to have a remarkable ability to sense patches of desert vegetation at long distance, and apparently obtain sufficient water from their diet of grasses. Greatly enlarged hooves and a long stride improve travel over the sandy and stony desert soils.

The sparsely vegetated subdesert and Sahelian steppes of north Africa are inhabited by the Scimitar oryx, another large white antelope of the same size as the addax, but with scimitar-shaped horns and russet markings. Similar habitats in the Arabian and Sinai peninsulas were once populated by the smaller Arabian oryx (see pp 572–573). The distinctively marked Beisa oryx inhabits the short-grass savannas and dry open bushland in seasonally arid areas of the Horn of Africa, and the closely related gemsbok is distributed over the dry plains and subdesert of the kalahari. These semidesert species supplement a basic diet of grass with browse: acacia pods, wild melons, cucumbers, tubers and the bulbs of succulents.

The arid environments of the oryx and addax have given rise to a remarkably tight social structure. The typical herd numbers less than 20 (60 in the gemsbok) and may contain several adult males in addition to females and young. The herd is closed to outsiders, horns providing females with the means to exclude competitors from scarce resources. Temporary aggregations of several hundred animals occur in areas where rainstorms have brought on the vegetation, but it is probable that groups typically inhabit conservative clan home ranges. Within the group, both bulls and cows are hierarchically organized; gener-

ally bulls rank over cows, but sometimes subadult bulls are dominated by high ranking cows. While the alpha bull of an oryx herd frequently reinforces his position over the other males through dominance encounters, he does not fight them and (in a complete reversal of the typical pattern among hoofed mammals) may actively prevent subordinate bulls from leaving.

The demands of a hierarchical society, coupled with the risk of conflict with long sabre-like horns, have given rise to the unique and highly ritualized oryx tournaments, which may involve many members of the herd running around in circles with sudden spurts of galloping and ritualized pacing interspersed with brief horn clashes.

Formerly protected by their inhospitable environment, addax and Scimitar oryx have become easy prey to mounted and motorized hunters with modern firearms. Sadly, the Scimitar oryx is now extinct north of the Sahara. Each year, deep boreholes for watering cattle on the edge of the Sahel are dug further north and each year the southern population of Scimitar oryx gets rarer, as cattle consume their traditional pastures. The Red Data Book states that the addax and Scimitar oryx are in real danger of total extinction in the very near future. The Arabian oryx was totally exterminated in the wild in the 1970s, but saved from extinction by the intervention of the Fauna and Flora Preservation Society and the Oman government (see pp 572–573). Addax and oryx are easily tamed and were domesticated by the ancient Egyptians. A domestic herd of oryx is now being run successfully on a cattle ranch in Kenya.

Sharing the same tribal name as the oryx,

▲ **Male and female Sable antelopes,** showing a pronounced sexual difference in coat color. The male is administering the foreleg kick, a courtship gesture.

▶ **Abundance in a National Park.** A mixed group of gemsbok and springbuck and a vast flock of Red-billed quelea in Etosha National Park, Namibia.

▼ **The impala clan.** The basic social unit of impala is the herd or clan of up to 100 females and young. The clan is frequently broken up into several groups which are unstable in composition. Males generally leave the herd before they reach breeding age. During the mating season, territorial males, as here, attempt to control groups of females and young that enter their territory.

the thickly-maned Roan and Sable antelopes inhabit the moister grasslands and open woodlands. Their tribal affiliation shows up in a preference for dry hillsides over fertile river valleys. Roan antelope in Kruger National Park, South Africa, move in herd ranges averaging 80sq km (31sq mi) with minimal range overlap. Their social organization is similar to that of oryx, with closed groups of 4–18 females, but with only one adult male per group. Aggregations of up to 150 occur in the dry season, when births are rare. Sable antelope particularly enjoy woodland and grassland near to water points and are the least conditioned by their dry-country origins. The herd home range may cover an area anywhere from 10–320sq km (3.8–124sq mi) in size on an annual basis, but is confined to as little as 2.5sq km (0.9sq mi) during the breeding season. The movements of Sable antelope at

this time are sufficiently localized and predictable for dominant males to relinquish their positions as masters of a mobile harem in preference for that of owner of a territory. In Zimbabwe, territories of 25–40ha (62–100 acres) have been held continuously for two or more years. During the two month rutting peak, breeding groups of 10–20 are rounded up by territorial males who turn escaping females back to their borders with loud snorts and vicious horn sweeps.

The last member of the tribe of horse-like antelope vanished in about 1800. Standing in the path of the pioneer columns, the bluebuck of the Cape Province (a close relative of the Roan antelope) fell prey to hunters and sportsmen before the concept of conservation had been born. Mounted specimens are in museums in Leiden, Paris, Stockholm, Vienna and Uppsala University.

MGM

THE 24 SPECIES OF GRAZING ANTELOPE

Abbreviations: HBL = head-body length. TL = tail length. SH = shoulder height. HL = horn length. wt = weight. Approximate nonmetric equivalents: 2.5cm = 1in; 1kg = 2.2lb.

E Endangered. EX Extinct. V Vulnerable. R Rare. * CITES listed.

Tribe Reduncini

Africa. Antelopes of wetlands, tall or tussock grassland. Horns on male only; long hair on sides and neck except in kob.

Genus *Redunca*

Light and graceful animals characterized by whistling and high bouncing jumps. Females of similar size to males. Glandular spot occurs beneath ears. 3 species.

Southern reedbuck

Redunca arundinum

Africa north to Tanzania and west to Angola. Southern savannas. Largest reedbuck. HBL 134–167cm; TL 26–27cm; SH 84–96cm; HL 30–45cm; wt 50–80kg. Coat: light buff to sandy brown; pale zone at base of horns; black and white markings on forelegs. Subspecies: 2.

Mountain reedbuck

Redunca fulvorufula

Cameroun, Ethiopia and E Africa, S Africa. Montane grasslands up to 5,000m. Smallest reedbuck. HBL 110–136cm; TL 20cm; SH 65–76cm; HL 14–38cm; wt 30kg. Coat: soft, woolly and gray. Prominent eyes and sockets. Subspecies: 3, each confined to its own highland area, including **Chanler's mountain reedbuck** (*R.f. chanleri*), Ethiopia and E Africa.

Bohor reedbuck

Redunca redunca

Senegal east to Sudan and south to Tanzania. Northern savannas. HBL 100–130cm; TL 18–20cm; SH male 75–89cm, female 69–76cm; HL 20–41cm; wt 45kg. Coat: light buff with strongly marked forelegs. Horns forward hooked. Subspecies: 7, including the Sudan form (*R.r. cottoni*) with long splayed horns.

Genus *Pelea*

A single species.

Gray rhebok†

Pelea capreolus
Gray rhebok, Vaal ribbok or rhebuck.

S Africa. High plateaux. SH 76cm; HL 20–29cm; wt 23kg. Gracefully built with long slender neck. Coat: brownish gray, soft and woolly. Tail short and bushy. No gland patch beneath the ears. Horns short, straight and almost vertical.

† This species is now recognized as not being a reduncine. Its exact affinities are still under debate.

Genus *Kobus*

Medium- to large-sized animals with a relatively heavy gait. Males have ridged horns and females are smaller than males. 5 species.

Waterbuck

Kobus ellipsiprymnus

S Africa north to Ethiopia and S Sudan, west to Senegal. Savanna and woodland near to permanent water. HBL 177–235cm; TL 33–40cm; SH 125cm; HL 55–99cm; wt 170–250kg. Big and shaggy with heavy gait. Coat: dark gray or reddish. Two taxonomic groups. The **Common waterbuck**, *ellipsiprymnus* group, with a white ellipse on the rump, is restricted to the east of the rift valley for most of its range extending from 6°N to 29.5°S. Subspecies: 4. The **Defassa waterbuck**, *defassa* group, with a white blaze on the rump is predominantly reddish in color. North and west of the rift valley. Subspecies: 9.

Kob

Kobus kob

Gambia east to Sudan and Ethiopia, Uganda. Low-lying flats and gently rolling hills close to permanent water. HBL 160–180cm; TL 10–15cm; SH male 90–100cm, female 82–92cm; HL 40–69cm; wt 77kg. Male robust and thick-necked with lyre-shaped horns; females more slender and graceful. Coat: reddish with white underside and throat chevron; legs with black markings. Subspecies: 10, including **Buffon's kob** (*K.k. kob*), Senegal to NW Nigeria; **White-eared kob** (*K.k. leucotis*), of S Sudan, whose males are predominantly black.

Lechwe V *

Kobus leche

Botswana, Zambia, SE Zaire. Floodplains and seasonally inundated grasslands. SH 99cm; HL 50–92cm; wt male 125kg, female 70kg. Coat: long and rough, bright chestnut to black with white underparts. Hooves long and pointed. Horns long, thin and lyre-shaped. Subspecies: 3. **Red lechwe** (*K.l. leche*), Botswana and Zambia. **Black lechwe** (*K.l. smithemani*), NE Zambia and SE Zaire. **Kafue Flats lechwe** (*K.l. robertsi*), Kafue River, Zambia.

Nile lechwe

Kobus megaceros
Nile or Mrs Gray's lechwe.

Sudan, W Ethiopia. Swamps in the vicinity of the White Nile, Sobat, Baro and Gilo Rivers. SH 94cm; HL 60–87cm; wt 86kg. Coat: long and rough, blackish chocolate with white underparts and white patch on shoulders; female uniformly reddish fawn. Hooves long, pointed and splayed. Horns long and spread out.

Puku

Kobus vardoni

S Zaire, Botswana, Angola, Zambia, Malawi, Tanzania. Margins of lakes, swamps and rivers, and on floodplains. HBL 126–142cm; TL 28–32cm; SH 77–83cm; HL 40–54cm; wt male 77kg, female 66kg. Coat: fairly long and shaggy, bright golden yellow. Horns short, thick and less lyre-shaped. Subspecies: 2. **puku** (*K.v. vardoni*) and **Senga kob** (*K.v. senganus*).

Tribe Alcelaphini

Africa. Grazers of open woodland, moist grassland and the zone between these two habitats. Characteristically with high population density. Except for impala, tribe characteristics include horns on both sexes, a long face, elevated shoulder and sloping hindquarters, preorbital glands and glands on forefeet only. The female is slightly smaller than the male. Impala have commonly been classed with the gazelles (Antelopini) although affinity with the kobs and reedbuck (Reduncini) has also been suggested. Evidence now suggests that impala are an early offshoot of the Alcelaphini, having diverged at the start of that tribe's history as an independent unit.

Genus *Beatragus*

A single species.

Hirola R

Beatragus hunteri
Hirola or Hunter's hartebeest.

E Kenya and S Somalia. Grassy plains between dry acacia bush and coastal forest. HBL 120–200cm; TL 30–45cm; SH 100–125cm; HL 55–72cm; wt 80kg. Coat: uniform sandy to reddish tawny, with white spectacles. Horns long and lyre-shaped on short bony base (pedicel). Female smaller with lighter horns.

Genus *Damaliscus*

Characterized by a long head but frontal region not drawn upward into a bony pedicel. Slope of body less exaggerated than that of hartebeest. Horns thick and ridged.

Topi *

Damaliscus lunatus
Topi, tsessebe, sassaby, tiangs damalisc, korrigum or Bastard hartebeest.

Senegal to W Sudan, E Africa through to S Africa. Green grassland of open savanna and swampy floodplains. HBL 170cm; TL 43cm; SH 124cm; HL 35–60cm; wt male up to 170kg, female 130kg. Coat: bold pattern of black patches (notably on face) on glossy mahogany red; fawn on lower part of legs. Subspecies: 7, including **tsessebe** (*D.l. lunatus*), from Zambia southwards; *D.l. jumela* and **topi** (*D.l. topi*) in NE Zaire and E Africa; **tiang** (*D.l. tiang*) in NW Kenya, W Ethiopia, and S Sudan; and **korrigum** (*D.l. korrigum*) from Senegal to W Sudan.

Bontebok

Damaliscus dorcas
Bontebok or blesbok.

S Africa. Open grassland. SH 84–99cm; HL 35–51cm; wt 59–100kg. Coat: Rich purplish chestnut brown, darker on the neck and hindquarters; white face patch, rump, belly and lower legs. Horns rather small in simple form. Subspecies: 2. **bontebok** (*D.d. dorcas*); **blesbok** (*D.d. phillipsi*).

Genus *Alcelaphus*

Characterized by a long narrow hammer-shaped head, heavy hooked and ridged horns and pronounced slope of back. 2 species.

Hartebeest

Alcelaphus buselaphus
Hartebeest or kongoni.

Coarse grassland and open woodland. Senegal to Somalia, E Africa to S Africa. HBL 195–200cm; TL 30cm; SH 112–130cm; HL 45–70cm; wt male 142–183kg, female 126–167kg. Coat: uniform sandy fawn to bright reddish, lighter on hindquarters, sometimes with black markings on legs. Frontal region of head drawn up into a bony pedicel. Horn shape diagnostic of races. Subspecies: 12, including **bubal** or **Northern hartebeest** (*A.b. buselaphus*) ranged N of Sahara, extinct; **Western hartebeest** (*A.b. major*), Senegal and

Guinea; **Lelwel hartebeest**
(*A.b. lelwel*), S Sudan, Ethiopia, N
Uganda and Kenya; **Tora hartebeest**
(*A.b. tora*) E, Sudan, Ethiopia;
Swayne's hartebeest (*A.b. swaynei*)
E, Ethiopia, Somalia; **Jackson's
hartebeest** (*A.b. jacksoni*). E Africa,
Rwanda; **Coke's hartebeest** (*A.b.
cokii*), Kenya, Tanzania; **Cape** or **Red
hartebeest** (*A.b. caama*), S Africa,
Namibia, Botswana, W Zimbabwe.

Lichtenstein's hartebeest

Alcelaphus lichtensteini

Tanzania, SE Zaire, Angola, Zambia,
Mozambique, Zimbabwe.
Open woodland. HBL 190cm;
TL 46cm; SH 124cm; HL 45–62cm;
wt male 160–205kg, female 165kg.
Coat: bright reddish with fawn flanks
and white hindquarters; dark stripe
down front legs. Frontal region of
skull does not form pedicel.

Genus *Connochaetes*

Characterized by massive head and
shoulders with mane on neck and
shoulders, beard under throat and
long tail reaching nearly to the
ground. Both sexes horned. 2 species.

White-tailed gnu

Connochaetes gnou
White-tailed gnu or Black wildebeest.

S Africa. Open grass veld. SH 115cm;
HL 53–74cm; wt 150–180kg. Coat:
dark brown to black; tail white. Face
covered by brush of stiff upward-
pointing hairs, tuft of hair between
front legs. Horns descending forwards
then pointing upwards. Name derived
from the hottentot 't' *gnu* which
describes the typical loud bellowing
snort.

Brindled gnu

Connochaetes taurinus
Brindled gnu or Blue wildebeest.

Northern S Africa to Kenya just south
of the equator. Moist grassland and
open woodland. HBL 194–209cm;
TL 45–56cm; SH 128–140cm;
HL 40–73cm; wt male 230kg, female
160kg. Bovine appearance. Coat:
slaty to dark gray; tail black. Horns
curving downwards laterally and
then pointing upwards and inwards.
Subspecies: 5, including **Blue
wildebeest** (*C.t. taurinus*), Zambia and
to the south and west; **Cookson's
wildebeest** (*C.t. cooksoni*), Luangwa
valley, Zambia; **White-bearded
wildebeest** (*C.t. mearnsi*), Tanzania,
Kenya.

Genus *Aepyceros*

A single species.

Impala

Aepyceros melampus

S Africa to Kenya, Namibia to
Mozambique. Open deciduous
woodland, especially near water.
HBL male 142cm, female 128cm;
TL 30cm; SH male 91cm; female
86cm; HL 49cm; wt male 80kg,
female 45kg. Straight backed, light
limbed, and graceful; the
quintessential antelope. Coat: light
mahogany with fawn flanks and
white undersurface of the belly; black
vertical stripes on tail and thighs, and
black glandular tuft on the fetlocks.
Horns on male only, lyre-shaped.
Subspecies: 6, including **Black-faced
impala** (*A.m. petersi*) E, SW Angola,
NW Namibia.

Tribe Hippotragini

Horse-like antelope of dry country
and savanna. Africa, Arabia. Females
only marginally smaller than males
and both sexes with well-developed
horns. 7 species (1 extinct).

Genus *Hippotragus*

Characterized by heavily ringed horns
curving backwards. Well-developed
mane of stiff hairs. 3 species.

Roan antelope [*]

Hippotragus equinus
Roan or Horse antelope.

Gambia to the Somali arid zone, C
Africa to S Africa, but rarely east of
the Rift Valley. Thinly treed
grasslands. HBL 190–240cm; TL
37–48cm; SH 126–145cm; HL
55–99cm; wt male 280kg, female
260kg. Coat: sandy fawn to dark
reddish with white underparts;
contrasted black and white face
markings. Ears long with tuft of hair
on tip. Subspecies: 6.

Sable antelope [*]

Hippotragus niger

C Africa from Kenya to S Africa,
Angola to Mozambique. Woodland
and woodland-grassland edges. HBL
197–210cm; TL 38–46cm;
SH 117–140cm; HL 50–164cm; wt
male 260kg, female 220kg. Coat:
adult females and young bulls rich
russet with pale underparts,
darkening to black; mature bulls
black with white bellies; calves
uniformly dun. Forehead narrow and
horns scythe-like. Subspecies: 4,
including **Giant sable antelope** (*H.n.
variani*), Angola.

Bluebuck [EX]

Hippotragus leucophaeus
Bluebuck or blaauwbok.

Formerly SW Cape Zone; extinct
about 1800. Woodland glades. SH
104cm; HL 34–62cm. Coat: Bluish
gray with white underparts and
insides of limbs (the velvety-bluish
sheen of the coat was much admired);
face markings indistinct or absent.
Hairs on neck mane directed forward,
tail short and tufted.

Genus *Oryx*

Characterized by short mane, hump
over shoulder and large hooves.

Scimitar oryx [E]

Oryx dammah
Scimitar or White oryx.

Formerly over most of N Africa,
presently extinct north of the Sahara;
confined to a narrow strip between
Mauritania and the Red Sea. Sahel
and semidesert. SH 119cm; HL
102–127cm; wt 204kg. Coat: pale
with neck and chest ruddy brown and
brownish markings on the face.
Scimitar-shaped horns.

Gemsbok

Oryx gazella
Gemsbok, oryx or Beisa oryx.

Discontinuous distribution. Namibia,
Angola, Botswana, Zimbabwe,
S Africa, and Tanzania north to the
Ethiopian coast. Seasonally arid
areas. HBL 153–170cm; TL 47cm;
SH 120cm; HL 65–110cm; wt male
200kg, female 162kg. Coat: fawn
with white underparts, black and
white markings on the head; black
line down throat and across flanks;
black tail. Horns straight and long.
Subspecies: 5, including **gemsbok**
(*O.g. gazella*) in SW Africa; **Beisa oryx**
(*O.g. beisa*) in the Horn of Africa; and
Fringe-eared oryx (*O.g. callotis*) in
Kenya and Tanzania.

Arabian oryx [E]

Oryx leucoryx
Arabian or White oryx.

Formerly Arabian peninsula, Sinai
peninsula. Presently reintroduced
into Oman. Stony semidesert. About
two-thirds of the size of *gemsbok*.
SH 81–102cm; HL 38–68cm; wt
65–75kg. Coat: white with black
markings on the face; legs dark
chocolate brown to black; tawny
colored line across flanks. Horns
nearly straight.

Genus *Addax*

A single species.

Addax [V]

Addax nasomaculatus

Formerly through entire Sahara.
Presently remnant populations in
Mauritania, Mali, Niger, Chad,
S Algeria, W Sudan. Sandy and stony
desert far from water. SH 100–110cm;
HL 76–109cm; wt 81–122kg. Coat:
grayish white with white rump, belly
and limbs; chestnut wig on forehead.
Tail long with black tip. Horns long
and spirally twisted.

MGM

The Arabian Oryx

A specialist for extremes

The desert environment of central Arabia is one of cruel extremes: summer shade temperatures peak at 48–50°C (118–122°F), while winter minima drop to 6–7°C (43–45°F) with strong cold winds. On every day of the year, the temperature varies by 20°C (36°F). Over large areas, rain may not fall for years and the vegetation lies dormant or in seed; natural surface water is short-lasting after rain. Sand storms can reduce visibility to a few meters for days on end. Resources of food, water, shade and shelter are sparse and scattered, making exacting demands on the animals living here. Of all the oryx species, gemsbok in the Kalahari, beisa in the East African Somali desert, and Scimitar-horned oryx in the Sahara, none equals the Arabian oryx as a desert specialist.

Sparse food has selected for a small size of 65–75kg (143–165lb), only a quarter of the weight of the gemsbok in the Kalahari which, although a barren landscape, is well-vegetated in comparison with central Arabia. Its color patterning has been simplified to produce a predominantly white coat which largely reflects incoming solar radiation. On winter mornings, the animals' coat has a suede look as they erect their hairs to absorb the sun's warmth, which is retained at night by the thicker winter coat. The blacker leg markings in winter also increase heat absorption.

The hooves are splayed and shovel-like, with their large surface areas fully in contact with the ground as an adaptation to sandy surfaces. The oryx is not a great runner, but it can walk for hours on end to reach a favorable habitat, and treks of 25–30km (15–19mi) in one night are not unusual. The oryx often combines trekking with the unavoidable chore of rumination, but if the herd encounters a patch of good grazing, they switch to feeding at once.

As the desert allows only low overall densities of oryx, the problems of finding a mate and reducing the risk of predation are solved by having only one herd type, containing all ages and approximately equal numbers of each sex. Such herds of 10–20 animals probably stay together for a considerable time.

Group stability is achieved through a linear dominance hierarchy in which there is a dominant bull to whom all other animals will defer; the second senior animal will be another bull or adult female who takes precedence over all except the dominant bull, and so on. Only the dominant male mates, so during his dominance, all calves are half-siblings. Newcomers meet with resistance in the form of threats, lunges and chases from established members of their own sex, and between males there can be severe fighting. If a newcomer persists in moving with the herd as a satellite, it is usually tolerated within 2–3 weeks, and is allowed to integrate progressively. Because both males and females feed in close proximity, every animal has to defend its food resources, and thus both sexes bear similar horns. The mixing of familiar individuals of each sex means that valuable energy is not used on the development of extravagant secondary sexual characteristics, and, thus, males and females are similar. The Arabian oryx expends less energy than other oryx in interactions between each other, and its musculature and tendon mass between the withers and skull is correspondingly reduced.

The low frequency of aggressive interactions allows animals to share scattered shade trees under which they may spend

eight of the daylight hours in the summer heat. Their small size allows them to creep under quite small acacia canopies for shade. These trees—like the oryx, natives of Africa—are also shorter in the desert. Under shade trees the oryx excavate scrapes with their fore-hooves so that they lie in cooler sand and reduce the surface area exposed to drying winds. They "fine-tune" their heat regulation carefully by seeking shade earlier on hot days and not venturing out until a cooler evening breeze blows. Through behavioral avoidance of excessive heat load, precious water is not lost by panting to cool down.

The members of a feeding herd spread out until neighbors may be 50–100m (165–330ft) apart, but constant visual checking, especially in undulating terrain, ensures that the animals keep in touch. Cohesion is helped by strong synchronization of activity within the herd. When the herd is trekking, a subdominant male leads, up to 100m (330ft) in front. Changes of direction when feeding can be initiated by any adult female. She will start in the new direction, then stop and look over her shoulder at the others until more or, gradually, all start to follow her.

Separation from the herd seems accidental. Singly oryx search for their herd and can recognize and follow fresh tracks in the sand. Moreover, as oryx are visible to the naked eye at 3km (2mi) in sunlight, the white coat may have evolved partially as a flag to assist herd-location in a open environment where merging with the environment is less necessary. Non-human predators such as the Arabian wolf and Striped hyena have never been abundant.

Historically, the Arabian oryx ranged through Arabia, up through Jordan and into Syria and Iraq. It had always been a prized trophy and source of meat for bedu tribesmen who, hunting on foot or from camel with a primitive rifle, were unlikely to have significantly depleted the populations. But from 1945, motorized hunting and automatic weapons caused a severe contraction of its range and numbers, and it became extinct in 1972, leaving a few in private collections in Arabia and the World Herd in the USA. This herd grew out of the 1962 Operation Oryx, organized by the far-sighted Fauna and Flora Preservation Society to ensure the survival of the species in captivity. By 1982, ecological and social conditions in central Oman were deemed right for the release of a carefully developed herd, with the long-term aim of re-establishing a viable population.   MSP

▲ **Peace in the shade.** Arabian oryx cope with their inhospitable surroundings by reducing unnecessary energy expenditure. They are extremely tolerant of each other, thus avoiding wasteful fights, and happily share the precious shade afforded by trees and bushes.

◄ **The patient trek** of Arabian oryx moving to a grazing area in late afternoon. They are highly organized and methodical in their movements, with a subdominant male leading the group and the dominant male at the rear, rounding up the calves.

GAZELLES AND DWARF ANTELOPES

Subfamily: Antilopinae
Thirty species in 12 genera.
Family: Bovidae.
Order: Artiodactyla.
Distribution: Africa, Middle East, Indian subcontinent, China.

Habitat: varied, from dense forest to desert and rocky outcrops.

Size: head-and-body length from 45–55cm (18–21.5in) in the Royal antelope to 145–172cm (57–68in) in the Dama gazelle; weight from 1.5–2.5kg (3.3–5.5lb) in the Royal antelope to 40–85kg (88–188lb) in the Dama gazelle.

Dwarf antelopes
Tribe: Neotragini.
Twelve species in 6 genera.
Species include: **Günther's dikdik** (*Madoqua guentheri*), **klipspringer** (*Oreotragus oreotragus*), **oribi** (*Ourebia ourebi*), **Pygmy antelope** (*Neotragus batesi*), **Royal antelope** (*Neotragus pygmaeus*), **steenbuck** (*Raphicerus campestris*).

Gazelles
Tribe: Antilopini.
Eighteen species in 6 genera.
Species include: **blackbuck** (*Antilope cervicapra*), **dibatag** (*Ammodorcas clarkei*), **Dorcas gazelle** or **jebeer** (*Gazella dorcas*), **gerenuk** (*Litocranius walleri*), **Goitered gazelle** (*Gazella subgutturosa*), **Grant's gazelle** (*Gazella granti*), **Red-fronted gazelle** (*Gazella rufifrons*), **Speke's gazelle** (*Gazella spekei*), **springbuck** (*Antidorcas marsupialis*), **Thomson's gazelle** (*Gazella thomsoni*).

► **Pronking springbuck** fleeing from a predator ABOVE. Pronking involves a sudden vertical leap in the middle of fast running.

► **Kirk's dikdik,** one of the smaller dwarf antelopes. Dwarf antelopes are well-endowed with scent-glands and the pre-orbital gland of this male dikdik is especially prominent.

THE gazelles and dwarf antelopes make up a tantalisingly contrasting group, containing some of the most abundant and the rarest, the most studied and the least known of the hoofed mammals. They occupy habitats from dense forest to desert and rocky outcrops, and their range spans three zoogeographic regions from the Cape to eastern China.

The dwarf antelope tribe (Neotragini) is very varied in form and habitat: from pygmy antelopes in the dense forests of central Africa to the oribi in the open grass plains adjacent to water; from the dikdiks in arid bush country to the klipspringer on steep rocky crags. The only common features of the tribe are their small size, with females 10–20 percent larger than males, and well-developed glands for scent marking, especially preorbital.

Their small size is associated with their diet. All the species except the oribi, which is a grazer, are "concentrate selectors," taking easily digested vegetation low in fiber, such as young green leaves of browse and grass, buds, fruit and fallen leaves. A smaller body size increases the metabolic requirement per kilogram of body-weight, so a small herbivore has to assimilate more food per kilogram of body-weight than a large her-bivore. Ruminants like the dwarf antelopes have a limit to the rate of food intake which is dictated by the length of time food stays in the rumen. In dwarf antelopes this limit is so low that they have to choose vegetation of a high nutritive quality. Small size is evidently a secondary adaptation since their gestation time is more typical of the larger hoofed mammals the size of the gazelles. The reduction in the extent of forest habitat since their evolution in the Miocene (about 12 million years ago) has meant that smaller size has been increasingly favored.

Unusually among hoofed mammals, female dwarf antelopes are larger than the males. It is advantageous for the females to be as large as possible, within the constraints of their habitats, since they have the added burden of raising young. However, many dwarf antelopes are territorial, and among some territorial species inter-male competition for females favors larger males. But in many dwarf antelopes, the males form a lifetime bond with one or a few females, and this reduces inter-male conflict and therefore precludes the need for larger size.

Territoriality has been favored in the evolution of dwarf antelopes: because the food items are so varied and scattered

through the habitat, it is most efficient for an animal to have an intimate knowledge of its home range and the distribution of resources therein, and to exclude competitors. A permanent bond between the individuals of a breeding group, and the demarcation of an exclusive territory, help to maintain such a system.

In order to demarcate and maintain these territories, dwarf antelopes have well-developed scent glands, whose secretions have become more important than visual displays and encounters. Secretions from the preorbital glands in front of the eyes are daubed repeatedly on particular stems, where a sticky mass accumulates. Pedal glands on the hooves mark the ground along frequently traveled pathways. Males also mark females in this way, thus reinforcing the bond.

The number of scent glands is highest in the oribi, which has six different gland sites, including one below the ear. Oribi have large territories (1sq km; 0.4sq mi) in open grassland, and their well-developed glands may be a response to the need to increase the amount of marking.

In dwarf antelopes dung and urine are deposited on particular sites, and the male has a characteristic posture and linked urination/defecation when adding to these piles. In most species, scent marking is done by both sexes. Typically, when a female defecates on these piles, the male will sniff, paw, and add his own contribution to hers immediately after. Oribi and suni adopt a ritualized posture and perform preorbital marking when neighboring territorial males meet; fights are very rare. Dikdiks "horn" the vegetation and raise their crests as a threat display.

The range of the dwarf antelopes has almost certainly been affected by disturbance to the habitat, since many of the species prefer the secondary growth that invades disturbed areas, notably from slash-and-burn cultivation. The pygmy antelopes, dikdiks, grysbucks and steenbuck all favor this habitat.

Those species that live in dense cover, such as the pygmy antelopes, have a crouched appearance, with an arched back and short neck, a body shape that is suited to rapid movement through the dense vegetation. The other species live in more open habitat, where detection of predators by sight is more important, and they have a more upright posture, with a long neck and a raised head. The exception is the klipspringer, which has an arched back, enabling it to stand with all four limbs together, to take advantage of small patches of level rock. Its physical appearance is unusual, having a thick coat of lightweight hollow-shafted hair to protect the body against the cold and physical damage from the bare rock in its exposed habitat of rocky crags. It has peg-like hooves each with a rubbery center and a hard outer ring to gain a sure purchase on the rock.

The proboscis of the dikdiks, which is most pronounced in the Günther's dikdik, is an adaptation for cooling. Venous blood is cooled by evaporation from the mucous membrane into the nasal cavity during normal breathing or under greater heat stress from nasal panting.

Dwarf antelopes generally rely on concealment to escape detection and thereby minimize predation. Their first response on detecting a predator is to freeze, and then, on the predator's closer approach to run away. Grysbucks tend to dash away and then suddenly drop down to lie hidden from sight. In tall grass, oribi lie motionless to escape detection, but, if the grass is short they flee, often with a stotting action like that of gazelles. The oribi and the dikdiks have a whistling alarm call.

Births occur throughout the year, but there are peaks that coincide with the vegetation flush that follows the early rains. In equatorial regions where there are two rainy seasons a year, two birth peaks occur. Young are born singly. Females become sexually mature at six months in the smaller species and ten months in the larger, and males become sexually mature at about fourteen months. The newborn lie out for the first few weeks after birth, and the mother will come to suckle it, calling with a soft bleat which the infant answers. When the young become sexually mature, they are increasingly threatened by the territorial male. The young responds with submissive displays such as lying down, and it eventually leaves the territory at 9–15 months old.

In contrast to the dwarf antelopes, gazelles (tribe Antilopini) are all very similar in shape, with only two exceptions: the gerenuk (see pp 582–583) and dibatag. They have long legs and necks and slender bodies, and most have ringed (annulated) S-shaped horns. They are generally pale fawn above and white beneath. The exception is the blackbuck of India, in which the males develop conspicuous black upperparts at about three years of age. The two-tone coloration of gazelles probably acts as countershading to obscure the animal's image to a predator and minimize detection.

Gazelles live in more open habitat than the dwarf antelopes and so rely more on visual signals. Some species, such as the Thomson's gazelle, Speke's gazelle, Red-fronted gazelle and the springbuck, have conspicuous black side-bands. Three functions have been suggested for these bands: they act as a visual signal to keep the herds together, they communicate alarm when all the members are fleeing, and they may break up the outline of individuals in a herd. All the species have white buttocks with at least some black on the tip of the tail, and dark bands that are more or less distinct, being most conspicuous in Grant's gazelle. Another feature of the gazelles is their stotting or pronking gait, seen when they are apparently playing or alarmed: bouncing along stiff-legged with all four limbs landing together. The possible functions of this are that it communicates alarm, gives the animal a better view of the predator, and also confuses or even intimidates it. Pronking is most pronounced in the springbuck, hence its name; it also erects the line of white hairs on its back at the same time.

The sense of sight and hearing are well-developed and are reflected in the large orbits and ear cavities of the skull. In contrast to the dwarf antelopes, most females possess horns, the exceptions being the Goitered gazelle, blackbuck, gerenuk and dibatag. One theory of horn evolution is that they have evolved in response to territoriality and increased inter-male conflict, and females have evolved horns to defend their food resources. Patchily distributed food resources can be defended, and this is particularly so for gazelles in the dry season or in winter, when the nutritive quality of the vegetation is generally poor, and patches of more nutritious grass and bushes occur.

The range of the tribe is very extensive, from the springbuck in the Cape to the blackbuck and Dorcas gazelle in India, and Przewalski's gazelle in eastern China. The species in north Africa, the Horn, and Asia inhabit arid and desert regions where their distribution is very patchy and they occur in low densities, with populations separated by geographical barriers of uninhabitable terrain, mountain ranges, and seas such as the Persian Gulf and the Red Sea. This has led to many variations in form. A good example is the Dorcas gazelle: in Morocco it is a pale-colored gazelle with relatively straight parallel horns and as it extends eastwards it goes through various discrete changes, becoming a reddish color, with a smaller body size and lyre-shaped horns in India.

Gazelles are mixed feeders, though they

► **Species of dwarf antelopes and gazelles.**
(1) Klipspringer (*Oreotragus oreotragus*) defacating on its territorial boundary.
(2) Dibatag (*Ammodorcas clarkei*) in the alarmed posture. (3) Slender-horned gazelle (*Gazella leptoceros*), the palest gazelle. (4) Oribi (*Ourebia ourebia*) scent marking a grass stem with its ear gland. (5) Kirk's dikdik (*Madoqua kirkii*) horning vegetation. (6) Royal antelope (*Neotragus pygmaeus*), the smallest antelope. (7) Steenbuck (*Raphicerus campestris*) scent marking with the pre-orbital gland. (8) Tibetan gazelle (*Procapra picticaudata*). (9) Springbuck (*Antidorcas marsupialis*) pronking. (10) Goitered gazelle (*Gazella subgutturosa*). (11) Dama gazelle (*Gazella dama*), the largest gazelle.
(12) Blackbuck (*Antilope cervicapra*) in territorial display gesture.

THE LIFE OF THE THOMSON'S GAZELLE

▲ **The typical stance** of fighting gazelles is displayed by these Thomson's gazelles: heads as close to the ground as possible and horns interlocked.

◄ **Mating.** The males establish territories during the breeding season and mate with receptive females who enter it.

▶ **A newborn Thomson's gazelle,** with the cord still visible. The young lie out for the first few weeks until they can run with the herd.

▷ **Meal fit for a cheetah.** The not uncommon end of a gazelle's life: one graceful creature devoured by another.

mainly browse, taking grass, herbs and woody plants to a greater or lesser degree, depending on their availability. Generally, they eat whatever is greenest, so in the early spring or rains when the grass flushes they will turn to browse. Thomson's gazelle is almost entirely a grazer, up to 90 percent of its food being grass. The gerenuk is a browser that feeds on young shoots and leaves of trees, particularly acacias (see pp582–583), up to a height of 2.6m (8.5ft). The dibatag also feeds in this manner and shows similar adaptations of body shape.

In all of the gazelles, the males establish territories, at least during the breeding season, from which they actively exclude other mature males while females are receptive. Four types of groupings occur: single territorial males; female groups with their recent offspring, usually associated with a territorial male; bachelor groups comprising nonbreeding males without territories; and mixed groups of all sexes and ages that are common outside the breeding season. There is a certain amount of mixing of the groups. Males mark their territories with urine and dung piles and, in most species, with secretions from their preorbital glands. Subordinate males will also add to the dung piles when they move through the territories, and territorial males will tolerate familiar subordinate males within their territories as long as they remain subordinate and do not approach the females with any intent. Threat displays start at the lowest level of intensity with head raised, chin up and horns lying along the back. This display reaches its most advanced form in the Grant's gazelle. Two males stand antiparallel with each other, chins up and heads turned away, and at the same instant both whip their heads around to face each other. The next level of intensity is a head-on approach with chin tucked in and horns vertical. The next level is head lowered to the ground and horns pointing towards the opponent.

Fights, when they develop, consist of opponents interlocking horns with their heads down on the ground and pushing and twisting. At any time, an individual can submit by moving away at a greater or lesser speed, depending on the closeness of the visitor's pursuit. Sparring bouts are very common between bachelor males; these do not result in change of status but doubtless provide experience in sizing up opponents' abilities. Intense, wounding fights do occur between neighboring territorial males. Territorial males will herd females with a chin-up display, but if females decide to leave a

territory the male will not stop them or attempt to retrieve them once they have crossed the boundary.

In contrast to the dwarf antelopes, the food items of the gazelles are more abundant and continuously distributed in the habitat, so that territories can support a greater density of animals. Males can therefore afford to maintain many females in their territories. Since there are thus fewer breeding males per female, there is increased inter-male conflict for the right to establish territories and breed. Increased size of the male confers an advantage and so evolution has favored males larger than females.

In the temperate and cold zones of their geographical range, gazelles breed seasonally, so that the births coincide with the vegetation flush in spring or early rains. Females go off on their own to give birth and fawns lie out for the first few weeks like the dwarf antelopes. Once the fawns can run sufficiently well they join the group.

Many species of gazelles have been greatly reduced in numbers and range by man. The species that inhabit North Africa, the Horn of Africa and Asia have suffered most because they live in arid habitats where their densities are less than 1 per sq km (2.6 per sq mi), and where they can easily be hunted from vehicles. Domestic sheep and goats also compete for the same food plants, and access to springs has been denied them by human activity. Several of these species are endangered. In Asia, cultivation has removed many areas of essential winter range, so that the vast winter aggregations, particularly of the Goitered gazelle and the blackbuck, are now rare or absent. The springbuck, though still common, no longer occurs in migrating herds of tens of thousands, due to excessive hunting and to barring their natural movements with fences for ranching of domestic stock.   BO'R

THE 30 SPECIES OF DWARF ANTELOPES AND GAZELLES

Abbreviations: HBL = head-and-body length. TL = tail length. SH = shoulder height. HL = horn length. WT = weight. Approximate nonmetric equivalents: 2.5cm = 1in; 1kg = 2.2lb.
[E] Endangered. [V] Vulnerable. [I] Indeterminate. [*] CITES listed.

Tribe Neotragini
Dwarf antelopes

Small delicate antelopes, females slightly larger than males. Horns short and straight, females hornless, except in a subspecies of klipspringer. Except for oribi, diet young leaves and buds, fruit roots and tubers, fallen leaves and green grass. Live singly or in small family groups. Males territorial. Gestation about 6 months. Territories marked with dung piles, preorbital and pedal glands, scent glands well-developed. Breeding throughout the year with birth peaks in early rains. Underparts white or buffish.

Genus *Neotragus*

Horns smooth, short, inclined backwards. Tail relatively long. Back arched, neck short. 3 species.

Royal antelope
Neotragus pygmaeus

Sierra Leone, Liberia, Ivory Coast and Ghana. Dense forest. Occurs singly or in pairs, shy and secretive. Little known. Smallest horned ungulate. HBL 45–55cm; TL 4–4.5cm; SH 20–28cm; HL 2.5–3cm; WT 1.5–2.5kg. Head and neck dark brown. Coat: back brown, becoming lighter and bright reddish on flanks and limbs, contrasting with white underparts; tail reddish on top and white underneath and at tip.

Pygmy antelope
Neotragus batesi

SE Nigeria, Cameroun, Gabon, Congo, W Uganda and Zaire. Dense forest. Predominantly solitary. Often move into plantations and recently disturbed land at night. Females have overlapping home ranges. HBL 50–58cm; TL 4.5–5cm; SH 24–33cm; HL 2–4cm; WT 3kg; Coat: shiny dark chestnut on the back, becoming lighter on the flanks; tail dark brown.

Suni
Neotragus moschatus

Patchily distributed from Zululand through Mozambique and Tanzania to Kenya. Coastal, riverine and montane forest with thick undergrowth. Occur singly or in small groups, with never more than one adult male per group. Males grate their horns on tree trunks as a territorial marker. HBL 58–62cm; TL 11–13cm; SH 30–41cm; HL 5–13cm; WT 4–6kg; Coat: dark brown, slightly freckled; transparent pink ears. Horns strongly annulated at base.

Genus *Madoqua*

Long, erectile hairs on forehead. Tail minute. Underparts white. Pairs of a male and female form lifelong territories. Larger groupings commonly occur at sites of food abundance between territories. 3 species.

Swayne's dikdik
Madoqua saltiana
Swayne's or Phillips' dikdik.

Horn of Africa. Arid evergreen scrub in foothills and outliers of Ethiopean mountains, particularly in disturbed or over-grazed areas with good thicket vegetation. HBL 52–67cm; TL 3.5–4.5cm; SH34.5–40.5cm; HL 4–9cm; WT 3–4kg. Coat: thick, back gray and speckled, flanks variable gray to reddish; legs and forehead and nose bright reddish; white ring round eye. Nose slightly elongated.

Günther's dikdik
Madoqua güntheri

N Uganda, eastwards through Kenya and Ethiopia to the Ogaden and Somalia. Semi-arid scrub. HBL 62–75cm; WT 3.5–5cm; SH 34–38cm; HL 4–9cm; WT 4–5.5kg. Coat: back and flanks speckled gray, reddish nose and forehead. Nose conspicuously elongated.

Kirk's dikdik
Madoqua kirkii
Kirk's or Damaraland dikdik.

Tanzania and southern half of Kenya; Namibia and Angola. The two ranges are separate. HBL 60–72cm; TL 4.5–5.5cm; SH 35–43cm; HL 4–9cm; WT 4.5–6kg. Coat: whitish ring round eye. Nose moderately elongated.

Genus *Oreotragus*

A single species.

Klipspringer
Oreotragus oreotragus

From the Cape of Angola, and up the eastern half of Africa to Ethiopia and E Sudan. Also two isolated massifs in Nigeria and Central African Republic. Well drained rocky outcrops. Gait a stilted bouncing motion. Seldom solitary; occur in small family groups. Aggregations occur at favourable feeding sites. Lack pedal glands. HBL 75–90cm; TL 6.5–10.5cm; SH 43–51cm; HL 6–16cm; WT 10–15kg. Coat: yellowish, speckled with gray, ears round, conspicuously bordered with black; black ring above hooves. Tail

minute. Back conspicuously arched. Fur is thick, coarse and brittle, loosely rooted and lightweight, giving the animal a stocky appearance, hooves peg-like. Horns smooth, nearly vertical.

Genus *Raphicerus*

Coat reddish, large white-lined ears, tail minute, horns smooth and vertical. Three species.

Steenbuck
Raphicerus campestris

From Angola, Zambia and Mozambique southwards to the Cape, and in Kenya and Tanzania. Open, lightly wooded plains. Usually seen singly. Have large home ranges. Preorbital gland apparently not used for territorial marking. HBL 70–95cm; TL 4–6cm; SH 45–60cm; HL 9–19cm; WT 10–15kg. Coat: reddish fawn; large white lined ears, shiny black nose. Horns smooth and vertical. Tail minute.

Sharpe's grysbuck
Raphicerus sharpei

Tanzania, Zambia, Mozambique, Zimbabwe. Woodland with low thicket or secondary growth. Mainly nocturnal and cryptic. HBL 61–75cm; TL 5–7cm; SH 45–60cm; HL 3–10cm; WT 7.5–11.5kg. Coat: Reddish brown, speckled with white on back and flanks; large white-lined ears. Horns small, smooth and vertical. Tail small. Back slightly arched.

Cape grysbuck
Raphicerus melanotis

Restricted to the southern Cape. Not as red as Sharpe's grysbuck, otherwise same.

Genus *Ourebia*

A single species.

Oribi
Ourebia ourebi

Range patchy and extensive. Eastern half of southern Africa, Zambia, Angola and Zaire, and from Tanzania northwards to Ethiopia, and westwards to Senegal. Grassy plains with only light bush, near water. Live mostly in pairs or small family groups. Aggregations do occur at favorable feeding sites. Scent glands and marking very well developed by males. Territories large. Commonly run with stotting gait. Grazer. HBL 92–140cm; TL 6–15cm; SH 54–67cm; HL 8–19cm; WT 14–21kg. Coat: reddish fawn back and flanks

contrasting conspicuously with white underparts; forehead and crown reddish brown; black glandular spot below ear. tail short with black tip. Ears large and narrow. Horns short, vertical, slightly annulated at base.

Genus *Dorcatragus*

A single species.

Beira [V]
Dorcatragus megalotis

Somalia and Ethiopia, bordering the Red Sea and the Gulf of Aden. Stony barren hills and mountains. Very rare and little known. HBL 70–85cm; TL 14–20cm; SH 52–65cm; HL 7.5–12.5cm; WT 15–26kg. Gazelle-like. Large for a neotragine. Gray, finely speckled on back and flanks. Distinct dark band on sides. Underparts yellowish. Ears very large. Tail long and white. Horns widely separated, curving slightly forward. Rubbery hooves adapted to their rocky habitat.

Tribe Antilopini
Gazelles

Slender body and long legs. Male larger than female. Fawn upperparts, white underparts, typically with gazelline facial markings of dark band on blaze, white band on either side, dark band from eye to muzzle and white round eye. Horns generally S-shaped, annulated. Usually both sexes have horns, though female horns are shorter and thinner. Tail black tipped. Populations comprise typically territorial males, at least during the breeding season, female groups with their offspring, and bachelor goups of non-territorial males. Will migrate in response to seasonal changes in vegetation and climate, forming large mixed-sex aggregations during the winter or dry seasons. In seasonal parts of their range, birth peaks occur to coincide with vegetation flush in spring or early rains. Mixed feeders, but mostly browse.

Genus *Gazella*

Subgenus *Nanger*
Large gazelles, white rump and buttocks, tail white with black tip. 3 species.

Dama gazelle
Gazella dama

Sahara, from Mauritania to Sudan. Desert. Occur singly or in small groups. In rainy season move north into Sahara, in dry season back to Sudan. Very rare and disappearing. HBL 145–172cm; TL 22–30cm; SH 88–108cm; HL 33–40cm; wt 40–85kg. Long neck and legs for a gazelle. Coat: neck and underparts reddish brown, sharply contrasting with white rump, underparts and head; white spot on neck. Horns sharply curved back at base, relatively short.

Soemmerring's gazelle
Gazella soemmerringi

Horn of Africa, northwards to Sudan. Bush and acacia steppe. Occurs in small groups of 5–20. HBL 122–150cm; TL 20–28cm; SH 78–88cm; HL 38–58cm; wt 30–55kg. Coat: Pale fawn on head, neck and underparts; facial markings very pronounced. Short neck, long head. Horns sharply curved inwards at tips.

Grant's gazelle
Gazella granti

Tanzania, Kenya, and parts of Ethiopia, Somalia and Sudan. From semidesert to open savanna. Live in small groups of up to 30. Preorbital gland not used. HBL 140–166cm; TL 20–28cm; SH 75–92cm; HL 45–81cm; wt 38–82kg. Coat: black pygal band. Heavily built. Horns long, variable according to race.

Subgenus *Gazella*
Small gazelles. White on underparts and buttocks not extending to rump. Gestation five and a half months. HBL 70–107cm; TL 15–26cm; SH 40–70cm; HL 25–43cm; wt 15–32kg. Except *G. thomsoni*, occur in small groups. 7 species.

Mountain gazelle
Gazella gazella

Arabian peninsula, Palestine. Extinct over much of its range. Semidesert and desert scrub in mountains and coastal foothills. Breed seasonally. Coat: upperparts fawn, pygal and flank bands distinct.

Dorcas gazelle [E]
Gazella dorcas
Dorcas gazelle or jebeer.

From Senegal to Morocco, and westwards through N Africa and Iran to India. Semi-desert plains. Coat: upperparts pale, pygal and flank bands indistinct.

Slender-horned gazelle [E]
Gazella leptoceros

Egypt eastwards into Algeria. Mountainous and sandy desert. Coat: upperparts very pale. Ears large. Hooves broadened. Horns long, only slightly curved.

Red-fronted gazelle
Gazella rufifrons

From Senegal in a narrow band running eastward to Sudan. Semidesert steppe. Coat: reddish upperparts, narrow black band on side with reddish shadow band below, contrasting with white underparts. Horns short, stout, only slightly curved.

Thomson's gazelle
Gazella thomsoni

Tanzania and Kenya, and an isolated population in southern Sudan. Open grassy plains. Very abundant. Occur in large herds of up to 200. Have aggregations of several thousand during migration. Mixed feeders, but predominantly grazers. Largest of the subgenus. Coat: upperparts bright fawn, broad conspicuous dark band on the side; well-pronounced facial markings. Horns long, only slightly curved; small and slender in females.

Speke's gazelle [I]
Gazella spekei

Horn of Africa. Bare stony steppe. Little known. Coat: upperparts pale fawn, broad dark side-band. Small swollen extensible protruberance on nose.

Edmi [E]
Gazella cuvieri

Morocco, N Algeria, Tunis. Semidesert steppe. Coat: upperparts dark gray-brown, dark side-band with shadow band below; facial markings pronounced.

Subgenus *Trachelocele*
A single species.

Goitered gazelle
Gazella subgutturosa

From Palestine and Arabia eastwards through Iran and Turkestan to E China. Semidesert and desert steppe. In Asia, form winter aggregations of several thousand at lower altitudes to avoid snow. In summer disperse; Females disperse further than males. Stocky body, relatively short legs. HBL 38–109cm; TL 12–18cm; SH 52–65cm; HL 32–45cm; wt 29–42kg.

Coat: upperparts pale, pygal and side-bands indistinct; facial markings not pronounced, fading to white with age. Horns arise close together and curved in at tips. Female mostly hornless. Male larynx forms conspicuous swelling.

Genus *Antilope*
A single species.

Blackbuck [*]
Antilope cervicapra

Indian subcontinent. From semidesert to open woodland. HBL 100–150cm; TL 10–17cm; SH 60–83cm; wt 25–45kg. Coat: adult males, upperparts and neck dark brown to black, contrasting with white chin, eye and underparts; immature males and females light fawn. Horns long and spirally twisted. Females hornless.

Genus *Procapra*
Gazelle-like. Pale fawn upperparts, white rump and buttocks. 3 species.

Tibetan gazelle
Procapra picticaudata

Most of Tibet. Plateau grassland and high altitude barren steppe. HBL 91–105cm; TL 2–10cm; SH 54–64cm; HL 28–40cm; wt 20–35kg. Face glands and inguinal glands absent. Horns S-shaped, not curving in at tips.

Przewalski's gazelle
Procapra przewalskii

China, from Nan Shan and Kukunor to Ordos Plateau. Semidesert steppe. Same size as Tibetan gazelle. Horns curve in at tip.

Mongolian gazelle
Procapra gutturosa

Most of Mongolia and Inner Mongolia. Dry steppe and semidesert. HBL 110–148cm; TL 5–12cm; SH 30–45cm; wt 28–40kg. Small preorbital glands and large inguinal glands present.

Genus *Antidorcas*
A single species.

Springbuck
Antidorcas marsupialis

Southern Africa west of the Drakensberg Mountains and northwards to Angola. Open arid plains. Very gregarious. Used to migrate in vast herds of tens of thousands. Have characteristic pronking gait in which white hairs on

back are erected. Mixed feeeders, taking predominantly grass. HBL 96–115cm; TL 20–30cm; SH 75–83cm; HL 35–48cm; wt 25–46kg. Coat: bright reddish fawn upperparts, dark side-band contrasting with white underparts; face white with dark band from eye to muzzle; buttocks and rump white, and line of white erectile hairs in fold of skin along lower back. Horns short, sharply curved in at tip.

Genus *Litocranius*
A single species.

Gerenuk
Litocranius walleri

Horn of Africa south to Tanzania. Desert to dry bush savana. Occur singly or in small groups. Browse on tall bushes by standing on hindlegs. Feeds delicately on leaves and young shoots. Independant of water. Do not form migratory aggregations. HBL 140–160cm; TL 22–35cm; SH 80–105cm; HL 32–44cm; wt 29–52kg. Coat: reddish brown upperparts, back distinctly dark, white around eye and underparts. Very long limbs and neck with small head and weak chin. Tail short with black tip. Horns stout at base, sharply curved forward at tips. Female hornless.

Genus *Ammodorcas*
A single species.

Dibatag [V]
Ammodorcas clarkei

Horn of Africa. Grassy plains and scrub desert. Occur singly or in small groups. Able to stand on hindlegs to browse. Hold tail erect in flight. Do not form migratory aggregations. HBL 152–168cm; TL 30–36cm; SH 80–88cm; HL 25–33cm; wt 23–32kg. Coat: upperparts dark reddish gray, contrasting with white underparts and buttocks; head with chestnut gazelline markings; tail long, thin and black. Long neck and limbs. Horns curve backwards at base then forwards at tip like a reedbuck. Females hornless.

BO'R

The Graceful Gerenuk

Survival in dry country

The delicate beauty of the gerenuk, with its extremely long, slender legs and neck, long ears and lyre-shaped horns, raises the question of how this seemingly frail creature is adapted to hot and dry country.

The gerenuk seems to have followed an evolutionary path similar to that of the giraffe, the long neck enabling it to reach higher up to gather food than other species.

Apart from size, the gerenuk differs from the giraffe, and from most other bovids, in that it habitually extends its reach even higher up by rising onto its hindlegs. In fact, all ungulates can rear up on their hindlegs—they do so when mating—but the gerenuck is unique in its maneuverability, sometimes stepping sideways round a tree while continuing to feed in a near-vertical stance. Minor adaptations in the skeleton and muscles of limbs and vertebral column facilitate this behavior, which enables the gerenuk to occupy an ecological niche different to those of its near relatives, the gazelles.

The gerenuk eats almost exclusively leaves, as well as some flowers and fruits, of a great variety of trees and shrubs, but no grass. Over 80 plant species are eaten by gerenuks in Tsavo National Park, Kenya. Their diet varies substantially between rainy and dry seasons, as the availability of different plant species changes too. Some of the plants are evergreens, with thick, rather hard leaves, coated with a thick cuticle which prevents evaporation of water. On evergreens, gerenuks are highly selective browsers, sniffing over a plant carefully before plucking small, tender leaves and shoots with their highly mobile lips. These and the narrow, pointed muzzle also permit them to pick the delicate leaves of acacias from among forbidding thorns.

The main advantage that the gerenuk derives from its highly selective feeding behavior is that it ingests only the juiciest, most nutritious plant parts. Probably, this is why gerenuks are independent of free water; even in captivity they hardly ever drink. This allows them to inhabit dry country that many other bovids would find too inhospitable. The gerenuk also has a less complicated digestive tract and weaker molar teeth, with a smaller grinding surface, than some of its relatives, eg Thomson's gazelle, that eat the more fibrous grass.

However, these benefits also have their costs: in the dry areas inhabited by the gerenuk, suitable food items are widely dispersed, particularly in the dry season; gathering enough to satisfy both water and energy requirements means spending considerable time searching for them. This has

two main consequences. Firstly, the individual gerenuk ought to minimize energy expenditure, so as to optimize its overall energy budget. A gerenuk can do this partly by avoiding strenuous activities such as fighting, long-distance, seasonal traveling, and partly by reducing energy losses in inclement weather. One way in which gerenuks—and some other gazelles—achieve this is to lie down in response to rainfall and/or strong wind. This reduces the surface area exposed to heat loss.

Secondly, gerenuk populations cannot achieve high biomass densities, compared to other species of hoofed mammals. In Tsavo National Park, where gerenuks are locally quite common, they contribute less than 0.5 percent to the total hoofed mammal biomass. This is also reflected in their social organization. Gerenuks do not form herds, but rather small groups of 1–5 individuals, although occasional aggregations of up to

▲ **Lacking the horns** of the male, the long pointed ears of the female gerenuk are her most distinctive feature.

▶ **Stilt walker.** The gerenuk is unique among antelopes in its ability to maneuver when at full stretch on its hindlegs. Like the giraffe, this ability to browse at heights allows the gerenuk to occupy a distinctive niche.

▼ **Courtship in the gerenuk** involves four stages: (1) the female raises her nose in a defensive gesture, while the male displays by a sideways presentation of the head; (2) the male marks the female from behind on the thigh; (3) the male performs the foreleg kick, tapping the females hindlegs with his foreleg; (4) the male performs the flehmen or lip-curl test, sampling the female's urine to determine whether she is ready to mate.

15 or 20 animals sometimes occur.

Even large groups rarely contain more than one adult male, as these are strictly territorial, excluding other males from their areas. Exclusion need not necessarily mean physical combat. Rather, the territorial system is maintained through frequent marking of dry twigs etc with the secretion of the scent glands in front of the eyes. Potential intruders apparently "respect" such marks or, at least, are able to assess the risk of progressing further. Such scent marking may be particularly important for male gerenuks, as their territories are comparatively large—2–4sq km (0.8–1.6sq mi).

Whereas other adult males are kept at a distance by the scent marks, a territorial male may tolerate subadult males near himself, as long as they acknowledge his superiority. Young males generally accompany their mothers before—at $1-1\frac{1}{2}$ years—they start seeking the company of peers with which they form small all-male groups. Eventually, these break up, as each male begins to look for a territory of his own. Associations between females and adult, territorial males are usually short-lived, unless both reside in the same small area.

Overall, the gerenuk, although a moderately large antelope, shows a number of traits more typical of small, forest-living bovids, such as highly selective feeding, sedentary life, occupancy of a resource-type territory coupled with extensive scent marking, a rather weak tendency to form groups, and partial reliance on "freezing" to avoid predators. These characteristics enable the gerenuk to live, and even thrive, in a habitat that would appear to be only marginal for a herbivore of this size.   WL

GOAT ANTELOPES

Subfamily: Caprinae
Twenty-six species in 13 genera.
Order: Artiodactyla.
Distribution: Asia, C Europe, N and C America, N Africa.

Habitat: steep terrain from hot deserts and most jungles to arctic barrens.

Saiga
Tribe: Saigini.
Two species: **saiga** (*Saiga tatarica*), **chiru** (*Pantholops hodgsoni*).

Rupicaprids
Tribe: Rupicaprini.
Five species in 4 genera.
Species include: **chamois** (*Rupicapra rupicapra*), **goral** (*Nemorhaedus goral*), **Mainland serow** (*Capricornis sumatrensis*), **Mountain goat** (*Oreamnos americanus*).

Musk ox
Tribe: Ovibonini.
Two species: **Musk ox** (*Ovibos moschatus*), **takin** (*Budorcas taxicolor*).

Caprids
Tribe: Caprini.
Seventeen species in 5 genera.
Species include: **argalis** (*Ovis ammon*), **Barbary sheep** (*Ammotragus lervia*), **Blue sheep** (*Pseudois nayaur*), **Himalayan tahr** (*Hemitragus jemlahicus*), **ibex** (*Capra ibex*), **mouflon** (*Ovis musimon*).

▶ **Species of goat antelopes** in order (left to right) of increasing body size as a response to increasing climatic severity. (1) Goral (*Nemorhaedus goral*). (2) Himalayan tahr (*Hemitragus jemlahicus*). (3) Urial (*Ovis orientalis*). (4) Barbary sheep (*Ammotragus lervia*). (5) Wild goat (*Capra aegagrus*). (6) Japanese serow (*Capricornis crispus*). (7) Chamois (*Rupicapra rupicapra*). (8) Ibex (*Capra ibex*). (9) Mountain goat (*Oreamnos americanus*). (10) Argalis (*Ovis ammon*). (11) Takin (*Budorcas taxicolor*). (12) Musk ox (*Ovibos moschatus*).

THE goat antelopes and their descendants (subfamily Caprinae) blossomed during the ice ages into a great diversity of species. They occupied extreme terrestrial environments such as hot deserts, arctic barrens, alpine plateaux, the steep, snow-covered cliffs at the edge of huge glaciers, and they survived in their evolutionary home, the humid tropics.

The basic goat-antelope body plan dates back to the mid-Tertiary period (35 million years ago), and is still reflected today in the serow of Southeast Asia. As the goat antelopes advanced into extreme climates, they abandoned the ancestral body plan, grew in size and diverged in appearance. The end products are giants such as the Musk ox from the arctic barrens, the giant rams from the highlands of central Asia, the long-horned ibex from the high alpine, and the bull-headed takin from the canyons of western China.

The goat antelopes and their kin are usually stocky, gregarious bovids that live on steep terrain. The primitive forms tend to have small horns and strongly patterned haircoats; the true sheep and goats are characterized by long, massive curving horns; the Musk oxen and their relatives are all large bodied with large horns. There is no consistent explanation for the very diverse body colors among these species, which vary from pure white, as in mountain goats and Dall's sheep, to pure black, as in the Musk oxen, serow or Stone's sheep. Gregarious forms tend to have strong frontal and rear markings. The females invariably are smaller in size and in the sheep and goats also have small horns.

At one extreme (the most primitive), there are so-called resource defenders, which live in small areas of highly productive and diverse habitats that supply all their needs year round and can be easily defended against members of the same species.

At the opposite extreme (the most recently evolved) are the grazers, which live in less productive and climatically more severe habitats. They are highly gregarious, roam over larger areas, and have retained very few attributes of their tropical ancestors. They carry large horns, engage in a variety of "sporting engagements" and the sexes may be strikingly different.

In goat antelopes we find not only species adapted to each of these extremes, but also to a range of habitats in between—the evolutionary phases can be not only predicted but also observed. In broad terms, rupicaprids are resource defenders and their descendants are grassland exploiters.

Resource defense imposes a number of demands on life-style. Such animals usually lead a solitary life and are territorial. They defend their small patches of food resources against other members of their own species. To defend the resources they have weapons capable of inflicting the greatest damage to their opponents, such as short, sharp, piercing horns. Most are dark in coloration, with male and female alike in armament and appearance. Since it is easier to defend a small than a large territory, a territory holder must make maximum use of resources available; so such species are small bodied with selective food habits.

The little that is known of the serow fits with this picture of a resource defender. Male and female look alike; both have short and sharp horns and large preorbital glands for scent marking. Serow are aggressive and attempt to fight off predators, even Asian black bears. Their teeth are typical of browsers, with lower crowns than in other rupicaprids. Their legs are less specialized for climbing than are those of their advanced

relatives. Serow are spectacular with long, tasseled ears, a long neck-mane and dark color. Their home ranges are well structured, with trails, "horn-rubs" on saplings and dung heaps. Serow are widely distributed throughout tropical and subtropical Southeast Asia and regionally vary greatly in color. They are largely confined to moist, shrubby or timber-covered rock outcrops, but have penetrated cold temperate zones, giving rise to the Japanese serow, which can tolerate heavy snowfall. There are also small island species, such as the Taiwan serow. The mountain goat may be a serow that evolved in a glacial region.

The two extreme forms of rupicaprids are the European chamois (see pp 590–591) and the American mountain goat. Both are adapted to cold, snowy mountains; both have glands on the back of the head. But here similarities end. The mountain goat is a massive, slow rock climber which only reverts to resource defense under severe winter conditions. It clings closely to cliffs and can live successfully even in the coastal ranges of Alaska and British Columbia with their high snowfalls. The chamois is relatively small bodied, gregarious, with a strikingly marked face and a bright rump patch. Only large males may usurp pockets of rich habitat in summer so as to fatten for the strenuous rut. In mountain goats, it is the adult females with a kid at heel which dominate males. In chamois, adult females also dominate males, but only young ones. The chamois' smaller size may be linked to its superlative dodging and evasive maneuvers in fighting; the mountain goat also evades but males have thick skin, especially thick on their rumps, for protection during fights.

In the goat antelopes, the break with resource defense correlates with exploitation of grasslands in open landscapes. Here, individuals band together in anonymous, but cooperative herds as a means of reducing predation. The larger the herd, the more secure each individual. Not only can many share in the task of vigilance and thereby permit each an adequate time for grazing, but animals in the center of the herd are virtually secure from predators. With grazing comes freedom from sedentary existence and an increase in tooth size to deal with the abrasive forage. Since grassland is productive, and all of it is available for feeding, a high density of grazers per unit area is possible. Several lines of goat antelopes are adapted to grazing, leading to the development of the caprids (sheep and goats), the Musk ox and probably the takin.

Among caprids, the most primitive form, related to the ancestral rupicaprids, is the tahr. This is intermediate in form and behavior between caprids and rupicaprids. It is closely wedded to cliffs and has broad food habits. It inhabits low latitudes in the Arabian peninsula, Nilgiri hills of southern India and the Himalayas, and has a very wide geographic distribution. The subspecies are divided by huge distances and oceans, which indicates their great age.

The Barbary sheep or aoudad from North Africa exemplifies the next stage in evolution. "New designs" arise at the periphery of the ancestor's range and only when colonizing a new habitat over a large area. Although a new species adapts to the environment it colonizes, changes in its external appearance arise, initially, as a consequence of social selection during colonization. The scarcity of colonizers prompts a high reproduction and social competition and rapid selection for larger bodies, stouter weapons and better bluffing abilities.

As expected, the Barbary sheep has a bigger head and horns than the tahr, yet it still retains a great diversity of combat forms, although head-clashing is clearly prevalent. In external appearance the Barbary sheep is close to goats, but biochemically it is closer to sheep.

The subsequent evolution of sheep and goats can be deciphered only in its broadest outlines. Both genera moved into cold mountains, increasing in body and horn size. However, there were also opportunistic local radiations which detract from a picture of orderly altitudinal and latitudinal progression. There must have been many hybridizations between populations during the ebb and flow of population movements, and there must have been many extinctions.

Goats are specialized for cliffs; sheep inhabit the open, rolling dry lands close to cliffs. These differences arise from different ways of dealing with predators: sheep escape by running and by clumping into cohesive herds, while goats are more specialized in putting obstacles and terrain with insecure footing in the way of pursuing predators. In external appearance, sheep and goat males differ considerably but females do not. In goats, the males carry long, scimitar-shaped, knobby horns as well as a chin beard. In sheep, males lack a chin beard and have massive curling horns; primitive sheep may have long hair on the cheeks and on the neck. Goats retain a tail and never have a large rump patch; sheep have short tails and large rump patches. Male goats spray themselves with urine and may consequently be quite malodorous. Where sheep and goats occur together they occupy different habitats, but this is not so when one occurs without the other. In the absence of sheep, some goats evolved horns not unlike those of sheep and lost their strong body odor. In the absence of goats, sheep move to cliffs and become ibex-like in body shape, such as American-type sheep. In both genera, as horns increase, the hair coat and body colors become less showy. Goats and sheep reach ever higher altitudes and latitudes, growing more grotesque and larger, specializing in combat delivering ever harder blows. Sheep had at least two great radiations: the first generated the 54/52 chromosome sheep (mouflon and American bighorn sheep), the second the giant argalis with 56 chromosomes. The extreme geographic forms, the bighorns and argalis, are remarkably similar in social behavior and appearance.

Asiatic and American sheep differ in ways of dealing with predators. The American sheep are ibex-like and cling to the vicinity of cliffs. Compared to Asiatic sheep, American sheep have much lower reproductive rates, for they do not bear twins, but they live longer. As populations, they cannot respond to decimation by a rapid build-up of numbers. They have a gestation period about one month longer than Asiatic sheep.

The goats had less success in geographic expansion than sheep. They stuck closely to cliffs, but along them they penetrated south as far as the Ethiopian highlands, and they moved west into Europe. In mountains without adjacent dry, grassy foothills goats apparently were more successful than sheep. The geographic expansion of goats and sheep is such that where these are found together primitive goats are paired with primitive sheep, and advanced goats with

▲ ► **The sheep from the goats.** Two species inhabiting the Rocky Mountains: an American bighorn ram ABOVE and a Mountain goat RIGHT. There could hardly be a greater contrast between two species, yet they are relatives, occupy the same mountains, endure similar weather, eat much the same food and escape from the same predators. The Mountain goat's hair is white and long; the sheep's is dun and short. The goat's horns are tiny, black and sharp; the sheep's are huge, light-colored and blunt. Goat fights are short, violent, bloody and rare; sheep fights are very long, ritualistic, mostly harmless and frequent. Sheep bang heads; goats do not. These differences are related to their very different lifestyles: goats are resource defenders, becoming ruthlessly territorial when food is scarce on the cliffs in deep snow, whereas sheep are migratory, moving seasonally across wide expanses of forest.

advanced sheep. The most primitive form is probably the markhor. The most advanced form is the ibex. These are stocky with thick, massive, knobby horns and a more uniform coat color. Species most distant from Asia Minor differ most from the wild goat. However, the evolutionary pattern of goats is complex, with some species assuming sheep-like shapes and behavior. Farthest from the Middle East and most sheep-like of all goats is the Blue sheep of Tibet and western China.

The Caprini are one radiation from the Rupicaprini that adapted to open landscapes and temperate climates. The Musk oxen radiation is another one, as is probably that of the takin. Both evolved frontal combat and head banging. With that comes hierarchical social structures, male groups and a high degree of gregariousness. The interpretation of these groups is difficult due to extinctions. Only the most highly evolved of the Musk oxen remains, and even it was

eradicated in Eurasia and survives only in the Arctic, North America and Greenland.

The takin is still a mystery animal, which, like the goats, has dispensed with a collection of odoriferous glands in favor of a general body odor. The takin's domain comprises steep cliffs covered in varying part with shrubs and trees and having a temperate climate. Like the overgrown rupicaprid it appears to be, it still clings to forest cover, but like an open-country form with gregarious tendencies it will also flock to large, open alpine meadows.

To early man, caprids were of little economic significance as game animals. In the Neolithic, some eight millenia ago, sheep and goats were domesticated in the Near East. These animals quickly rose to supreme importance. Their value lies in their ability to exploit land of low productivity which cannot be otherwise exploited by man, the ease with which these can be controlled with little manpower, and their numerous valuable products. Sheep and goats are raised for meat, wool, milk products and fur. Goats in particular are associated with the image of the poor man's cow, and are severely destructive of sparse vegetation in arid regions. Domestic sheep and goats arose in the same geographic region. The mouflon gave rise to Domestic sheep, as indicated by the similarity of their chromosomes; Domestic goats arose from the common Wild goat. Neither ibex nor advanced sheep contributed to the domestications, although such could have been possible; wild sheep and goats are quite easily habituated to humans and can even be tamed in the wild in the absence of hunting. Caprids are greatly renowned as game animals and the heads of the giant sheep and bighorns rate as the greatest trophies of Asia and North America. In modern times, caprids have had a mixed history. In America, bighorn sheep have not done well and can be maintained only by judicious care, limited hunting and artificial reintroduction into areas of former occupation. Desert bighorns in particular are susceptible to diseases of livestock and to disturbances. In Asia, due to the high reproductive rates of Old World wild sheep and goats, the situation is a happy one in some areas, but in others, such as the Himalayan region, it is desperate. Severe competition from domestic stock and overhunting has eliminated wild caprids from many areas. In Europe, the artificial spread of ibex is a great success, as is the management of mouflons and chamois. The ibex was rescued from the verge of extinction and, in Canada, so was the Musk ox. VG

THE 26 SPECIES OF GOAT ANTELOPES

Abbreviations: HTL = head-to-tail length. ht = height. wt = weight. Approximate nonmetric equivalents: 2.5cm = 1in; 1kg = 2.2lb.
E Endangered. V Vulnerable. * CITES listed.

Tribe Saigini

Formerly considered caprids but now normally classified as gazelles (subfamily Antilopinae). Like gazelles they have inflatable bags beside nostrils and carpal glands. Only males have horns. Inguinal glands are present.

Saiga
Saiga tatarica

N Caucasus, Kazakhstan, SW Mongolia, Sinkiang (China). Cold, arid steppe. Male HTL 123–146cm; ht 69–79cm; wt 32–51kg. Female HTL 108–125cm; ht 57–73cm; wt 21–41kg. Coat: sandy; horns amber and translucent. Head with grotesque proboscis that increases in size in males during rut, at which time hair tufts below eyes become impregnated with sticky smelly preorbital gland secretions; ears and tail short; neck with short mane. Twinning common. Population greatly expanded in recent years. Gestation: 145 days.

Chiru
Pantholops hodgsoni
Chiru or Tibetan antelope.

Tibet, Tsinghai, Sichuan (China), Ladak (N India). High Tibetan plateau. HTL 170cm; ht 90–100cm; wt up to 40kg in males. Coat: sandy, horns black, 50–72cm; rump patch large; face of male dark. Tail and ears short; nasal sacs when inflated size of pigeon eggs; auxillary glands present. Gestation: about 180 days.

Tribe Rupicaprini

The most primitive caprids. Defend resources using short sharp horns. Little difference between sexes in weight and appearance. Tail short. Strong affinity for steep slopes and shrub or tree cover. Food habits broad with emphasis on browse. Four teats. Young follow mother immediately after birth.

Mainland serow *
Capricornis sumatrensis

Tropical and subtropical E Asia. Male HTL unknown; ht up to 107cm; wt up to 140kg. Female HTL unknown; ht up to 90cm. wt up to 100kg. Coat: very variable, often dark, with long (40cm) white or brown neck mane; hair coarse, long and bristly; short beard from mouth to ears. Horns in male stouter than in female, 15–25cm; ears long, lance-shaped; large preorbital glands and foot glands.

Japanese serow
Capricornis crispus
Japanese or Taiwanese serow.

Japan, Taiwan. Particularly adapted to low temperatures and snowy climates in Japan. Dwarfed island form on Taiwan. Smaller and lighter colored than Mainland serow with conspicuous cheek beards in both sexes.

Goral *
Nemorhaedus goral
Goral, Red goral, Common goral.

N India and Burma to SE Siberia and south to Thailand. Very steep cliffs in dry climates; Manchurian form in moist, snowy climate. Male (Manchuria) HTL 106–117cm; ht 69–78cm; wt 28–42kg. Female HTL 106–118cm; ht 50–75cm; wt 22–35kg. Coat: highly variable, ranging from foxy red to dark gray to white; throat patch white; body hair long; neck mane short in male; horns black, up to 23cm in males, 20cm in females. Tail longer than in serow, with terminal tuft; tail longer in Manchurian form than in southern races. Rudimentary preorbital glands. Gestation: 250–260 days in Manchurian form, shorter in southern forms. 8 races from India, China, E Siberia. Endangered in Far East.

Chamois
Rupicapra rupicapra

NW Spain, Italy, Pyrenees to Caucasus, Carpathian and Tatra Mts; NE Turkey; Balkans; introduced to New Zealand. Moist alpine and subalpine, particularly open mountain. HTL 125–135cm; ht 70–80cm; wt 30–50kg (male), 24–42kg (female). Coat: in alpine form, pale brown in summer, dark brown/black with white rump patch and white to yellow facial stripes in late fall/winter. In southwestern form, reddish beige in summer, dark brown with wide yellowish patches on throat, neck sides, shoulders, hindlimbs in late fall/winter. Occipital glands present in both sexes. Tail short. Gestation 160–170 days. 9 subspecies.

Mountain goat
Oreamnos americanus

SE Alaska, S Yukon and SW Mackenzie to Oregon, Idaho and Montana; introduced in other N American mountain areas. Steep cliffs and edge of major glaciers in areas of high snowfall. Size very variable regionally: male HTL up to 175cm; ht up to 122cm; wt up to 140kg. Female HTL up to 145cm; ht up to 92cm; wt up to 57kg. In southern forms, wt to 70kg (male) and 57kg (female). Coat: yellowish-white, with long underwool and longer guard hairs that elongate into a stiff mane on neck and rump and into "pantaloons" on legs; molts in July; horns black, 15–25cm long, thicker in males. Body with massive legs and very large hooves. Tail short. Gestation: about 180 days.

Tribe Ovibonini

Grotesque giant rupicaprids from arctic and alpine environments, with long-haired, massive squat bodies carried on short, stout legs. Sexes similar in appearance, but females only 60 percent of males' weight. Horns relatively larger than in rupicaprids, but smaller than in caprids, curved for frontal attacks. Single young: gestation 8–8.5 months. Tail short. Four teats.

Takin
Budorcas taxicolor
Takin or Golden-fleeced cow.

W China, Butan, Burma. High alpine and subalpine bamboo forests on steep rugged terrain. Size variable: Male HTL 170–220cm; ht 100–130cm; wt up to 350kg, female up to 250kg. Coat: normally light, long and shaggy, golden in one race; face dark in bulls, but only nose dark in cows and calves. No localized skin glands, but oily, strong smelling substance with burning taste, secreted over whole body.

Musk ox
Ovibos moschatus

Alaska to Greenland. Arctic and tundra near glaciers; adapted to extreme cold; mixed feeder, restricted to areas with thin snow. Size variable, increasing from north to south: Male HTL (south) 245cm; ht 145cm; wt 350kg; male (Greenland) HTL 180cm; ht 110–130cm; wt up to 650kg (in captivity). Coat: black with light saddle and front, possibly bleaching in spring, dense and long, with hair strands up to 62cm. Foot glands present; preorbital glands large and secrete copiously in bull in rut.

Tribe Caprini

Advanced caprids with large differences in weight and external appearance between sexes. Specialized for frontal combat, with hierarchical social system in males. Spatial segregation of sexes outside the mating season. Horns with annual growth, much larger in males; length of display hair and horn mass inversely related. Teeth strongly molarized. Teats 2–4. Strong preference for open areas and grazing. Highly gregarious.

Genus *Hemitragus*

Most primitive and rupicaprid-like in form and behavior of the caprini. Both sexes horned. Those of males stouter, long and heavier, up to 44cm. Prefer steep, arid or cold cliffs. Males grow thick, heavy skin. Not malodorous, lack foot, groin, preorbital or tail glands. Females about 60 percent of male weight. Gestation: 180 days.

Himalayan tahr
Hemitragus jemlahicus

Himalayas. Introduced to New Zealand. HTL 130–170cm; ht 62–100cm; wt may exceed 108kg in males. Coat: variable over body from copper to black; hair relatively long with long-haired, shaggy neck ruff in males; haircoat shorter in females.

Nilgiri tahr V
Hemitragus hylocrius

S India. ht 62–100cm. Coat: dark, almost black, short-haired, with silver saddle in males; mane in male short and bristly; females grayish brown, with a white belly.

Arabian tahr E
Hemitragus jayakari

Oman. HTL, ht, wt: not known; is said to be the smallest species. Coat: brownish with dark dorsal stripe, black legs, white belly and long hair about jaw and withers.

Genus *Ammotragus*
A single species.

Barbary sheep *
Ammotragus lervia
Barbary sheep or aoudad.

N Africa. Mountains, particularly high desert. Male HTL 155–165cm; ht 90–100cm; wt up to 140kg; females 50 percent of male wt. Coat: short with long-haired neck ruffs and pantaloons on front legs in both sexes; beard and rump patch absent. Tail long, flat, naked on underside. Callouses on knees. Subcaudal glands present. Males do not smell offensively. Horns in males up to 84cm, in females up to 40cm. Ears slim and pointed. Gestation period 160 days. 4 subspecies.

Genus *Pseudois*

A single species.

Blue sheep

Pseudois nayaur
Blue sheep or bharal.

Himalayas, Tibet, E China. Slopes close to cliffs. Few measurements available: male HT 91cm; wt 60kg; females wt 40kg. Coat: bluish with light abdomen, legs strongly marked; lacks beard and shin glands; rump patch small; tail naked on underside. No callouses on knees. Horns curve rearward, cylindrical, up to 84cm long in males, tiny in females. Ears pointed and short. Gestation: 160 days. Longevity: average 9 years, but up to 24 years.

Genus *Capra*

Males with chin beards and strong smelling. No preorbital, groin, or foot glands; anal glands present. Both sexes have callouses on knees and long, flat tails with bare underside. Ears long, pointed, except in alpine forms. Rump patch small. Cliff-adapted jumpers, highly gregarious, which penetrate alpine and desert environments. Body size varies locally. Females 50–60 percent of male weight. In males, horns increase in length and weight with age. Female horns 20–25cm in all species. Breeding coat of male becomes more colorful with age in most species. Gestation: 150 days in small-bodied species, up to 170 days in large-bodied species. Twins common. Average life expectancy about 8 years. Many races, exact number unsettled.

Wild goat

Capra aegagrus
Wild goat or bezoar.

Greek Islands, Turkey, Iran, SW Afghanistan, Oman, Caucasus, Turkmenia, Pakistan and adjacent India; Domestic goat worldwide. Size highly variable: male HTL (Crete) unknown; HT unknown; wt 26–42kg; male (Persia) up to 90kg, female up to 45kg. Coat: old males very colorful compared to females or young males. Horns flat and scimitar-shaped, 58–126cm. 4 subspecies, including the Domestic goat (*C.a. hircus*).

Spanish goat

Capra pyrenaica
Spanish goat or Spanish ibex.

Pyrenees Mts. Male HTL 130–140cm; HT 65–70cm; wt 65–80kg. Female HTL 100–110cm; HT 70–75cm; wt 35–45kg. Coat: similar to Wild goat in color. Horns differ from those of ibex or goats, up to 75cm. 4 subspecies.

Ibex

Capra ibex

C Europe, Afghanistan and Kashmir to Mongolia and C China, N Ethiopia to Syria and Arabia. Extreme alpine of desert. Male HTL (Siberian) 115–170cm; HT 65–105cm; wt 80–100kg. Female HTL 130cm; HT 65–70cm; wt 30–50kg. Alpine ibex may be larger than this (wt up to 117kg); Nubian and Walia ibex are smaller. Coat: less colorful than in Wild goat; uniformly brown in alpine races; chin beards smaller than in goats. Horns massive and thick, but much more slender than in Caucasian turs; maximum length 85cm (Alpine), 128–143cm (Siberian).

Markhor [V] [*]

Capra falconeri

Afghanistan, N Pakistan, N India, Kashmir, S Uzbekistan, Tadzhistan. Woodlands low on mountain slopes. Male HTL 161–168cm; HT 86–100cm; wt 80–110kg. Female HTL 140–150cm; HT 65–70cm; wt 32–40kg. Coat: diagnostic due to long neck mane in male, pantaloons and strong markings; female does not have display hairs. Horns twisted, maximum length 82–143cm. The largest goat. Dentition more primitive than ibex and turs.

East Caucasian tur

Capra cylindricornis

E Caucasus Mts. Male HTL 130–150cm; HT 79–98cm; wt 65–100kg. Female HTL 120–140cm; HT 65–70cm; wt 45–55kg. Coat: uniformly dark brown in winter, red in summer; chin beard very short, up to 7cm. Horns cylindrical and sharply backward winding, as in *Pseudois*; maximum length 103cm.

West Caucasian tur

Capra caucasica

W Caucasus Mts. Size and form as East Caucasian tur except that horns are similar to those of alpine ibex, but more massive and curved. Skull form diagnostic and different from ibex. Chin beard up to 18cm.

Genus *Ovis*

Characterized by the presence of preorbital, foot and groin glands. Males do not have offensive smell. Tail short; rump patch small in primitive, large in advanced species. Females 60–70 percent of male weight; two teats. Horns present in both sexes, except a few mouflon populations where females lack horns; female horns normally very small. Horns increase in mass from urials to argalis sheep. In the latter horns form up to 13 percent of the male's body mass. In mouflons the horns may wind backwards, otherwise horns wind forwards. Horn mass in large mature rams: urials 5–9kg; Altai argalis 20–22kg, exceptionally more; Snow sheep about 6–8kg; Stone's and Dall's sheep 8–10kg; bighorns rarely in excess of 12kg. Horns used as sledge hammers in combat. All species prefer grazing. Highly gregarious, with sexes segregated except at mating season.

Urial [*]

Ovis orientalis

Kashmir to Iran, particularly rolling terrain and deserts. Male HTL 110–145cm; HT 88–100cm; wt 36–87kg. Coat: color variable, usually light brown; males with whitish cheek beards and light-colored long neck ruff; rump patch diffuse; tail thin and long for a sheep. Horns relatively light, forward winding. Long-legged, fleet-footed. Gestation: 150–160 days. Longevity: 6 years.

Argalis [*]

Ovis ammon

Pamir to Outer Mongolia and throughout Tibetan plateau. Cold, high alpine and cold desert habitats. Male HTL 180–200cm; HT 110–125cm; wt 95–140kg (180kg in Altai argalis); largest sheep. Coat: light brown with large white rump patch and white legs; size of neck ruff inversely related to size of horns; horns up to 190cm long and 50cm in circumference. Twinning common.

Mouflon

Ovis musimon

Asia Minor, Iran, Sardinia, Corsica, Cyprus; widely introduced in Europe. Cold and desert habitats. Male HTL 110–130cm; HT 65–75cm; wt 25–55kg. Coat: dark chestnut brown with light saddle; rump patch distinct; lacks a cheek bib but possesses dark ruff; tail short, broad and dark. Face of adults becomes lighter with age. Smallest wild sheep.

Snow sheep

Ovis nivicola
Snow sheep or Siberian bighorn

NE Siberia. Extreme alpine and arctic regions, particularly cliffs. Male HTL 162–178cm; HT 90–100cm; wt 90–120kg; females 60–65kg. Coat: dark brown with small, distinct rump patch and broad tail; some races light colored. 4 races.

Thinhorn sheep

Ovis dalli
Thinhorn, Stone's, Dall's or White sheep.

Alaska to N British Columbia. Extreme alpine and arctic regions, particularly cliffs. Male HTL 135–155cm; HT 93–102cm; wt 90–120kg. Coat: white in 2 subspecies, black or gray in the Stone's sheep; large, distinct rump patch in latter.

American bighorn sheep [*]

Ovis canadensis
American bighorn sheep or Mountain sheep.

SW Canada to W USA and N Mexico. Alpine to dry desert, particularly cliffs. Male HTL 168–186cm; HT 94–110cm; wt 57–140kg depending on locality; female 56–80kg. Coat: light to dark brown, lacks ruff or cheek beards; with large rump patches. Body stocky as in ibex. Gestation: 175 days. Longevity: average 9 years but to 24 years. 7 races.

VG

The Nimble Chamois

Social life of a mountain goat

The extraordinary agility of the chamois seems to defy gravity. Only a few hours after birth, a kid can nimbly follow its mother along narrow ledges or down precipitous screes. This agility enables them to survive predation from the eagles, wolves, lynxes and Brown bears that share their range.

Chamois live in mountainous country in Europe from the Cantabrians to the Caucasus and from the High Tatra to the Central Apennines. There are 10 subspecies, divided into two groups: the Southwestern races and the Alpine races. Some recent evidence suggests that these two groups should be recognized as separate species. In any case, the behavioral ecology of different chamois populations varies greatly and the species' social system is very flexible.

Some 400 of the Apennine chamois survive, all on a few mountains of the Abruzzo National Park, Central Italy, where they are strictly protected. The females with young live mainly in woodland during the cold months, but they move to Alpine meadows in late spring until the end of the fall.

Females are usually resident and tend to live in flocks, the largest group sizes being reached in late summer, prior to the rut. Males stay with the mother's flock until sexual maturity at 2–3 years old. Then, they live nomadically until they reach full maturity at 8–9 years old, when they become attached to a definite area. This may or may not lie within the females' range, and so not all males have access to females during the rut. However, should a female emigrate because of a high population density, she is likely to meet one such peripheral male and be fertilized.

Fully adult males live a solitary life in steep, rocky woodland throughout most of the year, only joining the female flocks occasionally, whereas young males mix readily with these. During the rut, the young males are chased away by older males, which defend their harems and begin spending more and more time with them on the Alpine meadows. The peak of the rut is reached in mid-November. Harem holders have to chase rivals off the harem, to test females for receptivity and prevent them from leaving: difficult, energy-consuming tasks which the harem holder has to undertake just before the winter. Despite the harem male's efforts, the females become skittish and restless when they come into heat, and the harem scatters. Thus, peripheral males have chances to mate too. This makes the advantage of being a harem-holder difficult to understand; it may lie in saving time, since courtship of a receptive female may last for hours, but a familiar male is likely to be accepted earlier than a stranger.

With the first heavy snowfalls, the chamois flocks split up and move to the woodland winter ranges. There they feed upon sparse, scattered food sources (buds, lichens, small grass patches) which would make living in a flock disadvantageous through increasing food competition. Heavy winter mortality may be suffered, especially

▲ **Winter quarters.** Chamois at 2,100m (6,900ft) in the Swiss Alps.

▶ **Out on the edge.** Chamois leaping down a rockface in the Abruzzo National Park, Italy.

◀ **Gestures of the chamois.** (1) "Side display" is an intimidatory stance used mainly by females and young males: the head is held high, the back is arched and the animal moves stiffly towards the receptor, which may move away in an alarmed "tail-up" posture (2). Grown-up males prefer to use a comparable behavior pattern where the neck stretches upwards and the long hair on withers and hind quarters is erected (3). The tail sticks between the rumps. Thus, an apparent increase of body size is achieved. Often subordinates creep away from displaying dominants in an inconspicuous, submissive, "low-stretching" attitude (4).

by the youngest age classes in relation to the duration of the snow cover. Also, the nomadic habits of young males may lead them to ecologically unsuitable areas.

Kids are born in May–June on rocky, secluded, inaccessible areas, where pregnant females have isolated themselves just before delivering. Large groups do not form again until mid-June. Normally, a mature female will give birth to a single kid per year.

Chamois are aggressive among themselves to the extent that potentially lethal fights are common. Normally, however, a vast repertoire of vocal and visual threat displays lessens the danger of direct forms of aggressiveness. In a fight, chamois show no inhibitions about goring rivals, hooking them in the throat, chest or abdomen, unless the loser is quick to lie flat on the ground, stretching forward its neck in an extreme submissive posture resembling an exaggeration of the suckling posture. Probably, infantile mimicry works to soothe the attacker's aggressiveness, but it is also the only posture which prevents the latter from goring efficiently; in fact the chamois horns are strongly crooked at their tops, so that the blows must be delivered with an upward motion to be effective. By lying down the loser simply prevents the winner from using its weapons! SL

SMALL HERBIVORES

RODENTS

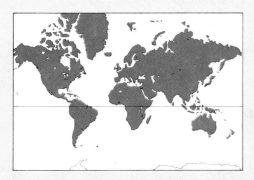

ORDER: RODENTIA
Thirty families: 389 genera: 1,702 species.

Squirrel-like rodents
Suborder: Sciuromorpha—sciuromorphs
Seven families: 65 genera: 377 species.

Beavers
Family: Castoridae
Two species in 1 genus.

Mountain beavers
Family: Aplodontidae
One species.

Squirrels
Family: Sciuridae
Two hundred and sixty-seven species in 49 genera.
Includes **Bryant's fox squirrel** (*Sciurus niger*), **flying squirrels** (subfamily *Petaurista*), **Gray squirrel** (*Sciurus carolinensis*), **ground squirrels** (genus Spermophilus), **marmots** (genus *Marmota*), **prairie dogs** (genus *Cynomys*), **Red squirrel** (*Sciurus vulgaris*), **Woodchuck** (*Marmota monax*).

Pocket gophers
Family: Geomyidae
Thirty-four species in 5 genera.

Scaly-tailed squirrels
Family: Anomaluridae
Seven species in 3 genera.

Pocket mice
Family: Heteromyidae
Sixty-five species in 5 genera.
Includes **Big-eared kangaroo rat** (*Dipodomys elephantinus*), **Kangaroo mice** (genus *Microdipodops*).

Springhare or springhass
Family: Pedetidae
One species.

Mouse-like rodents
Suborder: Myomorpha—myomorphs
Five families: 264 genera: 1,137 species.

Rats and mice
Family: Muridae
One thousand and eighty-two species in 241 genera and 15 subfamilies.
Includes: **Australian water rat** (*Hydromys chrysogaster*), **Brown lemmings** (genus *Lemmus*), **Collared lemming** or **Arctic lemming** (*Lemmus torquatus*), **Common vole** (*Microtus arvalis*), **cotton rats** (genus *Sigmodon*), **Golden hamster** (*Mesocricetus auratus*), **House mouse** (*Mus musculus*), **Multimammate rat** (*Praomys (Mastomys) natalensis*), **Nile rat** (*Arvicanthis niloticus*), **Norway lemming** (*Lemmus lemmus*), **Norway** or **Common** or **Brown rat** (*Rattus norvegicus*), **Roof rat** (*Rattus rattus*), **Saltmarsh**

Rodents have influenced history and human endeavor more than any other group of mammals. Nearly 40 percent of all mammal species belong to this one order, whose members live in almost every habitat, often in close association with man. Frequently this association is not in man's interest, for rodents consume prodigious quantities of his carefully stored food, and spread fatal diseases. It is said that rat-borne typhus has been a greater influence upon human destiny than any single person, and that in the last millennium rat-borne diseases have taken more lives than all wars and revolutions put together. Rodents have never been highly beneficial to man, though some larger species have been, and still are, sought for food in many parts of the world. Only a handful of species, such as guinea pigs and edible dormice, have been deliberately bred for food. Ironically, rats, mice and guinea pigs today play an inestimable role as "guinea pigs" in the testing of drugs and in biological research.

The reason for this ambivalent relationship is that rodents are as highly adaptable as man and like man, are opportunists. Our shared adaptability has guaranteed conflict of interest. In expanding our own world we invariably opened new habitats for our rodent competitors.

The success of the rodents is partly attributable to the fact that in evolutionary terms, they are quite young (between 26 and 38 million years), and populations retain large, untapped stocks of genetic variability. This variability is exposed to the selective forces of evolution rapidly since rodents produce many large litters each year. They are able to try out quickly new genetic combinations in the face of new environmental conditions (see pp664–65). A second facet of their success is that rodents have a very wide ranging diet.

Rodents occur in every habitat, from the high arctic tundra, where they live and breed under the snow (eg lemmings), to the hottest and driest of deserts (eg gerbils). Others glide from tree to tree (eg flying squirrels), seldom coming down to the ground, while others spend their entire lives in an underground network of burrows (eg mole-rats). Some have webbed feet and are semiaquatic, (eg muskrats), often undertaking complex engineering programs to regulate water levels (eg beavers), while others never touch a drop of water throughout their entire lives (eg gundis). Such species can derive their water requirements from fat reserves.

Rodents show less overall variation in body plan than do members of many mammalian orders. Most rodents are small, weighing 100g (3.5oz) or less. There are only a few large species of which the largest, the capybara, may weigh up to 66kg (146lb). All rodents have characteristic teeth, including a single pair of razor-sharp incisors. With these teeth the rodent can gnaw through the toughest of husks, pods

▶ **The expected build of a rodent.** This bush rat (*Rattus fuscipes*) would be recognizable as a rat to anyone familiar with the world's main rats, the Norway (or Common) and Roof rats, which are found wherever humans live. This rat lives along the eastern and southern coasts of Australia.

▼ **A highly adapted rat.** Many rodents are adapted for swimming, including this Australian water rat which is found in all nonarid areas of Australia. Its fur is seal-like and waterproof, its face is streamlined, its ears and eyes are small, its hindfeet webbed. It lives in marshes, swamps, backwaters and rivers, as do many of the world's rodents. It is also primarily carnivorous.

harvest mouse (*Reithrodontomys raviventris*),
Wood mouse (*Apodemus sylvaticus*).

Dormice
Families: Gliridae, Seleviniidae
Eleven species in 8 genera.

Jumping mice and birchmice
Family: Zapodidae
Fourteen species in 4 genera.

Jerboas
Family: Dipodidae
Thirty-one species in 11 genera.

Cavy-like rodents
Suborder: Caviomorpha—caviomorphs
Eighteen families: 60 genera: 188 species.

New World porcupines
Family: Erethizontidae
Ten species in 4 genera.

Cavies
Family: Caviidae
Fourteen species in 5 genera.

Capybara
Family: Hydrochoeridae
One species.

Coypu
Family: Myocastoridae
One species.

Hutias
Family: Capromyidae
Thirteen species in 4 genera.

Pacarana
Family: Dinomyidae
One species.

Pacas
Family: Agoutidae
Two species in 1 genus.

Agoutis and acouchis
Family: Dasyproctidae
Thirteen species in 2 genera.

Chinchilla rats
Family: Abrocomidae
Two species in 1 genus.

Spiny rats
Family: Echimyidae
Fifty-five species in 15 genera.

Chinchillas and viscachas
Family: Chinchillidae
Six species in 3 genera.

Degus or Octodonts
Family: Octodontidae
Eight species in 5 genera.

Tuco-tucos
Family: Ctenomyidae
Thirty-three species in 1 genus.

Cane rats
Family: Thryonomyidae
Two species in 1 genus.

African rock rat
Family: Petromyidae
One species.

Old World porcupines
Family: Hystricidae
Eleven species in 4 genera.

Gundis
Family: Ctenodactylidae
Five species in 4 genera.

African mole-rats
Family: Bathyergidae
Nine species in 5 genera.

and shells to secure the nutritious food contained within. The name "rodent" comes from the Latin *rodere*, which means to gnaw. Gnawing is facilitated by a sizable gap, called the diastema, immediately behind the incisors, into which the lips can be drawn, so sealing off the mouth from inedible fragments dislodged by the incisors. Rodents have no canine teeth, but they do possess a substantial battery of molar teeth by which all food is finely ground. These often massive and complexly structured teeth are traversed by convoluted layers of enamel. The pattern made by these layers is often of taxonomic significance. Most rodents have no more than 22 teeth, though one exception is the Silvery mole-rat from Central and East Africa which has 28. The Australian water rat has just 12. Since rodents feed on hard materials, the incisors have open roots and grow continuously throughout life; as much as several millimeters per week. They are constantly worn down by the action of their opposite number on the other jaw. If the teeth of rodents become misaligned so that they are not automatically worn down they will grow around and pierce the skull.

Most rodents are squat, compact creatures with short limbs and a tail. In South America, where there are no antelopes, several species have evolved long legs for a life on the grassy plains (eg maras, pacas and agoutis), and show some convergence towards the antelope body form. A very variable anatomical feature is the tail (see opposite).

The order is divided into three on the basis of the arrangement of jaw muscles. The principle jaw muscle is the masseter muscle which not only closes the lower jaw on the upper, but also pulls the lower jaw forward so allowing the gnawing action to occur. This action is unique. In the extinct Paleocene rodents the masseter was small and did not spread far onto the front of the skull. In the squirrel-like rodents (Sciuromorpha) the lateral masseter extends in front of the eye onto the snout; the deep masseter is short and used only in closing the jaw (see p604). In the cavy-like rodents (Caviomorpha) it is the deep masseter that extends forwards onto the snout to provide the gnawing action (see p684). Both the lateral and deep branches of the masseter are thrust forward in the mouse-like rodents (Myomorpha), providing the most effective gnawing action of all, with the result that they are the most successful in terms of distribution and number of species (see pp636–83).

Most rodents eat a range of plant products, from leaves to fruits, and will also eat small invertebrates, such as spiders and grasshoppers. A few are specialized carnivores; the Australian water rat feeds on small fish, frogs and mollusks and seldom eats plant material. To facilitate bacterial digestion of cellulose rodents have a relatively large cecum (appendix) which houses a dense bacterial flora (though this structure is relatively smaller in dormice). After the food has been softened in the stomach it passes down the large intestine and into the cecum. There the cellulose is split by bacteria into its digestible carbohydrate constituents, but absorption can only take place higher up the gut, in the stomach. Therefore rodents practise refection—reingesting the bacterially treated food taken directly from the anus. On its second visit to the stomach the carbohydrates are absorbed and the fecal pellet that eventually emerges is hard and dry. It is not known how rodents know which type of feces is being produced. The rodent's digestive system is very efficient, assimilating 80 percent of the ingested energy.

All members of at least three families (hamsters, pocket gophers, pocket mice) have cheek pouches. These are folds of skin, projecting inwards from the corner of the mouth, and are lined with fur. They may reach back to the shoulders. They can be everted for cleaning. Rodents with cheek pouches build up large stores—up to 90kg (198lb) in Common hamsters.

Rodents are intelligent and can master simple tasks for obtaining food. They can be readily conditioned, and easily learn to avoid fast-acting poisoned baits—a factor that makes them difficult pests to eradicate (see p602–3). Their sense of smell and their hearing are keenly developed. Nocturnal species have large eyes and all rodents have long, touch-sensitive whiskers (vibrissae).

In fact two thirds of all rodent species belong to just one family, the Muridae, with 1,011 species. Its members are distributed worldwide, including Australia and New Guinea, where it is the only terrestrial placental mammal family to be found (excluding modern introductions such as the rabbit). The second most numerous family is that of the squirrels (Sciuridae) with 268 species distributed throughout Eurasia, Africa and North and South America.

The fossil record of the rodents is pitifully sparse. Rodent remains are known from as far back as the late Paleocene era (57 million years ago), when all the main characteristics of the order had developed. The earliest

THE RODENT BODY PLAN

▼ **Skull of the Roof rat.** Clearly shown are the continuously growing, gnawing incisors and the chewing molars, with the gap (diastema) left by the absence of the canine and premolar teeth. All mouse-like rodents lack premolars but the squirrel- and cavy-like rodents have one or two on each side of the jaw.

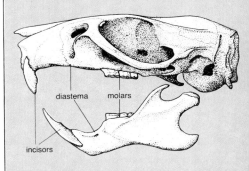

diastema   molars

incisors

▶ **Skeleton of the Roof rat.** This is typical of rodents with its squat form, short limbs, plantigrade gait (ie it walks on the soles of its feet) and long tail.

▶ **Evolutionary relationships** within the order Rodentia. The chart shows both the extent of the known fossil record (plum) and hypothetical lines of descent (pink).

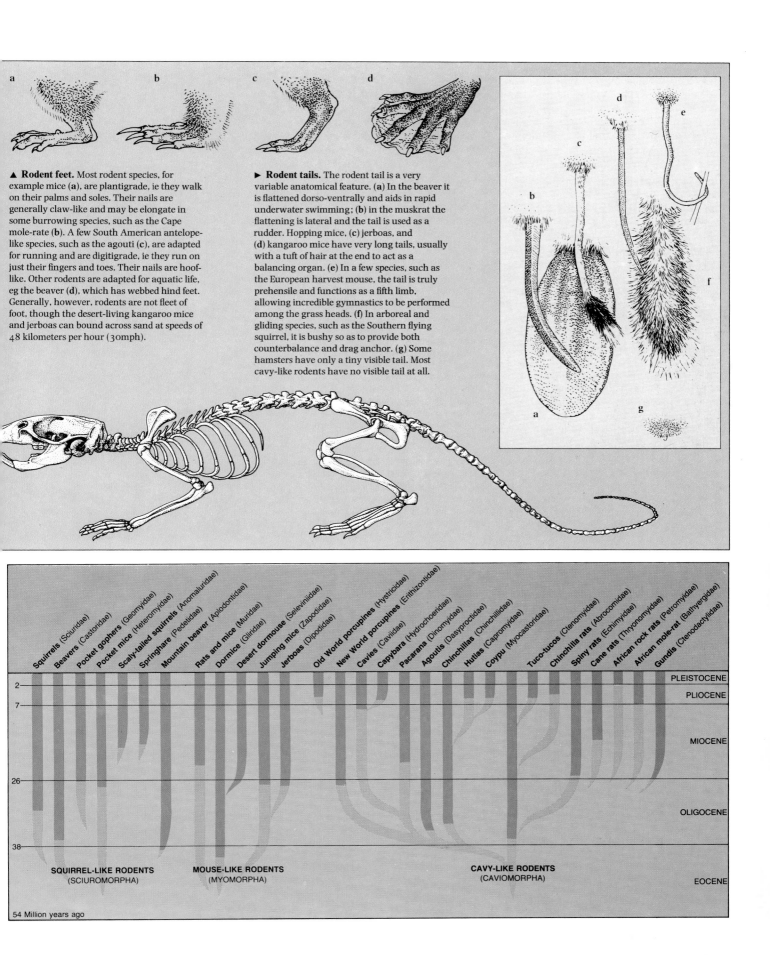

a **b** **c** **d**

▲ **Rodent feet.** Most rodent species, for example mice (**a**), are plantigrade, ie they walk on their palms and soles. Their nails are generally claw-like and may be elongate in some burrowing species, such as the Cape mole-rate (**b**). A few South American antelope-like species, such as the agouti (**c**), are adapted for running and are digitigrade, ie they run on just their fingers and toes. Their nails are hoof-like. Other rodents are adapted for aquatic life, eg the beaver (**d**), which has webbed hind feet. Generally, however, rodents are not fleet of foot, though the desert-living kangaroo mice and jerboas can bound across sand at speeds of 48 kilometers per hour (30mph).

▶ **Rodent tails.** The rodent tail is a very variable anatomical feature. (**a**) In the beaver it is flattened dorso-ventrally and aids in rapid underwater swimming; (**b**) in the muskrat the flattening is lateral and the tail is used as a rudder. Hopping mice, (**c**) jerboas, and (**d**) kangaroo mice have very long tails, usually with a tuft of hair at the end to act as a balancing organ. (**e**) In a few species, such as the European harvest mouse, the tail is truly prehensile and functions as a fifth limb, allowing incredible gymnastics to be performed among the grass heads. (**f**) In arboreal and gliding species, such as the Southern flying squirrel, it is bushy so as to provide both counterbalance and drag anchor. (**g**) Some hamsters have only a tiny visible tail. Most cavy-like rodents have no visible tail at all.

Squirrels (Sciuridae)
Beavers (Castoridae)
Pocket gophers (Geomyidae)
Pocket mice (Heteromyidae)
Scaly-tailed squirrels (Anomaluridae)
Springhare (Pedetidae)
Mountain beaver (Aplodontidae)
Rats and mice (Muridae)
Dormice (Gliridae)
Desert dormouse (Seleviniidae)
Jumping mice (Zapodidae)
Jerboas (Dipodidae)
Old World porcupines (Hystricidae)
New World porcupines (Erithizontidae)
Cavies (Caviidae)
Capybara (Hydrochoeridae)
Pacarana (Dinomyidae)
Agoutis (Dasyproctidae)
Chinchillas (Chinchillidae)
Hutias (Capromyidae)
Coypu (Myocastoridae)
Tuco-tucos (Ctenomyidae)
Chinchilla rats (Abrocomidae)
Spiny rats (Echimyidae)
Cane rats (Thryonomyidae)
African rock rats (Petromyidae)
African mole-rat (Bathyergidae)
Gundis (Ctenodactylidae)

PLEISTOCENE
2
PLIOCENE
7

MIOCENE

26

OLIGOCENE

38

SQUIRREL-LIKE RODENTS (SCIUROMORPHA)   MOUSE-LIKE RODENTS (MYOMORPHA)   CAVY-LIKE RODENTS (CAVIOMORPHA)

EOCENE

54 Million years ago

rodents belonged to the extinct sciuromorph family Paramyidae. During the Eocene era (54–38 million years ago) there was a rapid diversification of the rodents, and by the end of that epoch it seems that leaping, burrowing and running forms had evolved. At the Eocene–Oligocene boundary (38 million years ago) many families recognizable today occurred in North America, Europe and Asia, and during the Miocene (about 26 million years ago) the majority of present day families had arisen. Subsequently the most important evolutionary event was the appearance of the Muridae in the Pliocene era (7 million years ago) from Europe. About 2.5 million years ago, at the start of the Pleistocene era, they entered Australia, probably via Timor, and then underwent a rapid evolution. At the same time, when North and South America were united by a land bridge, murids invaded from the north with the result that there was an explosive radiation of New World rats and mice in South America (see p640).

The Role of Odor in Rodent Reproduction

Reproduction—from initial sexual attraction and the advertisement of sexual status through courtship, mating, the maintenance of pregnancy and the successful rearing of young—is influenced, if not actually controlled, by odor signals.

Male rats are attracted to the urine of females that are in the sexually receptive phase of the estrous cycle and sexually experienced males are more strongly attracted than naive males. Furthermore, if an experienced male is presented with the odor of a novel mature female alongside the odor of his mate he prefers the novel odor. Females, on the other hand, prefer the odor of their stud male to that of a stranger. The male's reproductive fitness is best suited by his seeking out, and impregnating, as many females as possible. The female needs to produce many healthy young, so her fitness is maximized by mating with the best quality stud male who has already proved himself. The otherwise solitary female Golden hamster must attract a male when she is sexually receptive. This she does by scent marking with strong-smelling vaginal secretions in the two days before her peak of receptivity. If no male arrives she ceases marking, to start again two days before the next peak.

In gregarious species such as the House mouse a dominant male can mate with 20 females in 6 hours if their cycles are synchronized. The odor of urine of adult sexually mature male rodents (eg mice, voles, deermice) accelerates not only the peak of female sexual receptivity but also the onset of sexual maturity in young females and brings sexually quiescent females into breeding condition. The effect is particularly strong from the urine of dominant males. Urine from castrated males has no such effect, so it would appear that the active ingredient—a pheromone—is made from, or dependent upon the presence of, the male sex hormone testosterone. Male urine has such a powerful effect that if a newly pregnant female mouse is exposed to the urine odor of a male who is a complete stranger to her she will resorb her litter and come rapidly into heat. If she then mates with the stranger she will become pregnant and carry the litter to term. The odor of the urine of females has either no effect upon the timing of the onset of sexual maturity in young females, or slightly retards it. If female mice are housed together in groups of 30 or more and males are absent the normal 4- or 5-day estrous cycles start to lengthen and the incidence of pseudopregnancy increases, indicating the power of the odor of female urine. However, the presence of the urine odor of an adult male will regularize all the lengthened cycles within 6–8 hours and the females will come into heat synchronously.

Female mice also produce a pheromone in the urine which has the effect of stimulating pheromone production in the male, but the female pheromone is not under the control of the sex glands (ovaries). It is not known what does control its production. A sexually quiescent female could stimulate pheromone production in a male, which would then bring her into sexual readiness.

It is thought that the reproductive success of the House mouse owes much to this system of pheromonal cuing, for mice would never be faced with a situation in which neither sex could stimulate or respond to the other because both stimulus and response were dependent upon the presence of fully functional sex glands in both sexes. Although only the House mouse has been studied in such detail, parts of the model have been discovered in other species and it may be of widespread occurrence.

About 8 days after giving birth, female rats start to produce a pheromone—an odor produced in the gut and broadcast via the feces—which inhibits *Wanderlust* in the young. It ceases to be produced when the young are 27 days old and almost weaned.

Finally some studies have involved a surgical removal of part of the brain which is involved with smell (the main and accessory olfactory bulbs). Removal of the bulbs in the Golden hamster, irrespective of previous sexual experience, bring an immediate cessation of all sexual behavior. In sexually experienced rats the operation has little effect, but in sexually naive rats the effect is as severe as in hamsters. Thus it appears that rats can learn to do without their sense of smell once they have gained some sexual experience.

DMS

▲ **Portable shade in the desert.** These African ground squirrels (*Xerus inauris*), which live in the Kalahari desert, provide themselves with shade by fluffing out their tails.

◄ **Threat displays** are very dramatic in some rodent species. (1) When slightly angry the Cape porcupine raises its quills and rattles the specialized hollow quills on its tail. If this fails to have the desired effect the hind feet are thumped on the ground in a war dance accompaniment to the rattling. Only if the threat persists will the porcupine turn its back on its opponent and charge backwards with its lethal spines at the ready. (2) Slightly less dramatic is the threat display of the Kenyan crested rat. This slow and solidly built rodent responds to danger by elevating a contrastingly colored crest of long hairs along its back, and in so doing exposes a glandular strip along the body. Special wick-like hairs lining the gland facilitate the rapid dissemination of a strong, unpleasant odor. (3) Finally, the little Norway lemming stands its ground in the face of danger and lifts its chin to expose the pale neck and cheek fur which contrast strongly with the dark upper fur.

With their high powers of reproduction and ability to invade all habitats, rodents are of great ecological importance. Occasionally their numbers become astoundingly high—almost 200,000 House mice per ha (80,000 per acre) were recorded in Central Valley in California in 1941–42—with densities of up to 1,200 per ha (500 per acre) for grassland species such as the Common vole being common. Apart from the immediate economic ruin that such hordes may bring to farmers, such densities have a profound effect upon the ecological balance of an entire region. Firstly, considerable damage is inflicted on vegetation resulting from the wholesale deaths of basic plants, from which it may take several years to recover. Secondly, the abundance of predators increases in response to the abundance of rodents, and when the rodents have gone they turn their attention to other prey. Such dramatic densities seem to occur only in unpredictable environments (but not all), such as the arctic tundra, and in marginal farming regions of arid zones. The numbers of tropical rodents, for the most part, rarely increase such that they become a nuisance to man.

A characteristic type of population explosion is seen in many rodents of the arctic tundra and taiga. In these areas population explosions occur with regularity, every 3–4 years, involving the Norway lemming in Europe and the Brown and Collared lemmings in North America. The population density builds up to a high level and then dramatically declines. A number of ideas have been put forward to explain the decline, but they are all too simplistic. For example, it was long held that the decline was brought about by disease (tularaemia or lemming fever which is found in many rodent populations), but it seems more likely that disease simply hastens the decline. Another suggestion was that the rodents become more aggressive at high density, which leads to a failure of courtship and reproduction. Others include the action of predators, and the impoverishment of the forage. Objective observation of lemming behavior at high density shows it to be adaptive, providing the lemmings with the best chances for survival (see p654).

Rodent cycles would not occur if rodents were not such prodigious breeders. Norway lemmings, for example, mate when they are about 4 weeks old and just newly weaned. Litters of about 6 young can be produced at 3–4 week intervals. Ecologists regard rodents as "r" selected, ie they are adapted for rapid reproductive turnover. Throughout mammals, species with short lives reproduce rapidly while long-lived species reproduce more slowly.

Although some species of rodents are extremely rare (eg the Big-eared kangaroo rat from Central California, *Dipodomys elephantinus*), few are listed in the *Red Data Book* as being endangered, only the Vancouver Island marmot, the Delmarva fox squirrel and the Saltmarsh harvest mouse (from marshes of southern California). A number of species are under threat, including six species of hutias (Family Capromyidae) from the islands of the West Indies, and in all these cases the danger comes from the draining of land and the clearing of formerly dense bush.

Rodents are highly social mammals, frequently living in huge aggregations. Prairie dog townships may contain more than 5,000 individuals. The solitary way of life appears to be restricted to those species that live in arid grasslands and deserts—hamsters and some desert mice—and to a few strictly territorial species such as the woodchuck. Their behavior lacks the ability to switch off aggression in others—they have no capacity for appeasement. Observation of a cage of mice for just a few moments will indicate the frequency of appeasement signals: a turning away of the head, a closing of the eyes. Fighting never occurs when one of a pair shows appeasement.

Rodent behavior is as adaptable as every other aspect of rodent biology. For example, individual House mice may be strongly territorial when population density is low, with adult males dribbling urine around their territories to saturate it with the owner's odor, but when density is high the territorial system breaks down in favor of a social hierarchy, or "pecking order." The consequences of both systems are the same: discrimination between the "haves" and the "have-nots" with regard to allocated limited resources. Dominant individuals display a confident deportment and show little appeasement. They are frequently the oldest and largest male members of the population. The blood level of the male sex hormone testosterone is higher in higher ranking individuals. Some species have extremely complex systems of social organization, for example Naked mole rats (see pp 710–11), but for the great bulk of mouse- and rat-sized species little is known.

Having alert and active senses, rodents communicate by sight, sound and smell. Some of the best known visual displays are seen in the arboreal and the day-active terrestrial species. Courtship display in tree squirrels may be readily observed in city parks in early spring. The male pursues the female through the trees, flicking his bushy tail forward over his body and head when he is stationary. The female goads him by running slightly ahead, but he responds by uttering infantile "contact-keeping" sounds similar to those infants produce to keep their mothers close. These stop the female, allowing the male to catch up. Threat displays are dramatic in some species (see pp 598–99).

Rodents make considerable use of vocalizations in their social communication. Occasionally, as in the threat of a lemming, the sound is audible (ie below 20kHz), but more often frequencies are far above the range of human hearing (at about 45kHz).

Rodents communicate extensively through odors which are produced by a variety of scent glands (see p 598). Males tend to produce more and stronger odors than females, and young males are afforded a measure of protection from paternal attack by smelling like their mothers until they are sexually mature.   DMS

Rodents as pests

Of the approximately 1,700 species of rodents only a handful are economically important, of which some occur worldwide. Recently the Food and Agrictulture Organization (FAO) of the United Nations estimated that each year rodent pests worldwide consume 42.5 million tonnes of food, worth 30 billion US dollars, equivalent to the gross national product of the world's 25 poorest countries. In addition, rodent pests are involved in the transmission of more than 20 pathogens, including plague (actually transmitted to man by the bite of the rat flea). Bubonic plague was responsible for the death of 25 million Europeans from the 14th century to the 17th.

The principal rodent pests are three species of the rats and mice family: the Norway, Common or Brown rat, the Roof rat and the House mouse. Squirrel-like pests include the European red and Gray squirrels which are, in terms of the cost of their damage, relatively minor pests (see pp 626–627). In 1974, for example, only 1 percent of the 48 percent of trees susceptible in the United Kingdom were damaged. Other pests include the Common hamster,

▲ **Dynamic reproductive power** is latent in the major rodent pests. Female Norway rats, for example, can breed when only two months old and can produce up to 11 young in one litter, and give birth again three or four weeks later.

▶ **A worldwide problem.** Vast populations of the major rodent pests occur worldwide, and can mobilize themselves with devastating results. From top to bottom: rat damage to orange in Egypt; a citrus tree stripped by rats in Cyprus; sugar cane damaged by the Hispid cotton rat. In the bottom picture of a sugar cane plantation all but the dark green areas was destroyed by rats.

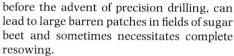

which is actually common in central Europe and is a pest on pasture land, and the gerbils which devastate agricultural crops in Africa. In North America, eastern and western Europe voles are prominent pests: they strip bark from trees, often causing death, and consume seedlings in forest plantations or fields. When vole populations peak (once every 3 or 4 years) there may be 2,000 voles per ha (almost 5,000 per acre). Other major rodent pests include the Cotton or Cane rat and *Holochilus braziliensis*, from, principally, Latin America, the Nile rat and the Multimammate rat of North and East Africa, *Rattus exulans* of the Pacific Islands, the *Bandicota* rats of the Indian subcontinent and Malaysian Peninsula and the prairie dogs, marmots and ground squirrels of the Mongolian and Californian grasslands.

The characteristics that enable these animals to become such major pests are a simple body plan, adapted to a variety of habitats and climates; high reproductive potential; opportunistic feeding behavior; and gnawing and burrowing habits.

In Britain the Norway rat may live in fields during the warm summer months where food is plentiful and they seldom reach economically important numbers. However, after harvest and with the onset of cold weather they move into buildings. Also in Britain, and some other western European countries, the wood mouse *Apodemus sylvaticus*, normally only a pest in the winter when it may nibble stored apples for example, has learnt to locate, probably by smell, pelleted sugar beet seed (now commonly precision drilled to avoid the need for hoeing out the extra production of seedling plants which used to be allowed for). This damage, possibly occurring unnoticed before the advent of precision drilling, can lead to large barren patches in fields of sugar beet and sometimes necessitates complete resowing.

Some rodents cause damage to agricultural crops, stored products and structural components of buildings etc. Such hazards result from rodents' feeding behavior or indirectly by their gnawing and burrowing. Although rodents usually consume about 10 percent of their body weight in food per day, much of the damage they do is not due to direct consumption. Three hundred rats in a grain store can eat 3 tonnes of grain in a year, but every 24 hours they also produce and contaminate the grain with 15,000 droppings, 3.5 litres (6 pints) of urine and countless hairs and greasy skin secretions. In sugar cane, rats may chew at the cane directly consuming only a part of it. The damage, however, may cause the cane to fall over (lodge) and the impact of the sun's rays will be reduced and harvesting impeded. The gnawed stem allows microorganisms to enter reducing the sugar content. Apart from the value of the lost crop, if the result is a loss in the sugar content of the crop of 6 percent that represents 6 percent of the investments in land preparation, fertilizers, pesticides, irrigation water, management, harvesting and processing. In Asia rodents routinely consume 5–30 percent of the rice crop and not infrequently devastate whole areas of rice fields, sometimes in excess of 10,000ha (25,000 acres). Rice field rats eat seedling rice and cut the shoots of rice as they ripen to consume the dough or milky stage of developing rice seedheads. A small amount of early damage may be compensated for by extra growth of the plant, but even if this happens the extra growth may ripen too late for harvesting. In oil palm in West Africa and especially Malaysia rats eat the hearts out of very young oil palm trees and consume the flesh of the oil palm fruit leading to losses in Malaysia alone of 5 percent of the final oil palm yield, valued at about 50 million US dollars per year.

Tropical crops damaged by rodents include coconuts, maize, coffee, field beans, citrus, melons, cocoa and dates. Each year rodents consume food equivalent to the total of all the world's cereal and potato harvest; a train 5,000km long, (about 3,000mi) as long as the Great Wall of China, would be needed to haul all this food.

Rodents also cause considerable losses of stored grain. For example, China produces about 320 million tonnes of grain a year and has to import a further 11 million tonnes to feed its 1,008 million population. Every year rodents, particularly the House mouse and Norway rat, are responsible for losses from this stored grain, amounting to at least 5–10 percent of the total, most from farm or village stores. The prevention of the loss, which could be achieved in a cost-effective manner, would eliminate the need for grain imports.

Structural damage attributable to rodents results, for example, from the animals burrowing into banks, sewers and under roads. The effects include subsidence, flooding or even soil erosion in many areas of the world. Gnawed electrical cables can cause fires, with their enormous economic impact. In the insulated walls of modern poultry units rodents will gnaw through electrical wires causing malfunctions of air-conditioning units, and subsequent severe economic losses. In Sudan, for example, a single power loss of 1 hour resulted in overheating of one unit causing the death of 8,000 birds to the value of 20,000 US dollars.

It is extremely difficult to put a value on the effects of diseases transmitted by rodents. There is little accurate information, partly due to a dearth of specialists in the field and partly because many governments do not wish to reveal the problems of disease in their countries in case it adversely affects their tourist industry.

Apart from plague, which persists in many African and Asian countries as well as in the USA (where wild mammals transmit the disease killing fewer than 10 people a year), murine typhus, food poisoning with *Salmonella*, Weils disease (Leptospirosis) and the West African disease Lassa fever, to mention just a few, are potentially fatal diseases transmitted by rats. Rats are probably responsible for more human deaths than all wars and revolutions. Difficult though it is to put an economic value on loss of life it is even more difficult to evaluate the economic consequences of debilitating chronic disease reducing the working efficiency of whole populations. One developing Asian country's capital was recently found to have 80 percent of its population seropositive for murine typhus, ie they had all, sometime, been in contact with the disease and had suffered from at least a mild form of it. Forty percent of people admitted to hospital in the capital city were diagnosed as having fever of unknown origin, at least some of these, maybe a majority, were probably suffering from murine typhus. The impact of this disease on the economy of the country is impossible to determine and the same country also suffered frequent out-

breaks of plague. Rats are involved in the transmission of both diseases.

To control the impact of rodents the simplest method is to reduce harborage and available food and water. This, however, is often impossible. At best such "good house-keeping" can prevent rodent populations building up, but it is seldom an effective method for reducing existing rodent populations. The same principles of management can be applied to fields, stores or domestic premises. In buildings, food and organic rubbish must be made inaccessible to rodents by blocking nooks and crannies where rodents may find refuge. They must be denied access to underground ducting (including pipe runs and sewers) as well as to air-conditioning ducts.

To reduce existing populations of rodents, predators—wild or domestic (such as cats or dogs)—have relatively little effect. Their role may lie in limiting population growth. It is a widely accepted principle that predators do not control, in absolute terms, their prey although the abundance of prey may effect the numbers of predators. Mongooses were introduced to the West Indies and Hawaiian Islands and Cobra snakes to Malaysian oil palm estates, both for controlling rats. The rats remain and the mongooses and cobra are now considered pests. Even the farm cat will not usually have a significant effect on rodent numbers: the reproductive rate of rodents keeps them ahead of the consumption rate of cats!

One of the simplest methods for combating small numbers of rodents is the use of traps. Few, however, are efficient: most just maim their victims. The most efficient method of control is by modern chemical rodenticides. The oldest kind of rodenticide, fast-acting, nonselective poison, appears in the earliest written record of chemical pest control. Aristotle, in 350 BC, described the use of the poison strychnine. However, fast-acting poisons (such as cyanide, strychnine, sodium monofluoroacetate (compound 1080), thallium sulfate and zinc phosphide) have various technical and ecological disadvantages, including long-lasting poison shyness in sublethally poisoned rodents and high hazards to other animals, and are not normally recommended today.

Since 1945, when warfarin was first synthesized, several anticoagulant rodenticides have been developed. These compounds decrease the blood's ability to clot, and lead to death by internal or external bleeding. These compounds have some selectivity as most have to be ingested by rodents over a number of days (5–10) for a

▲ **Rodents' impact on human history.** Numerous major outbreaks of rodent-borne disease in history have caused economic dislocation and social change on an enormous scale. Here victims of the Great Plague in London (1664–5) are buried. They died of *Pasteurella pestis*, transmitted by fleas carried by rats.

▶ **Rat in a grain store:** a Norway rat.

▼ **Communal destruction:** a night time scene in an Australian grain store.

lethal dose to be achieved; vitamin K_1, is used as an antidote in the case of accidental poisoning. Such compounds are best suited to controlling the Norway rat but used properly can result in total control of infestations of susceptible species. In some countries rodents, particularly the House mouse and Norway rat, developed genetic resistance to these "first generation" anticoagulants, notably in the United Kingdom, USA, Canada, Denmark and France, where the products have been used intensively for 10 years or more. This led to the development of "second generation" anticoagulants, compounds more potent than their predecessors. One of these is brodifacoum, which is usually toxic in single doses. But death is delayed and therefore, as with other anticoagulants, no poison shyness develops in sublethally poisoned animals. Unlike the first generation anticoagulants, repeated feeding is not necessary for the poison to take effect. Pulsed baiting with many small baits at about weekly intervals minimizes social interaction effects of rodent hierarchies and maximizes the exposure of individual rodents. This technique is not possible with the earlier anticoagulants. Second generation anticoagulants are revolutionizing rodent control in agriculture, which is, for the first time, practical and effective.

Apart from the choice of toxicant, the timing of rodent control and the coordinated execution of a planned campaign are important in serious control programs. The most effective time to control agricultural rodents, for example, is when little food is available to them and when population is low (probably just before breeding). A few "avant-garde" methods of rodent control are sometimes suggested (eg chemosterilants, ultrasonic sound, electromagnetism) but none can yet claim to be as effective as anticoagulant rodenticides.   ACD

SQUIRREL-LIKE RODENTS

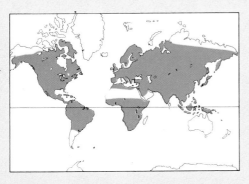

Suborder: Sciuromorpha
Seven families: 65 genera: 377 species.
Distribution: worldwide except for Australia region, Polynesia, Madagascar, southern S America, desert regions outside N America.

Habitat: very diverse, ranging from semiarid desert, flood plains, wetlands associated with streams, scrub, grassland, savanna to rain forest.

Size: head-body length from 5.6cm (2.2in) in the pocket mouse *Perognathus flavus* to 120cm (47in) in the beaver; weight from 10g in the African pigmy squirrel and *Perognathus flavus* to 30kg (66lb) in the beaver.

Beavers
Family: Castoridae
Two species in 1 genus.

Mountain beaver
Family: Aplodontidae
One species.

Squirrels
Family: Sciuridae
Two hundred and sixty-seven species in 49 genera.

Pocket gophers
Family: Geomyidae
Thirty-four species in 5 genera.

Scaly-tailed squirrels
Family: Anomaluridae
Seven species in 3 genera.

S QUIRRELS are predominantly seedeaters and are the dominant arboreal rodents in many parts of the world, but in the same family there are almost as many terrestrial species, including the ground squirrels of the open grasslands, also mostly seedeaters, and the more specialized and herbivorous marmots.

Athough they may be highly specialized in other respects, members of the squirrel family have a relatively primitive, unspecialized arrangement of the jaw muscles and therefore of the associated parts of the skull, in contrast to the mouse-like ("myomorph") and cavy-like ("caviomorph") rodents which have these areas specialized in ways not found in any other mammals. In squirrels the deep masseter muscle is short and direct, extending up from the mandible to terminate on the zygomatic arch. This feature is shared by some smaller groups of rodents, notably the Mountain beaver (Aplodontidae), the true beavers (Castoridae), the pocket gophers (Geomyidae) and the pocket mice (Heteromyidae), and has led to these families all being grouped with the squirrels in a suborder, Sciuromorpha. However, retention of a primitive character is not by itself an indication of close relationship and these groups are so different in other respects that it is fairly generally agreed that the suborder is not a natural group.

These families appear to have diverged from each other and from other rodents very early in the evolution of rodents and have very little in common other than the retention of the "sciuromorph" condition of the chewing apparatus.

A further primitive feature retained by these rodents is the presence of one or two premolar teeth in each row, giving four or five cheekteeth in each row instead of three as in the murids. GBC

▶ **The profile of a squirrel:** a distinctive head, long cylindrical body and bushy tail.

▼ **Distinguishing feature of squirrel-like rodents.** The lateral masseter muscle (green) extends in front of the eye onto the snout, moving the lower jaw forward during gnawing. The deep masseter muscle (blue) is short and used only in closing the jaw. Shown here is the skull of a marmot.

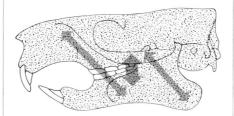

Pocket mice
Family: Heteromyidae
Sixty-five species in 5 genera.

Springhare or **Springhass**
Family: Pedetidae
One species.

Beaver

Red squirrel

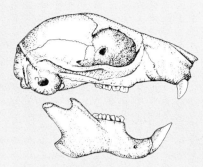

Red squirrel   4.5cm

Skulls and teeth of squirrel-like rodents.

Skulls of squirrels show few extreme
adaptations although those of the larger
ground squirrels, such as the marmots, are
more angular than those of the tree squirrels.
Most members of the squirrel family have
rather simple teeth, lacking the development
of either strongly projecting cusps or of sharp
enamel ridges as found in many other rodents.
In the beavers, however, there is a pattern of
ridges, adapted to their diet of bark and other
fibrous and abrasive vegetation and
convergent with that found in some unrelated
but ecologically similar rodents, such as the
coypu.

BEAVERS

Family: Castoridae

Two species in genus *Castor*.

Distribution: N America, Scandinavia, W and E Europe, C Asia, NW China.

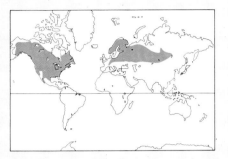

Habitat: semiaquatic wetlands associated with ponds, lakes, rivers and streams.

Size: head-body length 80–120cm (32–47in); tail length 25–50cm (10–20in); shoulder height 30–60cm (12–23in); weight 11–30kg (24–66lb). No difference between sexes.

Coat: yellowish brown to almost black; reddish brown is most common.

Gestation: about 105 days.

Longevity: 10–15 years.

North American beaver
Castor canadensis
North American or Canadian beaver.
Distribution: N America from Alaska E to Labrador, S to N Florida and Tamaulipas (Mexico). Introduced to Europe and Asia.

European beaver
Castor fiber
European or Asiatic beaver.
Distribution: NW and C Eurasia, in isolated populations from France E to Lake Baikal and Mongolia.

▶ **At home in the water.** ABOVE Of all rodents the beaver is one of the best-adapted for movement in water—torpedo-shaped with waterproof fur, commanding thrust from its flattened tail and webbed feet.

▶ **Kits are precious:** a female beaver produces only one small litter per year. This well-developed kit is being carried by its parent's front feet and mouth.

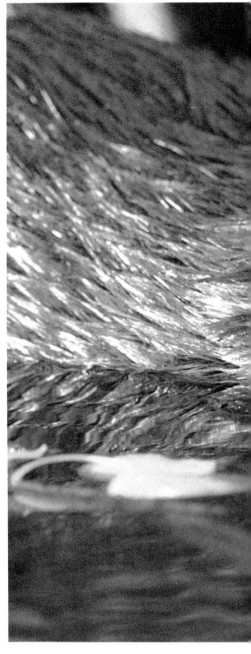

Few wild animals have had as much influence on world exploration, history and economics as the North American beaver. Much of the exploration of the New World developed from the fur trade—stimulated principally by the demand for beaver fur for making the felt hats that were very popular in the late 18th and early 19th centuries. Even wars were fought over access to beaver trapping areas, such as the French and Indian War (or the Seven Year's War) of 1754–63 in which the British defeated the French and gained control of northern North America.

Biologically, the beaver's flat, scaly tail, webbed hind feet, huge incisor teeth and unique internal anatomy of the throat and digestive tract are all distinctive. Beavers display a rich variety of constructional and behavioral activities, perhaps unsurpassed among mammals. Their dams promote ecological diversity, affect water quality and yields and have affected landscape evolution. Beavers also have tremendous inherent interest (testified by the quantity of literature on them), partly because they display many human traits, such as family units, complex communication systems, homes (lodges and burrows), food storage, and transportation networks (ponds and canals).

Beavers are the second heaviest rodent in the world—occasionally attaining a weight of over 30kg (66lb). They are adapted for a semiaquatic life, with a torpedo-shaped body and large webbed hind feet. The beaver's large tail—horizontally flattened and scaly—provides both steering and power; it can be flexed up and down to produce a rapid burst of speed. When the beaver dives its nose and ears close tight and a translucent membrane covers the eyes. The throat can be blocked by the back of the tongue, and the lips can close behind the incisors so the animal can gnaw and carry sticks underwater without choking.

On land the beaver is slow and awkward: it waddles clumsily on its large, pigeon-toed rear feet and on its small, short front legs and feet. Its posture—nose down, pelvis up—creates an impression of a walking wedge. If alarmed on land it will gallop or hop towards water.

The North American beaver can be found across most of the North American continent (its historic range), but only because populations were reestablished by State and federal wildlife agencies. In the eastern USA the animal came close to extirpation in the late 19th century. Populations of the North American beaver have been introduced in Finland, Russia (Karelian Isthmus, Amur basin and Kamchatka peninsula), and Poland.

The European beaver was once found throughout Eurasia but only isolated populations survive, in France (Rhone), Germany (Elbe) and Scandinavia, and in central Russia. The beavers in many of these groups are classified as subspecies, but some scientists consider the west European beaver a distinct species, *Castor albicus*, mainly because of differences in its skull.

The earliest direct ancestor of Eurasian beavers was probably *Steneofiber* from the Middle Oligocene (about 32 million years ago). The genus *Castor*, which originated in Europe during the Pliocene (7–3 million years ago), entered North America and because of its geographic isolation evolved into the present species. Thus the North

American beaver is considered to be younger and more advanced than the European beaver. During the Pleistocene, 10,000 years ago, these two species co-existed with giant forms which weighed 270–320kg (600–700lb), for example *Castoroides* in North America.

Beavers are herbivores whose diet varies with changes in season. In spring and summer they feed on nonwoody plants or plant parts, eg leaves, herbs, ferns, grasses and algae. In fall they favor woody items, taking them from a great variety of tree species but preferring aspen and willow. The digestion of woody plant material and cellulose is enhanced by microbial fermentation in the cecum and by the reingestion of the cecal contents.

Thanks to the adaptations of its teeth the beaver possesses a remarkable ability to cut down trees, for food and building materials. Like the incisor teeth of all living rodents the beaver's incisors are large and grow as fast as they are worn down by gnawing. They have a tough outer layer and a softer inner layer, forming a chisel-like edge which is kept sharp by rubbing or gnashing. Through them the beaver can exert enormous pressure when cutting.

During the fall, in northern climates, beavers gather woody stems as food for the winter. These are stored underwater, near a beaver's winter lodge or burrow. The water acts as a refrigerator, keeping the stems at a temperature around 0°C (32°F), preserving their nutritional value.

Beavers live in small, closed family units (often incorrectly termed "colonies") which usually consist of an adult pair and offspring from one year or more. An established family contains an adult pair, kits of the current year (up to 12 months old), yearlings born the previous year (12–24 months old), and possibly one or more subadults of either sex from previous breeding seasons (24 or more months old). The subadults generally do not breed.

The beaver's social system is unique among rodents. Each family occupies a discrete, individual territory. In northern latitudes male and female have a "monogamous" and long-term relationship. Their family life is exceptionally stable, being based on a low birth rate (one litter per year of 1–5 young in the European beaver and up to eight in the North American beaver), a high survival rate, parental care of a high standard and up to two years in the family for the development of adult behavior. The family is structured by a hierarchy in which adults dominate yearlings and yearlings

dominate kits, each conveying its own status by means of vocalizations, postures and gestures. Physical aggression is rare. Mating occurs during winter, in the water. Kits are born in the family lodge in late spring, soon after the yearlings have dispersed for distances usually less than 20km (12mi) but occasionally up to 250km (155mi). At birth they have a full coat of fur, open eyes and are able to move around inside the lodge. Within a few hours they are able to swim, but their small size and dense fur makes them too buoyant to submerge easily, so they are unable to swim down the passage from the lodge. Kits nurse for about 6 weeks and all members of the family share in bringing solid food to them with the male most actively supplying provisions. The kits grow rapidly but require many months of practice to perfect their ability to construct dams and lodges.

One of the ways by which beavers communicate is by depositing scent, often around the edges of water areas occupied by a family. The North American beaver scent marks on small mounds made of material dredged up from under water and placed on the bank whereas the European beaver marks directly on the ground. The scent, produced by castoreum from the castor sacs and secretions from the anal glands, is pungent and musty. All members of a beaver family participate, but the adult male marks most frequently. Scent marking is most intense in the spring and probably conveys information about the resident family to dispersing individuals and to adjacent families.

Beavers also communicate by striking their tails against the water (tail-slapping). This is done more often by adults than by kits, usually when they detect unusual stimuli. The slap is a warning to other members of the family who will move quickly to deep water. It may also frighten enemies.

Beaver construction activities modify their environment to provide greater security from predators, enhance environmental stability, and permit more efficient exploitation of food resources. Of the various construction activities in which beavers engage canal building is the least complex and was probably the first that beavers developed. Beavers use their forepaws to loosen mud and sediment from the bottom of shallow streams and marshy trails and push it to the side out of the way. Repeated digging and pushing of material creates canals that enable beavers to remain in water while moving between ponds or to feeding areas. This behavior occurs most frequently in summer when water levels are low and all members of the family participate.

Dams are built across streams to impound water, using mud, stones, sticks and branches. In northern climates the ponds created make the beavers' lodges more secure from predators and permit them to use more distant and larger food items. They must also be deep enough for the family to be able to swim under the ice from their lodge to the food cache. The sound of flowing water and visual cues stimulate dam building. Beavers use their small, agile forepaws to loosen mud, small stones and earth from the stream bottom and carry the material to the dam site. Sticks and branches are towed with the incisors to the dam. The branches support the dam and keep the mud

▶ **The towering jumble** of a beaver's dam—two or three meters high—conceals the careful, firm construction of mud, stones, sticks and branches.

▶ **Silent approach.** BELOW A beaver tows a branch to the base of its dam.

▼ **The beaver's lodge** or house is a large conical pile of mud and sticks, either on the shore or in the middle of a pond. The structure begins with beavers digging a burrow underwater into the bank (or into a slight mound adjacent to a waterway) with their forepaws. The burrow is extended upwards towards the surface as the water level rises behind the dam, and if it breaks through the ground the exposed hole is covered with sticks, branches and mud. Material continues to be added on top of the burrow while the living chamber is hollowed out above the water and within the mound of sticks and mud. In many cases the rising water isolates the lodge in the middle of the beaver pond. Most lodges are begun in later summer or early fall. Each fall in northern latitudes families add a thick coating of mud and sticks to the exterior of the lodge in which the family will spend the winter. Frozen mud provides insulation and prevents predators from digging into the lodge. All family members, except kits, help to build and maintain the lodges. Females are more active than males and the adult female is the most active lodge builder.

The Beaver's 29-hour Day

During most of the year beavers are active at night, rising at sunset and retiring to their lodges at sunrise. This regular daily cycle of activity is termed a circadian rhythm. In northern winters, however, where ponds freeze over, beavers stay in their lodges or under the ice because temperatures there remain near 0°C (32°F) while air temperatures are generally much lower. Activity above ice would require a very high production of energy.

In the dimly lit water world of the lodges and the surrounding water, light levels remain constant and low throughout the 24-hour day, so that sunrise and sunset are not apparent. In the absence of solar "cues" activity, recorded as noise and movement, is not synchronized with the solar day above. The circadian rhythm breaks down and so-called beaver days become longer. This may be an incidental consequence of the beaver's ability to live under the ice, but the biological significance of this change in activity is not clearly known. The beaver days vary in length from 26 to 29 hours, so during the winter beavers experience fewer "days" under the ice than occur above. This type of cycle is termed a free-running circadian rhythm.

and small sticks in place. Beavers keep adding mud and sticks to make the dam higher and longer; some may reach over 100m (330ft) long and 3m (10ft) high. Dam building activity is most intense during periods of high water, primarily spring and fall, although material may be added throughout the year. Adults and yearlings build dams, but female members of the family are more active than males and the adult female is most active.

Beavers are harvested for their fur and meat and are actively managed throughout most of their range. Annual harvests in Canada since the 1950s amounted to 200,000 to 600,000 pelts and during the 1970s in the USA 100,000 to 200,000. In Eurasia recent harvests from Nordic countries took less than 1,000 pelts per year and in Russia about 8,500 in 1980.   RAL/HEH

MOUNTAIN BEAVER

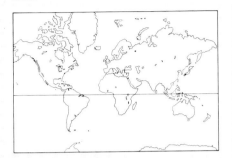

Aplodontia rufa
Sole member of family Aplodontidae.
Mountain beaver or beaver or boomer or
sewellel.
Distribution: USA and Canada along Pacific
coast.

Habitat: coniferous forest.

Size: head-body length 30–41cm
(12–16in); tail length 2.5–3.8cm
(1–1.5in); shoulder height
11.5–14cm (4.5–5.5in); weight
1–1.5kg (2.2–3.3lb). Sexes are
similar in size and shape, but
males weigh slightly more than
females.

Coat: young in their year of birth have grayish
fur, adults blackish to reddish brown, tawny
underneath.

Gestation: 28–30 days.

Longevity: 5–10 years.

M OUNTAIN beavers, the most primitive of
all present-day rodents, live only in
southwest Canada and along the west coast
of the USA, where they inhabit some of the
most productive coniferous forest lands in
North America. Not to be confused with the
flat-tailed stream beaver (*Castor*), Mountain
beavers are land animals that spend much of
their time in burrows, where they sleep, eat
defecate, fight and reproduce and do most of
their traveling. They are seldom seen outside zoos.

Mountain beavers are medium-sized,
bull-necked rodents with a round, robust
body and a moderately flat and broad head
with small black, beady eyes and long, stiff,
whitish whiskers (vibrissae). Not only are
their incisor teeth rootless and continuously
growing but also their premolars and
molars. Their ears are relatively small,
covered with short, soft, light-colored hair.
Their short legs give them a squat appearance; a short vestigial tail makes them look
tailless. The fur on their back presents a
sheen whereas white-to-translucent-tipped
long guard hairs give their flanks a grizzled
affect. A distinctive feature is a soft, furry
white spot under each ear.

Mountain beavers can be found at all
elevations from sea level to the upper limit of
trees and in areas with rainfall of 50–350cm
(20–138in) per year, and where winters are
wet, mild and snow-free, summers moist,
mild and cloudy. Their home burrows are
generally in areas with deep, well-drained
soils and abundant fleshy and woody plants.
In drier areas Mountain beavers are restricted to habitats on banks and to ditches that
are seasonally wet or have some free-
running water available for most of the
year.

The location of Mountain beavers is in
part explained by the fact that they cannot
adequately regulate the temperature of their
bodies and must therefore live in stable, cool,
moist environments. Nor can they effectively conserve body moisture or fat, which
prevents them from hibernating or spending
the summer in torpor. They need to consume considerable amounts of water and
food and to line their nests well for insulation. They can satisfy most of their requirements with items obtained within 30m
(about 100ft) of their nest. Water is obtained
mainly from succulent plants, dew or rain.

When defecating, Mountain beavers, unlike any other rodent or rabbit, extract fecal
pellets individually with their incisors and
toss them on piles in underground fecal
chambers. Like rodents and rabbits they also
reingest some of the pellets.

Mountain beavers are strictly vegetarians. They harvest leafy materials (such
as fronds of Sword fern, new branches of
salal and huckleberry, stems of Douglas fir
and vine maple, and clumps of grass or
sedge) and succulent, fleshy foods (such as
fiddle heads of Bracken fern, roots of False
dandelion and bleeding heart). These are
eaten immediately or stored underground.

Most food and nest items are gathered
above ground between dusk and dawn and
consumed underground. Decaying uneaten
food is abandoned or buried in blind chambers; dry, uneaten food is added to the nest.

It used to be thought that all Mountain
beavers were solitary animals, except during the breeding and rearing season. Recent
studies using radio tracking have demonstrated that this view is inadequate. Some
Mountain beavers spend short periods together, in all seasons. Neighbors, for
example, will share nests and food caches, or
a wandering beaver will stay a day or two in
another beaver's burrow system. Sometimes
a beaver's burrow will also be occupied, in
part if not in its entirety, by other animals,
eg salamanders, frogs or deer mice.

Unlike many rodents, Mountain beavers
have a low rate of reproduction. Most do not
mate until they are at least two years old
and females conceive only once a year, even

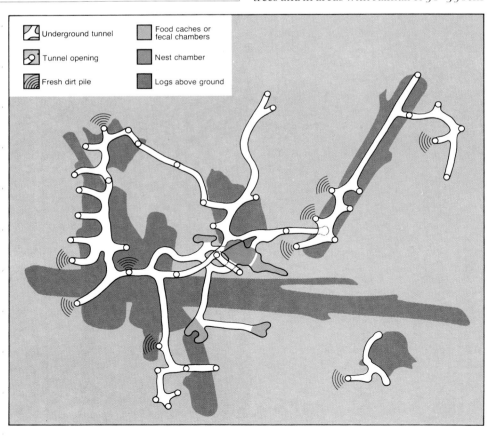

Underground tunnel
Tunnel opening
Fresh dirt pile
Food caches or
fecal chambers
Nest chamber
Logs above ground

▲ **Out foraging.** Mountain beavers rarely appear above ground in daylight. This exceptional picture, showing well the contrast between the sheen of the back and the grizzled flanks, offers a portrait of the Mountain beaver: plump-bodied, bull-necked, snouty.

◀ **The burrow system of a Mountain beaver.** Each consists of a single nest chamber and underground food caches and fecal chambers which are generally close to the nest. Most nests are about 1m (3.3ft) below ground in a well-drained, dome-shaped chamber, although some may occur 2m (6.6ft) or more below ground. Tunnels are generally 15–20cm (6–8in) in diameter and occur at various levels; those closest to the surface are used for travel; deep ones lead to the nest and food caches.

Burrow openings occur every 4–6m (13–20ft) or more depending on vegetative cover and number of animals occupying a particular area. Densities vary from 1 or 2 Mountain beavers per ha (2.5 acres) in poor habitat to 20 or more per ha in good habitat. Up to 75 per ha have been kill-trapped in reforested clear-out areas, but such densities are rare. Individual systems often connect.

if they lose a litter. Breeding is to some extent synchronized. Males are sexually active from about late December to early March, with aggressive older males doing most of the breeding. Conception normally occurs in January or February. Litters of 2–4 young are born in February, March or early April. Young are born blind, hairless and helpless in the nest. They are weaned when about 6–8 weeks old and continue to occupy the nest and burrow system with their mother until late summer or early fall. Then they are forced out, to seek sites for their own homes.

Once on its own a young beaver may establish a new burrow system or, more commonly, restore an abandoned one. New burrows may be within 100m (330ft) of the mother's burrow or up to 2km (1.2mi) away. The distance depends on population densities and on the quality of the land. Both sexes disperse in the same manner.

Dispersing young travel mainly above ground: they become very vulnerable to predators, for example owls, hawks and ravens, coyotes, bobcats and man. Those living near roads are liable to be killed by vehicles.

Mountain beavers are classed as non-game mammals, hence they are not managed like game (eg deer, elk) or like furbearing animals (eg beavers, muskrat). The damage they cause affects about 111,000ha (275,000 acres) of forest; they pose a tremendous threat to young conifers planted for timber. Almost all damage occurs while Mountain beavers are gathering food and nest materials. The result of their activities is the loss of timber worth several million US dollars every year.

Although some of the Mountain beavers' forest habitat has given way to urban development and agriculture, they range over about as extensive an area now as they did 200 to 300 years ago. They are probably more abundant now than they were in the early 20th century thanks to forest logging practices. Mountain beavers do not appear to be in any immediate danger of extermination from man or natural causes. JE

SQUIRRELS

Family: Sciuridae
Two hundred and sixty-seven species in 49 genera.
Distribution: worldwide apart from the Australian region and Polynesia, Madagascar, southern S America, desert regions (eg Sahara, Egypt, Arabia).

Habitat: varies from lush tropical rain forest to city parks and semiarid desert.

Size: ranges from head-body length 6.6–10cm (2.6–3.9in), tail length 5–8cm (2–3.1in), weight about 10g (0.35oz) in the African pygmy squirrel to head-body length 53–73cm (20.8–28.7in), tail length 13–16cm (5.1–6.3in), weight 4–8kg (8.8–17.6lb) in the Alpine marmot.

Gestation: about 40 days (ground squirrels and sousliks 21–28 days; marmots 33 days).

Longevity: varies considerably to maximum of 8–10 years (up to 16 in captivity).

Species and genera include: **African pygmy squirrel** (*Myosciurus pumilio*), **Alpine marmot** (*Marmota marmota*), **American red squirrel** (*Tamiasciurus hudsonicus*), **Arctic ground squirrel** (*Spermophilus undulatus*), **beautiful squirrels** (genus *Callosciurus*), **Black giant squirrel** (*Ratufa bicolor*), **chipmunks** (genus *Tamias*), **Cream giant squirrel** (*Ratufa affinis*), **Douglas pine squirrel** (*Tamiasciurus douglasii*), **flying squirrels** (genera *Aeromys, Belomys, Eupetaurus, Glaucomys, Hylopetes, Petaurista, Petinomys, Pteromys, Pteromyscus, Trogopterus*), **Fox squirrel** (*Sciurus niger*), **giant flying squirrels** (genus *Petaurista*), **giant squirrels** (genus *Ratufa*), **Gray squirrel** (*Sciurus carolinensis*), **ground squirrels** (genus *Spermophilus*), **Horse-tailed squirrel** (*Sundasciurus hippurus*), **Kaibab squirrel** (*Sciurus kaibabensis*), **Little souslik** (*Spermophilus pygmaeus*), **Long-clawed ground squirrel** (*Spermophilopsis leptodactylus*), **Long-nosed squirrel** (*Rhinosciurus laticaudatus*), **marmots** (genus *Marmota*), **Plantain squirrel** (*Callosciurus notatus*), **prairie dogs** (genus *Cynomys*), **Prevost's squirrel** (*Callosciurus prevosti*), **Red squirrel** (*Sciurus vulgaris*), **Russian flying squirrel** (*Pteromys volans*), **Siberian chipmunk** (*Tamias sibiricus*), **Slender squirrel** (*Sundasciurus tenuis*), **Southern flying squirrel** (*Glaucomys volans*), **Spotted giant flying squirrel** (*Petaurista Elegans*), **Sunda squirrels** (genus *Sundasciurus*), **Three-striped ground squirrel** (*Lariscus insignis*), **tree squirrels** (genus *Sciurus*), **woodchuck** (*Marmota monax*).

Like Squirrel Nutkin in Beatrix Potter's children's story *The Tale of Squirrel Nutkin*, squirrels have a popular image as attractive but cheeky opportunists. Their opportunism, however, is the basis of their success. Today squirrels are among the most widespread of mammals, and are found on every continent other than Australia and Antarctica. Because they are relatively unspecialized, squirrels have been able to evolve a wide variety of body forms and habits which fit them for life in a broad range of habitats, from lush tropical rain forests to rocky cliffs or semiarid deserts, from open prairies to town gardens. This successful family includes such varied forms as the ground-dwelling and burrowing marmots, ground squirrels, prairie dogs and chipmunks; the arboreal and day-active tree squirrels; and nocturnal flying squirrels.

Squirrels range in size from the tiny, mouse-like and arboreal African pygmy squirrel to the sturdy cat-sized marmots. Typically squirrels have a long, cylindrical body with a short or long bushy tail. Most squirrels have fine soft hair and in some species the coat is very thick and valued by man as fur. Many species molt twice a year and often sport a summer coat lighter in color. Male, female and young are similar in appearance, but even within species there can be considerable variation in color as for instance in the variable squirrels of Thailand where some populations are pure white, others pure black or red and yet others a combination of these and other colors.

Squirrels usually have large eyes, and tree and flying squirrels have large ears. Some species, eg the European red squirrel and the Kaibab squirrel, have conspicuous ear tufts. They have the usual arrangement of teeth in rodents, viz a single pair of chisel-shaped incisor teeth in each jaw and a large gap in front of the premolars because canines are absent. The incisors grow continuously and are worn back by use; the cheek teeth are rooted and have abrasive chewing surfaces. The lower jaw is quite movable, and some genera like chipmunks and ground squirrels have cheek pouches for carrying food.

Squirrels have sharp eyesight and wide vision but the ability of flying squirrels to see color is poor. In European red squirrels, marmots and prairie dogs the entire retina has the sensitivity that most mammals possess only within a small area of the retina (the foveal region). These squirrels can distinguish vertical objects particularly well, an ability important for tree-dwelling species, which are able to estimate distances between trees with great accuracy. Many species also have a well-developed sense of touch, thanks partly to touch-sensitive whiskers (vibrissae) on the head, feet and the outside of the legs.

Squirrels have short forelimbs, with a small thumb and four toes on the front feet, and longer hindlimbs, with four (woodchuck) or five toes on the hind feet (tree squirrels). Although the "thumb" is usually poorly developed it can have a very long claw (prairie dogs) or flat nail (marmots); all other toes have sharp claws. The soles of the feet have soft pads; in the desert-living long-clawed ground squirrels they are covered in hair which acts as an insulator to enable the animal to move over hot sand. These squirrels also have fringes of stiff hairs on the outside of the hind feet which push away the soft sand as they burrow. Among other adaptations are those of several of the terrestrial squirrels which nest and take refuge in underground burrows: they are heavy bodied with powerful forelimbs and large scraping claws for digging.

► **Pinned to the bark.** Tree squirrels, such as this Arizona gray squirrel, hold themselves head-down while waiting to make the next move.

▼ **A shrill screeching** means that this Long-tailed marmot has been disturbed or feels threatened.

Tree squirrels live and make their nests (dreys) in trees and are excellent at climbing and jumping; their toes, with their sharp claws, are well adapted for clinging to tree trunks. Squirrels descend tree trunks head first, sticking the claws of the outstretched hind feet into the bark to act as anchors. The tail serves as a balance when the squirrel runs and climbs and as a rudder when it jumps. The tail is also used as a flag to communicate social signals and is wrapped around the squirrel when the animal sleeps. On the ground, tree squirrels move in a sequence of graceful leaps, often pausing between jumps to raise their heads and look around. When feeding, all squirrels squat on their haunches holding the food between their front paws.

Tree squirrels are able to jump across the considerable gaps in the canopy. In a flying leap the legs are outspread and the body flattened, with the tail slightly curved so that a broad body surface is presented to the air. It is only a short step from such a leap to the controlled glides of the true "flying" squirrels. Like the other gliding mammals (the flying lemurs and flying phalangers) these squirrels have developed a furred flight membrane or patagium which extends along the sides of the body and acts like a parachute when the animal leaps. It extends from the hind legs to the front limbs and is bound in front by a thin rod of cartilage attached to the wrist. The bushy tail is free and acts like a rudder. The squirrel steers by changing the position of the limbs and tail and the tension in the muscular gliding membrane. Flying squirrels descend in long smooth curves, avoiding branches and trees, to land low on a tree trunk. As it lands the flying squirrel brakes, by turning its tail and body upward. The larger flying squirrels can glide for 100m (330ft) or more but the smaller forms cover much shorter distances. Gliding is an economical way to travel through the tall forest and enables the squirrel to escape quickly from a predator such as a marten. The flight membrane does have some drawbacks, however; when moving around in trees flying squirrels are less agile than tree squirrels and it is probably significant that the mammals that have adopted a gliding habit are active only at night when they are less conspicuous to keen-sighted birds of prey.

Most squirrels feed on nuts and seeds, fruits and other plant material, supplemented with a few insects, though they have been known to feed on reptiles and young birds. At certain times of year fungi and insects may be eaten in quantity. For grow-ing juvenile Gray squirrels in English woodland insects are an important source of protein in early summer when other nutritious foods are scarce. Several tropical squirrels have become totally insectivorous and the incisors of the long-nosed squirrel *Rhinosciurus laticaudatus* have become modified into forceps-like structures for grasping the insects that make up the animal's diet.

The Gray squirrel in England regularly strips bark (see pp626–627) and many other species will do so if food is short. Squirrels of the genus *Sundasciurus* specialize in feeding on bark and sap from the boles of trees in Malayan rainforest. Many species take large quantities of leaf matter, especially young leaves; the Giant flying squirrel is primarily a leaf-eater. Red squirrels feast on young conifer shoots, and thereby cause much damage in forestry plantations. Terrestrial squirrels feed mainly on low-growing plants and may affect the vegetation where they live. Prairie dogs feed in the immediate area around their burrows; they bite off all tall plants to open up a field of view. They feed on herbs and grasses—cropping plants and keeping vegetation low. The process induces changes in vegetation, encouraging the development of fast-growing plants. Prairie

▲ **Bushy-tailed mountain rodent.** The heavy Hoary marmot dwells in the mountainous areas of northern Canada and Alaska. Reserves of fat accumulated during summer for hibernation can amount to 20 percent of body weight.

◀ **Tucked away,** the gliding membrane of the Southern flying squirrel does not impede the animal's movement on tree trunks.

▼ **Membrane full spread,** a Giant flying squirrel glides between trees at night.

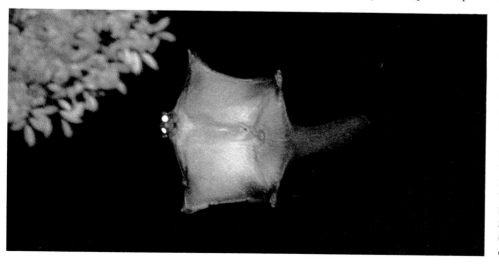

ing nuts. A squirrel grips a nut with both hands and its upper incisors while it gnaws a hole in the shell with the lower incisors which are then used as a lever to split the nutshell in half. Young squirrels also learn to distinguish between good and wormy nuts and discard the latter.

Because of the seasonal nature of flowering and fruiting in temperate forests the squirrels that live there must depend on different foods in different seasons but nuts and seeds are often cached, for use during the winter. From July, when the first fruits mature, until the following March Gray squirrels in English woodlands for instance depend on fresh and buried hazelnuts, acorns and beech mast. A poor mast crop can have serious consequences for squirrel populations, possibly preventing breeding the following spring or provoking large-scale migrations, over several kilometers, usually of younger subordinate animals. Even infant squirrels hoard and bury food and show instinctive innate burying behavior. Adult Red squirrels can smell and relocate pine cones buried 30cm (12in) below the surface. A squirrel does not necessarily find the hoard it buried—it might find another's. Many hoards are never retrieved and hence the burying of surplus food disperses seeds in temperate forests. This is not the case in tropical forests where the role of squirrels is more that of seed predators. There has been a long coevolution of squirrels with many cone- and mast-producing trees (eg pine, hazel, oak, beech, chestnut). Because their dentition enables them to gnaw through hard tissue and destroy the nutritive fruit embryo, squirrels are only successful seed dispersers where foods are seasonally superabundant and the animals make food caches. Gray squirrels carry acorns up to 30m (about 100ft) from a fruiting tree and bury them. Since many of these buried acorns are not found again they survive to germinate and eventually become new oak trees.

The Douglas pine squirrel has an equally important role for several conifers as it cuts unopened pine cones and caches them in damp places under logs or in hollow stumps in stores of 160 cones or more.

Many ground squirrels hibernate during the harsh winter months and prepare for this hibernation by laying up food supplies in their dens (tree squirrels do not hibernate though they may not emerge from their nests for a few days in bad weather). The Siberian chipmunk can carry up to 9g (0.3oz) of grain in its cheek pouches for a distance of over 1km (0.6m). It stores seeds,

dogs and most squirrels get all the water they require from the plant material they ingest but the European red squirrel in particular must stay close to sources of drinking water. Trappers take advantage of this fact and use water to attract squirrels into traps during a hot summer.

For most tree squirrels nuts and seeds are a major and highly nutritious part of the diet. One Red squirrel in Siberia took 190 pine cones in one day, eating the seeds and discarding the winged part. Squirrels have a special innate levering technique for open-

buds, acorns and mushrooms all stashed in different compartments adjoining the sleeping quarters of its underground burrow. One animal may store as much as 2–6kg (4.5–13lb) of food for its winter supplies. Similarly ground squirrels and sousliks become quite fat before hibernation, which lasts 5–7 months, but also store winter food supplies which are generally used only in spring when the animals reawaken. Marmots are also true hibernators. The entire marmot family unit (up to 15 animals) retreats into its den and sleeps huddled together throughout the winter. The last animal in, usually an adult male, plugs the entrance hole from the inside with hay, earth and stone. Marmots hibernate for 6 months or so and while they sleep their metabolism slows down and their heartbeat, body temperature and respiratory rate drop. When the temperature outside is below freezing the hibernating marmots have a body temperature of 4.5–7.5°C (40–45.5°F). Every 3–4 weeks the marmots wake to defecate and urinate and are much slimmer when they emerge in the spring, when they promptly begin to spring clean their burrows.

Some ground squirrels, eg the Arctic ground squirrel, the Little souslik and the Long-clawed ground squirrel, not only hibernate during the winter but also sleep during part of the summer when there is drought and the vegetation has withered away. They close off their dens with grass and sand and then settle down to sleep.

Most squirrels are sexually mature and able to breed when one year old, though marmots are not fully grown until they are two. Hibernating species usually mate a few days after the end of hibernation; marmots mate in May while still in their winter burrows and sexual activity is stimulated by odors from the anal glands of both sexes. Marmot families consist of both males and females—males are tolerant of each other even during the mating season. Gestation lasts about five weeks and the pregnant female closes off her living quarters with hay several days before the birth. Like all infant squirrels baby marmots are born naked, toothless and helpless with their eyes closed, but by six weeks of age they are fully-furred and sufficiently independent to be able to venture out of the burrow. While the young marmots play, other members of the family stand guard. In prairie dog societies too, both sexes are friendly towards the pups and take responsibility for them. The young

◄ ▲ **Representative species of squirrels.**
(1) A Southern flying squirrel (*Glaucomys volans*)
gliding from a nest hole in a tree trunk.
(2) Prevost's squirrel (*Callosciurus prevosti*).
(3) African pygmy squirrel (*Myosciurus pumilio*)
descending a tree head first. (4) An Abert or
Tassel-eared squirrel (*Sciurus aberti*). (5) An
American red squirrel (*Tamiasciurus hudsonicus*)
hanging by its hindlegs. (6) An Indian giant or
Malabar squirrel (*Ratufa indica*). (7) An Asiatic or
Siberian chipmunk (*Tasmias sibiricus*) with filled
cheek pouches. (8) Alpine marmot (*Marmota
marmota*) in vigilant upright posture giving
alarm whistle. (9) A Shrew-faced ground or
Long-nosed squirrel (*Rhinosciurus laticaudatus*)
foraging for termites. (10) Geoffrey's or Western
ground squirrel (*Xerus erythropus*) with its tail
arched and fluffed, an indication of anxiety.

marmots hibernate and live together in the parents' burrow for the next summer.

The situation described above where both male and female help care for the young is exceptional among squirrels; in most species parental care is the sole responsibility of the female. In spring male woodchucks and Red and Gray squirrels may travel considerable distances to mate. Gray and Red squirrels have mating chases with several males following a receptive female. After mating the male usually has no further association with the female who lines the breeding drey or den and rears the young alone. If she feels the young are threatened the mother moves them to another nest, carrying them by the scuff of the neck.

In all squirrel species gestation is short: 3–6 weeks. The young are always born naked, toothless and helpless with eyes closed. Litter sizes range from one or two (Spotted flying squirrel, Giant squirrel) to as many as nine or more (Gray squirrel); litter sizes may vary according to the age and condition of the mother. Many species breed only once a year (prairie dogs) but Gray and Red squirrels have two breeding seasons— in spring and summer—if conditions are favorable. If there has been little or no spring breeding because of winter food shortages large numbers of litters are produced in the summer to compensate.

The development of a young Red squirrel is fairly typical. At birth it will weigh only 8–12g (0.28–0.42oz) but it grows fast. Hair begins to appear at 10–13 days and it will be fully covered in hair by three weeks. The lower incisors appear at 22 days, the upper ones at 35. The eyes open after 30 days and the animal soon becomes more mobile and body cleaning and growing movements develop. It leaves the nest for the first time at 45 days, at which time it also takes its first solid food, and after 8 weeks is fully weaned and independent, though it remains near the mother and may still share her nest. Juvenile mortality is high with only about 25 percent of the young surviving to be one year old. Many fall to birds of prey or pine martens or die during migrations away from the maternal home range.

Development in young flying squirrels is slower than for many other mammals of similar size (eg cavies whose young are born fully furred and active); this is probably because of the hazards of nighttime arboreal activity and gliding. At birth infant flying squirrels already have a well-developed gliding membrane. Baby Southern flying squirrels are nursed for 60 or 70 days and are not active until that age.

Many of the terrestrial squirrels are very social animals. Alpine marmots live together in colonies ranging from 2 or 3 to 50 or more animals and one large colony may occupy an immense burrow system. Within the colony they scent mark their territory with substances secreted from the cheek glands and any intruder will be chased away by colony residents who gnash their teeth and call loudly. Once marmots have settled down and built their burrows they do not

▲ **The Red squirrel**–the archetypal squirrel. This species inhabits woodlands across Europe and Central Asia, but is hunted for its fur in the USSR.

▲ **The first summer** of a young prairie dog's life is a time of friendly relations with parents.

▷ **Half a prairie dog's life** (up to 15 months) OVERLEAF can be absorbed in growing to adult size.

▼ **The Black-tailed prairie dog** lives in colonies on plains and plateaus from southern Saskatchewan south to northern Mexico.

move far away from their home; in the mountains a family of marmots works about 0.25ha (0.6 acre).

Ground squirrels and sousliks also live socially in colonies with massive tunnel systems but some, like the European souslik, live singly within the colonies, each animal in its own den. Prairie dogs have an even more highly organized society: they live in social groups called coteries and several coteries make up "towns" which may cover areas of up to 65ha (160 acres). These interconnected burrows provide the prairie dogs with a refuge and a place to rear their young. Burrows may be surrounded by

crater-shaped mounds of earth, which help to prevent them flooding.

All animals from a coterie may use all parts of the coterie's burrow system. When coterie members meet they often touch noses; this is probably a form of identification or greeting. The dominant male usually maintains the coterie boundaries by challenging (with chases and calls) other prairie dogs from neighboring coteries. He may be helped by a subordinate coterie member, often his mate, who vocalizes at the intruder. Non-breeding coteries may consist of several males and females; one male usually dominates the other males but there is no clear dominance hierarchy among the females. In breeding coteries there is usually one adult male for every four adult females. Both sexes are friendly towards the young in their first summer and pups follow adults and play with them.

Relationships between coteries change with the season. In summer coterie boundaries are relaxed, and friendly contacts with neighboring coteries are common. By the fall coteries are exclusive and by December and January the dominant male is busy defending his boundaries against the neighbors. During the spring boundaries are gradually relaxed again, permitting cross-coterie contacts and inter-breeding. When

conditions become crowded adult prairie dogs emigrate, expanding the town's boundaries and leaving the old burrows to their offspring. This has good survival value in that older, more experienced animals leave to colonize new areas leaving the inexperienced young in familiar surroundings.

Something similar happens among American red squirrels where the female gives up her territory to her offspring. Although many Gray squirrel young emigrate from the area where they were born into areas of low population (if space allows), some establish home ranges within the original range of their mother, ie in territory with which they are already familiar and where they are likely to have a higher chance of survival. Gray squirrels are not territorial. Although they usually forage independently several squirrels will use the same area of woodland and often sleep together in the same dreys. Squirrels have overlapping home ranges with the male's ranges (4–6ha, 10–15 acres) larger than those of females (2ha or 5 acres) and one male's range will overlap those of several females.

There is a definite dominance hierarchy in Gray squirrels including both males and females; animals of low status are often forced to range widely and emigrate at times of food shortage. Animals taking part in these migrations are often the young of the year. It is these low status animals which cause the most damage in young plantations where they strip bark (see pp 626–627).

In fact the Gray squirrel pattern of social organization with larger male ranges overlapping smaller female ranges is typical of non-territorial squirrels and of many other arboreal mammals. Individual Gray squirrels do not have the social advantages of mutually cooperative food searching or protection against predators in the form of an efficient alarm system and cooperative resistance (compare prairie dogs). Instead they are widely spread over the available food resources in such a way that competition is minimal but each squirrel has social contact with its neighbors either directly or by scent marking trees (by gnawing under branches, roots, etc and depositing urine) and via them with the whole population. Such a social system allows efficient outbreeding with plenty of social feedback to influence population changes such as timing of reproduction, range shift and emigration. Little is known about social organization in flying squirrels though some are

reported to sleep in pairs in tree holes, but they too probably have overlapping home ranges.

Squirrel population densities vary considerably according to the species, habitat and number of species present. Red squirrels in pine forests in Scotland occur at densities of only 0.8 animals per ha (0.3 animals per acre) whereas Gray squirrels often reach densities of 5 or 6 animals per ha (about 2 animals per acre) in deciduous woodlands in England. Total squirrel densities in tropical forests are usually much lower (less than 2 animals per ha or 0.8 per acre) in spite of the great diversity of species and the year-round productivity of the forest.

Factors that limit the density of day-active squirrels are the spacing and timing of available food and competition for these sources. One limiting factor for flying squirrels must be the availability of suitable tree holes where they rest during the day. The low densities of tropical squirrels can be explained by competition among squirrels and the high densities of other arboreal mammals, especially primates, which not only compete for the same foods but also have a long-term influence on the ecology and evolution of the forest.

The main natural predators of squirrel populations are carnivores such as weasels, foxes, coyotes, bobcats, martens and birds of prey and owls, but also man, who kills squirrels because they are agricultural pests, or for sport or for fur. Some species may be threatened by habitat loss due to timber felling or change of land use.

Man, however, has had a long association with and affection for many species of squirrel. Red squirrels were kept as pets by Roman ladies, marmots were taught to perform at medieval fairs and prairie dogs and woodchucks feature in American Indian legends.

Red squirrels are prominent in many folk stories and fairy tales and were particularly important in Indian and Germanic religions and myths. The Red squirrel was holy to the Germanic god Donar because of its color and in Germany and England squirrels were once sacrificed at the feasts of spring and the winter solstice. In an Indian saga squirrel dries up the ocean with its tail.

In the United States Gray and Fox squirrels are prized as game animals though elsewhere squirrels are shot on account of the damage they cause in young tree plantations. The burrowing activity of ground squirrels and sousliks can improve the land. But by bringing up soil from lower depths they may be a nuisance in agricultural

Niche Separation in Tropical Tree Squirrels

For two or more species of mammal to live in the same habitat their use of resources must be sufficiently different to avoid the competitive exclusion of one species by another. Such differences in life-style as ground-living or tree-dwelling, active in daytime or nocturnal, insect-eating or fruit-eating are obvious means of ecological separation between squirrel species. Often, however, species occurring naturally in the same habitat appear to be utilizing the same food resources but closer study reveals that each occupies a somewhat different niche.

The situation is well illustrated by the squirrels found in the lowland forests of West Malaysia. Of the 25 Malayan species 11 are nocturnal and the rest active in daytime and the latter can be divided into terrestrial, arboreal and climbing categories with different species showing different use of the various forest strata. The Three-striped ground squirrel and Long-nosed squirrel feed on the ground or eat fallen wood whereas the Slender squirrel is most active on the tree trunks of the lower forest levels. The Plantain and Horse-tailed squirrels travel and feed mainly in the lower and middle forest levels but nest in the upper canopy. The three largest species live highest in the canopy.

The Malaysian squirrels show considerable divergence in choice of food when food is abundant but considerable overlap when it is scarce (all species then rely heavily on bark

and sap). None of the smaller forest squirrels apart from the Horse-tailed squirrel are seed specialists (unlike African or temperate forest species of comparable size). The Three-striped ground squirrel feeds on plant and insect material and the Long-nosed squirrel is an insectivore; these species overlap somewhat with the tree shrews rather than with the arboreal squirrels. The Sunda squirrels (*Sundasciurus* species) feed mainly on bark and sap and most of the beautiful squirrels (genus *Callosciurus*) are opportunistic feeders on a variety of plant material, supplemented in the smaller species with insects. The larger flying squirrels eat a higher proportion of leaves than the smaller species which take mainly fruit.

The three largest species of squirrels active by day show less clear ecological divergence than the smaller species of the lower forest levels. All three are fruit-eaters but the Cream giant squirrel uses more of the middle canopy levels and takes a significant proportion of leaves in its diet. The Black giant squirrel and Prevost's squirrel are often seen feeding together at the same fruit trees but Prevost's squirrel eats a smaller range of fruit. They also have different foraging patterns. While the larger giant squirrel is at an advantage in competitive situations its travel and basic metabolism are more expensive in energy, whereas Prevost's squirrel can afford to spend less time feeding each day and can travel further to food trees.

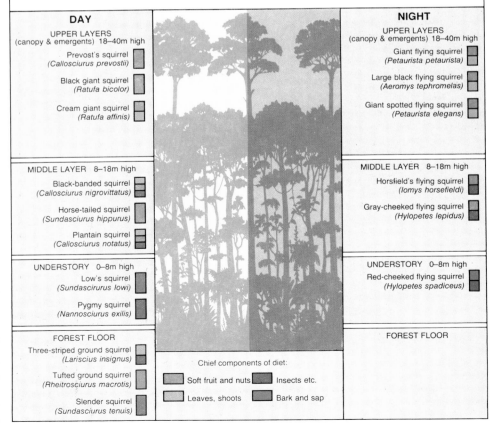

DAY

UPPER LAYERS (canopy & emergents) 18–40m high
Prevost's squirrel (*Callosciurus prevostii*)
Black giant squirrel (*Ratufa bicolor*)
Cream giant squirrel (*Ratufa affinis*)

MIDDLE LAYER 8–18m high
Black-banded squirrel (*Callosciurus nigrovittatus*)
Horse-tailed squirrel (*Sundasciurus hippurus*)
Plantain squirrel (*Callosciurus notatus*)

UNDERSTORY 0–8m high
Low's squirrel (*Sundasciurus lowi*)
Pygmy squirrel (*Nannosciurus exilis*)

FOREST FLOOR
Three-striped ground squirrel (*Lariscus insignis*)
Tufted ground squirrel (*Rheithrosciurus macrotis*)
Slender squirrel (*Sundasciurus tenuis*)

NIGHT

UPPER LAYERS (canopy & emergents) 18–40m high
Giant flying squirrel (*Petaurista petaurista*)
Large black flying squirrel (*Aeromys tephromelas*)
Giant spotted flying squirrel (*Petaurista elegans*)

MIDDLE LAYER 8–18m high
Horsfield's flying squirrel (*Iomys horsefieldi*)
Gray-cheeked flying squirrel (*Hylopetes lepidus*)

UNDERSTORY 0–8m high
Red-cheeked flying squirrel (*Hylopetes spadiceus*)

FOREST FLOOR

Chief components of diet:
Soft fruit and nuts
Leaves, shoots
Insects etc.
Bark and sap

▲ **Appealing but destructive.** The Least chipmunk, like other chipmunks, can be tamed and kept as a pet. Its natural habitat, however, is semiopen land where it can grub up newly planted corn seed in spring and invade granaries in fall.

▶ **Inverted heavyweight.** The Malabar squirrel and the other three species of giant squirrels (all of which live in Southeast Asia) can weigh up to 3kg (6.6lb). Yet for eating this is the characteristic position adopted: hung by the hind feet from a branch.

◀ **Strata use** by squirrels in a Bornean forest (see Box Feature).

areas and lead to the collapse of irrigation channels.

Those species of squirrels that live in colder climates are often hunted for their thick soft winter fur. Marmots play an important role in the Mongolian economy where they are hunted for fur and their meat. In the USSR sousliks, European flying squirrels and Red squirrels are important to the fur trade: the best furs come from the Taiga where the Red squirrels have dark gray winter coats. Gray squirrel tails are used in artists' brushes and marmot fat is used in the Alps as a remedy for chest and lung diseases.

Squirrels can also be carriers of disease. The fleas on the fur of sousliks can carry the plague bacillus and both sousliks and marmots may spread this disease. Golden-mantled ground squirrels are also carriers of bubonic plague and tularemia and marmots may carry Rocky mountain tick fever. For most people, however, squirrels are not considered as a source of disease but as a charming and amusing addition to our woodlands and prairies.   KM

The Role of Kinship

The annual round in Belding's ground squirrels

Belding's ground squirrels (*Spermophilus beldingi*) are meadow-dwelling rodents that inhabitat mountainous regions in the western United States. They are active above ground during the day and spend the night in subterranean burrows. While primarily vegetarian, and especially fond of flower heads and seeds (*Spermophilus* means "seed loving"), they also eat insects, birds' eggs, carrion and occasionally Belding young

A population of these animals located at Tioga Pass, high in the central Sierra Nevada of California (3,040m, 9,945ft) has been studied for 15 consecutive years. There ground squirrels are active only from May to October, hibernating the rest of the year. In the spring males emerge first, a few weeks before females; to reach the surface they must often tunnel through several meters of accumulated snow. Once the snow melts, females emerge and the annual cycle of social and reproductive behavior begins.

Females mate about a week after they emerge. During a single afternoon of sexual receptivity each female typically mates with 3–5 different males; studies of paternity have revealed that most litters (55–78 percent) are sired by more than one male. In the presence of receptive females, males threaten, chase and fight with each other; often they sustain physical injuries. The heaviest, oldest, most experienced males usually win such conflicts. They thereby remain near receptive females, enabling them to mate frequently. The majority of males, however, seldom or never copulate.

After mating each female digs a nest burrow in which she rears her litter. Gestation lasts about 24 days, lactation 27 days. The mean litter size is 5, with a range of 1–11. A typical nest burrow is 3–5m (10–16ft) long, 30–50cm (11.5–19.5in) below ground, and has at least two surface openings. Females shoulder the entire parental role. Indeed males often do not even interact with the young of the year because by the time weaned juveniles begin to emerge above ground in late-July or August some males are already reentering hibernation. The females begin to hibernate early in the fall and finally, when it begins to snow, the young of the year emerge to begin their first long, risky winter.

The 7–8 month hibernation period is a time of heavy mortality. Two-thirds of the juveniles and one-third of the adults perish during the winter. Most die because either they deplete their stores of body fat and freeze to death or else they are eaten by predators. Males typically live 2–3 years, compared to 3–4 years for females; a few

males survive 6 years but many females live at least 10 years. Males apparently die younger due to injuries incurred during fights over females and because males that are disabled by infected wounds are particularly susceptible to predators.

Dispersal also differs between the sexes: while females are sedentary from birth, males are relatively nomadic. Soon after they are weaned, juvenile males disperse from the area in which they were born. Typically they move 300–400m (975–1,300ft) and they rarely, if ever, return home or associate again with their maternal relatives. In mid-summer, when females are lactating, many adult males also emigrate. These dispersal episodes virtually preclude social interactions and inbreeding between males and their close kin. Females, in contrast, never disperse and they remain close to their natal burrow and interact with maternal kin throughout their lives.

The ground squirrels' population structure has set the stage for the evolution of nepotism (the favoring of kin). There are four major manifestations of nepotism among females. First, they seldom chase or fight with their close relatives—offspring, mother or sisters—when establishing nest burrows. Among kin, females thus obtain residences with minimum expenditure of energy and little danger of injury. Second, close relatives share portions of their nesting territories and permit each other access to food and burrows on such areas. Third, close kin join together to evict distant relatives or nonkin from each other's territories. Fourth,

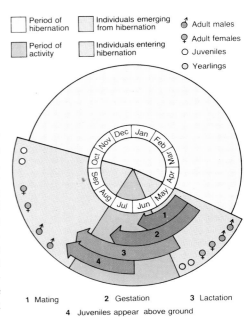

□ Period of hibernation	□ Individuals emerging from hibernation
▨ Period of activity	▨ Individuals entering hibernation

♂ Adult males
♀ Adult females
○ Juveniles
○ Yearlings

1 Mating　　　2 Gestation　　　3 Lactation
4 Juveniles appear above ground

▲ **The annual cycle** of hibernation, activity, and breeding in the Belding's ground squirrel at Tioga Pass, California.

▶ **Alarm call.** A female Belding's ground squirrel calls as a coyote approaches. The function of such calls is apparently to warn offspring and other close relatives of impending danger.

▶ **First day above ground.** BELOW These pups are about 27 days old and have just emerged from their natal burrow.

▼ **Grass for the nest,** carried by a female Belding's ground squirrel. To line one nest can require more than 50 loads of dry grass.

females give warning cries when predatory mammals appear. For example, at the sight of a badger, coyote or weasel some ground squirrels stand erect and give loud, staccato alarm calls. More callers than noncallers are attacked and killed by predators, so calling is self-sacrificial or "altruistic" behavior. However, not all individuals take the same risks. The most frequent callers are old (4–9 years), lactating, resident females with living offspring or sisters nearby. Males and young (1 year) or barren females call infrequently. In other words, callers behave as if they are trading the risks of exposure to predators for the safety and survival of dependent kin.

Cooperative defense of the nest burrow is another important manifestation of nepotism. During gestation and lactation females defend the area surrounding their nest burrow against intrusion by all but their close kin. Such territoriality helps protect the helpless pups from other Belding's ground squirrels: when territories are left unattended, even briefly, unrelated females or one-year-old males may arrive and kill pups. Yearling males typically eat the carcasses, so their infanticide may be motivated by hunger. A different stimulus causes females to become infanticidal. When all of a mother's own young are killed by coyotes or badgers, she often emigrates to a new, safer site. Upon arrival she attempts to kill young there. Only if she is successful is she able to settle. By removing juveniles (females) who are likely to remain in the preferred area, infanticidal females reduce future competition for a nest site. Mothers with close relatives as neighbors lose fewer offspring to infanticide than do females without neighboring kin. This is because groups of females detect marauders more quickly and expel them more rapidly than do individuals acting alone, and because pups are defended by their mother's relatives even when she is temporarily away from home.   PWS

The Squirrel's Devastation

Why do Gray squirrels strip bark?

Many mammals strip living bark from tree trunks. Some, such as deer, rabbits and domestic livestock, eat the tough outer bark, which tends to be rich in mineral nutrients, especially zinc. Other mammals, including some bears, dormice and several species of squirrel, remove and discard the outer bark and then scrape off and eat the soft, sap-containing tissues underneath. These are the tree's circulatory system, which transports water and dissolved minerals up from the roots, and sugars or other organic compounds down from the leaves in the crown of the tree. Although trees tolerate a limited amount of bark-stripping, and may eventually cover wounds with callus, the removal of bark right round the trunk cuts the flow of nutrients to and from the crown. When this happens low down, the whole tree will die, whereas a ring higher up may cause the upper part alone to succumb and the tree will become stunted. A tree not killed outright risks attack by insects or fungi on its exposed heartwood, and even wounds that heal leave inner crevices which spoil the tree's value as timber.

The Gray squirrel (*Sciurus carolinensis*) was introduced to Britain in 1876 from the deciduous woodlands of North America, where it occasionally causes damage by stripping bark from sugar maples. In Britain, Gray squirrels damage mainly sycamore (which is also a maple), beech and oak, although other trees are sometimes attacked. The native Red squirrel (*Sciurus vulgaris*), which has been replaced by the closely related Gray in most deciduous woods, also sometimes strips these species (as well as damaging conifers). Both squirrels strip bark from the stems and large side branches while the trees are still quite young, usually with a diameter less than about 20cm (8in) at the base and not more than 30 or 40 years old. On older trees the bark becomes too thick for easy removal, except on the smaller branches.

Squirrels are sometimes seen peeling fine strips of bark from the outer twigs of limes, thujas and some other trees, but this is taken as a soft lining for tree nests (dreys) or for dens in hollow trees. They do not hibernate during the winter but in Britain in the coldest months they usually forage for only 1–3 hours each day, conserving their body heat at other times in their well-insulated dreys and dens. Squirrels in these nests may also benefit from the shared warmth of several others: up to seven in one drey.

Because squirrel damage is costly and discourages the planting of native beech and oak woods, it is important to find effective ways to prevent bark-stripping. One solution is to kill squirrels in areas with vulnerable young trees. However, shooting and trapping can be expensive, and do not always remove enough squirrels to prevent damage. Poisoning with warfarin is more effective, but may harm other wildlife. Another approach has been to study the causes of the damage, in order to find cheaper solutions to the problem.

There have been many attempts to explain bark-stripping. One was that squirrels might have an uncontrolled gnawing reflex, rather like a dog's tendency to keep scratching with its hind leg after being

▶ **Destroyed in its prime.** Young trees with relatively thin bark, such as this sycamore, succumb most easily to the stripping activity of Gray squirrels.

▼ **Trees that survive,** especially oak, may well house squirrel dreys. Gray squirrels depend on oak trees for supplies of acorns, yet this does not exclude oaks from attack.

tickled behind an ear. Another suggestion was that gnawing might be necessary to prevent the incisor teeth growing too long. As in other rodents, squirrel incisor teeth grow continuously throughout life, and there are no hard nuts to chew in mid-summer, when most damage occurs. These theories can be discounted, because the squirrels do not simply gnaw but peel and eat. The theory that squirrels are marking territory boundaries must also be wrong, because although Gray squirrels share overlapping ranges in which they have a "pecking order" they do not defend specific territories. A fourth idea, that squirrels eat

sap to obtain water, leaves unexplained the fact that damage often occurs next to ponds or streams.

Any theory that attempts to explain bark-stripping must also show why the damage does not occur in all areas of young woodland, nor every year in the areas that are damaged. One possibility is that squirrels are short of food in particular areas or years during mid-summer: tree flowers and buds provide an abundance of food in spring, but food may then be less plentiful for several weeks before autumn nuts and mast are ripe. Another theory is that squirrels simply like the sweet sap.

Strange as it may seem, recent research strongly supports the last suggestion. Individual trees in a wood differ greatly in the thickness of their sappy tissue, even when their age is similar; the squirrels strip most bark from the trees with the thickest sap. Moreover, trees in some woods tend to be more sappy than in others; the squirrels cause most damage in the areas where the trees in general have thick sap. There is little bark-stripping where trees have little sap. Since sap thickness varies from year to year in the same area, this probably explains why the amount of bark-stripping is so variable. It also explains the forester's lament, "that they always ruin the best trees," since it is the trees with most sap, and thus with most circulating nutrients, that are growing most strongly.

Although the extent of damage in an area seems to depend mainly on the quality of its

trees, there is also evidence that aggressive behavior may sometimes trigger the bark-stripping. Spring breeding tends to be best in the worst-affected areas, so that gnawing-displays may be most likely there, leading to the discovery of the sweet sap. Gnawing in search of food may help squirrels to discover the sap too, because squirrels tend to have lost most weight in the areas with worst damage. Sap is not a rich food, however, and damage can be quite heavy and yet provide no more than one day's food for five or six squirrels. Nevertheless, whatever triggers the damage, it is probably least likely to start if there are few squirrels in an area.

An improved understanding of the damage indicates new ways of trying to prevent it. Where tree growth does not need to be rapid, in nature reserves for instance, we can try to regenerate beech without damage by making sure that the young trees grow densely, and therefore have a poor sap flow until they are old enough for their bark to resist squirrels. We may eventually be able to breed oak or beech trees with unpleasant-tasting sap, or to tempt squirrels away from them with artificial "sap" dispensers. Moreover, we can probably keep squirrel numbers down by planting vulnerable young trees well away from mature oaks, chestnuts, hazels and conifers, which provide abundant winter food for squirrels. In our existing woodlands, however, there is often no alternative to killing squirrels, so perhaps future research should concentrate on how best to prepare squirrel pie.   REK

POCKET GOPHERS

Family: Geomyidae
Thirty-four species in 5 genera divided into two tribes.
Distribution: N and C America, from C and SW Canada through the W and SE USA and Mexico to the Panama–Colombia border.

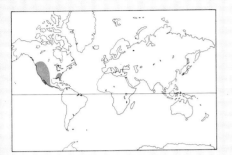

Habitat: friable soils in desert, scrub, grasslands, montane meadows and arid tropical lowlands.

Size: ranges from head-body length 12–22.5cm (4.7–8.9in) and weight 45–400g (1.6–14oz) in the genus *Thomomys* to head-body length 18–30cm (7–11.8in) and weight 300–900g (11–32oz) in the genus *Orthogeomys*. (Males are always larger than females, up to twice the weight.)

Gestation: 17–20 days in the genera *Thomomys* and *Geomys*.

Longevity: maximum 4 years (6 years recorded in captivity).

Tribe Geomyini

Eastern pocket gophers
Genus *Geomys*, 5 species including: **Plains pocket gopher** (*G. bursarius*).

Taltuzas
Genus *Orthogeomys*, 10 species.

Yellow and cinnamon pocket gophers
Genus *Pappogeomys*, 9 species including: **Yellow-faced pocket gopher** (*P. castanops*).

Michoacan pocket gopher
Zygogeomys trichopus.

Tribe Thomomyini

Western pocket gophers
Genus *Thomomys*, 9 species including **Valley pocket gopher** (*T. bottae*), **Northern pocket gopher** (*T. bulbivorus*).

POCKET gophers present a paradox: common to all species and genera is a basic body plan and a similar life cycle, both necessary for their life of digging, yet one of the most characteristic features of the family is an extensive variability in details of biology, both between species and within them.

In the United States the term "gopher" is used not only for the members of the family Geomyidae but is also applied to some ground squirrels (genus *Spermophilus*), to salamanders and to mud turtles; it suggests a way of life in which digging is important. "Pocket" refers to the external, fur-lined cheek pouches located on either side of the pocket gopher's mouth, a feature of the body which, among mammals, is only otherwise found in the closely related family Heteromyidae (pocket mice and kangaroo rats).

Highly modified for subterranean digging, pocket gophers have thickset, tubular bodies with little external evidence of a neck; short and powerful fore and hind limbs of approximately equal size; and a short, nearly naked tail, which is particularly sensitive to touch. Both upper and lower incisors project through the furred lips so that these teeth are exposed even when the mouth is closed. This adaptation enables pocket gophers to cut roots or dig with their teeth without dirt entering the mouth cavity. They excavate soil with the enlarged claws of their forefeet; the incisors are secondary implements. Populations or species living in harder soils tend to have more forward-pointing incisors and probably make more use of them for digging than do those living in looser soils.

Their skin fits loosely. It is usually clothed in short, thick fur through which are scattered hairs sensitive to touch. The loose skin enables individuals to execute tight turns in the constricted space of their burrows. Gophers are very agile and can move rapidly both backwards and forwards.

The pocket gopher's skull is massive and strongly ridged, with massive jaw muscles. Its incisors may be either smooth or grooved on their anterior surface, according to the genus. The four cheekteeth in each quarter of the jaw form a particularly effective battery for grinding tough and abrasive foodstuffs. All the cheekteeth grow continuously. Indeed the rate of growth of both incisors and foreclaws is amazingly rapid, from 0.5 to 1mm (0.02–0.04in) per day in the Valley pocket gopher.

Within the range of their distribution pocket gophers are ubiquitous in virtually all habitats with extensive patches of friable soil. The range of habitats in which they are found can be extreme: the Valley pocket gopher ranges from desert soils below sea level to alpine meadows well above the timberline, over 3,500m (11,500ft). In tropical latitudes populations of the same species may be found in mountain forest meadows or in arid tropical scrub, but few penetrate true tropical savanna.

Both genera and species are distributed contiguously. Except in very narrow zones of overlap only one kind of pocket gopher will be found in any particular area. This results from an apparent inability to subdivide the two-dimensional subterranean habitat (the fossorial niche). In areas where several species and/or genera meet the pattern of species distribution is mosaic-like. In the mountains of the western USA and Mexico species live in a succession of zones arranged according to altitude.

The two tribes of living pocket gophers represent major branches in the evolutionary divergence of their family. Modern genera first appear in the fossil record in the late Pliocene era (about 4 million years ago). The evolutionary history of the family can be characterized as successive attempts to invade the fossorial niche. Each successive group shows better developed adaptations for subterranean existence. The modern pocket gophers, all members of the subfamily Geomyinae (two other subfamilies are extinct), display the most strongly devel-

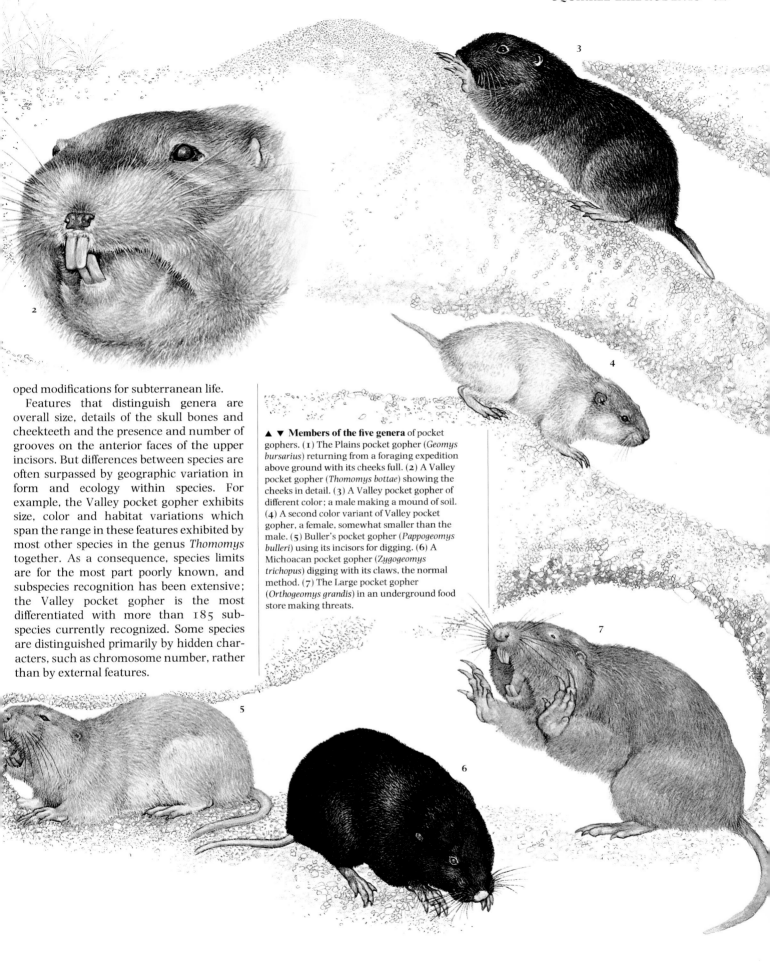

oped modifications for subterranean life.

Features that distinguish genera are overall size, details of the skull bones and cheekteeth and the presence and number of grooves on the anterior faces of the upper incisors. But differences between species are often surpassed by geographic variation in form and ecology within species. For example, the Valley pocket gopher exhibits size, color and habitat variations which span the range in these features exhibited by most other species in the genus *Thomomys* together. As a consequence, species limits are for the most part poorly known, and subspecies recognition has been extensive; the Valley pocket gopher is the most differentiated with more than 185 subspecies currently recognized. Some species are distinguished primarily by hidden characters, such as chromosome number, rather than by external features.

▲ ▼ **Members of the five genera** of pocket gophers. (1) The Plains pocket gopher (*Geomys bursarius*) returning from a foraging expedition above ground with its cheeks full. (2) A Valley pocket gopher (*Thomomys bottae*) showing the cheeks in detail. (3) A Valley pocket gopher of different color; a male making a mound of soil. (4) A second color variant of Valley pocket gopher, a female, somewhat smaller than the male. (5) Buller's pocket gopher (*Pappogeomys bulleri*) using its incisors for digging. (6) A Michoacan pocket gopher (*Zygogeomys trichopus*) digging with its claws, the normal method. (7) The Large pocket gopher (*Orthogeomys grandis*) in an underground food store making threats.

Pocket gophers eat a wide variety of plants. Above ground they take leafy vegetation from the vicinity of burrow openings; underground they devour succulent roots and tubers. They often prefer forbs and grasses, but diet shifts seasonally according to the availability of food and the gophers' needs for nutrition and water. For example, water-laden cactus plants may become a major dietary component during the hot and dry summer months in arid habitats. The external cheek pouches are filled with parts of plants by dextrous motions of the forepaws for transportation to cache storage areas in the burrow. The storage areas are usually sealed from the main tunnel system.

Males do not become sexually mature until the breeding season after their birth, ie when about one year old. Females in some populations, however, may start to breed during the season of their birth, when approximately 70 days old. For most species and populations the reproductive period is bounded by the changes of the seasons, but in its timing it varies geographically. In montane regions breeding follows the melting snow, but in coastal and desert valleys and in temperate grasslands it coincides with winter rainfall.

The number of litters per female may vary geographically within species. In the Valley pocket gopher many populations have but one litter per year while in others a given female may have several successive litters. Pocket gophers of this species, living in irrigated agricultural fields, may breed year-round though neighboring populations living amidst natural vegetation may have sharply delimited seasonal breeding. Females of Yellow-faced and Plains pocket gophers typically produce multiple litters each year. Litter size varies from *Pappogeomys* in which two is the commonest number of young to *Thomomys* which usually produces five. Some females, however, bear up to 10 young per pregnancy. The onset of breeding, length of season and number of litters are largely controlled by local environmental conditions, primarily temperature, moisture and vegetation.

Young are born completely dependent on their parents. Their cheek pouches do not open for about 24 days; eyes and ears open at 26 days. In Northern and Valley pocket gophers the molting from juvenile coats takes 100 days. Weaning in most gophers probably occurs by 40 days after birth, but dispersal from the mother's burrow system may not occur for about 60 days. The maximum longevity in nature is about 5 years, but the average life span of an adult is

considerably shorter, just over one year. Females live nearly twice as long as males. On average females survive for 56 weeks and males for 31 weeks in the Yellow-faced pocket gopher while in the Valley pocket gopher longevity is 4.5 years for females and 2.5 years for males. The ratio between adult numbers of the sexes varies geographically from near equality to a high preponderance of females (3 to 4 females for every male) in probably all species, although this observation is best documented for Valley, Northern, Plains and Yellow-faced pocket gophers. Such variation is related to geographic differences in population density, with males becoming proportionately less common as the population size increases. In part the proportional loss of adult males is thought to result from increased mortality due to male-male aggressive encounters over territorial space and mates. Virtually all adult females within a population will breed each season, but there may be high variance in male reproductive success. It is possible that some males never breed during their life span, and that most young in a given population are sired by relatively few.

Pocket gophers are solitary creatures, with individuals regardless of sex maintaining separate and contiguous territories. In the Valley pocket gopher the maximum territory size is about 250sq m (2,700sq ft). The size of territories varies in part according to the quality of habitat: high-quality habitat will support denser populations with smaller individual territory sizes than will habitats of poorer quality. Territorial boundaries break down momentarily during the

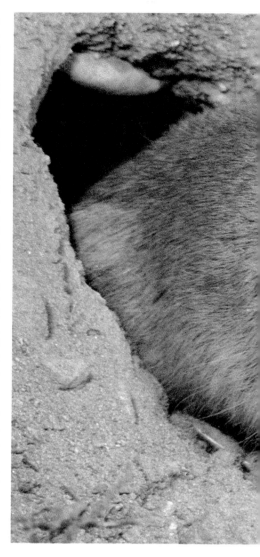

▲ **Out of the raw earth,** a Northern pocket gopher. Gophers prefer deep, sandy, easily crumbled, well-drained soils, but this species can live in a wider range of soils than any other. Gopher burrows are commonly 15m (about 50ft) long.

◄ **The digging machine.** This Valley pocket gopher displays well the contrast between gophers' small eyes and ears and their obtrusive claws.

tend to be evenly dispersed, occupying all available habitat. In poor-quality regions density is depressed, individuals tend to be clumped in distribution and their territories will shift in both size and position throughout the year as animals search for food.

Male and female young disperse from the mother's system at the same time, but in the Valley pocket gopher female young of the year initially move longer distances, establishing territories in which they may breed during the season of their birth. Male young tend to live in shallow systems in marginal and peripheral habitat until they disperse to establish territories just prior to the next breeding season. Despite their adaptations for digging, pocket gophers can and do move considerable distances and most dispersal happens above ground, on dark nights. Movements over shorter distances may take advantage of long tunnels just under the surface.

Individuals of both sexes are pugnacious and aggressive, and will fight if placed together in limited quarters. Individuals of both sexes in all species compete for parcels of land, which must contain all their requirements for long-term survival.

Pocket gophers serve a major role in soil dynamics. Their constant digging generates a vertical cycling of the soil, counteracting the packing effects of grazers, making the soil more porous and hence slowing the runoff of water, and providing increased aeration for plant growth. They can have profound effects on plant communities, often, through continual disturbance, creating soil conditions that favor the growth of herbaceous plants which are often preferred foods. The range and population density of many pocket gopher species have increased as native plant communities have been replaced by agricultural development. However, although pocket gophers can benefit agricultural interests by working the soil, they can cause extensive economic loss through their voracious herbivorous appetite and their digging activities. Dense populations can produce severe loss of crops and, particularly in the arid western region of North America, irrigation systems can rapidly be undermined by their extensive burrow systems. Millions of dollars have been spent on programs to control pocket gopher populations in the USA. However, most control efforts have been unable to deal effectively with the increase in population density and reproductive rate in pocket gopher populations which results from their invasion of the high-quality resources provided by agricultural habitat.   JLP

breeding season, at which time there occurs multiple occupancy of burrows by adult females and adult pairs or by females with young. Male territory size in the Valley pocket gopher averages more than twice that of females in both area and total burrow length. Each adult male territory may be contiguous with two or more female burrow systems. While each individual maintains a tunnel system for exclusive use, adjacent males and females may have common burrows and deep, common nesting chambers.

Densities in smaller species such as the Valley pocket gopher will generally not exceed 40 adults per ha (99 per acre) and will be as low as 7 per ha (17 per acre) in the large-bodied Yellow-faced pocket gopher. In high-quality habitat individual territories are stable in both size and position, with most individuals living their entire adult lives in very limited areas. In such conditions densities are high and individuals

SCALY-TAILED SQUIRRELS AND POCKET MICE

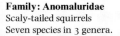

Family: Anomaluridae
Scaly-tailed squirrels
Seven species in 3 genera.
Distribution: W and C Africa.

Family: Heteromyidae
Pocket mice
Sixty-five species in 5 genera.
Distribution: N, C and northern S America.

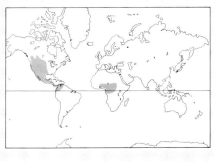

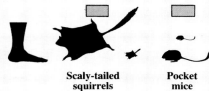

**Scaly-tailed
squirrels**　**Pocket
mice**

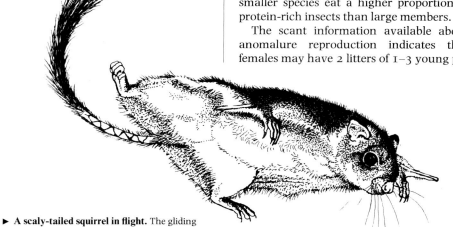

▶ **A scaly-tailed squirrel in flight.** The gliding membrane extends between the tail base and the hindlegs and from the hindlegs to the front limbs where it is attached to the upper arm and a gristle rod strung from the elbows to the neck. At sunset animals leave their roost trees and glide between trees in search of food for distances of up to 200m (approx 650ft). Pygmy scaly-tails have been reported moving as far as 6.5km (4mi) from their roost trees to eat the flesh of palm nuts.

THE tropical and subtropical forests of the Old World are inhabited by an interesting array of gliding mammals. In tropical Africa this niche is filled by members of the Anomaluridae, the gliding **scaly-tailed squirrels,** which apparently share only distant evolutionary relationships with true squirrels.

The scaly-tailed squirrels are squirrel-like in form with a relatively thin, short-furred tail whose underside contains an area of rough, overlapping scales near the base of the tail.

The ecology and behavior of scaly-tailed squirrels is poorly known although such species as Lord Derby's scaly-tailed flying squirrel and Zenker's flying squirrel are common and sometimes come into contact with humans. All species are probably nocturnal, and some spend their days in hollow trees where colonies of up to 100 Pygmy scaly-tailed squirrels have been found. Beecroft's anomalure apparently does not sleep in hollow trees but instead rests on the outside of tree trunks during the day where it relies on its cryptically colored fur for protection from predators. Lord Derby's scaly-tailed flying squirrel, the largest member of the family, eats a variety of plant products, including bark, fruits, leaves, flowers and green nuts, as well as insects. If anomalures are like other groups of rodents, smaller species eat a higher proportion of protein-rich insects than large members.

The scant information available about anomalure reproduction indicates that females may have 2 litters of 1–3 young per year. At birth babies are large, well-furred, active, and their eyes are open. Female pygmy scaly-tails apparently leave their colonies to bear their single young alone.

Except for Derby's anomalure, these rodents depend entirely on primary tropical forest for their existence. To the extent that African primary forests are being destroyed, these interesting but poorly-known rodents are endangered.

By day, summer conditions in the deserts in the southwestern USA are formidable. Surface temperatures soar to over 50°C (122°F), sparsely distributed plants are parched and dry and signs of mammalian life are minimal. As the sun sets, however, the sandy or gravelly desert floor comes alive with a high density of rodents (10–20 animals per ha, 25–49 per acre). Greatest diversity occurs among the **pocket mice** in which five or six species can coexist in the same barren habitat.

Contrast these hot, dry (or in winter, cold and apparently lifeless) desert conditions with the habitat in which species of the genus *Heteromys* can attain densities of up to 18 per ha (44 per acre): the tropical rain forests of Central and northern South America. Rich in rain and in vegetation, the rain forest is yet near bare in heteromyid species: only one species, Desmarest's spiny pocket mouse, occurs, at most sites.

The most likely explanation for this difference in species richness lies in the diversity and availability of seeds. In North American deserts, seeds of annual species can accumulate in the soil (to a depth of 2cm, 0.8in) in densities of up to 91,000 seeds per sq m (8,450 per sq ft). Patchily distributed by wind and water currents, small seeds, weighing about a milligram tend to accumulate in great numbers under shrubs and bushes and on the leeward sides of rocks whereas larger seeds occur in clumps in open areas between vegetation. This soil-seed matrix is the arena for the seed-gathering activities not only of nocturnal heteromyid rodents but also of a rather diverse array of seed-harvesting ants (about 50 species) and a modest diversity of seed-eating birds (four species, active by day).

Tropical forests are also rich in seeds, but many of those produced by tropical shrubs and trees are protected, chemically, against predation by seed-eating insects and their larvae, and by birds and mammals. This is especially true of large seeds weighing several grams, which considerably reduces the variety of seeds available to rodents. In effect, then, the tropical rain forest is a desert in the eyes of a seed-eating rodent whereas the actual desert is a "jungle" with regard to seed diversity and availability. Most of the 15–20 species of rodents inhabiting New World tropical forests are omnivorous or are fruit-eaters.

With large cheek pouches and a keen sense of smell, heteromyid rodents are admirably adapted for gathering seeds. Most of the time they are active outside their underground burrow systems is spent collecting

▲ **The benefits of hopping.** Kangaroo rats can range over wider areas than pocket mice and therefore can be more selective when foraging. They usually dig and glean seeds from one or two spots under a shrub and then rapidly hop to another shrub and dig again. Foraging away from cover exposes kangaroo rats, such as this Pacific kangaroo rat, to greater risk from such predators as owls, but their hearing—sensitive to low frequencies—enables kangaroo rats to detect predators more readily than other heteromyid rodents can.

Quadrupedal pocket mice are on the other hand almost "filter-feeders": they push slowly through loose soil, usually beneath bushes or shrubs, sorting and stuffing seeds into their cheek pouches as they encounter them. Not all seeds collected will be eaten or stored. For consumption or storage, pocket mice tend to select only seeds rich in calories.

seeds in various locations within their home ranges. Members of the two tropical genera (*Liomys* and *Heteromys*) search through the soil litter for seeds with which they fill their cheek pouches. Some of these will be buried in shallow pits scattered around the home range whereas others will be stored underground in special burrow chambers.

Breeding in desert heteromyids is strongly influenced by the flowering activities of winter plants which germinate only after at least 2.5cm (1in) of rain have fallen between late September and mid December. In dry years, seeds of these plants fail to germinate, and a new crop of seeds and leaves is not produced by the following April and May. In the face of a reduced food supply heteromyids do not breed, and their populations decline in size. In years following good winter rains most females produce two or more litters of up to five young, and populations increase rapidly in size.

This "boom or bust" pattern of resource availability also influences heteromyid social structure and levels of competition between species. When seed availability is low, seeds stored in burrow or surface caches become valuable and defended resources. Behavior becomes asocial in most species of arid-land heteromyids (including species of *Liomys*): adults occupy separate burrow systems (except for mothers and their young) and when two members of a species meet away from their burrows they engage in "boxing matches" and sand kicking fights. In the forest, in contrast, species of *Heteromys* are socially more tolerant; individuals have widely overlapping home ranges, they share burrow systems and are less likely to fight each other.

Experiments conducted in Arizona, USA, indicate that heteromyids not only compete for seeds among themselves but also with seed-harvesting ants. In one set of experiments heteromyid numbers and biomass increased 20–29 percent above levels on control plots after ant colonies had been removed from plots of 0.1ha (0.25 acre). Similarly, the total number of ant colonies increased 71 percent over control levels on plots from which heteromyids had been removed. Exclusion of large kangaroo rats from plots surrounded by "semipermeable" fences which permitted the immigration of small pocket mice resulted in a 3.5-fold increase in pocket mouse numbers after a period of eight months. Densities of small omnivorous rodents (eg *Peromyscus*, *Onychomys*) did not differ between experimental and control plots, which indicates that kangaroo rats influence only the abundance of pocket mice, not of all species of small rodents.

The diversity and availability of edible seeds is the key to the evolutionary success of heteromyid rodents. Seed availability affects foraging patterns, population dynamics and social behavior. In North American deserts, because seed production influences levels of competition among heteromyids, ants and probably other seed-eaters, there is a clear link between resources and the structure of an animal community. Thus the abundance and diversity of seed-eaters is directly related to plant productivity.   TEF

The 3 genera of scaly-tailed squirrels
Habitat: tropical and subtropical forests. Size: ranges from head-body length 6.8–7.9cm (2.7–3.1in), tail length 9.1–11.7cm (3.6–4.8in), weight 14–17.5g (0.5–0.6oz) in the Pygmy scaly-tailed squirrel (*Idiurus zenkeri*) to head-body length 27–37.9cm (10.6–15in), tail length 22–28.4cm (8.6–11.2in), weight 450–1,090g (16–38oz) in Lord Derby's scaly-tailed flying squirrel (*Anomalurus derbianus*). Gestation: unknown. Longevity: unknown (probably several years in Lord Derby's scaly-tailed flying squirrel).

Scaly-tailed flying squirrels
Genus *Anomalurus*.
Africa from Sierra Leone E to Uganda and S to N Zimbabwe. Four species including: **Beecroft's anomalure** (*A. beecrofti*), **Lord Derby's scaly-tailed flying squirrel** (*A. derbianus*).

Pygmy scaly-tailed squirrels
Genus *Idiurus*.
Africa from Cameroun SE to Lake Kivu in E Zaire. Two species including **Zenker's flying squirrel** (*I. zenkeri*).

Flightless scaly-tailed squirrel
Zenkerella insignis.
Cameroun.

The 5 genera of pocket mice
Habitat: semiarid to arid regions of N America for species in the genera *Perognathus*, *Microdipodops* and *Dipodomys*; tropical forests and grasslands for *Liomys* and *Heteromys*. Size: ranges from head-body length 5.6–6cm (2.2–2.4in), tail length 4.4–5.9cm (1.7–2.3in) and weight 10–15g (0.3–0.5oz) in *Perognathus flavus* to head-body length 12.5–16.2cm (4.9–6.4in), tail length 18–21.5cm (7.1–8.5in) and weight 83–138g (3.0–4.8oz) in *Dipodomys deserti*. Gestation: ranges from 25 days in *Liomys pictus* to 33 days in *Dipodomys nitratoides*. Longevity: most live only a few months, but up to 5 years have been recorded in the hibernating pocket mouse *Perognathus formosus*.

Kangaroo rats
Genus *Dipodomys*.
SW Canada and USA west of Missouri River to south C Mexico. Bipedal (hind legs long, front legs reduced). Twenty-two species.

Spiny pocket mice
Genus *Heteromys*.
Mexico, C America, northern S America. Quadrupedal. Eleven species including: **Desmarest's spiny pocket mouse** (*H. desmarestianus*).

Spiny pocket mice
Genus *Liomys*.
Mexico and C America S to C Panama. Quadrupal. Five species.

Kangaroo mice
Genus *Microdipodops*.
USA in S Oregon, Nevada, parts of California and Utah. Bipedal (hind legs long, front legs reduced). Two species.

Pocket mice
Genus *Perognathus*.
SE Canada, W USA south to C Mexico. Quadrupedal. Twenty-five species.

SPRINGHARE

Pedetes capensis
Springhare or springhass.
Sole member of family Pedetidae.
Distribution: Kenya, Tanzania, Angola,
Zimbabwe, Botswana, Namibia, South Africa.

Habitat: flood plains, fossil lake beds, savanna,
other sparsely vegetated arid and semiarid
habitats on or near sandy soils.

Size: head-body length
36–43cm (14–17in); tail
length 40–48cm (16–19in);
ear length 7cm (3in); hind
foot length 15cm (6in);
weight 3–4kg (6–9lb).
(Dimensions are similar for
both female and male.)

Coat: upper parts, lower half of ears, basal half
of tail yellow-brown, cinnamon or rufous
brown; upper half of ears, distal half of tail and
whiskers black; underparts and insides of legs
vary from white to light orange.

Gestation: about 77 days.

Longevity: unknown in wild; more than 14.5
years in captivity.

▼ **Springhare postures:** (1) standing,
(2) foraging on all fours. (3) leaping, (4) grooming.

SCATTERED through the arid lands of East and South Africa are numerous grass-covered flood plains and fossil lake beds. After the rainy season grass is superabundant but eventually, however, it is eaten away by large herbivores and for much of the year it is too short and too sparse for large grazers to forage efficiently. The result is an unused food supply or, to use the zoologist's concept, an empty niche. To fill it requires an animal small enough to use the grass efficiently yet large and mobile enought to travel to the grass from areas that can provide shelter from the weather and from predators. These are attributes of the springhare.

However the springhare still faces formidable problems. It is small enough to be killed by snakes, owls and mongooses, and large enough to be attractive to the largest predators, including lions and man. It seems sensible therefore, to interpret many of the animal's specialized physical and behavioral features as adaptations for an arid environment and for the efficient detection and avoidance of predators.

There is little evidence of the springhare's origins. Some people believe that its closest living relatives are the scaly-tailed flying squirrels (Anomaluridae). The springhare actually resembles a miniature kangaroo. Its hindlegs are very long and each foot pad has four toes, each equipped with a hoof-like nail. Its most frequent and rapid type of movement is hopping on both feet. Its tail is slightly longer than its body and helps to maintain balance while hopping. The front legs are only about a quarter of the length of the hindlegs. Its head is rabbit-like, with large ears and eyes and a protruding nose;

sight, hearing and smell are well-developed. When pursued by a predator a springhare can leap 3–4m (10–13ft). When captured it attempts to bite the predator with its large incisors and to rake it with the sharp nails of its powerful hind feet.

Though springhares are herbivorous, they occasionally eat mineral-rich soils and accidentally ingest insects. They are very selective grazers, preferring green grasses high in protein and water.

Springhares are active above ground at all times of night. Normally they forage within 250m (820ft) of their burrows but occasionally they travel as far as 400m (1,300ft). While foraging they are highly vulnerable to predators because they are completely exposed to detection and are far from the safety of their burrows. On nights with a full moon they appear to be particularly vulnerable and move only an average of 4m (13ft) onto the feeding area; by contrast on moonless nights they move on average 58m (190ft). When above ground springhares spend about 40 percent of their time in groups of two to six animals, pre-

sumably because a group is more efficient than an individual at detecting predators.

Springhares spend the hours of daylight in burrows located in well-drained, sandy soils. Burrows lie about 80cm (31in) deep, have 2–11 entrances and vary in length from 10m to 46m (33–151ft). Each burrow is occupied by one springhare or by a mother and an infant. Burrows provide considerable protection against the arid environment and against predators. Some predators, eg snakes and mongooses, can enter burrows however, so springhares often block entrances and passageways with soil after entering. The tunnels and openings in the burrow system provide many escape routes when predators do enter. The absence of chambers and nests within the burrow suggests that springhares do not rest consistently in any one location within the burrow—probably another precaution against predators.

In springhare populations the number of males equals the number of females. There is no breeding season and about 76 percent of the adult females may be pregnant at any one time. Adult females undertake about three pregnancies per year, each resulting in the birth of a single, large well-developed infant.

New-born are well furred and able to see and move about almost immediately, yet they are confined to the burrow and are completely dependent on milk until half grown when they can become completely active above ground. Immature springhares usually account for about 28 percent of all individuals active above ground.

Although the reproductive rate of springhares is surprisingly low, there are two distinctive advantages to be found in the springhare's reproductive strategy. First, the time and energy alloted to the female springhare for reproduction is funneled into a single infant. This results apparently in low infant and juvenile mortality. When the juvenile springhare first emerges from its burrow its feet are 97 percent and its ears 93 percent of their adult size: it is almost as capable of coping with predators and other environmental hazards as a fully grown adult. Second, in having to provide care and nutrition for only one infant the mother is subject to minimal strain. Females that can remain in good physical condition and avoid predators and disease during breeding are most likely to survive to breed again.

Springhares are generally common where they occur, even when they are frequently hunted by man. In the best habitats there may be more than 10 springhares per hectare (4 per acre). However, when arid, ecologically sensitive areas are overgrazed by domestic stock, as occurs in the Kalahari Desert, springhare densities decrease in response to decrease in the supply of food.

Springhares are of considerable importance to man as a source of food and skins. In Botswana springhares are the most prominent wild animal in the human diet. A single band of bushmen may kill more than 200 springhares in one year. However, the springhare can also be a significant pest to agriculture, feeding on a wide variety of crops including corn, peanuts, sweet potatoes and wheat. TMB

◄ **Caught.** An African bushman displays his spoil. No part of a dead Springhare goes to waste. Over 60 percent is eaten, including the eyes, brain and contents of the intestine. The skin is used to make bags, clothing and mats, the long sinew from the tail is used as thread; the fecal pellets are smoked.

▼ **As if a miniature kangaroo,** a Springhare sits upright, showing its short front limbs and powerful hindlegs.

MOUSE-LIKE RODENTS

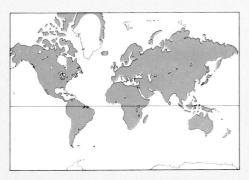

Suborder: Myomorpha
Five families: 264 genera: 1.137 species
Distribution: worldwide except for Antarctica.

Habitat: all terrestrial habitats except snow-covered mountain peaks and extreme high arctic.

Size: head-body length from 4cm (1.6in) in dwarf jerboas to 48cm (19in) in Cuming's slender-tailed cloud rat; weight from 6g (0.2oz) in the Pygmy mouse to 2kg (4.4lb) in Cuming's slender-tailed cloud rat.

Rats and mice
Family: Muridae
One thousand and eighty two species in 241 genera and 15 subfamilies.

New World rats and mice
Subfamily: Hesperomyinae
Three hundred and sixty-six species in 69 genera.

Voles and lemmings
Subfamily: Microtinae
One hundred and ten species in 18 genera.

Old World rats and mice
Subfamily: Murinae
Four hundred and eight species in 89 genera.

Blind mole-rats
Subfamily: Spalacinae
Eight species in 2 genera.

African pouched rats
Subfamily: Cricetomyinae
Five species in 3 genera

African swamp rats
Subfamily: Otomyinae
Thirteen species in 2 genera.

Crested rat
Subfamily: Lophiomyinae
One species.

African climbing mice
Subfamily: Dendromurinae
Twenty-one species in 10 genera.

Root or bamboo rats
Subfamily: Rhizomyinae
Six species in 3 genera.

Madagascan rats
Subfamily: Nesomyinae
Eleven species in 8 genera.

Oriental dormice
Subfamily: Platacanthomyinae
Two species in 2 genera.

MORE than a quarter of all species of mammals belong to the suborder of Mouse-like rodents (Myomorpha). They are very diverse—difficult to describe in terms of a typical member. However, the Norway rat and the House mouse are fairly representative of a very large proportion of the group, both in overall appearance and in the range of size encountered. Like these familiar examples the great majority of mouse-like rodents are small, terrestrial, prolific, nocturnal seedeaters. The justification for believing them to comprise a natural group, derived from a single ancestor separate from the other suborders of rodents, is debatable but lies mainly in two features: the structure of the chewing muscles of the jaw and the structure of the molar teeth.

The way in which the lateral masseter muscle, one of the principal muscles that close the mouth, passes from the lower jaw up through the orbit and from there forwards into the muzzle is not only unique among rodents but is not found in any other mammals. The structure of the teeth is more variable but there is never more than one premolar tooth in front of the molars.

The great majority of mouse-like rodents belong to the family Muridae, the mouse family. The minority groups are the dormice (family Gliridae) and the jerboas and jumping mice (families Dipodidae and Zapodidae). These represent early offshoots that have remained limited in number of species and also somewhat specialized, the dormice being arboreal and (in temperate regions) hibernating, the jerboas being adapted for the desert. The members of the mouse family (murids) have undergone more recent and much more extensive changes (adaptive radiation) beginning in the Miocene period, ie within the last 20 million years. Some of the resultant groups are specialists, for example the voles and lemmings which are adapted to feeding on grass and other tough but abundant vegetation. However, many have remained rather versatile generalists, feeding on seeds, buds and sometimes insects, all more nutritious but less abundant than grass.

The greatest proliferation and diversification of species that has ever taken place in the evolution of mammals has occurred in the mouse family, which has over 1,000 living species. Its members are found throughout the world, in almost every terrestrial habitat. They are often the dominant small mammals in these habitats. Those that most closely resemble the common ancestor of the group are probably the common mice and rats found in forest habitats world-wide, typified by the European Wood mouse and the very similar, although not very closely related, American Deer mouse. These are versatile animals, predominantly seedeaters but capable of using their seedeating teeth to exploit many other foods, such as buds and insects.

From such an ancestor many more specialized groups have arisen, capable of exploiting more difficult habitats and food sources. Most gerbils (subfamily Gerbillinae) have remained seedeaters but have adapted to hot arid conditions in Africa and Central Asia. The hamsters (subfamily Cricetinae) have adapted to colder arid conditions by perfecting the arts of food storage and hibernation; the voles and lemmings (sub-

▶ **The telling face of a rare rat.** The False water rat, a species belonging to the rat and mouse subfamily of Australian water rats and their allies, is seldom seen. A nocturnal creature, it lives on the edge of coastal swamps in northern Australia where it climbs trees rather than swims. Yet for all its obscurity its pointed face, bead eyes and bristling whiskers are instantly recognizable as those of a rat.

▼ **Distinguishing feature** of mouse-like rodents. Both lateral (green) and deep (blue) masseter muscles are thrust forward, providing very effective gnawing action, the deep masseter passing from the lower jaw through the orbit (eye socket) to the muzzle. Shown here is the skull of the muskrat.

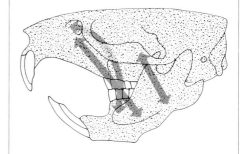

▷ **A more familiar mouse-like rodent.** OVERLEAF The Wood mouse, also known as the Common field or Long-tailed field mouse, can be found across Europe and Central Asia in a wide variety of habitats—woodlands, fields, hedgerows, gardens, even in buildings. Success is the result of versatility. The Wood mouse is extremely energetic, agile and omnivorous. If necessary it can swim.

Zokors or Central Asiatic mole-rats
Subfamily: Myospalacinae
Six species in 1 genus.

Australian water rats
Subfamily: Hydromyinae
Twenty species in 13 genera.

Hamsters
Subfamily: Cricetinae
Twenty-four species in 5 genera.

Gerbils
Subfamily: Gerbillinae
Eighty-one species in 15 genera.

Dormice
Families: Gliridae, Seleviniidae
Eleven species in 8 genera.

Jumping mice and birchmice
Family: Zapodidae
Fourteen species in 4 genera.

Jerboas
Family: Dipodidae
Thirty-one species in 11 genera.

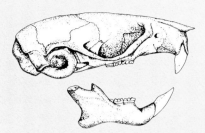

Harvest mouse 1.6cm

European hamster

Libyan jird

Harvest mouse

Norway lemming

Skull and teeth of mouse-like rodents.

Most mouse-like rodents have only three
cheekteeth in each row, but they vary greatly
in both their capacity for growth and the
complexity of the wearing surfaces. The most
primitive condition is probably that found in
the hamsters—low-crowned, with cusps
arranged in two longitudinal rows and no
growth after their initial eruption. The rats
and mice of the subfamily Murinae, typified by
the Harvest mouse, are similar but have three
rows of cusps. The gerbils have high-crowned
but mostly rooted teeth, in which the original
pattern of cusps is soon transformed by wear
into a series of transverse ridges. The voles
and lemmings take this adaptation to a tough,
abrasive diet—in this case mainly grass—even
further by having teeth that continue to grow
and develop roots only late in life or, more
commonly, not at all.

family Microtinae) and the superficially similar African swamp rats (subfamily Otomyinae) have cracked the problem of feeding on grass and similar tough herbage and have thereby opened up new possibilities for expansion by feeding on the vegetation that is dominant over very large areas.

Three independent groups, the blind mole-rats (subfamily Spalacinae), the zokors (subfamily Myospalacinae) and the root rats (subfamily Rhizomyinae) have taken to an underground life, emulating the more distantly related pocket gophers in America (family Geomyidae) and the African mole-rats (family Bathyergidae).

A different kind of diversification, and on a smaller scale, is shown by the rats and mice that succeeded in colonizing the island of Madagascar. There they found a whole variety of vacant habitats waiting to be occupied and therefore the Madagascan rodents (subfamily Nesomyinae), although probably derived from a single colonization, have diversified within that one island in much the same way as have their relatives that stayed behind on the African mainland, with a variety of vole-like, mouse-like and rat-like species.

The indigenous American members of the family (the New World rats and mice, subfamily Hesperomyinae), although very numerous as species (about 366), do not show quite the same degree of diversification as the Old World members. In particular there is only one underground species, the Brazilian shrew-mouse, and it is far less adapted to a mole-like existence than the various Old World mole-rats. This role in the Americas is played by other groups, the pocket gophers in North America and the tuco-tucos in South America.

If one looks at any one area, for example Europe, these specialized groups—mice, voles, hamsters, mole-rats—seem so different that it is tempting to treat them as separate families. This is often done, with considerable justification. However, it is more difficult when the problem is considered worldwide: the whole question of the interrelationships between these groups is still very uncertain.

The ancestral member of the mouse family probably had molar teeth similar to those found in many New World mice and hamsters. These have moderately low-crowned teeth with rounded cusps on the biting surface arranged in two longitudinal rows. The typical Old World mice and rats (subfamily Murinae) appear to have evolved from such an ancestor by developing a more complex arrangement of cusps, forming three rows, while retaining most of the other primitive characters. These two groups are often treated as separate families, the Cricetidae (so-called cricetine rodents) and Muridae (so-called murine rodents) respectively. On this basis several other groups like the gerbils and pouched rats are considered as derivations of, or are included in, the Cricetidae. Others, like the voles, African swamp rats and the various underground groups, have their molar teeth so modified for a rough diet, with high crowns and complex shearing ridges of hard enamel, that their origin is less certain, although most are generally believed to be derived from the cricetid rather than the murid pattern. The African climbing mice (subfamily Dendromurinae) have retained low-crowned, cusped teeth, with additional cusps as in the Murinae but forming a different pattern, suggesting an independent origin from the primitive "cricetid" plan.

These various dental adaptations are reflected in other ways also. By comparison with typical forest mice, species living in dense grassland tend to have shorter tails, legs, feet and ears, as in the voles and hamsters. However, those that live on even more open ground, such as the gerbils in dry steppes and semidesert, tend to retain the "mouse" characters of long hind feet and long tail, adapted to the very fast movement required to escape from predators in the absence of cover. They also retain large ears, and many have enormously enlarged bony *bullae* surrounding the inner ear, acting as resonators and enabling them to detect low frequency sounds carried over great distances and so recognize danger and to communicate over long distances in desert conditions where refuges may be far apart.

Some voles, for example those in the genera *Pitymys* and *Ellobius*, make extensive underground tunnels and in these species the extremities are reduced further and the eyes are very small. These features are also found in the tropical root rats (subfamily Rhizomyinae) and are taken to greater extremes in the more completely underground groups—the zokors (subfamily Myospalacinae) and the blind mole-rats (subfamily Spalacinae).

In ecological terms, most mouse-like rodents would be classified as "r-strategists," ie they are adapted for early and prolific reproduction rather than for long individual life spans ("k-strategists"). Although this applies in some degree to most rodents, the rats and mice show it more strongly and generally than, for example, their nearest relatives, the dormice and the jerboas.   GBC

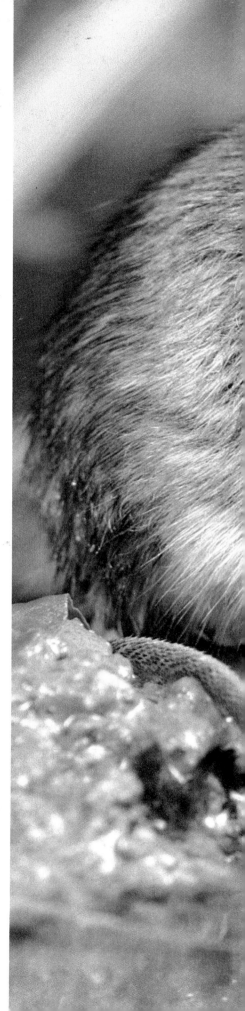

NEW WORLD RATS AND MICE

Subfamily: Hesperomyinae
Three hundred and sixty-six species in 69
genera combined into 13 tribes in 5 groups.
Family: Muridae.
Distribution: N and S America and adjacent
offshore islands.

Habitat: all terrestrial habitats (including
northern forests, tropical forest and savanna)
excluding snow-covered mountain peaks and
extreme high Arctic.

Size: ranges from head-body length 5–8.1cm
(2–3.2in), tail length 3.5–5.5cm (1.4–2.2in),
weight 7g (0.25oz) in the Pygmy mouse to
head-body length 16–28.7cm (6.3–11.3in), tail
length 7.6–16cm (3–6.3in), weight 700g (1lb
11oz) in the South American giant water rat.

Gestation: 20–50 days.
Longevity: maximum in wild 1 year; some live
up to 6 years in captivity.

▼ **A craving for sticks** is one of the most
distinctive features of the North American
woodrats. Most species, including *Neotoma
micropus* seen here, collect sticks and with them
build piles by their nests. A few desert living
species, however, use cactus instead.

THE New World rats and mice are an
array of 366 species divisible into 69
genera, at present distributed from the
northern forests of Canada south through
central Canada, the USA, Central and South
America to the tip of the continent. They
originated in North America where they
descended from the same kind of primitive
rodents as did the hamsters of Europe and
Asia and the pouched rats of Africa (their
nearest Old World relatives today). These
ancestors, cricetine rodents, first appeared
in the Old World in the Oligocene era (about
38 million years ago) and are found in North
America by the mid-Oligocene (about 33
million years ago). They were adapted to
forest environments, but as land dried during the succeeding Miocene era (26–8 million years ago) some became more terrestrial in their habits and developed into forms
recognizable as those of modern New World
rats and mice, the Hesperomyinae. In the
course of their evolution they occupied
many habitats similar to those occupied by
their Old World counterparts. In North
America, for example, there are harvest
mice of the genus *Reithrodontomys* that have
counterparts in the Old World in the mice of
the genus *Micromys*. The North American
wood mice of the genus *Peromyscus* have
counterparts in the Murinae genus *Apodemus*, the wood mice of the Old World.

The rats and mice of South America
developed in a similar way. After the land
bridge between North and South America
was formed during the Pliocene era (7–3
million years ago) several stocks of primitive
North American cricetine rodents, probably
animals equipped for climbing and in other
ways adapted for living in forests, moved
into South America where they underwent
an extensive radiation, in the first instance
for occupying spacious grassland habitats.
Subsequently they occupied many habitats,
some of which are not found occupied by
rodents in other parts of the world. A
particularly interesting feature of this radiation in South America is a consequence of
the absence in much of the continent of
members of the orders Insectivora (insectivores) and Lagomorpha (rabbits, hares and
pikas). Many South American rats and mice
have evolved in their forms and activities so
that they somewhat resemble shrews, water
shrews, moles and rabbits: the South American species include *Blarinomys breviceps*,
the Shrew mouse, the genus *Notiomys*, the
mole mice, the genus *Rheomys*, the fish-
eating mice, and the genus *Reithrodon*, the
rabbit rat.

This account to some extent simplifies
a very complicated evolutionary history.
Phases of change in South America have
occurred sporadically, often in response to
changes in environment. For example, during the Pleistocene era (2 million–20,000
years ago) zones of forest expanded and
contracted, sometimes creating isolated
groups which then developed along different
lines. The resulting mass of species and
genera creates considerable problems for the
natural historian trying to classify all the
animals belonging to the New World rats
and mice.

Though New World rats and mice include
a vast range of forms they are all small: the
largest living species are no longer, in head-
body length, than 30cm (about 1ft). Their
tails vary in length according to the extent of
adaptation for arboreal life. Highly arboreal
animals have long tails (usually they are
longer than half the animal's length), terrestrial forms the shortest. Most forms have a
brown back and white belly, but some
depart from this scheme and exhibit very
attractive coat colors, for example the Chinchilla mouse which has a strongly contrasting combination of buff to gray back and
white belly. The tails of most species have
few hairs, but the exceptional species have
well-haired tails, for example the Bushy
tailed woodrat. All exhibit the basic arrangement of teeth common to highly
evolved rodents, viz. three molars on each
side of the jaw separated, by a distinct gap
from a pair of incisors. The incisors grow
continuously. They have enamel on their
anterior surfaces which enables a sharp
cutting edge to be maintained.

In adapting for life in different environments New World rats and mice have

▲ **The unmatched deer mouse.** Of the 49 species of deer or white-footed mice of North and Central America *Peromyscus maniculatus* is most widespread. It can be found in alpine areas, forests, woodlands, grasslands and desert—anywhere except in moist habitats.

▶ **Rubbery droplets.** Mice reproduce rapidly, by a combination of short estrous cycle, short gestation and long breeding season. The average litter of mice is larger than the average mammal litter. Female deer mice produce on average 3.4 young per litter, in this as it were premature state, after a gestation of about three weeks. A similar period is required for weaning.

developed several modifications to the basic rat or mouse appearance. Burrowing forms have short necks, short ears, short tails and long claws. Forms adapted for an aquatic life often have webbed feet (for example the marsh rats of the genus *Holochilus*) or a fringe of hair on the hind feet which increases the surface areas of each foot (for example the fish-eating rats, genus *Daptomys*). In forms even more developed for aquatic life the external ear is reduced in size or even absent (as in the fish-eating rats of the genus *Anotomys*). Species that live in areas of semidesert or desert often have a light-colored back, and elsewhere the color of the back is often a shade that matches the surrounding soil. This enables exposed terrestrial New World rats and mice to reduce the threat they face from such predators as owls.

Above the generic level the classification of New World rats and mice is controversial, and subject to change. Genera can be combined to form 13 tribes, which in turn can be thought of as belonging to six groups, though the groups are of varying validity (see table).

The **White-footed mice** and their allies (tribe Peromyscini) consists of 10 genera, the genus *Peromyscus* containing the most species which are popularly known as white-footed mice or deer mice. In general they are nocturnal and eat seeds. Within the genus species show an array of adaptations. Head-body sizes range from 8–17cm (3–6.7in) and tail lengths from 4–20.5cm (1.6–8in). In terrestrial forms tails are shorter than the head-body length, in arboreal forms they are longer. Those that inhabit environments with great climatic fluctuations often produce many young. Species that occupy more stable habitats generally have low litter sizes and increased longevity, and also a relatively large brain. Such species include the California mouse, with a mean litter size of two and a brain weight equaling 2.9 percent of the body weight. An example of a species with a small brain and large average litter size is the Deer mouse, with a mean litter size of five and a brain weight that is 2.4 percent of the body weight.

Where several species of *Peromyscus* co-exist the habitat is subdivided in such a way that there is a segregation with respect of microhabitat. When three species co-exist, forming a guild, they are usually of three different sizes, with a small member having more versatile feeding habits and microhabitat requirements and often producing many young. The medium-sized member of the guild is usually somewhat more restricted in its habitat requirements and presumably takes a narrower range of food sizes. The largest member of the guild will show the lowest fecundity, the greatest longevity, will have the most restricted microhabitat requirements, and is often specialized for feeding on larger seeds, nuts and fruits. A typical example is the three species complex in California: *P. truei*, *P. maniculatus* and *P. californicus*.

Closely related to the genus *Peromyscus* is the Volcano mouse which is a burrow user and is quite terrestrial in its habits. It occurs at elevations of 2,600m to 4,300m (8,530–14,100ft) and there is a birth peak in July and August.

▲ **A home in the desert.** Desert woodrats that have access to rocks sometimes prefer to make their homes in crevices rather than in specially built nests. Woodrats are solitary: each has to keep watch for himself.

▶ **The Bushy-tailed woodrat,** whose range reaches into northwest Canada, is the northernmost of the woodrats. This may account for the animal's total insulation.

THE 6 GROUPS AND 13 TRIBES OF NEW WORLD RATS AND MICE

North American Neotomine-Peromyscine group

White-footed mice and their allies
Tribe Peromyscini.
Genera: **white-footed mice** or **deer mice** (*Peromyscus*), 49 species, from N Canada (except high Arctic) S through Mexico to Panama; species include **California mouse** (*P. californicus*), **White-footed mouse** (*P. Leucopus*). **Harvest mice** (*Reithrodontomys*), 19 species, from W Canada and USA S through Mexico to W Panama; species include **Western harvest mouse** (*R. megalotis*). *Habromys*, 4 species, from C Mexico S to El Salvador. *Podomys floridanus*, Florida peninsula. **Volcano mouse** (*Neotomodon alstoni*), montane areas of C Mexico. **Grasshopper mice** (*Onychomys*), 3 species, SW Canada, NW USA S to N C Mexico. *Osgoodomys bandaranus*, W C Mexico. *Isthmomys*, 2 species, Panama. *Megadontomys thomasi*, C Mexico. **Golden mouse** (*Ochrotomys nuttalli*), SW USA.

Woodrats and their allies
Tribe Neotomini.
Genera: **woodrats** (*Neotoma*), 19 species, USA to C Mexico. **Allen's woodrat** (*Hodomys alleni*), W C Mexico. **Magdalena rat** (*Xenomys nelsoni*), W C Mexico. **Diminutive woodrat** (*Nelsonia neotomodon*), C Mexico.

Pygmy mice and brown mice
Tribe Baiomyini.
Genera: **pygmy mice** (*Baiomys*), 2 species, from SW USA S to Nicaragua. **Brown mice** (*Scotinomys*), 2 species. Brazil, Bolivia, Argentina.

Central American climbing rats
Tribe Tylomyini.
Genera: **Central American climbing rats** (*Tylomys*), 7 species, S Mexico to E Panama. **Big-eared climbing rat** (*Ototylomys phyllotis*), Yucatan peninsula of Mexico S to Costa Rica.

Nyctomyine group

Vesper rats
Tribe Nyctomyini.
Genera: **Central American vesper rat** (*Nyctomys sumichrasti*), S Mexico S to C Panama. **Yucatan vesper rat** (*Otonyctomys hatti*), Yucatan peninsula of Mexico and adjoining areas of Mexico and Guatemala.

Thomasomyine-Oryzomyine group

Paramo rats and their relatives
Tribe Thomasomyini.
Genera: **paramo rats** (*Thomasomys*), 25 species, areas of high altitude from Colombia and Venezuela S to S Brazil and NE Argentina. **South American climbing rats** (*Rhipidomys*), 7 species, low elevations from extreme E Panama S across northern S America to C Brazil. **Colombian forest mouse** (*Chilomys instans*), high elevations in Andes in W Venezuela S to Colombia and Ecuador. *Aepomys*, 2 species, high elevations in Andes in Venezuela, Colombia, Ecuador. **Rio de Janeiro rice rat** (*Phaenomys ferrugineus*), vicinity of Rio de Janeiro.

Rice rats and their allies
Tribe Oryzomini.
Genera: **rice rats** (*Oryzomys*), 57 species, SE USA S through C America and N S America to Bolivia and C Brazil. **Brazilian spiny rat** (*Abrawayaomys ruschii*), SE Brazil. **Galapagos rice rats** (*Nesoryzomys*), 5 species, Galapagos archipelago of Ecuador. **Giant rice rats** (*Megalomys*), 2 species (recently extinct), West Indies (Martinique and Santa Lucia). **Ecuadorean spiny mouse** (*Scolomys melanops*), Ecuador. **False rice rats** (*Pseudoryzomys*), 2 species, Bolivia, E Brazil, N Argentina. **Red-nosed mouse** (*Wiedomys pyrrhorhinos*), E Brazil. **Bristly mice** (*Neacomys*), 3 species, E Panama across lowland S America to N Brazil. **South American water rats** (*Nectomys*), 2 species, lowland S America to NE Argentina. **Brazilian arboreal mouse** (*Rhagomys rufescens*), SE Brazil.

Akodontine-Oxymycterine group

South American field mice
Tribe Akodontini.
Genera: **South American field mice** (*Akodon*), 33 species, found in most of S America (from W Colombia to Argentina). *Bolomys*, 6 species, montane areas of SE Peru S to Paraguay and C Argentina. *Microxus*, 3 species, montane areas of Colombia, Venezuela, Ecuador, Peru. **Cane mice** (*Zygodontomys*), 3 species, Costa Rica and N S America.

Burrowing mice and their relatives
Tribe Oxymycterini.
Genera: **burrowing mice** (*Oxymycterus*), 9 species, SE Peru, W Bolivia E over much of Brazil and S to N Argentina. **Andean rat** (*Lexonus apicalis*), SE Peru and W Bolivia. **Shrew mouse** (*Blarinomys breviceps*), E C Brazil. **Mole mice** (*Notiomys*), 6 species, Argentina and Chile. **Mount Roraima mouse** (*Podoxymys roraimae*), at junction of Brazil, Venezuela, Guyana. *Juscelimomys candango*, vicinity of Brasilia.

Ichthyomyine group

Fish-eating rats and mice
Tribe Ichthyomyini.
Genera: **fish-eating rats** (*Ichthyomys*), 3 species, premontane habitats of Venezuela, Ecuador, Peru. *Daptomys*, 3 species, French Guiana, premontane areas of Peru and Venezuela. **Water mice** (*Rheomys*), 5 species, C Mexico S to Panama. **Ecuadorian fish-eating rat** (*Neusticomys monticolus*), Andes region of S Colombia and N Ecuador.

Sigmodontine-Phyllotine-Scapteromyine group

Cotton rats and marsh rats
Tribe Sigmodontini.
Genera: **marsh rats** (*Holochilus*), 4 species, most of lowland S America. **Cotton rats** (*Sigmodon*), 8 species in S USA, Mexico, C America, NE S America as far S as NE Brazil; species include **Hispid cotton rat** (*S. hispidus*).

Leaf-eared mice and their allies
Tribe Phyllotini.
Genera: *Graomys*, 3 species, Andes of Bolivia S to N Argentina and Paraguay. *Andalgalomys*, 2 species, Paraguay and NE Argentina. *Galenomys garleppi*, high altitudes in S Peru, W Bolivia, N Chile. *Auliscomys*, 4 species, mountains of Bolivia, Peru, Chile and Argentina. **Puna mouse** (*Punomys lamminus*), montane areas of S Peru. **Rabbit rat** (*Reithrodon physodes*), steppe and grasslands of Chile, Argentina, Uruguay. **Vesper mice** (*Calomys*), 8 species, most of lowland S America. **Chinchilla mouse** (*Chinchillula sahamae*), high elevations S Peru, W Bolivia, N Chile, Argentina. **Chilean rat** (*Irenomys tarsalis*), N Argentina, N Chile. **Andean mouse** (*Andinomys edax*), S Peru, N Chile. **Highland desert mouse** (*Eligmodontia typus*), S Peru, N Chile, Argentina. **Leaf-eared mice** (*Phyllotis*), 12 species, from NW Peru S to N Argentina and C Chile. **Patagonian chinchilla mice** (*Euneomys*), 4 species, temperate Chile and Argentina. **Andean swamp rat** (*Neotomys ebriosus*), Peru S to NW Argentina.

Southern water rats and their allies
Tribe Scapteromyini.
Genera: **red-nosed rats** (*Bibimys*), 3 species, SE Brazil W to NW Argentina. **Argentinean water rat** (*Scapteromys tumidus*), SE Brazil, Paraguay, E Argentina. **Giant South American water rats** (*Kunsia*), N Argentina, Bolivia, SE Brazil.

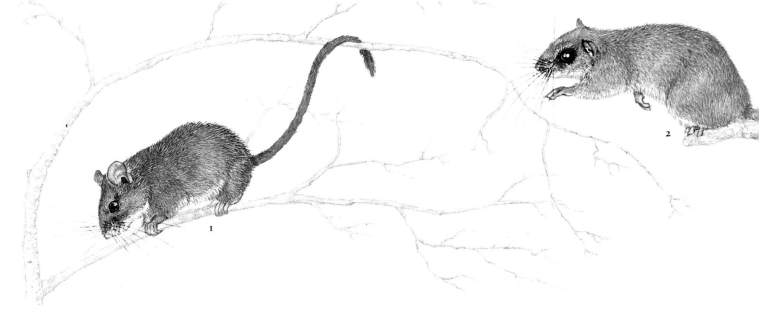

The size of harvest mice varies considerably from 6–14cm (2–5.7in) in head-body length, from 6.5–9.5cm (2.5–3.7in) in tail length. The North American species tend to be smaller than the Central American species, rarely weighing more than 15g (0.5oz). The Western harvest mouse is typical of the grassland areas of western North America. A nocturnal seed or grain eater it constructs globular nests approximately 24cm (9in) off the ground in tall grass.

Grasshopper mice are specialized forms that inhabit arid and semiarid habitats. In size they range from 9–13cm (3.5–5.1in) in head-body length, though their tail is rather short, 3–6cm (1.2–2.4in). These mice feed on insects and small vertebrates. They construct burrows in which they live as pairs during the breeding season. Litters disperse upon weaning and the duration of pairing in subsequent seasons is unknown. These rodents are well known for their high-pitched squeak (usually above 20kHz) which may function as a spacing call. Both sexes emit these calls, but males call more frequently than females. In nature the burrows are widely spaced.

The Golden mouse is confined to the moderately wet wooded habitats of the southeastern USA. The distinctive golden brown color of its back contrasts sharply with the white belly. This is an extremely arboreal form which builds a complex leafy nest in tangles of vines.

The **woodrats** and their allies (tribe Neotomini) are rat-sized rodents, varying in color from dark buff on the back to paler shades on the belly. In general they eat a wide range of foods but some species are highly adapted for feeding on the green parts of plants; indeed, *Neotoma stephensi* feeds almost entirely on the foliage of juniper trees. Many species are adapted for living in and around crevices or cracks in rocky outcrops, others construct burrows. But

whether burrowers or rock-dwellers all have the habit of creating mounds of sticks and other detritus in the vicinity of the nest hole or crack. Desert species often utilize such items as pads of spiny cacti. This habit and the transportation of materials to the mound have earned them the name "pack rats" in some parts of their range. Each stick nest tends to be inhabited by a single adult individual, but individuals may visit neighboring nests; in particular males apparently visit females when they are receptive.

The Magdalena rat occurs in an extremely restricted area of tropical deciduous forests in western Mexico in the states of Jalisco and Colima, where it may have an extended season of reproduction. It is a small nocturnal woodrat with excellent climbing ability.

The Diminutive woodrat is found in the mountainous areas of central and western Mexico where it is known to shelter in crevices of rock outcroppings at elevations exceeding 2,000m (6,500ft).

Pygmy mice and **brown mice** (tribe Baiomyini) are the smallest New World rodents. Pygmy mice have a relatively small home range (often less than 900sq m, about 9,700sq ft) compared with a larger seed-eating rodent such as *Peromyscus leucopus*, which has a home range of 1.2–2.8ha (2.9–6.9 acres). Pygmy mice are seedeaters which inhabit a grass nest, usually under a stone or log. They may be monogamous while pairing and rearing their young.

Brown mice are small subtropical mice which employ a high-pitched call apparently to demarcate territory. Males produce this call more frequently than females.

The **Central American climbing rats** (tribe Tylomyini; two genera) are associated with water edge forested habitat at lower elevations. They are extremely arboreal, strictly nocturnal and feed primarily on fruits, seeds and nuts. The **Big-eared climbing rat** is smal-

▲ **Representatives from six tribes** of New World rats and mice. (1) A South American climbing rat (genus *Rhipidomys*; tribe Thomasomyini). (2) A Central American vesper rat (genus *Nyctomys*; tribe Nyctomini). (3) A Central American climbing rat (genus *Tylomys*; tribe Tylomyini). (4) A pygmy mouse (genus *Baiomys*; tribe Baiomyini). (5) A white-footed or deer mouse (genus *Peromyscus*; tribe Peromyscini). (6) A woodrat or pack rat, carrying a bone (genus *Neotoma*; tribe Neotomini).

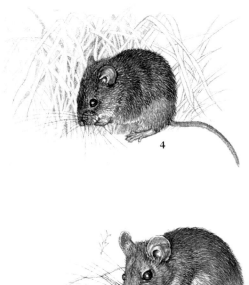

ler than the other species and again is also confined to lowland tropical forests. Although adapted for living in trees it also forages on the ground. It produces small litters of fully haired young whose eyes open after about six days. These well-developed young are a unique departure from the normal pattern found in New World rats and mice which usually produce hairless young whose eyes open after ten to twelve days. The gestation of the Big-eared climbing rat is approximately 50 days. Most New World rats and mice have gestation periods of approximately three weeks.

The tribe Nyctomyini contains just two species of **vesper mice**. The Central American vesper rat is a specialized nocturnal, arboreal fruit-eater which builds nests in trees and has a long tail and large eyes.

The Yucatan vesper rat, also highly arboreal, is a relict species in the Yucatan peninsula. It probably was once more broadly distributed under different climatic conditions but became isolated in the Yucatan during one of the drying cycles in the Pleistocene era (2 million–20,000 years ago).

Paramo rats and their allies (tribe Thomasomyini) are distributed throughout South America. In the Andes many species of paramo rats are adapted for life at elevations exceeding 4,000m (13,000ft). Otherwise they are almost always confined to forests or to forests along rivers. They are nocturnal and fruit-eating, but their biology is very poorly known. Litters of two to four young have been recorded, but in general their reproductive potential is considered to be quite low.

The South American climbing rats are likewise adapted for life in trees. They are also nocturnal, and feed upon fruits, seeds,

fungi and insects. Their litter size is small: for *R. mastacalis* two or three young per litter have been recorded.

Rice rats and their allies (tribe Oryzomini) are an assemblage with two tendencies: a retention of the long tail as an adaptation for a life spent partly in trees or, alternatively, the exploitation of moist habitats, culminating in semiaquatic adaptations. Many of the species are adapted to varying altitudes. For example, in northern Venezuela *Oryzomys albigularis* occurs above elevations of 1,000m (about 3,300ft); at lower elevations it is replaced by *O. capito*. When two species of rice rats occur in the same habitat one is often more adapted to life in trees than the other. Such is the case when *O. capito*, a terrestrial species, occurs with *O. bicolor*, a species adapted for climbing.

The South American water rat (one of the two species of *Nectomys*) is semiaquatic and the dominant aquatic rice rat over much of South America. *Oryzomys palustris*, found on the gulf coast of northern Mexico north to the coast of southern Maryland (USA), is also semiaquatic in its habits. This rice rat has a wide-ranging diet though at certain times of the year over 40 percent of its food may consist of snails and crustaceans. Across much of its range it has an extended breeding season from February to November, and is able to produce four young at 30-day intervals with the result that it can rapidly become a serious agricultural pest.

The small bristly mice or spiny rice rats, which have a distinctive spiny coat, are nocturnal and eat seeds. *Neacomys tenuipes*, in northern South America, can exhibit wide variations in population density.

Rice rats have excelled at colonizing islands in the Caribbean and the Galapagos Islands. Many island species of *Oryzomys* are currently threatened with extinction, due to human activities. The introduction of the Domestic house cat and murine rats and mice has had a severe impact on the Galapagos rice rats.

South American field mice (tribe Akodontini) are adapted for foraging on the ground and many are also excellent burrowers.

Akodon species have radiated to fill a variety of habitats. In general the species are omnivorous, including in their diets green vegetation, fruits, insects and seeds. Most species of *Akodon* are adapted to moderate to high elevations. *Akodon urichi* typifies the adaptability of the genus. Since it tends to be active both in daytime and at night it is terrestrial and eats an array of food items including fruits, seeds and insects.

Members of the genus *Bolomys* are closely allied to *Akodon* but are more specialized in adaptations for terrestrial existence. They have short ears, a short tail and a body form very similar to that of the Field vole. Members of the genus *Microxus* are similar to *Bolomys* in appearance but their eyes show even further reduction in size. They are strongly adapted for a terrestrial burrowing life as evidenced by a short neck, reduced external ear length, and reduction in eye size.

Cane mice are widely distributed in South America, taking the place of *Akodon* at low elevations in grasslands and bushlands. In grasslands it sometimes constructs runways which can be visible to the human observer. Cane mice eat a considerable quantity of seeds and do not seem to be specialized for processing green plant food. In grassland habitats subject to seasonal fluctuations in rainfall the cane mice may show vast oscillations in population density. In the llanos of Venezuela they show population explosions when productivity of the grasslands is exceptionally high, enabling them to harvest seeds and increase their production of young. Densities can vary from year to year from a high of 15 per ha (6 per acre) to a low of less than one per ha.

The **burrowing mice** and their relatives (tribe Oxymycterini) are closely allied to the South American field mice. They are distinguished by adaptations for burrowing habits and a trend within the species assemblage towards specialization for a more insectivorous food niche. Rodents specialized in this way show a reduction in the size of their molar teeth, and an elongate snout. Longer claws aid in excavating the soil for soil arthropods, larvae and termites.

Burrowing mice of the genus *Oxymycterus* are long-clawed with a short tail. Their molar teeth are weak, their snout long. These features correlate with eating insects. A grass nest is made in a burrow system where the young, two or three, are born. The small litter size of some species of

▼ **Representatives of seven tribes** of New World rats and mice. (**1**) A South American field mouse (genus *Akodon*; tribe Akodontini) grooming its tail. (**2**) A fish-eating rat (genus *Ichthyomys*; tribe Ichthyomyini). (**3**) The Argentinian water rat (*Scapteromys tumidus*; tribe Scapteromyini). (**4**) A cotton rat (genus *Sigmodon*; tribe Sigmodontini) attempting to remove an egg. (**5**) A mole mouse (genus *Notiomys*; tribe Oxymycterini). (**6**) A South American water rat (genus *Nectomys*; tribe Oryzomini). (**7**) A leaf-eared mouse (genus *Phyllotis*; tribe Phyllotini).

Oxymycterus may correlate with a lowered metabolic rate as an adaptation to termite feeding. Mammals specializing for feeding on termites often have a metabolic rate lower than would be expected, probably because the high chitin content of insects gives a lower return in net energy than other protein and carbohydrates. Lower metabolic rates in termite- and ant-feeding forms is an outcome of adjustment to the rate of energy return. A side effect of this is a reduced reproductive capacity, reflected in smaller litter sizes.

The Shrew mouse (one species), represents an extreme adaptation for a burrowing way of life. Its eyes are reduced in size and its ears are so short that they are hidden in the fur. It burrows under the litter of the forest floor and can construct a deep, sheltering burrow. Its molar teeth are very reduced in size, a characteristic that indicates adaptation for a diet of insects.

Mole mice are widely distributed in Argentina and Chile and exhibit an array of adaptations reflected in their exploitation of both semiarid steppes and wet forests. Some species are adapted to higher elevation forests, others to moderate elevations in central Argentina. They have extremely powerful claws which may exceed 0.7cm (0.3in) in length. The name mole mice derives from their habits of burrowing and spending most of their life underground.

The **fish-eating rats and mice** (tribe Ichthyomyini) represent an interesting group in that they have specialized for a semiaquatic life and have altered their diets so as to feed on aquatic insects, crustaceans, mollusks and fish.

Little is known of the details of the biology of fish-eating rats and mice other than that they occur on or near high elevation fresh water streams and exploit small crustaceans, arthropods and fish as their primary food sources.

Fish-eating rats of the genus *Ichthyomys* are among the most specialized of the genera. They have a head-body length that almost attains 33cm (9in), which is just exceeded by the tail. Their fur is short and thick, their eyes and ears are reduced in size, and their whiskers are stout. A fringe of hairs on the toes of the hind feet aids in swimming, and the toes are partially webbed.

They resemble a large water shrew or some of the fish-eating insectivores of West Africa and Madagascar.

Fish-eating rats of the genus *Daptomys* are similar to *Ichthyomys* and are distributed disjunctly in the mountain regions of Venezuela and Peru. The Ecuadorian fish-eating rat is the least specialized for aquatic life.

Water mice are slightly smaller than *Ichthyomys* and rarely exceed 19cm (7in) in head-body length. They occur in mountain streams of central America and Colombia and are known to feed on snails and possibly fish. In the webbing of their hind feet and in the hairs on the outer sides of their feet they are similar to *Ichthyomys*.

Cotton rats and **marsh rats** (tribe Sigmodontini) are united by a common feature, namely, folded patterns of enamel on the molars which when viewed from above tend to approximate to an "S" shape. They exhibit a range of adaptations; the species referred to as marsh rats are adapted for a semiaquatic life, whereas the cotton rats are terrestrial. Both groups, however, feed predominantly on herbaceous vegetation.

The genus *Holochilus* contains the web-footed marsh rats, of which two species (of four) are broadly distributed in South America. The underside of the tail has a fringe of hair, an adaptation to swimming. They build a grass nest near water, sufficiently high to prevent flooding, which may exceed 40cm (15.7in) in diameter. In the more southern parts of their range in temperate South America breeding tends to be confined to the spring and summer (ie September–December).

Cotton rats are broadly distributed from the southern USA to northern South America. In line with their adaptations for terrestrial life the tail is always considerably shorter than head-body length. Cotton rats are active during both day and night, and although they eat a wide range of food they consume a significant amount of herbs and grasses during the early phases of vegetational growth after the onset of rains. A striking feature of the reproduction of the Hispid cotton rat is that its young are born fully furred and their eyes open within 36 hours of their birth. It has a very high reproductive capacity and although it produces well-developed (precocial) young, the gestation period is only 27 days. The litter size is quite high, from five to eight, with 7.6 as an average. The female is receptive after giving birth and only lactates for approximately 10–15 days. Thus the turn-around time is very brief and a female can produce a litter every month during the breeding season. In agricultural regions this rat can become a serious pest.

Leaf-eared mice and their allies (tribe Phyllotini) are typified by the genera *Phyllotis* and *Calomys*. *Calomys* (vesper mice) includes a variety of species distributed over most of South America. They have large ears (as do *Phyllotis*) and feed primarily on

▲ **Equilibrium on a twig.** American harvest mice are nimble, agile rodents, adept at climbing. Their nests are usually found above ground, in grasses, low shrubs, or small trees; they are globular in shape, woven of grass.

▶ **A wolf among mice.** All three insect-eating species of grasshopper mice stand erect to utter repeated shrill sounds, each lasting about a second. They occur when one mouse detects another nearby, or just as a mouse is about to kill.

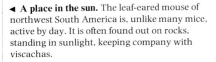

◀ **A place in the sun.** The leaf-eared mouse of northwest South America is, unlike many mice, active by day. It is often found out on rocks, standing in sunlight, keeping company with viscachas.

plant material; arthropods are an insignificant portion of their diet. The genus *Phyllotis* (leaf-eared mice) is composed of several species, most of which occur at high altitudes. They are often active in the day and may bask in the sun. They feed primarily on seeds and herbaceous plant material. The variation in form and the way in which several species of different size occur in the same habitat are reminiscent of *Peromyscus*, the deer mice (tribe Peromyscini).

The Rabbit rat is of moderate size and has thick fur adapted to the open country plains of temperate Chile, Argentina and Uruguay. It feeds primarily on herbaceous plant material and may be active throughout both day and night.

The Highland desert mouse is one of the few South American rodents specialized for semiarid habitats. Its hind feet are long and slender, resulting in a peculiar gait where the forelimbs simultaneously strike the ground followed by a power thrust from the hind legs where the forelimbs leave the ground. The kidneys of this species are very efficient at recovering water; indeed, it can exist for considerable periods of time without drinking, being able to derive its water as a by-product of its own metabolism.

Patagonian chinchilla mice are distributed in wooded areas, from central Argentina south to Cape Horn. The Puna mouse is found only in the altiplano of Peru. This rodent is the most vole-like in body form of any South American rodent. It is active both day and night and its diet is apparently confined to herbaceous vegetation. The Chilean rat is an inhabitant of humid temperate forests. This is an extremely arboreal species. It may be a link between the phyllotines and the oryzomyine rodents or rice rats.

The Andean marsh rat, occurs at high elevations near streams and appears to occupy a niche appropriate for a vole.

The **southern water rats** and their allies (tribe Scapteromyini) are adapted for burrowing in habitats by or near rivers. The Argentinean water rat is found near rivers, streams and marshes. It has extremely long claws and can construct extensive burrow systems.

The giant South American water rats prefer moist habitats and have considerable burrowing ability. *Kunsia tomentosus* is the largest living New World rat with a head-body length that may reach 28cm (11in) and a tail length of up to 16cm (6.3in).

Red-nosed rats are small burrowing forms allied to the larger genera, but whose biology and habits are poorly known.   JFE

VOLES AND LEMMINGS

Subfamily: Microtinae
One hundred and ten species in 18 genera.
Family: Muridae.
Distribution: N and C America, Eurasia, from
Arctic S to Himalayas, small relic population in
N Africa.

Habitat: burrowing species are common in
tundra, temperate grasslands, steppe, scrub,
open forest, rocks; 5 species are aquatic or
arboreal.

Size: most species head-body
length 10–11cm (4–4.5in),
tail length 3–4cm
(1.2–1.6in), weight 17–20g
(0.6–0.7oz).

Gestation: 16–28 days.

Longevity: 1–2 years.

► **On the nibble:** a Bank vole. Bank voles live
in Europe and Central Asia in woods and
scrubs, in banks and swamps, usually within a
home range of about 0.8ha (2 acres).

► **A summer scene in the arctic tundra.**
Collared lemmings (including these Arctic
lemmings) live along the arctic edge of
northern Europe, Siberia and North America.
In summer, coats are brown and heavy, but
with the approach of winter, coats turn white
and on the front paws the third and fourth
claws grow a second prong.

THE fascinating lemmings and voles that
belong to the Muridae subfamily Microtinae have two features of particular interest.
Firstly their populations expand and contract considerably, in line with cyclical
patterns. This has made them the most
studied subfamily of rodents (and the basis
of much of our understanding about the
population dynamics of small mammals).
Secondly, though they neither hibernate
like such larger mammals as ground squirrels, nor can rely on a thick layer of fat like
bears, most voles and lemmings live in
habitats covered by snow for much of the
year. They are able to survive thanks to their
ability to tunnel beneath the snow where
·they are insulated from extreme cold.

Vole and lemmings are small, thickset
rodents with bluntly rounded muzzles and
tails usually less than half the length of their
bodies. Only small sections of their limbs are
visible. Their eyes and ears tend to be small
and in lemmings the tail is usually very
short. Coat colors vary not only between
species but often within them. Lemmings'
coats are especially adapted for cold temperatures: they are long, thick and waterproof.
The Collared lemming is the only rodent
that molts to a complete white coat in
winter.

Some species display special anatomical
features: the claws of the first digit of the
Norway lemming are flattened and enlarged
for digging in the snow while each fall the
Collared lemming grows an extra big claw
on the third and fourth digits of its forelegs
and sheds them in spring. Muskrats have
long tails and small webbing between toes
which assist in swimming. The mole lemmings, adapted for digging, have a more
cylindrical shape than other species and
their incisors, used for excavating, protrude
extremely.

Adult males and females are usually the
same color and approximately the same size,
though the color of juveniles' coats may

differ from the adults'. Although most adult voles weigh less than 100g (3.5oz) the muskrat grows to over 1,400g (50oz). The size of the brain, in relation to body size, is lower than average for mammals.

Smell and hearing are important, well-developed senses, able to respond, respectively, to the secretions that are used to mark territory boundaries, indicate social status and perhaps to aid species recognition, and to vocalizations (each species has a characteristic suite of calls). Calls can be used for alarm, to threaten or as part of courtship and mating. Brandt's voles, which live in large colonies, sit up on the surface and whistle like prairie dogs.

Microtines, especially lemmings, are widely distributed in the tundra regions of the northern hemisphere where they are the dominant small mammal species. Their presence there is the result of recolonization since the retreat of the last glaciers. Voles are also found in temperate grasslands and in the forests of North America and Eurasia.

Because the Pleistocene era (2 million-10,000 years ago) has yielded a rich fossil record of microtine skulls, much of the taxonomy of this subfamily is based on the kind and structure of teeth. They are also used to distinguish the microtines from other rodents. All microtine teeth have flattened crowns and prisms of dentine surrounded by enamel. There are 12 molars (3 on each side of the upper and lower jaws) and 4 incisors. Species are differentiated by the particular pattern of the enamel of the molars. Dentition has not proved sufficient

for solving all taxonomic questions, and some difficulties remain both in delimitating species and defining genera. It is often possible to distinguish the species of live specimens by using general body size, coat color or length of tail. Subdivision of species into many subspecies (not listed here) reflects geographic variation in coat color and size in widely distributed species. *Microtus*, the largest genus (accounting for nearly 50 percent of the subfamily), is a heterogeneous collection of species.

Many species have widely overlapping ranges. For example, there are six species in the southern tundra and forest of the Yamal Peninsula in the USSR: two lemmings and four voles. Each can be differentiated by its habitat preference or diet. The ranges of Siberian lemmings and Collared lemmings overlap extensively but Collared lemmings prefer upland heaths and higher and drier tundra whereas Siberian lemmings are found in the wetter grass-sedge lowlands. In the northwest USA the Long-tailed vole is actively excluded from grasslands by the dominant Montane vole.

Voles and lemmings are herbivores, and usually eat the green parts of plants, but some species prefer bulbs, roots, mosses or even pine needles. The muskrat occasionally eats mussels and snails. Diet usually changes with the seasons and varies according to location, reflecting local abundances of plants. Species living in moist habitats, such as lemmings and the Tundra vole, prefer grasses and sedges while those species in drier habitats such as the Collared lemmings prefer herbs. But animals select their food to some extent; diets do not just simply mimic vegetation composition.

Voles and lemmings can be found foraging both day and night, although dawn and dusk might be preferred. They obtain food by grazing, or digging for roots; grass is often clipped and placed in piles in their runways. Some cache food in summer and fall, but in winter, when the snow cover insulates the animals that nest underneath the surface, food is obtained by burrowing and animals also feed on plants at ground surface. The Northern red-backed vole feeds on, among other items, berries, so in summer has to compete for them with birds. In winter, when the bushes are covered with snow, this vole can burrow to reach the berries. So only during the spring thaw is the animal critically short of food. In winter the Sagebrush vole utilizes the height of snow packs to forage on shrubs it normally cannot reach.

Most microtines, for example all species of

microtus, the Florida water rat and the Steppe lemming, have continuously growing molars and can chew more abrasive grasses and a greater quantity of grass than species with rooted molars. In tundra and grasslands voles and lemmings help the recycling of nutrients by eliminating waste products and clipping and then storing food below ground.

The life span of microtines is short. They reach sexual maturity at an early age and are very fertile. Mortality rates, however, are high: during the breeding season, of the animals alive one month only 70 percent are alive the next. The age of sexual maturity can vary considerably. The females of some species may become sexually mature only two or three weeks from birth. Males take longer, usually six to eight weeks. The Common vole has an extraordinarily fast development time. Females have been observed coming into heat while still suckling and may be only five weeks old at the birth of their first littler. In species that breed only during the summer, young born early will probably breed that summer while later litters may not become sexually mature until the following spring.

The length of the breeding season is highly variable but lemmings can breed in both summer and winter. Winter breeding is less common in voles, which tend to breed from late spring to fall. Voles in Mediterranean climates, on the other hand, breed in winter and spring, during the wet season, but not in summer. The breeding season may vary within species in different years or in different parts of the range. The muskrat will breed all year in the southern part of its range but only in summer elsewhere. Some species of meadow voles will breed in winter during the phase of population increase but not when numbers are declining.

In many species, such as the Montane, Field and Mexican voles, the presence of a male will induce ovulation in a female.

▶ ▼ **Representative species of voles and lemmings.** (1) A muskrat (*Ondata zibethicus*) sitting on its house of branches and twigs. (2) A Red-tree vole (*Phenacomys longicaudatus*), a highly specialized, tree-living vole. (3) A Southern (or Afghan) mole-vole (*Ellobius fuscocapillus*). (4) A Taiga vole (*Microtus xanothognathus*). (5) A Norway lemming (*Lemmus lemmus*). (6) A Meadow vole (*Microtus pennsylvanicus*), drumming an alarm signal with its hind foot. (7) A Collared (or Arctic) lemming (*Dicrostonyx torquatus*) in its winter coat with double foreclaws. (8) A Collared lemming wearing its summer coat with single foreclaws. (9) A European water vole (*Arvicola terrestris*).

3

6

The 3 tribes of voles and lemmings

Lemmings
Tribe Lemmini.
Four genera and 9 species in N America and Eurasia inhabiting tundra, taiga and spruce woods. Skull broad and massive, tail very short, hair long; 8 mammae. Genera:

Collared lemmings
Genus *Dicrostonyx*, 2 species including: **Collared** or **Arctic lemming** (*D. torquatus*).

Brown lemmings
Genus *Lemmus*, 4 species including: **Norway lemming** (*L. lemmus*).

Wood lemming
Myopus schisticolor.

Bog lemmings
Genus *Synaptomys*, 2 species including: **Northern bog lemming** (*S. borealis*).

Mole lemmings
Tribe Ellobii.
One genus and 2 species in C Asia inhabiting steppe. Form is modified for a subterranean life; coat color varies from ocher sand to browns and blacks; tail short; no ears; incisors protrude forwards. Species: **Northern mole-vole** (*Ellobius talpinus*): **Southern** or **Afghan mole-vole** (*E. Fuscocapillus*).

Voles and mole-voles
Tribe Microtini.
Thirteen genera and 99 species in N America, Europe, Asia and the Arctic. Main genera are:

Red-backed and bank voles
Genus *Clethrionomys*.
Japan, N Eurasia, N America, inhabiting forest, scrub and tundra. Back usually red, cheek teeth rooted in adults, skull weak; 8 mammae; seven species including: **Bank vole** (*C. glareolus*), **Northern red-backed vole** (*C. rutilus*).

Meadow voles
Genus *Microtus*.
N America, Eurasia, N Africa. Coat and size highly variable, molars rootless, skull weak; 4–8 mammae; burrows on surface and underground. Forty-four species including: **Brandt's vole** (*M. brandti*), **Common vole** (*M. arvalis*), **Field vole** (*M. agrestis*), **Meadow vole** (*M. pennsylvanicus*), **Mexican vole** (*M. mexicanus*), **Montane vole** (*M. montanus*), **Prairie vole** (*M. ochrogaster*), **Taiga vole** (*M. xanognathus*), **Townsend's vole** (*M. townsendii*), **Tundra vole** (*M. oeconomus*).

Pine voles
Genus *Pitymys*.
Not clearly distinct from *Microtus*. E and S USA, E Mexico, Eurasia. Fur soft, dense, mole-like; skull weak; 4–8 mammae; more given to digging than *Microtus*. Seventeen species including: **American pine vole** (*P. pinetorum*), **Mediterranean pine vole** (*P. duodecimcostatus*).

The other 10 genera of the tribe Microtini: **Mountain voles** (*Alticola*), 4 species in C Asia including: **Large-eared** or **High Mountain vole** (*A. macrotis*). **Water voles** (*Arvicola*), 3 species in N America and N Eurasia including: **European water vole** (*A. terrestris*). **Martino's snow vole** (*Dinaromys bogdanovi*), Yugoslavia. **Oriental voles** (*Eothenomys*), 11 species in E Asia including: **Père David's vole** (*E. melanogaster*). *Hyperacrius*, 2 species in Kashmir and the Punjab including: **True's vole** (*H. fertilis*). **Steppe lemmings** (*Lagarus*), 3 species in C Asia and W USA including: **Sagebrush vole** (*L. curtatus*). **Florida water rat** (*Neofiber alleni*), Florida. **Muskrat** (*Ondata zibethicus*), N America. *Phenacomys*, 4 species in W USA and Canada including: **Red-tree vole** (*P. longicaudus*). **Long-clawed mole-vole** (*Prometheomys schaposchnikowi*), Caucasus, USSR.

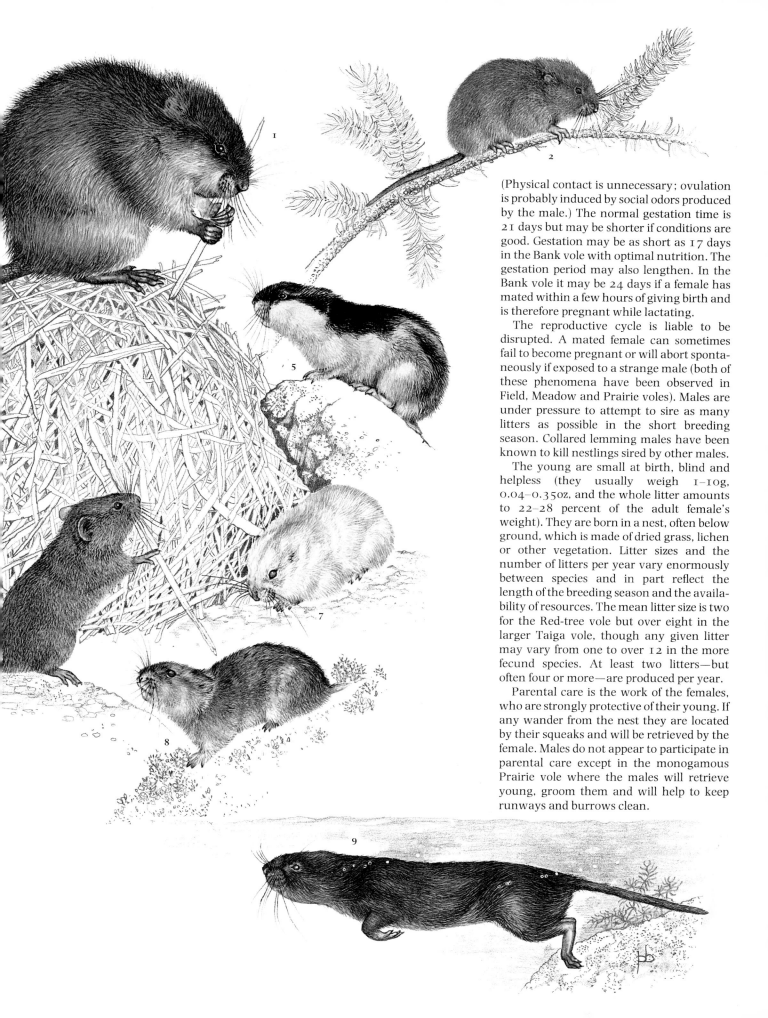

(Physical contact is unnecessary; ovulation is probably induced by social odors produced by the male.) The normal gestation time is 21 days but may be shorter if conditions are good. Gestation may be as short as 17 days in the Bank vole with optimal nutrition. The gestation period may also lengthen. In the Bank vole it may be 24 days if a female has mated within a few hours of giving birth and is therefore pregnant while lactating.

The reproductive cycle is liable to be disrupted. A mated female can sometimes fail to become pregnant or will abort spontaneously if exposed to a strange male (both of these phenomena have been observed in Field, Meadow and Prairie voles). Males are under pressure to attempt to sire as many litters as possible in the short breeding season. Collared lemming males have been known to kill nestlings sired by other males.

The young are small at birth, blind and helpless (they usually weigh 1–10g, 0.04–0.35oz, and the whole litter amounts to 22–28 percent of the adult female's weight). They are born in a nest, often below ground, which is made of dried grass, lichen or other vegetation. Litter sizes and the number of litters per year vary enormously between species and in part reflect the length of the breeding season and the availability of resources. The mean litter size is two for the Red-tree vole but over eight in the larger Taiga vole, though any given litter may vary from one to over 12 in the more fecund species. At least two litters—but often four or more—are produced per year.

Parental care is the work of the females, who are strongly protective of their young. If any wander from the nest they are located by their squeaks and will be retrieved by the female. Males do not appear to participate in parental care except in the monogamous Prairie vole where the males will retrieve young, groom them and will help to keep runways and burrows clean.

While the young are living in the nest they learn to recognize the scent and behavioral cues of their species. So if young Brown lemmings are fostered (artificially) with Collared lemmings they will be more aggressive towards their own species as adults than to their fostered species. The length of lactation varies, but usually lasts about three weeks and in some species is terminated by the female abandoning the nest.

Until recently microtines were considered to have a relatively simple social system. A decade of careful experimentation and observation has shown, however, that the two sexes and different species have different social behaviors and that these may change radically between seasons.

The key to understanding both the social relationships and the spatial organization of voles and lemmings is the spacing behavior of males and females. In the breeding season males and females form territories delimited by scent marks. Male and female territories may be separate (as in the Montane vole and the Collared lemming), overlap (as in the European water and Meadow voles) or several females may live in overlapping ranges within a single male territory (for example, Taiga and Field voles). Males form

hierarchies of dominance; subordinates may be excluded from breeding. They may act to exclude strange males from an area—to reduce the incidence of induced abortion. The Common, Sagebrush and Brandt's voles live in colonies. The animals build complicated burrows.

Although in most species males are promiscuous, a few are monogamous. In the Prairie vole the males and females form pair bonds while the Montane vole appears to be monogamous at low density, as a result of the spacing of males and females, but with no pair bond. At high density this species is polygynous (males mating with several females). Monogamy is favored when both adults are needed to defend the breeding territory from intruders.

The social system may vary seasonally. In the Taiga vole the young animals disperse and their territories break down late in summer. Groups of 5–10 unrelated animals then build a communal nest which is occupied throughout the winter by both males and females. Communal nesting or local aggregations of individuals is also observed in the Meadow, Gray and Northern red-backed voles. Huddling together reduces the energy requirements.

Dispersal (the movement from place of

▶ **Territories of seven female water voles** along a stretch of water in the Bure Marshes (Norfolk, England), approximately 0.5km (0.8mi) long, in (**a**) January and (**b**) March. In January individuals occupied overlapping and undefended territories. However, by March the females had extended their ranges which now did not overlap with those of other individuals and were defended against intruders. One individual was unable to find a territory and migrated; another new individual came in to occupy a territory.

▶ **Swimming without trouble.** BELOW Many populations of water voles, distributed across Western Europe, are able to swim, both on the surface and underwater, even along the bottoms of streams. A few develop hairy fringes on their feet, but in general there seem to be no specific adaptations for aquatic life.

▼ **Evidence for population cycles** can come in two forms: (**a**) indirectly from records kept by the Hudson Bay Company for trade in Arctic fox pelts. The main prey of this predator is lemmings, so when prey densities increase so do those of the Arctic fox; (**b**) directly from the numbers of voles trapped in field experiments (the figures refer to the numbers of the Red-gray vole trapped in northern USSR over a sample period of 100 days). Both graphs show clearly cyclical changes in population levels taking three to four years.

Population Cycles in Lemmings and Voles

Populations of many species of lemmings and voles fluctuate in regular patterns, loosely called "cycles," in which high populations recur at intervals of 3–4 years. These cycles are most famous in lemmings on the tundra but they also occur in a great variety of species of voles.

The cycles are generated by changes in the rates of reproduction and mortality. Reproduction is at maximum rate during the phase of increase, augmented by a breeding season of increased length (often involving winter breeding in lemmings) and by animals achieving sexual maturity at a very young age. When the population reaches its peak reproduction is curtailed by a shortening of the breeding season and an increase in the age of sexual maturity. At the same time juvenile mortality rises and the population stops growing; a large number of animals die of stress and food shortage. The next year the population is very low and can make up for this only with three or four years of reproduction.

There are two competing schools of thought. One suggests that a population's tendency is to increase but cycles are induced because high-density populations exhaust the supply of high-quality food, and so population

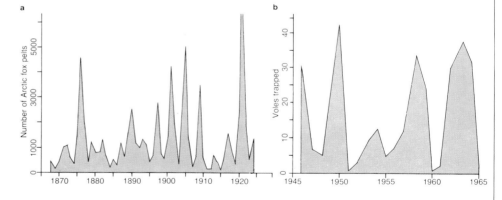

declines from malnutrition. Some herbs are toxic to voles. If a vole population exhausts its supply of palatable forage, individuals may be forced to feed on toxic plants or face starvation. An alternative view is that cycles arise because of changes of behavior within populations as individuals strive to increase their fitness in a fluctuating population. As populations grow individuals become more aggressive in defending their territories against intruders, and this increased aggression may inhibit reproduction and result in additional mortality from fighting or dispersal.

The two sexes often have different strategies. Males compete with other males for

access to reproductive females. Males may show infanticidal behavior. Females defend their nests from marauding individuals of both sexes and may also kill strange juveniles who wander into their territory. Intruder pressure increases with population density and at peak densities social strife can stop all reproduction and cause a decline in numbers.

Natural selection may thus favor large, aggressive individuals at high density because only these individuals will be able to hold a breeding territory under crowded conditions, and then favor small, docile individuals at low density when there is no one around to fight and the premium is on rapid reproduction.

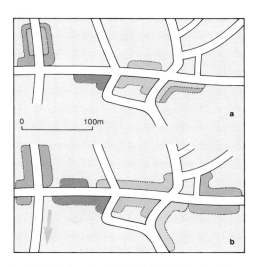

birth to place of breeding) is one of the most important aspects of microtine behavior and has been the subject of considerable research. It plays an important role in regulating population size and allows the animals to exploit efficiently the highly seasonal patchy nature of their habitat: strategies change according to population density. Norway lemmings move in summer into wet grassy meadows but in the fall move into deep mossy hillside habitats to overwinter (see pp656–657).

Dispersers differ from nondispersing animals in a number of ways. In the Meadow and Prairie voles animals that disperse on average are genetically different from permanent residents. Young males often disperse, forced to leave by the aggressive behavior of the dominant males. Additionally, in the Field, Prairie, Meadow, Tundra and European water voles many young pregnant females will disperse. This is quite unusual for mammals but it does enable voles to colonize rapidly vacant habitats.

The numbers of juvenile males and females are equal, except in the Wood lemming which has an unusual feature for mammals: some females are genetically programmed to have only daughters which is advantageous in an increasing population. In most species, however, there are usually more female adults than males, probably as a consequence of dispersal by males who are then more susceptible to predators. In the European water vole it is the females who usually disperse and also in this species there is a slight excess of males in the adult population. In Townsend's vole the survival and growth of juveniles are lower when the density of adult females is high.

Territory size varies but males usually have larger territories than females. In the Bank vole home ranges in one study were found to be 0.7ha (1.7 acres) for females and 0.8ha (2 acres) for males. Territory size decreases as population density rises. In the Prairie vole home range length drops from 25m (82ft) at low density to 10m (33ft) at high density.

Many species live in areas of little or no direct agricultural importance to man or in areas little changed by human habitation. These species are neither pests nor are they endangered. Nevertheless, species living in temperate areas can be serious agricultural pests and damage pastures, crops, orchards, vineyards and commercial forests. The American pine vole burrows in winter around the base of apple trees and chews on the roots or girdles the stems, resulting in a considerable loss in apple production. The extensive underground burrows of Brandt's vole can become a danger to grazing stock, while muskrats can damage irrigation ditches.

In addition to their status as pests many species harbor vectors of diseases such as plague (some *Microtus* and mole-lemmings) and sylvatic plague (Sagebrush vole). *Microtus* and water voles carry tularemia (an infectious disease which causes fever, chills and inflammation of the lymph glands) in eastern Europe and central Asia. The clearing of forest in these areas has increased the habitat for *Microtus* and also the incidence of tularemia.   CJK

Lemming Migrations

Their function in the life cycle of the Norway lemming

According to Scandinavian legend, every few years regimented masses of lemmings descend from birch woods and invade upland pastures where they destroy crops, foul wells and, with their decomposing bodies, infect the air, causing terror-stricken peasants to suffer from giddiness and jaundice. Driven by an irresistible compulsion, not pausing at obstacles—neither man nor beast, river nor ravine—they press on with their suicidal march to the sea. Another story tells of their origins: lemmings are spontaneously generated from foul matter in the clouds and fall to earth during storms and sudden rains. The Eskimos similarly considered them to be not of this world: the name for one North American species means "creatures from outer space."

The life cycle of Norway lemmings is not exceptional. They are active throughout the year. During winter they tunnel and build nests under the snow where they breed and remain safe from predators, except for an occasional attack by an ermine or weasel. The frequency with which they come to the surface depends on the depth and continuity of the snow cover and on the availability of food. With the coming of spring thaw, the lemmings' burrows are in danger of flooding and collapse so the animals are forced to move to higher ground in the alpine zone or lower down in parts of the birch-willow forest. Between May and August they spend much time in the safety of depressions and cavities in the ground or tunnel through the shallow layers of soil and vegetation. At these times of year their enemies include Snowy owls, skuas, ravens and other birds of prey. Well-marked paths and runways often betray the lemmings' presence even when they themselves are out of sight. In the fall, with the freezing of the ground and withering of the sedges, there is a seasonal movement back to sheltered places in the alpine zone. Lemmings are particularly vulnerable at this time: should freezing rain and frost blanket the vegetation with ice before the establishment of snow cover, the

difficulty in gathering food can be fatal.

The mass migrations that have made the Norway lemming famous usually begin in the summer or fall if there has been a period of rapid increase in numbers. In years of low density there may be 3–50 lemmings per ha (1.2–20 per acre) but in peak years 330 per ha (134 per acre) or more. Initially the migrations appear as a gradual movement from densely populated areas in mountain heaths downwards into the willow, birch and conifer forests. The lemmings appear to wander at random. However, as a migration continues groups of animals may be forced to coalesce by local topography. For instance, a large lake may block a migratory movement and thus halt the onward flow, or the lemmings may be caught in a funnel where two rivers meet. In such situations the continuous accumulations of animals become so great that a sort of mass panic ensues and the animals take to wreckless flight—upwards, over rivers, lakes and glaciers and occasionally into the sea.

Although the causes of mass migrations are far from certain, it is widely believed that they are triggered by overcrowding.

▲ **The first known representation of lemmings** appeared in 1555, a woodcut illustration in *Historia de gentibus septentrionalibus* ("History of the Northern Peoples") by the Swedish Catholic priest Olaus Magnus. Not only does the print allude to their origins in the clouds but shows a lemming migration.

▶ **The face of the Norway lemming.** This lemming's normal distribution covers the tundra region of Scandinavia and northwest Russia, in normal years. During "lemming years" the distribution expands considerably.

▶ **A watery grave.** BELOW The main cause of death on lemming migrations is drowning. All such deaths are in a sense accidental. Lemmings can swim well and normally take to water only when they feel they can cross successfully to a far shore. Changes in conditions, eg of the weather, make their destination to be not on this earth.

▼ **Norway lemmings are individualistic,** intolerant creatures. When numbers rise so does aggressive behavior, which is well developed. Here (1) two males box, (2) wrestle and (3) threaten.

1

2

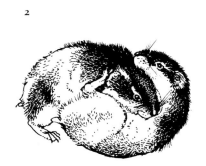

3

Females can breed in their third week and males as early as their first month; reproduction continues year round. Litters consist of five to eight young and may be produced every three or four weeks. In a short period lemmings can produce several generations and increase the overall population rapidly. Short winters without sudden thaws or freezes followed by an early spring and late fall provide favorable conditions for continuous breeding and a rapid increase in population density.

Lemmings are generally intolerant of one another and, apart from brief encounters for mating, lead solitary lives. According to one 19th-century naturalist, Robert Collett (1842–1912): "The enormous multitudes [during peak years] require increased space, and individuals ... cannot, on account of their disposition, bear the unaccustomed proximity of neighbours. Involuntarily the individuals are pressed out to the sides until the edge of the mountain is reached." A want of food is apparently not important in initiating mass migrations since enough food appears to be available even in areas where lemmings are most numerous. It is possible that during a peak year the number of aggressive interactions increases drastically and that this triggers migrations. This seems to be confirmed by reports that up to 80 percent of migrating Norway lemmings are young animals (and thus are likely to have been defeated by larger individuals).

"Lemming years" often correlate with marked increases in the number of other animals living in the same area, for example shrews and voles, capercaillies, and even certain butterflys (whose caterpillars strip entire birch forests of their leaves). But the essential feature of long-distance migrations appears to be a desire to ensure survival. The lemming species of Alaska and northern Canada (the Brown lemming and the Collared lemming) also engage in similar if less spectacular migrations. Although countless thousands of Norway lemmings may perish on their long journeys, the idea that these migrations always end in mass suicide in nonsense.   UWH

OLD WORLD RATS AND MICE

Subfamily: Murinae
Four hundred and eight species in 89 genera.
Family: Muridae.
Distribution: Europe, Asia, Africa (excluding
Madagascar), Australia; also found on many
offshore islands.

Habitat: grassland, forest, mountains.

Size: ranges from head-body length 4.5–8.2cm
(1.7–3.2in), tail length 2.8–6.5cm (1.1–2.5in),
weight about 6g (0.2oz) in the Pygmy mouse to
head-body length 48cm (19in), tail length
20–32cm (8–13in) in Cuming's slender-tailed
cloud rat.

Gestation: in most small species 20–30 days
(longer in species that give birth to precocious
young, eg Spiny mice 36–40 days); not known
for large species.

Longevity: small species live little over 1 year;
Bandicota indica (900g) 21 days.

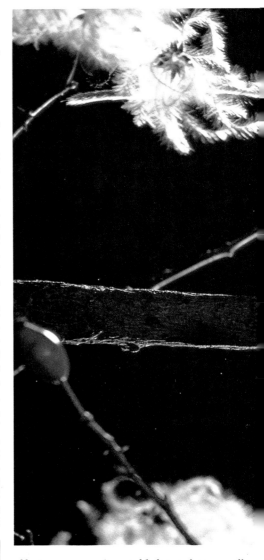

THE Old World rats and mice, or Murinae,
include 406 species distributed over the
major Old World land masses from immediately south of the Arctic circle to the tips of
the southern continents. If exuberant radiation of species, ability to survive, multiply
and adapt quickly are criteria of success,
then the Old World rats and mice must be
regarded as the most successful group of
mammals.

They probably originated in Southeast
Asia in the late Oligocene or early Miocene
(about 25–20 million years ago) from a
primitive (cricetine) stock. The earliest fossils (*Progonomys*), in a generally poor fossil
representation, are known from the late
Miocene (about 8–6 million years ago) of
Spain. Old World rats and mice are primarily a tropical group which have sent a few
hardy migrant species into temperate
Eurasia.

Their success lies in the combination of
features they have probably inherited and
adapted from a primitive, mouselike "archimurine." This is a hypothetical form, but
many features of existing species point to
such an ancestor, from which they are little
modified. The archimurine would have been
small, perhaps about 10cm (4in) long in
head and body with a scaly tail of similar
length. The appendages would have been
of moderate length thereby facilitating
subsequent elongation of the hind legs in
jumping forms and short robustness in the
forelimbs of burrowers. It would have a full
complement of five fingers and toes. The
sensory structures (ears, eyes, whiskers and

olfactory organs) would have been well
developed. Teeth would have consisted of
continuously growing, self-sharpening incisors and three elaborately rasped molar
teeth in each side of each upper and lower
jaw, with powerful jaw muscles for chewing
a wide range of foods and preparing material
for nests. The archimurine would have had
a short gestation period, would have produced several young per litter and therefore
would have multiplied quickly. With its
small size the archimurine could have
occupied a wide variety of microhabitats.
Evolution has produced a wide range of
adaptations, but only a few, if highly significant, lines of structural change.

Modifications to the tail have produced
organs with a wide range of different capabilities. It has become a long balancing
organ, with (as in the Australian hopping
mice) or without (as in the Wood mouse) a
pencil of hairs at its tip. It has become a

Species and genera include:

African creek rat (*Pelomys isseli*), African forest
rat (*Praomys jacksoni*), African grass rats (genus
Arvicanthis), African marsh rat (*Dasymys incomtus*),
African meadow rat (*Praomys fumatus*), African
soft-furred rats (genus *Praomys*), African swamp
rats (genus *Malacomys*), Australian hopping mice
(genus *Notomys*), Bushy-tailed cloud rat
(*Crateromys schadenbergi*), Chestnut rat (*Niviventer
fulvescens*), Climbing wood-mouse (*Praomys alleni*),
Cuming's slender-tailed cloud rat (*Phloeomys
cumingi*), Eastern small-toothed rat (*Macruromys
major*), Edwards long-footed rat (*Malacomys
edwardsi*), Four-striped grass mouse (*Rhabdomys
pumilio*), giant naked-tailed rats (genus *Uromys*),
Greater tree mouse (*Chiruromys forbesi*), harsh-
furred rats (genus *Lophuromys*), Harvest mouse
(*Micromys minutus*), Hind's bush rat (*Aethomys
hindei*), House mouse (*Mus musculus*), Larger
pygmy mouse (*Mus triton*), Lesser bandicoot rat
(*Bandicota bengalensis*), Lesser ranee mouse
(*Haeromys pusillus*), Long-footed rat (*Malacomys
longipes*), Mimic tree rat (*Xenuromys barbatus*), New
Guinea jumping mouse (*Lorentzimys nouhuysi*), Nile
rat (*Arvicanthis niloticus*), Norway, Brown or

Common rat (*Rattus norvegicus*), Old World rats
(genus *Rattus*), Oriental spiny rats (genus
Maxomys), Palm mouse (*Vandeleuria oleracea*),
Pencil-tailed tree mouse (*Chiropodomys gliroides*),
Peter's arboreal forest rat (*Thamnomys rutilans*),
Peter's striped mouse (*Hybomys univittatus*),
Polynesian rat (*Rattus exulans*), Punctated grass-
mouse (*Lemniscomys striatus*), Pygmy mouse (*Mus
minutoides*), Roof rat (*Rattus rattus*), Rough-tailed
giant rat (*Hyomys goliath*), Rufous-nosed rat
(*Oenomys hypoxanthus*), Short-tailed bandicoot rat
(*Nesokia indica*), small-toothed rats (genus
Macruromys), Smooth-tailed giant rat (*Mallomys
rothschildi,*), Speckled harsh-furred rat (*Lophuromys
flavopunctatus*), spiny mice (genus *Acomys*), Stick-
nest rat (*Leporillus conditor*), Striped grass mouse
(*Lemniscomys barbarus*), Temminck's striped mouse
(*Hybomys trivirgatus*), Tree rat (*Thamnomys
dolichurus*), Western small-toothed rat (*Macruromys
elegans*). Wood mouse (*Apodemus sylvaticus*).

▲ **The night shift.** Yellow-necked field mice—tree-dwelling nocturnal creatures distributed across much of Europe and Israel—are important for nature as distributors of seeds. This particular mouse is carrying a hazel nut, probably for food. Its rodent incisor teeth enable it both to transport the nut easily and to open it.

grasping organ to help in climbing, as in the Harvest mouse. It has become a sensory organ with numerous tactile hairs at the end furthest from the animal as well as being prehensile (as in the Greater tree mouse). In some it has become a bushy structure, as in the Bushy-tailed cloud rat. In some genera, for example spiny mice, as in some lizards, the tail is readily broken, either in its entirety or in part. Unlike lizards' tails it does not regenerate. In species where the proximal part of the tail is a dark color and the distal part white (for example the Smooth-tailed giant rat) the tail may serve as an organ of communication

Hands and feet show a similar range of adaptation. In climbing forms big toes are often opposable though sometimes relatively small (eg Palm mouse, Lesser ranee mouse). The hands and/or feet can be broadened to produce a firmer grip (eg Pencil-tailed tree mouse, Peter's arboreal

forest rat). In jumping forms the hind legs and feet may be much elongated (eg Australian hopping mice), while in species living in wet, marshy conditions the hind feet can be long and slightly splayed (eg African swamp rats), somewhat reminiscent of the webbed foot of a duck.

The claws are often modified to be short and recurved, for attaching to bark and other rough surfaces (eg Peter's arboreal forest rat) or to be large and strong, in burrowing forms (eg Lesser bandicoot rat). In some of the species with a small opposable digit the claw of this digit becomes small, flattened and nail-like (eg Pencil-tailed tree mouse).

Fur is important for insulation. In some species some hairs of the back are modified into short stiff spines (eg spiny mice) while in others it can be bristly (eg harsh-furred rats), shaggy (eg African marsh rat) or soft and woolly (eg African forest rat), with many intermediates. The function of spines is not known, although it is speculated that they deter predators.

The coat patterns of many species are dominated by medium to dark browns on the back and flanks. Others are designed to conceal an animal. In the Four-striped grass mouse and striped grass mice alternating stripes of dark brown and yellow-brown run along the length of the body so that during the day or evening they can blend with their environments. Mice that live in the deserts of Australia (eg Australian hopping mice) have a sandy brown fur. Many species have a white, gray or light-colored underside. This may be protective, because shadows can make the lighter underside look the same tone as the upper body.

Ears can range from the large, mobile and prominent (as in the Stick-nest rat) to the small and inconspicuous, well covered by surrounding hair (eg African marsh rat). In teeth there is considerable adaptation in the row of molars. In what is presumed to be the primitive condition there are three rows of three cusps on each upper molar tooth. The number of cusps is often much smaller, particularly in the third molar which is often small. The cusps may also coalesce to form transverse ridges. But the typical rounded cusps, although they wear with age, make excellent structures for chewing a wide variety of foods.

The adaptations of teeth have, at the extremes, resulted in the development of robustness and relatively large teeth (in Rusty-nosed rat, Nile rat) and in the reduction of the whole tooth row to a relatively small size, as in the Western small-toothed rat of New Guinea. The food of this rat probably requires little chewing. It possibly consists of soft fruit or small insects.

Murines are found throughout the Old World with considerable variations in the numbers of species in different parts of their range, though in examining their natural distribution the House mouse, Roof rat, Norway rat and Polynesian rat must be discounted as they have been inadvertently introduced in many parts of the world.

The north temperate region is poor in species with, in Europe, countries such as Norway, Great Britain and Poland having respectively as few as 2, 3 and 4 species each. In Africa the density of species is low from the north across the Sahara until the savanna is reached where the richness of species is considerable. Highest densities occur in the tropical rain forest and in adjacent regions of the Congo basin. This can be illustrated by reference to selected

locations. The desert around Khartoum, the arid savanna at Bandia, Senegal, the moist savanna in Rwenzori Park, Uganda, and the rainforests of Makokou, Gabon, support 0, 6, 9 and 13 species respectively. Zaire boasts 44 species, and Uganda 36.

Moving to the Orient, species are most numerous south of the Himalayas. India and Sri Lanka have about 35 species, Malaya 22. In the East Indies some islands are remarkably rich: about 41 species in New Guinea, 33 in Sulawesi (Celebes), 32 in the Philippines. Within the Philippines there has been a considerable development of native species with 10 of the 12 genera and 30 species, found only there (ie endemic); only two species of *Rattus* are found elsewhere. A notable feature is the presence of 10 large species having head-body lengths of about 20cm (8in) or more. The largest known murine is found in the Philippines, Cuming's slender-tailed cloud rat, over 40cm (16in) in head-body length. Just slightly shorter are the Pallid slender-tailed rat and the Bushy-tailed cloud rat. This high degree of endemism and the tendency to evolve large species is also found in the other island groups. In New Guinea there are 6 species with head-body lengths of more than 30cm (12in) (including the Smooth-tailed giant rat, the Rough-tailed giant rat, the Eastern small-toothed rat, the giant naked-tailed rats and the Mimic tree rat) and only one small species, the New Guinea jumping mouse, about the size of the House mouse. In Australia there are about 49 species of which approximately 75 percent are to be found in the eastern half of the continent and 55 percent in the west.

It has proved difficult to give an adequate and comprehensive explanation of the evolution and species richness of the murines. There are some pointers to the course evolution may have followed, based on structural affinities and ecological considerations. The murines are a structurally similar group and many of their minor modifications are clearly adaptive, so there are few characters that can be used to distinguish between, in terms of evolution, primitive and advanced conditions. In fact only the row of molar teeth has been used in this way: primitive dentition can be recognized in the presence of a large number of well-formed cusps. Divergences from this condition may represent specialization or advancement. Ecological considerations account for abundance and for the types of habitat preference a species may show. From this analysis two groups of genera have been recognized. The first contains the

▶ ▼ **Old World rats and mice.** (1) A spiny mouse (genus *Acomys*). (2) The Pencil-tailed tree mouse (*Chiropodomys gliroides*). (3) The African marsh rat (*Dasymys incomtus*). (4) The Harsh-furred mouse (*Lophuromys sikapusi*) eating an insect. (5) The Multimammate rat (*Praomys natalensis*). (6) The Four-striped grass mouse (*Rhabdomys pumilio*). (7) The Fawn-colored hopping mouse (*Notomys cervinus*). (8) The Smooth-tailed giant rat (*Mallomys rothschildi*). (**a**) Tail of the Bushy-tailed cloud rat (*Crateromys schadenbergi*). (**b**) Tail of the Greater tree mouse (*Chiruromys forbesi*). (**c**) Tail of the Harvest mouse (*Micromys minutus*). (**d**) Tail of a wood mouse (genus *Apodemus*). (**e**) Hindfoot of the Palm mouse (*Vandeleuria oleracea*); first and fifth digits opposable to provide grip for living in trees. (**f**) Hindfoot of Peter's arboreal forest rat (*Thamnomys rutilans*); has broad, short digits for providing grip. (**g**) Paw of the Lesser Bandicoot rat (*Bandicota bengalensis*) showing long, stout claws. (**h**) Hindfoot of an African swamp rat (genus *Malacomys*) showing long, splayed foot with digits adapted for walking in swampy terrain.

dominant genera (African soft-furred rats, Oriental spiny rats, Old World rats, giant naked-tailed rats, *Mus*, African grass rats and African marsh rats) which have been particularly successful, living in high populations in the best habitats. These are believed to have evolved slowly because they display relatively few changes from the primitive dental condition. The second group contains many of the remaining genera which are less successful, living in marginal habitats and often showing a combination of aberrant, primitive and specialized dental features.

The dominant genera (with the exception of the African marsh rat) contain more species than the peripheral genera and are constantly attempting to extend their range. Considerable numbers of new species have apparently arisen within what is now the range center of a dominant genus (eg soft-furred rats in central Africa and Old World rats in Southeast Asia). The reasons for this await explanation.

It is quite common for two or more species of murine to occur in the same habitat, particularly in the tropics. One of the more interesting and important aspects of studies is to explain the ecological roles assumed by each species in a particular habitat, and then to deduce the patterns of niche occupation and the limits of ecological adaptations by animals with a remarkably uniform basic structure. A particularly favorable habitat and one amenable to this type of study is regenerating tropical forest.

In Mayanja Forest, Uganda, 13 species were found in a recent study from a small area of about 4sq km (1.5sq mi). Certain species were of savanna origin and restricted to grassy rides, ie Rusty-bellied rat and Punctated grass mouse. Of the remaining 11 species all have forest and scrub as their typical habitat with the exception of the two smallest species, the Pygmy mouse and the Larger pygmy mouse, which are also found in grasslands and cultivated areas. Three species, the Tree rat, the Climbing wood-mouse and Peter's arboreal forest rat, seldom, if ever, come to the ground. The small Climbing wood mouse preferred a bushy type of habitat, being frequently found within the first 60cm (24in) off the ground. The two other arboreal species were strong branch runners and were able to exploit the upper as well as the lower levels of trees and bushes. All three species were found alongside a variety of plant species (in the case of the wood mouse, 37 were captured beside 19 different plants, with *Solanum* among the most favored). All species are herbivorous

Rats, Mice and Man

Some Old World rats and mice have a close detrimental association with man through consuming or spoiling his food and crops, damaging his property and carrying disease.

The most important species commensal with man are the Norway or Brown or Common rat, the Roof rat and the House mouse. Now of worldwide distribution, they originated from around the Caspian Sea, India and Turkestan respectively. While the Roof rat and the House mouse have been extending their ranges for many hundreds of years the Norway rat's progress has been appreciably slower, being unknown in the west before the 11th century. The Norway rat is well established in urban and rural situations in temperate regions, and it is the rodent of sewers. In the tropics it is mainly restricted to large cities and ports. The Roof rat is more successful in the tropics where towns and villages are often infested, though it cannot compete with the indigenous species in the field. In the absence of competitors in many Pacific, Atlantic and Caribbean islands, Roof rats are common in agricultural and natural habitats.

With even the solitary House mouse capable of considerable damage in one's home, the scale of mass outbreak damage is difficult to envisage, as when an Australian farmer recorded 28,000 dead mice on his veranda after one night's poisoning, and 70,000 were killed in a wheat yard in an afternoon. In addition to these cosmopolitan commensals there are the more localized Multimammate rat in Africa, the Polynesian rat in Asia and the Lesser bandicoot rat in India.

There are many rodent-borne diseases transmitted either through an intermediate host or directly. The Roof rat, along with other species, hosts the plague bacterium which is transmitted through the flea *Xenopsylla cheopsis*. The lassa fever virus of West Africa is transmitted through urine and feces of the Multimammate rat. Other diseases in which murines are involved include murine typhus, rat-bite fever and leptospirosis.

▲ **Tiny nimble-footed rodents:** the Old World harvest mouse. These are among the smallest of rodents and have a highly developed way of life in tall vegetation, for which they have equipped themselves well. Their legs and feet are highly flexible, their tail able to grasp. They use tall grasses or reeds as if they were a gymnasium, or a forest, building their nests of woven grass high above ground.

▲ **Peripheral rat.** ABOVE LEFT Most of the 63 species of rats belonging to the genus *Rattus* live well away from human habitation, including this bush rat from Australia.

and nocturnal and the two larger species construct elaborately woven nests of vegetation.

Two species are found on both the ground and in the vegetation up to 2m (6.5ft) above it. Of these the African forest rat was abundant and the Rufous-nosed rat much less common. The African forest rat lives in burrows (in which it builds its nest); their entrances are often situated at the bases of trees. It is nocturnal, feeding on a wide range of insect and plant foods. The Rufous-nosed rat is both nocturnal and active by day and constructs nests with downwardly

projecting entrances in the shrub layer. They are constructed of grass, on which this species is known to feed.

Of the 11 forest species, Peter's striped mouse, the Speckled harsh-furred rat, the Long-footed rat, the Pygmy mouse, the Larger pygmy mouse and Hind's bush rat are ground dwellers. Of these the striped mouse is a vegetarian, preferring the moister parts of the forest, the Harsh-furred rat an abundant species, predominantly predatory, favoring insects but also prepared to eat other types of flesh. The Long-footed rat is found in the vicinity of streams and swampy conditions; it is nocturnal and includes in its diet insects, slugs and even toads (a specimen in the laboratory constantly attempted to immerse itself in a bowl of water). The two small mice are omnivores and Hind's bush rat is a vegetarian species that inhabits scrub.

A further important feature, which could well account for dietary differences in these species, is the size ranges they occupy. The three mice are in the 5–25g (0.2–0.9oz) range with the Pygmy mouse rather smaller than the other two. The Rufous-nosed rat is in the 70–90g (2.5–3.2oz) range and the Long-footed rat and Hind's bush rat above this. The remaining species, the Tree rat, Peter's arboreal forest rat, African forest rat, Speckled harsh-furred rat and Peter's striped mouse, have weights between 35g and 60g (1.2–2.1oz).

Within the tropical forests there is a high precipitation, with rain falling in all months of the year. This results in continuous flowering, fruiting and herbaceous growth which is reflected in the breeding activity of the rats and mice. In Mayanja Forest the African forest rat and the Speckled harsh-furred rat were the only species obtained in sufficient numbers to permit the monthly examination of reproductive activity. The former bred throughout the year while in the latter the highest frequency of conception coincided with the wetter periods of March to May and October to December.

The foregoing account has attempted to highlight the species richness and adaptability of this group of mammals. In spite of their abundance and ubiquity in the Old World, particularly the tropics, the murines remain a poorly studied group. Exceptions include a few species of economic importance and the Palaearctic wood mouse. Many species are known only from small numbers in museums supported by the briefest information on their biology. There are undoubtedly endless opportunities for research on this fascinating and accessible group of mammals.   MJD

An Animal Weed

Man's relationship with the House mouse

At least since the beginning of recorded history the presence of the common House mouse has compelled man to form a view of its nature and constitution. The first written reference to the raising and protection of mice by man amounts to evidence of mouse worship in Pontis (Asia Minor) 1,400 years before the birth of Christ. Homer mentions Apollo Smintheus, a god of mice, about 1200 BC, and his worship was still popular at the time of Alexander the Great 900 years later. Mice were worshiped by the Teucrans of Crete, who attributed their victory over the Pontians to a God who caused mice to gnaw the leather straps on the shields of their enemies. A temple at Tenedos on the entrance to the Dardanelles (whose foundations still remain) was built in which mice were maintained at public expense. The mouse cult spread to other cities in Greece and continued as a local form of worship until the Turkish conquest in 1543. Pliny wrote "They (white mice) are not without certain natural properties with regard to the sympathy between them and the planets in their ascent ... Sooth-sayers have observed that it is a sign of prosperity if there be a store of white ones bred." Hippocrates recorded that he did not need to use mouse blood as a cure for warts in the same way as his colleagues because he had a magic stone with lumps on it which had proved to be an efficient remedy.

The other main center of ancient mouse culture was in the Far East. In China, albino mice were used as auguries, and Chinese government records show that 30 albino mice were caught in the wild between 307 and 1641 AD. A word for "spotted mouse" appears in the first Chinese dictionary, written in 1100 BC. Waltzing mice have been known since at least 80 BC (waltzing or dancing in mice is produced by a defect in the inner ear affecting balance, which is usually inherited). Clearly there has been a mouse fancy in China (and Japan) for many centuries which valued new and unusual forms. The Japanese had in their fancy such traits as albinism, non-agouti, chocolate, waltzing, dominant and recessive spotting, and other color variants still known to mouse breeders. Some of these varieties were brought to Europe in the mid-19th century by British traders.

In western Europe as early as 1664 the English scientist Robert Hooke used a mouse to study the effects of increased air pressure. William Harvey (1578–1657) used mice in his anatomical studies. The English chemist Joseph Priestly gives a delightful account in his *Experiments and Observations* (1775) of his experiments with mice, including how he trapped and maintained them. Half a century after this, a Genevan pharmacist named Louis Coladon bred large numbers of white and gray mice and obtained segregations in agreement with Mendelian expectation 36 years before Mendel published his work on peas. One of the earliest demonstrations of genetical segregation in animals after the rediscovery of Mendel's papers was made by L. Cuenot working in Paris with mice (1902).

The modern history of laboratory mouse breeding began in 1907 when an undergraduate at Harvard University, C.C. Little, began to study the inheritance of coat color under the supervision of W.E. Castle. Two years later, Little obtained a pair of "fancy" mice carrying alleles (variations of genes) for the recessively inherited traits dilution (*d*), brown (*b*), and non-agouti (*a*), and inbred their descendants brother to sister, selecting for vigorous animals. The recessive trait only appears when an animal inherits the *same* allele from both its parents (each parent contributes one allele of the pair in the offspring). If the trait appears when an animal has only one allele, it is said to be dominantly inherited. Inbreeding has the

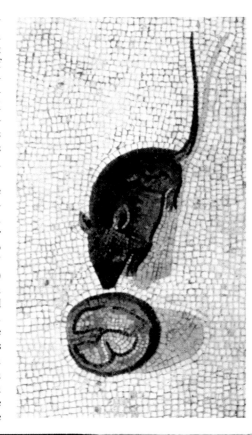

▲ **The variable House mouse.** Genetic diversity is exemplified in this nest of laboratory mice by the color variations in the young.

◄ **A delight in the mouse** is shown in this detail from a Roman mosaic.

▼ **Man's constant companions.** The wide distribution of the House mouse may be due to its characteristics or could be a historical development, resulting from its occurrences with Neolithic man (around 10,000 years ago) in the first cereal-growing sites in the Near East and subsequently spread with increasing cereal production.

effect that animals will share ancestors, and therefore the likelihood that they will get the same allele for a particular characteristic from both parents will be much greater than if they were unrelated. A strain which has been inbred will breed true for inherited characteristics. The first truly inbred strain (exclusively brother-sister) mated for over 20 generations) was Little's DBA strain (called after the three mutant genes it carried). Because all the animals in an inbred strain carry the same genes, the effects of different treatments (drugs, physical environment, etc) can be compared without any confusion being produced by variation due to genetic differences.

There are now many other inbred strains, most of them descended from mice caught in the northeastern USA. Laboratory workers have recently become increasingly interested in wild mice, because they carry many inherited variants not found in established inbred strains, and allow us to find out more about development and immunology.

The House mouse is an animal weed— quick to exploit opportunity and able to withstand local adversity and extinction without the species as a whole being harmed. This means that it must be able to

breed rapidly, tolerate a wide range of conditions, and adjust quickly to changes in environment. These traits are responsible for success in all but a few parts of the world where it is excluded by extreme environments (eg polar regions) or competition from other small mammals (eg Central Africa). Whether all *Mus* species are equally adaptable is unknown.

If mice adjusted to their environments exclusively genetically it would be easy to summarize the situation. However, mouse populations differ considerably in their genetical composition (genomes), and different genomes respond differently to similar environmental pressures. For example, the tail of the mouse seems to be a heat-regulating organ; generally, relative tail length is greater in mice reared in hot than in cold conditions but the genome is the same. Tail length decreases from about 95 percent of head and body length in southern England to less than 80 percent in Orkney—but mice from the Shetland and Faroe groups north of Orkney have tails as long as southern English animals.

House mice are among the more genetically variable mammals, with the most variable population known in Hawaii.   RJB

OTHER OLD WORLD RATS AND MICE

**Ninety-three species in 45 genera belonging to
10 subfamilies**
Distribution: Balkans, S USSR, Africa,
Madagascar, S India, China, New Guinea,
Australia.

Blind
mole-rats

African
pouched
rats

Crested
rat

African
climbing
mice

Root and
Bamboo
rats

Madagascan
rats

African
swamp
rats

Oriental
dormice

Zokors

Austral-
asian
water rats

THROUGHOUT the Old World there are small groups of rats and mice that cannot be included in the three major subfamilies of the muridae. Here they are grouped in 10 small subfamilies. Relationships amongst these are not clear and they are placed in a geographical sequence for convenience. Some of these rodents are superficially very similar to members of the major groups; eg the zokors can be considered as specialized voles. Two groups, the blind mole-rats and the bamboo rats, are more distinctive and are sometimes treated as separate families.

Of all the subterranean rodents the **blind mole-rats** (subfamily Spalacinae) show the most extreme adaptations to life underground and they are often treated as a separate family, Spalacidae. Their eyes are completely and permanently hidden under their skin and there are no detectable external ears or tail. The incisor teeth protrude so far that they are permanently outside the mouth and can be used for digging without the mouth having to be opened. A unique feature is the horizontal line of short, very stiff (presumably touch-sensitive) hairs on each side of the head. Most blind mole-rats are about 13–25cm (5–9in) long but in one species, the Giant mole-rat of southern Russia, they can reach 35cm (14in).

Blind mole-rats are found in dry but not desert habitats from the Balkans and southern Russia round the eastern Mediterranean as far west as Libya. Apart from being entirely vegetarian they live very much like the true moles (which are predators belonging to the order Insectivora). Each animal makes its own system of tunnels, which may reach as much as 350m (1,150ft) in length, throwing up heaps of soil. They feed especially on fleshy roots—bulbs and tubers —but also on whole plants. Although originally animals of the steppes they have adapted well to cultivation and are a considerable pest in crops of roots, grain and fruit.

Blind mole-rats breed in spring, with usually two or three in a litter which disperse away from the mother's tunnel system as soon as they are weaned, at about three weeks. There is sometimes a second litter later in the year.

As in many other burrowing mammals their limited movement has led to the evolution of many local forms which make the individual species very difficult to define but provides a bonanza for the study of genetics and the processes of evolution and species formation. The number of species that should be recognized and how they are classified are still very uncertain. Eight species are recognized here but they have been reduced to three elsewhere.

The five species of **African pouched rats** (subfamily Cricetomyinae) resemble hamsters in having a storage pouch opening from the inside of each cheek. The two short-tailed pouched rats resemble hamsters in general appearance but the other three species are rat-like, with long tail and large ears.

The giant pouched rats are among the largest of the murid rodents, reaching 40cm (16in) in head-body length. They are common throughout Africa south of the Sahara and in some areas are hunted for food. They feed on a very wide variety of items, including insects and snails as well as seeds and fruit. In addition to carrying food to underground storage chambers, the cheek pouches can be inflated with air as a threat display. The gestation period is about six weeks and litter size usually two or four.

The three large species are associated with peculiar blind, wingless earwigs of the genus *Hemimerus* which occur in their fur and in their nests where they probably share the rat's food.

African swamp rats (or vlei rats) (subfamily Otomyinae) are found throughout much of

▶ ▼ **Representative of minor subfamilies** of rats and mice. (**1**) The Spiny dormouse (*Platacanthomys lasiurus*). (**2**) *Macrotarsomys ingens* (one of the Madagascan rats). (**3**) The Crested rat (*Lophiomys imhausii*) with mane erect, showing a glandular patch. (**4**) The East Siberian Zokor (*Myospalax aspalax*), kicking back excavated soil with its hindfeet. (**5**) The Ehrenberg's mole-rat (*Nannospalax ehrenbergi*) showing the broad nose used for ramming soil and tactile hairs on the face. Note the absence of external eyes.

Africa south of the Sahara. Externally they can hardly be distinguished from some of the voles of the northern temperate region; like the voles of the genera *Microtus* and *Arvicola* they are medium-sized ground-living rodents with short ears and legs, small eyes, a short blunt muzzle and shaggy fur. The tail is shorter than the head and body but not as short as in most voles. The color, as in many voles, is usually a grizzled yellow brown. The various species are very similar and have not yet been adequately defined.

Like many voles most swamp rats live among thick grass, especially in wet areas, and they are particularly abundant in the alpine zone of many African mountains such as Mount Kenya and Kilimanjaro. Where ground vegetation is thick they are active by day as well as by night, making runways at ground level among the grass tussocks. They are adapted to feeding on grass and tough stems, having high-crowned molar teeth with multiple transverse ridges of hard enamel, although they are not ever-growing as in most voles.

Swamp rats have small litters, usually of one or two quite well-developed young.

The two species of whistling rats are very similar to the species of *Otomys* but they are lighter in color and live in dry country with sparse grass and low shrubs. They are sociable animals, active by day, and when they are alarmed outside their burrows they whistle loudly and stamp their feet before disappearing underground.

The **Crested rat** has so many peculiarities that it is placed in a subfamily of its own (Lophiomyinae) and it is not at all clear what are its nearest relatives. It is a large, dumpy, shaggy rodent with a bushy tail and tracts of long hair along each side of the back which can be erected. These are associated with specialized scent glands in the skin and the individual hairs of the crests have a unique lattice-like structure which probably serves to hold and disseminate the scent. These hair tracts can be suddenly parted to expose the bold striped pattern as well as the scent glands.

The skull is also unique in having a peculiar granular texture and in having the cavities occupied by the principal, temporal, chewing muscles roofed over by bone—a feature not found in any other rodent.

Crested rats are nocturnal and little is known of their way of life. They spend the day in burrow, rock crevice or hollow tree. They are competent climbers and feed on a variety of vegetable material. The stomach is unique among rodents in being divided into a number of complex chambers similar to those found in ruminant ungulates such as cattle and deer.

The majority of **African climbing mice** (subfamily Dendromurinae) are small agile mice with long tails and slender feet, adapted to climbing among trees, shrubs and long grass. Though they are confined to Africa south of the Sahara some of them closely resemble mice in other regions that show similar adaptations, such as the Eurasian harvest mouse (subfamily Murinae) and the North American harvest mice (subfamily Hesperomyinae). They are separated from these mainly by a unique pattern of cusps on the molar teeth and it is on the basis of this feature that some superficially very different rodents have been associated with them in the subfamily Dendromurinae.

Typical dendromurines, eg those of the genus *Dendromus*, are nocturnal and feed on grass seeds but are also considerable predators on small insects such as beetles and even young birds and lizards. Some species in other genera are suspected of being more completely insectivorous. In the genus *Dendromus* some species make compact globular nests of grass above ground, eg in bushes; others nest underground. Breeding is seasonal, with usually three to six, naked, blind nestlings in a litter. Among the other genera the most unusual are the fat mice. They make extensive burrows and during the dry season spend long periods underground in a state of torpor, after developing thick deposits of fat. Even during their active season fat mice become torpid, with reduced body temperature, during the day.

Many species of this subfamily are poorly

▲ **Apparatus for life underground.** The East African root rat rarely comes above ground—only for the occasional forage. It digs the underground burrows that provide its environment using its powerful incisor teeth.

▶ **Breeding mound of a blind mole-rat** (*Nannospalax ehrenbergi*). Each animal makes its own system of tunnels which may be as much as 350m (1,150ft) in length, throwing up heaps of soil. These mole-rats feed especially on fleshy roots—bulbs and tubers—but also on whole plants by pulling them down into their tunnels by the roots. They store food: their underground food storage chambers have been known to hold as much as 14kg (31lb) of assorted vegetables. The breeding nest is placed centrally within the mound.

The 10 subfamilies of Other Old World rats and mice

Blind mole-rats
Subfamily Spalacinae
8 species in 2 genera
Balkans, S Russia, E Mediterranean E N Africa, in dry (but not desert) habitats. Genera: *Spalax*, 5 species; *Nannospalax*, 3 species.

African pouched rats
Subfamily Cricetomyinae
5 species in 3 genera
Africa S of the Sahara, in savanna, dry woodland, tropical forest. Genera: **giant pouched rats** (*Cricetomys*), 2 species; **Lesser pouched rat** (*Beamys hindei*); **short-tailed pouched rats** (*Saccostomys*), 2 species.

African swamp rats
Subfamily Otomyinae
13 species in 2 genera
Africa S of the Sahara. Genera: **swamp and vlei rats** (*Otomys*), 11 species; **whistling rats** (*Parotomys*). 2 species.

Crested rat
Subfamily Lophiomyinae
1 species, *Lophiomys imhausi*, Kenya, Somalia, Ethiopia, E Sudan, in mountain forests between 1,200 and 2,700m.

African climbing mice
Subfamily Dendromurinae
21 species in 10 genera
Africa S of the Sahara, in savanna. Genera: **climbing mice** (*Dendromus*), 6 species; **Large Ethiopian climbing mouse** (*Megadendromus nikolausi*); **Dollman's tree mouse** (*Prionomys batesi*); **Link rat** (*Deomys ferrugineus*); *Dendroprionomys rousseloti*; Leimacomys buettneri; Malacothrix typica; **fat mice** (*Steatomys*), 6 species; **rock mice** (*Petromyscus*), 2 species; **Delany's swamp mouse** (*Delanymys brooksi*).

Root and bamboo rats
Subfamily Rhizomyinae
6 species in 3 genera
E Africa and SE Asia, in grassland and savanna and in forests respectively. Genera: **bamboo rats** (*Rhyzomys*), 3 species; **Bay bamboo rat** (*Cannomys badius*); **root rats** (*Tachyoryctes*), 2 species.

Madagascan rats
Subfamily Nesomyinae
11 species in 8 genera
Madagascar with 1 species in S Africa, in forest, scrub, grassland and marsh. Genera: *Macrotarsomys*, 2 species; *Nesomys rufus*; *Brachytarsomys albicauda*; *Eliurus*, 2 species; *Gymnuromys roberti*; *Hypogeomys antimena*; *Brachyuromys betsileoensis*; **White-tailed rat** (*Mystromys albicaudatus*).

Oriental dormice
Subfamily Platacanthomyinae
2 species in 2 genera
Spiny dormouse (*Platacanthomys lasiurus*), S India, in forested hills between 500 and 1,000m. **Chinese dormouse** (*Typhlomys cinereus*), S China and N Vietnam, in forest.

Zokors or Central Asiatic mole rats
Subfamily Myospalacinae
6 species in 1 genus (*Myospalax*)
China and Altai Mountains, underground.

Australian water rats
Subfamily Hydromyinae
20 species in 13 genera
Australia, New Guinea, and Philippines, in forest, rivers, marshes. Genera: *Hydromys*, 4 species including **Australian water rat** (*H. chrysogaster*); **Coarse-haired water rat** (*Parahydromys asper*); **Earless water rat** (*Crossomys moncktoni*); **False swamp rat** (*Xeromys myoides*); **shrew-rats** (*Rhynchomys*), 2 species; **striped rats** (*Chrotomys*), 2 species; **Luzon shrew-rat** (*Celaenomys siliceus*); **shrew-mice** (*Pseudohydromys*), 2 species; **Groove-toothed shrew-mouse** (*Microhydromys richardsoni*); **One-toothed shrew-mouse** (*Mayermys ellermani*); **Short-tailed shrew-mouse** (*Neohydromys fuscus*); **Long-footed hydromyine** (*Leptomys elegans*); *Paraleptomys*, 2 species.

known. Several distinctive new species have been discovered only during the last 20 years and it is likely that others remain to be found, especially arboreal forest species.

The genera *Petromyscus* and *Delanymys* have sometimes been separated from the others in a subfamily Petromyscinae.

Root rats (subfamily Rhizomyinae) are large rats adapted for burrowing and show many of the characteristics found in other burrowing rodents—short extremities, small eyes, large protruding incisor teeth and powerful neck muscles reflected in a broad angular skull. The **bamboo rats**, found in forests of Southeast Asia, show all these features in less extreme form than do the root rats of East Africa. They make extensive burrows in which they spend the day but they emerge at night and do at least some of their feeding above ground. The principal diet consists of the roots of bamboos and other plants, but above-ground shoots are also eaten. In spite of their size breeding is similar to the normal murid pattern with a short gestation of three to four weeks and three to five naked, blind young in a litter, though weaning is protracted and the young may stay with the mother for several months.

The **African root rats** are more subterranean than the bamboo rats but less so than African mole-rats (family Bathyergidae) or the blind mole-rats (subfamily Spalacinae). They make prominent "mole hills" in open country. Roots and tubers are stored underground. As in most mole-like animals each individual occupies its own system of tunnels. The gestation period is unusually long for a murid—between six and seven weeks.

It has long been debated whether the **Madagascan rats**, the ten or so indigenous rodents of the island of Madagascar, form a single, closely interrelated group, implying that they have evolved from the same colonizing species, or whether there have been multiple colonizations such that some of the present species are more closely related to mainland African rodents than to their fellows on Madagascar. The balance of evidence seems to favor the first alternative hence their inclusion here in a single subfamily, Nesomyinae, but the matter is by no means settled.

The inclusion of the South African white-tailed rat is also debatable, the implication being that it is the sole survivor on the African mainland of the stock that colonized Madagascar. The problem arises from the diversity of the Madagascan species, especially in dentition, coupled with the fact

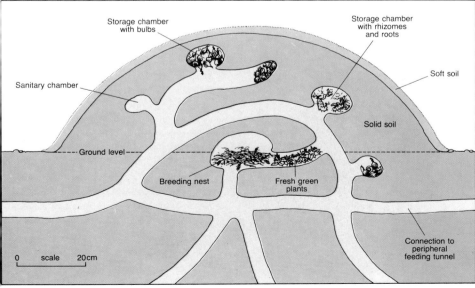

Storage chamber with bulbs

Storage chamber with rhizomes and roots

Soft soil

Sanitary chamber

Solid soil

Ground level

Breeding nest

Fresh green plants

Connection to peripheral feeding tunnel

0 scale 20cm

that none of them match very closely any of the non-Madagascan groups of murid rodents.

The group includes some species that are typical, small, agile mice with long tail, long slender hind feet and large eyes and ears (eg *Macrotarsomys bastardi* and *Eliurus minor*). *Nesomys rufus* is typically rat-like in its size and proportions while *Hypogeomys antimena* is rabbit-sized and makes deep burrows, although it forages for food on the surface.

The two species of *Brachyuromys* are remarkably vole-like in form, dentition and ecology. They live in wet grassland or marshes and are apparently adapted to feeding on grass. Externally they can only with difficulty be distinguished from Eurasian water voles.

The two species of **Oriental dormice** (subfamily Platacanthomyinae) have been considered to be closely related to the true dormice (family Gliridae) which they resemble externally and in the similar pattern of transverse ridges on the molar teeth, although there are only three molars on each row, not preceded by a premolar as in the true dormice. More recently opinion has swung towards treating them as aberrant members of the family Muridae (in its widest sense as used here). Whatever their affinities, they are very distinctive, arboreal mice with no very close relatives, and very little is known of their way of life. The Spiny dormouse is mainly a seed-eater and is a pest of pepper crops in numerous parts of southern India.

Zokors (subfamily Myospalacinae) are burrowing, vole-like rodents found in the steppes and open woodlands in much of China and west as far as the Altai Mountains. Although they live almost entirely underground they are less extremely adapted than the blind mole-rats. Both eyes and external ears are clearly visible, though tiny, and the tail is also distinct. Digging is done mainly by the very large claws of the front feet rather than the teeth. The front claws are so enlarged that when the animal moves it folds the fingers backwards and walks on the knuckles. The coat is a rather uniform grayish brown in most species and there is often a white streak on the forehead.

Like the blind mole-rats, zokors feed on roots, rhizomes and bulbs but they do occasionally emerge from their tunnels to collect food such as seeds from the surface. Massive underground stores of food are accumulated, enabling the animals to remain active all winter.

Breeding takes place in spring when one litter of up to six young is produced. Their

◆ ◄ **Representatives of minor subfamilies** of Old World rats and mice. (**1**) Brant's climbing mouse (*Dendromus mesomelas*). (**2**) A Savanna giant pouched rat (*Cricetomys gambianus*) with both pouches full of food. (**3**) A Vlei rat (*Otomys irroratus*) sitting in a grass runway (**4**) An East African mole-rat (*Tachyoryctes splendens*) burrowing with its incisors. (**5**) An Australian water rat (*Hydromys chrysogaster*) diving.

social organization is little known but the young appear to stay with the mother for a considerable time.

Australian water rats and their allies (subfamily Hydromyinae) are found mostly in New Guinea with a few species in Australia and in the Philippines. Although generally rat-like in superficial appearance the rodents in this group are diverse and it is not at all certain that they form a natural group more closely related to each other than to other murid rodents. The characteristic that appears to unite them is the simplification of the molar teeth, which lack the strong cusps or ridges found in most of the mouse-like rodents, and in some species the molars are also reduced in number, the extreme being seen in the One-toothed shrew-mouse which has only one small, simple molar in each row.

This adaptation is most likely to be related to a diet of fruit or soft-bodied invertebrates, but little information is available on the diet of most species. It is likely that all the members of this group have been derived from rats of the subfamily Murinae, mainly by the simplification of the molar teeth. However, it is quite possible that this could have happened more than once, so that, for example, the shrew-rats of the Philippines and the water rats of Australia and New Guinea may have arisen independently from separate groups of murine rats at different times.

Many of the species show few other peculiarities but two more specialized groups can be recognized. These are the water rats and the shrew-rats. The water rats, of which the best known, and largest, representative is the Australian water rat (weighing 650–1,250g, 23–44oz), have broad, webbed hind feet. The Australian water rat is a common and well-known animal, often seen by day (and often mistaken for a platypus) as it hunts underwater for prey such as frogs, fish, mollusks, insects, and crabs. They have been seen to bring mussels out of the water and leave them on a rock until the heat makes them open and easier to extract from the shell. These water rats breed in spring and summer, producing two or even three litters of usually four or five young after a gestation of about 35 days.

The most specialized of the water rats is the Earless water rat, which not only lacks external ears but has longitudinal fringes of long white hairs on the tail, forming a pattern that is remarkably similar to that found in the completely unrelated Elegant water shrew, *Nectogale elegans*, of the Himalayas.

Although not structurally specialized the little-known False water rat has the most specialized and unusual habitat, for it lives in the mangrove swamps of northern Australia. It climbs among the mangroves and does not appear to spend much time swimming.

The shrew-rats of the Philippines have long snouts with slender, protruding incisor teeth like delicate forceps, presumably adapted, as in the true shrews, for capturing insects and other invertebrates. However, the remaining teeth are small and flat-crowned, quite unlike the sharp-cusped batteries found in the true shrews. The two species of the genus *Rynchomys* show this adaptation in extreme form (and have sometimes been placed in a separate subfamily, the Rynchomyinae).

The two species of striped rats, also in the Philippines, are unique in the group in having a bold pattern consisting of a central bright buff stripe along the back, flanked on each side by a black stripe, producing a simplified version of the pattern seen in some chipmunks.

Of the remaining species the five shrew-mice of New Guinea are small smoky-gray mice living in mountain forest. They have more normal incisors than the shrew-rats but have very rudimentary molars. The three species of *Leptomys* and *Paraleptomys* are rat-sized and probably arboreal. GBC

HAMSTERS

Subfamily: Cricetinae
Twenty-four species in 5 genera.
Family: Muridae.
Distribution: Europe, Middle East, Russia,
China.

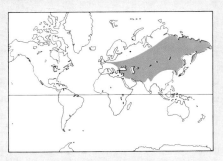

Habitat: arid or semiarid areas varying from
rocky mountain slopes and steppes to cultivated
fields.

Size: ranges from head-body length
5.3–10.2cm (2–4in), tail length 0.7–1.1cm
(0.3–0.4in), weight 50g (1.8oz) in the
Dzungarian hamster to head-body length
20–28cm (7.9–11in), weight 900g (32oz) (no
tail) in the Common or Black-bellied hamster.

Gestation: ranges from 15 days in the Golden
hamster to 37 days in the White-tailed rat.

Longevity: 2–3 years.

Mouse-like hamsters
Genus *Calomyscus*
Iran, Afghanistan, S Russia, Pakistan. Five
species including: **Mouse-like hamster**
(*C. bailwardi*).

Rat-like hamsters
Genus *Cricetulus*
SE Europe, Asia Minor, N Asia. Eleven species
including: **Korean gray rat** (*C. triton*).

Common hamster or Black-bellied hamster
Cricetus cricetus.
C Europe, Russia.

Golden hamsters
Genus *Mesocricetus*
E Europe, Middle East. Four species including:
Golden hamster (*M. auratus*).

Dwarf hamsters
Genus *Phodopus*
Siberia, Mongolia, N China. Three species
including: **Dzungarian hamster** (*P. sungorus*).

UNTIL the 1930s the Golden hamster was known only from one specimen found in 1839. However, in 1930 a female with 12 young was collected in Syria and taken to Israel. There the littermates bred and some descendants were taken to England in 1931 and to the USA in 1938 where they proliferated. Today the Golden hamster is one of the most familiar pets and laboratory animals in the West. The other hamster species are less well-known, though the Common hamster has been familiar for many years.

Most hamsters have small, compact, rounded bodies with short legs, thick fur and large ears, and prominent dark eyes, long whiskers and sharp claws. Most have cheek pouches which consist of loose folds of skin starting from between the prominent incisors and premolars and extending along the outside of the lower jaw. When hamsters forage they can push food into the pouches which then expand, enabling them to carry large quantities of food to the underground storage chamber—a very useful adaptation for animals that live in a habitat where food may occur irregularly but in great abundance. The paws of the front legs are modified hands, giving great dexterity to the manipulation of food. Hamsters also use a characteristic forward squeezing movement of the paws as a means of emptying their cheek pouches of food. Common hamsters are reputed to inflate their cheek pouches with air when crossing streams, presumably to create extra buoyancy.

Hamsters are mainly herbivorous. The Common hamster may hunt insects, lizards, frogs, mice, young birds and even snakes, but such prey contributes only a small amount to the diet. Normally hamsters eat seeds, shoots and root vegetables, including wheat, barley, millet, soybeans, peas, potatoes, carrots, beets as well as leaves and flowers. Small items (such as millet seeds) are carried to the hamster's burrow in its pouches, larger items (such as potatoes) in

its incisors. Food is either stored for the winter, eaten on returning to underground quarters, or, in undisturbed conditions, eaten above ground. One Korean gray rat managed to carry 42 soybeans in its pouches. The record for storage in a burrow probably goes to the Common hamster: chambers of this species have been found to contain as much as 90kg (198lb) of plant material collected by one hamster alone. Hamsters spend the winter in hibernation in their burrows, only waking on warmer days to eat food from their stores.

As children's pets, hamsters have a reputation for gentleness and docility. In the wild, however, they are solitary and exceptionally aggressive towards members of their own species. These characteristics may result from intense competition for patchy but locally abundant food resources, but may also serve to disperse population throughout a particular area or habitat. Large species, such as the Common hamster and the Korean gray rat, also behave aggressively towards other species, and have been known to attack dogs or even people when threatened. To defend itself from attack by predators the Korean gray rat may throw itself on its back and utter piercing screams.

Species studied in the laboratory have been shown to have acute hearing. They communicate with ultrasounds (high frequency sounds) as well as with squeaks audible to the human ear. Ultrasounds appear to be most important between males and females during mating and perhaps synchronize behavior. Their sense of smell is also acute. It has recently been shown that the Golden hamster can recognize individuals, probably from flank gland secretions, and that males can detect stages of a female's estrous cycle by odors, and even recognize a receptive female from the odors of her vaginal secretions.

Most hamsters have an impressive capacity for reproduction and become sexually mature soon after weaning (or even during it). Female Common hamsters become receptive to males at 43 days and can give birth at 59 days. Golden hamsters have slightly slower development and become sexually mature between 56 and 70 days. In the wild they probably breed only once, or occasionally twice, per year during the spring and summer months but in captivity they can breed year round. Their courtship is simple and brief, as befits animals that are in general solitary creatures and meet only to copulate. Odors and restrained movement suffice to indicate that the partners are

▲ **The Common hamster** was once widespread across Central Europe and the Soviet Union. Following the introduction of modern agricultural methods, however, its numbers—at least in Europe—have declined dramatically. Its front paws are extremely dextrous, making it easy to place food in its cheek pouches.

◄ **The Dzungarian hamster,** ABOVE, little known to natural historians, lives in Siberia, Mongolia and northern China. Although only a mere 5cm (2in) or so in length it displays all the adaptations of its larger relatives.

ready and willing to mate. Immature animals or females not in heat will either attack or be attacked by other individuals. As soon as a pair have copulated they separate and may never meet again. The female builds a nest for the young in her burrow from grass, wool and feathers and gives birth after about 16–20 days in the Common hamster. The young are born hairless and blind and are cared for by the female alone. During this time she may live off her food store in another section of the burrow. The young are weaned at about three weeks of age in the Golden hamster. In

the slowest developing species, the Mouse-like hamster, adult coloration and size may not be reached until six months old.

Hamsters are considered serious pests to agriculture in some areas. In some countries dogs are trained to kill them. Chinese peasants sometimes catch the large rat-like hamsters and eat them, and the Common hamster is trapped for its skin. Despite these pressures, none of the hamsters appears to be endangered at the moment, perhaps because most live in regions inhospitable to man and have such a high reproduction rate. JF

GERBILS

Subfamily: Gerbillinae
Eighty-one species in 15 genera.
Family: Muridae.
Distribution: Africa, Asian steppes from Turkey and SW Russia in the west to N China in the east.

Habitat: desert, savanna, steppe, rocks, cultivated land.

Size: ranges from head-body length 6.2–7.5cm (2.4–2.9in), tail length 7.2–9.5cm (2.8–3.7in), weight 8–11g (0.3–0.4oz) in Henley's (pygmy) gerbil to head-body length 15–20cm (5.9–7.9in), tail length 16–22cm (6.2–8.7in), weight 115–190g (4–6.7oz) in the Antelope rat or Indian red-footed gerbil.

Gestation: 21–28 days.

Longevity: usually 1–2 years.

To most people a gerbil is an attractive pet rodent with large dark eyes and a furry tail. The animal they have in mind, however, is the Mongolian gerbil, just one of the many species of gerbils, jirds and sandrats that belong to the largest group of rodents in Africa and Asia adapted to arid conditions.

The different genera of gerbils are very distinctive; most are either mouse-like or rat-like. Within genera, however, there are numerous ways in which gerbils differ from one another—in size, color, length of tail, the fashioning of the tail tuft, the color of nails. Variations can even be found within well-defined species. Given such complexities it is impossible to be certain about how many species exist and sometimes about the genus to which a species might belong.

Most gerbils live in arid climates in arid habitats, to both of which they have adapted in interesting ways. For any animal to survive and live it must not lose more water than it normally takes in. (Water loss usually occurs by evaporation from the skin, in air exhaled from the lungs, and by urination and defecation.) The predicament of the gerbil is that it has a large body-surface compared with its volume, and that it has to find ways of obtaining water and minimizing loss.

The gerbil therefore cannot afford to sweat and indeed cannot survive temperatures of about 45°C (113°F) for more than about two hours. During the day it lives below the surface, at a depth of about 50cm (20in), where the temperature is constant during both day and night, at about 20–25°C (68–77°F). Often the burrow entrance is blocked during the day. Most species are therefore nocturnal: the only gerbils that live on the surface in daytime

The 15 genera of gerbils

Ammodillus imbellis
N Kenya, Somalia, E Ethiopia inhabiting steppe and desert.

Brachiones przewalskii (**Przewalski's gerbil** or **jird**)
N China and Mongolian Republic inhabiting desert and steppe.

Desmodilliscus braueri (**Short-eared rat**)
Senegal E to S Sudan inhabiting desert, savanna and wooded grassland.

Desmodillus auricularis (**Namaqua gerbil** or **Cape short-toed gerbil**)
S Africa inhabiting desert, savanna, steppe.

Dipodillus, 3 species
W Sahara and African countries bordering the Mediterranean inhabiting desert and semidesert. Species include **Simon's dipodil** (*D. simoni*).

Gerbillurus, 4 species
S Africa inhabiting savanna and desert. Species include **Brush-tailed gerbil** (*G. vallinus*).

Gerbillus (**smaller gerbils**) 34 species
N Africa, Middle East, Iran, Afghanistan, W India, inhabiting desert, semidesert and coastal areas. Species include **Hairy-footed gerbil** (*G. latastei*), **Pallid gerbil** (*G. perpallidus*), **Wagner's gerbil** (*G. dasyurus*).

Meriones (**jirds**), 14 species
N Africa, Turkey, Middle East, Iran, Afghanistan, NW India, Mongolia, N China inhabiting savanna, desert and steppe. Species include **Indian desert jird** (*M. hurrianae*), **King jird** (*M. rex*), **Mongolian gerbil** (*M. unguiculatus*).

Microdillus peeli
N Kenya, Somalia, E Ethiopia inhabiting desert and steppe.

Pachyuromys duprasi (**Fat-tailed gerbil**)
N African countries bordering on the Sahara, inhabiting desert and semidesert.

Psammomys, 2 species
N African countries bordering on the Sahara inhabiting desert and semidesert. Species include **Fat sand rat** or **Fat jird** (*P. obesus*).

Rhombomys opimus (**Great gerbil**)
Afghanistan, SW USSR, Mongolia and N China inhabiting steppe and desert.

Sekeetamys calurus (**Bushy-tailed gerbil**)
E Egypt, S Israel, Jordan, Saudi Arabia inhabiting desert.

Tatera (**larger gerbils**), 9 species
S and E Africa, SW Asia, inhabiting savanna, steppe and desert.

Taterillus, 7 species
Senegal E to S Sudan and S to Tanzania inhabiting desert, savanna and wooded grassland. Species include: **Emin's gerbil** (*T. emini*), **Harrington's gerbil** (*T. harringtoni*).

The genera are grouped as follows:
Gerbillus group: *Ammodillus, Dipodillus, Gerbillus, Microdillus, Pachyuromys, Sekeetamys*
Meriones group: *Brachiones, Meriones, Psammomys, Rhombomys*
Tatera group: *Desmodilliscus, Desmodillus, Gerbillurus, Tatera, Taterillus*

▲ **The digestive system of the female gerbil** has to extract sufficient liquid from her food to meet the requirements of herself and her young. These are born in a pre-mature condition, and may have to be suckled for about three weeks. The litter of this Mongolian gerbil is of average size.

are the "northern" species, for example the Great gerbil or the Mongolian jird, though some jirds that live further south emerge during the day in winter.

Often in the dry world of the gerbil the only food available is dry seeds. The animal's nocturnal activity enables it to make the most of these. By the time it emerges it finds the seeds permeated with dew, and by taking them back to its burrow where the humidity is relatively high it can improve

the water content further. Water is extracted efficiently from the food by the gerbil's digestive system, thus minimizing water loss in the feces, and retained by the efficient kidneys which produce only a few drops of concentrated urine.

Other features of the gerbil seem to be adaptations to reduce the risk of being captured and killed by predators. All gerbils take the color of the ground on which they live—even subpopulations of a single species living in different habitats: gerbils found on dark lava sands are dark brown whereas members of the same species living on red sand are red. The purpose of the match seems to be to conceal the gerbil from flying predators, though its effectiveness might seem to be compromised by the tail ending in a tuft of contrasting color. This probably acts as a decoy: it distracts a predator and if caught will come away from the gerbil's body with part of or the entire tail.

Another special structure of the gerbil is a particularly large middle ear, which is largest in those species that live in open habitats. It enables the animal to hear sounds of low frequency, such as the wing beats of an owl. The eyes of the gerbil are also positioned so that they give it a wide field of vision. This may also help the animal to become aware of predators.

The geographical range of gerbils can be divided in three major areas. The first is Africa south of the Sahara, mainly savanna and steppes, but also the Namib and Kalahari deserts. Here the temperature does not fall below freezing in the winter. The second area includes the "hot" deserts and semidesert regions along the Tropic of Cancer in North Africa and Southwest Asia plus Som-

alia, east Ethiopia and north Kenya. The third covers the deserts, semideserts and steppes in Central Asia where winter temperatures fall below freezing.

To a limited extent the genera of gerbils can be formed into groups that inhabit each of these three areas. Gerbils grouped with the *Tatera* species occur in the first area, except for the Antelope rat. The *Gerbillus*-type gerbils occur in the second area, but only the *Meriones* group live in the third area, though some also live in the second area.

Gerbils are basically vegetarians, eating various parts of plants—seeds, fruits, leaves, stems, roots, bulbs etc—but many species will eat anything they encounter, including insects, snails, reptiles and even other small rodents. Some species are very specialized, living on one type of food. The nocturnal *Gerbillus* species often search for wind-blown seeds in deserts. The large Fat sand rat, the Great gerbil and the Antelope rat are basically herbivorous. The Fat sand rat is so specialized that it only occurs where it can find salty succulent plants. The Antelope rat depends on fresh food all year round, so it tends to occur near irrigated crop fields. Wagner's gerbil has such a liking for snails

▼ **Representative species of gerbils.** (1) A South African pygmy gerbil (*Gerbillurus paeba*) grooming its muzzle and spreading secretions. (2) A Tamarisk gerbil (*Meriones tamariscinus*) exposing its ventral gland. (3) A Libyan jird (*Meriones libycus*) making an attack. (4) A Cape short-eared gerbil (*Desmodillus auricularis*) making a submissive crouch. (5) A Great gerbil (*Rhombomys opimus*) with a heap of sand and feces or urine. (6) *Gerbillus gerbillus* (one of the smaller gerbils) marking sand with secretions from its ventral gland. (7) A female Mongolian gerbil (*Meriones unguiculatus*) with hair raised darting away from a male (part of the mating sequence). (8) A Fat sand rat (*Psammomys obesus*) holding and sniffing a ball of sand and urine.

The Communal Life of the Mongolian Gerbil

Mongolian gerbils live in large social groups which at their largest, in summer, consist of 1–3 adult males, 2–7 adult females and several subadults and juveniles; their home is a single burrow. Detailed studies have demonstrated that they engage in various activities as a group. For example they all collect food to hoard for the winter. They also spend the winter together in their burrow. Their integrity as a community seems, under normal circumstances, to be jealously guarded. Strange gerbils—and other animals—are chased off.

The interesting question that arises from this situation is: who are the parents of the subadults and juveniles? It is not evident from the behavior of males and females within the community, even though they have been observed to form pairs within the groups studied.

They may be the offspring of young adults that migrated from another burrow, but in theory this is unlikely. If these animals were to leave late in the summer to establish their own burrow community elsewhere they would be vulnerable to predators, they would suffer the effects of bad weather, and possibly they would have to contend with other gerbils into whose territories they might wander (when population densities are high there may be as many as 50 burrows per ha, 20 per acre). Moreover they would lose the food they had helped to collect for winter. The most serious objection, however, is that this pattern would perpetuate inbreeding and produce genetic problems.

The solution to the problem has come from the study of animals in captivity. This showed firstly that communal groups do remain stable and territorial, but when females are in heat they leave their own territories and visit neighboring communities to mate. They then return to their own burrows where their offspring will eventually grow up under the protection and care of not their mother and father but of their mother and their uncles.

3

4

8

in its diet that it virtually threatens the existence of local snail populations: big piles of empty shells are found outside this gerbil's burrows. Most gerbils in fact take the precaution of carrying their food back to their burrows and consuming it there. Species that live in areas with cold winters must hoard in order to survive. One Mongolian gerbil was found to have hidden away 20kg (44lb) of seeds in its burrow. The Great gerbil not only hoards plants in its burrow but constructs large stacks outside. Some have been found measuring a meter in height and three in length (3 × 10ft).

As yet the social organization of gerbils has been little studied. Species that live in authentic deserts, whatever genera and groups they belong to, tend to lead solitary lives, though burrows are often found to be close enough together that colonies could exist. Perhaps because the supply of food cannot be guaranteed in such an environment each animal fends for itself. By con-

trast species from savannas, where food is more abundant, are more social: there have been reports of stable pairs being formed and even of family structures emerging.

The most complex social arrangements have been found in those species within the *Meriones* group that live in areas with cold winters. Groups larger than families gather in single burrows, perhaps to keep each other warm, but also perhaps to guard food supplies. The best-known example is the Great gerbil of the Central Asian steppes which lives in large colonies composed of numerous subgroups which themselves have developed from male-female pairs. A similar social structure is found among Mongolian gerbils, but the *Meriones* species that occur in the hot and cooler areas of North Africa and Asia are reported to be solitary in the hot climates but social in the cooler.

Likewise reproduction seems to be linked to climate and food. Desert-dwelling species, like most desert rodents, give birth after the rainy season. Gerbils in areas where fresh food is available may reproduce all the year round, a female perhaps giving birth to two or three litters a year.

The number of young born can vary between one and 12, but the mean lies between three and five according to species. The young appear in a pre-mature state—hairless, eyes closed, unable to regulate their body temperatures. For about 20 days they then depend on their mother's milk. Where there is a breeding season only those born early within it become sexually mature so as to breed in the same season (when aged about two months). Those born later become sexually mature after about six months and breed during the next season.

Most gerbils live in areas of the world uninhabited by man. When the two do come into contact, in the Asian steppes and in India for example, the natural activities of the gerbils make them man's natural enemy. When collecting food to hoard for winter they will pilfer from crops. When burrowing they can cause great damage to pasture, and to irrigation channels, the embankments of road and railways, even to the foundations of buildings. They also carry the fleas that transmit deadly diseases, such as plague and the skin disease Leishmaniasis. Though they serve mankind in medical research, man is keen to eradicate the gerbil when it interferes with his own life. Many gerbil burrows are destroyed, by gasing and plowing, even though some of them may have been used by generations of gerbils for hundreds of years.   GA

DORMICE

Families: Gliridae, Seleviniidae
Eleven species in 8 genera.
Distribution: Europe, Africa, Turkey, Asia,
Japan.

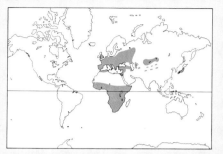

Habitat: wooded and rocky areas, steppe,
gardens.

Size: head-body length
6.1–19cm
(2.4–7.5in); tail
length 4–16.5cm
(1.6–6.5in); weight
15–200g (0.5–7oz).

Gestation: 21–32 days.
Longevity: 3–6 years in wild.

Family Gliridae
Species include: **African dormouse** or **Black
and white dormouse** (*Graphiurus murinus*);
Common dormouse or **Hazel mouse**
(*Muscardinus avellanarius*); **Edible dormouse** or
Fat dormouse or **Squirrel-tailed dormouse**
(*Glis glis*); **Forest dormouse** or **Tree dormouse**
(*Dryomys nitedula*); **Garden dormouse** or
Orchard dormouse (*Eliomys quercinus*);
Japanese dormouse (*Glirulus japonicus*);
Mouse-like dormouse or **Asiatic dormouse**
(*Myomimus personatus*).

Family Seleviniidae
Desert dormouse (*Selevinia betpakdalensis*).

THE dormice or Gliridae originated at least as early as the Eocene era (60–40 million years ago). In the Pleistocene era (2 million–10,000 years ago) giant forms lived on some Mediterranean islands. Today dormice are the intermediates, in form and behavior, between mice and squirrels. Key features of dormice are their accumulations of fat and their long hibernation period (about seven months in most European species). The Romans fattened dormice in a special enclosure (Latin *glirarium*) while the French have a phrase "To sleep like a dormouse," similar to the English phrase "To sleep like a log."

Dormice are very agile. Most species are adapted to climbing but some also live on the ground (eg Garden and Forest dormice). The Mouse-like dormouse is the only dormouse that lives only on the ground. The four digits of the forefeet and the five digits of the hind feet have short, curved claws. The underside of each foot is bare with a cushion-like covering. The tail is usually bushy and often long, and in some species (Fat dormouse, Common dormouse, Garden dormouse, Forest dormouse, African dormouse) it can come away when seized by other dormice or predators. Hearing is particularly well-developed, as is the ability to vocalize. Fat, Common, Garden and African dormice make use of clicks, whistles and growlings in a wide range of behavior—antagonistic, sexual, explorative, playful.

Dormice are the only rodents that do not have a cecum, which indicates a diet with little cellulose. Analysis of the contents of the stomachs of dormice has shown that they are omnivores whose diet varies according to season and between species according to region. The Edible and Common dormice are the most vegetarian, eating quantities of fruits, nuts, seeds and buds. Garden, Forest and African dormice are the most carnivorous—their diets include insects, spiders, earthworms, small vertebrates, but also eggs and fruit. In France 40–80 percent of the diet of the Garden dormouse consists of insects, according to region and season. But there is also another factor at work. In summer the Garden dormouse eats mainly insects and fruit, in fall little except fruit, even though the supply of insects at this time of year is plentiful. The change in the content of the diet is part of the preparation for entering hibernation; the intake of protein is reduced, sleep is induced.

In Europe dormice hibernate from October to April with the precise length varying between species and according to region.

During the second half of the hibernation period they sometimes wake intermittently, signs of the onset of the hormone activity that stimulates sexual activity. Dormice begin to mate as soon as they emerge from hibernation, females giving birth from May onwards through to October according to age. (Not all dormice that have recently become sexually mature participate in mating.) The Edible and Garden dormice produce one litter each per year but Common and Forest dormice can produce up to three. Vocalizations play an important part in mating. In the Edible dormouse the male emits calls as he follows the female, in the Garden dormouse the female uses whistles to attract the male. Just before she is due to give birth the female goes into hiding and builds a nest, usually globular in shape and located off the ground, in a hole in a tree or in the crook of a branch for example. Materials used include leaves, grass and moss. The Garden and Edible dormice use hairs and feathers as lining materials. The female Garden dormouse scent marks the

▷ **Clinging tight.** OVERLEAF Most Garden dormice in fact live in forests across central Europe, though some inhabit shrubs and crevices in rocks.

◀ **Coming down,** a Common dormouse. This richly colored species lives in thickets and areas of secondary growth in forests. It is particularly fond of nut trees.

◀ **Looking out,** BELOW LEFT, an Edible dormouse. This squirrel-like dormouse also lives in trees, but also inhabits burrows. It has a great liking for fruit.

▼ **Curled up,** a Common dormouse in hibernation. For the long winter sleep this dormouse resorts to a nest, either in a tree stump, or amidst debris on the ground, or in a burrow. The length of hibernation is related to the climate, but can last for as long as nine months.

area around the nest and defends it.

Female dormice give birth to between two and nine young, with four being the average litter size in about all species. The young are born naked and blind. In the first week after birth they become able to discriminate between smells, though the exchange of saliva between mother and young appears to be the means whereby mother and offspring learn to recognize each other. This may also aid the transition from a milk diet to a solid food one. At about 18 days young become able to hear and at about the same time their eyes open. They become independent after about one month to six weeks. Young dormice then grow rapidly until the time for hibernation approaches, when their development slows. Sexual maturity is reached about one year after birth, towards the end of or after the first hibernation.

Dormice populations are usually less dense than those of many other rodents. There are normally between 0.1 and 10 dormice per ha (0.04–4 per acre). They live in small groups, half of which are normally juveniles, and each group occupies a home range, the main axis of which can vary from 100m (330ft) in the Garden dormouse to 200m (660ft) in the Edible dormouse. In urban areas the radiotracking of Garden dormice has indicated that their home range is of an elliptic shape and related to the availability of food. In fall the home range is about 1,000sq m (about 10,800sq ft). A recent study of the social organization of the Garden dormouse has revealed significant changes in behavior in the active period between hibernations. In the spring, when Garden dormice are emerging from hibernation, males form themselves into groups in which there is a clearcut division between dominant and subordinate animals. As the groups are formed some males are forced to disperse. Once this has happened, although groups remain cohesive, behavior within them becomes more relaxed so that by the end of the summer the groups have a family character. In the fall social structure includes all categories of age and sex. Despite the high rate of renewal of its members a colony can continue to exist for many years.

The Desert dormouse, which is placed in a family of its own, occurs in deserts to the west and north of Lake Balkhash in eastern Kazakhstan, Central Asia. It has exceptionally dense, soft fur, a naked tail, small ears, and sheds the upper layers of skin when it molts. It eats invertebrates, such as insects and spiders and is mostly active in twilight and night. It probably hibernates in cold weather. CB

JUMPING MICE, BIRCHMICE AND JERBOAS

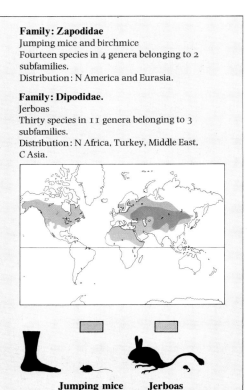

Family: Zapodidae
Jumping mice and birchmice
Fourteen species in 4 genera belonging to 2 subfamilies.
Distribution: N America and Eurasia.

Family: Dipodidae.
Jerboas
Thirty species in 11 genera belonging to 3 subfamilies.
Distribution: N Africa, Turkey, Middle East, C Asia.

Jumping mice
and birchmice   Jerboas

▼ **Mouse on stilts,** a Meadow jumping mouse. Its most important food is grass seeds, many of which are picked from the ground, but some are pulled from grass stalks. To eat timothy seeds, for example, the mouse may climb up the stalk and cut off the seed heads. At other times the mouse will reach as high as it can, cut off the stalk, pull the top portion down, cut it off again, etc, until the seed head is reached. The stem, leaf stalk and uneaten seeds are usually left in a criss-cross pile of match-length parts. Another favorite seed is that of the touch-me-not. These seeds taste like walnut, but have bright turquoise endosperm which turns the entire contents of the stomach brilliant blue.

THE name "jumping mouse" is something of a misnomer. All **jumping mice** are equipped for jumping, with long back feet, and long tails to help them maintain their balance in the air, but the most common species, those belonging to the genus *Zapus*, are more likely to crawl under vegetation or to run by making a series of short hops rather than long leaps. However, the Woodland jumping mouse often moves by bounding 1.5–3m (5–10ft) at a time. In addition to the feet and tail, the outstanding characters of jumping mice are their colorful fur and their grooved upper incisors. The function of the groove is unknown: it may improve the efficiency of the teeth as cutting tools, or it may just strengthen them.

Jumping mice are not burrowers. They live on the surface of the ground, though their nests may be underground or in a hollow log or other protected places, and the hibernating nest is often at the end of a burrow in a bank or other raised area. For the most part, however, jumping mice hide by day in clumps of vegetation. They also usually travel about in thick herbaceous cover, though they will use runways or sometimes burrows of other species when present.

The habitat of jumping mice varies, but lush grassy or weedy meadows are the preferred habitat of Meadow jumping mice, although they are often quite abundant in wooded areas, in patches of heavy vegetation (especially of touch-me-not, *Impatiens*), particularly in areas where there are no Woodland jumping mice. Woodland jumping mice usually occur in woods, almost never in open areas, and are most abundant in wooded areas with heavy ground cover. Moisture is often mentioned as a factor favorable to jumping mice, but it seems more important as a factor favoring the development of lush vegetation rather than as a factor directly favoring the mice.

Jumping mice are profound hibernators, hibernating for 6–9 months of the year according to species, locality and elevation. The Meadow jumping mouse in the eastern USA usually hibernates from about October to late April. Individuals that hibernate successfully put on about 6–10g (0.21–0.35oz) of fat in the two weeks prior to entering hibernation. This they do by sleeping for increasingly longer periods until they attain deep hibernation with their body temperature just a little above freezing. Their heart rate, breathing rate and all bodily functions drop to low levels. However, the animals wake about every two weeks, perhaps urinate, then go back to sleep. In the spring the males appear above ground about two weeks before the females. Of the animals active in the fall, only about a third—the larger ones—are apparently able to put on the layer of fat, enter hibernation and awaken in the spring. The remainder—young individuals or those unable to put on adequate fat—apparently perish during the winter retreat.

Jumping mice give birth to their young in a nest of grass or leaves either underground, in a clump of vegetation, in a log or in some other protected place. Gestation takes about 17 or 18 days, or up to 24 if the female is lactating. Each litter contains about 4–7 young. Litters may be produced at any time between May and September, but most enter the world in June and August. Most females probably produce one litter per year.

Meadow jumping mice eat many things, but seeds, especially those from grasses, are the most important food. The seeds eaten change with availability.

The major animal foods eaten by jumping mice are moth larvae (primarily cutworms) and ground and snout beetles. Also important in the diet is the subterranean fungus *Endogone*. This forms about 12 percent of the diet (by volume) in the meadow jumping mice, and about 35 percent of the diet in the Woodland jumping mouse.

Birchmice differ from jumping mice in having scarcely enlarged hind feet and upper incisors without grooves. Moreover their legs and tail are shorter than those of jumping mice, yet they travel by jumping and climb into bushes using their outer toes to hold on to vegetation and their tails for partial support. Birchmice, also unlike jumping mice, dig shallow burrows and make nests of herbaceous vegetation underground.

Like jumping mice, birchmice are active primarily by night. They can eat extremely large amounts of food at one time and can also spend long periods without eating. Their main foods are seeds, berries and insects. Birchmice hibernate in their underground nests for about half of the year. It has been suggested that *Sicista betulina* spends the summer in meadows but hibernates in forest. Gestation probably lasts about 18–24 days and parental care for another four weeks. Studies of *S. betulina* in Poland have shown that one litter a year is produced and that any female produces only two litters during her lifetime.   JOW

Jerboas are small animals built for jumping. Their hind limbs are elongated—at least four times as long as their front legs—and in

▲ **The long-legged burrow-dweller.** The burrows used by desert jerboas can run as deep as 2m (about 6ft). At this depth the animals are insulated against fluctuations in outside temperature. Burrows can be complex, with passages off the main chambers from which the animals can "burst" through the soil to the surface when disturbed or threatened. There is usually one jerboa to a burrow, but the desert jerboas (as here) are fairly sociable and live in loose colonies with two or three animals often using the same nest. In spite of having long hind legs desert jerboas are adept at burrowing, using their short front feet and incisor teeth for digging, and their hind feet for throwing away the sand.

most species the three main foot bones are fused into a single "cannon bone" for greater strength (in the subfamilies Dipodinae and Euchoreutinae, but not in the Cardiocraniinae). The outer toes on the hind feet are small in size and do not touch the ground in species with five toes. In other species the outer toes are absent, so there are three toes on each hind foot. One species, *Allactaga tetradactylus*, has four toes. Jerboas living in sandy areas have tufts of hairs on the undersides of the feet which serve as snowshoes on soft sand and help them to maintain traction and to kick sand backwards when burrowing. These jerboas also have tufts of hair to help keep sand out of the ears.

Some jerboas, those belonging to the genus *Jaculus* for example, have a fold of skin which can be pulled forward over the nostrils when burrowing. Jerboas use their long tails as props when standing upright and as balancing organs when jumping. Jerboas are nocturnal and have large eyes.

The well-developed jumping ability of jerboas enables them to escape from predators as well as to move about, though they also move by slow hops. Otherwise only the hind legs are used in moving; the animal walks on its hind legs. The front feet then can be used for gathering food. Jumps of 1.5–3m (5–10ft) are used when the animal moves rapidly. Desert jerboas, *Jaculus*, can jump vertically to nearly 1m (3ft).

Jerboas feed primarily on seeds, but sometimes also on succulent vegetation. In some areas they may be a pest to growers of melons. They also eat insects, and one species, *Allactaga sibirica*, feeds primarily on beetles and beetle larvae. One individual of *Salpingotus* in captivity ate only invertebrates. In *Dipus* all individuals in a population emerge for their nightly forays at about the same time, and move by long leaps to their feeding grounds, which may be some distance away. There they feed on plants, especially those with milky juices, but they also smell out underground sprouts and insect larvae in underground galls (gallnuts). Like pocket mice, jerboas do not drink water; they manufacture "metabolic water" from food.

Some jerboas hibernate during the winter, surviving off their body fat. Also, some species enter torpor during hot or dry periods. They are generally quiet, but when handled will sometimes shriek or make grunting noises. Some species have been known to tap with a hind foot when inside their burrows.

In northern species mating first occurs shortly after the emergence from hibernation, but most female jerboas probably breed at least twice in a season, producing litters of between two and six young.

There are four kinds of burrows used by various jerboas, depending on their habits and habitats: temporary summer day burrows for hiding during the day, temporary summer night burrows for hiding during nightly forays, permanent summer burrows used as living quarters and for producing young, and permanent winter burrows for hibernation. The two temporary burrows are simple tubes, which are in length respectively 20–50cm (8–20in) and 10–20cm 4–8in).

The permanent summer burrows have secondary chambers for food storage, and the permanent winter burrows are at least 22cm (9in) below the surface and also have secondary chambers. Some species build a mound at the entrance, and some provide one or more accessory exits. The Comb-toed jerboa lives in sand dunes where it burrows into the protected side of the dunes. Most of the burrows that have been dug up consisted of single passages. This is one species from which tapping sounds have been recorded from within the burrow. JOW

Jumping mice and birchmice

Subfamily Zapodinae
Jumping mice

Four species in three genera. Distribution: N America with one species in China (*Eozapus setchuanus*). Habitat: meadows, moors, steppes, thickets, woods. Size: head-body length 7.6–11cm (3–4.3in), tail length 15–16.5cm (5.9–6.5in), weight up to 28g (1oz). Gestation: about 17–21 days in *Zapus* and *Napaeozapus*. Longevity: probably one or two years. Species: *Eozapus setchuanus*. *Napaeozapus insignis* (**Woodland jumping mouse**). *Zapus hudsonius* (**Meadow jumping mouse**), *Z. princeps*.
(The animal sometimes classified separately as *Z. trinotatus* is here considered to be synonymous with *Z. princeps*.)

Subfamily Sicistinae
Birchmice

Nine species in one genus. Distribution: Eurasia. Habitat: mainly birch forests but other habitats also. Size: head-body length 5–9cm (1.9–3.5in), tail length 6.5–10cm (2.6–3.9in), weight up to 28g (1oz). Gestation: 18–24 days in *Sicista betulina*. Longevity: probably less than a year. Species of the genus *Sicista* include *S. betulina* (N Eurasia), *S. caucasica* (W Caucasus and Armenia), *S. concolor* (China), *S. subtilis* (USSR and E Europe).

Jerboas

Habitat: desert, semidesert, steppe, including patches of bare ground. Size: head-body length 4–26cm (1.6–10in), tail length 7–30cm (2.7–11in), hind foot length 2–4cm (0.8–10in). Gestation: 25–42 days. Longevity: probably less than two years.

Subfamily Cardiocraniinae
Genera: **Five-toed dwarf jerboa** (*Cardiocranius paradoxus*), W China, Mongolia. *Salpingotulus michaelis* Pakistan. **Three-toed dwarf jerboas** (Genus *Salpingotus*) Asia in deserts. Three species.

Subfamily Dipodinae
Genera: **four-** and **five-toed jerboas** (Genus *Allactaga*) NE Africa (Libyan desert), Arabian peninsula, C Asia. Eleven species. **Lesser five-toed jerboa** (*Alactagulus pumilio*) S European Russia in clay and salt deserts and prairies. **Feather-footed jerboa** (*Dipus sagitta*) China and USSR. **Desert jerboas** (Genus *Jaculus*) N Africa, Russia, Iran, Afghanistan, Pakistan in various habitats (including desert, rocky areas, meadows). Five species. **Comb-toed jerboa** (*Paradipus ctenodactylus*) USSR in dry, sandy deserts. **Fat-tailed jerboas** (Genus *Pygeretmus*) USSR in salt and clay deserts. Three species. **Thick-tailed three-toed jerboa** (*Stylodipus telum*) Russia, Mongolia, China, in clay and gravel deserts.

Subfamily Euchoreutinae
Genera: **Long-eared jerboa** (*Euchoreutes naso*) China and Mongolia in sandy areas.

CAVY-LIKE RODENTS

Suborder: Caviomorpha
Eighteen families: 60 genera: 188 species.
Distribution: America, Africa, Asia.

Habitat: desert, grassland, savanna, forest.

Size: head-body length from 17cm (6.8in) in gundis to 134cm (53in) in the capybara; weight from 175g (6.2oz) in Speke's gundi to 64kg (141lb) in the capybara.

New World porcupines
Family: Erethizontidae
Ten species in 4 genera.

Cavies
Family: Caviidae
Fourteen species in 5 genera.

Capybara
Family: Hydrochoeridae
One species.

Coypu
Family: Myocastoridae
One species.

Hutias
Family: Capromyidae
Thirteen species in 4 genera.

Pacarana
Family: Dinomyidae
One species.

Pacas
Family: Agoutidae
Two species in 1 genus.

Agoutis and acouchis
Family: Dasyproctidae
Thirteen species in 2 genera.

Chinchilla rats
Family: Abrocomidae
Two species in 1 genus.

Spiny rats
Family: Echimyidae
Fifty-five species in 15 genera.

Chinchillas and viscachas
Family: Chinchillidae
Six species in 3 genera.

Degus or Octodonts
Family: Octodontidae
Eight species in 5 genera.

Tuco-tucos
Family: Ctenomyidae
Thirty-three species in 1 genus.

Cane rats
Family: Thryonomyidae
Two species in 1 genus.

THE familiar Guinea pig is a representative of a large group of South American rodents usually referred to as the caviomorph rodents and formally classified in a suborder, Caviomorpha. Most are large rodents, confined to South and Central America. Although they are very diverse in external appearance and are generally classified in separate families, they share sufficient characteristics to make it virtually certain that they constitute a natural, inter-related group.

Externally many caviomorphs have large heads, plump bodies, slender legs and short tails, as in the guinea pigs, the agoutis and the giant capybara, the largest of all rodents at over a meter (about 3ft) in length. Others, however, eg some spiny rats of the family Echimyidae, come very close in general appearance to the common rats and mice.

Internally the most distinctive character uniting these rodents is the form of the masseter jaw muscles, one branch of which extends forwards through a massive opening in the anterior root of the bony zygomatic arch to attach on the side of the rostrum. At its other end it is attached to a characteristic outward-projecting flange of the lower jaw. Caviomorphs are also characterized by producing small litters after a long gestation period, resulting in well-developed young. Guinea pigs, for example, usually have two or three young after a gestation of 50–75 days, compared with seven or eight young after only 21–24 days in the Norway (murid) rat.

This group is also frequently referred to as the hystricomorph rodents, implying that the various Old World porcupines, family Hystricidae, are closely related to the American "caviomorphs." Although they share all the caviomorph features mentioned above there has been considerable debate as to whether such features indicate a common ancestry or that they have evolved independently in two groups—a problem related to the question as to whether the caviomorphs reached South America from North America or from Africa. The African cane rats (family Thryonomyidae) are closely related to the Hystricidae but some other families, namely the gundis (Ctenodactylidae) and the rock rat (Petromyidae) are much more doubtfully related, showing only some of the "caviomorph" characters.

Most of the American caviomorphs are terrestrial and herbivorous but a minority, the porcupines (Erethizontidae) are arboreal and one group, the tuco-tucos (Ctenomyidae) are burrowers.   GBC

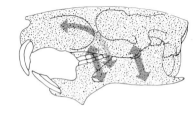

▲ **Distinguishing feature of cavy-like rodents.** The deep masseter muscle (blue) provides the gnawing action, extending forward through an opening in the zygomatic arch to attach to the muzzle. The lateral masseter (green) is only used in closing the jaw.

▶ **The face of a true cavy-like rodent.** Like most South American members of this suborder the degu has a plump, well-furred body and a large head.

▼ **A distant relative:** an Old World porcupine (Cape porcupine). Even porcupine young are born with quills.

African rock rat
Family: Petromyidae
One species.

Old World porcupines
Family: Hystricidae
Eleven species in 4 genera.

Gundis
Family: Ctenodactylidae
Five species in 4 genera.

African mole-rats
Family: Bathyergidae
Nine species in 5 genera.

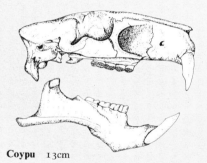

Coypu 13cm

Coypu

Mara

Capybara

Tuco-tuco

Skulls and teeth of cavy-like rodents.

Most cavy-like rodents have rather angular skulls and very strongly developed incisor teeth. The wearing surfaces of the four cheekteeth show enormous variation in pattern and complexity amongst the different species. Those of the coypu are typical of a large group of herbivorous species, including the agoutis and the American porcupines, and are closely paralleled in the Old World porcupines and cane rats. The teeth of the mara and of the capybara, although superficially very different in degree of complexity, resemble each other in being ever-growing as in the unrelated but also grass-eating voles and rabbits. At the other extreme the tuco-tucos have surprisingly simple cheekteeth considering that they feed mainly on roots and tubers.

NEW WORLD PORCUPINES

Family: Erethizontidae
Eleven species in 5 genera.
Distribution: N America (except SE USA),
S Mexico, C America, N S America.

Habitat: forest areas, open grasslands, desert,
canyon.

Size: ranges from head-body length 30cm (12in)
and weight 900g (32oz) in the prehensile-
tailed porcupines to head-body length 86cm
(34in) and weight 18kg (40lb) in the North
American porcupine.

Gestation: 210 days in the North American
porcupine.

Longevity: up to 17 years in the North
American porcupine.

Prehensile-tailed porcupines
Genus *Coendou*
S Panama, Andes from NW Colombia to N
Argentina, NW Brazil; forest areas. Two
species: **Prehensile-tailed porcupine**
(*C. prehensilis*), **South American tree porcupine**
(*C. bicolor*).

Upper Amazonian porcupine
Echinoprocta rufescens
C Colombia; forest areas.

North American porcupine
Erethizon dorsatum
Alaska, Canada, USA (except extreme SW, SE
and Gulf coast states), N Mexico; forest areas.

South American tree porcupines
Genus *Sphiggurus*
S Mexico, C America, S America as far S as
N Argentina; forest areas. Six species including:
Mexican tree porcupine (*S. mexicanus*), **South
American tree porcupine** (*S. spinosus*).

Thinned-spined porcupine
Chaetomys subspinosus
E and N Brazil, forest areas.

NEW World porcupines strongly resemble Old World porcupines, but New World porcupines are arboreal in habits, unlike Old World porcupines which are terrestrial. For such heavy-bodied animals they are excellent climbers with well-developed claws and unfurred soles on their large feet. The soles consist of pads and creases which increase the supporting surface and the gripping power of the feet. Individual genera have further modifications to improve their climbing abilities. The prehensile-tailed porcupines and the South American tree porcupines—the most arboreal genera—have smaller first digits on their hind feet than the other genera, but they are incorporated in the footpads which increases the width and the gripping power of the pads.

The same genera also have long spineless tails for grasping. Their tips form upward curls and have a hard skin (callus) on the upper surface. In the prehensile-tailed porcupines the tail contributes 9 percent of the total body weight; nearly half of the weight of the tail is composed of muscle fibers.

New World porcupines are very near-sighted, and have keen senses of touch, hearing and smell. They produce a variety of sounds—moans, whines, grunts, coughs, sniffs, shrieks, barks and wails. All porcupines have large brains and appear to have good memories.

In habits porcupines range from the North American porcupine, which is semi-arboreal, to prehensile-tailed and South American tree porcupines, which are specialized, arboreal feeders. All forms spend much of their time in trees, but even tree porcupines are known to come to the ground to feed and to move from one tree to another. In winter North American porcupines feed on conifer needles and on the bark of a variety of trees, except Red maple, White cedar and hemlock. During the summer these porcupines feed more frequently on the ground and select roots, stems, leaves, berries, seeds, nuts and flowers. In the spring they often come out from forested areas into meadows to feed on grasses in the evening hours. They will eat bark at all times of the year, and can be destructive to forest plantations. Prehensile-tailed and South American tree porcupines feed more on leaves and have many characteristics of an arboreal leaf-eater. However, they are also reported to feed on tender stems, fruits, seeds, roots, tubers, insects and even small reptiles.

In the North American porcupines the female reaches sexual maturity when about 18 months old. The estrous cycle is 29 days,

and these animals may have more than one period of estrus in a year. They have a vaginal closure membrane so females form a copulatory plug. The gestation period averages 210 days, and in both North American and prehensile-tailed porcupines usually one young is produced (rarely twins). The weight of the precocial newborn is about 400g (14oz) in prehensile-tailed porcupines and 600g (21oz) in North American porcupines. Lactation continues for 56 days, but the animals also feed on their own after the first few days. Porcupine young are born with their eyes open and are able to walk. They exhibit typical defensive reactions and within a few days are able to climb trees. These characteristics probably explain why infant mortality is very low. Porcupines grow for three or four years before they reach adult body size.

The home range of North American porcupines in summer averages 14.6ha (36

▲ **Sustained on a branch.** The Prehensile-tailed porcupines live mainly in the middle and upper layers of forests in Central and South America, only descending to the ground to eat. Although their claws are large, firm and stiff, they do not pin themselves to or cling to trees but rely for adhesion on their weight, a firm hold with their claws and, as seen here on the far left, their tail which can be coiled around branches and has a callus pad to provide grip.

▶ **In search of food,** a South American tree porcupine. More time is spent on the ground in summer than in winter. The prehensile-tailed porcupines are more herbivorous than other genera, eating large quantities of leaves, roots, stems, blossoms and fruit. Others have more developed tastes for insects and small reptiles.

acres). In winter, however, they do not range great distances—they stay close to their preferred trees and shelters. Prehensile-tailed porcupines can have larger ranges, though these vary from 8 to 38ha (20–94 acres). They are reported to move to a new tree each night, usually 200–400m (660–1,300ft) away, but occasionally up to 700m (2,300ft). Prehensile-tailed porcupines in South Guyana are known to reach densities of 50–100 individuals per sq km (130–260 per sq mi). They have daily rest sites, in trees, 6–10m (20–33ft) above the ground, usually on a horizontal branch. These porcupines are nocturnal, change locations each night and occasionally move on the ground during the day. Male prehensile-tailed porcupines are reported to have ranges of up to four times as large as those of females.

Porcupines in general are not endangered, and the North American porcupine can in fact be a pest. The fisher (a species of marten) has been reintroduced to some areas of North America to help control porcupines, one of its preferred prey. The fisher is adept at flipping the North American porcupine over so that its soft and generally unquilled chest and belly are exposed. The fisher attacks this area, killing and eating the porcupine from below. One study found that porcupines declined by 76 percent in an area of northern Michigan (USA) following the introduction of the fisher. Prehensile-tailed porcupines are frequently used for biomedical research, which contributes to the problem of conserving the genus, but the main threat is habitat destruction. In Brazil prehensile-tailed porcupines have been affected by the loss of the Atlantic forest, and the South American tree porcupine is included on the list of endangered species published by the Brazilian Academy of Sciences. One species of porcupine may have become extinct in historic times: *Sphiggurus pallidus*, reported in the mid-19th century in the West Indies, where no porcupines now occur.   CAW

▶ **Lord of the conifer forest.** This is the North American porcupine, found in forests across most of Canada, the USA and northern Mexico, but which is mainly terrestrial. It has relatively poor eyesight, cannot jump, moves slowly and clumsily, but frequently climbs trees to enormous heights, in search of food—twigs, leaves, berries, nuts. Its small intestine digests cellulose efficiently.

▼ **Almost a primate**—a North American porcupine in Alaska.

CAVIES

Family: Caviidae
Fourteen species in 5 genera.
Distribution: S America (mara in C and S Argentina).

Habitat: open areas in forests, semiarid thorn scrub, arboreal savanna, Chaco, pampas, high altitude puna, desert (scrub desert and grasslands for mara).

Size: in small cavies ranges from head-body length 22cm (8.7in), weight 300g (10.7oz) in the genus *Microcavia* to head-body length 38cm (15in), weight 1kg (2.2lb) in the genus *Kerodon*: in mara head-body length 50–75cm (19.7–30in), tail length 4.5cm (1.8in), weight 8–9kg (17.6–19.8lb).

Gestation: in small cavies varies from 50 days in genera *Galea* and *Microcavia* to 75 days in genus *Kerodon*; in mara 90 days.

Longevity: in small cavies 3–4 years (up to 8 in captivity); in mara maximum 15 years.

Guinea pigs and cavies
Genus *Cavia*.
S America, in the full range of habitats.
Coat: grayish or brownish agouti; domesticated forms vary. Five species: *C. aperea*, *C. fulgida*. *C. nana*, *C. porcellus* (**Domestic guinea pig**), *C. tschudii*.

Mara or Patagonian hare
Genus *Dolichotis*.
S America (C and S Argentina), occurring in open scrub desert and grasslands. Head and body are brown, rump is dark (almost black) with prominent white fringe round the base; belly is white. Two species: *D. patagonum* and *D. salinicolum*.

Genus *Galea*
S. America, in the full range of habitats.
Coat: medium to light brown agouti with grayish-white underparts. Three species: *G. flavidens*, *G. musteloides*, *G. spixii*.

Rock cavy
Kerodon rupestris.
NE Brazil, occurring in rocky outcrops in thorn-scrub.
Coat: gray, grizzled with white and black; throat is white, the belly yellow-white, the rump and backs of the thighs reddish.

Desert cavies
Genus *Microcavia*.
Argentina and Bolivia, in arid regions.
Coat: a coarse dark agouti, brown to grayish.
Three species: *M. australis*, *M. niata*, *M. shiptoni*.

Most people are familiar with cavies, but under a different and somewhat misleading name: the Guinea pig. "Guinea" refers to Guyana, a country where cavies occur in the wild; and the short, squat body gives this rodent a piggish appearance, at least to the imaginative eye. The pork-like quality of the flesh doubtless contributes to the use of the name.

The Guinea pig of pet stores and laboratories has little in common with its wild cousins besides a shared evolutionary history. The Domestic guinea pig was being raised for food by the Incas when the conquistadors arrived in Peru in the 1530s and is now found the world over with one exception: it no longer occurs in the wild. Wild cavies share the same squat body form as the Guinea pig, but their simple external appearance belies their ecological adaptability. Cavies are among the most abundant and widespread of all South American rodents.

All cavies (except the mara or Patagonian cavy, for which see pp 694–695) share a basic form and structure. The body is short and robust, the head is large, contributing about one-third of the total head-body length. The eyes are fairly large and alert, the ears large but close to the head. The fur is coarse and easily shed when the animal is handled. There is no tail. The forefeet are strong and flat, usually with four digits with sharp claws. The hind feet, with three clawed digits, are elongated markedly. They walk on their soles with the heels touching the ground. The incisors are short, and the cheekteeth, which are arranged in rows that converge towards the front of the mouth, have the shape of prisms and are evergrowing. Both sexes are alike, apart from each possessing certain specialized glands.

Cavies are very vocal, making a variety of chirps, squeaks, burbles and squeals. One genus, *Kerodon*, emits a piercing alarm whistle when frightened. *Galea* rapidly drum their hind feet on the ground when anxious.

Cavies first appeared in the mid-Miocene era of South America. Since their appearance, some 20 million years ago, the family Caviidae has undergone an extensive adaptive radiation, reaching peak diversity 5–2 million years ago, during the Pliocene (when there were 11 genera), then declining in the number of species to present levels (5 genera) during the Pleistocene (about 1 million years ago).

The 12 remaining species of the subfamily Caviinae (all cavies except *Dolichotis* species) are widely distributed throughout South America. All 12 species are to a degree specialized for exploiting open habitats. Cavies can be found in grasslands and scrub forests from Venezuela to the Straits of Magellan. But each cavy genus has also evolved to be able to exist in a slightly different habitat. *Cavia* is the genus most restricted to grasslands. In Argentina *Cavia aperea* is restricted to the humid pampas in the northeastern provinces. *Microcavia australis* is the desert specialist and is found throughout the arid Monte and Patagonian deserts of Argentina. Other *Microcavia* species, *M. niata* and *M. shiptoni*, occur in the arid high-altitude puna (subalpine zone) of Bolivia and Argentina. The specialized genus *Kerodon* is found only in rock outcrops called *lajeiros* which dot the countryside in the arid thorn scrub, or *caatinga*, of northeastern Brazil. *Galea* seem to be the jacks-of-all-trades of the cavies. *Galea* species are

▲ **Hutch in the wild.** Cavy genera occupy a variety of habitats. These include rocky outcrops in which the Rock cavy is common.

▶ **Grazing stock for the table.** The Domestic guinea pig has been bred by South American Indians for its meat for at least 3,000 years.

◀ **Rock cavy mating procedure.** An adult male blocks the path of a female (1), passes under her chin (2) and begins to mount (3).

▷ **Elegance attained.** OVERLEAF Selective breeding has produced Domestic guinea pigs of numerous strains. As pets they are bred for beauty, though in South America some have joined their ancestors in the wild.

found in all of the above habitats, and it is the only genus that occurs along with the other three genera.

Regardless of habitat, all cavies are herbivorous. *Galea* and *Cavia* feed on herbs and grasses. *Microcavia* and *Kerodon* seem to prefer leaves—both genera are active climbers. *Kerodon* are especially surprising because they lack both claws and a tail, two adaptations usually associated with life in trees. The sight of an 800g (28oz) guinea pig scooting along a pencil-thin branch high in a tree is quite striking.

All cavies become sexually mature early, at between one and three months of age. Of the species studied to date, only *Microcavia* show marked seasonality in reproduction. The gestation period in cavies is fairly long for rodents, varying from 50 days (*Galea* and *Microcavia*) to 60 (*Cavia aperea*) and 75 (*Kerodon*). Litter sizes are small, averaging about three for *Galea* and *Microcavia*, two for *Cavia* and 1.5 for *Kerodon*. Young are born highly precocial. Males contribute little obvious parental care. The male generally ignores the female and her young once the litter is born. Whether or not the male defends resources needed by the female is unclear. This seems to be the case with *Kerodon* but not with *Galea*.

With cavies being so similar in form and in diet and reproductive biology, what are the differences between species? Some of the most interesting originate in social behavior and in adaptations, in very subtle ways, to the environment.

Three species of cavies have been studied in northeastern Argentina: *Microcavia australis*, *Galea musteloides* and *Cavia aperea*. *Cavia* and *Microcavia* never occur in the same area, *Cavia* preferring moist grasslands and *Microcavia* more arid habitats. *Galea* occurs with both genera. Competition between *Galea* and *Microcavia* seems to be minimized by different foraging tactics: *Microcavia* is more of a browser, and arboreal. The degree to which *Cavia* and *Galea* interact within the same areas is unknown. Home-range sizes are known only for *Microcavia*, on average 3,200sq m (approx. 34,500sq ft), and for *Galea* 1,300sq m (approx. 14,000sq ft).

All three of these cavies have a similar social structure: mating is promiscuous, no male-female bonds are formed, and there is no permanent social group. But there are some subtle differences. *Microcavia australis*, the species most adapted to arid regions with limited resources, has the highest level of amicable behavior among individuals. *Cavia aperea* occurs in habitats that have a high productivity of grasses and herbaceous vegetation. Although food resources are abundant, *Cavia* is the most aggressive of the cavy genera (adults are especially aggressive towards juveniles, causing them to disperse early). *Galea* cavies live in areas where resources are intermediate in abundance. They have a social structure with moderate levels of adult-juvenile aggression.

Two species of cavies coexist in northeastern Brazil: *Kerodon rupestris* and *Galea spixii*. *Galea spixii* is similar to the Argentine *Galea* in morphology, color, ecology and behavior. They inhabit thorn forests, are grazers and have a noncohesive social organization. *Kerodon rupestris* (the Rock cavy) is markedly different from all the other small cavy species. It is larger and leaner, and has a face that is almost dog-like. All small cavies except *Kerodon* have sharply clawed digits; *Kerodon* have nails growing from under the skin with a single grooming claw on the inside hind toes. Their feet are extensively padded. The modifications of the feet facilitate movement on slick rock surfaces. They are strikingly agile as they leap from boulder to boulder, executing graceful mid-air twists and turns. They are also exceptional climbers, and forage almost exclusively on leaves in trees. There is little competition for resources with *Galea*.

Perhaps the most interesting difference between *Kerodon* and *Galea* is behavioral. *Galea*, like the Argentine cavies, inhabit a relatively homogeneous habitat: the thorn scrub forest. *Galea spixii* also has a promiscuous mating system. *Kerodon* inhabit isolated patches of boulders, many of which can be defended by a single male, and have what appears to be a harem-based mating system: one in which a single male has exclusive access to two or more females. The clumped distribution of the boulder piles allows for the males to be able to do this. This makes *Kerodon* similar to the unrelated hyraxes of eastern Africa. Hyraxes live in kopjes, a kind of rock outcrop similar to the ones occupied by *Kerodon*.

Most cavy species can be found in altered and disturbed habitat, and some do well in and among human habitation. The one exception is *Kerodon*, a unique mammal, which has a patchy distribution throughout its range: Rock cavies are hunted extensively, and are declining in numbers. This specialized rodent is in desperate need of protection. Because these rodents occur in such a rare and patchily distributed habitat, large areas will be needed to assure the existence of *Kerodon*; indeed two research reserves have been set aside in Brazil.　TEL

Life-long Partners

Colonial breeding in the monagamous mara

Dawn broke across the Patagonian thornscrub as a large female rodent, with the long ears of a hare and the body and long legs of a small antelope, cautiously approached a den, followed closely by her mate. They were the first pair to arrive, and so walked directly to the mouth of the den. At the burrow's entrance the female made a shrill, whistling call, and almost immediately eight pups of various sizes burst out. The youngsters were hungry, not having nursed since the previous night, and all thronged around the female, trying to suckle. Under this onslaught she jumped and twirled to dislodge the melee of unwelcome mouths which sought her nipples. The female sniffed each carefully, lunging at and chasing off those that were not her own. Finally, she managed to select her own two offspring from the hoard and led them 10m (33ft) away from the den to a site where, despite intermittent harassment from other hungry infants, they would be nursed on and off for an hour or more.

In the meantime her mate sat alert nearby. If another adult pair had approached the den while his female was there, coming to tend their own pups, he would have made a vigorous display directly in front of his mate. If the newcomers had not moved away he would have dashed towards them, with his head held low and neck outstretched, and chased them off. The second pair would then have waited, alert or grazing, at a distance of 20–30m (about 65–100ft). When the original pair had left the area the new pair would then approach the den, to collect their own pups.

The animal being observed was the mara or Patagonian cavy, an 8kg (17.6lb) hare-like day-active cavy, *Dolichotis patagonum*. (The behavior of the only other member of the subfamily Dolichotinae, the Salt desert cavy, *Dolichotis salinicolum*, is unknown in the wild.) A fundamental element of the mara's social system is the monogamous pair bond. Certainly in captivity, and probably in the wild too, the bond between two animals lasts for life. The drive that impels males to bond with females is so strong that it can lead to "cradle snatching"—adult bachelor males attaching themselves to females while the latter are still infants. Contact between paired animals is maintained primarily by the male who closely follows the female wherever she goes, discouraging approaches from other maras by policing a moving area around her of about 30m (about 100ft) in diameter. In contrast, females appear much less concerned about the whereabouts of their mates. While

foraging, members of a pair maintain contact by means of a low grumble which can be heard only a few meters away.

Monogamy is not common in mammals and in the mara several factors probably combine to favor this social system. Monogamy typically occurs in species where there are opportunities for both parents to care for the young, yet in maras virtually all direct care of the offspring is undertaken by the mother. However, the male does make a considerable indirect investment. Due to the high amount of energy a female uses in bearing and nursing her young she has to spend a far greater proportion of the day feeding than the male—time during which her head is lowered and her vigilance for predators impaired. On the other hand, the male spends a larger proportion of each day scanning and is thus able to warn the female and offspring of danger. Also, by defending the female against the approaches of other maras, he ensures uninterrupted time for her to feed and to care for his young. Furthermore, female maras are sexually receptive only for a few hours twice a year;

▲ **A congregation of maras.** As many as 50–100 maras will associate for a period on the dried-out bed of a shallow lake in the Patagonian thornscrub and at the same time preserve their monogamous pair bonds within the crowd.

▶ **Mother and pups.** A female mara will normally give birth at one time to a maximum of three well-developed (precocial) young. She will nurse them for an hour or more once or twice a day for up to four months.

▶ **A long-legged rodent.** The mara, about 45cm (18in) high, can walk, gallop or run. It has been known to run at speeds up to about 45km per hour (28mph) over long distances.

▼ **Males fight hard** to ward off the challenges of other maras. A male rarely manages to usurp another male's female and even then he will, within a few hours, return to his own mate and allow the deprived male to rejoin his female.

in Patagonia females mate in June or July and then come into heat again in September or October, about 5 hours after giving birth, so a male must stay with his female to ensure he is with her when she is receptive.

Mara pairs generally avoid each other and outside the breeding season it is rare to see pairs within 30m (about 100ft) of each other. Then their home ranges are about 40ha (96 acres). Perhaps the avoidance between pairs is an adaptation to the species' eating habits. Maras feed primarily on short grasses and herbs which are sparsely, but quite evenly, distributed in dry scrub desert. So far, detail of their spatial organization is unknown, but there is at least some overlap in the movements of neighboring pairs. Furthermore, there are some circumstances when, if there is an abundance of food, maras will aggregate. In the Patagonian desert there are shallow lakes, 100m to several km in diameter, which contain water for only a few months of the year. When dry, they are sometimes carpeted with short grasses which maras relish. At these times, towards the end of the breeding season (January to March), up to about 100 maras will congregate.

The strikingly cohesive monogamy of maras is noteworthy in its contrast with, and persistence throughout, the breeding season when up to 15 pairs become at least

superficially colonial by depositing their young at a communal den. The dens are dug by the females and not subsequently entered by adults. The same den sites are often used for three or more years in a row. Each female gives birth to one to three young at the mouth of the den; the pups soon crawl inside to safety. Although the pups are well-developed, moving about and grazing within 24 hours of birth, they remain in the vicinity of the den for up to four months, and are nursed by their mother once or twice a day during this period. Around the den an uneasy truce prevails amongst the pairs whose visits coincide. What social bonds unite them is unknown. The number of pairs of maras breeding at a den varies from 1 to at least 15 and may depend on habitat. Pairs come and go around the den all day and in general at the larger dens at least one pair is always in attendance there.

Even when 20 or more young are kept in a creche, cohabiting amicably, the monogamous bond remains the salient feature of the social system. Each female sniffs the infants seeking to suckle and they respond by proferring their anal regions to the female's nose. Infants clambering to reach one female may differ by at least one month in age. A female's rejection of a usurper can involve a bite and violent shaking. Despite each female's efforts to nurse only her own progeny, interlopers occasionally secure an illicit drink. Although females may thus be engaging in communal nursing they rarely seem to do so as willing collaborators. Indeed, the bombardment of its mother by other youngsters is, to a suckling pup, disadvantageous as the female will interrupt nursing to pursue the pestilent pups.

The reasons why normally unsociable pairs of maras keep their young in a communal creche instead of using separate dens are unknown, but it may be that the larger the number of infants in one place the lower the likelihood that any one will fall victim to a predator; indeed, the more individuals at a den (both adults and young) the more pairs of eyes there are to detect danger. Furthermore, some pairs travel as much as 2km (1.2mi) from their home range to the communal den, so the opportunities for shared surveillance of the young may diminish the demands on each pair for protracted attendance at the den. The unusual breeding system of the monogamous mara may thus be a compromise, conferring on the pups benefits derived from coloniality, in an environment wherein association between pairs is otherwise apparently disadvantageous.   ABT/DWM

CAPYBARA

Hydrochoerus hydrochaeris
Family: Hydrochoeridae.
Distribution: S America, east of the Andes from
Panama to NE Argentina.

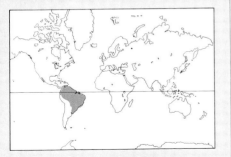

Habitat: open grassland, always near water;
also found in a variety of other habitats
including tropical rain forest.

Size: head-body length
106–134cm (42–53in),
shoulder height 50–62cm
(20–24in), weight 35–64kg
(77–141lb) for males, 37–66kg (81.6–146lb)
for females.

Coat: light brown, consisting of short,
abundant hairs in young; adults have long,
sparse, bristle-like hairs of variable color,
usually brown to reddish.

Gestation: 150 days.
Longevity: 5–10 years.

Subspecies: *H.h. hydrochaeris*, the most
widespread, replaced by *H.h. dabbenei* in
Paraguay and NW Venezuela and by
H.h. uruguayensis in Uruguay and E Argentina;
H.h. isthmius from NW Venezuela, N Colombia
and Panama is the smallest of subspecies and is
sometimes considered a separate species.

CAPYBARAS are the largest living rodents. They are found only in South America where they live in groups near water. The first European naturalists to visit South America called them Water pigs or Orinoco hogs, though they are neither pigs nor totally aquatic. Although their scientific name, *Hydrochoerus*, means water pig, their nearest relatives are the cavies. Capybaras are the largest living rodents, but the smallest members of their subfamily, Hydrochoerinae. Some extinct forms were twice as long and probably eight times as heavy as modern capybaras, ie they were heavier than the largest modern North American Grizzly bear.

Capybaras are ponderous, barrel-shaped animals. They have no tail and their front legs are shorter than their back legs. Their slightly webbed toes, four in the front feet and three in the back, make them good swimmers, able to stay under water for up to 5 minutes. Their skin is very tough and covered by long, sparse, bristle-like hairs. Their nostrils, eyes and ears are situated near the top of their large, blunt head and hence protrude out of the water when the animal swims. Two pairs of large, typically rodent incisors allow them to eat very short grasses which they grind up with their molar teeth. There are four molars on each side of each lower jaw. The fourth molar is characteristic of the subfamily in being as long as the other three.

They have two kinds of scent glands. One, highly developed in males and almost non-existent in females, is located on top of the snout and called the morrillo (meaning hillock in Spanish). It is a dark, oval-shaped, naked protrusion that produces a copious white, sticky secretion. Both sexes also produce odors from two glandular pockets located on each side of the anus. Male anal glands are filled with easily detachable hairs abundantly coated with layers of hard crystalline calcium salts. Female anal pockets also have hairs but theirs are not detachable and are coated in a greasy secretion rather than with crystalline layers. The proportions of each chemical present in the secretions of individual capybaras are different, potentially providing a means of individual recognition via each personal "olfactory fingerprint."

Capybaras have several distinct vocalizations. Infants and young constantly emit a guttural purr, probably to maintain a contact with their mother or other members of the group. This sound is also produced by losers in aggressive interactions, possibly as an appeasement signal to the opponent. Another vocalization, the alarm bark, is given by the first member of the group to detect a predator. This coughing sound is often repeated several times and the reaction of nearby animals may be to stand alert or to rush into the water.

Capybaras are exclusively herbivorous, feeding mainly on grasses that grow in or near water. They are very efficient grazers and can crop the short dry grasses left at the end of the tropical dry season. Usually they spend the morning resting, then bathe during the hot midday hours; in the late

▲ The head of the world's largest rodent. Because the eyes and ears are small the male's morrillo gland assumes prominence. The smallness of eyes and ears is probably an adaptation for life underwater. The animal's English name is derived from the word used by Guaran-speaking South American Indians. It means "master of the grasses."

◄ Lazing by a lake. Capybaras live either in groups averaging 10 in number or in temporary larger aggregations, containing up to 100 individuals, composed of the smaller groups. The situation varies according to season.

afternoon and early evening they graze. At night they alternate rest periods with feeding bouts. Never do they sleep for long periods; rather they doze in short bouts throughout the day.

In the wet season capybaras live in groups of up to 40 animals, but 10 is the average adult group size. Pairs with or without offspring and solitary males are also seen. Solitary males attempt to insinuate themselves into groups, but are rebuffed by the group's males. In the dry season, groups coalesce around the remaining pools to form large temporary aggregations of up to 100 animals. When the wet season returns these large aggregations split up, probably into the original groups that formed them.

Groups of capybaras tend to be closed units where little variation in core membership is observed. A typical group is composed of a dominant male (often recognizable by his large morrillo), one or more females, several infants and young and one or more subordinate males. Among the males there is a hierarchy of dominance, maintained by aggressive interactions consisting mainly of simple chases. Dominant males repeatedly shepherd their subordinates to the periphery of the group but fights are rarely seen. Females are much more tolerant of each other and the details of their social relationships, hierarchical or otherwise, are unknown. Peripheral males may have more fluid affiliation with groups.

Capybaras are found in a wide variety of habitats, ranging from open grasslands to tropical rain forest. Groups may occupy an area varying in size from 2 to 200ha, (4–494 acres) with 10–20ha (24.7–49 acres) being most common. Each home range is used mainly, but not exclusively, by one group. Particularly in the dry season, but at other times as well, two or more groups may be seen grazing side by side. Density of capybaras in some areas may be as high as two individuals per ha (5 per acre) but lower densities (eg less than 1 per ha) are more frequent.

Capybaras reach sexual maturity at 18 months. In Venezuela and Colombia they appear to breed year round with a marked peak at the beginning of the wet season in May. In Brazil, in more temperate areas, they probably breed just once a year. When a female becomes sexually receptive a male will start a sexual pursuit which may last for an hour or more. The female will walk in and out of the water, repeatedly pausing while the male follows closely behind. The mating will take place in the water. The female stops and the male clambers on her back, sometimes thrusting her under water with his weight. As is usual in rodents, copulation lasts only a few seconds but each sexual pursuit typically involves several mountings.

One hundred and fifty days later up to seven babies are born; four is the average litter size. To give birth the female leaves her

group and walks to nearby cover. Her young are born a few hours later—precocial, able to eat grass within their first week. A few hours after the birth the mother rejoins her group, the young following as soon as they become mobile, three or four days later. Within the group the young appear to suckle indiscriminately from any lactating female; females will nurse young other than their own. The young in a group spend most of their time within a tight-knit creche, moving between nursing females. When they are active they constantly emit a churring purr.

Capybara infants tire quickly and are therefore very vulnerable to predators. They

▲ **Mating** is an aquatic activity, as in hippopotamuses. Unlike the young of the latter, however, capybaras are born on land. For capybaras water is a place of refuge.

▶ **Scent marking.** ABOVE A male deposits its white sticky secretion from its morrillo.

▶ **Capybara young.** Capybaras are born after a long gestation of over five months. Even though they emerge in a well-developed (precocial) condition and can eat soon after birth they require over a year before sexual maturity is attained.

Farming Capybara

In Venezuela there has been consumer demand for capybara meat at least since Roman Catholic missionary monks classified it as legitimate lenten fare in the early 16th century, along with terrapins. The similar amphibious habits of these two species presumably misled the monks who supposed that capybaras have an affinity with fish. Today, because of their size, tasty meat, valuable leather and a high reproductive rate capybaras are candidates for both ranching and intensive husbandry.

It has been calculated that where the savannas are irrigated to mollify the effects of the dry season, the optimal capybara population for farming is 1.5–3 animals per ha (about 7.4 per acre), which can yield 27kg per ha (147lb per acre) per annum. Those ranches that are licensed to take 30–35 percent of the population at one annual harvest can sustain yields of about 1 capybara per 2ha (or 1 animal per 0.8 acre) in good habitat. An annual cull takes place in February, when reproduction is at a minimum and the animals congregate around

waterholes. Horsemen herd the capybaras which are then surrounded by a cordon of cowboys on foot. An experienced slaughterman then selects adults over 35kg (77lb), excluding pregnant females, and kills them with a blow from a heavy club (see illustration). The average victim weighs 44.2kg (97.4lb) of which 39 percent (17.3kg; 38lb) is dressed meat. These otherwise unmanaged wild populations thus yield annually over 8kg of meat per ha (3lb per acre).

In spite of this yield, farmers feared that large populations of capybaras would compete with domestic stock. In fact because capybaras selectively graze on short vegetation near water they do not compete significantly with cattle (which take more of taller, dry forage) except near wetter, low-lying habitats. There capybaras are actually much more efficient at digesting the plant material than are cattle and horses. So ranching capybara in their natural habitat appears to be, both biologically and economically, a viable adjunct to cattle ranching.

have most to fear from vultures and feral or semiferal dogs which prey almost exclusively on young. Cayman, foxes and other predators may also take young capybaras. Jaguar and smaller cats were certainly important predators in the past, though today they are nearly extinct in most of Venezuela and Colombia. In some areas of Brazil, however, jaguars seize capybaras in substantial numbers.

When a predator approaches a group the first animal to detect it will emit an alarm bark. The normal reaction of other group members is to stand alert but if the danger is very close, or the caller keeps barking, they will all rush into the water where they will form a close aggregation with young in the center and the adults facing outward.

Capybara populations have dropped so substantially in Colombia that c. 1980 the government prohibited capybara hunting. In Venezuela they have been killed since colonial times in areas devoted to cattle ranching. In 1953 the hunting of capybara there became subject to legal regulation and controlled, but to little effect until 1968 when, after a five year moritorium, a management plan was devised, based on a study of the species biology and ecology. Since then, of the annually censused population in licensed ranches with populations of over 400 capybaras, 30–35 percent are harvested every year. This has apparently resulted in local stabilization of capybara populations. Capybaras are not threatened at present but control on hunting and harvesting must remain if population levels are to be maintained.   DWM/EH

OTHER CAVY-LIKE RODENTS

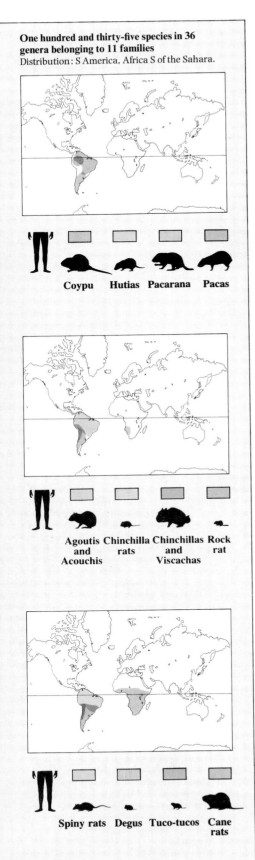

One hundred and thirty-five species in 36 genera belonging to 11 families
Distribution: S America, Africa S of the Sahara.

Coypu   Hutias   Pacarana   Pacas

Agoutis and Acouchis   Chinchilla rats   Chinchillas and Viscachas   Rock rat

Spiny rats   Degus   Tuco-tucos   Cane rats

THE families assembled in this chapter are a disparate assemblage of ancient rodent fauna of South America, a few of which have migrated into Central and North America since the two continents were rejoined in the Pliocene era (7–3 million years ago). The diversity they show is in marked contrast to the relative homogeneity of the more recent rodents.

In this group there are small rodents and large rodents; some are covered with barbed spines, others have soft silky fur. Some species are common and widespread, others known only from a few museum specimens. They inhabit forests and grasslands, water and rocky deserts, coastal plains and high mountains; some are solitary, others colonial. Many species are eaten by humans, others prized for their fur; some are pests, others carry the diseases of man and domestic animals.

The larger species, such as agoutis, paca, pacarana and viscacha, are prey for the large and medium-sized species of carnivore (jaguar, ocelot, pampas cat, maned wolf, bush dog, foxes etc). They are herbivorous and may be considered as the South American ecological equivalent of the vast array of ungulate herbivores which are so important in the African ecosystems. It is thought that these rodents radiated into this role as the primitive native herbivores became extinct and before the arrival of the new fauna from the north.

The **coypu** is a large robust rodent, weighing up to 10kg (22lb). It lives in burrows in river banks, feeds on water plants and is an expert swimmer. Families of up to 10 young are recorded—most of the female's mammae are situated in a row high on the side to enable her to feed the precocial young while swimming.

The coypu has small rounded ears and webbed hind feet. Its fur is darkish brown-yellow and the tip of its muzzle white. Its outer hair is long and coarse, covering the thick soft underfur known in the fur trade as nutria—a word corrupted from the Spanish word for otter.

Hutias are found only in the West Indies, living in forests and plantations and eating not only vegetation but sometimes small animals such as lizards. Hutias are robust, short-legged rats ranging in length from 20 to about 60cm (8–24in) and weighing up to 7kg (15lb). Their fur is rough but with a soft underfur. Two species of the genus *Plagiodontia* are known only from subfossil remains found in caves and kitchen middens. Three other genera, *Hexalobodon*, *Aphaetreus* and *Isolobodon*, are known from similar subfossil bones and are thought to have become extinct within historical time. Several other species have recently become extinct thanks to human destructiveness. Among living representatives of the group, *Capromys* is a forest dweller weighing up to 7kg (15lb) and is hunted for its flesh in Cuba, and *Geocapromys* is a short-tailed nocturnal form which lives on leaves, bark and twigs and weighs up to 2kg (4.4lb). Little is known of the biology of these animals and several species are thought to be in peril from the Burmese mongoose which has been introduced to the West Indies.

The **pacarana** is a slow-moving, robust animal, weighing up to 10–15kg (22–33lb), which resembles a spineless porcupine. It is coarse haired, black or brown with two more or less continuous white stripes on each side. It has broad, heavily clawed feet and uses its forepaws to hold food while eating. Its tail is about one quarter of its head-body length which may reach 80cm (31in). A forest-dwelling species, seldom encountered, about which little is known,

◀ **Coypu afloat.** Coypus are probably the most aquatic of cavy-like rodents. They have webbed feet, spend most of their waking hours in water and live in burrows bored in river banks.

▼ **Coypu abroad.** Coypu fur, labeled as "nutria" fur, has been popular since the early 19th century. Consequently coypu farms have been established on every continent except Antarctica, but in many countries animals have become feral, including this inhabitant of East Africa. Some have also become pests.

this inoffensive herbivore is prey for jaguar, ocelot and other medium-sized carnivores and is hunted for food by man. It has a remarkably long gestation period, which can vary from 220 days to 280.

About 10 species of **agouti** and two of **acouchi** make up the family Dasyproctidae, all large rodents sometimes considered to belong to the same family as the pacas. They are relatively common animals although secretive: they hide in burrows and become nocturnal in areas where they are disturbed. The coat of agoutis is orange to brown or blackish above, yellowish to white below, with a contrasting rump color. Acouchis are reddish to blackish green above, yellowish below, with a bright color on the head. Agoutis live mainly on fallen fruits. They are attracted to the sound of ripe fruits hitting the ground. When food is abundant they carefully bury some for use in time of scarcity. This behavior is important in dispersing the seeds of many species of forest trees. Acouchis, which have relatively long tails, are rarely seen and their biology in the wild is almost unknown. These animals are hunted for food and preyed upon by a variety of carnivores.

Chinchilla rats are soft-furred rats with short tails. *Abrocoma cinerea* is smallest, with a head-body length of 15–20cm (6–8in) and tail length 6–15cm (2.5–6in), with comparable figures for *A. bennetti*, 20–25cm (8–10in) and 13–18cm (5–7in), respectively. Both have soft dense underfur, overlain with long, fine guardhairs; coat color is silver gray or brown above, white or brown below. Their pelts are occasionally sold but are of much poorer quality than the pelts of true chinchillas. They live in tunnels and crevices in colonies from sea level to the high Andes of southwestern South America. *Abrocoma bennetti* is distinguished by having more ribs than any other rodent (17 pairs).

Spiny rats comprise some 15 genera and about 56 species and are a peculiar assemblage of robust, medium-sized, herbivorous rodents, most of which have a spiny or bristly coat. (Some species however are soft furred.) Some are very common and widespread while others are extremely rare; some have short tails, others long tails (both types having a tendency to become lost), and some are arboreal, others burrowing. The taxonomy of some genera is poorly understood and the numbers of species are tentative.

The body form in this family is correlated to life style. Robust, short-tailed forms (eg *Clyomys, Carterodon* and *Euryzygomatomys*) are burrowing savanna species; relatively

slender, long-tailed forms are arboreal (eg *Thrinacodus, Dactylomys, Kannabateomys*). Of the intermediate forms *Proechimys, Isothrix, Hoplomys* and *Cercomys* are more or less terrestrial and *Mesomys, Louchothrix, Echimys* and *Diplomys* are mostly arboreal. Of the many genera only two have produced any number of species: *Proechimys* and *Echimys*.

Chinchillas and **viscachas** are soft-furred animals with long, strong hind legs, large ears and tails up to one third the length of the body. Their teeth are characteristically divided into transverse plates. The Plains viscacha inhabits pampas and shrub in Argentina. It has soft underfur overlain with stiffish guardhairs and is brown or gray above, white below, with a distinctive black and white face. Plains viscachas are sexually dimorphic with males at 8kg (18lb) almost twice the size of females. Mountain viscachas occur in the Andes of Peru, Bolivia, Argentina and Chile and have thick soft fur, gray or brown above, whitish or grayish below. Chinchillas, which weigh only up to 0.8kg (1.8lb), inhabit the same regions as mountain viscachas. They have very dense fur, bluish gray above, yellowish below. All species are subject to pressure from human hunting—chinchillas have been hunted for their valuable fur to near extinction, mountain viscachas are prized for both food and fur and the Plains viscachas compete for grazing with domestic animals, ten viscachas eating as much as a

sheep. In addition they destroy pasture with their acidic urine, and so undermine the pampas that men, horses and cattle are often injured by falling into their concealed tunnels. The animals collect a variety of materials (bones, sticks and stones) lying loose in their surroundings and heap them in piles above the entrances to their burrows.

Degus (or **octodonts**) occur in southern South America from sea level to about 3,000 (10,000ft). The name octodonts refers to the worn enamel surface of their teeth which forms a pattern in the shape of a figure eight. All are robust rodents, with a head-body length of 12–20cm (5–8in) and tail length of 4–18cm (1.6–7in), and a long silky coat. Degus and choz choz are gray to brown above, creamy yellow or white below. White corucos are brown or black and Rock rats dark brown all over. Viscacha rats are buffy above, whitish below, with a particularly bushy tail up to 18cm (7in) long. Most are adapted to digging, particularly the Rock rats, or to living in rock crevices. Corucos have broad well-developed incisors, used in burrowing. They eat bulbs, tubers, bark and cactus.

Tuco-tucos comprise one genus with about 33 species, which inhabit sandy soils from the altoplano of Peru to Tierra del fuego in a wide variety of vegetation types. They are fossorial (adapted for digging), building extensive burrow systems and may be compared to the pocket gophers of North

The 12 families of other Cavy-like rodents

Coypu
Family: Myocastoridae.
1 species, *Myocastor coypus*, S Brazil, Paraguay, Uruguay, Bolivia, Argentina, Chile; feral populations in N America, N Asia, E Africa, Europe.

Hutias
Family: Capromyidae.
12 species in 4 genera.
W Indies. Genera: **Cuban hutias** (*Capromys*), 9 species; **Bahaman and Jamaican hutias** (*Geocapromys*), 2 species; *Isolobodon*, 3 species, recently extinct; **Hispaniolan hutias** (*Plagiodontia*), 1 living and 1 recently extinct species.

Pacarana
Family: Dinomyidae.
1 species, *Dinomys branicki*, Colombia, Ecuador, Peru, Brazil, Bolivia, lower slopes of Andes.

Pacas
Family: Agoutidae.
2 species in 1 genus.
Mexico S to S Brazil.

Agoutis and Acouchis
Family: Dasyproctidae.

13 species in 2 genera.
Mexico S to S Brazil. Genera: **agoutis** (*Dasyprocta*), 11 species; **acouchis** (*Myoprocta*), 2 species.

Chinchilla rats
Family: Abrocomidae.
2 species in 1 genus.
Altoplano Peru, SW Bolivia, Chile, NW Argentina.
Species: *Abrocoma bennetti, A. cinerea.*

Spiny rats
Family: Echimyidae.
56 species in 15 genera.
Nicaragua S to Peru, Bolivia, Paraguay, S Brazil.
Genera: **spiny rats** (*Proechimys*), 22 species; **spiny rats** (*Echimys*), 10 species; **Guiara** (*Euryzygomatomys spinosus*); **Owl's rat** (*Carterodon sulcidens*); **Thickspined rat** (*Hoplomys gymnurus*); *Clyomys laticeps*; **Rabudos** (*Cercomys cunicularis*); *Mesomys*, 4 species; *Lonchothrix emiliae*; **toros** (*Isothrix*), 3 species; **soft-furred spiny rats** (*Diplomys*), 3 species; **corocoro** (*Dactylomys*), 3 species; *Kannabateomys amblyonyx*; Thrinacodus, 2 species.

Chinchillas and Viscachas
Family: Chinchillidae.
6 species in 3 genera.

Peru, Argentina, Chile, Bolivia. Genera: **Plains viscacha** (*Lagostomus maximus*); **mountain viscachas** (*Lagidium*), 3 species; **chinchillas** (*Chinchilla*), 2 species.

Degus or **Octodonts**
Family: Octodontidae.
8 species in 5 genera.
Peru, Bolivia, Argentina, Chile, mainly in Andes (some along coast). Genera: **Rock rat** (*Aconaemys fuscus*); **degus** (*Octodon*), 3 species; **Choz choz** (*Octodontomys gliroides*); **viscacha rats** (*Octomys*), 2 species; **Coruros** (*Spalacopus cyanus*).

Tuco-tucos
Family: Ctenomyidae.
33 species in 1 genus (*Ctenomys*).
Peru S to Tierra del Fuego.

Cane rats
Family: Thryonomyidae.
2 species in 1 genus (*Thryonomys*).
Africa S of the Sahara.

African rock rat
Family: Petromyidae.
1 species, *Petromus typicus*, W South Africa N to SW Angola.

▲ **Representatives of 10 families** of other cavy-like rodents. (**1**) A chinchilla (family Chinchillidae). (**2**) An American spiny rat climbing (family Echimyidae). (**3**) A hutia sunning itself (family Capromyidae). (**4**) A cane rat (family Thryonomyidae). (**5**) A Rock rat (family Petromyidae). (**6**) A pacarana (*Dinomys branickii*) feeding. (**7**) A chinchilla rat (*Abrocoma bennetti*). (**8**) A paca (*Agouti paca*). (**9**) A tuco-tuco (*Ctenomys opimus*) excavating with its incisors. (**10**) A degu (*Octodon degus*).

stored in cells in the tunnels but such stores are often left to decay.

Cane rats and the African **rock rat** are African rodents sufficiently distinctive to be placed in families of their own. They appear to be related to the African porcupines which in turn may have some affinities with the cavy-like rodents of the Neotropics.

There are probably only two species of cane rat although many varieties have been described. They are robust rats weighing up to 9kg (20lb), occasionally more. Head and body length ranges from 35cm to 60cm (14–24in), tail length from 7cm to 25cm (3–10in).

Cane rats prefer to live near water and eat a variety of vegetation, especially grasses, and as their name suggests they can be pests of plantations. They are prey for leopard, mongoose and python in addition to being hunted for food by man in many parts of their range. Their coat is coarse and bristly, the bristles being flattened and grooved along their length. There is no underfur. Coat color is brown speckled with yellow or gray above, buffy white below. Preferred habitats are reed beds, marshes and the margins of lakes and rivers. In some parts of their range cane rats breed all year round but most young are born between June and August with two to four in a litter. Of the two species *Thryonomys swinderianus* is the larger, attaining weights of up to 9kg (20lb) and head and body length of up to 60cm (24in); *T. gregorianus* may occasionally reach 7kg (15lb) and head and body length of 50cm (20in). The latter is said to be less aquatic than *T. swinderianus*.

The single species that belongs to the family Petromyidae is known as the African rock rat or dassie. It has a flattened skull and very flexible ribs which allow it to squeeze into narrow rock crevices. Its color is very variable and mimics the color of the rocks amongst which the animal lives. It is particularly active at dawn and dusk, and lives on vegetable matter such as leaves, berries and seeds. Head and body length varies from 14cm to 20cm (5.5–8in), tail length from 13cm to 18cm (5–7in). Its coat is long and soft but there is no recognizable underfur. Shades of gray, yellow and buff predominate. Its preferred habitat is rocky hillsides of southwestern South Africa and the animal is found only where there are rocks for shelter. Its tail has hairs scattered throughout its length and long white hairs at the tip. Rock rats breed only once per year and produce two young to a litter, and thus have a relatively slow reproduction rate for small mammals in the tropics.   IRB

America. Their hind feet bear strong bristle fringes, their ears are small, their claws strong and their incisors well developed. Their size ranges from head-body length 15–25cm (6–10in), tail length 6–11cm (2.4–4.3in) and they weigh up to 0.7kg (1.6lb). Their coat is very variable, ranging from gray or buff to brown and reddish brown above, lighter below; their hairs can be long or short, usually dense, but never bristly. The name "tuco-tucos" refers to their calls.

Their shallow burrows may have several entrances and by opening and plugging tunnels as necessary *Clenomys* can regulate the burrow temperature, which is normally maintained at about 20–22°C (68–72°F). Feeding is often accomplished by pulling roots down into a tunnel. Food is often

OLD WORLD PORCUPINES

Family: Hystricidae
Eleven species in 4 genera and 2 subfamilies.
Distribution: Africa, Asia.

Habitat: varies from dense forest to semidesert.

Size: ranges from head-body length of 37–47cm
(14.6–18.5in) and weight 1.5–3.5kg (3.3–7.7lb)
in the brush-tailed porcupines to head-body
length 60–83cm (23.6–32.7in) and weight
13–27kg (28.6–59.4lb) in the crested porcupines.

Gestation: 90 days for the Indian porcupine,
93–4 days for the Cape porcupine, 100–10
days for the African brush-tailed porcupine,
105 days for the Himalayan porcupine, 112
days for the African porcupine.

Longevity: approximately 21 years recorded for
the crested porcupines (in captivity).

Brush-tailed porcupines
Genus *Atherurus*.
C Africa and Asia; forests; brown to dark brown
bristles cover most of the body; some single
color quills on the back. Two species: **African
brush-tailed porcupine** (*A. africanus*); **Asiatic
brush-tailed porcupine** (*A. macrourus*).

Indonesian porcupines
Genus *Thecurus*.
Coat: dark brown in front, black on posterior;
body densely covered with flattened flexible
spines; quills have a white base and tip with central
parts black; rattling quills on the tail are hollow.
Three species: **Bornean porcupine** (*T. crassispinis*);
Phillipine porcupine (*T. pumilis*); **Sumatran
porcupine** (*T. sumatrae*).

Bornean long-tailed porcupine
Trichys lipura.
Body covered with brownish flexible bristles;
head and underparts hairy.

Crested porcupines
Genus *Hystrix*.
Africa, India, SE Asia, Sumatra, Java and
neighboring islands, S Europe; recently
introduced to Great Britain; hair on back of
long, stout, cylindrical black and white erectile
spines and quills; body covered with black
bristles; grayish crest well developed. Five
species: **African porcupine** (*H. cristata*); **Cape
porcupine** (*H. africaeaustralis*); **Himalayan
porcupine** (*H. hodgsoni*); **Indian porcupine**
(*H. indica*); **Malayan porcupine** (*H. brachyura*).

Porcupines have a peculiar appearance, due to parts of their bodies being covered with quills and spines. It is often thought that they are related to hedgehogs or pigs. Their closest relatives, however, are guinea pigs, chinchillas, capybaras, agoutis, viscachas and cane rats. Many of these have in common an extraordinary appearance, and are well known, at least to zoologists, for their unusual ways of solving problems involved in reproduction.

The Old World porcupines belong to two distinct subfamilies: Atherurinae and Hystricinae. The brush-tailed porcupines (of the subfamily Atherurinae) have long, slender tails which end in a tuft of white, long, stiff hairs. In the genus *Atherurus* these hairs have hollow expanded sections at intervals which produce a rustling sound when the tail is shaken. Their elongated bodies and short legs are covered with short, flat, chocolate-colored, sharp bristles, with only a few long quills on the back. Crested porcupines (subfamily Hystricinae), on the other hand, have short tails surrounded by an array of cylindrical, stout, sharp quills with the tip of the tail being armed with a cluster of hollow open-ended quills with a narrow stalk-like base, which produce a characteristic rattling sound when the tail is shaken, serving as a warning signal when the animal is annoyed by an intruder. The posterior two thirds of the upper parts and flanks of the body are covered with black and white spines, up to 50cm (20in) in length, and sharp cylindrical quills which cannot be projected at the enemy but which are on contact easily detached.

When aggressive, porcupines erect their spines and quills, stamp their hind feet, rattle their quills and make a grunting noise. If threatened further they will turn their rump towards an intruder and defend themselves further by quickly running sideways or backwards into the enemy. If the quills penetrate the skin of the enemy they become stuck and detach. Although not poisonous, embedded quills often cause septic wounds which can prove fatal for their natural predators, eg lions and leopards.

The remainder of the stout body is covered with flat, coarse, black bristles. Most species have a crest that can be erected, extending from the top of the head to the shoulders, consisting of long, coarse hair. Their heads are blunt and exceptionally broad across the nostrils. Their small pig-like eyes are set far back on the sides of the head. The sexes look alike. The female's mammary glands are situated on the side of the body.

Most Old World porcupines are vegetarians. In their natural habitats they feed on the roots, bulbs, fruit and berries of a variety of plants. When they enter cultivated areas they will eat such crops as groundnuts, potatoes, pumpkins, melons and maize. African porcupines are reported as able to feed on plant species known to be poisonous to cattle. Porcupines manipulate their food with their front feet and while eating hold it against the ground. Their chisel-like incisors enable them to gnaw effectively. Porcupines are solitary feeders but groups of two or three individuals, comprising one or two adults and offspring, have been observed on occasion. Porcupine shelters often contain accumulations of bones, carried in for gnawing, either for sharpening teeth or as a source of phosphates. Brush-tail porcupines are active tree-climbers and feed on a variety of fruits.

Detailed information on the reproductive biology of porcupines is only available for the Cape porcupine. Sexual behavior leading to copulation is normally initiated by a female who, after having approached a male, or after being approached by a male, will take up the sexual posture, in which she raises her rump and tail and holds the rest of her body close to the surface. The male mounts the female by standing bipedally behind the female with his forepaws resting on her back. Thrusting only occurs after intromission, which only occurs when the female is in heat (every 28–36 days), when the vaginal closure membrane becomes perforated. Sexual behavior without intromission is exhibited during all stages of the sexual cycle.

The young are born in grass-lined chambers which form part of an underground burrow system. At birth they are unusually precocial: fully furred, eyes open, with bristles and heralds of future quills already present. These harden within a few hours after birth. They weigh 300–330g (10.6–11.6oz) and start to nibble on solids between nine and 14 days. They begin to feed at four to six weeks but are nursed for 13–19 weeks, when they weigh 3.5–4.7kg (7.7–10.4lb). Litter size varies but 60 percent of all births produce one young and 30 percent produce twins. Three is normally the maximum number. Females produce only one litter per year and although not seasonal when kept in captivity they breed only in summer in their natural habitats. In a colony of porcupines all members protect the young, adult males playing an important role by being particularly aggressive towards invaders. Sexual maturity is at-

▲ **On a desert dune,** a crested porcupine. These porcupines have amazing versatility, in some ways resembling that of some mouse-like rodents. They can live in deserts, steppe, rocky areas and forest. They shelter in existing holes and crevices or dig their own burrows. They are also successful in dealing with predators. Such factors probably enable them to live long, often for 12–15 years.

▼ **Amidst grass in Malaysia:** an Asiatic brush-tailed porcupine. This is a nocturnal species which lives in groups in forests.

tained at an age of two years.

All evidence suggests that Cape porcupines live in colonies comprising at least an adult pair and their consecutive litters, with as many as six to eight individuals occupying a burrow system. Females as well as males will be aggressive towards strange males and females, irrespective of the sexual state of the female. Both will also accompany young up to the age of six or seven months when they go out foraging.

Population density in the semiarid regions of South Africa varies from one to 29 individuals per sq km (75 per sq mi) with approximately 40 percent of populations comprising individuals less than one year old.

No information is available about territoriality or the size of home ranges but porcupines have been reported to forage up to 16km (10mi) from their burrows per night. They move along well-defined tracks, almost exclusively by night. Porcupines are catholic in their habitat requirements, providing they have shelter to lie in during the day. They take refuge in crevices in rocks, in

caves or in abandoned aardvark holes in the ground, which they modify by further digging to suite their own requirements.

Porcupines are often reported as a menace to crop-producing farmers. They are destroyed by various methods. Their flesh is enjoyed by indigenous people throughout Africa and the killing of porcupines, whenever the opportunity arises, has apparently become a favored pastime. They still, however, occur in great numbers throughout Africa, thanks to the near absence of their natural predators (lions, leopards, hyenas) over much of their range and also because of the increase in food production through the cultivation of agricultural crops. African porcupines also carry fleas, which are responsible for the spread of bubonic plague, and ticks, which spread babesiasis, rickettsiasis and theilerioses. Brush-tailed porcupines are known to be hosts of the malarial parasite *Plasmodium atheruri*. In spite of being killed in large numbers there is no reason to believe that porcupines are endangered.   RVA

GUNDIS

Family: Ctenodactylidae
Five species in 4 genera.
Distribution: N Africa.

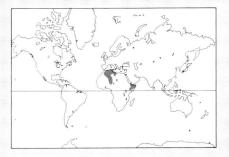

Habitat: rock outcrops in deserts.

Size: ranges from
head-body length
17–18cm (6.8–7.2in),
tail length 2.8–3.2cm
(1.1–1.3in), weight 178–195g (6.3–6.9oz) in
the Felou gundi to head-body length
17.2–17.8cm (6.9–7.1in), tail length
5.2–5.6cm (2–2.2in), weight 175–180g
(6.2–6.3oz) in Speke's gundi.

Gestation: 56 days in the Desert or Sahara
gundi (unknown for other species).

Longevity: 3–4 years (10 years recorded for
Speke's gundi in captivity).

North African gundi
Ctenodactylus gundi
SE Morocco, N Algeria, Tunisia, Libya,
occurring in arid rock outcrops.

Desert gundi
Ctenodactylus vali
Desert gundi or Sahara gundi.
SE Morocco, NW Algeria, Libya, occurring in
desert rock outcrops.

Mzab gundi
Massoutiera mzabi
Mzab gundi or Lataste's gundi.
Algeria, Niger, Chad, occurring in desert and
mountain rock outcrops.

Felou gundi
Felovia vae
SW Mali, Mauritania, occurring in arid and
semiarid rock outcrops.

Speke's gundi
Pectinator spekei
Speke's gundi or East African gundi.
Ethiopia, Somalia, N Kenya occurring in arid
and semiarid rock outcrops.

▶ **Mzab gundi.** ABOVE An extraordinary
feature of this gundi is that its ears are flat and
immovable.

▶ **Speke's gundi,** from East Africa, has a range
of well-developed vocalizations.

THE first gundi was found in Tripoli in
1774 and called the gundi-mouse
(gundi is the North African name). In the
mid-19th century the explorer John Speke
shot gundis in the coastal hills of Somalia,
and later French naturalists found three
more species; skins and skulls began to
arrive in museums. But no attempt was
made to study the ecology of the animal.
Some authors said gundis were nocturnal;
others, diurnal. Some said they dug burrows, others said they did not. Some said
they made nests. Some heard them whistling; others, chirping like birds—and there
were fantastic tales about them combing
themselves in the moonlight with their hind
feet. The family name is Ctenodactylidae
which means comb-toes. In 1908 two
French doctors isolated a protozoan parasite
(now known to occur in almost every
mammal) from the spleen of a North African
gundi and called it *Toxoplasma gondii*.

Gundis have short legs, short tails, flat
ears, big eyes and long whiskers. Crouched
on a rock in the sun with the wind blowing
through their soft fur they look like powder
puffs.

The North African and the Desert gundi
have tiny wispy tails but the other three
have fans which they use as balancers.
Speke's gundi has the largest and most
elaborate fan which it uses in social displays.
Gundis also have rows of stiff bristles on the
two inner toes of each hind foot which stand
out white against the dark claws. They use
the combs for scratching. Sharp claws adapted to gripping rocks would destroy the soft
fur coat that insulates them from extremes
of heat and cold. The rapid circular scratch
of the rump with the combed instep is
characteristic of gundis.

The gundi's big eyes convinced some
authors that the animal was nocturnal. In
fact the gundi is adapted to popping out of
sunlight into dark rock shelters. Equally the
gundi can flatten its ribs to squeeze into a
crack in the rocks.

Gundis are herbivores: they eat the
leaves, stalks, flowers and seeds of almost
any desert plant, including grass and acacia.
Their incisors lack the hard orange enamel
that is typical of most rodents. Gundis are
not, therefore, great gnawers. Food is scarce
in the desert and gundis must forage over
long distances—sometimes as much as 1km
(0.6mi) a morning. Regular foraging is
essential because gundis do not store food.
Home range size varies from a few square
meters to 3sq km (1.9sq mi).

Foraging over long distances generates
body heat which can be dangerous on a hot

desert day. It is unusual for small desert
mammals to be active in daytime but gundis
behave a bit like lizards. In the early morning they sunbathe until the temperature
rises above 20°C (68°F) and then forage for
food. After a quick feed they flatten themselves again on the warm rocks. Thus they
make use of the sun to keep their bodies
warm and to speed digestion. It is an
economical way of making the most of
scarce food. By the time the temperature has
reached 32°C (90°F), the gundis have taken
shelter from the sun under the rocks and
will not come out again until the temperature drops in the afternoon. When long
foraging expeditions are necessary gundis
alternate feeding in the sun and cooling off
in the shade. In extreme drought gundis eat
at dawn when plants contain most moisture. They obtain all the water they need
from plants; their kidneys have long tubules
for absorbing water. Their urine can be
concentrated if plants dry out completely.

But this emergency response can only be sustained for a limited period.

Gundis are gregarious, living in colonies that vary in density from the Mzab gundi's 0.3 per ha (0.12 per acre) to over 100 per ha (40 per acre) for Speke's gundi. Density is related to the food supply and the terrain. Within colonies there are family territories occupied by a male and female and juveniles or by several females and offspring. Gundis do not make nests and the "home shelter" is often temporary. Characteristically a shelter retains the day's heat through a cold night and provides cool draughts on a hot day. In winter, gundis pile on top of one another for warmth, with juveniles shielded from the crush by their mother or draped in the soft fur at the back of her neck.

Each species of gundi has its own repertoire of sounds, varying from the infrequent chirp of the Mzab gundi to the complex chirps and chuckles and whistles of Speke's gundi. In the dry desert air and the rocky terrain their low-pitched alert calls carry well. Short sharp calls warn of predatory birds; gundis within range will disappear under the rocks. Longer calls signify ground predators and inform the predator it has been spotted. The Felou gundi's harsh chee-chee will continue as long as the predator prowls around. Long complex chirps and whistles can be a form of greeting or recognition. The *Ctenodactylus* species—whose ranges overlap—produce the most different sounds: the North African gundi chirps, the Desert gundi whistles. Thus members can recognize their own species. All gundis thump with their hind feet when alarmed. Their flat ears give good all-round hearing and a smooth outline for maneuvering among rocks. The bony ear capsules of the skull are huge, like those of many other desert rodents. The acute hearing is important for picking up weak low-frequency sounds of predators—sliding snake or flapping hawk—and for finding parked young. Right from the start, young are left in rock shelters while the mother forages. The young are born fully furred and open-eyed. The noise they set up—a continuous chirruping—helps the mother to home in on the temporary shelter.

The young have few opportunities to suckle: from the mother's first foraging expedition onwards they are weaned on chewed leaves. (They are fully weaned after about 4 weeks.) The mother has four nipples—the average litter size is two—two on her flanks and two on her chest. But a gundi has little milk to spare in the dry heat of the desert. WG

AFRICAN MOLE-RATS

Family: Bathyergidae
Nine species in 5 genera.
Distribution: Africa S of the Sahara.

Habitat: underground in different types of soil
and sand.

Size: ranges from head-body length 9–12cm
(3.5–4.7in), weight 30–60g (1–2.1oz) in the
Naked mole-rat to head-body length 30cm
(11.8in), weight 750–1,800g (26–63oz) in the
genus *Bathyergus*.

Gestation: 70 days in the Naked mole-rat;
unknown for all other genera.

Longevity: unknown (captive Naked mole-rats
have lived for over 10 years).

Subfamily Bathyerginae

Dune mole-rats
Genus *Bathyergus*.
Sandy coastal soils of S Africa; the largest mole-
rats. Two species including: **Cape dune mole-
rat** (*B. suillus*).

Subfamily Georychinae

Common mole-rats
Genus *Cryptomys*.
W, C and S Africa; the most widespread genus.
Three species including: **Common mole-rat**
(*C. hottentotus*).

Cape mole-rat
Georychus capensis.
Cape Province of the Republic of S Africa, along
the coast from the SW to the E.

Silvery mole-rats
Genus *Heliophobius*.
C and E Africa. Two species including: **Silvery
mole-rate** (*H. argenteocinereus*).

Naked mole-rat
Heterocephalus glaber.
Arid regions of Ethiopia, Somalia and Kenya.

▶ **Shaped like a cylinder,** ᴀʙᴏᴠᴇ, the Common
mole-rat. Members of its genus may be the
inhabitants of the longest constructed and
maintained burrows of any animal (up to
350m, 1,150ft).

▶ **Rodent digging power,** the head of a Cape
mole-rat.

THE mole-rat is a rat-like rodent that has
become totally conditioned to life underground and has assumed a mole-like
existence. Consequently its anatomy and
life-style are distinctive if not unusual: it
digs out an extensive system of semiperma-
nent burrows for foraging—to which are
attached sleeping areas and rooms for storing food—and throws up the soil it excavates as "mole hills." Most rodents of comparable size grow rapidly and live for only a
couple of years. Naked mole-rats, however,
take over a year to attain adult size and can
live for several years. All these features must
have contributed to the evolution of their
highly structured social system.

Many of the physical features of the mole-
rat are designed for its life underground
where boring through soil and pushing it up
as molehills requires considerable power
and energy. Efficiency in effort is essential.
Mole-rats have cylindrical bodies with short
limbs so as to fit as compactly as possible
within the diameter of a burrow. Their loose
skin helps them to turn within a confined
space: a mole-rat can almost somersault
within its skin as it turns. Mole-rats can also
move rapidly backwards with ease and so
when moving in a burrow tend to shunt to
and fro without turning round.

All genera, except dune mole-rats, use
chisel-like incisors protruding out of the
mouth cavity for digging. To prevent soil
from entering the mouth while digging
there are well-haired lip-folds behind the
incisors. Dune mole-rats dig with the long

senses of smell and hearing are well developed and their noses and ears are modified on the outside so as to cope with the problems of living in a sandy environment: the nostrils can be closed during digging while the protruding parts of the ears (pinnae) have been lost.

Mole-rats are vegetarians and obtain their food by digging foraging tunnels. These enable them to find and collect roots, storage organs (geophytes) and even the aerial portions of plants without having to come above ground. The large Dune mole-rat, less well equipped, cannot afford to be a specialist feeder and lives mainly on whatever it encounters—grass, herbs and geophytes. On the other hand, the small Common and Naked mole-rats are more selective and live entirely on geophytes and roots. This may be possible because they displace less soil as they dig and, being social, share the cost of digging and finding food between members of the colony. Foraging burrows can be very extensive: one system containing 10 adults and 3 young Common mole-rats ran for 1km (0.6mi). Burrow length depends on the number, ages and conditions of mole-rats in a system and on the abundance of food items. Apart from providing nest, food storage and toilet areas, the entire burrow system is dug in search of food.

Burrowing activity normally increases just after the rainy season, when the soil is soft, moist and easily worked. It appears that food sources located at this time are exploited at a later date. Small food items are stored (by Common and Cape mole-rats for example), larger items are left growing *in situ* and are gradually hollowed out by the mole-rats, thus ensuring a constantly fresh and growing food supply. In some areas where Naked mole-rats occur, tubers may weigh as much as 50kg (110lb). When feeding, the mole-rat holds small items with its forefeet, shakes them free of soil, cuts them up into small pieces with its incisors and then chews these with its cheekteeth.

In southwestern Cape Province, South Africa, differences in diet, burrow diameter and depth, and perhaps social organization, enable three genera to occupy different niches within the same geographical area—indeed in the same field: *Bathyergus*, *Georychus*, and *Cryptomys*. This sympatry is unusual for burrowing mammals, where the normal pattern is for one species to occupy one area exclusively.

The social behavior of three genera (*Bathyergus*, *Georychus* and *Heliophobius*) follows the normal pattern for subterranean mammals: they are solitary. Nothing is

claws on their forefeet and so are less efficient at mining. Moreover their body size is larger. These disadvantages seem to restrict them to areas with easily dug sandy soil. This difference in digging method is reflected in the teeth of mole-rats: the incisors of tooth-diggers have roots that extend back behind the row of cheekteeth.

When a mole-rat is tunneling it pulls the soil under its body with its forefeet. Then, with the body weight supported by the forefeet, both hind feet are brought forward, collect the soil and kick it behind the animal. Once a pile of soil has accumulated the mole-rat reverses along the burrow with the soil and pushes it up a side branch to the surface. In all but the Naked mole-rats solid cores of soil are forced out onto the surface to form a molehill. Naked mole-rats kick a fine spray of soil out of an open hole—an "active" hole looks like an erupting volcano. A number of Naked mole-rats cooperate in digging, one animal excavating, a number transporting soil and another individual kicking it out of the hole. In all mole-rats the hind feet are fringed by stiff hairs which help hold the soil down during digging movements.

Because mole-rats live in complete darkness for most of their lives their eyes are small. (Interestingly, the Cape mole-rat, which spends some time on the surface, has eyes larger than those of other mole-rats.) Evidence derived from experiments suggests that the eyes are sightless: only their surface is used, to detect air currents which would indicate damage to the burrow system. Indeed if damage occurs they rapidly repair it. Touch is important in finding their way round the burrow system; many genera have long touch-sensitive hairs scattered over their bodies (in the Naked mole-rat these are the only remaining hairs). Their

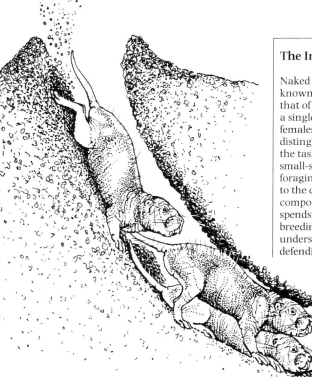

The Insect-like Rodent

Naked mole-rats are the only mammals known to have a colony structure similar to that of social insects. Within each colony only a single pair breeds; the remaining males and females belong to castes which are distinguishable by differences in size and by the tasks they perform. More numerous, small-sized, working-caste mole-rats dig the foraging burrows and carry food and nesting to the communal nest. A nonworking caste, composed of fewer, large-sized individuals, spends most of the time in the nest with the breeding female: its role is not yet clearly understood but it may be involved in defending the colony.

energies of colony members along specific avenues (some finding food, others bearing young) these mole-rats can survive in areas where single or pairs of mole-rats cannot.

Experiments have shown that nonbreeding members of Naked mole-rat colonies are not sterile. They can found new colonies and can also replace the breeding animals if they die. In this latter case, if the colony is an established one and otherwise undisturbed several females initially show signs of sexual activity, but one will grow rapidly and become sexually dominant within a few weeks of the death of the former breeding female. Usually no fighting occurs, suggesting that a hierarchy exists within the colony and that the successor is already determined.

If, therefore, the nonbreeding mole-rats are

known of mating behavior but their young disperse soon after weaning. In *Georychus* the dispersing young appear to burrow away from the parent system and block up the linking burrows; this probably also occurs in the other solitary genera and would ensure that the young are protected from predators during this otherwise very vulnerable phase in their life history. If forcibly kept together in captivity, levels of aggression build up until litter mates will kill each other. The Common mole-rat occurs in pairs or small groups about which little is known, while Naked mole-rat colonies may contain over 80 individuals, but here only a pair of mole-rats breed (see box).

Because they live in a well-protected, safe environment, mole-rats, unlike most rodents, have little to fear in the way of predators. Perhaps as a consequence their litters contain few young, usually between two and five. There are exceptions: the Cape mole-rat produces up to 10 young and the Naked mole-rat has given birth to as many as 27 in captivity, though the average litter size is 12. It must be remembered, however, that the Cape mole-rat tends to surface more often than other species and is therefore more liable to be preyed upon, and so may produce more young to compensate for exceptional loss, and that in a Naked mole-rat colony only one female breeds.

All genera, except Naked mole-rats, appear to be seasonal breeders, having one or two litters during the breeding season. In captivity the breeding female Naked mole-

The young born to the colony are cared for by all the mole-rats but suckled only by the breeding female. During weaning (from three weeks old), in addition to eating food brought to the nest the young beg feces from colony members: among other things this appears to be important in providing nutrients for the young mole-rats. Once weaned they join the worker caste, but whereas some individuals appear to remain as workers throughout their lives, others eventually grow larger and become nonworkers. It is therefore likely that a colony is composed of a number of closely related litters—many with the same parents. As with social insects, this relatedness is probably an important factor in the evolution and maintenance of a social structure in which some individuals in the colony never breed. By caring for closely related mole-rats which share their genetic make-up, the nonbreeding individuals ensure the survival and passing on of their own genetic characteristics, even when they themselves do not actually breed.

This type of social system often evolves in animals in which an individual or pair has a poor chance of surviving on its own and of successfully rearing young. Under these circumstances it pays them to team up with other individuals and for some animals not to breed but to devote all their energies to ensuring that young born to their close relatives survive. This appears to be true for Naked mole-rats which live in arid regions and have to bore through very hard soil to find their food. By joining forces with a number of individuals and by channeling the

not sterile, why do they normally not breed? After all, individual mole-rats cannot appreciate the long-term evolutionary significance of not breeding, so what prevents them from breeding? It appears that the breeding female suppresses reproduction in her colony. As with the social insects, this control seems to be largely chemical (pheromones) produced by the breeding female. In captive colonies, with a history of many litters surviving to weaning, the whole colony is affected by the reproductive state of the breeding female. For example, just before a litter is born all the colony members develop teats and look like females: male hormone levels drop and some females come close to breeding condition. This strongly suggests that the colony is responding to chemical stimuli produced by the breeding female: in this case, the stimuli seem to prime the colony to receive and care for young which are not their own. The chemicals used and the ways in which the colony is exposed to them are currently under investigation.

Two sites of control are strongly implicated by experimental evidence. The first is through chemicals in the urine of the breeding female which is deposited in the communal toilet area and becomes spread over other colony members. The second site of possible control is in the communal nest where there is very close body contact. This would be an ideal situation for transmitting chemical signals to the colony.

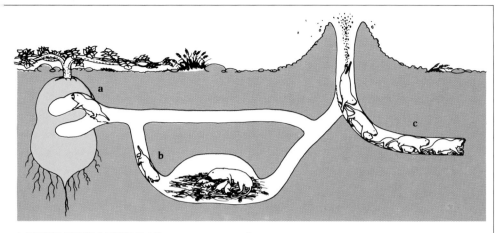

THE NAKED MOLE-RAT UNDERGROUND ORGANIZATION

▲ **Cross-section of a burrow system.** On the left (**a**) Naked mole-rats hollow out a growing tuber; in the center (**b**) is the main chamber which is occupied by the breeding female, subsidiary adults and young; on the right (**c**) a digging chain is at work.

◀ **A digging chain of Naked mole-rats.** The front mole-rat excavates with its teeth and pushes the soil backwards. The animal behind pulls it behind him and then moves backwards, holding himself close to the tunnel floor, until it can pass the soil to the animal responsible for dispersal. It then returns to the front, straddling the line of mole-rats pushing soil away.

▶ **The appearance of Naked mole rats** varies according to function. From top to bottom: (**i**) breeding female; (**ii**) nonworking adult; (**iii**) working adult.

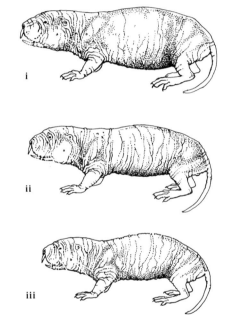

▼ **A young Naked mole-rat** in the huddle.

rat produces a litter every 80 days (just over 4 litters a year).

The only predators known to be able to pursue mole-rats underground are snakes. Field evidence suggests that the Mole snake (*Pseudaspis cana*) is attracted to the smell of freshly thrown up soil and will penetrate the burrow system via a new molehill. This may also be true of the Eastern beaked snake (*Rhamphiophis oxyrhunchus rostratus*) which has been seen preying on Naked mole-rats as they kick soil out of the burrow. Other predators take mole-rats when they venture above ground or are working very near the surface; mole-rat skulls are not uncommon in the pellets of birds of prey.

In addition to protecting the mole-rat against predators, the underground environment provides a uniform humid microclimate. This and the high mosture content of the mole-rats' food precludes the necessity of having to drink free water. The burrow temperature remains stable throughout the day, often in stark contrast to the surface temperature. In Naked mole-rat country, for example, surface temperatures of over 60°C (140°F) have been recorded while burrows (20cm (8in), below ground) remained a steady 29–30°C (84–86°F). In response to this stable burrow temperature Naked mole-rats have almost lost the ability to regulate their body temperature which consequently remains close to that of the burrow. If they need to alter their body temperature they huddle together when cold and take refuge in cooler areas within the system if they overheat when, for example, digging close to the surface. Their naked skin permits a rapid transfer of heat between them and the environment. Because Naked mole-rats have low body temperatures their metabolic rates are also low and such things as their need for food and their rate of growth are less than normal.

Though inconspicuous animals, mole-rats can cause considerable damage to human property. They have the ability to chew through underground cables and to undermine roadways, and sometimes devour root crops. Even their hills of excavated soil can damage the blades of harvesting machines, not to mention garden lawns and golf courses. Recent studies, however, have shown that mole-rats are important agents in soil drainage and soil turnover (a single Cape dune mole-rat may throw up as much as 500kg (1,100lb) of soil each month). They may play a role is dispersing geophytes (plants with underground storage organs) and may eat plants that are poisonous to farm livestock. JUMJ

LAGOMORPHS

ORDER: LAGOMORPHA
Two families; 11 genera; 58 species.

Pikas
Family: Ochotonidae–ochotonids
Fourteen species in 1 genus (*Ochotona*).

Rabbits and hares
Family: Leporidae–leporids
Forty-four species in 10 genera.
Includes: **Antelope jack-rabbit** (*Lepus alleni*),
Black-tailed jack-rabbit (*L. californicus*),
European hare (*L. europaeus*), **Arctic hare**
(*L. timidus*), **Snowshoe hare** (*L. americanus*),

Sumatran hare (*Nesolagus netscheri*), **European
rabbit** (*Oryctolagus cuniculus*), **Volcano rabbit**
(*Romerolagus diazi*), **Swamp rabbit** (*Sylvilagus
aquaticus*), **Forest rabbit** (*S. brasiliensis*), **Marsh
rabbit** (*S. palustris*), **Eastern cottontail**
(*S. floridanus*).

WHAT do the mad "March hare" and the diminutive rock coney or pika have in common; and how do they relate to the pestilential European rabbit and the rare Volcano rabbit? They are all lagomorphs, which literally means "hare-shaped."

Lagomorphs occur throughout the world either as native species or introduced by man. The order contains two families: the small rodent like pikas (family Ochotonidae) which weigh less than 0.5kg (1.1lb); and the rabbits and hares (family Leporidae), the largest of which may weigh over 5kg (11lb). Pikas are small with short front and hind legs and are well adapted for living in rocky habitats where many species are found; the tail is virtually absent (at least externally) and their ears are short and rounded. Rabbits and hares, on the other hand, have a more flattened body, and elongate hindlimbs adapted for running at speed over open ground. Their ears are long and mobile and the tail is usually a small "powder puff."

Lagomorphs, whether hares or pikas, have characteristics which give them a similar appearance. Their fur is usually long and soft and their feet, unlike those of many rodents, are fully furred; their ears are large (even pikas have ears larger than most rodents) and their eyes are set high on their heads and look sideways giving them a wide field of vision. The nose has slit like nostrils which can be opened and closed by a fold of skin above—thus rabbits are often said to "wink" their noses. In stature lagomorphs have weak but flexible necks which enable them to turn their heads more than most rodents and when resting they usually tuck their heads back into their shoulders. Lagomorphs have only one external opening (cloaca) for both anus and urethra.

Lagomorphs are herbivores feeding on grasses, leaves and bark, as well as seeds and roots, even though some have been reported as eating snails and insects—even ants in the case of the Eastern cottontail. The lagomorph digestive system is highly modi-fied for coping with large quantities of vegetation; particularly they eat some of their feces (see diagram).

Lagomorphs were originally classified as rodents (order Rodentia) because of their gnawing incisors and herbivorous diet. In 1912 J. W. Gridley recognized several distinctive features which set them apart from rodents and he created the new order Lagomorpha. Unlike rodents, lagomorphs possess a second pair of small, peg like incisors behind the long constantly growing pair in the upper jaw. This gives rabbits and hares a dental formula of I2/1, C0/0, P3/2, M3/3; pikas have one fewer upper molar on each side. Serological studies indicate that lagomorphs are no more closely related to rodents than to any other mammal order.

The oldest known lagomorphs occur in the late Eocene (50 million years ago) in Asia and North America, but the more primitive Asian forms may pre-date the first North American records. Despite probable Asian origins, the family Leporidae underwent most of its Oligocene and early Miocene development in North America (38–20 million years ago). The first pikas appeared in Asia in the middle Oligocene and spread to North America and Europe in the Pliocene (7–2 million years ago). Pikas seem to have peaked in distribution and diversity during the Miocene (26.7 million years ago) and since declined while the rabbits and hares have maintained a widespread distribution since the Pliocene. Evolutionary changes in both families have been conservative.

Rabbits and hares have nevertheless occupied a remarkably wide range of habitats; these include the cold snow-covered arctic habitat of the Arctic and Snowshoe hares, the semi-desert habitat of the Antelope and Black-tailed jack rabbits, the grassland habitat of the European hare, the tropical forest of the South American Forest rabbit, the tropical mountain forest of the Sumatran hare and the swamps of the

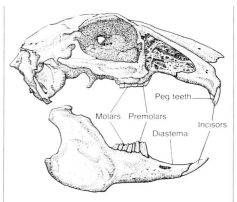

▲ **Rabbit skull.** Lagomorphs have long, constantly growing incisors, as do rodents, but lagomorphs differ in having two pairs of upper incisors, the back non-functional ones known as peg teeth. There is a gap (diastema) between the incisors and premolars. The dental formula of rabbits and hares (family Leporidae) is I2/1, C0/0, P3/2, M3/3, with the pikas (family Ochotonidae) having one less upper molar in each jaw.

▶ **Motionless in grass,** a young European hare (leveret) hides in a form awaiting its mother's return. European hare leverets are dispersed to such separate forms about three days after birth.

▶ **Double digestion.** The lagomorph digestive system is highly modified for coping with large quantities of vegetation. The gut has a large blind-ending sac (the cecum) between the large and small intestines which contains a bacterial flora to aid the digestion of cellulose. Many products of the digestion in the cecum can pass directly into the blood stream, but others such as the important vitamin B_{12} would be lost if lagomorphs did not eat some of the feces (refection) and so pass them through their gut twice. As a consequence lagomorphs have two kinds of feces. Firstly, soft black viscous cecal pellets which appear during the day in nocturnal species and during the night in species active in the daytime. These are usually eaten directly from the anus and stored in the stomach, to be mixed later with further food taken from the alimentary mass. Secondly round hard feces which are passed normally.

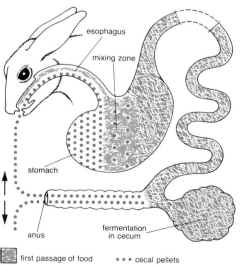

esophagus

mixing zone

stomach

anus

fermentation in cecum

first passage of food | • • • cecal pellets

alimentary mass | • • • hard feces

North American Marsh and Swamp rabbits. Pikas, on the other hand, are highly adapted to their alpine habitat of North America and Asia, and steppes in Asia.

Lagomorphs represent staple prey to a range of medium-sized mammalian and bird, and larger predators. In species like the European and cottontail rabbits, predation, disease and climatic factors are the primary agents responsible for mortality rates often in excess of 90 percent among young of the year. High mortality rates are counteracted by a well recognized high reproductive capacity. Most species reach sexual maturity relatively early (three months in female European rabbits). The gestation period is short: 40 days in *Lepus* species and about 30 days in all other members of the order; litter size is often large. Other features of lagomorph reproduction which minimize inter-birth interval for females include the phenomenon of induced "ovulation", where eggs are shed in response to copulation rather than only a cyclic basis; and "postpartum estrus" where a female is able to conceive immediately after giving birth. Female lagomorphs are also capable of "resorbing" embryos under adverse conditions, for example climatic or social stress; this clearly reduces the energetic loss of premature pregnancy termination for females. There is also evidence that some species, like the European hare, are capable of conceiving a second litter before birth of the last young—"superfetation"; explanations as to just how a female is able to support embryos at different stages of development present reproductive physiologists with something of a challenge!

JAC/ES

RABBITS AND HARES

Family: Leporidae
Forty-four species in 10 genera.
Distribution: Americas, Europe, Asia, Africa;
introduced in Australia, New Zealand and
other islands.

Habitat: wide ranging, from seashore to upper
mountainous regions, from arctic tundra to city
center, from dry desert to the swamp, from
agricultural landscape to forest.

Size: ranges from head-body length 25–29cm
(10–11.4in), weight 0.3kg (0.66lb) for Pygmy
rabbit, and tail length 1.5cm (0.6in) (Amami
rabbit) to head-body length 50–76cm
(20–30in), weight 2.5–5kg (5.5–11lb) and tail
length 7–12cm (2.8–4.7in) in European hare.
Ear length ranges from about 4.3cm (1.7in) in
Sumatran hare to 18–20cm (7–8in) in the
Antelope jackrabbit; ears longer than wide.
Hind limbs longer than forelimbs. Y-shaped
naked groove extends from the upper lip to
and around nose ("hare-lip").

Coat: usually thick and soft, but coarse in some
forms; hair on the ears shorter and thinner; tail
well furred or even bushy; feet hairy on both
surfaces; coloration ranges through reddish
brown, brown, buff, gray, or white, belly often
covered with lighter or pure white hair. One
species striped, and arctic forms change into
white for winter.

Gestation: usually longer in hares (Mountain
hare 50 days) than in rabbits (European rabbit
28–33 days).

Longevity: average less than 1 year in the wild;
physiological maximum age of European hare
is estimated 12–13 years; oldest recorded age in
the wild is 12.5 years.

Species include: **Bushman hare** (*Bunolagus
monticularis*); **Hispid hare** (*Caprolagus hispidus*);
Antelope jackrabbit (*Lepus alleni*); **Black-
tailed jackrabbit** (*L. californicus*); **Snowshoe
hare** (*L. americanus*); **European hare**
(*L. europaeus*); **Arctic hare** (*L. timidus*);
Sumatran hare (*Nesolagus netscheri*); **European
rabbit** (*Oryctolagus cuniculus*); **Amami rabbit**
(*Pentalagus furnessi*); **Volcano rabbit**
(*Romerolagus diazi*); **Forest rabbit** (*Sylvilagus
brasiliensis*); **Eastern cottontail** (*S. floridanus*).

THERE are few mammals so involved with
man as the European rabbit. Domestication of the species started probably in
Roman times in northern Africa or Italy;
today there are over 50 established strains of
domestic rabbit all selectively bred from this
one species. There have been many deliberate worldwide introductions of the European rabbit to countries like Australia by
19th century empire-builders (seeking to
establish the fauna and hunting from
home), and to scattered islands throughout
the oceans as a food source for shipwrecked
sailors. Most of the successful mainland
invasions have been by wild-type stock;
colonies of domesticated rabbits only managed to survive on islands without heavy
predation pressures. In the absence of those
natural selection pressures (predators, climate and disease) operating in their
ancestral home, many of these introduced
populations exploded to reach the pest proportions familiar in Australia in the first half
of this century.

Yet the European rabbit is just one species
of a family (Leporidae) of 44 species, some of
which have similar but less notorious in-
volvement with man, and some whose
numbers can be counted possibly in
hundreds rather than tens of millions.

The family Leporidae can be broadly
divided into two groups: the jackrabbits and
hares (genus *Lepus*); and the rabbits in the
remaining 9 genera. Although many of the
latter genera include species popularly
known as "hares", for example the Hispid
hare, they look more like rabbits.

All rabbits and hares are adapted for
quick movement and flight from danger.
Their hind legs are long and adapted for
running—in full flight some of the larger
hares can reach speeds of 80km/h (50mph).
Their ears are large and mobile, to facilitate
the detection of approaching danger. Unlike
those of pikas (the other family of the order
Lagomorpha), the eyes of rabbits and hares
are large and adapted to their twilight and
night-time activities.

The forefeet have five toes and the hindfeet four, all equipped with strong claws.
The lower surface of the feet is covered with
long, thick brush-like hairs which provide a
better grip and cushion when moving on
hard ground. This is especially true in the
Snowshoe hare where it also helps the
animal to run on top of snow.

Many of the differences between hares
and rabbits are associated with specializations for a burrowing life (most rabbits) or
improved running ability (hares). A few
hare species are known to burrow in extreme climates, eg the Black-tailed jackrabbit digs short burrows to avoid high summer
temperatures in the desert. Snowshoe and
Arctic hares may burrow into snow. In
Scotland, Arctic hare populations often dig
burrows 1–2m (3.3–6.6ft) long which are
used as bolt holes by young—but rarely used
by adults. The same species is known to dig
complex 7m (23ft) tunnels in Khatanga
(Russia), but doesn't appear to burrow in
other countries, for example the Swiss Alps
or Ireland. Forms—surface depressions in
the ground or vegetation—are more commonly used as resting up sites by hares.
These may be well-established sites used by
successive generations or temporary refuges
occupied for a few hours. The underground
network of tunnels and chambers dug by
rabbits show considerable variety in form,
depending on species and within species
according to habitat (see pp724–725).

The occurrence of rabbits and hares in
different climatic conditions is reflected in

8

Representative species of rabbits and hares.
(1) The Greater red rockhare (*Pronolagus crassicaudatus*) in an alert scanning posture.
(2) The Hispid hare (*Caprolagus hispidus*), sitting among cuttings and pellets. (3) The European hare (*Lepus europaeus*), boxing. (4) The Volcano rabbit (*Romerolagus diazi*) reingesting pellets, sitting amongst vegetation of *zacatón* grasses.
(5) A male Eastern cottontail (*Sylvilagus floridanus*) in an alert posture. (6) The Sumatran hare (*Nesolagus netscheri*), grooming its muzzle and spreading scent. (7) The European rabbit (*Oryctolagus cuniculus*), a dominant male rubbing his chin. (8) An Amami rabbit (*Pentalagus furnessi*) digging its burrow. (9) A Bushman hare (*Bunolagus monticularis*) in an alert posture. (10) A Bunyoro rabbit (*Poelagus marjorita*) hopping.

their molting physiology. Different species undergo one to three regular seasonal changes of fur, in some cases resulting in sharp differences between summer and winter coloration. Molting is triggered by temperature and light intensity. The Arctic hare never attains a white winter coat in Ireland but retains it for five months of the year in the European part of the USSR and for seven months in some northern Asian regions.

Young hares (leverets) are born, at a form site at a more advanced stage of development than rabbit young (kittens). The latter are generally born at carefully constructed nests (of hair and grass) within the warren or special shallower breeding "stop" sites. Leverets are covered with fur at birth and their eyes are open; rabbit kittens are naked and their eyes do not open for several days (10 in the European rabbit). Those species studied so far appear to suckle their litters for only one brief period, typically under 5 minutes, once every 24 hours. (The milk of female European rabbits, for example, is known to be highly nutritious and with 10 percent fat and 15 percent protein it is richer than goat or cow milk.) For female European rabbits this will be the only contact between

mother and young until weaning (around 3 weeks of age); those litters born at breeding stop nests, for instance, will be re-enclosed into their soil-covered tunnel between each nursing. European hare leverets are dispersed to separate forms about 3 days after birth, but meet up daily at a specific location at a set hour (around sunset) for less than 3 minutes nursing. Radio-tracked litters of Snowshoe hare showed a similar behavior.

Puberty is attained in approximately 3–5 months in the European rabbit and about 1 year in most hare species; however some species of hares take two years.

The relationship between climate and reproduction is illustrated by the New World rabbits in the genus *Sylvilagus*. They show a direct correlation between latitude and litter size; species or subspecies in the north produce the largest litters which are generally correlated with the shortest breeding season. There is also a relationship between latitude and the length of the gestation period—rabbits in northern latitudes have the shortest gestation periods. The advantage of a short gestation is that the maximum number of young can be produced during the period when the weather is most

▲ **Big ears**—a Black-tailed jackrabbit in its hot, arid habitat. It is active at night, the day being spent in the shade. Its ability to regulate the flow of blood through the massive ears controls the intake or loss of heat according to environmental conditions.

▶ **The most abundant** and best-studied of cottontails, ABOVE, the Eastern cottontail, is found in almost all types of habitat and mature woodland. Young (kittens) are born on the surface, are blind and have only a sparse covering of hair.

▶ **Seeking shade under a cactus,** a Desert cottontail rests during the daytime heat. Most cottontails are active at night or twilight, but may be seen during the day at any time.

▷ **Jackrabbit in snow.** OVERLEAF As is common to several species from northern regions, the coat molts in the fall and is replaced by a thicker, paler (often white) one which aids heat conservation and camouflage in the snowy wastes.

suitable. Conversely, it is advantageous for rabbits to have longer gestation periods in southern localities because young rabbits born more fully developed are better able to avoid predators and fend for themselves.

The potentially high reproductive capacity of rabbits and hares means populations can have a rapid rate of increase. Population cycles have been observed in some species, particularly in the Snowshoe hare (see pp 722–723).

Rabbit social behavior varies from the highly territorial, breeding groups of the European rabbit to (see pp 724–725) the non-territorial Eastern cottontail.

The only auditory signals known for most species are the characteristic foot thumps—made in alarm or aggression—and the distress screams made by captured young and adults. Some species like the Volcano rabbit are reported to use a variety of calls. Scent signals seem to play a predominant role in the communication systems of most rabbits and hares.

Because of their abundance, rabbits and hares are widely interwoven into man's economy. They appear frequently in the art of ancient civilizations, for example of the Central American Mayans, and were regarded as prime sporting animals for the "hunt" by Europeans. To the Romans, roast rabbit meat and embryos—laurices—were a delicacy, with the result that they kept rabbits and hares in walled enclosures—"leporaria." The good quality of its light, soft meat made the European rabbit a valued source of food. During the Middle Ages rabbits were kept in the enclosures belonging to monasteries and many of today's populations of wild European rabbits are descendants of escapees. Rabbit meat remains an important source of protein but is less popular in, for example, the United Kingdom since myxomatosis.

Modern agriculture involving deforestation and grazing by livestock, and the elimination of predators, created conditions (open grassland, cultivated crops and few predators) favorable for rabbits (the European rabbit is preyed upon by over 40 vertebrate species in its ancestral home in the Iberian peninsula, but much fewer in the rest of Europe). In early postglacial times the European hare inhabited the open steppes of eastern Europe and Asia, and it is likely that few occurred in the limited open habitats of central and western Europe. Advancing agriculture later replaced the open forests with a patch-work of cultivated fields, meadows, pastures and isolated woodlands and hedgerows, and a profusion of "new" wild and cultivated plants emerged—ideal hare habitat. The European hare rapidly moved into this habitat and is now found throughout Europe.

In Australia, introduced European rabbits caused enormous damage to both virgin habitats and cultivated areas between 1900–1959. The European hare is regarded as a pest in Australia and New Zealand where it has been introduced. This species was also introduced to Argentina (from Germany) in 1888 at Rosaria Santa Fe province. Since then it has spread throughout the country from Patagonia to the subtropical north covering an area of around 2.7 million sq km (1 million sq mi). Densest populations occur in the "humid pampas" area where 5–10 million hares are caught each year with nearly 15,000 tonnes of meat exported to Europe in 1977.

In the United States the Black-tailed jackrabbit of California is the most serious pest of cultivated crops, while the Snowshoe hare and cottontails cause greatest disruption of the reforestation program through their destruction of tree seedlings.

Large-scale rabbit drives were a popular method of control at the beginning of the 20th century, but modern control programs use a combination of techniques including

poison baits, buffer crops, repellents, exclusion fencing, shooting and trapping. Control of European rabbits in Australia was achieved by the introduction of the virus disease myxomatosis. This is harmless to its natural host the Forest rabbit or tapiti of South America and to other members of the genus *Sylvilagus*, but is virulent and deadly to the European rabbit. Following the introduction of myxomatosis into the wild rabbits of Australia in 1951–52, massive numbers died, relieving the country of a devastating pest. A similar decline occurred in the United Kingdom and other European countries. But now both in Europe and Australia rabbits are beginning to develop a level of immunity to the disease and their numbers have again increased.

In North America the transplantation of eastern forms to the west coast has caused changes in the genetic make-up of resident forms. For example, with a decreasing population of the native subspecies of the Eastern cottontail, wildlife agencies, hunt clubs and private individuals started, in the 1920s, a massive cottontail importation program into Kansas, Missouri, Texas and Pennsylvania. The scheme failed in that the annual harvest remained low, but one incidental consequence is that the native subspecies has been replaced by its hybrid with the imported one, which is colonizing new habitats.

While some rabbits and hares have reached pest status others are in danger of extinction. In general these are relict species, often the only members of their genus, specialized to a restricted threatened habitat. The Sumatran hare, of which only 20 animals have ever been recorded, comes from remote mountainous regions of Sumatra. The Hispid hare found in scattered pockets of sal forests in northern India and Bangladesh is losing the fight to compete with humans who are either burning its habitat to improve grazing or using its food plants as thatch. The Bushman hare inhabits riverside habitats that are rapidly being cultivated. The Volcano rabbit is restricted to the volcanic slopes near Mexico City, that is within a 30 minute drive of 17 million people. As well as habitat destruction this endangered species suffers the pressures of tourism and hunting. The Amami rabbit is found only on two heavily forested islands in the Amami Island group of Japan. Incessant logging has reduced the population to about 5,000 and its endangered status is supported by its creation in Japan as a "special natural monument."

JAC/ES

THE 44 SPECIES OF RABBITS AND HARES

Abbreviations: HBL = head-body-length; TL = tail length; EL = ear length; HFL = hind foot length; WT = weight. Approximate measure equivalents: 1cm = 0.4in; 1kg = 2.2lb.

* CITES listed; E Endangered.

Genus *Bunolagus*

Bushman hare E
Bunolagus monticularis
Bushman or River hare.

Central Cape Province (S Africa).
Dense riverine scrub (not the
mountainous situations often
attributed). Now extremely rare.
Coat: reddish, similar to Red
rockhares with bushy tail.

Genus *Caprolagus*

Hispid hare E *
Caprolagus hispidus
Hispid hare or Assam rabbit.

Uttar Pradesh to Assam; Tripura
(India), Mymensingh and Dacca on
the western bank of river
Brahmaputra (Bangladesh). Sub-
Himalayan sal forest where grasses
grow up to 3.5m in height during the
monsoon months; occasionally
cultivated areas. HBL 476mm; TL
53mm; EL 70mm; HFL 98mm; WT
2.5kg. Coat: coarse and bristly;
upperside appears brown from
intermingling of black and brownish
white hair; underside brownish white
with chest slightly darker; tail brown
throughout, paler below. Claws
straight and strong. Inhabits burrows
which are not of its own making.
Seldom leaves forest shelter.

Genus *Lepus*
Hares
Most inhabit open grassy areas, but:
Snowshoe hare occurs in boreal
forests; European hare occasionally
forests; Arctic hare prefers forested
areas to open country; Cape hare
prefers open areas, occasionally
evergreen forests. Rely on well-
developed running ability to escape
from danger instead of seeking cover:
also on camouflage by flattening on
vegetation. HBL 400–760mm; TL
35–120mm; WT 1.3–5kg. Coat:
usually reddish brown, yellowish
brown or grayish brown above,
lighter or pure white below; ear tips
black edged with a significant black
area on the exterior in most species;
in some species the upperside of the
tail is black. Indian hare has a black
nape. Species inhabiting snowy
winter climate often molt into a white
winter coat, while others change from
a brownish summer coat into a
grayish winter coat. Diet: usually
grasses and herbs, but cultivated
plants, twigs, bark of woody plants

are the staple food if others are not
available. Usually solitary, but
European hare more social. Habitat
type has a marked effect on home-
range size within each species, but
differences also occur between
species, eg from 4–20ha in Arctic
hares to over 300ha in European
hares. Individuals may defend the
area within 1–2m of forms but home
ranges generally overlap and feeding
areas are often communal. Most live
on the surface, but some species, eg
Snowshoe and Arctic hares dig
burrows while others may hide in
holes or tunnels not of their making.
Breed throughout the year in
southern species; northern species
produce 2–4 litters during spring and
summer. Litter size from 1–9.
Gestation up to 50 days in Arctic
hare, other species shorter.
Vocalization: deep grumbling; shrill
calls given in pain. Twenty-one
species.

Antelope jackrabbit
Lepus alleni

S New Mexico, S Arizona to N Nayarit
(Mexico), Tiburon Is. Locally
common. Avoid dehydration in hot
desert by feeding on cactus and
yucca.

Black-tailed jackrabbit
Lepus californicus

Mexico, Oregon, Washington,
S Idaho, E Colorado, S Dakota,
W Missouri, NW Arkansas, Arizona,
N Mexico. Locally common.

White-sided jackrabbit
Lepus callotis

SE Arizona, SW New Mexico and
Oaxaco (Mexico). Locally common,
but declining.

Tehuantepec jackrabbit
Lepus flavigularis

Restricted to sand dune forest on
shores of salt water lagoons on
nothern rim of Gulf of Tehuantepec,
(Mexico). Nocturnal.

Black jackrabbit
Lepus insularis

Espiritu Santo Is (Mexico).

White-tailed jackrabbit
Lepus townsendii

S British Columbia, S Alberta,
SW Ontario, SW Wisconsin, Kansas,
N New Mexico, Nevada, E California.
Locally common.

Snowshoe hare
Lepus americanus

Alaska, coast of Hudson Bay,
Newfoundland, S Appalachians,
S Michigan, N Dakota, N New Mexico,
Utah, E California. Locally common.

Japanese hare
Lepus brachyurus

Honshu, Shikoku, Kyushu, (Japan).
Locally common.

Cape hare
Lepus capensis

Africa, S Spain (?), Mongolia,
W China, Tibet, Iran, Arabia. Locally
common.

European hare
Lepus europaeus
European or Brown hare.

S Sweden, S Finland, Great Britain
(introduced in Ireland), Europe south
to N Iraq and Iran, W Siberia. Locally
common but declining.

Savanna hare
Lepus crawshayi

S Africa, Kenya, S Sudan; relict
populations in NE Sahara. Locally
common.

Manchurian hare
Lepus mandshuricus

Manchuria, N Korea, E Siberia. Range
decreasing.

Ethiopian hare
Lepus starkei

Ethiopia.

Indian hare
Lepus nigricollis
Indian or Black-naped hare.

Pakistan, India, Sri Lanka (introduced
into Java and Mauritius).

Woolly hare
Lepus oiostolus

Tibetan plateau and adjacent areas.

Burmese hare
Lepus peguensis

Burma to Indochina and Hainan
(China).

Scrub hare
Lepus saxatilis

S Africa, Namibia.

Chinese hare
Lepus sinensis

SE China, Taiwan, S Korea.

Arctic hare
Lepus timidus
Arctic, Mountain or Blue hare.

Alaska, Labrador, Greenland,
Scandinavia, N USSR to Siberia,
Hokkaido, Sikhoto Alin Mts, Altai,
N Tien Shan, N Ukraine, Lithuania.
Locally common. Isolated populations
in the Alps and Ireland.

African savanna hare
Lepus whytei

Malawi. Locally common.

Yarkand hare
Lepus yarkandensis

SW Sinkiang (China). Rare.

Genus *Nesolagus*

Sumatran hare *
Nesolagus netscheri
Sumatran hare or Sumatran short-eared rabbit.

W Sumatra (1°–4°S) between
600–1,400m in Barisan range.
Primary mountain forest.
HBL 368–393mm; TL 17mm;
EL 43–45mm. Coat: variable; body
from buffy to gray, the rump bright
rusty with broad dark stripes from the
muzzle to the tail, from the ear to the
chin, curving from the shoulder to the
rump, across the upper part of the
hind legs, and around the base of the
hind foot. Diet: juicy stalks and leaves.
Strictly nocturnal; spends the day in
burrows or in holes (not of its own
making). Very rare—only one
specimen recorded in last decade.

Genus *Oryctolagus*

European rabbit
Oryctolagus cuniculus

Endemic on the Iberian Peninsula and
NW Africa; introduced in rest of
W Europe 2000 years ago, and to
Australia, New Zealand, S America
and some islands. Opportunistic,
having colonized habitats from stony
deserts to subalpine valleys; also
found in fields, parks and gardens,
rarely reaching altitudes of over
600m. Very common. HBL:
380–500mm; TL 45–75mm; EL
65–85mm; HFL 85–110mm; WT
1.5–3kg. Coat: grayish with a fine
mixture of black and light brown tips
of the hair above; nape reddish-
yellowish brown; tail white below;
underside light gray; inner surface of
the legs buffy gray; total black is not
rare. All strains of domesticated rabbit
derived from this species. Colonial

organization associated with warren systems. Diet: grass and herbs, roots and the bark of trees and shrubs, cultivated plants. Breeds from February to August/September in N Europe; 3–5 litters with 5–6 young, occasionally up to 12; gestation 28–33 days; young naked at birth; weight about 40–45g, eyes open when about 10 days old. Longevity: about 10 years in wild. Vocalization: shrill calls are given in pain or fear.

Genus *Pentalagus*

Amami rabbit [E]
Pentalagus furnessi
Amami or Ryukyu rabbit (erroneously).
Two of the Amami Islands (Japan). Dense forests. HBL 430–510mm; EL 45mm. Coat: thick and woolly, dark brown above, more reddish below. Claws are unusually long for rabbits at 10–20mm. Eyes small. Nocturnal. Digs burrows. 1–3 young are born naked in a short tunnel; two breeding seasons.

Genus *Poelagus*

Bunyoro rabbit
Poelagus marjorita
Bunyoro rabbit or Uganda grass hare.
S Sudan, NW Uganda, NE Zaire, Central African Republic, Angola. Savanna and forest. Locally common. HBL 440–500mm; TL 45–50mm; wt 2–3kg; EL 60–65mm. Coat: stiffer than that of any other African leporid; grizzled brown and yellowish above, becoming more yellow on the sides and white on the under parts; nape reddish yellow; tail brownish yellow above and white below. Ears small; hind legs short. Nocturnal. While resting, hides in vegetation. Young reared in burrows and less precocious than those of true hares. Said to grind teeth when disturbed.

Genus *Pronolagus*
Red rockhares
HBL 350–500mm; TL 50–100mm; HFL 75–100mm; EL 60–100mm; wt 2–2.5kg. Coat: thick and woolly, including that on the feet, reddish. Inhabits rocky grassland, shelters in crevices. Nocturnal, feeding on grass and herbs. Utters shrill vocal calls even when they are not in pain. Three species.

Greater red rockhare
Pronolagus crassicaudatus
S Africa.

Jameson's red rockhare
Pronolagus randensis
S Africa, E Botswana, Zimbabwe, Namibia.

Smith's red rockhare
Pronolagus rupestris
South Africa to Kenya.

Genus *Romerolagus*

Volcano rabbit [E] [*]
Romerolagus diazi
Volcano rabbit, teporingo or zacatuche.
Restricted to two volcanic sierras (Ajusco and Iztaccihuatl-Popocatepetl ranges) close to Mexico City. Habitat unique "zacaton" (principally *Epicampes*, *Festuca* and *Muhlenbergia*) grass layer of open pine forest at 2,800–4,000m. Smallest leporid. HBL 270–357mm; wt 400–500g; EL 40–44mm. Features include short ears, legs and feet, articulation between collar and breast bones and no visible tail. Coat: dark brown above, dark brownish gray below. Lives in warren-based groups of 2–5 animals. Breeding season December to July; gestation 39–40 days; average litter 2. Mainly active in daytime, sometimes at night. Vocal like pikas.

Genus *Sylvilagus*
Cottontails
HBL 250–450mm; TL 25–60mm; wt 0.4–2.3kg; smallest Pygmy rabbit, biggest Swamp rabbit. Most species common. Coat: mostly speckled grayish brown to reddish brown above; undersides white or buffy white; tail brown above and white below ("cottontail"): Forest rabbit and Marsh rabbit have dark tails. Molts once a year, except Forest and Marsh rabbits. Ears medium sized (about 55mm) and same color as the upper side; nape often reddish, but may be black. Range extends from S Canada to Argentina and Paraguay and a great diversity of habitats is occupied. Distributions of some species overlap. Most preferred habitat open or brushy land or scrubby clearings in forest areas, but also cultivated areas or even parks. Various species frequent forests,

marshes, swamps, sand beaches, or deserts. Diet: mainly herbaceous plants, but in winter also bark and twigs. Only Pygmy rabbit digs burrows; others occupy burrows made by other animals or inhabit available shelter or hide in vegetation. Not colonial, but some species form social hierarchies in breeding groups. Active in daytime or night. Not territorial; overlapping stable home ranges of a few hectares. Vocalization rare. Longevity: 10 years (in captivity). Thirteen species. Most locally common.

Swamp rabbit
Sylvilagus aquaticus
E Texas, E Oklahoma, Alabama, NW–S Carolina, S Illinois. A strong swimmer. Gestation 39–40 days; eyes open at 2–3 days.

Desert cottontail
Sylvilagus audubonii
Desert or Audubon's cottontail.
C Montana, SW–N Dakota, NC Utah, C Nevada and N and C California (USA), and Baja California and C Sinaloa, NE Puebla, W Veracruz, (Mexico).

Brush rabbit
Sylvilagus bachmani
W Oregon to Baja California, Cascade–Sierra Nevada Ranges. Average 5 litters per year; gestation 24–30 days; covered in hair at birth.

Forest rabbit
Sylvilagus brasiliensis
Forest rabbit or tapiti.
S Tamaulipas (Mexico) to Peru, Bolivia, N Argentina, S Brazil, Venezuela. Average litter size 2; gestation about 42 days.

Mexican cottontail
Sylvilagus cunicularius
S Sinaloa to E Oaxaca and Veracruz (Mexico).

Eastern cottontail
Sylvilagus floridanus
Venezuela through disjunct parts of C America to NW Arizona, S Saskatchewan, SC Quebec, Michigan, Massachusetts, Florida. Very common. Gestation 26–28 days; young naked at birth.

Tres Marías cottontail
Sylvilagus graysoni
Maria Madre Is, Maria Magdalena Is. (Tres Marías Is, Nayarit, Mexico).

Pygmy rabbit
Sylvilagus idahoensis
SW Oregon to EC California, SW Utah, N to SE Montana; isolated populations in WC Washington.

Omilteme cottontail
Sylvilagus insonus
Sierra Madre del Sur, C Guerrero (Mexico).

Brush rabbit
Sylvilagus mansuetus
Known only from San Jose Island, Gulf of California. Often regarded as subspecies of *S. bachmani*.

Mountain cottontail
Sylvilagus nuttalli
Mountain or Nuttall's cottontail.
Intermountain area of N America from S British Columbia to S Saskatchewan, S to E California, NW Nevada, C Arizona, NW New Mexico.

Marsh rabbit
Sylvilagus palustris
Florida to S Virginia on the coastal plain. Strong swimmer.

New England cottontail
Sylvilagus transitionalis
S Maine to N Alabama. Distinguished from overlapping Eastern cottontail by presence of gray mottled cheeks, black spot between eyes and absence of black saddle and white forehead.

ES

The Ten-year Cycle

Population fluctuations in the Snowshoe hare

Animal populations rarely, if ever, remain constant from year to year. Populations of the great majority of species fluctuate irregularly or unpredictably. An exception is the Snowshoe hare, whose populations in the boreal forest of North America undergo remarkably regular fluctuations which peak every 8–11 years. This is a persistent fluctuation, documented in fur-trade records for over two centuries and now widely known as "the 10-year cycle."

In addition to its regularity, the cycle is unusual in two other respects; firstly it is broadly synchronized over a vast mid-continental area from Alaska to Newfoundland, where regional Snowshoe hare peaks seldom differ by more than three years; secondly, the amplitude of change in numbers from a cyclic high to a low may be more than 100-fold.

There has been much speculation about what causes the Snowshoe hare cycle, but only recently have long-term field studies begun to provide the solution. It is now known that certain demographic events are consistently associated with each phase of the cycle. Thus declines from peak densities are initiated by markedly lower survival of young hares overwinter, and by sharp decreases in birth rates. These conditions persist for three or four years, and the population continues to contract. The survival of adult hares also declines during this phase of the cycle. The onset of the next cyclic increase in numbers is brought about by greatly improved rates of survival and birth. We know too that growth rates of young hares are highest during the increase phase of the cycle, and that overwinter weight losses are significantly lower at that time.

If Snowshoe populations fluctuate cyclically because of a pattern of changing survival and birth rates the next obvious question is what causes the latter? Acceptable explanations must account for the above-noted trends in juvenile growth rates and for weight losses overwinter, and for the cycle's synchrony between regions. There is currently no general consensus among biologists as to the ultimate cause of the 10-year cycle, but one tenable explanation runs as follows.

The cycle is repeatedly generated intrinsically when peak Snowshoe hare populations exceed their winter food supply of woody browse and resulting malnutrition triggers a population decline. As hare numbers fall, the ratio of predators to hares increases, as does the impact of predation on the hare population. This extends the cyclic

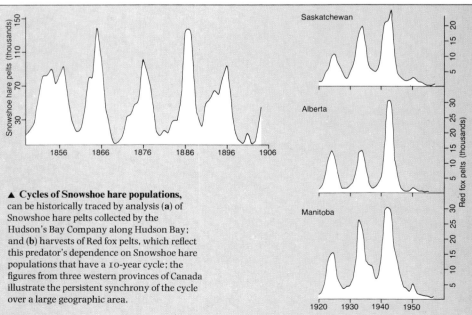

▲ **Cycles of Snowshoe hare populations,** can be historically traced by analysis (**a**) of Snowshoe hare pelts collected by the Hudson's Bay Company along Hudson Bay; and (**b**) harvests of Red fox pelts, which reflect this predator's dependence on Snowshoe hare populations that have a 10-year cycle; the figures from three western provinces of Canada illustrate the persistent synchrony of the cycle over a large geographic area.

► **Girdled by marauding hares.** Aspen trees in Alaska from which the bark has been chewed away by Snowshoe hares during winter. The thick snow cover enables the hares to reach far up the trunk. Snowshoe hares require such dense, low-growing woody vegetation for cover and winter food, while in the summer they eat a great variety of herbaceous plants.

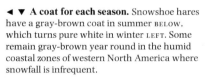

◄ ▼ **A coat for each season.** Snowshoe hares have a gray-brown coat in summer BELOW, which turns pure white in winter LEFT. Some remain gray-brown year round in the humid coastal zones of western North America where snowfall is infrequent.

decline beyond the period of winter food shortage, and drives the hare population still lower. The resulting scarcity of hares then causes predator populations to drop to low levels. Now largely free of predation, and with winter food once more abundant, the Snowshoe population begins another cyclic increase. Inter-regional synchrony is caused by mild winters that moderate mortality and thereby delay declines of malnourished hare populations, while permitting others that are lagging to reach critical peak densities. Such synchrony is reinforced by highly mobile predators responding to local differences in hare abundance.

Among the more notable ecological consequences of the Snowshoe hare cycle are: firstly, the impact of hare browsing on plant species composition and succession—such unpalatable shrubs as the honey suckles (Caprifoliaceae) increase whereas forest succession from aspen (*Populus tremuloides*) to White spruce (*Picea glauca*) is slowed; secondly a direct and overriding effect on predator populations; thirdly, an indirect effect on the alternate prey of hare predators. The Snowshoe hare's significant influence on other components of the boreal forest ecosystem stems from the amplitude of its fluctuations, and the densities of 400 to 2,400 per sq km (1,000–6,000 per sq mi) commonly attained during cyclic peaks. While plant ecologists and physiologists have only recently acknowledged the impact of hare cycles on vegetation, the destabilizing effect on numbers of lynx, Red fox, coyote, North American marten, fisher and other predatory mammals has long been recognized by the fur trade. Populations of birds of prey likewise respond to fluctuating hare abundance, and together with the mammalian predators may thereby influence such alternate prey as Ruffed, Spruce and Sharp-tailed grouse.

Not all Snowshoe hare populations have a 10-year cycle. Those that do not are found where suitable habitat (a dense understory of woody shrubs and saplings) is highly fragmented or island-like. This fragmentation exists naturally in the mountain ranges and along the southern limit of Snowshoe distribution. In these regions predators have a greater diversity of prey, and hence more stable populations. Accordingly, the absence of cyclic fluctuations among Snowshoe hares has been ascribed to their being held in check by sustained predation, especially on individuals dispersing from habitat fragments. There is evidently a parallel between the Snowshoe hare in North America and the Arctic hare in Soviet Eurasia, for the latter also has a 10-year cycle within continuous habitat of the taiga, but exhibits irregular short-term fluctuations to the south as habitats become increasingly disjunct.

Population explosions are also seen in many rodents of the arctic tundra and taiga where they occur with regularity every 3–4 years, as in the lemmings. **LBK**

Habitat and Behavior

Adaptable females and opportunist males in rabbit societies

In Europe throughout the Middle Ages rabbits were farmed successfully for their meat and fur. They appeared to be tolerant of crowding and bred profusely in their small enclosures (called warrens) and lived together in underground burrows. Naturally enough they came to be regarded as a classic example of a mammal that lives in groups. The first systematic research into the nature of their social habits was conducted on similar enclosed populations in both Australia and England during the 1950s. The resulting observations were interpreted as showing that rabbits form mixed-sex breeding groups containing 6–10 adults, and defend exclusive group territories. Since then, however, some studies of wild populations have cast doubt on the general validity of these conclusions.

Two long-term studies in England have now shown that the European rabbit's habitat can have a profound effect on its social organization and behavior. One study took place on chalk downland, the other on coastal sand-dunes. At the former, the burrows in which rabbits typically take refuge for more than half of each day are clustered together in tight groups (also, confusingly, called warrens), which are themselves randomly distributed over the down. Adult females (does), who do most burrow excavation, rarely attempted more than the expansion of an existing burrow system in

the hard chalky soil; completely new warrens hardly ever appeared. At the sand-dune site, by contrast, new burrows were continually being dug as others were collapsing or falling into disuse. Burrows were never found in the flat "slacks," which lie between the dunes and tend to flood easily, but even on the higher ground they were not clustered into the easily defined warrens found on the downland. Burrow entrances more than 5m (16ft) apart on the dunes were rarely connected by a tunnel and many had just a single entrance at the surface.

Does usually give birth and nurse their young in underground nesting chambers situated within pre-existing burrow systems. Thus, where space underground is in short supply natural selection should favor females that compete for and then defend some burrows. Not surprisingly there is some good evidence for such competition at the downland site. Of the disputes between adult females observed there, over 70 percent took place within 5m (16ft) of a burrow entrance habitually used by one of the contestants. There was also a direct relationship between the size of a warren and the number of adult females refuging in it: the larger the warren the more females lived there. Thus a group of females sharing a warren and feeding in extensively overlapping ranges around it are best regarded as reluctant partners in an uneasy alliance.

▲ ► **Problems of burrowing.** ABOVE European rabbits inhabiting sites with soft soil, for example sand dunes, have little difficulty digging new burrows, even overnight. BELOW RIGHT On hard soils, for example chalk, excavations are major endeavors so new burrows rarely appear. These differences considerably affect the social systems of rabbits on the two habitats; on sand dunes rabbits spread themselves out while those on chalk-land center on long-established burrow systems.

◄ **Group-living and social behavior** in European rabbits. Comparison of (**a**) dune-land and (**b**) chalk-land social organization.

On chalk-land rabbits have clustered burrows with females living as reluctant partners around each cluster; fights often occur between females and their home ranges overlap considerably within each group, but not with those of adjacent groups.

On dune-land rabbit burrows are not clustered and are randomly distributed, although they do not occur in the slacks which are prone to flooding. Females move freely between burrows and there is little fighting between individuals, and home ranges overlap less than on chalk-land.

In both habitats males have larger territories which overlap those of several females.

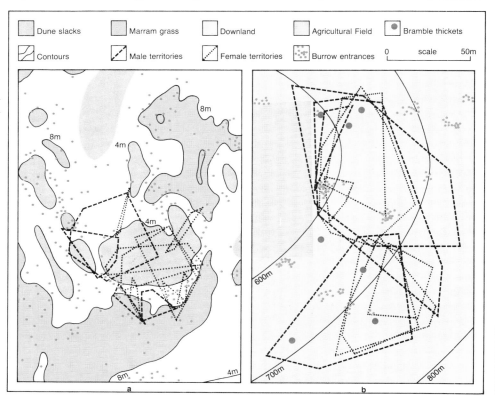

| Dune slacks | Marram grass | Downland | Agricultural Field | Bramble thickets |
| Contours | Male territories | Female territories | Burrow entrances | 0 scale 50m |

cent of interactions with does and refuged in the available burrows without hindrance from them. Males' home ranges were on average about twice the size of those of neighboring females with which they had an extensive overlap. Consequently these bucks could have been acquiring information, largely on the basis of scent rather than direct encounters, about the reproductive state of numerous females. Bucks did not seem to defend strict territories exclusive of all other males at either site. However, the frequent aggressive interactions observed between males, whether or not they were escorting does, may be best interpreted as attempts by them to curtail each other's use of space and access to females.

The behavior of bucks following females around can be regarded as "mate-guarding." Each female is usually accompanied by only one male, although males at the duneland site have been seen with up to three different females at different times over the course of a few days. If a second male approached a male-female pair he was promptly rebuffed by a look or a few paces in his direction by the consorting male. Occasionally energetic chases occurred. in which male antagonists jumped clear of the ground and attempted to rip each other with their claws as they passed in mid-air. Guarding seems to be effective: successful takeovers of paired females by approaching males were very rarely seen.

Despite these efforts by particular bucks to monopolize proximity to particular females, mating in rabbits is promiscuous. A recent Australian study involved the genetic typing (using blood proteins) of all potential parents in a population together with their weaned young. The resulting analysis showed that at least 16 percent of the young were not fathered by the male known from direct observation to be the usual escort of their mother. So, while promiscuous matings can yield offspring, the female's brief period in heat apparently makes it advantageous for bucks to try to monopolize sexual access to one female, or at most a few.

The dune- and downland studies in England have shown that, contrary to popular belief, rabbits are not always group-living animals. Sometimes females have to live together in order to make use of a limited supply of nesting sites, but this habit is not the automatic consequence of evolution. The evident flexibility of female social behavior contrasts with the similar behavior of the bucks in these two populations. As expected from the theory of sexual selection, male rabbits are sexual opportunists.   PJG/DPC

At the duneland site burrows were never apparently in short supply; new ones often appeared overnight. Here, as expected, aggressive interactions between females were rare and not obviously related to burrow possession. Some does even moved round a series of "home burrows" during single breeding seasons without incurring attacks from local residents. This evidence of rather fluid refuging conventions, together with analyses showing that females' home ranges were not bunched together in superimposed clusters, suggests that adult does did not form groups in the dunes. Their burrowing activity was not constrained by the sandy substrate there, so they spread themselves out over the habitat.

What of the males? Buck rabbits, like many other male mammals, are unable to contribute directly to parental care. Consequently their reproductive success will reflect how many matings they have achieved with receptive females. Females come into a period of heat for 12–24 hours about every seventh day or soon after giving birth. Males apparently monitor female condition closely: adult does were "escorted" by single males for about a quarter of their time above ground in the breeding season on both study areas. Bucks won over 90 per-

PIKAS

Family: Ochotonidae
Fourteen species belonging to the genus
Ochotona.
Distribution: N America, E Europe, Middle East,
Asia N of Himalayas.

Habitat: rocky slopes on mountains, steppe and
semidesert, ranging in height from sea level to
6,100m (20,000ft).

Size: ranges from head-body
length 18.3cm (7.2in),
weight 75–210g (2.6–7.4oz)
in the Steppe pika to
head-body length 20.2cm (8in), weight
190–290g (6.7–10.2oz) in the Afghan pika.
Tail length 0.5–2cm (0.2–8in); ear length
1.2–3.6cm (0.5–1.4in).

Gestation: known for the Afghan pika, 25 days,
and the North American pika, 30.5 days.

Longevity: 7 years for the North American
pika, 4 years for the Asian pikas.

Species include: **Afghan pika** (*O. rufescens*);
Collared pika (*O. collaris*); **Daurian pika**
(*O. daurica*); **Large-eared pika** (*O. macrotis*);
Mongolian pika (*O. pallasi*); **North American
pika** (*O. princeps*); **Northern pika**
(*O. hyperborea*); **Red pika** (*O. rutila*); **Royle's
pika** (*O. roylei*); **Steppe pika** (*O. pusilla*).

Pikas are small lagomorphs with relatively large rounded ears and short limbs. They almost look like cavies (Guinea pigs). The tail is slight and barely visible. Pikas are little known because they live high up in the mountains or below ground in deserts.

The word pika and the generic name *Ochotona* are derived from vernacular terms used by Mongol peoples. There are probably 14 species of pikas, 12 distributed in Eurasia, two in North America. However, zoologists are unable to agree on an exact number of species.

Each species has a particular preference in habitat. Most prefer slopes covered in rock debris (talus) or rock slides on mountains. These rock dwellers move in crevices between loose rocks or among slide-rocks, lava flows or even the stone walls of houses. Of the rock dwellers the Afghan pika and the Mongolian pika also live in areas without rocks, in burrows. The Steppe pika and the Daurian pika inhabit steppes by constructing burrow systems, though their feet and nails are not specialized for burrowing. Species probably do not coexist in the same habitat. In Mongolia, the Daurian pika occupies steppes whereas the Mongolian pika inhabits rocky terrain. On high mountains, two rock dwellers segregate their habitats vertically. In the Nepal Himalayas the Large-eared pika lives at higher altitudes than Royle's pika, and in Tien-Shan higher than the Red pika. In Nepal the distribution boundaries of the two pikas border each other.

Pikas are usually active by day and do not hibernate. They are very lively and agile, but often sit hunched up on rocks for long periods. Most species are most active in the morning and late afternoon, with some activity at night. Remarkable differences are found in two Himalayan species: Royle's pika is active at dawn and dusk, but the Large-eared pika, living at altitudes over 4,000m (13,000ft) in Nepal, is active only in the hours around midday. The activity rhythms of these species are thought to result from living where there are favorable temperatures.

Pikas appear to utilize whatever plants are available near to their burrows or at the edges of their rock slides. They eat the leaves, stalks and flowers of grasses, sedges, shrub twigs, lichens and mosses. Pikas cannot grasp plants with their forepaws, so they eat grasses or twigs from the cut end. Pikas, like other lagomorphs, produce two different types of feces: small spherical pellets like pepper seeds, and a soft, dark-greenish excrement. The latter, having high energy value (particularly in B vitamins), is reingested either directly from the anus or after being dropped.

During summer and fall pikas spend much time collecting plants for winter food. They cut down green plants which rock dwellers carry to traditional places under rocks while burrow dwellers with dry habitats construct haypiles in the open near their burrows or between shrubs. Weights of haypiles vary but sometimes reach 6kg (13lb). Moreover, the Mongolian pika carries pebbles, 3–5cm (1–2in) in diameter, in its mouth and places them near its burrow, probably to prevent hay from being scattered by the wind. This hay-gathering

underground or running. In the mating season males of two North American species (the North American and Collared pikas) and of the Northern pika frequently give successive calls to declare their possession of territory. From summer to fall these pikas, both sexes, frequently give short calls, accompanied by the development of hay-gathering behavior. In contrast to these pikas two of the Himalayan species rarely utter even weak sounds and virtually abstain from hay-gathering. Parallel development between food storage and vocalization from summer to fall seems to indicate a possible causal relationship between these behavior patterns in evolution.

There are two different pika reproduction strategies, which are linked with habitat preferences. For typical rock dwellers the mean litter size is less than five and the young first breed as yearlings, but for burrow dwellers, and the Afghan pika and the Mongolian pika with their burrowing ability, the litter size is larger and the young born first breed in the summer of their first year. The relatively low reproductive capacity in the typical rock dwellers distributed at high altitudes or high latitudes may be related to the short favorable reproductive season. On the other hand, in the latter high reproductive capacity group, the densities greatly fluctuate (for example, from 2–3 to 70–80 animals per ha (0.8–1.2 to 28–32 per acre) for the Daurian pika) and the dispersed young may extend their habitats by burrowing.

Pikas always move alone. However, in the Northern pika a male and female hold the possession of a definite territory throughout the year. Each resident will react to intruders of the same sex, though females confine themselves to their territories and male residents often trespass into adjacent territories. The haypiles stored in each territory are consumed by the pair during the winter. In the mating season the two North American species possess a pair territory similar to that of the Northern pika, but in the hay-gathering season their pair territories are divided into two solitary territories, each on average covering 709sq m (7,630sq ft). Each protects the hay within its own "hay territory" from intruders of both sexes (see pp 728–729). The social patterns of some Himalayan species from fall to winter are basically similar to the pair territoriality of the Northern pika. Thus the phenomenon of individual "hay territoriality" seems to be superimposed on the social organization of both North American species to ensure a supply of winter food. TKa

▲ ◄ **More like rats than rabbits** (their closest relatives), pikas have rounded ears, short limbs and no visible tail. The Collared pikas shown here (ABOVE, ABOVE LEFT) occur on rocky outcrops in Alaska and northwestern Canada. Their name is derived from the grayish patches below the cheek and around the neck. They spend about half of their time on the surface sitting on prominent rocks.

◄ **The Large-eared pika of Nepal** and surrounding areas occurs at heights between 2300–6100m (7500–20,000ft). It has the largest ears of all pikas.

behavior is common to all species except the two Himalayan ones (Royle's pika and the Large-eared pika). In winter, in addition to eating their hay, pikas make tunnels in the snow to reach and gnaw the bark of apple trees, aspens and young conifers. Mongolian herders and antelopes steal the hay of Daurian pikas in winter. To compensate, this pika and Royle's pika steal wheat and fried wheat cakes from houses.

Most pikas give distinctive calls while perching on rocks, and occasionally while

Family Bonds and Friendly Neighbors
The social organization of the North American pika

Two North American pikas darted into and out of sight on a rock-strewn slope (talus), one in pursuit of the other. The chase continued, onto an adjoining meadow then into the dense cover of a nearby spruce forest. When next seen, dashing back towards the talus, they were being chased by a weasel. One pika was caught and death came quickly less than 1m (about 3ft) from the talus and safety. Immediately all pikas in the vicinity, with one exception, broke into a chorus of consecutive short calls, the sounds that pikas utter when alarmed by predators. The dead pika had initiated the chase, but the object of his aggression had escaped the weasel. He now surveyed the talus from a prominent rock perched in silence.

Most accounts of the natural history of pikas have emphasized their individual territoriality, but recent work in the Rocky Mountains of Colorado has revealed in detail their social organization. For example, adjacent territories are normally occupied by pikas of the opposite sex. Male and female neighbors overlap each other's home ranges more and have centers of activity that are closer to each other than are the ranges or activity centers of nearest neighbors of like sex. The possession and juxtapositions of territories tend to be stable from year to year. Pikas in North America can live up to 6 years, and the appearance and whereabouts of vacancies on the talus are unpredictable. For a pika, trying to secure a vacancy is like entering a lottery where in part an animal's sex determines whether it will have

▲ ► **Collecting in the hay.** A characteristic (almost frantic) activity of pikas in late summer is the harvesting of vegetation (RIGHT) to store in haypiles on the talus, in part to serve as food over winter. Most stores of hay are located under overhanging rocks (ABOVE).

◄ **Singing pika.** Pikas have two characteristic vocalizations: the short call and the long call (or song). Long calls (a series of squeeks lasting up to 30 seconds) are given by males primarily during the breeding season. Short calls normally contain one-to-two note squeeks and may be given: from rocky promontories either before or after movement by the pika; in response to calls or movement by a nearby pika; while chasing another pika or while being chased; and when predators are active.

males. The pika killed by a weasel had forayed from his home territory to chase an unfamiliar, immigrant adult male. Juveniles that move away from their natal home range are similarly attacked by residents.

Affiliative behavior is seen in pairs of neighboring males and females, who are not only frequently tolerant of each other but engage in duets of short calls. Such behavior is rarely seen between neighbors of similar sex or between non-neighbor heterosexual pairs.

Adults treat their offspring as they do their neighbors of the opposite sex. Some aggression is directed toward juveniles, but also frequent expressions of social tolerance. Most juveniles remain on the home ranges of their parents throughout the summer.

Ecological constraints have apparently led to a monogamous mating system in pikas. Although males cannot contribute directly to the raising of young (hence they are not monogamous because of their need to assist a single female to raise her young), they still primarily associate with a single neighboring female. Polygyny evolves when males can monopolize sufficient resources to attract several females or when they can directly defend several females. The essentially linear reach of vegetation at the base of the talus precludes resource defense polygyny. Males cannot defend groups of females because they are dispersed and held apart by their mutual antagonism.

Juveniles of both sexes are likely to be repelled should they disperse and attempt to colonize an occupied talus. As a result juveniles normally settle close to their site of birth. This "philopatric" settlement may lead to incestuous matings and contribute to the low genetic variability found in pika populations.

The close association among male-female pairs and the close relatedness of neighbors may underlie the evolution of cooperative behavior patterns in pikas. First, attacks on intruders by residents may be an expression of indirect paternal care: if adults can successfully repel immigrants they may increase the probability of settlement of their offspring should a local site become available for colonization. Second—returning to the opening account—the alarm calls given by both sexes when the weasel struck the resident pika served to warn close kin—note that the unrelated immigrant was the only pika that did not call. Uncontested, the newcomer immediately moved across the talus to claim the slain pika's territory, half-completed haypile, and access to a neighboring female.

a winning ticket; territories are almost always claimed by a member of the same sex as the previous occupant.

The behavior pattern that may sustain this pattern of occupancy based on sex is apparently a compromise between aggressive and affiliative tendencies. Although all pikas are pugnacious when defending territories, females are less aggressive to neighboring males and more aggressive to females. Male residents rarely exhibit aggression toward each other, simply because they rarely come into contact, but they apparently avoid each other by using scent marking and vocalizations. Males, however, vigorously attack unfamiliar (immigrant)

ATS

ELEPHANT-SHREWS

ORDER: MACROSCELIDEA

Fifteen species in 2 subfamilies and 4 genera.
Family: Macroscelididae.
Distribution: N Africa, E, C and S Africa.

Habitat: varies considerably, including
montane and lowland forest, savanna, steppe,
desert.

Size: varies from the Short-eared elephant-
shrew with head-body length 10.4–11.5cm
(4.1–4.5in), tail length 11.5–13cm (4.5–5in),
weight about 45g (1.6oz) to the Golden-
rumped elephant-shrew with a head-body
length of 27–29.4cm (11–12in), tail length
23–25.5cm (9.5–10.5in), weight about 540g
(19oz).

Gestation: 57–65 days in the Rufous elephant-
shrew, about 42 days in the Golden-rumped
elephant-shrew.

Longevity: 2½ years in the Rufous elephant-
shrew (5½ recorded in captivity), 4 years in the
Golden-rumped elephant-shrew.

Species: **Golden-rumped elephant-shrew**
(*Rhynchocyon chrysopygus*), **Black and rufous
elephant-shrew** (*R. petersi*), **Chequered
elephant-shrew** (*R. cirnei*), **Short-snouted
elephant-shrew** (*Elephantulus brachyrhynchus*),
Cape elephant-shrew (*E. edwardi*), **Dusky-
footed elephant-shrew** (*E. fuscipes*), **Dusky
elephant-shrew** (*E. fuscus*), **Bushveld elephant-
shrew** (*E. intufi*), **Eastern rock elephant-shrew**
(*E. myurus*), **Somali elephant-shrew** (*E. revoili*),
North African elephant-shrew (*E. rozeti*),
Rufous elephant-shrew (*E. rufescens*), **Western
rock elephant-shrew** (*E. rupestris*), **Short-eared
elephant-shrew** (*Macroscelides proboscideus*),
Four-toed elephant-shrew (*Petrodromus
tetradactylus*).

Aɴʏoɴᴇ unfamiliar with elephant-
shrews might assume that they are
large versions of true shrews—the small
gray mammals with little beady eyes,
pointed snouts and short legs that barely lift
their bellies from the ground. Armchair
naturalists, never having seen elephant-
shrews in the wild, referred to them as
jumping shrews because they thought that
their long rear legs were used for hopping.
Field naturalists in Africa called them
elephant-shrews because they have long
snouts like elephants and eat invertebrates
like shrews. Although elephant-shrews do
have long snouts to forage for invertebrates,
their similarity to true shrews, ends there.
But with large eyes and long legs resembling
those of small antelope, a trunk-like nose,
high-crowned cheek teeth similar to those of
a herbivore, and a long rat-like tail they
have been shuffled from one taxonomic
group to another. At times they have been
included in the insectivores (insect-eaters),
classed as a type of ungulate, the Menotyphla, which once also included the tree
shrews, and placed in their own order, the
Macroscelidea. Most systematics now agree
that elephant-shrews are indeed unique and
belong in their own order, but what is
therefore their evolutionary relationship
with other mammals?

Recently discovered fossil material and a
reinterpretation of dental and foot morphology suggest that rabbits and hares
(Lagomorpha) and elephant-shrews had a
common Asian ancestor in the Cretaceous
era, about 100 million years ago. The
elephant-shrews became isolated in Africa,
and by the late Oligocene (about 30 million
years ago) they occurred in several diverse
forms that included small insectivorous
forms (Macroscelidinae), small herbivorous
species (Mylomygalinae), weighing about
50g (1.8oz) and resembling grass-eating
rodents, and large plant-eaters (Myo-
hyracinae), weighing about 500g (18oz)
that were so ungulate-like that they were
initially misidentified as hyraxes. Today all
that remains of these ancient groups are two
well-defined insectivorous subfamilies, both
still restricted to Africa: the giant elephant-
shrews (Rhynchocyoninae) and the small
elephant-shrews (Macroscelidinae). The
other subfamilies mysteriously died out by
the Pleistocene (2 million years ago).

Elephant-shrews are widespread in
Africa, occupying habitats as diverse as the
Namib Desert in southwest Africa, the
steppes and savannas of East Africa, the
mountain and lowland forests of central
Africa, and semiarid habitats of extreme
northwestern Africa. Their absence from
west Africa has never been adequately explained. Nowhere are elephant-shrews particularly common, and despite being active
above ground during the day and in the
evening they are difficult to see. The small
species have the size of a mouse and are
cryptic in behavior, the larger and more
colorful giant elephant-shrews are usually
only heard as they bound noisily away into
the forest. Both are very secretive.

All species are strictly terrestrial. Despite
the diversity of habitats and the difference in
size between the smallest and largest
species, there is little variation in social
organization. Individuals of the Golden-
rumped, Four-toed, Short-eared, Rufous
and Western rock elephant-shrews live as
loosely associated monogamous pairs on
contiguous home ranges that are defended
against neighboring pairs. As in most mono-

▶ **Representative species of elephant shrews.**
(1) Rufous elephant-shrew (*Elephantulus
rufescens*) foraging for insects. (2) Chequered
elephant-shrew (*Rhynchocyon cirnei*) scent
marking with anal glands. (3) Short-eared
elephant shrew (*Macroscelides proboscideus*)
clearing trail. (4) North African elephant-
shrew (*E. rozeti*) face washing at burrow
entrance. (5) Four-toed elephant shrew
(*Petrodamus tetradactylus*) extruding tongue
after insects. (6) Black and rufous elephant
shrew (*R. petersi*) tearing prey with teeth and
claws. (7) Golden-rumped elephant-shrew
(*R. chrysopygus*) stalking before chase. (8) Tail
of Four-toed elephant-shrew showing the
knobbed bristles uniquely found along the
bottom of the tail of some races. One of the
earliest suggestions for the occurrence was that
they result from scorching in the frequent
brush fires. Another idea was that the bristled
tail is used as a broom to sweep clean its trails.
More recently it has been proposed that they
are used to detect ground vibrations, such as
other foot-drumming elephant shrews and
approaching predators. During aggressive and
sexual encounters, this species depresses its tail
to the ground and lashes it from side to side,
dragging the bristles across the substrate.
Perhaps the animals are scent-marking during
these encounters, and the knobs act as swabs to
spread scent-bearing sebum from the large
glands at the base of each bristle.

gamous mammals the sexes are similar in size and appearance, with the exception of the larger canine teeth of the male giant elephant-shrews.

In territorial encounters, visual signals are important but all forms have scent glands that are used to mark their home ranges. These are located on the bottom of the tail in several species of the genus *Elephantulus*, on the soles of the feet in the Rufous elephant-shrew, on the chest of the Dusky-footed, Rufous and Somali elephant-shrews, and just behind the anus in the giant elephant-shrews. Vocal communication is unimportant, though sounds are created by the Four-toed elephant-shrew and some species of *Elephantulus* by drumming their rear feet; *Rhynchocyon* species slap their tails on the ground. When captured the giant elephant-shrews emit a sharp, highpitched scream. When handled they are surprisingly gentle, rarely attempting to bite despite having well-developed canine teeth.

Most species breed year-round. With a gestation of about 45 days for the giant forms and about 60 days for the smaller species; several litters per year are usually produced. Litters normally contain one or two young, born in a well-developed state with a coat pattern similar to that of adults. The North African elephant-shrew and Chequered elephant-shrew may produce three young per litter. The young of giant elephant-shrews require more care than those of the smaller species. They are confined to the nest for several days before they can accompany their mother.

The Four-toed elephant-shrew, Rufous elephant-shrew and Short-eared elephant-shrew clear and maintain complex trail networks to enable them to traverse their territories easily and quickly. Other species may create trails in habitats where vegetation and surface litter is particularly dense. The Short-eared, Western rock and Bushveld elephant-shrews dig short, shallow burrows in sandy substrates, but where the ground is too hard they use abandoned rodent burrows. The Eastern rock elephant-shrew is restricted to rocky areas where it shelters among cracks and crevasses. The most unusual sheltering habits are found in the Four-toed and Rufous elephant-shrews, which use neither burrow nor shelter but spend their entire lives relatively exposed on their trail systems, much as small antelopes do. Their distinct black and white facial pattern probably serves to disrupt the contour of their large black eye, thus camouflaging them from predators. The giant elephant-shrews are more typical of small mammals, in that they spend each night in a leaf nest on the forest floor.

Elephant-shrews spend much of their active hours feeding on invertebrates, and the small species also eat plant matter, especially small fleshy fruits and seeds. The giant forms search for their prey as small coatis or pigs do, using their proboscis-like noses to probe in thick leaf litter and the long claws on their forefeet to excavate in the soil. The small elephant-shrews normally feed by gleaning small invertebrates from the surface of the soil, leaves and twigs. All species have long tongues which will extend to the tips of their noses and are used to flick small items of prey into their mouths.

Elephant-shrews are of little economic importance to man, though along the coast of Kenya the Golden-rumped and Four-toed elephant-shrews are snared and eaten. Recently the Rufous elephant-shrew has been successfully exhibited and bred in numerous zoological gardens, especially in the United

The Trail System of the Rufous Elephant-shrew

Rufous elephant-shrews inhabiting the densely wooded savannas of Tsavo in Kenya are distributed as male-female pairs on territories that vary in size from 1,600 to 4,500sq m (about 17,200–48,400sq ft). Although monogamous, individuals spend little time together. Members of a territorial pair cooperate to the extent that they share common boundaries, but when defending their area they behave as individuals, with females only showing aggression towards other females and males towards males. This system of monogamy, characterized by limited cooperation between the sexes, is also found in several small antelopes, such as the dikdik and klipspringer.

There are also similarities between elephant-shrews and some small ungulates in predator avoidance. Camouflage is important in eluding initial detection, but if this fails elephant-shrews use their long legs to flee, swiftly, and outdistance pursuing predators. But how can a 58g (2oz) elephant-shrew, that stands only 6cm (2.4in) high at the shoulder manage to escape the numerous birds of prey, snakes, mongooses, and cats that also inhabit the Tsavo woodlands?

The answer lies partly in the use of a complex network of crisscrossing trails, which each pair builds, maintains and defends. The trails allow the elephant-shrews to take full advantage of their running abilities, providing they are kept immaculately clean. Just a single twig could break an elephant-shrew's flight, with disastrous consequence. Every day individuals of a pair separately traverse much of their trail network, removing accumulated leaves and twigs with swift side-strikes of the forefeet. Paths that are used infrequently are composed of a series of small bare oval patches on the sandy soil, on which an animal lands as it bounds along the trail. Paths that are heavily used become continuous bare channels through the litter.

The Rufous elephant-shrew produces highly precocial and independent young. Since only the female can nurse the young a male can do little to assist directly. Why then is the elephant-shrew monogamous? Part of the answer relates to their system of paths. Males spend nearly twice as much time trail-cleaning as females. Although this indirect help is not as dramatic as the direct cooperation of male wolves and marmosets, it is just as vital to the elephant-shrew's reproductive success, since without paths its ungulate-like habits would be completely ineffective.

States. Giant elephant-shrews are exceedingly difficult to keep in captivity and they have never been bred. Most elephant-shrews are fairly widespread and occupy habitats that have little or no agricultural potential. There is little immediate danger of their numbers being reduced by habitat destruction. The forest-dwelling giant elephant-shrews, however, face severe habitat depletion. This is especially true for the Golden-rumped and Black and rufous elephant-shrews, which occupy small isolated patches of forest that are quickly being cleared for subsistence farming, exotic tree plantations and tourist developments. It would be an incredible loss if these unique, colorful mammals were to disappear, after more than 30 million years of evolution in Africa, just because a few square kilometers of forest habitat could not be preserved.

GBR

▲ **Aggressive encounters.** Rufous elephant-shrews visibly mark their territories by creating small piles of dung in areas where the paths of two adjoining pairs meet. Occasionally aggressive encounters occur in these territorial arenas. Two animals of the same sex face one another and while slowly walking in opposite directions they stand high on their long legs and accentuate their white feet, much like small mechanical toys. If one of the animals does not retreat, a fight usually develops and the loser is routed from the area.

▶ **Long nose, big eyes** and small size leave little doubt as to why elephant-shrews got their name. However, they are not shrews but separated in a group of their own. Shown here is the Rufous elephant shrew which inhabits wooded savannas of East Africa.

Escape and Protection

The tactics and adaptations of the Golden-rumped elephant-shrew

The sun was just setting when a Golden-rumped elephant-shrew approached an indistinct pile of leaves, about 1m (3ft) wide on the forest floor. The animal paused at the edge of the low mound for 15 seconds, sniffing, listening and watching for the least irregularity. When nothing unusual was sensed, it quietly slipped under the leaves. The leaf nest shuddered for a few seconds as the elephant-shrew arranged itself for the night, then everything was still.

At about the same time the animal's mate was retreating for the night into a similar nest located on the other side of their home range. As this elephant-shrew prepared to enter its nest, a twig snapped. The animal froze, and then quietly left the area for a third nest, which it eventually entered, but not before dusk had fallen.

Every evening, within a few minutes of sunset, these pairs of elephant-shrews separately approach and cautiously enter one of a dozen or more nests they have constructed throughout their home range. Each evening a different nest is used to discourage forest predators such as leopards and eagle-owls from learning that elphant-shrews are always found in a nest.

This is just one of several ways by which Golden-rumped elephant-shrews have learnt to avoid predators. The problem they face is considerable. During the day they spend over 75 percent of their time exposed while foraging in leaf litter on the forest floor. They are the prey of Black mambas, Forest cobras and harrier eagles. To prevent capture by such enemies the Golden-rumped elephant-shrew has developed tactics that involve not only its ability to run fast but also its distinct coat pattern and flashy coloration.

Golden-rumped elephant-shrews can bound across open forest floor at speeds above 25 kilometers per hour (16 miles per hour)—about as fast as an average person can run. Because they are relatively small, they also can pass easily through patches of undergrowth, leaving behind larger terrestrial and aerial predators. Despite speed and agility, however, they are still vulnerable to ambush by sit-and-wait predators, such as the Southern banded harrier eagle. Most small terrestrial mammals have coats or skins with cryptic colors, acting as camouflage. However, the forest floor along the coast of Kenya where the Golden-rumped

▲ **Foraging elephant-shrew.** The Golden-rumped elephant-shrew has a small mouth located far behind the top of its snout, which makes it difficult to ingest large prey items. Small invertebrates are eaten by flicking them into the mouth with a long extensible tongue.

◄ **Daily activities** of the Golden-rumped elephant-shrew. (**a**) Nest-building occurs mainly in the early morning hours, when dead leaves are moist with dew and make little noise. Predators are less likely to be attracted by nest-building activity in the morning. Weathered nests are nearly indistinguishable from the surrounding forest floor. The elephant-shrews use a sleeping posture, with their heads tucked back under their chest. (**b**) The elephant-shrews in the Arabuko-Sokoke forest of coastal Kenya feed mainly on beetles, centipedes, termites, cockroaches, ants, spiders and earthworms, in decreasing order of importance. (**c**) Elephant-shrews chase intruders from their territory using a halfbound gait.

elephant-shrew lives is relatively open so camouflage would be ineffective.

This elephant-shrew's tactic is to "invite" predators to take notice. It has a rump patch that is so visible that a waiting predator will discover a foraging elephant-shrew while it is too far away for making successful ambush. The initial action of a predator detecting an elephant-shrew, such as rapidly turning its head or shifting its weight from one leg to another, may result in enough motion or sound to reveal its presence. By inducing a predator to disclose prematurely its presence or intent to attack, a surprise ambush can be averted. An elephant-shrew that has discovered a predator outside its flight distance does not bound away but pauses and then, with its tail, slaps the leaf litter every few seconds. The sharp sound produced by this behavior probably communicates a message to the predator: "I know you are there, but you are outside my

flight distance, and I can probably outrun you if you attack." Through experience the predator has probably learned that when it hears this signal it is probably futile to attempt a pursuit because the animal is on guard and can easily escape. But what happens, however, if an elephant-shrew unwittingly forages under, for example, an eagle that is perched within its flight distance? As the bird swoops to make its kill the elephant-shrew takes flight across the forest floor towards the nearest cover, noisily pounding the leaf litter with its rear legs as it bounds away. Only speed and agility will save it in this situation.

The Golden-rumped elephant-shrew is monogamous, but pairs spend only about 20 percent of their time in visual contact with each other. The rest they spend resting or foraging alone. So for most of the time they must communicate with scent or sound. The distinct sound of an elephant-shrew tail-slapping or bounding across the forest floot can be heard over a large area of a pair's 1.5ha (3.7 acres) territory. These sounds not only signal to a predator that it has been discovered, but also communicate to an elephant-shrew's mate and young that an intruder has been detected.

Each pair of elephant-shrews defends its territorial boundaries against neighbors and wandering subadults in search of their own territories. During an aggressive encounter a resident will pursue an intruder in a high-speed chase through the forest. If the intruder is not fast enough it will be gashed by the long canines of the resident. These attacks between elephant-shrews might be thought of as a special type of predator–prey interaction, and reveal yet another way in which this animal's coloration may serve to avoid successful predation. The skin under the animal's rump patch is up to three times thicker than the skin on the middle of the back. The golden color of the rump probably serves as a target to discourage attacks on such vital parts of the body as the head and flanks. The toughness of the rump makes it best suited to take attacks. Deflective marks are common in many invertebrates and have been shown to be effective in foiling predators. The distinct eye spots on the wings of some butterflies attract the predatory attacks of birds, allowing the insects to escape relatively unharmed. The yellow rump (and the white tip on the black tail) of the elephant-shrew may serve a similar function by also attracting the talons of an eagle or a striking snake to the rump, thus improving the chances of making a successful escape.   GBR

PIALS

INSECTIVORES

ORDER: INSECTIVORA
Six families: 60 genera; 345 species.

Tenrecs
Family: Tenrecidae
Thirty-three species in 11 genera.
Includes **Aquatic tenrec** (*Limnogale mergulus*),
Common tenrec (*Tenrec ecaudatus*) **Giant otter
shrew** (*Potamogale velox*).

Solenodons
Family: Solenodontidae
Two species in a single genus.
Includes **Hispaniola solenodon** (*Solenodon
paradoxurus*).

Hedgehogs and moonrats
Family: Erinaceidae
Seventeen species in 8 genera.
Includes **European hedgehog** (*Erinaceus
europaeus*).

Shrews
Family: Soricidae
Two hundred and forty-six species in 21
genera. Includes **Eurasian water shrew**
(*Neomys fodiens*) **European common shrew**
(*Sorex araneus*) **Vagrant shrew** (*Sorex vagrans*).

Golden moles
Family: Chrysochloridae
Eighteen species in 7 genera.
Includes **Giant golden mole** (*Chrysopalax
trevelyani*).

Moles and desmans
Family: Talpidae
Twenty-nine species in 12 genera.
Includes **European mole** (*Talpa europaea*),
Pyrenean desman (*Galemys pyrenaicus*),
Russian desman (*Desmana moschata*). **Star-
nosed mole** (*Cordylura cristata*).

Aᴌ of the approximately 345 species of insectivores are small animals (none are larger than rabbits) with long, narrow snouts that are usually very mobile. Most use a walk or run as their normal style of movement, although some are swimmers and/or burrowers. Body shapes vary widely—from the streamlined form of the otter shrews to the short, fat body of hedgehogs and moles. All walk with the soles and heels on the ground (plantigrade). In general, the limbs are short, with five digits on each foot. The eyes and ears are relatively small and external signs of both may be absent. Most members of the order are solitary and nocturnal, feeding mainly on invertebrates, especially insects, as the name of the group suggests.

The insectivores are often divided into three suborders to emphasize the relationships between the families. The Tenrecomorpha comprises tenrecs and golden moles; the hedgehogs and moonrats (the latter considered the most primitive of the living insectivores) are placed in the Erinaceomorpha; and the Soricomorpha consists of shrews, moles and solenodons.

Although the order as a whole is very widely distributed, only three families can be said to be widespread. These are the Erinaceidae (hedgehogs and moonrats), Talpidae (moles and desmans), and Soricidae (shrews) which between them account for almost all of the worldwide distribution. The other three have very limited distributions indeed! The Solenodontidae (solenodons) are found only on the Caribbean islands of Hispaniola and Cuba. The Tenrecidae (tenrecs) are also found mainly on islands—Madagascar and the Comores in the Indian Ocean—with some members of the family (the otter shrews) found only in the wet regions of Central Africa. Because of their differences in distribution, lifestyle and habitat, the otter shrews were at various times considered to be in a separate family, the Potamogalidae, although their teeth indicate that they are true tenrecs. The golden moles occur only in the drier parts of southern Africa.

As a group, the insectivores are generally considered to be the most primitive of living placental mammals and therefore representative of the ancestral mammals from which modern mammals are derived. This was not the original purpose of the grouping. The term "insectivore" was first used in a system of classification produced in 1816 to describe hedgehogs, shrews and Old-World moles (all primarily insect-eaters). The order soon became a "rag-bag" into which fell any animal that did not fit neatly into the other orders of mammalian classification. In 1817, the naturalist Cuvier added the American moles, tenrecs, golden moles and desmans. Forty years later, tree shrews, elephant shrews, and colugos were included. All were new discoveries in need of classification but none looked much like any other members of the group.

Confronted in 1866 by an order Insectivora containing a number of very different animals, the taxonomist Haeckel subdivided it into two distinct groups that he called Menotyphla and Lipotyphla. Menotyphlans (tree shrews, elephant shrews and colugos) were distinguished by the presence of a cecum (the human appendix) at the beginning of the large intestine; lipotyphlans (moles, golden moles, tenrecs and shrews) by its absence. Menotyphlans also differ greatly from lipotyphlans in external ap-

◄ **Leafy setting** for a Eurasian water shrew. Although they sometimes forage on land, they are usually found near water.

▼ **The best-known** of the insectivores is the European hedgehog. It is also the only insectivore to have a favorable relationship with human beings.

pearance: large eyes and long legs are only two of the more obvious characters. The colugos are so different that the new order Dermoptera was created for them as early as 1872. In 1926 the anatomist Le Gros Clarke suggested that the tree shrews are more similar to lemur-like primates than to insectivores, but the most modern view is that tree shrews comprise a separate order, the Scandentia. The elephant shrews also cannot be readily assigned to any existing order, so they have become the sole family in the new order Macroscelidea. Modern analyses (of skull features in particular) show that the remaining, lipotyphlan members of the Insectivora are probably descended from a common ancestor. This conclusion does not apply to the fossil members of the Insectivora, which includes a vast assortment of early mammals and remains very much a "waste-basket" group. Many of these early forms are known only from fossil fragments and teeth; they are assigned to the Insectivora largely as a matter of convenience, having insectivore affinities and no clear links with anything else.

Not all of the insectivores are primitive mammals. Most living insectivores have evolved specializations of form and behavior which mask some of the truly primitive characters they possess. "Primitive" characters are those features which probably would have been found in those animals' ancestors. These are contrasted with "derived" (or advanced) characters, found in animals which have developed structures and habits not found in their ancestors. The cecum is a primitive character, and its lack is therefore a derived character, a feature of the Insectivora as it now stands. There are, however, a number of characters considered to be primitive which are more commonly found in the Insectivora than in other mammalian orders. These include relatively small brains, with few wrinkles to increase the surface area, primitive teeth, with incisors, canines and molars easily distinguishable, and primitive features of the auditory bones and collar bones. Other primitive characteristics shared by some or all insectivores are testes that do not descend into a scrotal sac, a flat-footed (plantigrade) gait, and possession of a cloaca, a common chamber into which the genital, urinary and fecal passages empty (*cloaca*: from the Latin for sewer). Some of these primitive features, such as the cloaca and abdominal testes, are also characteristic of the marsupials, but insectivores, like all Eutherian (placental) mammals, are distinguished by the possession of the chorioallantoic placenta which permits the young to develop fully within the womb.

Many insectivores have acquired extremely specialized features such as the spines of the hedgehogs and tenrecs, the poisonous saliva of the solenodons and some shrews, and the adaptations for burrowing found in many insectivore families. A number of shrew and tenrec species are thought to have developed a system of echolocation similar to that used by bats.

If all these derived characters are ignored, it is possible to produce a picture of an early mammal, but only in the most general terms. They would have been shy animals, running along the ground in the leaf litter but capable of climbing trees or shrubs. Small and active, about the size of a modern mouse or shrew (the largest known fossil is about the size of a Eurasian badger), they probably fed mainly on insects; some may have been scavengers. They would have looked much like modern shrews, with small eyes and a long, pointed snout with perhaps a few long sensory hairs or true whiskers. A dense coat of short fur would have covered all of the body except the ears and soles of the paws. Perhaps they had a dun-colored coat, with a stripe of darker color running through the eye and along the side of the body—a common pattern, found even on reptiles and amphibians. The development of the ability to regulate body temperature, combined with the warm mammalian coat, meant that the early mammals could be active at night when the dinosaurs (their competitors and predators) were largely inactive due to lower air temperatures.

From this basic stock, two slightly different forms are believed to have developed,

THE INSECTIVORE BODY PLAN

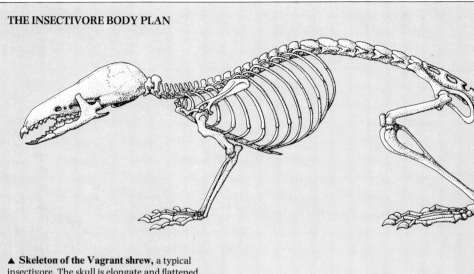

▲ **Skeleton of the Vagrant shrew,** a typical insectivore. The skull is elongate and flattened. Typical characteristics of insectivores include a small brain case, and the absence of a zygomatic arch (cheek bone), in all except hedgehogs and moles, or auditory bullae (bony) prominences around the ear opening). The teeth of shrews are well differentiated into molars, premolars and canines, with pincer-like front incisors. The dental formula of the Vagrant shrew is I3/1, C1/1, P3/1, M3/3 = 32. The teeth are partially colored by a brownish-red pigment.

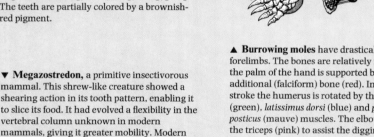

▼ **Megazostredon,** a primitive insectivorous mammal. This shrew-like creature showed a shearing action in its tooth pattern, enabling it to slice its food. It had evolved a flexibility in the vertebral column unknown in modern mammals, giving it greater mobility. Modern insectivores have added specialized adaptations to the primitive mammalian pattern but many retain such features as the cloaca (a common chamber into which the genital, urinary and fecal tracts empty).

▲ **Burrowing moles** have drastically modified forelimbs. The bones are relatively massive and the palm of the hand is supported by an additional (falciform) bone (red). In the digging stroke the humerus is rotated by the *teres major* (green), *latissimus dorsi* (blue) and *pectoralis posticus* (mauve) muscles. The elbow is flexed by the triceps (pink) to assist the digging stroke.

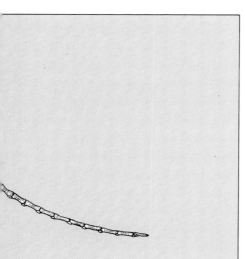

▼ **Skulls of insectivores.** Unlike shrews, the cheek bone (red) of hedgehogs is fully formed. The front incisors are enlarged and the molars are adapted to an omnivorous rather than an insectivorous diet. The dental formula is I2–3/3, C1/1, P3–4/2–4, M3/3 = 36–44. The Common tenrec has a long, tapered snout and, in the adult male, long canines, the tips of the bottom pair fitting into pits in front of the upper ones (red). The dental formula is I2/3, C1/1, P2/3, M3/3 = 38. Solenodons have an unusual cartilaginous snout which articulates with the skull via a "ball-and-socket" joint. Solenodons produce a toxic saliva which is released from a gland at the base of the second lower incisor. The dental formula is I3/3, C1/1, P3/3, M = 40.

Hedgehog 5.5cm

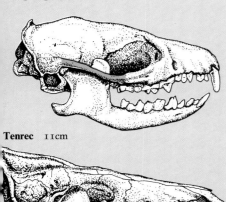

Tenrec 11cm

Solenodon 8.5cm

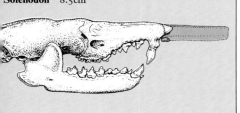

known today only from teeth and fragments of bone dating from the late Cretaceous (80 million years ago). These two groups are characterized mainly by very different teeth. From the evidence available, it appears that one group, the Paleaoryctoidea, eventually gave rise to the creodonts, a type of early carnivore, while the Leptictoidea were thought to have produced the modern insectivores. Recent research on leptictoid fossils suggests instead that most of them were less closely related to the Insectivora, and were perhaps "dead-end" offshoots from the main branch of insectivore evolution.

Despite the evolutionary relationships between the families, insectivores have little in common other than their apparent primitiveness. It is perhaps this diversity which is responsible for the success of the three larger families. The family Talpidae contains both the true moles, which spend practically all of their lives in their subterranean burrows, and the desmans, which spend much of their time in the water and construct burrows in stream banks only for shelter. Similarly, the mainly terrestrial family Tenrecidae also includes the rice tenrecs, which are said to burrow in the banks of rice paddies, while the Aquatic tenrec leads a semi-aquatic life along stream banks and the shores of lakes and marshes. The Soricidae show adaptations to every type of habitat—in addition to the "standard" shrew, running along the ground, there are also species, like the American short-tailed shrew, that are reported to burrow like miniature moles, and those like the European water shrew that swim much like the tenrecid otter shrews.

The order Insectivora is rich in examples of convergent evolution (where animals not closely related have adapted in similar ways to fit the demands of a specific habitat or way of life). Moles and golden moles, for example, are not closely related within the order yet both have adopted similar burrowing lifestyles and even look much alike. According to the fossil record, moles developed from an animal which resembled a shrew, whereas golden moles appear to be more closely related to the tenrecs. The similarity between golden moles and the Marsupial mole of Australia is even more remarkable. In this case, the lineages have been separate for 70 million years; one is placental, the other pouched—yet the golden mole is more similar in appearance to the Marsupial mole (even in the texture of its fur) than to the true mole. The eyes of both Marsupial and golden moles are covered by skin—the eyelids have fused, whereas the minute eyes

of the true moles are still functional. Comparisons can also be made between the otters (order Carnivora, family Mustelidae) and the otter shrews, members of the Tenrecidae. Hedgehogs are well known for their spiny coat and their ability to curl up into a ball when threatened. Several genera of tenrecs have also independently developed a similar coat of spines, and some members of three genera have acquired the ability to curl up as well.

The ability to curl up, combined with a dense coat of spines, is an obvious deterrent to predators. Some non-spiny species (moles and shrews for example) have strongly distasteful secretions from skin glands which may have a similar effect. Many of the other specializations are more likely to be the result of competition for food. If two or more species compete for food or some other resource, then either the worst competitor will become (at least locally) extinct, or all will evolve to specialize on different aspects of the resource and reduce the competition. Moles, for example, may have developed their burrowing life-style to avoid both predation and competition with surface-dwelling insectivores. Some species of both shrews and moles have become semi-aquatic, probably to exploit a different source of invertebrate food.

The tenrecs are thought to have been one of the first mammalian groups to arrive on Madagascar. Like the Australian marsupials, they provide a fascinating example of adaptive radiation, having evolved, in the absence of competition with an established fauna, a variety of forms which use most of the available habitats.

Insectivores rely heavily on their sense of smell to locate their prey, as would be expected from the relatively large center of smell in the brain. Invertebrates are the main food—most are thought to feed on insects and earthworms, although some of those associated with water also eat mollusks and possibly fish. Where food habits have been closely studied, the animals appear to eat almost anything organic they find or can catch, and many will attempt to kill prey which are substantially larger than themselves: hedgehogs, for example, can kill chickens; water shrews can kill frogs. The poisonous saliva of solenodons and some shrews may have evolved to enable these animals to catch larger prey than their body size would normally permit. The poison acts mainly on the nervous system, paralyzing the victim.

Shrews, in particular, have a reputation for gluttony, consuming more than their own body-weight daily, but recent studies indicate that their food is less nutritious than the dry seeds eaten by rodents of a similar size, so they need to consume a correspondingly greater bulk. High food requirements also result from the high metabolic rates of the shrews, which in the case of some northern shrews may be an adaptation to a highly seasonal climate. All but the otter shrews and some true shrews are active mainly at night, dawn and dusk.

Reported litter sizes vary widely in those species that have been studied. Females of the genus *Tenrec* have been caught with up to 25 developing embryos, of which probably only 12–16 survive to birth. Other tenrec species, and moles, hedgehogs, and shrews have litters of 2–10; solenodons have only one or two young per litter. The timing of the breeding season is governed mainly by food availability and thus the young of animals living in arid areas are usually born in the rainy season (winter), while temperate species breed in the spring and summer. Some tropical species (the moonrats, for example) are thought to breed throughout the year.

Many insectivore species are known only from a few museum specimens, and some have not subsequently been found in the original locality. In general, insectivores do not lend themselves to field study; most are nocturnal, secretive creatures of retiring habits. Shrews, for example, are relatively common but rarely seen in the wild. They are also often difficult to catch and keep in captivity, and many occur in areas where field research is difficult. It is not surprising, therefore, that very little is actually known about the social biology of insectivores. Those few species that have been the subject of research (mainly shrews, some hedgehogs and tenrecs) are solitary, with little communication between adults except at breeding times. Studies of European hedgehogs indicate that individuals do not defend territories, although shrews do.

Shrews are territorial in that their home ranges—about 500sq m (5,400sq ft) in the European common shrew—are mutually exclusive. When shrews meet they behave aggressively towards each other. Moles, such as the European mole, are even less sociable and, once the juveniles have been expelled from a female's burrow system, she will not normally tolerate another mole in her home burrow except for a few hours in the spring when she is ready to mate. Moles usually do not trespass into each others' tunnel systems, though they will quickly take over a territory if its occupant is

removed. It is likely that scent plays an important role in keeping the animals apart and they are probably circumspect about entering unfamiliar places smelling of other moles. This pattern of mutual avoidance is probably common among some other insectivore groups too. Mole home ranges are essentially linear, being constrained by tunnel walls. The burrow system acts as a pitfall trap which collects soil invertebrates as prey and in poor soils, longer burrow systems are needed to supply sufficient food. Thus a mole's linear home range may vary from 30–120m (100–400ft) or more, depending on soil type and food density. Surface activity, especially among dispersing juveniles may extend this home range considerably. Hedgehogs, such as the European hedgehog, which forage over an area from 1–5ha (2.5–12.5 acres) in a night, may use a home range of up to 30ha (74 acres) in a season; males travel further and have larger ranges than females. Unlike some small rodents, insectivores do not normally seem to make use of a three dimensional home range by climbing into bushes and trees. In desmans, the home range is linear, along a stream edge; in tenrecs it is an area around the burrow or den; but for many species there is no detailed ecological information at all.

A number of species, including the tenrecs and solenodons, are now facing a major problem: the introduction by man of new competitors and predators, such as rats, mice, cats and dogs, which are effective generalists and can thrive almost anywhere. These competitors also have high reproductive rates, bearing several litters in a year and can thus quickly build a sizeable population which overwhelms native species before they have any chance to adapt to the changed situation. Solenodons were the principal small carnivores in the Antilles until the 17th century, when the Spaniards brought with them dogs, cats and rats. Mongooses were introduced in the late 19th century to combat Black rats. The latter two proved to be highly effective as competitors for food and all apparently added solenodons to their diet. These pressures, combined with the clearance of jungle for cultivation, are now proving too great for the solenodons. *Nesophontes* (a solenodon-like insectivore from the Caribbean and the sole member of the family Nesophontidae) is thought to have become extinct in the last 50 years, and the two species in the Solenodontidae are unlikely to survive much longer in the wild.

The Russians are at present trying to preserve the remnants of their Russian

▲ **Insectivores in the Middle Ages.** TOP Hedgehogs, one encrusted with fruit, depicted in a bestiary from about 1300AD. The incorrect notion that hedgehogs deliberately carry fruit on their spines derives from the Roman author Pliny and has been handed down ever since. ABOVE The mole, depicted in *Le Bestiaire d'Amore* by Richard de Fournivall; early 14th century. According to this author: "The mole is blind with eyes under his heart, but he makes up for it by his acute hearing. So with all of the senses: if one is defective there is always another which surmounts it. Also the mole lies purely in earth, being one of those animals which live only in one element."

▶ **A rash of mole-hills** ABOVE on an upland pasture in Herefordshire, England, shows just how much of a nuisance moles can be, turning a significant proportion of the pasture into spoil heaps.

▶ **The fate of moles.** For centuries, moles have been trapped, both to reduce their effects on farmland and for their pelts. Now, eradication campaigns are still waged when they become serious pests but their pelts are no longer valued.

(especially in Madagascar and the West Indies) and the smaller ones are probably rarely noticed.

Most insectivores are too small and too scarce to be any use for food, and equally are unlikely to be serious economic pests. Many are, however, extremely abundant, and it has been suggested that if all shrews were to disappear suddenly, the number of insect pests in fields and gardens would increase noticeably. Other insectivores are more likely to become accidental victims of human activities than to be blamed as pests or praised as allies.

Some species have been economically exploited. The Russian desmans have a dense lustrous coat and, in the past, numbers of these animals were caught in nets and traps. They were also chased from their burrows and shot or clubbed. A century ago, a fortunate hunter might obtain 40 desman skins during the month-long hunting season at the time of the spring thaw. Despite their value, desman pelts were never a major commodity and the "industry" seems to have been opportunistic.

Mole-skins have been used for hats and trimming since Roman times. In the 17th and 18th centuries many were used in Germany for purses, caps and other things. Fashions are fickle and although moles were caught in sufficient numbers in Germany at the end of the 19th century to cause fears of extinction, the market had dwindled by 1914. Demand from America in the 1920s is said to have caused the export of some 12 million moles from Europe and, even as late as the 1950s, a million mole-skins a year were being trapped in Britain. Mole-skins are graded and valued according to size and whether or not they show signs of molt. Until the late 19th century many English parishes employed a professional mole catcher to rid the fields of troublesome moles. Some of these men claimed to kill a thousand or more moles each winter. However, mole-skins are no longer an economic proposition; they are worth comparatively little and cheaper synthetic substitutes are readily available.

A large proportion of insectivore species are found only in tropical countries where the emphasis is on development and exploitation of resources rather than preservation of wildlife. Few, if any, of the nature reserves have been designed with insectivores in mind; they are generally unprepossessing animals and, with the exception of a few species like the hedgehogs, a threat of extinction is unlikely to receive wide public attention. AW

desman population. This was once a staple of the fur trade, with tens of thousands of skins exported annually to western Europe, but hunting, combined with pollution of their aquatic habitat and competition from introduced coypus and muskrats, has drastically reduced the population. The small population of Pyrenean desmans in the Pyrenees is threatened by water pollution. Several species of golden mole are also threatened by various changes in their habitat.

The only insectivore which has a favorable relationship with man is the hedgehog. The folklore of both Europe and Asia contains tales of hedgehogs and although they are regarded as pests by some English gamekeepers, householders in the urban areas of western Europe provide them with bowls of bread and milk. Elsewhere in the world, the larger insectivores are eaten occasionally

TENRECS

Family: Tenrecidae
Thirty-four species in 11 genera.
Distribution: Madagascar, with one species
introduced to the Comoros, Mascarenes and the
Seychelles; W and C Africa.

Habitat: wide-ranging, from semi-arid to rain
forest, including mountains, rivers and human
settlements.

Size: ranges from head-body length 43mm
(1.7in), tail length 45mm (1.8in) and weight
5g (0.18oz) in *Microgale parvula* to head-body
length 250–390mm (10–15in), tail length
5–10mm (0.2–0.4in) and weight 500–1,500g
(18–53oz) in the Common tenrec.

Gestation: relatively uniform within the
Oryzoryctinae and Tenrecinae where known
(50–64 days); unknown in Potamogalinae.

Longevity: up to 6 years.

Coat: soft-furred to spiny, brown to gray to
contrasted streaks.

Tenrecs
Subfamily Oryzoryctinae
Twenty-five species in 4 Madagascan genera,
including **Aquatic tenrec** (*Limnogale mergulus*),
rice tenrecs (3 species), genus *Oryzoryctes*;
long-tailed tenrecs (20 species), genus
Microgale; **Large-eared tenrec** (*Geogale aurita*).

Subfamily Tenrecinae
Six species in 5 Madagascan genera: **Greater
hedgehog tenrec** (*Setifer setosus*), **Lesser
hedgehog tenrec** (*Echinops telfairi*), *Dasogale
fontoynonti* (possibly based on mistaken
identification), **Common tenrec** (*Tenrec
ecaudatus*), **streaked tenrecs** (*Hemicentetes
semispinosus* and *H. nigriceps*).

Otter shrews
Subfamily Potamogalinae
Three species in 2 African genera: **Giant otter
shrew** (*Potamogale velox*), **Ruwenzori least
otter shrew** (*Micropotamogale ruwenzorii*),
Mount Nimba least otter shrew (*M. lamottei*).

Tenrecs and otter shrews are remarkable in having a greater diversity than any other living family of insectivores. Yet this has been achieved in virtual isolation, since tenrecs themselves are confined to Madagascar, and the otter shrews to West and central Africa. Tenrecs retain characters which were perhaps more widespread among early placental mammals. These conservative features include a low and variable body temperature (see box), retention of a common opening for the urogenital and anal tracts (the cloaca) and undescended testes in the male. Some of the family comprise a more conspicuous part of the native fauna than temperate zone insectivores, being either an important source of food, relatively large and bold, or conspicuously colored. Even though detailed studies of the family are few and restricted to a handful of species, they provide an excellent basis for the elucidation of mammalian evolution.

The earliest fossils date from Kenyan Miocene deposits (about 25 million years ago), but by then the Tenrecidae were well differentiated, and had probably long been part of the African fauna. The tenrecs were among the first mammals to colonize Madagascar. The only survivors of this ancient group on the African mainland, the otter shrews (Potamogalinae), are sometimes regarded as a separate family.

Eyesight is generally poor in these largely nocturnal species, but the whiskers are sensitive and smell and hearing are well developed. Vocalizations range from hissing and grunting to twittering and echolocation clicks. The brain is relatively small and the number of teeth ranges from 32–42.

The aquatic Tenrecidae are active creatures of streams, rivers, lakes and swamps. The Giant otter shrew and Mount Nimba least otter shrew are confined to forest, but the Ruwenzori least otter shrew and the Aquatic tenrec are less restricted. All four species have a sleek, elegant body form with a distinctive, flattened head which allows the ears, eyes and nostrils to project above the surface while most of the body remains submerged. Stout whiskers radiate from around the muzzle, providing a means of locating prey. The fur is dense and soft, and frequent grooming ensures that it is waterproof, and traps insulating air during dives. Grooming is accomplished by means of the two fused toes on each hindfoot which act as combs. All otter shrews and the Aquatic

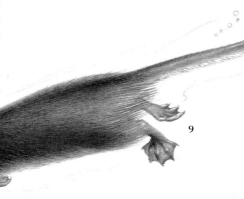

often figures as part fish and part mammal in African folklore, giving rise to such names as "transformed fish." The unmistakable deep, laterally flattened tail, which tapers to a point, is the main source of such beliefs even though it is covered by fine, short hair, but the animal's proficiency in water no doubt plays a part. Sinuous thrusts of the powerful tail extending up the lower part of the body provide the propulsion for swimming, giving rise to startling speeds and great agility. Although most of the active hours are spent in the water, the agility extends to dry-land foraging. The Boulou of southern Cameroun call the Giant otter shrew the *jes*: a person is said to be like a *jes* if he flares up in anger but just as rapidly calms down.

The long-tailed and Large-eared tenrecs are shrew-like, and the former have the least modified body plan within the Tenrecidae. Evergreen forest and wetter areas of the

▲ **Species of tenrecs.** (**1**) Lesser hedgehog tenrec (*Echinops telfairi*). (**2**) Common tenrec (*Tenrec ecaudatus*). (**3**) Streaked tenrec (*Hemicentetes nigriceps*). (**4**) Long-tailed tenrec (*Microgale melanorrachis*). (**5**) Greater hedgehog tenrec (*Setifer setosus*). (**6**) Rice tenrec (*Oryzoryctes tetradactylus*) (**7**) Giant otter shrew (*Potamogale velox*). (**8**) Ruwenzori least otter shrew (*Micropotamogale ruwenzorii*). (**9**) Aquatic tenrec (*Limnogale mergulus*).

tenrec have a chocolate brown back; the Aquatic tenrec has a gray belly and otter shrews have white bellies. The Madagascan Aquatic tenrec shows strong convergence with the least otter shrews, with a rat-size body and a tail approximately the same length. The Ruwenzori least otter shrew and the Aquatic tenrec have webbed feet, which probably provide most of the propulsion in the water, and their tails are slightly compressed laterally, providing each with an effective rudder and additional propulsion.

The Mount Nimba least otter shrew is probably the least aquatic, having no webbing and a rounded tail. However, all are probably agile both in water and on land. The Giant otter shrew, which is among the most specialized of the aquatic insectivores,

central plateau of Madagascar are the primary habitats for long-tailed tenrecs, and only one species extends into the deciduous forests of the drier western region. These tenrecs have filled semi-arboreal and terrestrial niches. The longest-tailed species, with relatively long hindlegs, can climb and probably spring among branches, while on the ground live jumpers and runners, together with short-legged semi-burrowing species. The Large-eared tenrec is also semi-burrowing in its western woodland habitat, and is apparently closely related to one of the oldest fossil species.

The rice tenrecs with their mole-like velvet fur, reduced ears and eyes, and relatively large forefeet fill Madagascar's burrowing insectivore niche. In undisturbed areas of northern and western Madagascar, these tenrecs burrow through the humus layers in a manner similar to the North American shrew mole, but the extensive cultivation of rice provides new habitats for them.

The subfamily Tenrecinae comprise some of the most fascinating and bizarre insectivores. The tail has been lost or greatly reduced, and varying degrees of spininess are linked with elaborate and striking defensive strategies. Both the Greater hedgehog tenrec and its smaller semi-arboreal counterpart, the Lesser hedgehog tenrec, can form a nearly impregnable spiny ball when threatened, closely resembling the Old World hedgehogs. Continued provocation may also lead to them advancing, gaping, hissing and head-bucking, the latter being common to all Tenrecinae. The brown adult Common tenrec, qualifying as the largest living insectivore, is the least spiny, but it combines a lateral open-mouthed slashing bite with head-bucking which can drive spines concentrated on the neck into an offender. A fully grown male with a gape of 10cm (4in) has canines measuring up to 1.5cm (0.6in) and the bite is powered by the massively developed masseter (jaw) muscles. A pad of thickened skin on the male's mid-back also provides some protection. The black-and-white striped offspring relies less on biting but uses numerous barbed, detachable spines to great effect in head-bucking. Common tenrecs have better eyesight than most Tenrecidae, but may also detect disturbances through long sensitive hairs on the back. Disturbed young can communicate their alarm through stridulation, which involves rubbing together stiff quills on the mid-back to produce an audible signal. Streaked tenrecs are remarkably similar to juvenile Common tenrecs in coloration, size and possession of a stridulating organ. Like juvenile Common tenrecs, they forage in groups, and their main defense involves scattering and hiding under cover. When cornered, they advance, bucking violently, their spines bristling.

Tenrecs and otter shrews are opportunistic feeders, taking a wide variety of invertebrates, as well as some vertebrates and vegetable matter. Otter shrews scour the water, stream bed and banks with their sensitive whiskers, snapping up prey and carrying it up to the bank if caught in the water. Crustacea are the main prey, including crabs of up to 5–7cm (2–3in) across the carapace. Rice tenrecs probably encounter most of their invertebrate prey in underground burrows or surface runs, but also consume vegetable matter. Fruit supple-

ments the invertebrate diet of the more omnivorous species such as the Common and hedgehog tenrecs. Common tenrecs are also large enough to take reptiles, amphibians and even small mammals. Prey are detected by sweeping whiskers from side to side, and by smell and sound. Similarly, semi-arboreal Lesser hedgehog tenrecs and long-tailed tenrecs perhaps encounter and eat lizards and nestling birds. The streaked tenrecs, one of which is active during daytime, have delicate teeth and elongated fine snouts for feeding on earthworms.

Tenrec reproduction is diverse and includes several features peculiar to the family. Where known, ovarian processes differ from those in other mammals in that no fluid-filled cavity, or antrum, develops in the maturing ovarian follicle. Spermatozoa also penetrate developing follicles and fertilize the egg before ovulation; this is known in only one other mammal, the Short-tailed shrew.

Most births occur during the wet season, coinciding with maximum invertebrate numbers, and the offspring are born in a relatively undeveloped state. Litter size varies from two in the Giant otter shrew and some Oryzoryctinae to an extraordinary maximum of 32 in the Common tenrec, and apparently reflects survival affected by the stability of the environment. For example, oryzoryctines inhabiting the comparatively stable high rain-forest regions are apparently long-lived and bear small litters. Similarly, average litter size of Common tenrecs inhabiting relatively seasonal woodland/savanna regions with fluctuating climatic conditions is 20, compared to 15 in rain-forest regions and 10 in Seychelles rain

▲ **The striped coat** of a young Common tenrec is a form of camouflage enabling it to accompany its mother on daylight foraging trips.

◄ **Very like a hedgehog,** this ball of spines is in fact the Lesser hedgehog tenrec, a species native to Madagascar.

forests within 5° of the equator. Variation in weight within the litter can reach 200–275 percent in Common and hedgehog tenrecs.

The Common tenrec feeds her offspring from up to 29 nipples, the most recorded among mammals. Nutritional demands of lactation are so great in this species that the mother and offspring must extend foraging beyond their normal nocturnal regime into the relatively dangerous daylight hours. This accounts for the striped camouflage coloration of juveniles, which only become more strictly nocturnal at the approach of the molt to the adult coat. Moreover, adult females have a darker brown coat than adult males, presumably because it affords better protection for daylight feeding throughout their brief spell of lactation. The striking similarity between juvenile Common tenrecs and adult streaked tenrecs suggests that a striped coat associated with daylight foraging has been an important factor in the evolution towards modern streaked tenrecs.

Rain-forest streaked tenrecs form multigenerational family groups comprising the most complex social groupings among insectivores. Young mature rapidly and can breed at 35 days after birth, so that each group may produce several litters in a season. The group, of up to 18 animals, probably consists of three related generations. They forage together, in subgroups, or alone, but when together they stridulate almost continuously. Stridulation seems to be primarily a device to keep mother and young together as they search for prey.

The primary means of communication among the Tenrecidae is through scent. Otter shrews regularly deposit feces either in or near their burrows and under sheltered banks. Marking by tenrecs includes cloacal dragging, rubbing secretions from eyeglands and manual depositing of neck-gland secretions. Common tenrecs cover 0.5–2ha (1.2–5 acres) per night, although receptive females reduce this to about 200sq m (2,150sq ft) to facilitate location by males. Giant otter shrews may range along 800m (0.5mi) of their streams in a night.

Common tenrecs have been a source of food since ancient times, but are not endangered by this hunting. Undoubtedly, some rain-forest tenrecs are under threat as Madagascar is rapidly being deforested, but some species thrive around human settlements. Forest destruction is also reducing the range of the Giant otter shrew and perhaps also of the Mount Nimba least otter shrew and the Ruwenzori least otter shrew.

MEN

Tenrec Body Temperature

Body temperature is relatively low among tenrecs, with a range of 30–35°C (86–95°F) during activity. The Large-eared tenrec and members of the Tenrecinae enter seasonal hypothermia, or torpor, during dry or cool periods of the year when foraging is difficult. This hypothermia ranges from irregular spells of a few days which are opportunistic, to continuous periods lasting six months; then it is integral to the animal's physiological and behavioral cycles, as in the Common tenrec in the hot, humid rain forests of the Seychelles within 5° of the equator. So finely arranged are the cycles of hypothermia, activity and reproduction that the Common tenrec must complete such physiological changes as activation of the testis or ovary while still torpid, since breeding begins within days of commencing activity.

The Giant otter shrew, some Oryzoryctinae and the Tenrecinae save energy at any time of year because body temperature falls close to air temperature during daily rest. In this way, the animals save energy otherwise used to keep the body at a higher, constant

temperature. Interactions between these fluctuations in body temperature and reproduction in the Tenrecidae are unique. For example, during comparable periods of activity in the Common tenrec, body temperatures of breeding males are on average 0.6°C (1.1°F) lower than those of nonbreeding males. This is because sperm production or storage can only occur below normal body temperature. Other mammals have either an elaborate mechanism for cooling the reproductive organs or, in a few rare cases, tolerate high temperatures. Normally, thermoregulation improves during pregnancy, but female Common tenrecs, and no doubt others in the family, continue with their regular fluctuations in body temperature dependent on activity or rest, regardless of pregnancy. This probably accounts for variations in recorded gestation lengths, as the fetuses could not develop at a constant rate if so cooled during maternal rest. Although torpor during pregnancy occurs among bats, it is well regulated, and the type found in tenrecs is not known elsewhere.

SOLENODONS

Family: Solenodontidae
Two species in a single genus.

Hispaniola solenodon [E]
Solenodon paradoxurus
Distribution: Hispaniola.
Habitat: forest, now restricted to remote
regions; nocturnal.

Size: head-and-body length 284–328mm
(11–13in), tail length 222–254mm (8.5–10in),
weight 700–1,000g (25–35oz).

Coat: forehead black, back grizzled gray-brown,
white spot on the nape, yellowish flanks; tail
gray except for white at base and tip.

Cuban solenodon [E]
Solenodon cubanus
Distribution: Cuba.
Habitat, activity and size as for the Hispaniola
solenodon, except that the tail is slightly
shorter.

Coat: finer and longer than in the Hispaniola
solenodon, dark gray except for pale yellow
head and mid-belly.

[E] Endangered.

▶ **A threatened species,** the primitive
Hispaniola solenodon falls prey to carnivores
and suffers competition from rodents
introduced to its sole location on the island of
Hispaniola.

The extraordinary solenodons of Cuba and Hispaniola face a real and immediate threat to their survival. They are so rare and restricted that the key to their survival lies in prompt governmental effort to provide adequate management of forest reserves in remote mountainous regions. Without such efforts, these distinctive ancient Antillean insectivores are likely to follow the West Indian shrews, the Nesophontidae, which may have declined to extinction upon the arrival of Europeans.

The solenodons are among the largest living insectivores, resembling, to some extent, large, well-built shrews. The most distinctive feature is the greatly elongated snout, extending well beyond the length of the jaw. In the Hispaniola solenodon, the remarkable flexibility and mobility of the cartilaginous appendage stems from its attachment to the skull by means of a unique ball-and-socket joint. The snout of the Cuban solenodon is also highly flexible but lacks the round articulating bone. Solenodons possess 40 teeth, and the front upper incisors project below the upper lip. The Hispaniola solenodon secretes toxic saliva and this probably occurs also in the Cuban species. Each limb has five toes, and the forelimbs are particularly well developed, bearing long, stout claws which are sharp. Only the hindfeet are employed in self-cleaning and can reach most of the body surface by virtue of unusually flexible hip joints. Only the rump and the base of the tail cannot be reached, but because these areas are hairless they require little attention. The tail is stiff and muscular and possibly plays a role in balancing.

As in most nocturnal terrestrial insectivores, brain size is relatively small, and the sense of touch is highly developed, while smell and hearing are also important. Vocalizations include puffs, twitters, chirps, squeaks and clicks; the clicks comprise pure high-frequency tones similar to those found among shrews, and probably provide a crude means of echolocation. Scent marking is probably important, as evidenced by the presence of anal scent glands, while contact perhaps plays a role in some situations.

In addition to the living genus, solenodons are known from North American middle and late Oligocene deposits (about 32–26 million years ago). Their affinities are difficult to ascertain owing to their long isolation, but their closest allies are probably the true shrews (Soricidae), or the Afro-Madagascan tenrecs. Some mammalogists have also considered the extinct West Indian nesophontid shrews to be within the Solenodontidae.

Solenodons were among the dominant carnivores on Cuba and Hispaniola before Europeans arrived with their alien predators, and were probably only occasionally eaten themselves by boas and birds of prey. Soil and litter invertebrates constitute a large part of the diet, including beetles, crickets and various insect larvae, together with millipedes, earthworms and termites. Vertebrate remains which have appeared in feces may have been the result of scavenging carrion, but solenodons are large enough to take small vertebrates such as amphibians, reptiles and perhaps small birds. Solenodons are capable of climbing near-vertical surfaces, but spend most time foraging on the ground. The flexible snout is used to investigate cracks and crevices, while the massive claws are used to expose the prey under rocks, the bark of fallen branches and in the

soil. A solenodon may lunge at prey and pin it to the ground with the claws and toes of the forefeet, while simultaneously scooping up the prey with the lower jaw. Occasionally the prey is pinned to the ground only by the cartilaginous nose, and must be held there as the solenodon advances. These advances take the form of rapid bursts to prevent the prey's escape, and a maneuvering of the lower jaw into a scoop position. Once it is caught, the prey is presumably immobilized by the toxic saliva.

The natural history of solenodons is characterized by a long life span and low reproductive weight, with a litter size of 1–2, as a consequence of having been among the dominant predators in pre-Columbian times. The frequency and timing of reproduction in the wild is not known, but receptivity lasts less than one day and recurs at approximately 10-day intervals. Events leading up to mating involve scent marking by both sexes, soft calling and frequent body contacts. In captivity, the scent marking involves marking projections in the female's cage with anal drags and also defecating and urinating in locations previously used by the female. The young are born in a nesting burrow, and they remain with the mother for an extended period of several months, which is exceptionally long among insectivores. During the first two months, each young solenodon may accompany the mother on foraging excursions by hanging onto her greatly elongated teats by the mouth. Solenodons are the only insectivores which practice teat transport, and carrying the offspring in the mouth is more widespread within this order. Initially, the offspring are simply dragged along, but as they grow they are able to walk with the mother, pausing when she stops. Teat transport would undoubtedly be useful if nursing solenodons change burrow sites regularly. More advanced offspring continue to follow the mother, learning food preferences from her by licking her mouth as she feeds, and getting to know routes around the nest burrow. The mother-offspring tie is the only enduring social grouping among solenodons; adults are otherwise solitary.

There are no accurate estimates of solenodon numbers in Cuba or Hispaniola, although the Cuban solenodon appears to be the rarer species. The low reproductive rate is one factor in the decline in solenodon abundance, but more important factors accounting for their rarity are habitat destruction and predation by introduced carnivores, against which solenodons have no defense. Mongooses and feral cats are the main predators on Cuba, whereas dogs decimate solenodon populations in the vicinity of settlements on Hispaniola. There is little hope for the Hispaniola solenodon in Haiti, the nation comprising the western half of Hispaniola, but protected areas of dense forest now exist in remote regions of the neighboring Dominican Republic, and on Cuba. These require prompt, efficient management to ensure the solenodon's survival. Such is the pressure for new land which accompanies the human population explosion on these islands that the solenodons' survival may ultimately rest upon the efforts of zoos. MEN

▼ **The solenodon's snout** is a unique feature, a cartilaginous appendage extending well beyond the jaw. In the Hispaniola solenodon, but not the Cuban solenodon, the snout articulates with the skull by means of a ball-and-socket joint, and is used to investigate cracks during foraging and to pin down prey.

HEDGEHOGS

Family: Erinaceidae
About seventeen species in 8 genera.
Distribution: Africa, Europe, and Asia north to
limits of deciduous forest but absent from
Madagascar, Sri Lanka, and Japan. Introduced
to New Zealand.

Size: ranges from head-body length 10–15cm (4–6in), tail length 1–3cm (0.4–1.2in) and weight 40–60g (1.4–2oz) in the Lesser moonrat to head-body length 26–45cm (10–18in), tail length 20–21cm (7.8–8.3in) and weight 1,000–1,400g (2.2–3lb) in the Greater moonrat.

Habitat: wooded or cultivated land (including urban gardens), tropical rain forest, steppe, and desert.

Gestation: known only for European hedgehog (34–49 days) and Long-eared hedgehog (37 days).

Longevity: up to 6 or 8 years (10 in captivity).

Moonrats or gymnures
Subfamily Echinosoricinae
Five genera with 5 species: **Greater moonrat** (*Echinosorex gymnurus*), **Lesser moonrat** (*Hylomys suillus*), **Hainan moonrat** (*Neohylomys hainanensis*), **Mindanao moonrat** (*Podogymnura truei*) [v], **Shrew-hedgehog** (*Neotetracus sinensis*).
Southeast Asia, China.

Hedgehogs
Subfamily Erinaceinae
Twelve species in 3 genera, including **Western European hedgehog** (*Erinaceus europaeus*), W and N Europe; **long-eared hedgehogs** (*Hemiechinus* spp), Asia and N Africa; **desert hedgehogs** (*Paraechinus* spp), Asia and N Africa.

[v] Vulnerable.

▶ **Grist to the mill.** Hedgehogs will eat almost any invertebrate prey. Here a South African hedgehog is devouring a grasshopper.

▶ **A white moonrat, foraging.** The Greater moonrat is usually black with whitish head and shoulders but some individuals are white all over.

THE hedgehogs are among the most familiar small mammals in Europe, distinguished by their thick coat of spines and ability to curl up into a ball when threatened. Hedgehogs have always enjoyed a close and rather friendly relationship with man. They figure prominently in folk tales and in many places are treated as free-ranging pets. Nevertheless, only the European species have been studied closely.

Even less is known about the moonrats or gymnures (sometimes considered to be spineless hedgehogs), which are found in Southeast Asia and China. These five species are very diverse in appearance: the Greater moonrat is the size of a rabbit, with coarse black hair and white markings on the face and shoulders; all the others are smaller and have soft fine hair (rather like large shrews). All of them have very elongated snouts.

With the exception of the spines and their associated musculature, the body plan of hedgehogs and moonrats is very primitive. The eyes and ears are well developed, the snout is long and pointed and extends a long way in front of the mouth. The front of the skull is quite blunt, however, so that the tip of the snout is unsupported and capable of great mobility. The teeth vary from 36 to 44 in number, usually with both upper and lower first incisors (the "front" teeth) larger than the others. In the hedgehogs, the first upper incisors are separated by a wide gap into which the blunt, forward-projecting lower incisors can fit (when a hedgehog closes its jaws on an insect, the lower incisors act as a scoop and push upward between the upper teeth to impale it). The rest of the teeth are either pointed (incisors, canines, and first premolars) or broad with sharp cusps (last premolars and molars).

All species but one have five toes on each foot, the exception being the Four-toed hedgehog which has only four on the hind-feet. The hedgehogs have powerful legs and strong claws and are good at digging (those species occupying dry areas often live in burrows). The five species of moonrats look and behave rather like large species of shrews, tending to move abruptly and quickly. Hedgehogs can run at up to 2m/sec (6.5ft/sec), but usually move with a slow shambling walk.

Among the five species of moonrat, two species, the Mindanao moonrat and the Hainan moonrat, are quite distinct. Of the other three species, all of which have similar geographic ranges and apparently occupy fairly similar habitats, the Greater moonrat is readily distinguished by its large size and striking coloration. Despite the superficial

similarity of the Lesser moonrat and the Shrew-hedgehog, the latter has fewer teeth.

About 12 species of hedgehogs are usually recognized, divided into three genera on the basis of shape and pattern of spines and length of ears, as well as skull morphology. The long-eared hedgehogs do not have a narrow spineless tract on the crown of the head while the other two genera do. Desert hedgehogs differ from woodland hedgehogs in having longitudinal grooves on their spines. Differences between the species within a genus often involve the color of the spines and the length of the toes or ears. In general, the division of a hedgehog genus into species rests upon their geographical ranges. Even this criterion fails when considering the two species of European hedgehogs: it is unclear whether there are one

two, or perhaps even 12 species present. The problem is that hedgehogs vary considerably in size, shape and color, and that groups of individuals from different parts of Europe tend to be quite different in appearance. Animals from Spain are paler in color, those from Crete and Britain are smaller, those from the western parts of Europe have mostly brown underparts while eastern European animals tend to be white on the breast. It is not known whether these subgroups are sufficiently and consistently different in appearance to be considered as subspecies. Even the division into eastern and western species of European hedgehog accepted here is not based on an inability to interbreed: hybrids have been produced in captivity and intermediate forms are occasionally found in areas where the two species meet. Nevertheless, the two are considered to be separate species because there are minor differences in the skull and, perhaps more importantly, there are differences in the appearance of their chromosomes.

Like other insectivores, hedgehogs and moonrats eat a wide variety of prey. European hedgehogs (in both Europe and New Zealand, where they have been introduced) eat virtually any available invertebrate. The major prey is the earthworm, followed by beetles, earwigs, slugs, millipedes and caterpillars. In addition, hedgehogs will scavenge the remains of any animal found dead and take eggs and young from birds' nests. Despite a reputation for killing frogs and mice (among others) and for destroying the eggs of ground-nesting birds, it is unlikely that they are a significant predator on any vertebrates. In contrast, the Daurian hedgehog of the Gobi desert has been reported to feed mainly on small rodents, perhaps because there is little else to eat in that area. The Greater moonrat enters the water quite readily, and eats crustaceans, mollusks and even fish. Little is known of the diets of the other species of moonrats, or of most of the non-European hedgehogs.

Hedgehogs will eat seeds, berries or fallen fruit. Stomach contents or droppings frequently reveal abundant plant matter, especially grasses and leaves, largely undigested.

There is no information on breeding for the Hainan and Mindanao moonrats, and little for the other three species. Both the Greater and Lesser moonrats breed all year round in the Tropics, while the Shrewhedgehog may have either a long breeding season (April to September) or two breeding seasons: one in April/May and the other in

▼ **Species of hedgehogs and moonrats.**
(1) Desert hedgehog (*Paraechinus aethiopicus*).
(2) Algerian hedgehog (*Erinaceus algirus*).
(3) Shrew-hedgehog (*Neotetracus sinensis*).
(4) Lesser moonrat (*Hylomys suillus*).
(5) Long-eared hedgehog (*Hemiechinus auritus*).
(6) Mindanao moonrat (*Podogymnura truei*).
(7) Greater moonrat (*Echinosorex gymnurus*).

August/September. Greater and Lesser moonrats have two or three young in a litter, the Shrew-hedgehog has four or five.

Most of the hedgehogs in tropical climates breed throughout the year, but those in desert or semidesert (long-eared and desert hedgehogs) breed only once between July and September. In the more temperate areas, where food is abundant but only seasonally available, hedgehogs may breed once or twice a year between May and September, depending on the weather. Litter sizes vary from 1–10 (usually 4–7) in woodland and long-eared hedgehogs and from 1–5 in desert hedgehogs.

In woodland hedgehogs, courtship occurs practically whenever a male encounters a female. The female will stand still, with partly erect spines. The male moves slowly around her, usually brushing against her, and frequently reversing direction. The female also turns so that she faces toward the male and, if unreceptive, will erect the spines on her forehead and butt at the male's flank. The female, and sometimes the male, make loud snorting noises throughout. Continued butts from an unreceptive female do not deter the male from circling her, often for hours.

Contrary to one old wives' tale, hedgehogs do not mate face to face. The female's vaginal opening is placed far back, immediately in front of the anus, while the male's penis is very long and located a long way forward. The female stands with spines flat and back depressed while the male mounts. Because the flattened spines on the female's back are very slippery, the male holds himself in position by grasping the spines on her shoulder in his teeth. Copulation may occur several times in succession before the two animals part company. There is no pair-bond formed and the male shows no paternal behavior.

During gestation, the female prepares a nest in which to give birth. The young are born naked with eyes and ears closed. Although spines are present at birth, they lie just beneath the skin, which is engorged with large amounts of fluid. This prevents the spines from piercing the skin and damaging the mother's birth canal. The fluid beneath the skin is rapidly resorbed after birth, allowing the skin to contract, which then forces the 150 or so white spines through the skin. Within 36 hours these are supplemented by the sprouting of additional, darkly pigmented spines. Young European hedgehogs grow quite rapidly; at 14 days their hair has started to grow, the eyes are beginning to open, and the animal

can partially roll up. At 21 days the first teeth have appeared and the young start to leave the nest with the mother in order to learn to fend for themselves. Weaning occurs at 6–7 weeks of age, after which the young either leave the nest or are driven off by the mother. The development of the young of desert hedgehogs is similar to that of woodland hedgehogs. Long-eared hedgehogs, however, appear to produce smaller young—5–6cm (2–2.4in) long, 3–4g (0.1–0.14oz), as opposed to 6–10cm (2.4–4in), 12–25g (0.4–0.9oz) in European hedgehogs—which mature much faster than the other two genera. The eyes are reputedly open at 1 week and by 2–3 weeks the young can eat solid food.

In the European hedgehogs, from mating to weaning takes about 12–13 weeks and so, if the food supply is adequate and breeding has occurred early enough in the year, a second litter is possible. The young of first litters have an advantage over those of

▲ **A spineless hedgehog.** Hedgehogs are born with the spines beneath the skin, to avoid damaging the mother's birth canal.

▷ **The Hedgehog family.** OVERLEAF The European hedgehog has litters of usually 4–7 young. At three weeks old the young accompany the mother and at 6–7 weeks they are weaned.

Spines and Curling in Hedgehogs

The most distinctive feature of hedgehogs is their spines. An average adult carries about 5,000, each about 2–3cm (1in) long, with a needle-sharp point. Each creamy white spine usually has a subterminal band of black or brown. Spines are actually modified hair, and along the animal's sides where they give way to true hair, thin spines or thick stiff hairs can often be found which may show the transition from one to the other. To minimize weight without losing strength, each spine is filled with many small air-filled chambers separated by thin plates. Towards its base, each spine narrows to a thin angled flexible neck and then widens again into a small ball which is embedded in the skin. This arrangement transforms any pressure exerted along the spine (from a blow, or from falling and landing on the spines, for example) into a bending of the thin flexible part rather than driving the base of the spine into the hedgehog's body. Connected to the base of each spine is a small muscle which is used to pull the spine erect. Normally, the muscles are relaxed and the spines laid flat along the back. If threatened, a hedgehog will often not immediately roll up but will first simply erect the spines and wait for the danger to pass. When erected, the spines stick out at a variety of different angles, criss-crossing over one another and supporting each other to create a virtually impenetrable barrier.

Hedgehogs are additionally protected by their ability to curl up into a ball. This is achieved by the presence of a rather larger skin than is necessary to cover the body, beneath which lies a powerful muscle (the *panniculus carnosus*) covering the back. The skin musculature is more strongly developed

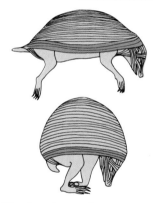

around its edges than at the center (where it forms a circular band, the *orbicularis* muscle) and is only very loosely connected to the body beneath. When the *orbicularis* contracts, it acts like the drawstring around the opening of a bag, forcing the contents deeper into the bag as the string is drawn tighter. When a hedgehog starts to curl up, two small muscles first pull the skin and underlying circular muscle forward over the head and down over the rump. Then the circular *orbicularis* muscle contracts, the head and hindquarters are forced together, and the spine-covered skin of the back and sides is drawn tightly over the unprotected underparts. So effective is this that on a fully curled hedgehog the spines that formerly covered its flanks and the top of the head are brought together to block the small hole (smaller than the width of a finger) which corresponds to the now-closed opening of the bag. As the skin is pulled tightly over the body, the small muscles which erect the spines are automatically stretched, and the spines erected, so that the tighter the hedgehog curls the spinier it becomes.

the second, in an additional 12–13 weeks preparation for winter and hibernation. The young of a second litter have only a month or two (when food supplies are already dwindling) in which to build up a store of fat. Up to 75 percent of the animals born in a summer may die due to an inability to prepare for winter.

The Greater moonrat is probably nocturnal, resting during the day among the roots of trees, under logs, or in empty holes. It is usually found near water. Both the Lesser moonrat and the shrew-hedgehog are also nocturnal.

Long-eared hedgehogs and desert hedgehogs both dig and live in burrows. These are 40–50cm (16–20in) deep and each is occupied by a single individual throughout the year (except when females have their young with them). This restriction to a single nest-site suggests that they may be territorial.

Woodland hedgehogs do not burrow but usually build nests of grass and dead leaves in tangled undergrowth. Such nests can be built quickly (compared to digging a burrow), and a single individual may occupy a given nest for only a few days and then move on and either build a new nest or find an abandoned one. For this reason, in open habitats (the only ones that have been studied) the home range of the Western European hedgehog is large for its body-weight—20–35ha (50–87 acres) for males and 9–15ha (22–37 acres) for females. The males move further and faster than females—up to 3km (1.9mi) per night versus 1km (0.6m) per night—probably because they need to look for females as well as for food.

All hedgehogs are capable of undergoing periods of dormancy (hibernation) during which their body temperature is allowed to drop to a level close to that of the surrounding air (heterothermy). This reduces the hedgehog's energy needs to a very low level (oxygen requirement declines from about 550ml/kg/h to about 10ml/kg/h) and allows it to survive periods of up to several months when food is scarce. Tropical hedgehogs do not hibernate, since food availability is not seasonal, but the two desert genera and the more temperate species usually hibernate during the winter (in northern Europe, October or November until March or April). Hibernation is dictated by the conditions and is not a species trait—tropical hedgehogs will hibernate if exposed to cold and low food levels, while European hedgehogs do not hibernate if food is available throughout the year. Hedgehogs intro-

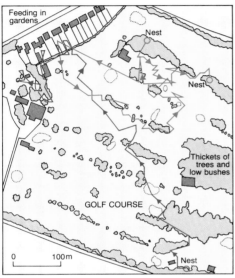

▲ **Hedgehogs and food bowls.** TOP Many people in Europe put out bowls of food in their garden for hedgehogs to visit at night. The animals do so, often with great regularity. This supplementary food may play an important role in achieving the weight needed to survive hibernation. Householders assume that "their" hedgehogs nest nearby and travel no further to reach the food than necessary. In fact, the animals may nest several hundred meters away, and although they move their daytime resting site frequently in summer, they make no attempt to live nearer to regular sources of supplementary food. Nor do they necessarily visit food bowls every night, even if available. Not all hedgehogs that could do so actually visit the food bowls. All this suggests that provision of large amounts of supplementary food does not significantly distort hedgehog behavior. One male (BLUE) was observed to travel over a 1km (0.6mi) round trip each night from its golf-course nest to a series of gardens to feed. It was blind. That it could find its way regularly and so easily indicates that sight was not a crucial sense. A female (RED) was tracked for 5¼ hours, during which she traveled 1.4km (0.9mi), foraging extensively both on the golf course and in the gardens.

duced into New Zealand from Europe hibernate only for very short periods during the southern winter or may not hibernate at all in the warmer areas.

Small supplies of energy are required to keep the body "ticking over" during hibernation and to provide the boost needed to bring the body temperature back up to normal when waking up. Most hibernators do not "sleep" continuously but will arouse periodically, once every two weeks for instance, stay awake for a day or so and then go back into hibernation again. The source of this energy is fat laid down during the summer. Thus European hedgehogs may almost double their weight during the summertime, mostly by laying down a thick layer of fat. Besides providing an energy store, the fat acts as an insulating layer. A specially built, well-insulated hibernation nest also helps to keep the animal warm, even when the external temperature is below freezing point.

Being largely solitary animals, hedgehogs and moonrats do not need an elaborate communication system. Visual signals are apparently absent and, although their hearing is acute, auditory signals are not widely used. Hedgehogs make snuffling and snorting noises while courting, and may hiss when angry or upset. Young hedgehogs separated from their mother make a twittering whistle. Probably the most important mode of communication is by odor. All hedgehogs and moonrats have a well-developed sense of smell and all appear to use odors in one way or another to obtain information about other members of their species. All also undoubtedly find their food largely by sense of smell.

Little is known about the moonrats. Both the Greater and Lesser moonrats have well-developed anal scent glands whose secretions smell to humans like rotten garlic or onions, noticeable from several meters. Captive animals use these glands to scent mark around the entrance to their nests. Hedgehogs also have anal scent glands, but they are not well developed and are not used in any overt way to leave marks. They do, however, leave trails of scent behind them as they walk along—probably just a generalized body scent, although it may contain information about the sex of the animal. Male hedgehogs have been observed to cross such a trail left by a female and immediately turn and follow it for 50m (165ft) or more until catching up with her and initiating courtship.

An unusual behavior of hedgehogs is self-anointing. The hedgehog suddenly pro-

duces copious foamy saliva which it smears all over the spines on its back and flanks. One opinion is that this behavior is elicited by strong-tasting substances—in experiments, it is elicited by cigarette ends, bookbinder's glue, toadskin, leather, creosote and cat food. The function of self-anointing is unknown but suggestions include: 1) it may produce strong odors which act as a sexual attractant; 2) it may reduce parasites on the skin; 3) it may be an attempt to clean the spines; 4) it may make the spines distasteful and thus deter predators.

Unlike the shy moonrats, hedgehogs rely on their spines for protection and so wander where there is no protective cover; if disturbed, most hedgehogs simply erect their spines and freeze until the disturbing factor has gone. Throughout Europe, the hedgehog has capitalized on an amicable relationship with people, and colonized urban areas. Gardens are a good habitat for a hedgehog, providing a mosaic of foraging places—lawns, flower beds, vegetable patches, and compost heaps, together with places in which to nest—hedges, rubbish or junk piles, or under garden sheds.

Hedgehogs play a small role in controlling insect pests. They carry large numbers of parasites among their spines, particularly fleas, ticks and ringworm (a fungus), but these all tend to be species which specialize on hedgehogs and are not easily transmitted to man or other animals. Throughout Europe, large numbers of hedgehogs are killed on the roads every year, but there is no evidence that this adversely affects their population size.

▲ **Hedgehog and hen's egg.** This European hedgehog seems to be making a not particularly urgent assault on a hen's egg. They do take birds' eggs and young when available.

◄ **Self-anointing.** This European hedgehog is coating its spines with saliva; this is flicked onto them by its long tongue, a strange, unexplained piece of behavior.

Hedgehogs figure prominently in folklore: over 2,000 years ago, the Roman author Pliny recorded that hedgehogs carry fruit impaled on their spines rather than in their mouths, and this story has been handed down through the ages. The Chinese tell the same tale. Undoubtedly, the spines of hedgehogs do occasionally catch and pick up things (usually dead leaves or twigs) and it is possible to impale soft fruit such as apples or grapes on the spines. However, European hedgehogs usually eat their food where they find it and do not normally carry it off somewhere else. Nor, in fact, would it be easy for them either to impale the fruit themselves or to remove it once it was attached. Other common folk tales are that hedgehogs suck milk from sleeping cows (possible, but unlikely—they may lap up milk that has oozed from a full udder), and that they are immune to snake bites (they may have some resistance, but in most cases a snake striking at the spine-covered back is

more likely to hurt itself than the hedgehog).

Future prospects are mixed for the hedgehog family. Those species which have formed benign relationships with man are not under any threat. Among the others, the Mindanao moonrat is already very rare and may be threatened with extinction because of forest clearance in the Philippines. The Hainan moonrat may be similarly affected but no reports on its status have been made since it was first discovered in 1959. The other moonrat species have much less restricted ranges but all live in forested areas in countries where increasing population and economic pressures have resulted in clearing the jungle for lumber and to provide agricultural land. As their habitat disappears, all three species of moonrat are likely to become rarer. The tropical hedgehogs may soon be similarly affected, but at the moment no hedgehog species is considered to be in danger of extinction.

AW

SHREWS

Family: Soricidae

Two hundred and forty-six species in 22 genera.

Distribution: Eurasia, Africa, North America, northern South America.

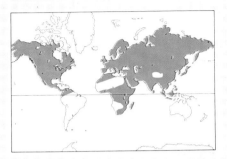

Habitat: forest, woodland, grassland, desert; terrestrial, but some species partially aquatic.

Size: head-to-tail length from 3.5–4.8cm (1.3–2.0in) in the Pygmy white-toothed shrew, the smallest living terrestrial mammal, to 26.5–29cm (10.5–11.5in) in the African forest shrew; weight from 2g (0.07oz) in the Pygmy white-toothed shrew to 35g (1.2oz) in the African forest shrew.

Gestation: 13–24 days.

Longevity: probably 12–18 months.

► **Refection.** Some shrews, like this Eurasian pygmy shrew, obtain certain nutrients by licking the rectum, which during the process projects from the anus. It is thought that some nutrients that would be lost in normal digestion are obtained in this way.

▼ **The smallest living terrestrial mammal** is the Pygmy white-toothed shrew. Its large ears are prominent here.

"IT is a ravening beast, feigning itself gentle and tame, but being touched it biteth deep, and poisoneth deadly. It beareth a cruel mind, desiring to hurt anything, neither is there any creature it loveth." So wrote Edward Topsell of the European common shrew in his *History of Four-footed Beasts* published in 1607. His characterization highlights the curiously ironical place of shrews in popular folklore. There can be few mammal groups which are less of an imposition on man yet few which have been attributed with such an unfavorable disposition—the words "shrewd," "shrewish" and "shrew" were coined originally to describe a rascally or villainous character, although their meaning has changed somewhat over the centuries. Shrews have been seen as poisoners of horses, scavengers of raven flesh, a cause of lameness in livestock, a certain cure (when burned, powdered and mingled with goose grease) for swelling and as talismans against their own and other evils.

In reality, shrews are small, secretive mammals superficially rather mouse-like but with characteristically long, pointed noses. They are typically terrestrial, foraging in and under the litter in woods, and the vegetation mat beneath herbage. Some, however, are aquatic. Ecologically, they are important in breaking down animal tissue and returning raw materials to the soil. The eyes are small, sometimes hidden in the fur, and vision appears to be poor. Hearing and smell, however, are acute. Even so, the external ears are reduced and difficult to discern in some species. The foot is not specialized except in the species which regularly enter water. The Tibetan water shrew is the only species with webbed feet. Other "aquatic" species, like the European water shrew, have feet, toes, fingers and tail fringed with stiff hairs. These hairs increase the surface area, aid in propulsion and trap air bubbles so the shrew can "run" on the surface of the water. In most genera, the genital and urinary systems have a common external opening, but in long-tailed shrews the openings are separate. In mouse-shrews an intermediate situation exists.

The first set of teeth are shed or resorbed during embryonic development so that shrews are born with their final set. One cause of death, at least in the European common shrew, is starvation due to wearing down of the teeth. The fact that teeth are not replaced makes tooth wear a useful index of age. Analyses of tooth wear in the American short-tailed shrew, water shrew and the European common shrew, water shrew and pygmy shrew have shown that there are usually two generations present during the summer and one during the winter.

A skeletal feature unique to shrews and found only in the Armored shrew, an African species, is the possession of interlocking lateral, dorsal and ventral spines on the vertebrae. Along with the large number of facets for articulation, the spines create an exceptionally sturdy vertebral column. There are reliable reports of the Armored shrew surviving the pressure of a full-grown man standing on it.

Although they mainly use touch, shrews do communicate vocally. The most conspicuous vocalizations are the characteristically high-pitched screams and twitterings used in disputes with members of their own species. Shrews are habitually solitary and many species defend feeding territories from which they oust intruders. There is some evidence in the European common shrew that the pitch and intensity of these vocalizations transmit information about the competitive ability of residents and intruders, and help in settling disputes quickly and without injury. In addition, some species, at least of the genera *Sorex* and *Blarina*, may use ultrasound, which seems to be generated in the larynx and could provide a crude means of echolocation.

Shrews are widely distributed geographically, but genera vary enormously in their distribution and ecology. Some genera, like *Sorex*, *Neomys*, *Blarina*, *Cryptotis* and *Suncus*, are distributed on a continental scale, while others, like *Podihik*, *Anourosorex*, *Feroculus* and *Diplomesodon*, are so restricted that they occur only on particular islands or in certain, often unique, mainland habitats. Ten genera comprise only a single species each and the Sri Lanka shrew and the African forest shrew are known from only two and three specimens respectively.

While shrews are known from the Eocene (54 million years ago) onwards, knowledge of the evolutionary lineages of present forms is extremely thin. Part of the reason is undoubtedly the small size of the bones and teeth, which are easily overlooked. Shrews are found from the Oligocene (38 million years ago) onwards in North America, from the Eocene (54 million years ago) onwards in Europe, from the Pliocene (7 million years ago) onwards in Asia, and from the Miocene (26 million years ago) onwards in Africa. The best records seem to be for the existing European species. The European pygmy shrew has the oldest fossil record, extending back to the beginning of the Pleistocene (2 million years ago). At least two main lines of shrews died out in Europe towards the end of this period, possibly as a result of the arrival of more *Crocidura* species from North Africa and Asia. The presence of five co-existing species of *Sorex* in Finland suggests that, in the absence of *Crocidura*, abundance and diversity of shrews may be greater.

At the end of the Pliocene, at least four species of shrew were present in western Europe and probably also in the British Isles: the European pygmy shrew, which has remained virtually unchanged to the present day, *Sorex magaritodon*, *S. runtonensis* and *S. kennardi*. From the fossil evidence, it seems that the White-toothed or Musk shrew has only recently invaded from the east, as has the Lesser white-toothed shrew from the eastern Mediterranean region. An intriguing subspecies of the latter, the Scilly shrew, is restricted in its distribution to the Scilly Isles. Studies suggest that the Scilly landmass was close to a tongue of ice derived from the Irish and Welsh ice sheets during the last glacial maximum. One possibility is that it was introduced by Iron-age or earlier traders from France and northern Spain who came to Cornwall in search of the tin that was mined there.

In general, it appears that the gross form of shrews has changed very little since the early Tertiary. The only overall change seems to have been a slight reduction in size. The European common shrew and the Pygmy shrew, for instance, were slightly larger in the Pleistocene than they are today. In this sense, shrews represent a primitive stage in mammalian evolution. However, the existing species exhibit many non-primitive specializations and so cannot

▲ **A shrew's delight** is a long, juicy earthworm. Their reputation for voracity stems from their small size and active life with consequent high metabolic rate.

▶ **Plastered with air bubbles,** a European water shrew forages underwater. These shrews feed on small fishes, aquatic invertebrates and small frogs.

The 22 Genera of Shrews

Genus *Sorex*
Fifty-two species in N Eurasia, N America, including **American water shrew** (*S. palustris*), **European common shrew** (*S. araneus*), **European pygmy shrew** (*S. minutus*), **Masked shrew** (*S. cinereus*), **Trowbridge's shrew** (*S. trowbridgii*); tundra, grassland, woodland.

Genus *Microsorex*
Two species: **American pygmy shrews**; N America; forest.

Genus *Soriculus*
Nine species: **mountain shrews**; Himalayas, China; montane forest.

Genus *Neomys*
Three species in N Eurasia, including **Eurasian water shrew** (*N. fodiens*); forest, woodland, grassland, streams, marshes.

Genus *Blarina*
Three species in eastern N America, including **American short-tailed shrew** (*B. brevicauda*); forest grassland.

Genus *Blarinella*
One species: **Chinese short-tailed shrew** (*B. quadraticauda*); S China; montane forest.

Genus *Cryptotis*
Thirteen species in E USA, Ecuador, Surinam, including **Lesser short-tailed shrew** (*C. parva*); forest, grassland.

Genus *Notiosorex*
Two species in SW USA and Mexico, including **Desert shrew** (*N. crawfordi*); semidesert scrub, montane.

Genus *Megasorex*
One species: **Giant Mexican shrew** (*M. gigas*); SW Mexico; forest.

Genus *Crocidura*
One hundred and seventeen species in Eurasia and Africa, including **African forest shrew** (*C. odorata*); forest to semidesert.

Genus *Suncus*
Fifteen species in Africa, including **Pygmy white-toothed shrew** (*S. etruscus*); forest, scrub, savanna.

Genus *Podihik*
(considered by some to be synonymous with *Suncus*) One species: **Sri Lanka shrew** (*P. kura*); north-central Sri Lanka; possibly aquatic.

Genus *Feroculus*
One species: **Kelaart's long-clawed shrew** (*F. feroculus*); Sri Lanka; montane forest.

Genus *Solisorex*
One species: **Pearson's long-clawed shrew** (*S. pearsoni*); Sri Lanka; montane.

Genus *Paracrocidura*
One species: *P. schoutedeni*; Cameroun.

Genus *Sylvisorex*
Seven species: Africa; forest, grassland.

Genus *Myosorex*
Ten species: **mouse shrews**; C and S Africa; forest.

Genus *Diplomesodon*
One species: **Piebald shrew** (*D. pulchellum*); Russian Turkestan; desert.

Genus *Anourosorex*
One species: **mole-shrew** (*A. squamipes*); S China to N Thailand, Taiwan; montane, forest.

Genus *Chimarrogale*
Three species: **oriental water shrews**; E Asia; montane streams.

Genus *Nectogale*
One species: **Tibetan water shrew** (*N. elegans*); Sikkim to Shenshi; montane streams.

Genus *Scutisorex*
One species: **Armored shrew** (*S. somereni*); C Africa; forest.

themselves be regarded as primitive mammals.

A feature, at least of the European common shrew, which has interesting evolutionary implications is variations in the chromosome complement. Mammalian cells usually have a constant number of chromosomes which are typical of the species. In the European common shrew the number varies from 21–27 in males and 20–25 in females, due to the fusion of certain small chromosomes (so-called Robertsonian chromosome mutations). There are also multiple sex chromosomes: males are generally XYY and females, as in other mammals, XX. This variation makes possible 27 different chromosomal arrangements of which 19 have so far been found. While variation in chromosome numbers is known in invertebrates, the European common shrew is the first example among mammals.

Shrews are very active and consume large amounts of food for their size. Studies of several species have shown that their metabolic rate is higher than that of rodents of comparable size. Shrews cope with increased demands for food and water primarily by living in habitats where these are abundant. However, their generally small size enables them to utilize thermally protected microhabitats. At least one species, the Desert shrew, has mastered the physiological problems of living in a hot, arid climate by lowering its metabolic rate in a similar way to other desert-dwelling mammals. Several species, mainly from the subfamily Crocidurinae, but including the Desert shrew, are capable of lowering their body temperature in response to food deprivation. However, similar adaptations are found in "higher" placental mammals and any differences appear to be in degree rather than in kind—a function of the shrews' small size rather than their "primitive" place in the mammal hierarchy. Interestingly, the subfamily Soricinae ("hot shrews") have higher metabolic rates than subfamily Crocidurinae ("cold shrews"). This difference may be related to their different origins, northern and tropical respectively. Digestion in shrews is fairly rapid and the gut may be emptied in three hours. Since they carry available reserves for only an hour or two, frequent feeding is imperative. For this reason, shrews are active throughout the day and night. Adult specimens of the European common shrew are often encountered dead in the open during the fall. These are generally old individuals that have bred the previous summer and whose teeth are worn.

The apparent cause of death is starvation, and the carcasses remain uneaten owing to the presence (in common with other species) of skin glands whose secretion renders shrews unpalatable to most carnivores.

Those species whose foraging behavior and dietary habits have been studied have turned out to be opportunists with little specialization. They are mainly insectivorous and carnivorous (also taking carrion), but some also eat seeds, nuts and other plant material. The mode of foraging used by shrews depends on their habits. They may use runways, particularly those created by voles or other small mammals, or hunt on the vegetation surface. The direction taken during foraging appears to be arbitrary, and shrews tend to scurry about haphazardly until they come across prey. Shrews can detect a very wide size range of prey, from 10cm (4in) earthworms to 1–2mm nematodes. Even small mites and the animal's own external parasites are not exempt. When prey are buried (as are some insect larvae and pupae), foraging involves digging or furrowing. Shrews excavate holes using the nose and forefeet and then lift the head with prey seized in the jaws. Where the ground is soft, the nose may be pushed into it and the body propelled forward by the hindfeet.

The bite of some shrews is venomous. The salivary glands of the American short-tailed shrew, for instance, produce enough poison to kill by intravenous injection about 200 mice. The poison acts to kill or paralyze before ingestion, and may be particularly important in helping to subdue large prey like fish and newts which water shrews are known to take.

Some shrew species, and possibly all, show refection. In the European common shrew the animal curls up and begins to lick the anus, sometimes gripping the hindlimbs with the forefeet to maintain position. After a few seconds, abdominal contractions cause the rectum to extrude and the end is then nibbled and licked for some minutes before being withdrawn. It appears that refection does not start until the intestine is free of feces. If the shrew is killed, the stomach and first few centimeters of the intestine can be seen to be filled with a milky fluid containing fat globules and partially digested food. It may be that shrews obtain trace elements and vitamins B and K in this way. Food-caching is known to occur in the European common shrew, the European water shrew and the American short-tailed shrew, and appears to be a means of securing short-term food supplies when competition is fierce or food is temporarily very abundant.

From the species studied, ovulation in shrews appears to be induced by copulation and females are receptive for only a matter of hours during each cycle. The European common shrew does not usually breed until the year after birth and only females have been recorded breeding in their first year. This is more common when the population density is high (up to 35 percent of individuals) than when it is low (around 2 percent of individuals). It is also common generally in the American short-tailed shrew. The breeding season is usually from March to November in northern temperate species, but throughout the year in the Tropics. The number of litters varies between one and 10 a year and the newborn are naked and blind. In the European common shrew some females mate one day after birth and are pregnant and lactating at the same time, suggesting that lactation and gestation are of similar duration (otherwise the second litter might be born before the first is weaned). The mammary glands begin to regress some days before the birth of the second litter. The milk supply is therefore adjusted to demand. The main determinant of weaning appears to be food-shortage. Pregnancy and suckling last for 13–19 days in the European common shrew and up to 21–24 days in the European water shrew. A peculiar feature of parent/offspring relationships in the genera Crocidura and Suncus is so-called "caravanning." When mature enough to leave the nest, the young form a line. Each animal grips the rump of the one in front and the foremost grips that of the mother. The grip is quite tenacious and the whole caravan can be lifted off the ground intact by picking up the mother.

Observations of mating in the European common shrew have shown that the male tends to approach the female warily, usually in a jerky series of approaches and retreats.

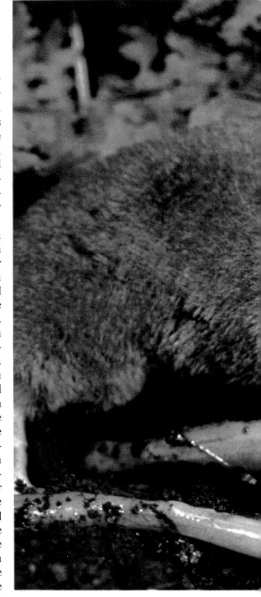

▲ **An efficient killer,** thanks to its venomous salivary glands, this American short-tailed shrew is killing a frog.

▶ **Territorial confrontation** between two American short-tailed shrews. This species, like several of the shrews, digs tunnel systems which form the basis of defended territories.

▼ **Caravanning in shrews.** Young shrews get a guided tour of their terrain by holding onto each other, led by their mother. They continue to hang on even if the mother is picked up.

Non-receptive females are very aggressive and emit high-pitched vocalizations. On the final approach, the male sniffs and licks the female's genitals and sometimes seizes the female and drags her into cover. He may attempt mating before the female goes into the mating posture and the penis is sometimes extruded on unsuccessful mounting attempts. When the female lifts her hindquarters, the male seizes the scruff of her neck and copulates. Penetration normally lasts for about 10 seconds.

Several species dig tunnel systems and these may be the focal point of defended territories. The American short-tailed shrew, the Lesser short-tailed shrew, the European common shrew and the European water shrew have been observed digging. Observations in captivity of the Smoky shrew, the Masked shrew, Trowbridge's shrew and the European pygmy shrew suggest that they do not burrow. In the European water shrew, the tunnel system is important in squeezing water from the fur and it also seems to be important in maintaining fur condition in the European common shrew. In captivity, European common shrews often cache food in their tunnels. Tunnel systems may also be important in avoiding predators and usually have more than one entrance/exit. Nests of grass and other plant material are usually built in a chamber off the tunnel system and shrews spend most of their sleeping and resting time there.

In the European common shrew, intruders onto feeding territories are attacked and driven off, and experiments have shown that the resident has a strong advantage. However, the resident advantage is greater when intruders have lower fighting ability and when food availability is low. When there is little food, the cost of intrusion to the resident is likely to be high and residents fight more vigorously. When two animals are tricked, in an experiment, into thinking they are both owners of a territory, the winner is the one who experiences the lower food availability. Winner and loser can therefore be alternated according to feeding experience. Although all shrew species appear to be solitary except at mating, there is evidence that the European water shrew may sometimes move about in small groups, probably families, and there is at least one account of an apparent mass migration by this species involving many hundreds of animals. The American least shrew is known to be colonial and there are reports of individuals cooperating in the digging of tunnels. CJB

GOLDEN MOLES

Family: Chrysochloridae
Eighteen species in 7 genera.
Distribution: Africa south of the Sahara.

Size: ranges from head-body length 70–85mm (2.7–3.3in) in Grant's desert golden mole to head-body length 198–235mm (7.8–9in) in the Giant golden mole.

Habitat: almost exclusively burrowing in grassveld forest, river-banks, swampy areas, mountains, desert and semidesert.

Gestation and longevity: unknown.

Genus *Chrysospalax*
Two species, including **Giant golden mole** (*C. trevelyani*) R. Distribution: forests in E Cape Province, Transkei, Ciskei. Coat: dark reddish brown, long, coarse and less glossy than other genera.

Genus *Cryptochloris*
Two species, including **De Winton's golden mole** (*C. wintoni*), with pale fawn fur; **Van Zyl's golden mole** (*C. zyli*), with darker brown fur. Distribution: sandy arid regions in SW Cape and Little Namaqualand.

Genus *Chrysochloris*
Three species, including **Stuhlmann's golden mole** (*C. stuhlmanni*). Distribution: mountains in C and E Africa. Coat: dark gray-brown, more golden on belly

Genus *Eremitalpa*
Grant's desert golden mole (*E. granti*). Distribution: sandy desert and semidesert of SW Cape, Little Namaqualand and Namib desert. Coat: silky grayish yellow, seasonally variable length.

Genus *Amblysomus*
Four species, including **Hottentot golden mole** (*A. hottentotus*). Distribution: S and E Cape, Natal, Swaziland, N and E Orange Free State, SE and E Transvaal. Coat: dark brown to rich reddish brown, more golden on belly.

Genus *Chlorotalpa*
Five species, including **Sclater's golden mole** (*C. sclateri*). Distribution: Cameroon to Cape Province.

Genus *Calcochloris*
Yellow golden mole (*C. obtusirostris*). Distribution: Zululand to Mozambique and SE Zimbabwe.

R Rare.

THE iridescent sheen of coppery green, blue, purple or bronze on the fur of most golden moles and the silvery yellow fur of Grant's desert golden mole probably gave the family its common name. Golden moles are known as far back as the Lower Miocene (about 25 million years ago); climatic modifications may be responsible for their present discontinuous distribution, but they have special adaptations for a burrowing mode of life in a wide geographic range of terrestrial habitats.

Golden moles are solitary burrowing insectivores with compact streamlined bodies, short limbs and no visible tail. The backward-set fur is moisture repellent, remaining sleek and dry in muddy situations, and a dense woolly undercoat provides insulation. The skin is thick and tough, particularly on the head. The eyes have been almost lost and are covered with hairy skin; the ear openings are covered by fur and the nostrils are protected by a leathery pad which assists with soil excavations. A muscular head and shoulders push and pack the soil and the strong forelimbs are equipped with digging claws. Of the four claws, the second and third are elongated, and the third claw is extremely powerful. The first and fourth are usually rudimentary, but the sand burrowers *Crysochloris* and *Eremitalpa* have the first claw almost as long as the second, and *Eremitalpa* also has a well developed fourth claw.

The sense of touch in golden moles is mainly used for detecting food, but smell is used during occasional exploratory forages on the surface in search of food, new burrow systems, or females for breeding. The ear ossicles are disproportionately large, giving great sensitivity to vibrations, which trigger rapid locomotion unerringly towards an open burrow entrance when on the surface or a bolt-hole when underground. Golden moles have an ability to orientate and when parts of burrow systems are damaged by flooding or other mechanical means they are joined together again by new tunnels constructed in the same places, linking up with the new burrow entrances.

Golden moles burrow just beneath the surface, forming soil ridges; desert-dwelling species "swim" through the sand just below the surface leaving U-shaped ridges. Stuhlmann's golden mole makes shallow burrows in sphagnum overlying peaty, swampy areas, or in rich humus and among dense roots. Similar damp mossy, peaty areas occur for the Hottentot golden moles in the Drakensberg mountains. Giant golden moles also use swampy areas and

▶ **A golden mole,** foraging on the surface. Golden moles spend more time burrowing in search of food than in any other activity. Food, consisting mainly of earthworms, insect larvae, slugs, snails, crickets and spiders, is eaten in the burrows immediately or carried along tunnels some distance from the site of capture before being eaten, but it is not cached. Desert golden moles seize legless lizards on the surface and drag them into the sand before eating them. Golden moles themselves are prey to snakes, owls and other birds of prey, otters, genets, mongooses and jackals.

▼ **Burrowing in the Hottentot golden mole.** A burrow system may extend to 95cm (37in) in depth and contain 240m (800ft) of burrows. Construction and occupation is influenced mainly by rainfall. Surface foraging for soil invertebrates amounts to 72 percent of total activity in the summer when rainfall is high. With hindfeet braced hard against tunnel walls, the mole loosens the soil with alternate or synchronized forward and downward pick-like strokes of the front claws. Displaced soil is tightly packed against the burrow walls and pushed up with the snout, head and shoulders to form ridges on the surface. To produce mole-hills, loose soil in deep tunnels is kicked behind with the hindfeet, swept into heaps with the snout, pushed along the tunnel and out onto the surface with thrusts of the head and shoulders. The burrow entrance is sealed with a compacted soil plug. Walls of the cylindrical burrows—4–6cm (1.5–2.4in) in diameter—are smoothed and compacted by the mole "waddling" back and forth with back arched against the ceiling, head down, stamping the floor with its hindfeet and snout. Bolt-holes are used by the moles when alarmed or as sleeping sites. While they are small, young golden moles live in a spherical nest of grass or leaves. The nest chamber may be a modified bolt-hole or sleeping site.

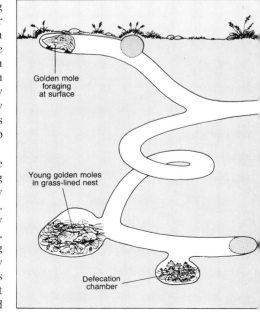

Golden mole foraging at surface

Young golden moles in grass-lined nest

Defecation chamber

scrutiny of tunnel walls by smell. Fighting occurs between individuals of the same sex and sometimes between male and female. Hottentot golden moles tolerate herbivorous mole-rats in the same burrow systems, and in the Drakensberg mountains golden mole burrows open into burrows of the ice-rat.

Courtship in Hottentot golden moles involves much chirruping vocalization, head-bobbing and foot stamping in the male, and grasshopper-like rasping and prolonged squeals with mouth wide open in the female. Both sexes have a single external urogenital opening. There appears to be no distinct breeding season. Testes are abdominal and sexually mature males may have enlarged or regressed testes throughout the year. Whether this is cyclic or what triggers breeding condition is not known. Females have two teats on the abdomen and two in the groin region. Pregnancy, lactation and the birth of 1–3 naked young—with a head-body length 47mm (1.9in); weight 4.5g (0.16oz)—also occurs throughout the year. Eviction from the maternal burrow system occurs when the young moles are 35–45g (1.2–1.6oz) in weight.

Species of *Amblysomus* inhabit areas where soil temperatures range from 0.8°C (33°F) in the Drakensberg mountains to 32°C (90°F) on the Natal coast. They are physiologically adapted to withstand unfavorable environmental conditions of extremes of temperature and scarcity of food. Although thermoregulation is poor because of low body temperature (33.5°C; 92°F), a metabolic rate of 2 percent higher than expected for their size (probably due to their carnivorous diet), and a high rate of heat loss from the body at the thermoneutral range of 23–33°C (73–91°F), *Amblysomus* can vary their body temperature between 27° and 37.5°C (47° and 53°F) in normal situations, or reduce it to that of their surroundings and become torpid, thereby conserving energy. At higher altitudes, *Amblysomus* tend to be heavier and have lower metabolic rates than those nearer the coast.

The versatility of body mass and metabolism, and a thermoneutral range wider than is usual for small mammals, may largely account for the ability of the golden moles to survive in a wide climatic range of habitats, hence their long evolutionary history. But, sadly, the Giant golden mole, which has a more restricted distribution, is now in grave danger of extinction. Its forest range is split into isolated localities, with large parts in the Transkei and Ciskei being destroyed by domestic livestock or converted for economic development.   MAK

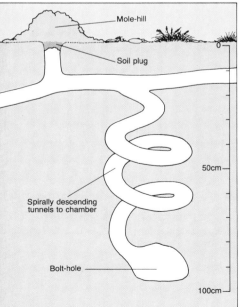

concentrate their burrows round the bases of bushes and trees. All except desert moles also excavate deeper burrows in which soil is deposited on the surface as molehills or in disused tunnels. This usually occurs when the surface becomes hard and dry. Hottentot golden moles may burrow 72m (41ft) on average every 24 hours. Average sustained burrowing lasts about 44 minutes, separated by inactive periods lasting about 2.6 hours. Grant's desert golden mole may leave 4.8km (3mi) of surface tracks in one night.

Food supply influences territorial behavior. Hottentot golden mole burrow systems are more numerous in the summer when food is more abundant, and a certain amount of home range overlap is tolerated. Burrow systems are larger and more aggressively defended in less fertile areas. A neighboring burrow system may be taken over by an individual as an extension of its home range. Occupancy is detected by

MOLES AND DESMANS

Family: Talpidae
Twenty-nine species in 12 genera.
Distribution: Europe, Asia and N America.

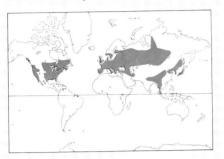

Habitat: Moles largely subterranean, usually under forests and grasslands but also under heaths. Desmans aquatic in lakes and rivers. Shrew moles construct true tunnels but forage in the litter layer.

Size: ranges from head-body length 24–75mm (1–3in), tail length 24–75mm (1–3in), weight under 12g (0.40z) in the shrew moles to head-body length 180–215mm (7–8.5in), tail length 170–215mm (6.5–8.5in), weight about 550g (19.50z) in the Russian desman.

Coat: desman fur is two-layered with short, dense waterproof underfur and oily guard hairs; stiff hairs enlarge the paws and tail for swimming. Moles have short fur of uniform length which will lie in any direction during tunneling. Shrew moles have guard hairs and underfur directed backwards. Moles are usually uniformly brownish black or gray, desmans brown or reddish on the back, merging to gray below.

Gestation: unknown in shrew moles (but greater than 15 days) and desmans; 30 days in European and 42 days in Eastern American mole.

Longevity: largely unknown. Up to 3 years in European and 4–5 years in Hairy-tailed mole.

Species include: **European mole** (*Talpa europaea*), **Mediterranean mole** (*Talpa caeca*), **Pyrenean desman** (*Galemys pyrenaicus*) [v], **Russian desman** (*Desmana moschata*) [v], **Star-nosed mole** (*Condylura cristata*), **Townsend mole** (*Scapanus townsendi*).

[v] Vulnerable.

Iₙ 1702 William of Orange, the King of England and Scotland, was out riding when his horse tripped on a mole-hill, throwing and killing its regal rider. This was to the delight of the Jacobites who henceforth drank a toast: "The little gentleman in black velvet."

Most people in Europe and North America have seen mole-hills, but very few have seen the gentleman in question or his relatives. Their subterranean way of life makes moles very difficult to study and it is only in the last three decades that detailed information on their biology has been gathered. The other major group within the family, the desmans, have a way of life as different as it is possible to imagine—they are aquatic. Again, little is known about many aspects of their biology, due to their relative inaccessibility in the mountain streams of the Pyrenees and in the vastness of Russia.

Moles and desmans have elongated and cylindrical bodies. The muzzle is long, tubular and naked, apart from sensory whiskers. It is highly mobile and extends beyond the lower lip. In the Star-nosed mole the nose is divided at the end into a naked fringe of 22 mobile and fleshy tentacles. The penis is directed to the rear and there is no scrotum.

The eyes are minute but structurally complete. They are hidden within the coat and in some species, including the Mediterranean mole, they are covered by skin. The eyes are sensitive to changes in light level but provide little visual acuity. There are no external ears except in the Asiatic shrew-moles. Both moles and desmans rely to a

Territorial Behavior in the European Mole

European moles spend almost their whole lives underground, in their tunnels, which makes detailed study of their social behavior very difficult. However recent radio-tracking studies have shown that although several neighboring moles each inhabit their own tunnel systems, territories do overlap to a small extent. It is not known, as yet, whether tunnels in the area of overlap are shared or whether they remain separate, running between each other in the soil column. There is much evidence, however, that moles are well aware of the presence and activities of their neighbors. For example, during any particular activity period, neighbors forage in non-adjacent parts of their territories, avoiding contact. It is a testimony to the efficiency of this system of avoidance that conflict between established neighbors has not been recorded.

When a mole dies, or is trapped and removed from its territory, neighboring animals very quickly detect its absence and invade the vacated area, sometimes within two hours. For example, after several weeks of observation one radio-tagged animal was trapped and removed; within 12 hours a neighboring mole was spending his morning activity period foraging in the vacated territory, and his afternoon shift in his own territory. This continued for several days until a third male invaded the area and forced the second back to his former territorial limits. Clearly, within mole populations not all individuals are equal.

In other cases, a vacated territory may be shared among several neighbors. Such was the case with a group of four moles who occupied neighboring territories until one was removed. Within a matter of hours, the others had enlarged their territories to incorporate the vacated area. These enlarged territories were retained for at least several weeks.

Moles probably advertise their presence and tenure of an area by scent marking. Both sexes possess preputial glands which produce a highly odorous secretion that accumulates on the fur of the abdomen and which is deposited on the floor of the tunnels, as well as at latrines. The scent is highly volatile and must be renewed regularly if the mole is to maintain its claim to ownership of the territory. In the absence of scent the territory is quickly invaded.

▲ Species of moles and desmans.
(1) European mole (*Talpa europaea*). (2) American shrew-mole (*Neurotrichus gibbsi*). (3) Lesser Japanese shrew-mole (*Urotrichus pilirostris*). (4) Star-nosed mole (*Condylura cristata*). (5) Pyrenean desman (*Galemys pyrenaicus*).

great extent on touch. The muzzle is richly endowed with projections called Eimer's organs, which are probably sensitive to touch. Various parts of the body, including the muzzle and tail, and the legs of desmans, are supplied with sensory whiskers.

Desmans are adapted for swimming. The tail is long, flattened and broadened by a fringe of stiff hairs. The legs and feet are proportionally long and powerful. The toes are webbed, the fingers half-webbed, and both are fringed with stiff hairs. The nostrils and ears are opened and closed by valves. The fur is waterproof. When swimming, the hindlegs provide the main propulsive force.

In moles, the forelimbs are adapted for digging. The hands are turned permanently outwards, like a pair of oars. They are large, almost circular and equipped with five large and strong claws. The teeth are un-

specialized and typical of the Insectivora.

The moles and desmans probably originated in Europe and spread from there to North America and more recently to Asia. Moles are found in North America, Europe and Asia, shrew-moles in North America and Asia and desmans in Europe only. The distribution map is accurate for North America and Europe but grossly incomplete for Asia, where it simply reflects collecting effort. Some species were formerly much more widely distributed, for example fossils of the Russian desman are to be found throughout Europe and in Britain.

Moles dig permanent tunnels and obtain most of their food from soil animals which fall into them. When digging new deep tunnels they brace themselves with their hindfeet and then dig with the forefeet which are alternately thrust into the soil and moved sideways and backwards. Periodically, they dig a vertical shaft to the surface and push up the soil to make the familiar mole-hill. These tunnels range in depth from a few centimeters to 100cm (40in). Probably 90 percent of the diet is foraged from the permanent tunnels, the rest from casual digging. The diet of moles consists largely of earthworms, beetle and fly larvae and, when available, slugs. The European mole, which consumes 40–50g (1.4–1.8oz) of earthworms per day, stores earthworms with their heads bitten off, near to its nest in October and November.

Desmans obtain nearly all their food from water, particularly aquatic insects such as stone-fly and caddis-fly larvae, fresh water gammarid shrimps and snails. The Russian desman also takes larger prey such as fish and amphibians.

Desmans are largely nocturnal but they often have a short active period during the day. In contrast, most of the moles are active both day and night. Until recently it was thought that European moles always had three active periods per day, alternating with periods of rest in the nest. Recent studies with moles fitted with radio-transmitters reveal a rather more complex picture. In the winter, both males and females show three activity periods, each of about four hours, separated by a rest of about four hours in the nest. At this time they almost always leave the nest at sunrise. Females maintain this pattern for the rest of the year except for a period in summer when they are lactating. Then they return to the nest much more frequently in order to feed their young.

Males are more complex. In spring they start to seek out receptive females and remain away from the nest for days at a time, snatching cat-naps in their tunnels. In the summer they return to their winter routine but in September they display just two activity periods per day.

Details of the life cycle are known only for a few species, in particular the European mole. Little is known about the desmans and shrew-moles. In general, moles have a short breeding season, and produce a single litter of 2–7 young each year. Lactation lasts for about a month. The males take no part in the care of the young, and the young normally breed in the year following their birth.

In Britain, European moles mate in March to May and the young are born in May or June. The time of breeding varies with the latitude: the same species is pregnant in mid-February in northern Italy, in March in southern England and not till May or June in northeast Scotland. This suggests that length of daylight controls breeding, which may seem strange for a subterranean animal. However, moles do come to the surface, for example to collect grass and

other materials for their nests (one nest, near to a licensed hotel, was made entirely of potato chip bags!).

The average litter size in Britain is 3.7 but it can be higher in continental Europe—an average of 5.7 has been reported for Russia. One litter per year is the norm, but there are records of pregnant animals in September in England and in October in Germany. Sperm production lasts for only two months but sufficient spermatozoa may be stored to allow the insemination of females coming into this late or second period of heat.

The young are born in the nest. They are naked at birth, have fur at 14 days and open their eyes at 22 days. Lactation lasts for 4–5 weeks and the young leave the nest weighing 60g (2oz) or more some 35 days after birth.

After leaving the nest, the young leave their mother's territory and move overground to seek an unoccupied area. At this time many of them are killed by predators and by cars. With an average litter size of about four the numbers of animals present in May must be reduced by 66 percent by death or emigration if the population is to remain stable.

Little is known about the population density and social organization of desmans beyond the following snippets. Pyrenean desmans appear to be solitary and to inhabit small permanent home ranges which they

▲ **Breaking cover,** a European mole emerges from the middle of a mole-hill. The relatively huge outward facing forelimbs are well to the fore here.

◀ ▼ **The Star-nosed mole** has a unique nose, divided into many fleshy tentacles. LEFT Eating a worm caught in a pool of water. BELOW a close-up of the coral-like tentacles.

scent mark with latrines. There is some evidence that Russian desmans are at times nomadic as a result of unpredictable water levels. They may be social, since as many as eight adults have been found in one burrow.

There is much more information on the moles, although most comes from the European mole. Populations of European moles have a sex ratio of 1:1. Unlike those of small rodents, populations appear to be relatively stable.

Most mole species are solitary and territorial, with individuals defending all or the greater part of their home range. The Star-nosed mole is exceptional in that during the winter a male and female may live together. In other species, males and females meet only briefly, to mate.

Population densities vary from species to species and from habitat to habitat. Average numbers per hectare are 5–25 for the European mole (2–10 per acre), 5–12 for the Eastern American mole (2–5 per acre), 5–25 for the Star-nosed mole (2–10 per acre) and 5–28 for the Hairy-tailed mole (2–11 per acre).

Radio-tracking studies have shown that territory sizes in the European mole vary between the sexes, from habitat to habitat, and from season to season. The habitat and sex differences reflect differences in food supply and the greater energy demands of the larger male.

Female territory sizes remain similar throughout the year but in the spring males increase theirs dramatically as they seek out receptive partners.

Normally, moles remain within their territories for their whole lives. However radio-tracking has revealed that during hot, dry weather some moles leave their range and travel as far as 1km (0.6mi) to drink at streams. This necessitates crossing the territories of up to 10 other moles!

The Russian desman is widely trapped for its lustrous fur, and most of the skins in Western museums have come from the fur trade. To increase production the species was introduced to the river Dnepr.

All moles, particularly those in North America and Europe, are regarded as pests by farmers, gardeners and, above all, by golf-course green-keepers! Moles disturb the roots of young plants, causing them to wilt and die, soil from mole hills contaminates silage, and the stones they bring to the surface cause damage to cutting machinery. On the positive side, they have on occasion brought worked flints and Roman tesserae to the surface.

Nowadays, moles are usually controlled by poisoning. In Britain the most widely used is strychnine, but it has the disadvantage of being unacceptably cruel and endangers other animals besides moles: although available only by government permit, a disturbing amount is diverted to killing other forms of "vermin," including birds of prey.

In the past, moles were trapped on a massive scale in Britain, by professional trappers and agricultural workers who sold the skins to be made into breeches, waistcoats and ladies' coats. After World War I the market collapsed due to changes in fashion and imports from Russia. There was a similar trade in mole-skins in North America, particularly from the Townsend mole.

The two extant species of desmans are both endangered species: the Russian desman because of overhunting for its fur, the Pyrenean desman from over-zealous scientific collecting and from habitat destruction, particularly the damming of mountain streams. The formulation of management plans is severely hampered by a lack of knowledge of both species' basic biology.

MLG

EDENTATES

ORDERS: EDENTATA AND PHOLIDOTA
Five families: 14 genera: 36 species.

Anteaters
Family: Myrmecophagidae
Order: Edentata.
Four species in 3 genera.

Sloths
Families: Megalonychidae and Bradypodidae
Order: Edentata.
Five species in 2 genera.

Armadillos
Family: Dasypodidae
Order: Edentata.
Twenty species in 8 genera.

Pangolins
Family: Manidae
Order: Pholidota.
Seven species of the genus *Manis*.

"EDENTATE" means "without teeth," but only the anteaters are strictly toothless: sloths and armadillos have very simple, rootless molars which grow throughout life. In the 18th century, pangolins were also included in the Edentata on the grounds that they lack teeth, but they are now considered to belong to a separate order: the Pholidota. The diet of edentates ranges from an almost exclusive reliance on ants and termites in the anteaters, through a wide range of insects, fungi, tubers and carrion in the armadillos, to plants in the sloths.

By the early Tertiary, 60 million years ago (the beginning of the "Age of Mammals"), the ancestral edentates had already diverged into two quite distinct lines. The first, comprising small, armorless animals of the suborder Palaeanodonta, rapidly became extinct. But the other, the suborder Xenarthra, was on the brink of a spectacular radiation that was later to produce some of the most distinctive and bizarre of all the New World mammals. The four families, 13 genera and 29 species of living edentates are descended from the early xenarthrans and, although highly specialized, they retain many common ancestral features.

The living and recently extinct edentates are distinguished from all other mammals (including the pangolins) by additional articulations between the lumbar vertebrae, which are termed xenarthrales. These bony elements provide lumbar reinforcement for digging, and are especially important for the armadillos. The living edentates also differ from most mammals in having rather simple skulls, with no canines, incisors or premolars, and in having a double posterior vena cava vein (single in other mammals), which returns blood to the heart from the hindquarters of the body. Females have a primitive divided womb only a step removed from the double womb of marsupials, and a common urinary and genital duct, while males have internal testes, and a small penis with no glans.

Despite these unifying characteristics, the extinct edentates differed greatly in size and appearance from their modern relatives and, in terms of numbers of genera, were ten times as diverse. The rise and fall of these early forms is closely linked to the fact that throughout the Tertiary, from 65 million years ago, South America was a huge, isolated island. At the beginning of this epoch, ancestral edentates shared the continent only with early marsupials and other primitive mammals, and flourished in the virtual absence of competition. By the early Oligocene (38 million years ago), three families of giant ground sloths had emerged, with some species growing to the size of modern elephants. In their heyday during the late Miocene, 30 million years later, ground sloths appeared in the West Indies and southern North America, apparently having rafted across the sea barriers as waif immigrants. Four families of armored, armadillo-like edentates were contemporary with the ground sloths for much of the Oligocene. The largest species, *Glyptodon*, achieved a length of 5m (16.5ft) and carried a rigid 3m (10ft) shell on its back, while the related *Doedicurus* had a massive tail with the tip armored like a medieval mace. Although *Glyptodon* and the giant ground sloths survived until historical times—and are spoken of in the legends of the Tehuelche and Araucan Indians of Patagonia—only the smaller tree sloths, anteaters and armadillos persisted to the present day.

The extinct edentates are believed to have been ponderous, unspecialized herbivores that inhabited scrubby savannas. They were probably outcompeted and preyed upon by the new and sophisticated northern invaders. In contrast, the success of the living edentates was due to their occupation of relatively narrow niches, which allowed little space for the less specialized newcomers. The anteaters and leaf-eating sloths, for example, have very specialized diets. To cope with the low energy contents

▶ **Prehistoric edentates.** The edentates produced three major groups: "shelled" forms (Loricata), including the extinct glyptodons and living armadillos; "hairy" forms (Pilosa), including the extinct Giant ground sloth and living tree sloths; and the anteaters (Vermilingua). (1) The Giant ground sloth (*Megatherium*), from the Pleistocene of South America, was up to 6m (20ft) long. (2) *Eomanis waldi*, a small armored pangolin from the Eocene of Germany. (3) *Glyptodon panochthus*, a giant shelled form from the Pleistocene of South America. (4) Giant anteater (*Scelidotherium*), from the Pleistocene of South America.

2

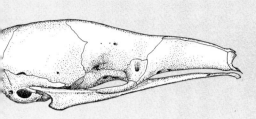

Southern tamandua   12cm

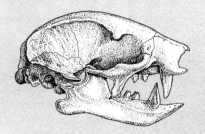

Linné's two-toed sloth   12cm

Skulls of edentates

Although Edentata means "without teeth", only the anteaters, such as the Southern tamandua, are completely toothless. Both sloths and armadillos are equipped with a series of uniform peg-shaped cheek or grinding teeth (premolars and molars). These lack an enamel covering and have a single so-called "open root" which allows continuous growth. True incisor and canine teeth are absent in all edentates, but sloths have enlarged canine-like premolars.

of their foods, both groups evolved metabolic rates that are only 33–60 percent of those expected for their body-weights, and variable but low — 32.7–35°C (91–95°F)— body temperatures that burn fewer calories. Armadillos eat a wide range of foods, but are specialized for a partly subterranean way of life; they also have low body temperatures— 33–35.5°C (91.5–96°F) — and rates of metabolism (29–57 percent of expected) to avoid overheating in their closed burrows. Lacking similarly sluggish metabolisms, the invading mammals were not able fully to exploit these habitats, and competition with the edentates was probably minimal.

As consequences of specializing and slowing their metabolisms, the sloths and anteaters use energy frugally, and generally move slowly over small home ranges. Females attain sexual maturity at 2–3 years of age and breed only once a year thereafter. They produce small precocious litters (usually one young), and invest much time and energy in weaning and post-weaning care. Defense against predators is passive, and primarily dependent on cryptic camouflage. While anteaters, and occasionally sloths,

may try to flee from an assailant, they more often stand their ground and strike out with their claws. Sloths are reputedly able to survive the most severe injuries: bite wounds and deep scars rarely become infected, and heal completely within weeks. Armadillos show similar trends towards economizing their use of energy, but these are not as marked as in their ant- and plant-eating relatives. The armadillos are less constrained because of their more varied and energy-rich diets, and the ability (of some species) to store fat and enter torpor.

The social lives of edentates are probably dominated by the sense of smell. All species produce odoriferous secretions from anal glands, which are used to mark paths, trees or conspicuous objects; these probably advertise the presence, status and possibly the sexual condition of the marking individual. Scent marks may also serve as territorial markers, and allow individuals priority of access to scarce resources, such as food.

With their lack of teeth, long, sticky tongues, and taste for ants, the pangolins exploit a niche equivalent to that occupied by the South American anteaters. These similarities suggest that selection for the ant-eating habit has acted in parallel in both the Old and New World.   CRD

ANTEATERS

Family: Myrmecophagidae
Four species in 3 genera.
Order: Edentata.
Distribution: southern Mexico, C America
south to Paraguay and northern Argentina;
Trinidad.
Habitat: savanna, parkland, thorn scrub, steppe;
deciduous, montane and tropical rain forest.

Size: from head-body length 16–22cm (6–9in),
tail length 16–25cm (6–10in) and weight
300–500g (10.6–17.6oz) in the Silky anteater
to head-body length 100–120cm (39–47in),
tail length 70–90cm (27.5–35.5in) and weight
20–39kg (44–86lb) in the Giant anteater. Male
anteaters are 10–20 percent heavier than
females.

Coat: coarse, stiff, dense; gray with black and
white shoulder stripe (Giant anteater) or light
fawn-dark brown with variable patches of black
or reddish-brown from shoulders to rump
(tamanduas); soft, silky gray-yellowish-orange
with darker mid-dorsal stripe (Silky anteater).

Gestation: 130–150 days (tamanduas); 190
days (Giant anteater).

Longevity: Unknown in the wild; 26 years
(captive Giant anteater); at least 9½ years
(captive Northern tamandua).

Species: **Giant anteater** (*Myrmecophaga
tridactyla*) v ✴, **Northern tamandua**
(*Tamandua mexicana*), **Southern tamandua**
(*Tamandua tetradactyla*) ✴, **Silky anteater**
(*Cyclopes didactylus*).

v Vulnerable. ✴ CITES listed.

► **Piggy-back anteaters.** ABOVE The young of
the Giant anteater may ride on its mother's
back for up to a year, although weaning takes
place at six months. The coloration of the
young is so similar to that of its mother that the
two animals appear to merge.

► **A Giant anteater** investigating a termite
mound.

WITH its long, elongated snout, diagonal
black and white shoulder stripe and
long, bushy tail, the Giant anteater differs
strikingly in appearance from all other New
World mammals. The three smaller species
of anteaters have prehensile tails, naked on
the undersides, with shorter snouts than
their larger relative, but all lack teeth and
specialize in eating ants and termites.

The anteaters have characteristically
long, narrow tongues with minute,
backward-pointing spines, which are
covered in sticky saliva when the animals
are feeding. All have five fingers and four or
five toes, although the fifth finger of the
Giant anteater and the first, fourth and fifth
fingers of the Silky, or Two-toed anteater are
much reduced and not visible externally.
The three middle fingers of tamanduas and
the second and third fingers of the Giant and
Silky anteaters bear sharp, powerful claws,
4–10cm (1.5–4in) long, which are used in
defense and for opening the nests of ants and
termites. To protect these claws, anteaters
walk on their knuckles and the sides of their
hands with the claws tucked inwards; this
produces an awkward limping gait.

The Giant anteater is distributed widely in
Central America and South America east of
the Andes, extending as far south as
Uruguay and northwestern Argentina. This
species specializes in eating large ground-
dwelling ants, such as carpenter ants in the
Venezuelan *llanos*, and usually ignores termites, leaf-eating ants, army ants and other
species with large jaws. Prey colonies are
located by scent, opened but not demolished
by the powerful claws and cropped on
average for about a minute; as few as 140
ants (0.5–1 percent of the daily requirement) may be taken from a single colony on
any day. Since a Giant anteater travels at
some 14m (46ft) a minute it can revisit
many prey colonies within its home range at
regular intervals, and so avoid overexploit-
ation of its food. Detection by smell of its
preferred prey is probably excellent: experiments have shown that captive animals can
discriminate an odor for which they have
been positively conditioned, even when it
exists at four thousandths of the concentration of a neutral odor. The tongue of the
Giant anteater can be pushed out a remarkable 60cm (24in) up to 150 times a
minute to obtain its prey, and the sheath
containing the tongue and tongue-
retracting muscles is anchored on the
breast-bone. The salivary glands secrete
enormous quantities of viscous saliva onto
the tongue, and trapped ants are drawn
back and masticated first by horny papillae

on the roof of the mouth and sides of the
cheeks, then by the muscular stomach. This
species evidently obtains most of its water
from ants, although the tongue may be used
on occasion to lap up free water, and juicy
fruits and larvae are sometimes picked up by
the lips.

The abundance of the Giant anteater
varies with habitat and perhaps corresponds
with the distribution of its preferred prey. In
tropical forest on Barro Colorado Island,
Panama, and in the undisturbed grasslands
of the southeastern highlands of Brazil,
populations may attain densities of 1–2
animals per sq km (5–10 per sq mi); in the
mixed deciduous forests and semi-arid *llanos*
of Venezuela, populations may be ten times
as sparse, with individuals occupying correspondingly large home ranges of 2,500ha
(6,200 acres).

In the southern part of its geographical
range, the Giant anteater is believed to breed

in the fall (March–May), but in captivity individuals also mate in spring (August–October). Courtship has not been described in this species; however, since adults are normally solitary, breeding individuals probably come together just before copulation and part shortly afterwards. The female gives birth standing upright, using her tail as the third leg of a tripod, and licks the precocious youngster after it has crawled up through the fur onto her back. Females usually bear only one young and suckle it for some six months.

Although they are capable of a slow, galloping movement at the age of just one month, the young emit short, shrill whistles if left on their own, and they do not feed independently of the mother until fully grown at the age of two years. Adults are normally silent, but bellow if provoked. Both sexes respond to the smell of their own saliva and can produce secretions from an anal gland; these odors are perhaps used in communication.

Giant anteaters have relatively slow rates of metabolism and one of the lowest recorded body temperatures—32.7°C (91°F)—of any terrestrial mammal. Although they do not burrow, they dig out shallow depressions with their claws and lie asleep in these for 14–15 hours a day; they are covered by their great fanlike tail and warned of predators, such as pumas and jaguars, by their keen hearing. Normally docile, threatened anteaters swiftly rear up on their hindlegs and slash at an adversary with their curved claws. An embrace in the immensely strong forearms of an anteater is as fearsome as its claws, and dogs, large cats, and even humans have succumbed.

The two species of tamanduas, or collared anteaters, are rather less than half the size of the Giant anteater, with distributions that generally overlap that of the larger species.

Both tamanduas are clothed in dense, bristly hair and, in marked contrast to the Giant anteater, have naked prehensile tails which assist in climbing. The ears of tamanduas are relatively large and point out from the head, suggesting a keen sense of hearing, but as in their large relative the eyes are small and their vision is probably poor. In both species the mouth opening is only the diameter of a pencil—half that of the Giant anteater—but the rounded tongue can be extended some 40cm (16in).

Although the background color of the tamanduas is fawn or brown, the northern species always has a black area that runs along the back, developing into a collar in front of the shoulders and a second, wider, band around the middle of the body. This "vest" is also present in the Southern tamandua, but only in individuals in the southern part of the range, remote from the northern species, southeast of the Amazon basin. The vest is absent in Southern tamanduas inhabiting northern Brazil and Venezuela across to a line west of the Andes; here the geographical distributions of the two species abut and the differences in their coat colors are most striking. Uniform gold, brown, black and partially vested Southern tamanduas occur throughout the Amazon basin. This confusing variation has contributed to this species being split into some 13 subspecies, in contrast to the more constant Northern tamandua with five subspecies and the Giant anteater with only three.

Tamanduas inhabit savanna, thorn scrub and a wide range of wet and dry forest habitats, where their mottled coloration provides effective camouflage. Northern tamanduas spend about half their time in trees in the tropical forest of Barro Colorado Island, whereas the Southern tamandua, in various habitats in Venezuela, has been estimated to spend 13–64 percent of its time aloft. Both species are nocturnal and may use tree hollows as sleeping quarters during the daylight hours. The tamanduas move in a clumsy, stiff-legged fashion on the ground, and are unable to gallop like their giant relative.

Because of their small size and arboreal habits, Northern tamanduas on Barro Colorado Island occupy home ranges of only 50–140ha (125–345 acres), less than one-sixth the size of those used by the terrestrial Giant anteater. However, in the more open *llanos* of Venezuela, Southern tamanduas may occupy larger ranges of 350–400ha (845–990 acres), with densities of perhaps three animals per sq km (7.5 per sq mi).

Tamanduas specialize in eating termites and ants and detect them by scent; however, they are repelled by leaf-eating ants, army ants and other species capable of marshalling chemical defenses. The Northern tamandua can detect and avoid soldiers of the Niggerhead termite, but will readily feed on the vulnerable workers and reproductive castes of the same species. The tamanduas feed to a larger extent on termites, especially arboreal termites, than the Giant anteater, although the methods of feeding and processing of prey are similar in all three species. Both tamanduas are also reputed to eat bees and their honey, and will take fruit and even meat in captivity.

Tamanduas probably mate in the fall and give birth to a single young in spring. Unlike the Giant anteaters, however, the offspring do not at first resemble the parents and vary in color from almost white to black. The mother carries the young on her back for an indeterminate time, setting it down only occasionally on the safety of a branch while she feeds. Tamanduas are normally solitary, but may communicate by hissing and by the release of a powerful and—to the human nose—unpleasant odor, evidently produced by the anal gland. This habit has earned them the nickname of "stinker of the forest."

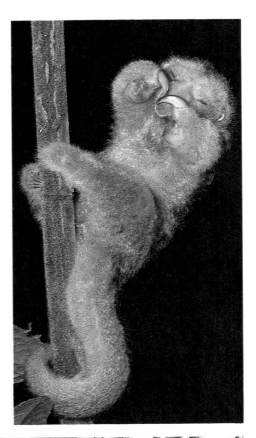

▲ **The gold, brown and white patterning** of the Southern tamandua is effective camouflage in its scrub and forest habitats.

◄ **Like a soft toy** with nylon fur, ABOVE a Silky anteater clings to a tree in a defensive posture, claws in front of its face.

◄ **Arboreal anteater.** The Northern tamandua is nocturnal and spends much of the daytime asleep in trees.

Although tamanduas defend themselves in much the same way as the Giant anteater, they are more likely to flee or hiss at an assailant than their larger relative. The attitude of the vested tamandua hugging its adversary has led to the further nickname "Dominus vobiscum" because of its supposed likeness to a priest at the altar.

Because of its nocturnal and strongly arboreal habits, the squirrel-sized Silky anteater is seen more rarely than its three larger relatives. Yet, it is distributed widely as seven subspecies from southern Mexico through Central America to northern Peru on the western slopes of the Andes, and east of the cordillera throughout most of the Amazon basin. In the northern part of this range, the short silky fur is a uniform golden yellow, but it becomes progressively grayer, with a darker mid-dorsal stripe, in the south. Although the Silky anteater is restricted to tropical forest, it is said to favor certain kinds of trees such as the Silk cotton tree; this species produces seeds embedded in massive balls of soft, silverish fibers which resemble, and camouflage, the anteater from owls, Harpy eagles and other predators. The Silky anteater has small ears, but a larger mouth and eyes than its relatives. It uses a peculiar joint in the sole of its hindfoot, which allows the claws to be doubled back under the foot, and its strongly prehensile tail as aids in grasping branches.

The Silky anteater rarely eats termites, but specializes instead on ants which live in the stems of lianas and on tree branches. In captivity, it will also take beetles and fruits. In keeping with its low body temperature—33°C (91.4°F)—and relatively low metabolic rate, the Silky anteater moves slowly and with deliberation; however, nightly bouts of activity last about four hours, and are punctuated by rests of only a few minutes. During the day, the animal sleeps curled up on a branch with its tail wrapped around its feet; usually no two days are spent in the same tree.

Nothing is known of mating and courtship in the Silky anteater but, some time after birth, the single young is fed on semi-digested insects that are regurgitated from the stomachs of the mother and father. Both parents carry the young for an indeterminate time although, as in the tamanduas, the young may occasionally be left alone in a leaf nest in a tree hollow while the mother feeds. Adults utter soft whistling sounds and have a cheek or facial gland, but it is not known if this has any function in communication.

Although some small use is made of tamandua skin in local leather industries, anteaters are of little commercial value and they are seldom used for food. Nevertheless, the spectacular Giant anteater is sought by trophy hunters and live animal dealers, and has been exterminated in many areas of Peru and Brazil. Large numbers of tamanduas die on roads in settled areas or are killed by locals and their dogs for "sport," while Silky anteaters are collected, at least in Peru, for the live animal trade. Unfortunately, because all anteaters have a very stable and predictable diet, the widespread and rampant destruction of habitat and associated prey species in many parts of South America may present a still greater threat to their survival than overhunting.

CRD

SLOTHS

Families: Megalonychidae (two-toed sloths), and **Bradypodidae** (three-toed sloths)
Five species in 2 genera.
Order: Edentata.
Distribution: Nicaragua through Colombia, Venezuela and the Guianas to north-central Brazil and nothern Peru (two-toed sloths); Honduras through Colombia, Venezuela and the Guianas to Bolivia, Paraguay and northern Argentina; on the west to coastal Ecuador (three-toed sloths).

Habitat: lowland and upland tropical forest; montane forest to 2,100m (7,000ft) (Hoffmann's two-toed sloth only).

Size: from head-body length 56–60cm (18–24in), tail length 6–7cm (2.4–2.8in) and weight 3.5–4.5kg (7.7–9.4lb) in the three-toed sloths to head-body length 58–70cm (23–28in), tail absent and weight 4–8kg (8.8–17.6lb) in the two-toed sloths.

Coat: Stiff, coarse, grayish-brown to beige, with a greenish cast provided by growth of blue-green algae on hairs. Two-toed sloths with darker face and hair to 15cm (6in); three-toed sloths with darker face and hair to 6cm (2.4in), light fur on shoulders; Maned sloth with dark hair on head and neck.

Gestation: 6 months (Linné's two-toed sloth, three-toed sloths); 11.5 months (Hoffmann's two-toed sloth).

Longevity: 12 years (at least 31 in captivity).

Two-toed sloths
Two species of genus *Choloepus*: **Hoffmann's two-toed sloth** (*Choloepus hoffmanni*) [*], **Linné's two-toed sloth** (*Choloepus didactylus*).

Three-toed sloths
Three species of genus *Bradypus*: **Brown-throated three-toed sloth** (*Bradypus variegatus*) [*], **Pale-throated three-toed sloth** (*Bradypus tridactylus*), **Maned sloth** (*Bradypus torquatus*) [E].

[E] Endangered. [*] CITES listed.

ALTHOUGH sloths are renowned for their almost glacial slowness of movement, they are the most spectacularly successful large mammals in Central and tropical South America. On Barro Colorado Island, Panama, two species—the Brown-throated three-toed sloth and Hoffmann's two-toed sloth—account for two-thirds of the biomass and half of the energy consumption of all terrestrial mammals, while in Surinam they comprise at least a quarter of the total mammalian biomass. Success has come from specializing in an arboreal, leaf-eating way of life to such a remarkable extent that the effects of competitors and predators are scarcely perceptible.

The sloths have rounded heads and flat-tened faces, with small ears hidden in the fur; they are distinguished from other tree-dwelling mammals by their simple teeth (five upper molars, four lower), and their highly modified hands and feet which terminate in long—8–10cm (3–4in)—curved claws. Sloths have short, fine underfur and an overcoat of longer and coarser hairs which, in moist conditions, is suffused with green. This color derives from the presence of two species of blue-green algae that grow in longitudinal grooves in the hairs, and help to camouflage animals in the tree canopy. All species have extremely large many-compartmental stomachs, which contain cellulose-digesting bacteria. A full stomach may account for almost a third of the body-weight of a sloth, and meals may be digested there for more than a month before passing completely into the relatively short intestine. Feces and urine are passed only once a week, at habitual sites at the bases of trees.

The sloths are grouped into two distinct genera and families, which can be distinguished most easily by the numbers of fingers: those of genus *Choloepus* have two fingers and those of genus *Bradypus* have three. Unfortunately, despite the fact that both genera have three toes, the two-fingered

▲ **Gathering moss,** or rather algae, a Brown-throated three-toed sloth clings to a tree in the rain forest of Panama. The algal growth on the hair seems appropriate in view of the extreme sluggishness of the animals. In fact the algae serves as camouflage.

◄ **Tree hanger.** ABOVE A Brown-throated three-toed sloth sunbathing in Panamanian rain forest.

◄ **Wet-look sloth.** Hoffman's two-toed sloth in the rain forest of Central America.

▷ **Cradled by its mother,** OVERLEAF a young Brown-throated three-toed sloth peers through the foliage. Young are carried for up to nine months and feed from this position.

forms are known as two-toed and the three-fingered forms as three-toed sloths.

While representatives of both families occur together in tropical forests throughout much of Central and South America, sloths within the same genus occupy more or less exclusive geographical ranges. These closely related species differ little (10 percent) in body-weight and have such similar habits that they are apparently unable to coexist.

Where two-toed and three-toed sloths occur together, the two-toed form is 25 percent heavier than its relative and it uses the forest in different ways. In lowland tropical forest on Barro Colorado Island, the Brown-throated three-toed sloth achieves a density of 8.5 animals per ha (3.5 per acre),

over three times that of the larger Hoffmann's two-toed sloth. The smaller species is sporadically active for over 10 hours out of 24, compared with 7.6 hours for the two-toed sloth and, unlike its nocturnal relative it is active both day and night. Three-toed sloths maintain overlapping home ranges of 6.6ha (16.3 acres), three times those of the larger species. Despite their apparent alacrity, however, only 11 percent of three-toed sloths travel further than 38m (125ft) in a day, and some 40 percent remain in the same tree on two consecutive nights; the three-toed sloths, by contrast, changes trees four times as often.

Both two-toed and three-toed sloths maintain low but variable body temperatures—30–34°C (86–93°F)—which fall during the cooler hours of the night, during wet weather and when the animals are inactive. Such labile body temperatures help to conserve energy: sloths have metabolic rates that are only 40–45 percent of those expected for their body-weights as well as reduced muscles (about half the relative weight for most terrestrial mammals), and so cannot afford to keep warm by shivering. Both species frequent trees with exposed crowns and regulate their body temperatures by moving in and out of the sun.

Sloths are believed to breed throughout the year, but in Guyana births of the Pale-throated three-toed sloth occur only after the rainy season, between July and September. The single young, weighing 300–400g (10.5–14oz), is born above ground and is helped to a teat by the mother. The young of all species cease nursing at about a month, but may begin to take leaves even earlier. They are carried by the mother alone for six to nine months and feed on leaves they can reach from this position; they utter bleats or pure-toned whistles if separated. After weaning, the young inherit a portion of the home range left vacant by the mother, as well as her taste for leaves. A consequence of "inheriting" preferences for different tree species is that several sloths can occupy a similar home range without competing for food or space; this will tend to maximize their numbers at the expense of howler monkeys and other leaf-eating rivals in the forest canopy. Two-toed sloths may not reach sexual maturity until the age of three years (females) or 4–5 years (males).

Adult sloths are usually solitary, and patterns of communication are poorly known. However, males are thought to advertise their presence by wiping secretions from an anal gland onto branches, and the pungent-smelling dung middens conceivably act as trysting places. Three-toed sloths produce shrill "ai-ai" whistles through the nostrils, while two-toed sloths hiss if disturbed.

Oviedo y Valdés, one of the first Spanish chroniclers of the Central American region in the 16th century, wrote that he had never seen an uglier or more useless creature than the sloth. Fortunately, little commercial value has since been attached to these animals, although large numbers, especially of two-toed sloths, are hunted locally for their meat in many parts of South America. The Maned sloth of southeastern Brazil is considered rare due to the destruction of its coastal rain-forest habitat, and the fortunes of all five species depend on the future of the tropical forests.   CRD

ARMADILLOS

Family: Dasypodidae
Twenty species in 8 genera.
Order: Edentata.
Distribution: Florida, Georgia and South
Carolina west to Kansas; eastern Mexico, C and
S America to Strait of Magellan; Trinidad and
Tobago, Grenada, Margarita.
Habitat: savanna, pampas, arid desert, thorn
scrub; deciduous, cloud and rain forest.

Size: from head-body length 12.5–15cm
(5–6in), tail length 2.5–3cm 1–1.2in) and
weight 80–100g (2.8–3.5oz) in the Lesser fairy
armadillo, to head-body length 75–100cm
(30–39in), tail length 45–50cm (18–20in) and
weight 45–60kg (99–132lb) in the Giant armadillo.

Coat: broad shield of pale pink or yellowish
dark brown armor (scute plates) over shoulders
and pelvis, with varying numbers of flexible
half rings over middle of back; some species
with white-dark brown hairs between the scute
plates.
Gestation: 60–65 days (hairy and Yellow
armadillos) to 120 days (prolonged by delayed
implantation) (Common long-nosed armadillo).
Longevity: 12–15 years (to 19 in captivity).

Long-nosed armadillos
Six species of genus *Dasypus*, including
Common long-nosed armadillo
(*D. novemcinctus*), **Southern lesser long-nosed
armadillo** (*D. hybridus*) **Brazilian lesser long-
nosed armadillo** (*D. septemcinctus*).

Naked-tailed armadillos
Four species of genus *Cabassous*, including
Southern naked-tailed armadillo
(*C. unicinctus*).

Hairy armadillos
Three species of genus *Chaetophractus*, including
Andean hairy armadillo (*C. nationi*),
Screaming hairy armadillo (*C. vellerosus*),
Larger hairy armadillo (*C. villosus*).

Three-banded armadillos
Two species of genus *Tolypeutes*, including
Southern three-banded armadillo (*T. matacus*).

Fairy armadillos
Two species of genus *Chlamyphorus*, including
Greater fairy armadillo (*C. retusus*).
Other species: **Yellow armadillo** (*Euphractus
sexcinctus*), **pichi** (*Zaedyus pichiy*), **Giant
armadillo** (*Priodontes maximus*) v *.

v Vulnerable. * CITES listed.

CLAD in a suit of hard and bony body
plates, the Giant armadillo is the most
heavily armored of modern mammals.
Though itself a dwarf when compared with
some of its enormous ancestors—whose 3m
(10ft) long body shells were used as roofs
or tombs by the early South American
Indians—it is the most conspicuous of the
twenty species in the armadillo family,
which includes the frequently encountered
Common long-nosed armadillo as well as
the rare Andean hairy species.

Armadillo armor develops from the skin
and is composed of strong bony plates, or
scutes, overlaid by horn. There are usually
broad and rigid shields over the shoulders
and hips, with varying numbers of bands
over the middle of the back which are
connected to the flexible underlying skin.
The tail, top of the head and outer surfaces of
the limbs are also armored, but the undersurface is covered only with soft, hairy skin.
To protect this vulnerable area, most species
are able to withdraw the limbs under the hip
and shoulder shields and sit tightly on the
ground, while some, such as the three-
banded armadillos, can roll up into a ball.
Armadillos have rather flattened heads with
a long extendable tongue, and small but
upright ears. Most species have 14–18 teeth
in each jaw, but the Giant armadillo, with a
total of 80–100 small and vestigeal teeth,
has more than almost any other mammal.
While the hindlimbs always bear five clawed
digits, the forelimbs are powerful and terminate in three, four or five curved claws
which are used for digging. Armadillos often
excavate labyrinthine burrow systems, but
also dig to obtain food and evade predators.
All species are nocturnal.

The most widespread species is the Common long-nosed or Nine-banded armadillo,
which occurs in six slightly different subspecific forms from the southern United
States through Central America to Uruguay
and Argentina. The population density of
this species varies throughout its range from
only 0.36 animals per sq km (0.94 per sq mi)
in the dry *cerrado* of Brazil, to about 50 per
sq km (130 per sq mi) in the coastal prairies
of Texas; it may be more abundant still in
the tropical forests of Panama.

The success of the Common long-nosed
armadillo is probably due to its flexible diet
and reproductive behavior, which allow it to
exploit all but the most arid habitats.
Although possessing relatively weak dentition and jaws, this species detects small
vertebrates and insects with its keen sense of
smell, and digs them from the soil and litter
with manic speed. To keep the scent of the

prey while digging, the animal places its
nose flush to the soil, and holds its breath for
up to six minutes to avoid inhaling the dust.
When digging, the armadillo loosens the soil
with its nose and forefeet and piles it up
underneath the body. Then, pausing momentarily to balance on its forefeet and tail, it
kicks reflexively with the hindfeet to scatter
the soil behind it. This armadillo also takes a
wide range of fungi, tubers, fallen fruits and
carrion when available, but in the southern
part of its range it has developed a taste for
termites and ants. Using its long, sticky
tongues to extract these prey from their nests,
it may eat up to 40,000 ants at a sitting.

No less flexible in its reproductive
behavior, the Common long-nosed armadillo mates in July and August (summer) in
the United States, but from November to
January (summer) in South America. However, implantation is delayed for about 14
weeks, so that after the four-month gestation period, the young—usually single-sex
quadruplets—are born in time for the spring

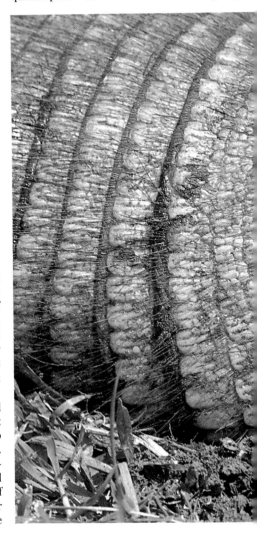

◄ **Full frontal.** The pichi, a small armadillo of the pampas of Argentina and Patagonia displays the beautiful radial patterns on its armored carapace.

▼ **The Larger hairy armadillo, feeding.** They often obtain maggots from beneath carcasses and sometimes burrow into the carcasses themselves.

flush in invertebrate food. At birth, the young are covered in a soft, leathery skin which hardens within a few weeks.

Unlike their common relative, the five other species of long-nosed armadillos are more specialized in their habits and have restricted geographical ranges which show no subspecific variation. The large Greater long-nosed armadillo, for example, occurs only in the rain forests of the Orinoco and Amazon basins, while the smaller Brazilian, Northern and Southern lesser long-nosed are restricted to savanna, forest edges and thorn scrub in parts of northern and central South America. The rare Hairy long-nosed armadillo occupies a still smaller range on the high Andean slopes of Peru at altitudes of 2,400–3,200m (7,900–10,500ft). These species seem to be less omnivorous than the Common long-nosed armadillo, and feed largely on termites and ants. Except for the Southern lesser long-nosed armadillo, which bears 8 or even 12 young, all long-nosed armadillos have litters of four. However, young of the Common long-nosed armadillo are weaned in only a few weeks and reach sexual maturity at the age of six months to a year (twice as long in all the other species).

While the long-nosed armadillos are most often seen alone, adult females show the strongest tendency to maintain exclusive home-range areas. In contrast, the home ranges of male Common long-nosed armadillos overlap freely and may encompass the ranges of several females; male ranges are 50 percent larger than those of females and range in size from 1.8ha (4.4 acres) on the coastal prairies of Texas to 10.8ha (26.6 acres) in Florida. The home ranges of both sexes include up to twelve different den sites (the average is 4.5 per animal on sandy soil, and 8.5 per animal on coastal prairie), which are usually used by a single individual on different nights. The burrows are 0.6—4.6m (24–180in) long, 18cm (7in) wide and up to 2m (79in) deep with two or more entrances. Grass, weeds and debris are raked in to line the one or two nest chambers, and often clog and hide the burrow entrances from view. Records of four or more animals in a nest are usually of mothers and their young or of newly independent all-brother or all-sister groups.

The spacing behavior of armadillos, especially of adult females, is probably regulated by olfactory communication and aggression, and to a lesser extent by the more poorly developed senses of sight and hearing. The long-nosed armadillos have glands on the ears, eye-lids and soles of the feet, as well as a bean-shaped pair of anal glands which produce a yellow nauseous secretion. Adults often sniff each other's anal regions on meeting, and it is likely that this strong odor advertises the presence of an individual to others; spacing behavior may be reinforced by the deposition of droppings and urine along paths or on prominent objects. The secretion may also deter predators, although it cannot be squirted, as in pangolins or skunks. Armadillos sometimes resolve territorial disputes by kicking, chasing and fighting but, except for a high-pitched squealing, the grunts and buzzes uttered during these squabbles differ little from the sounds made while foraging.

Along with the closely related Giant armadillo, the four smaller species of naked-tailed armadillos range widely from Central America to northern Argentina and Uruguay. These species have 10–13 movable bands across the back, short snouts and five long, powerful claws on the forefeet; that on the third digit is especially large. Although the geographical ranges of naked-tailed armadillos generally do not overlap, individuals are often found in savanna, thorn forest and tropical forest habitats with the Giant armadillo and other more distantly related species. Never abundant, the maximum density estimated for the Southern naked-tailed armadillo is only 1.2 per sq km (3.1 per sq mi) in the Venezuelan *llanos*; but the Giant armadillo in optimum lowland forest habitat in Surinam is only half as abundant as this. Using their long claws, the naked-tailed armadillos dig more efficiently than their long-nosed relatives, but run less swiftly on the surface of the ground. Their burrows are often near or underneath termite hills, and sparse observations of feeding suggest that these insects form 90–98 percent of the diet. They appear to breed at any time in captivity, and give birth to one or two blind, helpless young weighing 100–115g (3.5–4oz).

The armor of the three hairy armadillos is broader and flatter than in most other species, and is sparsely covered by brownish hairs, 3–4cm (1.2–1.6in) which grow between the bony scute plates. Until recently, they were classified with the more modestly haired Yellow armadillo and pichi as six-banded armadillos (the actual number of bands varies from six to eight), from which they differ principally in size. The hairy armadillos occur in desert, temperate grassland and forest in southern South America, and can tolerate cool and very dry conditions; indeed, the rare Andean Hairy armadillo occurs only in the high, bleak *puna* of

An Armored Invader

Although primarily an inhabitant of the tropical forests and warm, moist grasslands of South America, the Common long-nosed armadillo has extended its range south to the cooler regions of Uruguay and Argentina, and north into the southern United States. Formerly, this armadillo occurred no further north than Mexico, but in 1880 it crossed the Rio Grande and entered Texas. As the armadillo gradually advanced northwards, its taste for insect pests was noticed and it was successfully introduced to several southern crop growing states. In Florida, several armadillos escaped from zoos or private owners in the 1920s, and these also established wild populations. By 1939, specimens were being reported from all southeastern states, and today the species extends as far north as central Oklahoma and possibly New Mexico.

The opportunistic feeding and reproductive habits of the Common long-nosed armadillo have no doubt assisted the advance, but its

ability to excavate deep burrows in the most compacted soils has also been important. Unlike other armadillos, the Common long-nosed armadillo is able to swim across water barriers, inflating its stomach and intestine with air for bouyancy. Since the breath can be held for several minutes, this species is also able to trot across shorter river crossings.

The southward advance of the Common long-nosed armadillos is probably limited by competition from the Three-banded and Hairy armadillos, and by its inability to tolerate aridity.

In the United States, cold has slowed the northward advance. Unlike the more cold-adapted forms, the body temperature can be dropped only 2.5°C (4.5°F) (from 34.5°C; 94.1°F) to reduce the losses, while frosts reduce the abundance of insect food which is needed to make up the deficit. Like its giant ancestors, the armadillo may have to evolve a larger body with more insulating fat if it is to advance further north.

Bolivia. The small pichi, half the size of the hairy armadillos, occupies open pampas country in Patagonia, south to the Strait of Magellan, while the Yellow armadillo, evidently the most abundant of the six-banded forms, occurs in Brazilian savanna and forest at densities of up to 2.9 per sq km (7.5 per sq mi).

The hairy armadillos conserve heat and moisture in their dry, harsh environments by maintaining low but variable body temperatures—for example, 24.0–35.2°C (75.2–95.3°F)—in the pichi—and by spending the coldest times of day underground. Alone among armadillos, the pichi is able to

▲ **A Brazilian lesser long-nosed armadillo,** resting in a surface burrow.

◄ **The soft, pink leathery skin** of this young Common long-nosed armadillo will soon harden into a tough carapace.

◄ **Like a puzzle ball,** the Southern three-banded armadillo leaves no chinks for predators when it curls up.

enter torpor in winter, but all of the hairy species can shiver to produce heat and shift their daily activity cycle to forage, as required, in the warm mid-day sun or in the cool, moist evenings. In the northern Monte Desert of Argentina, where annual rainfall is only 13cm (5in), the Screaming hairy armadillo obtains moisture in summer from a broad diet of insects (45 percent), plants and tubers (29 percent) and rodents and lizards (26 percent), foraging over an area of 3.4ha (8.4 acres). Hairy armadillos are believed to produce up to two litters of two young each year.

Sharing the geographical ranges of the hairy armadillos, but differing in its preference for ants and termites, the Southern three-banded armadillo has adapted to dry forest and savanna habitats by evolving very thick armor. Air trapped beneath the shell reduces heat loss so effectively that the animal is often active on the coldest of winter days and nights. Unlike the hairy armadillos, it seldom digs burrows, and sleeps under bushes; in southern Brazil it may reach densities of seven animals per sq km (18 per sq mi). It has the ability,

unique among armadillos, to roll into a complete ball, presenting only its thick armor to would-be predators. The single young are born between October and January and, although blind, they are able to walk and close their shells within hours.

Fairy armadillos are the smallest members of the armadillo family, and are known only from the sandy soil regions of southern South America. They have very flexible body armor, and five powerfully clawed digits for digging. Largely subterranean, these species are believed to surface once a day to feed on insects.

The long-nosed, six-banded and Southern three-banded armadillos are hunted for meat throughout Central and South America but, except for the very rare and geographically restricted forms such as the Brazilian three-banded armadillo, they do not appear to be seriously declining. However, the Giant armadillo has disappeared from much of its former range in Brazil, Peru and elsewhere, and is currently considered endangered, while loss of habitat may reduce populations of the rare and vulnerable fairy armadillos.   CRD

PANGOLINS

Family: Manidae

Seven species of genus *Manis*.
Order: Pholidota.
Distribution: Senegal to Uganda, Angola, western Kenya; south to Zambia and northern Mozambique; Sudan, Chad, Ethiopia to Namibia and South Africa. Peninsular India, Sri Lanka, Nepal and southern China to Taiwan and Hainan; south through Thailand, Burma, Laos, Peninsular Malaysia, Java, Sumatra, Kalimantan and offshore islands.

Habitat: forest to open savanna.

Size: from head-body length 30–35cm (12–14in), tail length 55–65cm (22–26in) and weight 1.2–2.0kg (2.6–4.4lb) in the Long-tailed pangolin, to head-body length 75–85cm (30–33in), tail length 65–80cm (25–31in) and weight 25–33kg (55–73lb) in the Giant pangolin. Males 10–50 percent heavier than females in most species, up to 90 percent heavier in Indian pangolin.

Coat: Horny, overlapping scales on head, body, outer surfaces of limbs, and tail, varying in color from light yellowish-brown through olive to dark brown. Scales of young Chinese pangolin purple-brown. Undersurface hairs white to dark brown.

Gestation: 139 days (Cape pangolin). Longevity: at least 13 years (captive Indian pangolin).

Four African species: **Giant pangolin** (*M. gigantea*), **Cape pangolin** (*M. temmincki*), **Small-scaled tree pangolin** (*M. tricuspis*), **Long-tailed pangolin** (*M. tetradactyla*); Three Asian species: **Indian pangolin** (*M. crassicaudata*), **Chinese pangolin** (*M. pentadactyla*), **Malayan pangolin** (*M. javanica*).

▶ **Overlapping brown scales,** ABOVE, give the pangolins a unique, animal-artichoke appearance. This is the Small-scaled tree pangolin, an African species.

▶ **A Cape pangolin** drinking at Etosha National Park, Namibia. Cape pangolins sleep by day in burrows up to 6m (20ft) deep. They feed mostly on four genera of termites but when in competition with other termite-eaters such as the aardwolf and aardvark, they may take ants.

THE pangolins are distinguished from all other Old World mammals by their unique covering of horny body scales, which overlap like shingles on a roof. The scales, which grow from the thick underlying skin, protect every part of the body except the underside and inner surfaces of the limbs, and are shed and replaced periodically. With the body curled tightly into a ball, the scales form a shield that is impregnable to all but the larger cats and hyenas, with their bone-crunching jaws.

Pangolins specialize in eating ants and termites and, like their South American counterparts, they probe the nests of their prey with a long, narrow tongue. In the largest species, the Giant pangolin, the strap-like tongue can be pushed out 36–40cm (14–16in), although it is 70cm (27.5in) in total length, and is housed in a sheath that extends to an attachment point on the pelvis. Viscous saliva is secreted onto the tongue by an enormous salivary gland—360–400cu cm (22–24.5cu in)—which sits in a recess in the chest. The simple skull lacks teeth and chewing muscles; captured ants are ground up in the specialized, horny stomach. Pangolins have a small, conical head with a reduced or absent outer ear, and an elongate body that tapers to a stout tail. Thick lids protect their eyes from the bites of ants, and special muscles close the nostrils during feeding. The limbs are short but powerful and terminate in five clawed digits; the three middle claws on the forefoot are 55–75mm (2.2–2.9in) long and curved.

In Africa, two of the four species of pangolins are principally arboreal, and inhabit the rain-forest belt from Senegal to the Great Rift Valley. While the common Small-scaled tree pangolin occupies home ranges of 20–30ha (49–74 acres) in the lower strata of the forest, the smaller Long-tailed pangolin is more restricted to the forest canopy. Here, moving often by day to avoid its relative, the smaller species seeks the soft, hanging nests of ants and termites (preferring arboreal species), or attacks the columns that move among the leaves. Both species have a slender but strongly prehensile tail (the 46 or 47 tail vertebrae of the Long-tailed pangolin are a mammalian record), with a short, bare patch at the tip which contains a sensory pad. They are able to scale vertical tree trunks by gaining a purchase with the fore-claws and then drawing up the hindfeet just behind them; the jagged edges of the tail-scales provided additional support. These arboreal pangolins sleep aloft, curled up among epiphytes (plants growing on trees) or in the fork of a branch.

The terrestrial African pangolins are larger than their tree-dwelling relatives, and occur in a spectrum of habitats from forest to open savanna. They use their powerful claws to demolish the nests of ground termites and ants. The Giant pangolin, with a stomach capacity of 2 liters, may take 200,000 ants a night, weighing over 700g (25oz).

To protect their digging claws, the terrestrial pangolins walk slowly on the outer edges of their forefeet with the claws tucked up underneath; the curious, shuffling gait this produces has been likened to that of a "perambulating artichoke." However, all species can move more swiftly—up to 5km/h (3mph)—by rearing up and running on their hindlegs, using the tail as a brace.

The three Asian species of pangolins are less well known than their African counterparts, and can be distinguished from them

by the presence of hair at the bases of the body-scales. Intermediate in size between the African species, the Asian pangolins are nocturnal and usually terrestrial, but can climb with great agility. They inhabit grasslands, subtropical thorn forest, rain forest and barren hilly areas almost devoid of vegetation, but are nowhere abundant. The geographical range of the Chinese pangolin is said to approximate that of its preferred prey species, the subterranean termites.

Although pangolins are usually solitary, their social life is dominated by the sense of smell. Individuals advertise their presence by scattering feces along the tracks of their home ranges, and by marking trees with urine and a pungent secretion from an anal gland. These odors may communicate dominance and sexual status, and possibly facilitate individual recognition. The vocal expressions of pangolins are limited to puffs and hisses; however, these probably serve no social function.

Pangolins usually bear one young weighing 200–500g (7–18oz), although two and even three young have been reported in the Asian species. In the arboreal species, the young clings to the mother's tail soon after birth, and may be carried in this fashion until weaned at the age of three months. Young of the terrestrial species are born underground with small, soft scales, and are first carried outside on the mother's tail at the age of 2–4 weeks. In all species, births usually occur between November and March; sexual maturity is at two years.

In Africa, large numbers of pangolins are killed for their meat and scales by the native peoples, and the future of one species, the Cape pangolin, is seriously endangered. In Asia, powdered scales are believed to have medicinal and aphrodisiac qualities, and the animals are hunted indiscriminately. Unless controlled, the population densities and ranges of the three Asian pangolins will continue to dwindle.   CRD

BATS

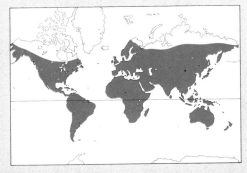

ORDER: CHIROPTERA

Nineteen families; 187 genera; 951 species.
Distribution: worldwide except Arctic,
Antarctic and highest mountains.
Habitat: highly diverse.

Size: weight and wingspan range from 1.5g
and 15cm in Kitti's hog-nosed bat to 1.5kg and
2m in flying foxes (*Pteropus* species).

Coat: variable, but mostly browns, grays,
yellows, reds and blacks.
Gestation: variable, and with delayed
implantation can range from 3 to 10 months in
a single species.
Longevity: maximum 30 years but average 4–5
years.

Suborder Megachiroptera

Flying foxes
Family Pteropopidae
Forty-four genera and 173* species in Old
World including: **Straw-colored flying fox**
(*Eidolon helvum*); **rousettes** (*Rousettus* species);
Rodriguez flying fox (*Pteropus rodricensis*);
Samoan flying fox (*P. samoensis*); **Hammer-
headed bat** (*Hypsignathus monstrosus*);
Franquet's flying fox (*Epomops franqueti*);
Dawn bat (*Eonycteris spelaea*); **long-tongued
fruit bats** (*Macroglossus* species).

Suborder Microchiroptera

Mouse-tailed bats
Family Rhinopomatidae
One genus and 3 species in Old World
including: **Greater mouse-tailed bat**
(*Rhinopoma microphyllum*).

Sheath-tailed bats
Family Emballonuridae
Thirteen genera and 50* species in Old and
New Worlds including: **Sac-winged bat**
(*Saccopteryx bilineata*).

Hog-nosed bat
Family Craseonycteridae
One species in Old World: **Kitti's hog-nosed bat**
(*Craseonycteris thonglongyai*).

Slit-faced bats
Family Nycteridae
One genus (*Nycteris*) and 11* species in Old
World.

False vampire bats
Family Megadermatidae
Four genera and 5 species in Old World
including: **Greater false vampire** (*Megaderma
lyra*). (Note: the New World False vampire bat
belongs to the family Phyllostomatidae.)

Horseshoe bats
Family Rhinolophidae
One genus and 69* species in Old World
including: **Greater horseshoe bat** (*Rhinolophus
ferrumequinum*).

Leaf-nosed bats
Family Hipposideridae
Nine genera and 61* species in Old World.

Leaf-chinned bats
Family Mormoopidae
Two genera (*Pteronotus* and *Mormoops*) and 8 species in New World.

Bulldog bats
Family Noctilionidae
One genus (*Noctilio*) and 2 species in New World.

Short-tailed bats
Family Mystacinidae
One genus (*Mystacina*) and 2 species in Old World.

Spear-nosed bats
Family Phyllostomatidae
Forty-seven genera and 140* species in New World including: **Greater spear-nosed bat** (*Phyllostomus hastatus*); **Fringe-lipped bat** (*Trachops cirrhosus*); **False vampire** (*Vampyrum spectrum*) (note: not to be confused with the Old World false vampires, family Megadermatidae); **Pallas' long-tongued bat** (*Glossophaga soricina*); **Mexican long-nosed bat** (*Leptonycteris nivalis*); **Geoffroy's long-nosed bat** (*Anoura geoffroyi*); **Great stripe-faced bat** (*Vampyrodes caraccioloi*); **Great fruit-eating bat** (*Artibeus lituratus*).

Vampire bats
Family Desmodontidae
Three genera and 3 species in New World including: **Common vampire** (*Desmodus rotundus*).

Funnel-eared bats
Family Natalidae
One genus (*Natalus*) and 8 species in New World.

Thumbless bats
Family Furipteridae
Two genera and 2 species (*Furipteus horrens* and *Amorphochilus schnabli*) in New World.

Disk-winged bats
Family Thyropteridae
One genus (*Thyroptera*) and 2 species in New World.

Sucker-footed bat
Family Myzopodidae
One species (*Myzopoda aurita*) in Old World.

Common or vesper bats
Family Vespertilionidae
Forty-two genera and 319* species in Old and New Worlds including: **Gray bat** (*Myotis grisescens*); **Little brown bat** (*M. lucifugus*); **Large mouse-eared bat** (*M. myotis*); **Natterer's bat** (*M. nattereri*); **Yuma myotis** (*M. yumanensis*); **Fish-eating bat** (*Pizonyx vivesi*); **Common pipistrelle** (*Pipistrellus pipistrellus*); **Eastern pipistrelle** (*P. subflavus*); **Leisler's bat** (*Nyctalus leisleri*); **Noctule bat** (*N. noctula*); **Big brown bat** (*Eptesicus fuscus*); **Bamboo bat** (*Tylonycteris pachypus*); **Hoary bat** (*Lasiurus cinereus*); **Red bat** (*L. borealis*); **Brown long-eared bat** (*Plecotus auritus*); **Schreiber's bent-winged bat** (*Miniopterus schreibersi*); **Painted bat** (*Kerivoula picta*).

Free-tailed bats
Family Molossidae
Twelve genera and 91* species in Old and New Worlds including: **Mexican free-tailed bat** (*Tadarida brasilensis*); **Wrinkle-lipped bat** (*T. plicata*); **Naked bat** (*Cheiromeles torquatus*).

* Number of species changing as research continues.

Nearly one quarter of mammalian species are bats. Apart from birds, they are the only other vertebrates capable of sustained flight. They have exploited all major land habitats with the exception of the polar regions, highest mountains, and some remote islands, particularly in the eastern Pacific. On New Zealand, Hawaii, the Azores and many oceanic islands, bats are the only indigenous mammals and, like birds, their mobility allows them readily to investigate and colonize new areas if roosts and food are available.

In Europe, the Leisler's bat long ago reached the Azores in the north Atlantic, and the Hoary bat from the Americas similarly colonized the Hawaiian Islands with minimum distances from the mainland of 1,500 and 3,700km (930 and 2300mi) respectively. Both species are narrow-winged, fast-flying bats that are migratory over at least part of their current ranges. Most bats are only active at night, but island species in the absence of birds of prey are often also active by day, and a few bats of most species will occasionally fly during daytime.

Flying, especially at night, poses problems of obstacle avoidance and navigation, but facilitates finding food which may be patchily distributed in space and time. In general, birds solved this problem by evolving superb eyesight, but their hearing is average and sense of smell very poor. Although some bats, such as Old World flying foxes, have excellent sight, most rely upon highly acute hearing which, with often complex sound production, enables bats to navigate, feed and locate roosts by echolocation. Many bats, particularly the fruit-eating species, have a keen sense of smell. The light-gathering capability of megachiropterans' eyes is enhanced by numerous projections from the rods (monochrome receptors).

Bats (order Chiroptera) are separated into the suborders Megachiroptera and Microchiroptera. The megachiropterans comprise a single family, the flying foxes, and live in the Old World tropics and subtropics from

◀ ▼ **Bat heads.** Bats exhibit a wide variety of head shapes. LEFT. The fruit-eating flying foxes, such as the Common long-tongued fruit bat (*Macroglossus minimus*) have a dog-like face and generally small simple ears, and characteristically large eyes which are the primary sense in navigation.

BELOW At the other extreme many insect-eating species, such as the Narrow-eared leaf-nosed bat (*Hipposideros stenotis*) have extraordinary, even grotesque, faces, often with huge ears and with elaborate growths (nose leaves) around nostrils or mouths, associated with echolocation.

THE BAT BODY PLAN

▶ **Body plan** of a typical bat with a simple nose.

▼ **Bat tails.** Major variations in tail shape of bats; (**a**) free tail (free-tailed bat—*Tadarida*); (**b**) mouse tail (mouse-tailed bat—*Rhinopoma*); (**c**) full membrane (mouse-eared bat—*Myotos*); (**d**) sheath tail (Old World sheath-tailed bat—*Emballonura*); (**e**) short tail (tube-nosed fruit bat—*Nyctimene*); (**f**) tail lacking (flying fox—*Pteropus*).

Africa to the Cook Islands in the Pacific. They include the largest bats, for example the Samoan flying fox with a reported wingspan of 2m (6.6ft) and weight 1.5kg (3lb), but some are tiny, such as the long-tongued fruit bats (*Macroglossus* species) (wingspan 30cm/12in, weight 15g/0.5oz). Microchiropterans occur throughout the world and are grouped into 18 families. They include the smallest bat (and mammal), Kitti's hog-nosed bat (wingspan 15cm/6in, weight as little as 1.5g/0.05oz), and species as large as the New World False vampire bat (wingspan up to 1m/3.3ft, weight 200g/7oz).

Bats have wings that flap, a character which separates them from all other mammals. Even so-called flying mammals such as flying squirrels and colugos which possess expanded flaps of skin are not able to undertake powered flight—they just glide. The membrane (or patagium) consists in bats of skin, sandwiching bundles of elastic tissue and muscle fiber, and is supported by the finger bones, arms and legs. The muscle fibers keep the wing tensioned in flight and gather it up while at rest. Holes heal within a few weeks—even broken finger bones mend quickly.

The wing pattern is essentially similar in all bat species, but differences in shape reflect the variety of ecological niches and feeding behavior exhibited by bats. The upper arm (humerus), is shorter than the forearm (radius) and the second forearm bone, the ulna, is more or less reduced to a sliver of bone. All bats have a clawed thumb, although in two species of smoky bats the thumbs are functionless. Bats mostly use thumbs for moving around roosts but some, especially the fruit-eating bats, hold and manipulate food with them. Flying foxes of the genus *Pteropus* have particularly large thumbs and claws which are also used for

▶ **Evolution of bats.** Evolution chart showing the relationships between present-day families and their grouping into superfamilies and orders. The fossil record of bats is exceptionally poor with only 30 fossil genera discovered. There are no known early flying foxes (Megachiroptera), so it is not known whether the two suborders evolved independently or from a common ancester.

Bats have some similarity with both insectivores and primates, and it is often stated that bats probably evolved from an ancestral shrew-like insectivore. In the absence of any evidence such speculation is pointless and may be misleading.

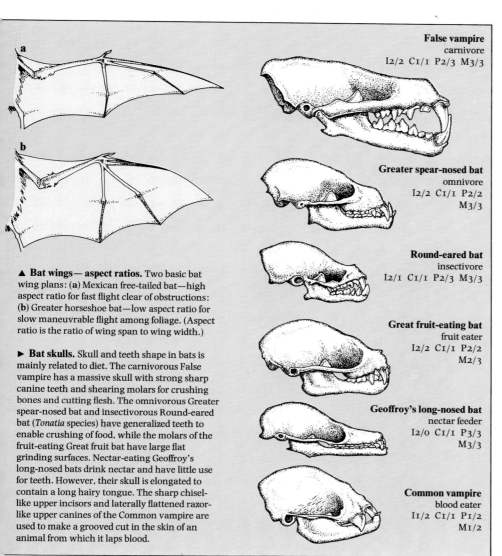

▲ Bat wings— aspect ratios. Two basic bat wing plans: (**a**) Mexican free-tailed bat—high aspect ratio for fast flight clear of obstructions: (**b**) Greater horseshoe bat—low aspect ratio for slow maneuvrable flight among foliage. (Aspect ratio is the ratio of wing span to wing width.)

► Bat skulls. Skull and teeth shape in bats is mainly related to diet. The carnivorous False vampire has a massive skull with strong sharp canine teeth and shearing molars for crushing bones and cutting flesh. The omnivorous Greater spear-nosed bat and insectivorous Round-eared bat (*Tonatia* species) have generalized teeth to enable crushing of food, while the molars of the fruit-eating Great fruit bat have large flat grinding surfaces. Nectar-eating Geoffroy's long-nosed bats drink nectar and have little use for teeth. However, their skull is elongated to contain a long hairy tongue. The sharp chisel-like upper incisors and laterally flattened razor-like upper canines of the Common vampire are used to make a grooved cut in the skin of an animal from which it laps blood.

False vampire
carnivore
I2/2 C1/1 P2/3 M3/3

Greater spear-nosed bat
omnivore
I2/2 C1/1 P2/2
M3/3

Round-eared bat
insectivore
I2/1 C1/1 P2/3 M3/3

Great fruit-eating bat
fruit eater
I2/2 C1/1 P2/2
M2/3

Geoffroy's long-nosed bat
nectar feeder
I2/0 C1/1 P3/3
M3/3

Common vampire
blood eater
I1/2 C1/1 P1/2
M1/2

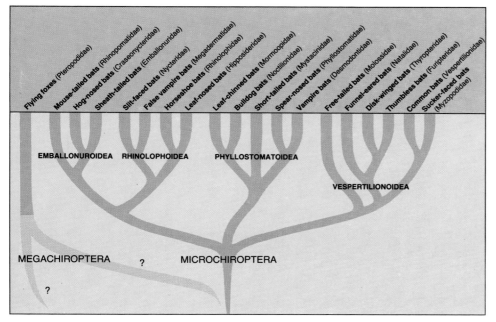

fighting. The thumb serves as an attachment for the propatagium (that part of the wing membrane in front of the forearm), which can be broad, especially in some of the flying foxes and slow-flying bats.

In all bats the second digit is relatively short, and for most flying foxes it terminates in a small claw, but this is absent in the Microchiroptera. The third digit is the longest and extends to the wing tip; the ratio of its length relative to the fifth digit, which is a measure of the wing width, characterizes the flight pattern (see diagram). Bats with the third digit about 1⅓ times longer than the fifth have short, broad wings, low aspect ratios and are generally slow flyers. (Aspect ratios are the ratio of wing span to average wing width). Bats with the third digit about twice as long as the fifth have long thin wings, high aspect ratios and fly rapidly. Most free-tailed bats that have high aspect ratios such as *Tadarida*, generally roost well above the ground so that when they take off they can fall 2–3m (6.6–10ft) to gain enough speed to fly. They fly faster than most other species (36–55km/h or 22.5–34mph) but they are not very maneuverable and normally fly clear of trees or other obstructions. By contrast, the horseshoe bats are mostly slow-flying, with low aspect ratios. They are highly maneuverable and may even hover or turn in a space no larger than the wing span. These bats generally fly less than 26km/h (16mph).

Some of the long-winged bats fold the wing tips when at rest. Most extreme in this respect is the Naked bat of Southeast Asia, heaviest of all microchiropterans, which, after folding its wings, tucks the ends into pouches beneath the wings that join in the middle of the back. In this way, the large 60cm (24in) wingspan presents no encumbrance while moving around the roost.

Tails in bats are extremely variable, in size ranging from absent, as in some flying foxes and Kitti's hog-nosed bat, to long and thin in mouse-tailed bats (*Rhinopoma* species). The tail membrane (or uropatagium) is absent in some bats, such as flying foxes of the genus *Pteropus*, or very small, as in the mouse-tailed bats, but can be large and supported by the tail as in the slit-faced bats (*Nycteris* species). The latter are unique among mammals in having a T-shaped bone at the tail tip, the function of which is unknown.

The tail membrane may be used to aid maneuverability, as it is a conspicuous feature of most insectivorous species. In some species, eg the Natterer's bat, insects are caught in the tail before transfer to the mouth. The bats that catch insects with the wing may transfer the prey to the tail which serves as a holding pouch until the bat lands at a perch to devour the food.

The legs, which support the uropatagium that extends between them as well as the plagiopatagium (the main part of the membrane between the body and fifth finger), are generally weak. They project sideways and backward and the knee bends back rather than forward as in other mammals. Together with the feet with their five clawed toes of equal length, the hind limbs function as "clothes hanger hooks." Some bats, such as the large flying foxes and horseshoe bats, cannot walk on all fours (quadrupedally), but others, such as many of the common bats, can scurry very rapidly around their roosts, or while chasing prey on the ground. Most agile is the Common vampire, which has especially strong legs and long thumbs enabling it to run and leap very quickly. The two short-tailed bats (*Mystacina* species) also run freely, climb with agility, and excavate burrows, aided by wings that roll up out of the way and by talons on the claws of thumbs and toes.

Bats spend much of their time at roosts washing and grooming, often hanging by one foot while the other vigorously combs all parts of the body. Although the feet show very little variation, fishing bats, such as *Noctilio* species, have long laterally flattened claws and toes. These are drawn through water to gaff fish which are lifted to the mouth prior to the bat landing to eat.

Evolution

Bats were originally tropical animals, though some have now adapted to living in temperate climates. Unlike other mammalian orders the fossil record of bats is poor. About 30 fossil genera have been described compared with 187 current genera, and excepting five genera they are attributable to at least 10 of the 19 current families. The five oldest fossils are not referrable to modern families but nevertheless do not vary significantly from the variety of body forms living today. The earliest fossil is of a virtually complete skeleton recovered from what was a lake bed in Wyoming about 50 million years ago in the Eocene epoch of the Tertiary period. This species, named *Icaronycteris index*, was so well preserved that membranes were visible and remains were found in the area of its stomach indicating that it was probably insectivorous. This fossil indicates that bats were fully developed early in the Eocene epoch (54–38 million years ago) and were

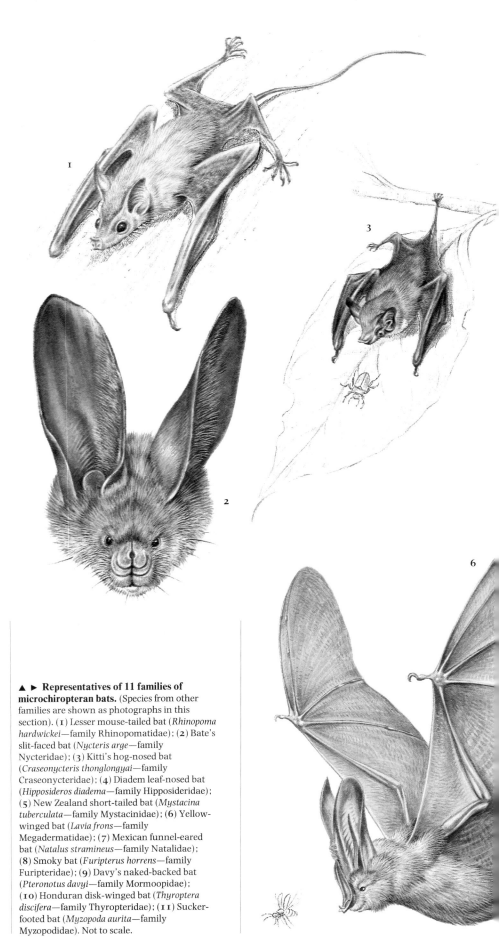

▲ ► **Representatives of 11 families of microchiropteran bats.** (Species from other families are shown as photographs in this section). (**1**) Lesser mouse-tailed bat (*Rhinopoma hardwickei*—family Rhinopomatidae); (**2**) Bate's slit-faced bat (*Nycteris arge*—family Nycteridae); (**3**) Kitti's hog-nosed bat (*Craseonycteris thonglongyai*—family Craseonycteridae); (**4**) Diadem leaf-nosed bat (*Hipposideros diadema*—family Hipposideridae); (**5**) New Zealand short-tailed bat (*Mystacina tuberculata*—family Mystacinidae); (**6**) Yellow-winged bat (*Lavia frons*—family Megadermatidae); (**7**) Mexican funnel-eared bat (*Natalus stramineus*—family Natalidae); (**8**) Smoky bat (*Furipterus horrens*—family Furipteridae); (**9**) Davy's naked-backed bat (*Pteronotus davyi*—family Mormoopidae); (**10**) Honduran disk-winged bat (*Thyroptera discifera*—family Thyropteridae); (**11**) Sucker-footed bat (*Myzopoda aurita*—family Myzopodidae). Not to scale.

contemporary with other mammals, including rodents, insectivores and primates.

Classification

Modern classification of bats suffers primarily from the lack of adequate data. About 100 species have been caught less than 20 times and about 20 species are recognized on the basis of a single specimen. New species are constantly being found and described as collecting techniques improve. At least eight species are known to have become recently extinct, some probably as a direct result of man's recent influence on habitats. Even in western Europe, three new species have been described in the last 25 years and others are suspected. As recently as 1974 a new family was erected to contain the newly discovered Kitti's hog-nosed bat from Thailand. Existing taxonomy is based mainly on external form (morphology) but results of modern biochemical techniques are beginning to demand reassessment. Ecological research is also providing clues to separate sibling species that were previously overlooked. Some species are clearly related while others may have characters intermediate between two presently recognized groups. As information improves some regrouping is necessary. For example, the rare short-tailed bats of New Zealand have recently been separated into two species, *Mystacina tuberculata* and *M. robusta*.

The variety of body forms that is the basis of our current classification reflects the adaptations each species has made in response to the ecological niche in which it has evolved and helped to differentiate. Only three of the 19 bat families have representatives in both the Old and New Worlds, although there are bats occupying similar niches in both areas. Those families that have representatives throughout the world are the sheath-tailed bats, common bats and free-tailed bats. These include about half of all bats and are almost exclusively insectivorous.

In the Old World, the fruit-eating niche is occupied mostly by one family, the flying foxes, whereas in the New World the niche is occupied by the spear-nosed bats. This family is the most diverse of all, embracing fruit-eaters, carnivores and insectivores, with the closely related vampires being blood-eaters (see pp812–813).

About 250 species of spear-nosed bats and flying foxes are important to over 130 genera of plants because they pollinate and/or disperse their seeds. In the New World alone over 500 plant species are pollinated by bats (see pp814–815). Bat-adapted plants often have large white flowers that show up in the dark, or smell strongly so as to attract bats, and all produce copious quantities of nectar and pollen. Nectar flow is synchronized with bat activity, in some plants, such as bananas, beginning at dusk and continuing for over half the night, while in a Passion flower nectar flow begins after midnight and stops shortly after dawn.

Over 650 species eat insects, representing members of all families except the flying foxes, and even a few of these take some insects, but perhaps only accidentally while eating ripe fruit.

The three species of vampires consume blood. Probably about 10 bats in the fisherman and common bat families catch fish, but none exclusively, as they also take various Arthropoda (insects etc). Similarly, about 15 bats, mainly of the false vampire and spear-nosed families, are carnivores, but some of these species also take insects and even fruit. There are no truly herbivorous species, nor marine species, although the two bulldog or fisherman bats roost and feed along shorelines.

Echolocation

For 150 years biologists have marveled at the ability of bats to fly and catch insects in the darkness, even if deliberately blinded. It was an Italian, Lazzaro Spallanzani, who in 1793 first discovered that bats were disoriented when they could not hear but that blinded bats could still avoid obstacles. In 1920 the English physiologist Hartridge suggested that bats navigated, and located and captured prey using their sense of hearing. In the late 1930s the invention of a microphone sensitive to high frequencies enabled Donald Griffin in the United States to discover in 1938 that bats produce ultrasonic sounds. The term "ultra-sonic" means sounds of higher frequency than is audible to humans.

Before describing what sounds bats produce and how they use them, it is important to understand the descriptive terminology. When an object vibrates it causes pressure changes in the surrounding air. The ear may intercept these pressure changes and, through the eardrum and middle ear, transfer them to the inner ear or cochlea. This has sensitive cells that selectively respond by sending signals to the brain, which interprets them as sound. The number of vibrations per second is termed "frequency" and is measured in hertz (Hz). Humans can perceive sounds from 20Hz to 20,000Hz, while bats' sensitivity ranges from less than

▶ **Echolocating prey.** Sonograms showing search, approach and terminal phases of the hunt in two species of bat.

(a) The North American Big brown bat, produces frequency modulated (FM) calls (see text) steeply sweeping from 70–30kHz. While foraging the bat emits 5–6 pulses per second, each of about 10 milliseconds (msec) duration until an insect is located. Immediately the pulse rate increases, duration shortens, with the frequency sweep starting at a lower frequency. As an insect is caught (or just missed) the repetition rate peaks at 200 per second, with each pulse lasting about 1msec.

(b) Hunting horseshoe bats produce their long (average 50msec) constant frequency (CF) calls (see text) at a rate of 10 per second. They often feed among dense foliage. A problem facing a bat is how to distinguish fluttering insect wings from leaves and twigs oscillating in the wind. While foliage produces a random background scatter of echoes the insect with a relatively constant rapid wing-beat frequency will appear like a flashing light to a bat using a CF component. As the bat closes on the insect the CF component of each pulse is suppressed in amplitude and reduced to under 10msec while the amplified terminal FM sweep is used for critical location and capture of the prey.

▶ **Sonar hunt**—a Greater horseshoe swoops on a moth. Such a battle is not necessarily one-sided. Some moths have evolved listening membranes that detect the bat's sonar pulses, giving the moth opportunity to escape. To counter this some tropical bats only send out signals at wavelengths that cannot be detected by the moths. In the last resort moths may dive away from the bat at the very last moment.

▼ **Sound collectors**—the external ears of bats using ultrasound to navigate and hunt prey as with this Lesser long-eared bat (*Nyctophilus geoffroyi*), are enlarged and folded into complex shapes. The separate lobe seen inside the front of the ear is known as a tragus.

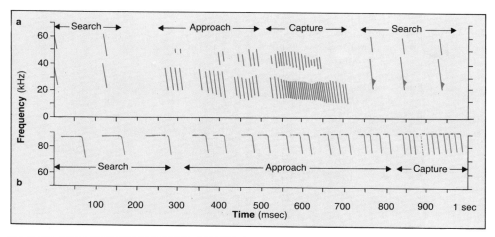

100Hz to 200,000Hz (normally written as 200kHz).

Sound vibrations travel through air in pressure "waves" and the distances between successive peaks, termed the wavelength, is measured in meters. The higher the frequency, the shorter the wavelength. Also sounds vary in intensity (from loud or quiet), and this reflects the energy or "amplitude" of each wave. The intensity of sound is usually recorded in decibels (dB).

Most animals produce simultaneous sounds comprising a number of frequencies and of differing amplitudes. Some sounds are "harmonics" of particular frequencies, that is, they are double, treble or quadruple etc, the lowest frequency. This latter is termed the fundamental (or base) frequency, with others being called the second, third or fourth etc harmonic.

Probably all microchiropterans use ultrasound which they produce with their larynxes. A single species, such as the high, fast-flying Noctule bat in Europe, produces different emissions while migrating, cruising looking for food, chasing and catching food, and when flying or feeding in close company with other bats. During high migrating flight loud low-frequency pulses at one-second intervals are used, presumably to keep it in contact with the ground. If prey is detected pulses lasting under 5 milliseconds (msec) are produced sweeping down through a frequency range of over 40kHz and up to 200 per second in the terminal phase of the chase. Individuals flying in a group alter their frequencies slightly so that they can more easily detect their own echoes.

Sounds are emitted through the open mouth or nostrils depending on species. Those bats with elaborate noses, including horsehoes, leaf-nosed, slit-faced, false vampires and spear-nosed bats, and some common bats like the long-eared *Plecotus*, emit sounds through the nose. The nose-leaf, acting as a transducer, may modify, direct and focus the sound, producing a more concentrated beam. In order to scan an area the head is moved from side to side. The shape of the nose-leaf is constantly modified to accommodate the changing needs.

Most species use pulses that sweep down through a range of frequencies—so-called frequency-modulated (FM) calls. They can be produced as a shallow sweep of long duration, or a steep sweep of short duration. It is believed the steep FM pulses improve object discrimination and that this can be further refined by producing harmonics. Some bats, eg *Nyctalus* species, suppress the fundamental while accentuating a harmonic.

Leaf-nosed and horseshoe bats, as well as at least one of the leaf-chinned bats, are known to emit pure constant frequency (CF) pulses, terminating in a short FM sweep. Bats of other microchiropteran families also produce a CF pulse, usually while traveling at high altitude well away from obstructions. (See diagram and caption for how bats find prey by echolocation.)

At sea level sound travels at 340m per second. A stationary observer listening to a passing train perceives higher frequencies as it approaches and lower frequencies when it is moving away – the so-called Doppler effect. It results because the number of sound waves arriving per second increases with oncoming vehicles and decreases as they go away. A bat emitting sounds and the prey reflecting echoes are moving independently, therefore the echo will be heard by the bat at a different frequency from the emission. Bats primarily using FM already listen to a wide range of frequencies but for species using pure-tone CF the echo may return at a frequency to which their ears are less sensitive. To compensate for this Doppler shift, bats like the Greater horseshoe lower the frequency of their CF emission so that the returning echoes arrive at their maximum hearing acuity of 83kHz.

Members of only one megachiropteran genus, *Rousettus*, produce echolocation sounds. Most flying foxes roost in trees and navigate only by sight. However, rousette bats usually roost in caves and their echolocation enables them to navigate out of the cave, and thereafter they rely on sight. Unlike other echolocation bats, rousettes produce sound pulses with their tongues.

The unsophisticated sounds embrace a wide band (5–100kHz), but include mostly high-amplitude long wavelengths (low frequencies) which are best for long-range orientation. The sounds are audible to humans as metallic clicks.

Reproduction

Reproductive behavior is known in detail for only a few bat species but even so a variety of systems are found. For many, particularly temperate bat species, food availability varies over the year. Because of the high energy demands of producing milk lactation) to feed the young it is crucial that birth coincides with a period of consistently abundant food. Migration and hibernation also limit the optimum mating season. Probably most bats produce one offspring per litter and one litter per year, but a number have twins and the Red bat in North America averages three. In northern Europe, pipistrelles produce a single offspring, but in more southerly areas twins are common. Since among recorded births twins are more frequent in well-fed captive bats than in the same species in the wild, twinning is probably related to better nutrition.

Sexual maturity is usually attained within 12 months from birth. For example, many pipistrelles and the Little brown bat give birth at the end of their first year although in the latter species males do not breed until their second. At the edge of their range in Britain male Greater horseshoe bats mostly mature in their second year; however, females are normally four years old before their first pregnancy (exceptionally seven or more years) and some do not breed each year for the first few years after

▲ **Crowded crèche.** Masses of newly-born, naked Schreiber's bent-winged bats in the roof of a nursery cave. Such nurseries can contain up to 3,000 young per square meter (about 300/square foot) which are nursed and reared to independence by their mothers. In Australia young are born in December and disperse, sometimes over hundreds of kilometers, during February and March. Nursery caves are selected for their high temperature and humidity, and have been used for thousands of years.

▶ **Blind as a bat**—but only until their eyes open. A female flying fox and newly born young. Most bats have only one young at a time.

◀ **Dawn bat roost in Indonesian cave.** Anything from a few dozen to tens of thousands of Dawn bats roost in such limestone caves. These bats breed throughout the year and at any time more than 50 percent of the females are either pregnant or nursing young.

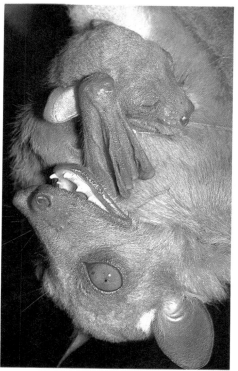

maturity. Almost certainly the poor breeding success of this insectivorous species is due to Britain's fickle climate, which greatly affects insect abundance. Greater horseshoe bats that go into hibernation with a relatively low body weight usually fail to breed. Also cold, wet, windy weather can result in insects becoming unavailable, and during lactation females need a continuous food supply if the baby is to survive. Lack of insects for several days will cause mothers to abandon their young.

In equatorial forests food supply is often relatively constant throughout the year, but bats still breed once a year, although not necessarily all at the same time. Some tropical species come into heat (estrus) several times a year, often with an estrus immediately after birth, which can result in two or even three litters per year. Such is the case with some insectivorous common bats (Myotis species) and the nectar-eating Pallas' long-tongued bat.

As far as is known, most bats are not selective in mate choice, are promiscuous and do not form pair bonds. A few species,

such as the Hammer-headed bat in Africa (see pp 816–817), form leks where the adult males gather in an area and advertise themselves by calling to attract females, who then select their mate. Male European Noctule bats, and the related Leisler's bat, occupy and probably defend a roost site throughout the fall, from which they repeatedly fly out during the night, calling loudly for a few minutes before returning to the roost, until a female is attracted. Up to 18 mature females may be with an individual male at any one time but no harem structure exists. Male Greater spear-nosed bats hold harems of females (see pp 796–797).

In cool temperate species the female's receptive period begins in the fall and continues during hibernation. The male sperm is formed in summer and mating begins in the fall shortly after lactation and weaning of the season's young. Mating takes place at roosts and may occur at times throughout hibernation, with some females inseminated several times. During hibernation, some aroused males fly along cave passages, landing beside torpid females, which they awaken and copulate with, to the accompaniment of loud vocalizations.

Both males and inseminated females store viable sperm for up to seven months. This facility is unique amongst mammals and is mostly found in the common and horseshoe bats, occuring in both tropical and temperate species. It appears to ensure that females in temperate areas can be ready for ovulation and fertilization as soon as conditions become favorable in spring.

Delayed implantation of the fertilized ovum has been recorded in at least two genera. In the Straw-colored flying fox of Africa, blastocyst implantation is delayed three months so that births coincide with the onset of the wettest season when the maximum amount of fruit is available.

Some populations of the widely distributed insectivorous Schreiber's bent-winged bat also exhibit delayed implantation. In tropical areas, development may proceed without delay after mating and fertilization. With increase in latitudes, implantation is delayed for an increasing period. In Europe, fertilization occurs in the fall but, because of hibernation, development does not begin until spring. Depending on how long the previous season's young are suckled, gestation in this species can vary from three to 10 months.

Among temperate bats with body temperatures that are variable, the length of gestation is variable, depending on the weather and availability of food. During cold weather when insects cannot be caught, the bats enter torpor and fetal development ceases. As a result these bats can time births precisely to coincide with maximum food availability. Fetal growth rates in bats are slower than any other mammalian order, although those of some primates are similar. Gestation, which may last 40 to 240 days, has a low metabolic cost for the pregnant bat.

Most temperate species form maternity or nursery colonies which consist almost exclusively of adult females. Such clustering reduces heat loss and hence energy costs to each individual. Males usually roost some distance away, so avoiding competition for food.

In some species the young are born with the mother hanging upside down, but in others the mother turns head uppermost and catches the baby in the interfemoral membrane. Suckling may begin within a few minutes of birth. Most species, and especially the insectivorous bats which require maximum maneuverability, leave

their offspring at roost while feeding. When young are carried it is usually only when changing roosts. The young of most small species develop very quickly and fly within about 20 days; however, it may be three months before the larger flying foxes take their first flights. Vampires are the slowest developers, being suckled for up to nine months. Although maximum body dimensions are achieved within a few weeks after weaning, maximum weight may be reached only after several years.

Greater horseshoe bats of northerly latitudes do not reach their maximum weight for nine years, by which time females tend to breed every year rather than in alternate years. This probably reflects the increasing skill of the individual in finding food and roost sites adjacent to the best feeding grounds. Hibernating bats from lower latitudes with longer summers have more time from the end of weaning to the beginning of hibernation in which to accumulate food reserves. Bats that have a relatively low weight in early winter often fail to produce young the following year.

Maximum longevity for the Little brown bat and probably many others is in excess of 30 years, but very few bats in any population will achieve that age. Average lifespan is often about 4–5 years.

Energy Conservation and Hibernation
Animals have differing ways of surviving seasons when food is sparse or absent. The herds of antelope on the African plains, for example, migrate hundreds of kilometers to find grass. Other species survive periods when food is scarce by laying in stores for the winter or by hibernating. Many bat species fall into this latter category.

Harem Life in Greater spear-nosed bats

In Trinidad in the West Indies, Greater spear-nosed bats roost by day within caves in clusters of 10–100 individuals. Each cluster is either a harem consisting of one adult male and many adult females (an average harem consists of about 18 females), or a group of "bachelor" males.

The membership of a harem and its roosting location are very stable. The same adult females roost together for many years, perhaps for life (10 years or longer) and harem males can retain a harem for over three years. Bachelor group membership is less stable than that of harems, due at least in part to the occasional replacement of a harem male by a bachelor. Females within a harem and males within a bachelor group appear amicable, but throughout the year harem males vigorously defend their females from intrusion by other harem males or bachelors.

The resident harem males father most, if not all, of the pups born into their harems. Because of the large size of harems and the potentially long tenure of harem males, some males father over 50 pups during their reproductive life-span, while many bachelor males father none. Clearly, the reproductive advantages to a male obtaining a harem are enormous.

Neither the membership nor the stability of female harem groups depends on the harem males. The basic social organization in Greater spear-nosed bats appears to result from a male attaching himself to an existing female group and attempting to exclude all other males from access to these females.

Stable associations of females are common in mammalian social systems, and in many cases, for example lions, African elephants, Black-tailed prairie dogs and Belding's ground squirrels, the females within a group are relatives. In Greater spear-nosed bats, females are not generally related because all juveniles

◀ **Up in the rafters,** a group of female Long-eared bats, with young that are left in the roost at night when mothers leave to forage.

▼ **Cluster of Greater spear-nosed bats** with nursing young. In Trinidad mating takes place in the day-roosting groups between October and December and each female gives birth to a single young; most bats are born within a few days of each other in early April. A male which resides with a harem during the previous mating season (October–December) fathers most bats born to his harem females. If this male is displaced between the mating season and the birth of young, the new harem male does not kill or interfere with the young. Since females give birth only once a year, such behavior would not hasten a female's ability to mate and reproduce with the new male. (See boxed feature.) (The colored rings were used to identify individuals.)

A number of physiological changes occur during hibernation which allow the body temperatures to be reduced and energy stores to be eked out. (A similar function is served by the daytime torpor of many temperate bats in summer—discussed towards the end of this section). Several groups of animals hibernate, for example bears (Carnivora), squirrels and dormice (Rodentia), hedgehogs (Insectivora), but none to the degree of many bats. The body temperatures of most mammal hibernators fall less than 10°C (18°F) from the normal active temperature, whereas the core body temperatures of some hibernating bats drop to slightly below 0°C (32°F). The lowest such temperature recorded is −5°C (23°F) for the Red bat.

In the fall, temperate bats rapidly gain weight as they accumulate food reserves, mostly in the form of subcutaneous fat, which can account for up to one-third of the total body mass at the beginning of hibernation. The change from summer to winter habit is sudden and may be triggered by an inter-relationship of daylength, temperature and food availability, combined with the body mass an individual bat has achieved. Generally old adult females are first to begin hibernation, followed in succession by adult males and juveniles.

It is not known how bats choose their hibernation sites (hibernacula). Their individual choice is crucial to their eventual survival through to spring. The lower the body temperature they can tolerate the longer their energy stores will last. However, low temperatures may have disadvantages, such as increased susceptibility to disease. Each species has its preferred range of temperatures. For example, the Brown long-eared bat of Europe and the Red bat of North America hibernating in hollow trees survive variable temperatures down to slightly below 0°C (32°F) while the European Greater horseshoe bat prefers the warmer 7–12°C/45–54°F) and more stable temperatures found in caves and mines.

Tree-holes and similarly exposed and poorly insulated sites are chosen by Noctule bats and many other hibernating bats which often gather in large clusters. They must prevent themselves freezing and use energy to maintain a warmer temperature with minimal cost to each individual. In cold weather, single roosting bats under the same conditions would need to move to a better insulated or warmer site to maintain the same level of energy consumption. In areas with very cold winters, for example central and eastern Canada and northeastern Europe (where the January isotherm is below −5°C/28°F), few bats hibernate in hollow trees but bats migrate south in the winter to places with less extreme climates where clusters of up to 1,000 bats are known in large trees.

Caves, mines and fortifications are used by many bats which prefer less variable temperatures. Some, like the Greater horseshoe bat, appear to select sites very precisely according not only to temperature but also to the quantity of their individual energy store. Old adult females weighing 26g (0.9oz) in November select temperatures of 11.5°C (52°F), while in April they are found in roosts of 8.5°C (47°F) and weighing 21g (0.75oz). Comparable figures for first year females are 22g (0.8oz) at 10.5°C (51°F) in November but 16g (0.6oz) and 6.0°C (43°F) in April.

disperse prior to their first birthdays. Young females born in different parental groups in the same cave and in different caves assemble to form new stable groups; rarely, young females may join established harems.

It is not certain why these stable groups of females form. Normally they travel alone and independently of one another to their foraging areas, but occasionally they "swarm" at a large patch of food, suggesting that cooperation may occur on the foraging grounds. This could involve sharing food or information about the location of food, or defending food from members of other groups. Whatever benefits these females obtain from living in groups, it is apparent that kin-selection plays no role in determining group membership or group stability.   GFG

It is important for all hibernating bats that humidity is high, usually over 90 percent, to prevent excess evaporative losses which would necessitate more frequent awakenings to drink. This is particularly important for species like horseshoe bats that hang in exposed sites wrapped in their wings. The other hibernating bats fold their wings at their sides and often seek crevices where evaporation will be negligible.

Individual bats, such as horseshoe bats, return to exactly the same roost each winter and for a given individual five or more precise sites are used in succession depending on the changing temperature needs throughout hibernation.

Bats do not hibernate continuously but periodically awake, sometimes actually flying to a new site, and others remaining *in situ*. Why they wake is puzzling, but a simple explanation is that they need to eliminate surplus water and waste products, which are toxic to tissues. Biologists have long noted that bats urinate shortly after being disturbed and this is part of the process of reestablishing a physiological balance (homeostasis). Some bats awake approximately every ten days, while others may go as long as 90 days. However, in the wild one cannot be sure that a bat that appears to have not moved in 90 days has not woken, urinated and returned to torpor without moving. Periods of torpor tend to be longer early in hibernation when warmer sites are selected but as the winter progresses and food reserves become depleted cooler roosts are preferred. It might be expected that bats choosing cooler sites would awake less frequently than the same species occupying warmer roosts. However, while this may be so in early winter, many bats in cooler areas awake with increasing frequency towards the end of hibernation.

Some cave-roosting species characteristically hibernate singly, or occasionally in small groups. Other species, such as the Little brown and Gray bats of North America and the Large mouse-eared and Shreiber's bent-winged bats of Europe and Asia, form clusters numbering tens or hundreds of thousands of individuals. These aggregations may have bats packed in densities of over 3,000 per square meter, (270/sq ft). The purpose of these dense gatherings is not understood because the temperatures within the clusters are often similar to those of individuals of the same species which are roosting separately. However, these clustered animals are often heavier at the end of hibernation than comparable bats that have roosted singly.

▲ **Daytime camp** of Spectacled fruit bats. Only large fruit bats roost in such exposed sites, sometimes stripping away leaves to improve vision. These bats are at the mercy of the elements and wrap their wings tightly around themselves when cold or wet, or hang with flapping outstretched wings when hot.

◄ **In cold storage.** The dew covering of this hibernating Daubenton's bat (*Myotis daubentoni*) indicates that its body temperature has dropped to that of its very humid surroundings in a cold cave.

▼ **Not clustered for warmth,** a hibernating group of Little brown bats in a cave roof. The temperature within this cluster would be similar to that of a solitary bat, so the reason for hibernating together is not known.

In contrast to hibernation, many temperate bats enter periods of torpor in their day roosts during summer when there is no overriding need to maintain a higher rate of metabolism. Vesper or common bats, and horseshoe bats can tolerate by far the widest range of body temperatures. These insectivorous bats which do not attempt to maintain a more or less constant temperature when living in temperate climates are termed heterotherms as distinct from homeotherms. For example in man, a homeotherm, normal temperature fluctuation is within 2°C (3.6°F) of about 37°C (98.6°F). Bats that hibernate often have active temperatures around 38–40°C (100.4–104°F) and up to 42°C (107.6°F) in flight, but they can allow their temperature to drop about 30°C (86°F) during digestion (often taking less than one hour), and subsequently down to the temperature of their surroundings. Corresponding heartbeat rates range from over 1,000 per minute to less than 20.

This ability to lower temperatures, and hence save energy, is particularly important for bats that live in cool temperate climates and depend primarily on flying insects, because insect abundance, even in summer, is variable from night to night. On wet, windy, cool nights insects will not fly and hence some bats will not even attempt to leave their day roost. Males generally become torpid at any time during the year but adult females in late pregnancy do not do so as they need to maintain higher metabolic rates so that the fetal development continues, ensuring young are born at the time of year when food is most plentiful.

Bats that hibernate may become torpid at any time in summer, especially in cold weather when food is absent. However, torpor in summer is less extreme than torpor in hibernation. The physiological differences

between summer and winter torpor are not clearly understood. Some species in the tropics may enter a period of summer torpor (or aestivation). Like hibernators, they put on food reserves when food is plentiful then enter a type of torpor when food is sparse. Often their body temperature is about 30°C (86°F), much higher than a true hibernator.

Ecology

Bats occupy niches in all habitats except polar or the highest alpine regions and the oceans. Most are insectivorous, but there is a wide range of diets: insects, caught in flight and at rest; other arthropods, including scorpions, woodlice and shrimps; vertebrates, including mice, other bats, lizards, amphibians (see pp810–811) and fish, and blood of mammals or birds, as well as fruits, flowers, pollen, nectar (see pp814–815) and some foliage. While most bats specialize on a relatively narrow diet range, with none more limited than the Common vampire, which feeds throughout its life on the blood of mostly one breed of cattle (see pp812–813), some, like the Greater spearnosed bat, are omnivorous, feeding on vertebrates, insects and fruit (see pp796–797).

Nearly all bats feed at night and roost during the day at a variety of sites depending on species. In cool temperate regions it is advantageous to select roosts large enough to contain a large number of bats, so that each bat minimizes heat and evaporative water losses. Fine tuning of requirements may be achieved by bats moving round the roost either throughout the day or seasonally and by spacing themselves at varying distances. For example, a Yuma myotis in California moves down from the warmer air at the top of the roost when temperatures reach about 38°C (100.4°F), then flies off if temperatures exceed 40°C (104°F).

There are three main types of roost (caves; holes or crevices; and the open) and each species tends to specialize in one type. Caves insulate against climatic changes, but are unevenly distributed, although generally large numbers of bats may be accommodated in a single site. Tree and other cavities are more widely distributed but can accommodate smaller numbers and are more exposed to climatic changes. Only the large flying foxes hang in exposed camps from tree branches which are often deliberately stripped of leaves so as to improve visual observation within the colony. These bats are at the mercy of the elements and wrap their wings tightly around themselves when cold or wet, or hang with flapping outstretched wings when hot.

Social organization in colonies is poorly known (but see pp796–797). Colonial roosts have the advantage of energy conservation, but at night the bats have to fly farther in search of food. For example, cave colonies of the insectivorous Mexican free-tailed bat in the southwestern United States can total about 50 million individuals, which consume at least 250,000kg (550,000lb) of insects nightly, collected from many hundreds of square miles. They fly to feeding areas in tens or hundreds and undertake group display flights similar to those of flocking birds like starlings. These are thought to attract other bats to good feeding areas.

For some unknown reason Natterer's bats roosting in bat boxes in Germany invariably emerge and return in a specific order, with the earliest bats out, the last to return.

Colonies of the small, insectivorous, Sac-winged bat in Central America maintain annual home ranges that encompass a variety of habitats. Insect abundance is very patchy, occuring at any one time over those plants which are in flower. Thus the size of each bat colony's home range is correlated with the distribution of vegetation types. This species forms year-round harems which roost on the side of trees with large buttresses. Up to five harems may be found on one tree, with much movement of females between adjacent harems. Territorial adult males entice females by elaborate displays involving vocalizations and flashing of the species' wing-sacs. The complex vocalizations include audible (to humans) "songs" lasting 5–10 minutes with repeated phrases. Intruding males will be pursued and even attacked. The territorial males similarly defend dense patches of insects from other males but allow their harem of up to eight females to feed. Large patches of food may be divided amongst several males with their harems.

Some bats feed on plants, mostly trees and large cacti, for nectar, pollen or fruit and such plants appear to have two flowering and fruiting patterns. There are those that flower and fruit in short seasons ("big bang" types) and others that produce small quantities over many months ("steady state" types). "Big bang" types tend to be visited by a large range of generalist bat species, while "steady state" plant species often have a species of pollinating bat that is specific to them. "Big bang" plants produce huge quantities of flowers, an adaptation which compensates for the fact that visiting bats destroy many flowers, while "steady state" plants are far more efficiently pollinated

and produce smaller quantities of flowers.

A typical "big bang" tree is the durian (*Durio zibethinus*) of Southeast Asia, which is pollinated by the cave-roosting Dawn bat. Throughout the year this species of bat visits at least 31 flower species. Dawn bats forage over an area of at least 40km (25mi) radius in flocks whose members may benefit from collective scanning of a wide area.

A typical "steady-state" plant is the Passion flower (*Passiflora mucronata*), which produces flowers at a rate of one per branch per night over several months that are pollinated primarily by Pallas' long-tongued bat. In its search for the widely dispersed flowers, the bat flies to and fro along a beat sipping nectar several times each night from the same flower, effectively bringing about cross-pollination.

Among the large array of fruit-eating spear-nosed bats in Central and South America, competition between species for limited food resources appears to be reduced by adjustment of feeding times. Four species of spear-nosed fruit bats (three *Artibeus* and a *Vampyrodes*) visit the same species of fig tree at different times and this probably reduces conflict between bat species. Territorial defense of fruiting trees does not appear to occur, although there is often much squabbling over fruits when large numbers of bats of the same and different species are present. Flying foxes tend to land in trees and consume fruit *in situ*, while spear-nosed fruit bats usually pluck fruits and fly to a safe perch to eat them. Predators, such as snakes and carnivores, often gather in trees with ripening fruit, knowing that bats will come to feed.

Typically bats are gregarious at roosts, forming mixed-sex groups for much of the year, but adult females often segregate for birth and weaning. Many of the insectivorous species from temperate regions are in this category, such as the Eastern pipistrelle, Big brown bat and Little brown myotis in North America, and the Common pipistrelle, Large mouse-eared bat, Greater horseshoe bat and Schreiber's bent-winged bat in Europe. In some species both sexes remain separate in single-sex groups except when mating, the extreme examples of this being species that are solitary roosters, for example the flying fox *Epomops franqueti* in tropical Africa, and the small insectivorous Red bat in North America. Some bats form harems which may be more or less permanent throughout the year, for example the Sac-winged bat or Greater spear-nosed bat in the neotropics (see pp796–797), and bamboo bats (*Tylonycteris* species) in South-

▶ **A second to live**—a Greater False vampire bat (*Megaderma lyra*) swoops on an unsuspecting mouse. False vampires are truely carnivorous feeding on rodents, birds, frogs, lizards, fish, and even other bats, as well as insects and spiders. They hunt among trees and undergrowth flying close to the ground. Captured prey is sometimes taken back to the roost—a hollow tree, cave or building—to be eaten.

▼ **Sipping nectar**—a Gray-headed flying fox (*Pteropus poliocephalus*) feeds from a eucalyptus flower.

The present figure of over 130 genera of plants, representing many hundreds of species, known to be bat-adapted will undoubtedly increase as research proceeds and many more bat/plant interactions are discovered. Since bats often form half the mammalian species in rain forests, their survival is vital to the ecology of these and to the well-being of other habitats. Even though some species form enormous colonies which tend to make people consider the species' numbers to be limitless, the number of colonies is normally small.

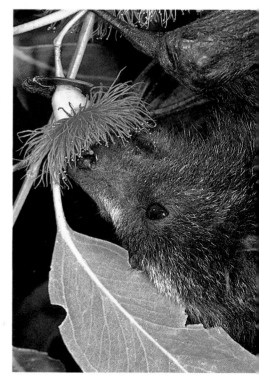

east Asia; others form harems just for mating, for example the Noctule bat in Europe. Others apparently occur in monogamous family units, for example the New World False vampire and the Painted bat from Asia.

Myths and Conservation

Europeans have long regarded bats with superstitious fear. Why some people fear bats is not known, but poorly understood animals are often feared. Bats are generally secretive by day and emerge at night, traditionally a time when the Devil is at work. In European medieval art the Devil is often depicted with bat wings. In more ancient civilizations, notably the Persians and Chinese, bats symbolized longevity and happiness and highly stylized bats decorate all kinds of objects from buildings to furniture, fabrics and porcelain. Superstitions like "blind as a bat" and "bats becoming inextricably stuck in a woman's hair" are now disappearing due to education programs, mostly in the popular press. Indeed bat conservation projects all contain a large element of education.

Bat populations are declining rapidly worldwide and several species have recently become extinct. Bats that form large colonies, especially in the tropics, provide valuable sources of high quality meat, and their guano (excrement) is used and traded as fertilizer.

In Africa, Southeast Asia and islands in the Indian and Pacific oceans, the large fruit-eating flying foxes of the family Pteropodidae have been caught in small numbers for their meat for hundreds of years. Recently an increase in accessibility to habitats and in numbers of firearms have resulted in over-exploitation and now many species are threatened with extinction. On the western Pacific island of Guam the flying fox, *Pteropus tokudae*, has become extinct and a larger relative, *Pteropus mariannus*, which formerly lived throughout the Mariana island chain, is very vulnerable. It is extinct on several islands but survives on uninhabited well-vegetated volcanic islands, although even here it is sought after by raiding hunters. Other flying foxes on Yap and Samoa are similarly close to extinction, and some of their meat is traded to Guam. Hunting is very inefficient because several bats may be hit by shot simultaneously, some injured bats flying away before dying, and hence only some of those hit are usually recovered.

Thirty years ago thousands of the endemic Rodriguez flying foxes occupied Rodrigues Island east of (and a dependency of)

Mauritius. Now only about 200 remain due to deforestation and consequent removal of roosting sites and food. The few remaining hectares of forest are critically threatened by encroaching agriculture and collection of firewood. These bats have always suffered periodic natural disasters since Rodrigues is in the Indian Ocean cyclone belt and every three years or so a large proportion of bats die by being blown out to sea, or subsequently by starvation since all the food is stripped off the trees. A breeding colony has been established by the Jersey Wildlife Trust in the English Channel Islands and survival of the species seems assured.

▶ **Silent hunter of Australian nights,** the Australian ghost bat (*Macroderma gigas*) has two notable statistics to its name. Firstly it is the largest microchiropteran bat; and secondly it is listed as vulnerable to extinction in the Red Data Book due to habitat loss—the fate of many bats with restricted distribution.

▶ **Survival in question.** BELOW Island species such as this Seychelles fruit bat (*Pteropus seychellensis*) are very sensitive to habitat disturbance and hunting.

▼ **Whirlwind of bats** leaving a cave at dusk in Java. Cave-dwelling bats are particularly susceptible to destruction through blockage of, or trapping at, narrow entrances.

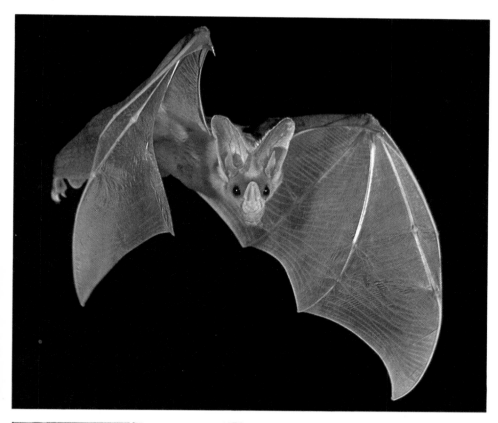

a single small entrance through which all bats must pass. In Thailand, the small insectivorous Wrinkle-lipped bat forms cave colonies of over one million animals, from which local people have been gathering guano weekly for over 200 years to sell to farmers as fertilizer. Hunters now set nets around the cave entrances to catch, for meat, the fruit bats that roost in the same cave, at the same time killing large numbers of Wrinkle-lipped bats which are discarded. In consequence, guano production has declined and a local livelihood is threated, as well as the bats.

Similar population losses have occurred in many countries worldwide, for example of the Mexican free-tailed bat in the southwest United States. Colonies up to 50 million were formerly estimated and more that 100,000 tonnes of guano have been extracted from one cave alone, but in this species declines in numbers appear to be related to the use of agricultural pesticides rather than to hunting.

In the United States fear of bats has been deliberately generated by the multi-million dollar pest-control industry to attract business for the destruction of bats in buildings. In reality, the anti-coagulant pesticide used in houses to kill bats is a more serious health hazard to humans than the bats themselves. The bats do not present a significant health hazard unless people pick up sick bats. In temperate regions the major causes of population decline appear to be loss of habitat and deliberate and accidental killing. Bats that are disturbed regularly while in hibernation will die through starvation and many roost sites such as caves and hollow trees have been blocked or felled, and habitats have been severely modified mainly through changes in agricultural and forestry practice, resulting in a reduction in the size of insect populations. Remedial timber treatments in buildings used by bats are known to cause the deaths of such colonies. To counter the decline in bat populations worldwide, education programs are being conducted by Bat Conservation International, an organization which gathers information on the status, distribution and threats, and implements research surveys and education programs (see pp 808–809).

Bats are now protected by legislation in all European and many other countries; some important roost sites and feeding habitats are being specially designated. In general there is a severe lack of detailed knowledge about the ecological requirements of most bat species, and hence it is difficult to plan appropriate conservation measures. RES

Because of their importance for pollination and seed dispersal, the decline of some flying foxes populations is having serious effects on crops, and tropical forest and savanna habitats. For example, in Malaysia, colonies of the cave-dwelling Dawn bat in the Batu caves, near Kuala Lumpur, are threatened with destruction because of limestone quarrying and by over-exploitation by hunters using fishing nets to catch them. Formerly these bats ranged up to 40km (25mi) to find nectar and pollen, particularly from durian, each hectare of which produced $10,000-worth of fruit annually. Since the decline in bat numbers durian production is failing.

In West Africa the Straw-colored flying fox seasonally forms clusters of one million bats. In the past a few bats were speared for food but modern shotguns can kill up to 30 per shot. Agricultural departments in several countries have suggested that these bats are a pest and should be thinned out or killed completely. However, recent research has demonstrated that this species is the most important disperser of seed in rain forests and savannas. For example, it disperses seed of iroko, a hardwood tree in West Africa which is the basis of a $100m-a-year industry.

Cave-dwelling bats are particularly vulnerable to destruction because often there is

THE 19 FAMILIES OF BATS

Abbreviations: HBL = head-and-body length; TL = tail length; FL = forearm length; WS = wingspan; WT = weight. Approximate nonmetric equivalents: 2.5cm = 1in; 28g = 1oz.
* contains species CITES-listed; E contains species endangered.

Despite their abundance, bats have not been studied in as much detail as most other mammalian orders. In recent years, partly as a result of technological advances which facilitate the study of nocturnal animals, an upsurge in interest has begun rapidly to improve our knowledge. New species are being discovered and others reclassified. However, various authorities recognize differing numbers of families, genera and species, and substantial changes will continue. Undoubtedly there are many new species awaiting discovery, especially in tropical forests. Unfortunately, with the loss of forests, species may become extinct before discovery.

With such a diverse order of mammals it is difficult in a few words to give a true flavor of the variety of shapes, sizes and life-styles. Diagnostic characters for each family, together with their size range and ecological niche, are given.

Suborder Megachiroptera

Flying foxes * E
Family Pteropodidae
Flying foxes or Old World fruit bats.

Over 173 species in 44 genera. Old World tropics and subtropics from Africa to E Asia and Australasia including many islands in the Indian and Pacific oceans east to the Cook Islands. A few species reach warmer temperate regions north to Turkey and Syria and to the extreme south of Africa and SE Australia.
Size ranges from small to the largest bats, having wingspans approaching 2m: HBL 5–40cm; tail absent; FL 3.7–23cm; WS 30–200cm; wt 15–1.500g.
Most species with dog-like faces, large eyes and conspicuous widely separated simple ears. Most have a claw on the second digit in addition to the thumb and males are generally larger than females. The majority do not navigate by echolocation but instead use their excellent eyesight. Members of the widespread genus *Rousettus* use poorly developed echolocation emissions in addition to eyesight, especially to locate their roosts, which are often in caves.
Coat: drab brown but a few species are brightly colored, eg **Rodriguez flying fox** (*Pteropus rodricensis*) varies from black to silver, yellow, orange and red. **Tube-nosed bats** (genus *Nyctimene*) are brightly colored with speckled membranes and a dorsal stripe; cryptic coloration helps avoid predation while roosting among foliage.
A few species have secondary sexual characters which develop in males for use in attracting females, eg tufts of light-colored or white hair emanating from glandular patches on shoulders. Males attract females by singing loudly and flashing the hair tufts. Diet primarily plant material, chiefly soft ripe fruits but also flowers, nectar and pollen which are taken by some smaller species (eg *Macroglossus* species) by their long tongue bearing bristle-like papillae; some may eat insects. May eat leaves at times of extreme food shortage. Flying foxes are essential to the pollination of many plants and to dispersing seeds. Some of the medium-sized and larger fruit bats are highly gregarious, forming colonies which may exceed one million individuals. The very large **Common flying fox** (*Pteropus vampyrus*) of SE Asia and Indonesia and the **Indian flying fox** (*P. giganteus*) of the Indian subcontinent often roost in the same sites throughout the year, whereas an ecologically comparable species in Africa, the **Straw-colored flying fox** (*Eidolon helvum*) usually migrates seasonally. The largest colonies, formed by the cave-dwelling **rousette fruit bats** (*Rousettus* species), number several million; these bats, which feed primarily on nectar and pollen, must fly considerable distances nightly to find sufficient food for survival.

Suborder Microchiroptera

Mouse-tailed bats
Family Rhinopomatidae
Mouse-tailed, rat-tailed or long-tailed bats.

Three species in 1 genus (*Rhinopoma*). N Africa to the S Sudan, Middle East, India and SE Asia, Sumatra. Arid or savanna areas, also agricultural and disturbed habitats.
Size small: HBL 5–8cm; TL 6cm; FL 5–7cm; wt 10–25g.
Tail exceptional being about as long as the head and body length and entirely free of membrane. Ears connected across the forehead with a distinct but simple tragus. Small noseleaf looking like a pig's muzzle. Insectivorous. Roosts of many thousands known, normally in caves or rock crevices but have been associated for several thousand years with the pyramids, temples, palaces and other man-made structures. Do not hibernate but accumulate large quantities of fat seasonally and become torpid during cold dry weather when insects unavailable.

Sheath-tailed bats
Family Emballonuridae
Sheath-tailed or sac-winged bats.

Fifty species in 13 genera. Worldwide including many islands in the Indian and Pacific Oceans. Include some of the world's smallest bats with others medium-sized: HBL 3.5–10cm; TL 0.5–1.3cm; FL 3–8cm; wt 3–40g.
Tails short, projecting on the upperside out of the tail membrane so that the tip is free but ensheathed. Many species have small glandular wing sacs (larger in males) found in the propatagium and open to the upper surface; purpose unknown, but may attract females. Other species have large throat glands whose secretions often smell strongly. Coat mostly drab brown or black, some white and a few have cryptic patterns and tufts of hair. One of the smallest bat species—the monotypic **Probosis bat** (*Rynchonycteris naso*) of Latin America—has grizzled fur with two light curved lines on the back and tufts of hair on the forearms. Insectivorous. Roost in all kinds of sites from hollow trees to buildings and caves as well as in the open. Roost holding their wings at about 45° to the body, appearing like dead leaves. Some highly colonial, eg **tomb bats** (*Taphozous* species); others solitary. Most occur in the tropics but some in cooler areas where they become torpid and may hibernate.

Kitti's hog-nosed bat * E
Family Craseonycteridae
Hog-nosed or Butterfly bat.

Single species (*Craseonycteris thonglongyai*), first described in 1974. W Thailand. Bamboo forests and teak plantations (natural vegetation had been removed mostly by the 1950s). The world's smallest bat and mammal: HBL 2.9–3.3cm; tail absent; FL 2.1–2.6cm; WS 15–17cm; wt 1.5–3g.
Tailess, but interfemoral membrane stretched between its rather thin legs. Ears relatively large with a tragus (prominence in front of outer ear opening). Nose glandular and pig-like in appearance. Upperparts brown to reddish or gray, underside paler, wings darker. Insectivorous and colonial, forming small roosting groups in caves. Total world population is thought to be about 200 bats.

Slit-faced bats
Family Nycteridae
Slit-faced, hollow-faced or hispid bats.

Eleven species in one genus (*Nycteris*). Africa and adjacent Asia, E Mediterranean and Red Sea, *N. javanica* in Malaysia and Indonesia. Arid areas as well as rain forests. Medium sized: HBL 4–8cm; TL 4–8cm; FL 3.5–6cm; wt 10–30g.
Complex nose-leaf divided by a groove containing nostrils towards the muzzle tip and a deep pit between the eyes. Ears large and tragus small; unique amongst mammals in having a tail that is T-shaped at tip.
Coat: long, usually rich brown to grayish.
Insectivorous, including arthropods such as scorpions and spiders. Roost in caves, rock cliffs and animal burrows as well as trees and buildings where they often roost singly. Sometimes small groups occur with other species. Have several periods of estrus (heat) and give birth twice per year.

False vampire bats * E
Family Megadermatidae
False vampire or yellow-winged bats.

Five species in 4 genera. Old World tropics from C Africa, through India and SE Asia to the Philippines and Australia. Among the larger microchiropteran bats with the **Australian ghost bat** (*Macroderma gigas*), the largest: HBL 6.5–14cm; tail absent; FL 5–12cm; wt 20–200g.
Ears very large and erect joined over the forehead; tragus divided. Eyes large. Nose-leaf prominent, largest in the African **Yellow-winged bat** (*Lavia frons*).
Coat: drab but very variable, ranging from bluish-gray to brown and whitish. Yellow-winged bat very colorful with ears and wings yellowish-orange and the fur usually bluish-gray to olive-green. Diet variable, including small vertebrates, bats, small mammals, birds, reptiles, amphibia and fish as well as insects and spiders; legs, feathers and wings often litter the ground beneath perches. Color linked to their favorite roost sites. Drab-colored species mostly roost in caves, some solitary and others in small colonies. The Yellow-winged bat hides among foliage in bushes or trees where it can watch for passing food, mostly insects which it will catch and eat by day or night. Hunt rather like flycatchers, returning to a favorite perch after each foray.

Horseshoe bats [E]
Family Rhinolophidae

About 69 species in one genus (*Rhinolophus*).
Old World especially in the tropics; a few species in temperate Europe, Asia and Japan.
Mostly small: HBL 3.5–11cm; TL 2.5–4.5cm; FL 3.5–7cm; wt 4–40g.
Vernacular name derives from the horseshoe-shaped front part of the complex nose-leaf. Nostrils open central to the horseshoe and a sella projects forward with a generally pointed lancet running lengthwise. Ears usually large, pointed and always without a tragus. Like the flying foxes, the heads of these bats face downwards (ventrally) whereas most bats look along their long axis. Hind limbs poorly developed, unable to walk quadrupedally. Broad wings makes them among the most maneuverable of species in flight. Coat: very variable color, from yellow through red to dark brown, gray and black.
Insectivorous, catching their food close to or from the ground. Roost mainly in caves or mines but also hollow trees and buildings. At roost fold their wings around themselves. Some species solitary but most are gregarious, sometimes forming huge colonies of many thousands. Several species hibernate and some aestivate.

Leaf-nosed bats
Family Hipposideridae
Leaf-nosed or trident bats.

At least 61 species in 9 genera.
Old World tropics from Africa through SE Asia to the Philippines, Solomons and Australia.
Closely related to the horseshoe bats but their size-range much larger: HBL 2.5–14cm; tail mostly absent but may be up to 6cm; FL 3–11.5cm; wt 4–120g.
Nose-leaf lacks a well-defined horseshoe and the lancet is a transverse leaf often with three points; no sella arising from the center. Ears generally large and pointed with no tragus. Mostly drab grays and browns but a few species brightly colored, orange or yellowish-gold, occasionally with whitish fur patches.
Diet: insects and other arthropods, but larger species such as the **Great round-leaf bat** (*Hipposideros armiger*) may opportunistically take small vertebrates as they do readily in captivity.
Roost mainly in caves but use all

kinds of shelter, including animal burrows. A few species apparently roost singly but all may be colonial and some form huge colonies numbering hundreds of thousands and possibly millions in large caves. However, each bat usually roosts slightly apart, at wing-tip distance, rather than forming dense clusters like many other species. Some species regularly roost in mixed-species groups and are often associated with a number of species from different families.

Leaf-chinned bats
Family Mormoopidae
Leaf-chinned, naked-backed or ghost-faced bats.

Eight species in 2 genera.
Extreme SW USA through C America and Caribbean south to central S Brazil.
Small to medium sized: HBL 4–7.7cm; TL 1.5–3.5cm; FL 3.5–6.5cm; wt 7–25g.
Lack a nose-leaf but have leaf-like development of the lips so that a dish-shape can be created. Several species apparently have naked backs but this is due to the wing membranes joining at the upper midline. Ears small with a tragus. Tail projects slightly beyond the end of the interfemoral membrane. Fur short and dense, reddish-brown to brownish-gray.
Insectivorous, feeding low, near to, or over water. Primarily roost in caves where at least half a million may occur, often producing large quantities of guano which is sometimes mined for fertilizer. Large colonies occur at over 3,300m in the Andes. Do not hibernate.

Bulldog bats
Family Noctilionidae
Bulldog or fisherman bats.

Two species in one genus (*Noctilio*).
Latin America from Mexico to Argentina.
Fishing bulldog bat (*Noctilio leporinus*): HBL 9.5–14cm; FL 7–9.5cm; wt 70g. **Southern bulldog bat** (*N. albiventris*): HBL 7cm; FL 5.4–7cm; wt 15–25g.
Fishing bulldog bat has short orange or yellowish fur which sheds water readily. This species is highly adapted for catching and eating fish, the most characteristic feature being huge feet on long legs with incredibly sharp claws. The toes are highly flattened laterally so as to present minimal resistance when pulled through water while attempting to gaff fish. Fish quickly transferred to the mouth where a combination of long thin

canines and large jowl-like upper lips, which form internal pouches, secure the slippery fish. Fish are thought to be caught by the bat echolocating ripples as the fish break the surface. Fish up to 8cm long are taken. Fishing is generally undertaken in pools, slow-flowing rivers or sheltered coastal lagoons, but bats have been seen over open water areas. They also take insects, especially if fish are difficult to find. The smaller species is primarily insectivorous.

Short-tailed bats [E]
Family Mystacinidae

Two species (*Mystacina tuberculata* and *M. robusta*). New Zealand and adjacent islands.
Small bats: HBL 6cm; TL 1.8cm; FL 3.6–4.9cm; wt 7–35g.
Thumb and toe claws have extra projection or talon, unique among bats, which may aid running, climbing, or burrowing for food or excavating roost sites. Fur mole-like, gray, dense and velvety. Wings with thick membranes, and fold or roll up to facilitate movement on ground and digging. Tail dorsally perforates the upperside of the interfemoral membrane, as in the Emballonuridae. Ears simple and separate, with a long tragus. The tongue is partly extensible with papillae at its tip. Omnivorous, eating fruit, nectar and pollen as well as insects and other arthropods. Scurry through leaf-litter looking for animal food but fly weakly in search of plant diet. Roost in caves, rock crevices, seabird burrows or specially excavated burrows in decaying trees. Colonial but little is known of ecology. These two species and a vespertilionid bat are the only indigenous mammals in New Zealand.

Spear-nosed bats
Family Phyllostomatidae
Spear-nosed or New World leaf-nosed bats.

About 140 species in 47 genera.
New World from extreme SW USA throughout C America and Caribbean south to N Argentina.
Generally robust animals with sizes ranging from small to the largest American bat, the **False vampire**, (*Vampyrum spectrum*): HBL 4–13.5cm; tail absent or 0.4–5.5cm; FL 3–11cm; wt 7–200g. Wingspan up to 1m in *Vampyrum*.
Most species have a spear-shaped nose-leaf but five have none, or a more complex shape. Ears usually simple but may be very large and a tragus is invariably present. Apart from one more or less white bat, the

Honduran white bat, (*Ectophylla alba*), others are brown, gray or black, occasionally with hair tufts which are red or white and associated with glands producing oily secretions. Several species have longitudinal whitish lines on the face and/or body. Diet mainly insects, but a few carnivorous or omnivorous, eating small bats and other mammals, birds, reptiles and amphibians; and many feed on fruit, pollen and nectar, aided by the presence of a long tongue bearing bristly papillae in many species, eg **Spear-nosed long-tongued bat** (*Glossophaga soricina*), **Geoffroy's long-nosed bat** (*Anoura geoffroyi*) and **Mexican long-nosed bat** (*Leptonycteris nivalis*).
Roost sites variable including caves, mines, culverts, tree-hollows, animal burrows, termite nests and among foliage. A few species of the genera *Uroderma* (**tent-building bats**) and *Artibeus* make shelters by biting through leaf-ribs and hence forming "tents." Generally form small aggregations but some live in colonies of many hundreds. None hibernate. One or two species in the USA and high in the Andes may aestivate.

Vampire bats
Family Desmodontidae

Three species in 3 genera.
Mexico to N Argentina.
Medium-sized bats: HBL 6.5–9cm; FL 5–6.5cm; wt 15–45g.
External tail lacking. Muzzles appear swollen and glandular, giving the impression of a nose-leaf. Fur is grizzled, being shades of brown, and one species, the **White-winged vampire** (*Diaemus youngi*), has white wing-tips and edges. Teeth highly specialized. The **Common vampire** (*Desmodus rotundus*), has 22 teeth, of which only the 6 chisel-like incisors and 4 razor-like canines play any part in feeding. A small sliver of skin about 3 × 8mm is removed, usually from an area devoid of hair or feathers. The blood that flows, aided by anticoagulants in the saliva, is lapped with the tongue which has two lateral grooves that narrow and widen during feeding. The Common vampire feeds on mammalian blood, mostly from domesticated species (see p812–813); the other two much rarer species apparently prefer birds. Colonial, but groups tend to be small, usually much less than 100. Some authorities include this family in the Phyllostomatidae.

CONTINUED ▶

Funnel-eared bats
Family Natalidae
Funnel-eared or long-legged bats.

Eight species in one genus.
N Mexico through Colombia and Brazil, and including many E Caribbean islands.
Small and delicate: HBL 3.5–5.5cm; TL 5–6cm; FL 2.5–4.2cm; wt 4–10g.
Lightly built with long slender wings, legs, and with tails longer than the head and body. Ears funnel-shaped and large with a short triangular tragus. Nose simple. Colors variable although rather drab from brown to reddish, yellowish and gray.
Little is known about these bats, but they occur in small groups or larger colonies mostly in caves, but also in hollow trees and elsewhere. Apparently feed on tiny insects which are caught using the very high frequency ultra-sounds of up to 170kHz. There is some evidence that the bats aestivate but they do not hibernate.

Thumbless bats
Family Furipteridae
Thumbless or smoky bats.

Two species in 2 genera.
Tropical S America from Panama and Trinidad to Peru and Brazil.
Small bats: HBL 3.5–6cm; TL 2.4–3.6cm; FL 3–4cm; wt 3–5g.
Thumbs virtually absent (actually present but functionless). Ears funnel-shaped as in funnel-eared bats but broader, covering the eyes. Coat brown to gray. Virtually nothing is known of these bats but some have been found roosting in caves and they feed on small insects.

◀ **Tropical tent-makers.** These tent-building bats (*Uroderma bilobatum*) bite through the ribs and veins of leaves so that they bend over giving protection from sun, wind, rain and predators. The bitten spots also afford good footholds.

Disk-winged bats
Family Thyropteridae
Disk-winged or New World sucker-footed bats.

Two species in one genus (*Thryroptera*).
Hondurus to Peru and Brazil; tropical forest.
Small bats: HBL 3.4–5.2cm; TL 2.5–3.3cm; FL 2.7–3.8cm; wt 4–5g.
Wrists, ankles and functional thumb have disk suckers borne on short stalks; just one sucker able to support the weight of the bat roosting in smooth furled leaves such as those of *Heliconia* or bananas. Ears funnel-shaped with a tragus. Coat: drab reddish-browns or blackish, and whitish eventually. Diet: insects. Roost head upmost, singly or in small groups; little else known of their ecology.

Sucker-footed bat
Family Myzopodidae

One species (*Myzopoda aurita*). Madagascar. Very rare, in palm forests.
Small to medium sized: HBL 5.5–6cm; TL 4.5–5cm; FL 4.5–5cm; wt ?
Suction discs present on wrists and ankles (as in disk-winged bats) but they are not on stalks and do not appear to be as efficient in support. Fossils have been found in mainland Africa. Diet: assumed to be insects but nothing is known of its ecology.

Common or vesper bats
Family Vespertilionidae

At least 319 species in 42 genera.
Worldwide except polar regions and remote islands; genus *Myotis*, containing over 60 species, probably more widely distributed than any mammalian genus except *Homo*.
Mostly small but a few are medium to large: HBL 3–10.5cm; TL 2.5–7.5cm; FL 2.2–8cm; wt 4–60g.
Most have simple muzzles but about 10 species in 3 genera have a slight nose-leaf and 12 species in two genera have tubular nostrils. Tails substantial, extending to or slightly beyond the end of the interfemoral membrane. Ears normally separate and range from small to enormous, being most extreme in the **long-eared bats** (*Plecotus* species), where they approach the head-and-body length. Bats of five genera have wing and/or foot disks to aid grip on smooth leaves or bamboo. Coat mostly rather drab browns and grays but several genera bright yellow, red or orange, sometimes with white patches. The **Painted bat** (*Kerivoula picta*) has scarlet or orange fur, black

membranes and the finger bones picked out in orange. This cryptic coloration allows the bats to hide among flowers and foliage and is similar to the **butterfly bats** (*Glauconycteris* species), which have white spots and stripes in the cream to black fur with wings elaborately pigmented to look like dead leaves. Small clusters of *K. argentata* look like the hanging mud nests of wasps. Diet mainly insects and other arthropods, but a few species eg *Pizonyx vivesi* eat fish and others may be partially carnivorous. Roost sites include hollow trees and crevices in trees and rocks, amongst flowers and foliage, in bird nests and animal burrows. Majority cave-dwelling but have adapted to living in man-made structures including buildings, tunnels, wells, and culverts. Recently the **Northern serotine** (*Eptesicus nilssoni*) has been able to colonize treeless areas north of the Arctic circle in Lapland due to the building of permanent houses. Among the most adapted bats for exploiting a specialized roost are the tiny flat-headed sucker-footed **bamboo bats** (*Tylonycteris* species) of SE Asia, which gain access to bamboo stems through the small oval flight hole of a beetle whose larva has fed on the bamboo.
Some species roost singly or in small groups, while others form colonies which may total over one million. Colonies generally occupy different roost sites seasonally and some species are known to migrate over 1,000km either to find suitable sites for hibernation or to find adequate food. Species inhabiting temperate areas generally hibernate while some in the tropics and subtropics aestivate. Marked bats in the wild have lived over 30 years.

Free-tailed bats
Family Molossidae

At least 91 species in 12 genera.
Warm or tropical areas from C USA, south to C Argentina; S Europe and Africa, east to Korea, the Solomons and Australasia.
Most are small- to medium-sized bats, with two large species, the naked bats of genus *Cheiromeles* of which *C. torquatus* is the heaviest microchiropteran: HBL 4–12cm; TL 1.5–8cm; FL 2.7–9cm; wt 8–230g.
Robust bats with a large proportion of the thick tail projecting beyond the tail membrane. Membranes leathery and wings long and narrow, facilitating fast flight. Ears with a tragus usually joined across the forehead and directed forward,

perhaps partly for streamlining. Fur usually short and sleek, drab brown, gray or black, but some are reddish with whitish patches, eg in *Xiphonycteris spurrelli*. The large naked bats have very thick black skin with wings joining at the mid-line of the back, and large skin pouches laterally into which the wing tips are tucked while at rest. Although essentially naked, they do have some short hairs, particularly associated with a throat sac, which many species possess, and the outer toe characteristically has short stiff bristles used in grooming. All *Molossus* species are insectivorous, catching their food in flight. Some members of several genera tend only to form small groups, others, especially of the genus *Tadarida* (about 40 species), form the largest colonies of any warm-blooded vertebrate. Nursery colonies of the **Mexican free-tailed bat** (*T. brasiliensis*) in SW USA have totalled about 50 million individuals and it is possible larger aggregations exist in tropical forest areas. The largest colonies occur in caves, but other species utilize all kinds of sites from rock clefts to hollow trees, among foliage and in buildings. Occasionally *Molossus* species appear to burrow and roost in rotten timber, and naked bats can be found in earth holes. Fly faster than members of any other family and may travel several hundred kilometers each night. Some species, eg *T. brasiliensis*, migrate over 1,300 km from summer to winter sites. None hibernate.

A Problem of Conservation

The Greater horseshoe bat

In the last hundred years the Greater horseshoe bat has rapidly declined in abundance and distribution throughout its range in Europe and Asia from Britain to Japan. It is already extinct in northern parts of Europe and in some areas of the eastern Mediterranean. In Britain, colonies of several thousand were common a century ago; now the total population is only about 2,200. The causes of decline are many, but include vandalistic killing, destruction of roosts, disturbance of roosts by tourists visiting caves, and sport caving. Recently large colonies have been killed by insecticides used in timber treatments in buildings.

Recent research has been designed to ascertain the timing and rates of declines, identification and assessment of the likely causes, particularly giving emphasis to developing a conservation strategy to prevent extinction. A major problem of working on such an endangered species is to avoid contributing to a further reduction—the Greater horseshoe (*Rhinolophus ferrumequinum*) is a sensitive species whose behavior patterns easily change as a result of disturbance. Two chief lines of research were followed relating to roosting and feeding behavior.

Greater horseshoe bats hibernate in sites with humidities close to saturation, so as to minimize water losses due to evaporation. Draught-free, dark, humid areas with slight seasonal temperature fluctuations are always used, each bat choosing the best temperatures according to its own body mass. In the fall, the heaviest bats occupy the warmest sites up to 12°C (54°F), while low-weight animals in the late winter selected about 6°C (43°F). First-year, low-weight Greater horseshoes lose about 30 percent of their body mass while hibernating, while adults, which weigh 15–20 percent more, only lose 20 percent. It would be expected that older, and therefore experienced, animals would use less energy in hibernation than the young bats, but the older bats select the warmest sites where metabolic rates are higher. However older animals, particularly females, awake from torpor less frequently than juveniles, resulting in considerable energy saving as consumption increases 100-fold during arousal.

Energy is also conserved in summer by clustering, especially after feeding, so that each bat spends less energy keeping warm during digestion. With the seasonally differing environmental needs of Greater horseshoes, a large number of roost sites are required by the colony, perhaps as many as 200. Individual bats will fly more than 60km (37mi) between roosts, even in very cold weather, to find hibernation sites.

In summer, pregnant females gather, often with non-breeding adults and immature individuals (including a few males), in a maternity roost. These sites traditionally were in caves but since the 98 percent-plus decline in population over the last 10 years in Britain, maternity roosts are now sited in buildings. This change is due to the advantage of using a roost whose temperature is warmer than a cave. In southern Britain, cave temperatures are typically about 10°C (50°F), whereas roofs heated by the sun exceed 30°C (86°F). Originally cave maternity colonies were in the apex of dome-shaped areas where the large cluster comprising many thousands of bats collectively raised the temperature to about 28–30°C (82–86°F) so as to ensure continual development of the fetus at minimal cost to the mother. With the declining size of clusters each bat uses more energy to keep warm and eventually none remains for reproduction.

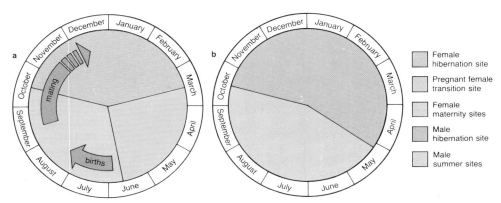

Female hibernation site

Pregnant female transition site

Female maternity sites

Male hibernation site

Male summer sites

▲ **Year in the life of horseshoe bats.** Adult males and females generally lead separate lives apart from mating. Then, it is thought, females visit a male at his traditional roost, although males are known occasionally to visit females at the maternity roost for unknown reasons. Juveniles and immatures tend to associate more with mature females than males.

(a) Females. On leaving hibernation sites (at which the oldest females tend to be solitary) pregnant females gather in increasing numbers at transition roosts before moving to maternity sites where birth and mating occurs. Most juveniles remain in the maternity roost during the summer.

(b) Males tend to occupy the same traditional site both during winter hibernation and during the summer.

▶ **Wrapped against the cold,** a Greater horseshoe hangs from a cave roof with its wings closed around it. This prevents excess water loss. By changing the degree of wrapping the bats are able to critically adjust their immediate surroundings.

▼ **A rare sight,** Greater horseshoes clustered in a cave roof.

Adult males, often with immature males, usually occupy traditional roosts for most of the year spaced out in all areas of the colony's annual range, which varies from 700–2,500sq km (270–965sq mi). Periodically these males fly to the maternity roost but do not stay. The purpose of this is unknown, but perhaps they are trying to attract females for mating when they are ready. By remaining dispersed from the maternity roost the males avoid competition for food. Sperm formation occurs in summer and by September only one male remains in each roost site. It is assumed that other males are driven off, because the same male remains often for many years. It is not definitely known where mating occurs but it is likely that females visit the males in these traditional sites as soon as they are in heat.

Analysis of fecal pellets, collection of insects at feeding sites and radio-tracking techniques have been used to study food and feeding behavior. In spring, when nights tend to be cold and windy, bats immediately enter woodland and feed preferentially on large (1g/0.04oz) cockchafer beetles. After a hot day in summer the bats fly direct to woodland, then, seemingly realizing that more insects would be available over pasture, return and forage close to the ground. If the weather turns cold again the bats first fly over pasture, covering a large area field by field, presumably in search of and remembering the previous night's dense patches of food, but finding none return quickly to woodland. Because of the high energy cost of flying, it is important that adequate food is found quickly. Sometimes bats will either fail to leave the roost or return quickly when insects are absent.

In midsummer, lactating females forage over permanent pasture which has the highest density of insects. Large noctuid moths form the major prey, but prior to hibernation the large Dor beetle becomes the dominant food.

Radio tracking has revealed many previously unsuspected facets of the behavior and roost selection of Greater horseshoes. Further work is required, but it appears that woodland and adjacent old pasture are the two key habitats for this species and both these habitats have disappeared rapidly in the last 50 years. For the Greater horseshoe bats to survive, a large number of different undisturbed roosts are required adjacent to large tracts of woodland interspersed with permanent pasture. This survival recipe will be difficult to realize, and it appears that further population declines are inevitable.

RES

To Eat or be Eaten

Predatory habits of a frog-eating bat

The Fringe-lipped bat lives in lowland trop-ical forests throughout most of Latin Amer-ica. It feeds upon a variety of prey ranging from insects to lizards, but prefers frogs. Specialized hearing adaptations permit these bats to locate and identify the calls of courting male frogs.

Some of the most elaborate and striking displays by animals are those employed by males to attract and court females. Familiar examples include the elaborate plumes of male peacocks, the huge antlers of the extinct Irish elk, and the brilliant flashing of fireflies. According to the theory of sexual selection which Darwin first proposed in *On the Origin of Species* and developed more fully in *The Descent of Man, and Selection in relation to Sex*, these displays have evolved under the influence of female choice—the more elabo-rate the male's display the more likely it is to attract a female. But Darwin and others realized that there was a limit to which these displays could evolve because of their "costs," which include increased energy expenditure and greater risk of predation: a male peacock, for example, with a longer tail might attract more females but even if he had the energy necessary for growth, a tail which was too cumbersome for flight would make him an easy victim for predators.

Despite the importance of this idea, there are only a few cases where males advertising for mates have been proven to incur in-creased risk of predation.

Like most bats, the Fringe-lipped bat (*Trachops cirrhosus*) is nocturnal. A hall-mark of the Panamanian jungle at night is the almost constant cacophony of calling male frogs which devote much time and energy to attracting mates. Male Tungara frogs, for example, sometimes call more than 7,000 times in a single night! But the calls do not only attract female frogs. Bats are well known for their ability to hear the very high frequency sounds they use for echolocation, but the Fringe-lipped bat also exploits the relatively low-frequency sounds made by male frogs. The bats simply home in on these calls for a quick meal. At one pond in Panama, several of these bats caught between them an average of 6.6 frogs per hour from a chorus of only 250 frogs.

Male frogs can increase their ability to attract females by making themselves more conspicuous. Depending on the species, they can do this by making more intense calls, calling more frequently, and by producing calls that contain more notes. If bat preda-tion is an important counter-selective force on the evolution of frog vocalizations, then by behaving in ways which attract more females, male frogs also would increase their risk of predation. And that is exactly what happens. In one experiment bats were placed in a flight cage and frog calls were broadcast from two speakers. In all cases bats were attracted to the calls known to be more alluring to females. Bats and female frogs alike selected the more intense calls over the less intense ones, the calls with a faster repetition rate over the ones repeated at a slower rate, and the calls that had more notes over calls with fewer notes. But how do those attractive male frogs avoid being eaten? In further experiments it was shown that bats rarely catch silent frogs and on all but the darkest nights frogs recognize ap-proaching bats by sight and stop calling immediately. Also frogs either ignore smal-ler insect-eating bats or start calling again quicker. Therefore, the frogs' vocalizations probably have evolved as a compromise between two opposing selective forces. Female choice selects for calls that make the males more conspicuous, while predation selects for calls that make the males less conspicuous.

Not all frogs are subject to predation by the Fringe-lipped bat. A number of species, such as toads, secrete chemicals that are highly toxic to most predators, including the Fringe-lipped bat. All species of frogs, includ-ing the poisonous ones, have their own distinctive, species-specific call. These calls enable female frogs to select mates of the correct species. But, once again, the bat is able to exploit information in the call origin-ally intended for females. The Fringe-lipped bat is able to distinguish the calls of different species and use this information to avoid poisonous frogs. Furthermore, not all non-poisonous frogs are appropriate prey. Fringe-lipped bats are of medium size, usu-ally weighing up to 30g (1oz). Some jungle frogs are larger, and one, the South Amer-ican bullfrog, is known to turn the tables and eat bats. Fringe-lipped bats recognize the calls of such large frogs, as they do those of poisonous species, and avoid them.

Studies of bat–frog interactions demons-trate the important role that predation can have in the evolution of an animal's com-munication system. As more studies of this nature are conducted, it should become clear that only through an understanding of complex relationships among species, can we hope to fully grasp the dynamics of the process of evolution. MJR/MDT

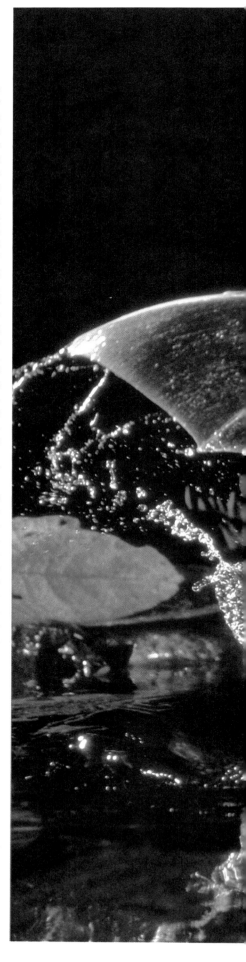

▶ **Successful hunt**—a Fringe-lipped bat catching a mud puddle frog in a Panamanian pond.

A Myth Exploded

Hunting behavior of vampire bats

No other species has contributed so much to the misunderstanding, even fear, of bats than the Common vampire bat (*Desmodus rotundus*). These bats feed almost exclusively on the blood of domestic stock, for example horses, cattle, burros, goats, pigs etc, with an occasional blood meal from a wild host; rarely do they attack humans.

On a foraging flight, a Common vampire will alight either on the ground near, or directly on, a potential host. It painlessly inflicts a small—3mm (0.12in) diameter, 1–2mm (0.04–0.08in) deep—wound on the hide of the host with its razor-sharp incisors. Its saliva contains anti-coagulants which keep blood flowing freely from the wounds. Common vampires do not "suck" blood; rather, they make use of capillary action and dart their tongues quickly in and out of the wounds. In one feeding (lasting $8\frac{1}{2}$–18 minutes in the case of cattle), a vampire may ingest up to 40 percent of its own body weight; although this is quite a load for the bat, it is rather an insignificant blood loss for the host animal. (In captivity, a single bat will consume 15–20g/0.5–0.7oz each night.) Nevertheless, several bats may feed from the same wound, causing more severe blood loss and possibly weakening of the victim. Some bats also transmit diseases such as rabies to the hosts—a serious threat to livestock in Central and South America.

The Common vampire is active only during the darkest hours of the night, and avoids moonlit periods. This might be a tactic to avoid predation by other nocturnal predators, such as owls, but is more probably related to the activity pattern of its most abundant host—domestic cattle. In the tropics, cattle are active, and therefore sensitive to bat attacks, during moonlit periods of the night; during the dark hours, they bed down in tight clusters.

Exactly how vampires locate their potential hosts remains a mystery. Relative to other bat species they have good eyesight and a well-developed sense of smell. Given the relative volumes of the different brain structures, and the ease with which vampires can be trained in conditioning experiments, learning may play an important role in their daily lives. Vampires often change daytime roosts (hollow trees or caves) to those nearest their preferred cattle herds.

Vampires, like many other bat species, often use riverbeds as "flyways" (flight corridors) to move from one area to another within their home ranges. At one study site, the incidence of animals bearing fresh bites in a population of about 1,200 domestic cattle, decreased from 2–8 animals each

▲ **Hopping to its prey.** Common vampires often land near to their prey then hop and leap forward on the ground. Sight and smell are probably its major senses used to locate prey.

◄ **Razor sharp incisor teeth** and a grotesque head leave little doubt as to why vampires have a bad name.

▼ **Lapping blood** from the head of a pig, a vampire bat takes a meal. Domesticated animals are the major prey of vampire bats.

night at the riverbed to nil at about 2km (1.2mi) on either side of such a river.

Vampires are very efficient at finding prey and most accomplish this within three hours of leaving their daytime roosts. However, males and females show some behavioral differences: although both sexes are equally active throughout the night, females, especially those pregnant or lactating, feed earlier in the evening and appear to give feeding a higher priority than males.

Vampires appear to be selective in their choice of hosts. Within mixed herds of tropical Zebu (Brahman) and Brown Swiss cattle breeds in Costa Rica, vampires preferred members of the Brown Swiss over Zebu animals, calves over their cows and cows in heat (estrus) over non-estrous cows and over bulls! The explanation for these preferences probably lies in the animals' accessibility to the vampires. When a mixed herd of Brown Swiss and Zebu beds down for the night in a tight cluster, the Brown Swiss are most often to be found on the edge of the herd and are thus more easily approached by the vampires. One should not forget that these bats are feeding on animals 10,000 times their own weight—to attempt to secure a blood meal from a host amidst a densely packed herd of such animals is certainly dangerous for the bat. Calves remain bedded down a greater proportion of the night than their cows, and members of herds with both cows and calves bed down with greater spaces between individuals than do members of pure cow herds. Both these factors increase the calves' exposure to vampire attack. Cows in heat are also found

on the perimeter of densely-packed herds. It has also been noted that Brown Swiss are more docile and do not react as vigorously to vampire bites as Zebu.

The onset of the rainy season also heralds changes in vampire/cattle relationships. Normally, vampires inflict their feeding wounds on the neck-shoulder region. However, during the rainy season (wettest month September) there is a notable increase in the number of bites found on the cow's flanks, above the hooves and in the anal region. Furthermore, more animals are bitten during the wet season than in the dry season (lowest rainfall in February), and the degree of preference for Brown Swiss, although still significant, slackens. These changes can also be related to accessibility, since during the rainy season, members of a herd bed down farther apart than during the dry months. This effectively increases the number of animals in a herd exposed to vampire attack, increases the area of a host's body exposed to such attacks, and lessens the importance of Brown Swiss perimeter animals as vampire targets.

It appears that the Common vampire bat is an extremely adaptable species that has almost completely switched over to hosts associated with civilization (domestic forms) over the past 400 years. Due to the elimination of its former natural (wild) hosts and the tremendous increase in domestic herds in many areas throughout its distribution range, the Common vampire has been forced by man to adopt new hunting strategies in order to survive, which it has done with success.   DCT

Unlikely Partners

Mexican long-nosed bats feed on the nectar of desert plants

Bees, butterflies and hummingbirds are familiar nectar feeders and pollinators of plants; the flowers they serve possess elaborate devices to advertise and deliver nectar and pollen. In a similar manner certain bats and plants have evolved together to become unlikely partners in pursuit of food and sex.

New World nectar-feeding bats belong to the subfamily Glossophaginae (family Phyllostomatidae), which comprises 13 genera. The group is basically tropical, but the Mexican long-nosed bat (*Leptonycteris nivalis*) is a nomadic species which follows sequentially blooming plants northward into the desert of Sonora state and summers in Arizona. Here the bats feed from flowers of the giant saguaro cactus and agave.

These small—20–24g (0.7–0.85oz)—bats need a tremendous energy input—their in-flight heart rates may exceed 700 beats per minute. Unlike other bats they lack the ability to conserve energy through a daily lowering of metabolism (torpor). Nor do they store fat or hibernate. Without food, they would starve to death in two days.

As they forage for their summer food plants, Mexican long-nosed bats form flocks which feed at successive plants. Generally flocks contain at least 25 bats comprising adult females and their young (both male and female) of the year. The groups appear to have no social structure while foraging and tests show that there is no difference in food intake dependent on sex, age or weight. There is a lack of antagonistic interactions which is surprising in view of the constant bickering observed in other bat colonies, although the intense mutual grooming which takes place during intermittent roosts every 15–20 minutes may serve a conciliatory function.

While foraging the bats circle above a plant and take turns swooping down over the flower to feed. After feeding for several minutes at one plant, a bat from the flock may move to an adjacent plant and all the other bats follow immediately with no further passes around the original plant. No individual bat "leads" consistently, but the first bat to leave the plant is always the one that most recently visited the flowers.

The groups are truly cooperative, rather than merely congregating at abundant food, and their degree of cohesiveness is impressive. In field experiments, when many of the plants were covered so food was drastically reduced, flocking was tightly maintained, and there was no change in the lack of aggression nor emergence of dominance. When single bats were released from a cage to feed on wild flowers, they were seemingly

reluctant to leave the flock. They returned and fluttered in front of their captive flock-mates and made repeated sallies to and from the cage as if soliciting company.

The greatest advantage of flocking for these nectar bats is increased foraging efficiency. Because the bats live on such a tight energy budget this benefit of cooperation is critical. The Sonoran desert food plants are only available patchily in both space and time, and the flocking bats make initial energy savings since many eyes search the environment more efficiently than a single pair. The discovery made by one bat soon becomes communal knowledge. This is likewise true for locating especially rich spots within a patch of plants.

One of the primary decisions for a bat is how long to stay with one patch of flowers before moving on to the next. As the flocks work *Agave* inflorescences, they randomly deplete the nectar in the flowers; as foraging continues, the chances are increased that a bat will visit empty, unrewarding flowers. At a certain point, it will cost the bat more energy to circle around the inflorescence to find a full flower than it would cost to switch to the next plant.

Since no one bat "leads" the flock in switching from one plant to another, each bat must be an equally good decision-maker. The question arises of whether, for the calories the bats invest in feeding behaviors, they are netting the highest possible caloric reward from the flower population—that is, are they switching patches at the right time? A computer simulation, considering bat flight metabolism, flight speed and natural plant spacing and nectar data could only improve on the bats' efficiency by one part in 10,000! It has been calculated that the Mexican long-nosed bat uses 0.002 kilo-calories (kcal) in its four-second flight around a plant and obtains 0.9kcal from feeding—a profit factor of 45.

In making such good decisions the bats assess such variables as levels of nectar (perhaps using their tongue as a dipstick), nectar concentration, number of empty flowers experienced, etc. They also form expectations. If they happen to feed from a particularly rich patch early in the evening and then switch to an average patch, they will tend to abandon the average one in a very short time. Evidently in comparison with recent experience, they "see" the patch as a poor one.

Young bats are apparently not good decision-makers, so performance appears to be improved through learning. Young bats foraging alone, which some do after the main group has returned to winter in Mexico, tend to spend shorter periods of time feeding at each patch of food, often leaving behind nectar. Thus these "impetuous youths" waste more of their energy travelling between plants than is necessary.

Unfortunately, nectar-feeding bat populations in Texas and Arizona have been decimated over the last 30 years. At best, a few thousand Mexican long-nosed bats now make the summer movement to the United States. A few decades ago, a single cave in Arizona was the maternity site for 20,000 individuals. Fears and supersitions have hindered attempts to conserve bats. General habitat destruction may be partially responsible. Another factor in northern Mexico may be "moonshine" pulque and tequila operations. Agaves have long been a local source of food and drink. In central Mexico, the plants are managed and replanted by the large tequila factories, but in Northern Mexico, "cottage" operations thrive. To make pulque and tequila the plants are harvested (and effectively destroyed) before they flower, leaving huge areas without flowering plants.

In United States agave populations which nectar bats no longer visit, plant reproduction is down to 1/3,000th of that in areas that are still visited by bats. Herbarium studies of fruiting capsules from agaves show a decline in the number of viable seeds over the last 30 years, paralleling the decline in bat pollinators. When agave populations diminish, the decline of the remaining bat populations is hastened. In Arizona, the organ pipe cactus and saguaro are also declining. Since these large, nectar-rich plants provide shelter and food for numerous other animals, the whole community is threatened. Individual animal species do not exist in a vacuum, and destruction even of one bat species can have far-reaching implications for the balance of nature, in consequence threatening the survival of further plants and animals.   DJH

◄ **Nectar mop.** The tongue of a *Leptonycteris* bat can be extended almost the length of its body and is tipped with fleshy bristles.

▼ **Quick feeders**—*Leptonycteris* feeding from a century plant. Since bats cannot hover, they spend only a fraction of a second feeding during each pass.

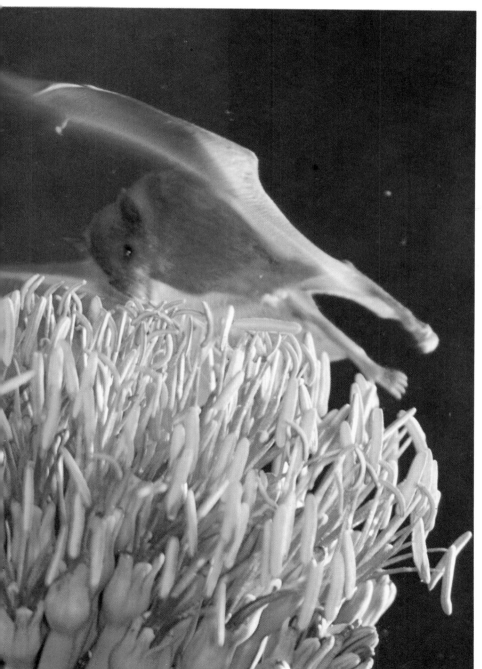

Buzzing Bats

The lek mating system of Hammer-headed bats

Hammer-headed bats are one of the few mammalian species which practice lek mating. A lek is an aggregation of displaying males to which females come solely for the purpose of mating. Females usually visit a lek, examine a number of males and then select one with whom to mate. Females are remarkably unanimous in this choice so that only a few males do all the mating. Females undertake all the parental care in lek species. Other forms with lek behavior include the Uganda kob among mammals and some birds, frogs, fish and insects.

Hammer-headed bats (*Hypsignathus monstrosus*) occur in tropical forests from Senegal, through the Congo basin to western Uganda; they form leks and mate during each of two—June–August and January–early March—annual dry seasons. A Hammer-headed bat lek nearly always borders a waterway, and varies from 0.7–1.5km (0.4–0.9mi) in length. Males are spaced about 10–15m (33–49ft) apart along the site and the array is usually about two males (range 1–4) deep. In Gabon, there are calling males on the sites for about $3\frac{1}{2}$ months each dry season. Early and late in the season, only a few males call. The number increases rapidly to a peak in July and February, and then declines more slowly towards the end of the dry season. Each night at sunset during the mating period, males leave their day roosts and fly directly to traditional lek sites. At the lek they hang in the foliage of the canopy edge and emit a loud metallic call while flapping their wings at twice the call rate. Early in the mating season, there is usually some fighting for calling territories. By the time females begin visiting the lek—and they start to do so before they are ready to mate—most territorial squabbles have been settled and there is little subsequent interference between males.

Typical leks contain 30–150 displaying males, each calling at one to four times a second and flapping its wings furiously. Females fly along the male assembly and periodically hover before a particular male. This causes the male to perform a staccato variation of its call and to tuck its wings close against its body. Females will make repeated visits on the same night to a decreasing number of males, each time eliciting a "staccato buzz." Finally, selection is complete, the female lands by the male of her choice, and mating is accomplished in 20–30 seconds. Females usually terminate mating with several squeals, and then fly off.

The importance of display in enabling a male to breed has obviously favored a heavy investment in the equipment it uses to advertise itself. Males are twice as large as females—425g compared to 250g (15oz/8.8oz)—have an enormous bony larynx which fills their chest cavity, and have a bizarre head with enlarged cheek pouches, inflated nasal cavities, and a funnel-like mouth. The larynx and associated head structures are all specializations for producing the loud call.

Females can first mate at an age of six months and reach adult size at about nine months of age. They can thus produce their first offspring (only one young is born at a time) as yearlings. Females come into heat immediately after birth (post-partum estrus) and thus can produce two successive young each year. In fact, many of the females mating during any dry season are carrying newborn young conceived at the last mating or lek period. As with many lek species, males mature later.

Despite early reports to the contrary, these bats are entirely fruit-eaters. The fruit of several species of *Anthocleista* and figs form the major part of their diets in Gabon. Females and juvenile males appear to feed more on the easily located and closer (1–4km/0.6–2.5mi from the roost) but less profitable *Anthocleista* on which fruits ripen

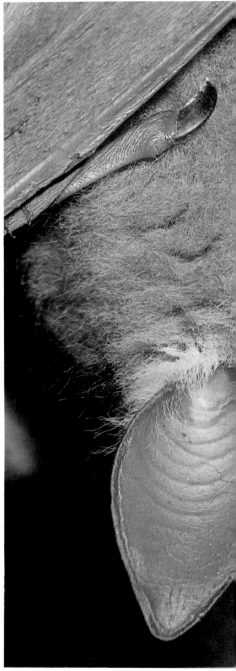

slowly and a few at a time. By contrast adult males fly 10km (6.2mi) or more to find the less predictable but more profitable patches of ripe figs where large numbers of fruit are available on a single trip, but only for a short while. This extra effort by males presumably pays off by providing more energy for vigorous display. It has the cost that, if unsuccessful, males may starve. The effects of variable food levels and the high energetic outlays during display may explain the higher parasitic loads (primarily hemosporidians in the blood cells) in adult males and the higher

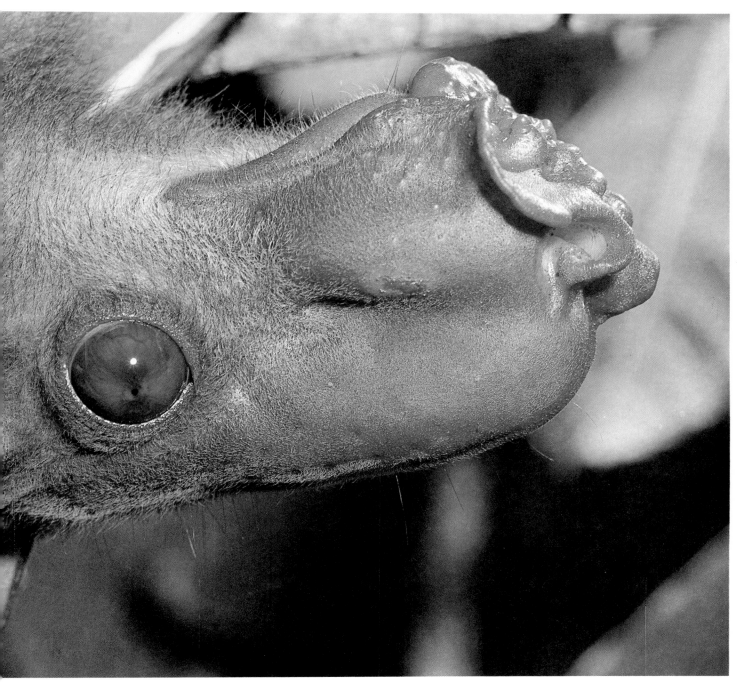

◀ ▲ **Genteel females, grotesque males**—sexual differences in Hammer-headed bats. LEFT female; ABOVE male. Young males quickly become heavier than females of identical age, but continue to have female-like heads until yearlings. During the ensuing six months, males become mature, complete development of the enormous larynx, and develop the bizarre head shape of an adult. They are then ready to join leks for mating and, on average, live long enough to attend 3–4 consecutive lek seasons.

annual mortality rates of adult males. It is also reflected by the abandonment of display by all males, even though females may be visiting for mating, following days of colder than average weather.

Lek mating is often considered a "default" mating system, adopted when males cannot provide parental care, defend resources which females require, or defend groups of females. It is easy to understand that the expensive and chancy business of self-advertisement in competition with other males might be undesirable for males—

unless it is absolutely necessary. Hammer-headed bats do appear to fit these generalizations: there is little males could do to assist itinerant females with young, females rarely form groups and those formed are at best transient aggregations, and neither roosts nor food sources are defensible, the latter because fig trees are widely dispersed and come randomly and unpredictably into fruit. The costs of display are certainly significant, yet males are committed both physically and behaviorally to this system.

JWB

MONOTREMES

ORDER: MONOTREMATA
Two families; 3 genera; 3 species.

Echidnas   Platypus

⊽ Vulnerable.

Family Tachyglossidae

Short-beaked echidna
Tachyglossus aculeatus
Short-beaked or Common echidna (or spiny anteater).
Distribution: Australia, Tasmania, New Guinea.
Habitat: almost all types, semi-arid to alpine.
Size: head-body length 30–45cm (12–18in);
weight 2.5–8kg (5.5–17.6lb). Males 25 percent
larger than females.
Coat: black to light brown, with spines on back
and sides; long narrow snout without hair.
Gestation: about 14 days.
Longevity: not known in wild (extremely long-
lived in captivity—up to 49 years).

WHILE the description of monotremes as "the egg-laying mammals" clearly distinguishes them from any other living animals, it exaggerates the significance of egg-laying in the group. The overall pattern of reproduction is clearly mammalian, with only a brief, vestigial period of development of the young within the egg. The eggs are soft-shelled and hatch after 10 days, whereupon the young remain (in a pouch in the echidnas) dependent on the mother's milk for 3–4 months in the platypus and about six months in echidnas. Nonetheless, the term "egg-laying mammal" has long been synonymous with "reptile-like" or "primitive mammal," regardless of the fact that monotremes possess all the major mammalian features: a well-developed fur coat, mammary glands, a single bone in the lower jaw and three bones (incus, stapes and malleus) in the middle ear. Monotremes are also endothermic; their body temperature, although variable in echidnas, remains above environmental temperatures during winter.

Monotremes have separate uteri entering a common urino-genital passage joined to a cloaca, into which the gut and excretory systems also enter. The one common opening to the outside of the body gave the name to the group to which the animals are now known to belong—the order Monotremata ("one-holed creatures").

Although clearly mammals, monotremes are highly specialized ones, particularly in regard to feeding. The platypus is a semi-aquatic carnivorous mammal that feeds on invertebrates living on the bottom of freshwater streams. Echidnas are terrestrial carnivorous mammals, specializing in ants and termites (Short-beaked or Common echidna) or noncolonial insects and earthworms (Long-beaked echidna). Such diets require grinding rather than cutting or tearing, and monotremes lack teeth as adults. In the platypus, teeth actually start to develop and may even serve as grinding surfaces in the

very young, but the teeth never fully develop and regress to be replaced by horny grinding plates at the back of the jaws. Reduction of teeth is common among ant-eating mammals, and echidnas never develop teeth, nor are their grinding surfaces part of the jaw. A pad of horny spines on the back of the tongue grinds against similar

spines on the palate, so breaking up the food.

In the platypus the elongation of the front of the skull and the lower jaw to form a bill-like structure is also a foraging specialization. The bill is covered with shiny black skin which contains many sensory nerve endings. Echidnas have a snout which is based on exactly the same modifications of

Long-beaked echidna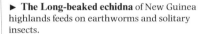

Zaglossus bruijni
Long-beaked or Long-nosed echidna (or spiny anteater).
Distribution and habitat: mountains of New Guinea.
Size: head-body length 45–90cm (18–35in); weight 5–10kg (11–22lb).
Coat: brown or black; spines present but usually hidden by fur except on sides; Spines shorter and fewer than Short-beaked echidna; very long snout, curved downwards.
Gestation: unknown.
Longevity: not known in wild (to 30 years in captivity).

Family Ornithorhynchidae

Platypus

Ornithorhynchus anatinus
Platypus or duckbill.
Distribution: E Australia from Cooktown in Queensland to Tasmania. Introduced in Kangaroo Island, S Australia.
Habitat: most streams, rivers and some lakes which have permanent water and banks suitable for burrows.
Size: lengths and weights vary from area to area, and weights change with season. Male: head-body length 45–60cm (17.7–23.6in); bill length average 5.8cm (2.3in); tail length 10.5–15.2cm (4.1–6in); weight 1–2.4kg (2.2–5.3lb). Female: head-body length 39–55cm (15.4–21.7in); bill length average 5.2cm (2in); tail length 8.5–13cm (3.3–5.1in); weight 0.7–1.6kg (1.5–3.5lb).
Coat: dark brown back, silver to light brown underside with rusty-brown midline, especially in young animals which have lightest fur. Short, dense fur (about 1cm/0.4in depth). Light patch below eye/ear groove.
Gestation: not known (probably 2–3 weeks).
Incubation: not known (probably about 10 days).
Longevity: 10 or more years (17-plus in captivity).

▶ **The Long-beaked echidna** of New Guinea highlands feeds on earthworms and solitary insects.

▼ **From snowy regions to deserts**, the smaller Short-beaked echidna makes its home where there is a plentiful supply of termites and ants on which it feeds.

skull and jaws but is relatively smaller and is cylindrical in shape. The mouth of an echidna is at the tip of this snout and can only be opened enough to allow passage of the cylindrical tongue.

Monotremes are one of the two groups of venomous mammals (the other is certain shrews). In echidnas, the structures which produce and deliver the venom are not functional, though present. It is only the male platypus that actually produces and is capable of delivering the venom. The venom-producing gland is located behind the knee and is connected by a duct to a horny spur on the back of the ankle. This spur can be erected from a fold of covering skin and is hollow to deliver the venom, which causes agonizing pain in man, and can kill a dog. Venom is delivered by a forceful jab of the hindlimbs. Because the venom gland enlarges at the beginning of the breeding season, it has long been assumed to be connected with mating behavior. The marked increase in aggressive use of the spurs observed between males in the breeding season may serve to decide spatial relationships in the limited river habitat. However, that does not explain why

in echidnas the system is present but non-functional. The spur in male echidnas makes it possible to distinguish them from females, which is otherwise difficult in monotremes since the testes never descend from the abdomen. However the echidnas' venom duct and gland are degenerate, and the male cannot erect the spur. If it is pushed from under its protective sheath of skin, few echidnas can even retract the spur. It may be that the venom system in monotremes originated as a defense against some predator long since extinct. Today adult monotremes have few, if any, predators. Dingos occasionally prey on echidnas, but dingos are themselves a relatively recent arrival in Australia.

The platypus is confined to eastern Australia and Tasmania, while the Long-beaked echidna is confined to New Guinea, and the Short-beaked echidna is found in almost all habitats in all of these regions. But these distributions are relatively recent, as there are Pleistocene fossils of, for example, Long-beaked echidnas at numerous sites in Australia and Tasmania. Fossil monotremes from the Pleistocene epoch (after about 2 million years ago) are much the same as the living types. There is a platypus fossil from the mid-Miocene (about 10 million years ago), but it, too, is much like the living platypus, although the adult may have had functional teeth. The logical assumption is that monotremes are an old, specialized derivative of a very early mammal stock that has survived only by being isolated in the Australian region. But there is little direct evidence to show what that ancestral stock was. Australia and South America are assumed to have some special relationship because of their marsupial faunas, but no monotreme fossils have been found in South America, or indeed elsewhere. It has recently been suggested that modern monotremes are descendants of a long-extinct (and much wider spread, although unknown from Australia) group of early mammals known as multituberculates, but there is little supporting evidence. So the zoogeography and evolution of monotremes remain an enigma.   MLA

Echidnas are readily recognized by their covering of long spines (shorter in the Long-beaked species). Fur is present between the spines. In the Long-beaked echidna and the Tasmanian form of the Short-beaked echidna the fur may be longer than the spines. The spiny coat provides an excellent defense. When suddenly disturbed on hard ground, an echidna curls up into a spiky ball; if disturbed on soft soil it may rapidly dig straight down, like a sinking ship, until all that can be seen are the spines of the well-protected back. By using its powerful limbs and erecting its spines, an echidna can wedge itself securely in a rock crevice or hollow log.

Echidna spines are individual hairs and are anchored in a thick layer of muscle (panniculus carnosus) in the skin. The spines obscure the short, blunt tail and the rather large ear openings, which are vertical slits just behind the eyes. Spines are lacking on the underside and limbs. The snout is naked and the small mouth and relatively large nostrils are located at the tip. Echidnas walk with a distinctive rolling gait, although the body is held well above the ground.

Males can be distinguished from females by the presence of a horny spur on the ankle of the hind limb. Males are larger than females within a given population. Yearling Short-beaked echidnas usually weigh less than 1kg (2.2lb), but beyond that there is no way of determining age.

Echidnas have small, bulging eyes. Although they appear to be competent at making visual discriminations in laboratory

studies, in most natural habitats vision is probably not important in detecting food or danger. Their hearing is very good, and echidnas hear a person approaching and take cover long before they can be seen. In locating prey, usually by rooting through the forest litter or undergrowth, the sense of smell is used. When food items are located they are rapidly taken in by the long, thin, highly flexible tongue, which Short-beaked echidnas can extend up to 18cm (7in) from the tip of the snout. The tongue is lubricated by a sticky secretion produced by the very large salivary glands. The ants and termites which form the bulk of the Short-beaked echidna's diet are available throughout the year. One variation of foraging strategy occurs during the early months of spring (August–September), when Short-beaked echidnas attack the mounds of the meat ant *Iridomyrmex detectus* to feed on the fat-laden females: this is done regardless of spirited defense by the stinging workers, although the mounds are prudently avoided the rest of the year when males and females have left.

Short-beaked echidnas are essentially solitary animals, inhabiting a home range the size of which varies according to the environment. In wet forest with abundant food the home range area is about 50ha (124 acres). The home range appears to change little, and within it there is no fixed shelter site. Echidnas take shelter in hollow logs, under piles of rubble and brush and under various thick clumps of vegetation when inactive. Occasionally they dig shallow burrows as long as 1.2m (4ft), which may be reused. A female incubating an egg or suckling young has a fixed burrow. The home ranges of several individuals overlap.

▲ **Digging into an ant's nest** TOP, a Short-beaked echidna searches for food.

▲ **Near-buried sleeper** ABOVE, this Short-beaked echidna retires from the heat of the Australian summer. Disturbed on soft soil, the echidna will dig down, disappearing like a "sinking ship."

▶ **Unmistakable beak-tip nostrils** ABOVE, small protruding eyes and digging foreclaws of a Short-beaked echidna. The echidna draws termites and ants into its mouth on its long saliva-coated tongue.

▶ **Snorkel swimmer** BELOW, a Short-beaked echidna demonstrates the species' ability to cross most types of terrain.

During the mating season (June–August on mainland Australia) the female leaves a scent track by everting the cloaca, the wall of which contains numerous glands. This presumably attracts males in overlapping ranges. The only time echidnas are observed in groups is during the mating season, and on the only occasion when the sex of individuals in such a group was determined, the group consisted of one female being followed by a line of five males. In captivity echidnas kept in spacious accommodation do not form any sort of groups and are mutually tolerant. If kept in confined over-crowded quarters echidnas may form a size-related dominance order, but this does not seem to be a natural behavior.

The chief periods of activity are related to environmental temperature. Short-beaked echidnas are usually active at dusk and dawn, but during the hot summer echidnas are nocturnal and during cold periods they may be active during the middle of the day.

They avoid rain and will remain inactive for days if rain continues. Echidnas do not hibernate. It should be noted, in view of reports that echidnas hibernate, that the mating season is mid-winter in the southern hemisphere.

The female pouch is barely detectable for most of the year. Before the breeding season, folds of skin and muscle on each side of the abdomen enlarge to form an incomplete pouch with milk patches at the front end. There are no teats. The single egg is laid into the pouch by extension of the cloaca while the female lies on her back. After about 10 days the young hatches, using an egg tooth and a horny caruncle at the tip of the snout. The single young remains in the pouch until the spines begin to erupt. For a further six months or so the young continues to suckle. The young become fully independent and move out to occupy their own home ranges about one year after being born.

The Short-beaked echidna is widespread

and common (although little seen) on mainland Australia and Tasmania. However its status in New Guinea is uncertain.

The Long-beaked echidna and several other similar genera were once distributed throughout Australia but disappeared by the late Pleistocene (10,000 years ago). Today there is only one species, restricted to the New Guinea highlands. It is likely that the disappearance of Long-beaked echidnas from Australia, while Short-beaked echidnas have remained, is related to climatic changes in Australia and to differences in diet. Long-beaked echidnas feed largely, if not solely, on earthworms. Their tongue is equipped with horny spines located in a groove that runs from the tip about one third of the way back. Worms are hooked by these spines when the long tongue is extended. It is necessary to take the worm into the long snout by either the head or tail, and if necessary the forepaws are used to hold the worm while the beak is positioned. Other details of the biology of Long-beaked echidnas are unknown although most of the information given above for Short-beaked echidnas applies to both species.

Like the Short-beaked echidna, which also occurs in New Guinea, the status of this animal is uncertain and zoologists who have visited New Guinea in recent times are concerned that Long-beaked echidnas are widely hunted for food.   MLA

Ever since the first **platypus** (a dried skin) arrived in Britain from the Australian colonies around 1798, this animal has been surrounded by controversy. Initially it was thought that the specimen was a fake, that a taxidermist had stitched together the beak of a duck and body parts of a mammal! Even when it was found to be real, the species was not accepted as a mammal. It did have fur but was found to have a reproductive tract similar to birds and reptiles. Because of this arrangement of the reproductive system it was suggested that the species laid eggs, instead of bearing its young live as did all mammals known at that time. But this suggestion (now known to have been correct) was discredited when it was found that the animal had the essential characteristic from which the class Mammalia takes its name—mammary glands.

Controversy did not end there however, and because the platypus has some skeletal features similar to reptiles (especially the pectoral or shoulder girdle) and because it does actually lay eggs, the species has been described as a "primitive mammal" or even a "furred reptile." As late as 1973 it was

thought that the animal could not regulate its body temperature in the precise manner of most mammals, and that it became hypothermic when swimming in water and frequently had to retire to its burrow to warm up. This seemed strange for an animal which must obtain all of its food in water at all seasons of the year, and recent studies have shown that the species is truly mammalian in this respect also.

The platypus is smaller than most people think. The females are significantly smaller than males and the young are about 85 percent of adult size when they first become independent. The animal is streamlined, with a covering of dense waterproof fur over all of its body except the feet and the bill. The bill looks a bit like that of a duck with the nostrils on top just back from the tip, but it is soft and pliable. It is well supplied with nerves and is used by the animal in locating food and in finding its way around underwater, as the eyes and ears are both closed when diving. Behind the bill are two internal cheek pouches opening from the mouth. These contain horny ridges which functionally replace the teeth lost by the young soon after they emerge from the burrows. The pouches are used to store food while it is being chewed and sorted.

The limbs are very short and held close to the body. The hindfeet are only partially webbed, being used in water only as rudders while the forefeet have large webs and are the main organs of propulsion. The webs of the front feet are turned back when the animal is walking or burrowing to expose large broad nails. The rear ankles of the males bear a horny spur which is hollow and connected by a duct to a venom gland in the thigh (see p819). The tail is broad and

experience water temperatures close to, and air temperatures well below, freezing in winter. When the platypus is exposed to such cold conditions it can increase its metabolic rate to produce sufficient extra heat to maintain its body temperature around the normal level of 32°C (89.6°F), lower than in many mammals, but certainly much warmer than that of reptiles in cold conditions. Good fur and tissue insulation help the animal to conserve body heat, and their burrows also provide a microclimate that moderates the extremes of outside temperature in both winter and summer.

Although mating is reputed to occur earlier in northern Australia than in the south, it occurs sometime in the spring (August–October). Mating apparently takes place in water after initial approaches by the female, followed by chasing and grasping of her tail by the male. Two (occasionally one or three) eggs are laid. When hatched the young are fed on milk, which they suck from the fur of the mother around the openings of the mammary glands (there is no pouch) for 3–4 months while they are confined to the special breeding burrow. This burrow is longer and more complex than the burrows normally inhabited, as it may be up to 30m (100ft) long and be branched with one or more nesting chambers. The young emerge from these burrows in late summer (late January–early March in New South Wales). They continue to take milk from their mothers for some time.

Although normally two eggs are laid, it is not known how many young are successfully weaned each year. Not all females breed each year, and new recruits to the population probably do not breed until they are at least two years of age. In spite of this low reproductive rate, the platypus has returned from near extinction in certain areas since its protection and the cessation of hunting at the turn of the century. This indicates that the reproductive strategy of having only a few young, but looking after them well, is effective in this long-lived species.

The platypus owes its success to its occupation of an ecological niche which has been a perennial one, even in the driest continent of the world. By the same token, because the platypus is such a highly specialized mammal it is extremely susceptible to the effects of changes in its habitat. Changes wrought by man in Australia have so far only affected individual populations of platypuses, but care and consideration for the environment must be maintained if this unique species is to remain. TRG

▲ **Scouring the water-bottom** for insect larvae, the platypus uses its touch-sensitive, pliable bill as its "ears and eyes" (the real ones are closed underwater).

◄ **"Beak of a duck, body of a mammal."** The glistening, streamlined surface presented by the long guard hairs of the platypus's coat conceals the thick, dry underfur that insulates the body from water temperatures which may in winter drop below zero in certain localities. The platypus can also increase the metabolic rate of its body chemistry so as to maintain normal body temperatures around 32°C (89.6°F).

flat and is employed as a fat-storage area.

The food of the platypus is almost entirely made up of bottom-dwelling invertebrates, particularly the young stages (larvae) of insects. Two introduced species of trout feed on the same sort of food and are possible competitors of the platypus. However, both the fish and the platypus are common in many rivers, so that it can be assumed that they do not compete seriously. Recent studies in one river system have shown that the trout eat more of the swimming species of invertebrates, while the platypus feed almost exclusively on those inhabiting the bottom of the river. Waterfowl may also overlap in their diets with platypuses, but most consume plant material which does not appear to be eaten by the platypus.

Certain areas occupied by platypuses

MARSUPIALS

ORDER: MARSUPIALIA
Eighteen families; 76 genera; 266 species.

Polyprotodonts
Suborder Polyprotodonta

American opossums
Family Didelphidae
Seventy-five species in 11 genera.

Monito del monte
Family Microbiotheriidae
One species, *Dromiciops australis*.

Shrew or rat opossums
Family Caenolestidae
Seven species in 3 genera.

Quolls, dunnarts and marsupial mice
Family Dasyuridae
Fifty-two species in 18 genera.

Numbat
Family Myrmecobiidae
One species, *Myrmecobius fasciatus*.

Thylacine or Tasmanian tiger
Family Thylacinidae
One species, *Thylacinus cynocephalus*.

Marsupial mole
Family Notoryctidae
One species, *Notoryctes typhlops*.

THE first marsupials brought from South America to Europe in the 16th century were considered zoological curiosities. It was not until the exploration and settlement of Australia from 1770 onwards that a sufficient variety of species was collected for European zoologists to realize that, rather than a few aberrant rodents, they were dealing with a natural group of mammals which shared a most unusual mode of reproduction.

Fossil evidence suggests that the ancestors of modern marsupials and placental mammals became distinct about 100 million years ago. The first undoubted marsupial fossils are known from the late Cretaceous of North America, about 75 million years ago, mainly from the family Didelphidae. Although there was at first some development in North America, marsupials later declined there as placental mammals increased in diversity, and they became extinct in North America about 15–20 million years ago. A few didelphid marsupials had spread to Europe by about 50 million years ago, but they were not conspicuously successful, and disappeared about 20–25 million years ago.

South America, on the other hand, had a considerable diversity of marsupial fossil forms only slightly later than the earliest known fossil marsupials in North America. This diversity persisted for the more than 60 million years that South America was isolated from North America. Seven families of living and fossil marsupials are known from South America, where they filled the omnivore and carnivore niches. About 2–5 million years ago, a land connection to North America was again established, and more placental mammals reached South America, including carnivores such as the jaguar. In the face of such competition, the large carnivorous marsupials disappeared, but the small omnivores have persisted successfully to the present day. Some marsupials moved north to recolonize North America,

most notably the Virgina or Common opossum.

The earliest marsupials found in Australia so far are from some 23 million years ago, at which time most modern families and forms were clearly established. There is no clear fossil evidence to establish whether marsupials originated in North America, South America or Australia. Extremely primitive marsupials are known from the Cretaceous in both South and North America, making them a more likely point of origin than Australia, with no fossil record at this time. The lack of fossil marsupials in Europe, Africa and Asia makes the most likely route of the spread from South America to Australia via Antarctica. At this time the three southern continents were still united in the land mass known as Gondwanaland.

▼ **Fossil marsupials** include many instances of gigantism and flesh-eaters. In South America, the hyena-like *Borhyaena* (1) preyed on marsupial and placental mammals alike during the Miocene (26–7 million years ago); the jaguar-sized Pliocene genus *Thylacosmilus* (2) was similar to the saber-toothed cats living at the same time in other continents; when North and South America rejoined 2–5 million years ago, placental carnivores moved south, and forced large marsupials into extinction. In Australia, and related to today's kangaroos, was the enormous Pleistocene short-headed browser *Procoptodon goliah* (3) standing 10ft (3m) high. The seven-foot long *Diprotodon* (4) may have survived until 20,000 years ago and the "marsupial lion" *Thylacoleo carnifex* (5) only died out some 30,000 years ago.

Bandicoots
Family Peramelidae
Seventeen species in 7 genera.

Rabbit-eared bandicoots or bilbies
Family Thylacomyidae
Two species in 1 genus.

Diprotodonts
Suborder Diprotodonta

Cuscuses and brushtails
Family Phalangeridae
Fourteen species in 3 genera.

Pygmy possums
Family Burramyidae
Seven species in 4 genera.

Ringtail possums
Family Pseudocheiridae
Sixteen species in 2 genera.

Gliders
Family Petauridae
Seven species in 3 genera.

Kangaroos and wallabies
Family Macropodidae
Fifty species in 11 genera.

Rat kangaroos
Family Potoroidae
Ten species in 5 genera.

Koala
Family Phascolarctidae
One species Phascolarctos cinereus.

Wombats
Family Vombatidae
Three species in 2 genera.

Honey possum
Family Tarsipedidae
One species, Tarsipes rostratus.

Gradually, Gondwanaland broke up as each continental piece was moved over the surface of the earth by convection currents deep within the earth. This breakup began 135 million years ago with the separation of India, followed by Africa, Madagascar and New Zealand. Australia and Antarctica remained connected until about 45 million years ago and South America and Antarctica separated about 30 million years ago. Forty-five million years ago the climate of the southern land mass was much more congenial than it is now—Antarctica supported forests of Southern beech. The first (and so far the only) fossil land mammal to be found in Antarctica was a marsupial of the extinct didelphoid family Polydolopidae, in rocks about 40 million years old, which indicates that marsupials did exist in Antarctica at about the time when Australia became isolated.

The Australian crustal plate (including the southern part of New Guinea) gradually drifted northward for some 30 million years before reaching its present latitude, during which time its plants and animals were isolated from other continents, until Southeast Asia was approached about 10–15 million years ago. It is this long isolation which allowed the extensive development of marsupials in Australia in the absence of any competition from placental mammals.

Early Australian fossil marsupials of 20–15 million years ago include a preponderance of arboreal and browsing terrestrial forms. From the mid-Miocene (15 million years ago) extensive forests became confined to coastal areas and savanna woodlands took over large areas of the interior. The changes in climate and vegetation were matched by an increase in terrestrial grazing marsupials. Kangaroos formed an increasingly significant part of the fauna, and today are the most numerous marsupials over much of inland Australia. As recently as 30,000 years ago, there existed many now extinct giant herbivores, such as the 3m (9.8ft) tall browsing kangaroo Procoptodon and Diprotodon, the largest marsupial that ever lived, the size and shape of a living rhinoceros, a member of the now extinct family Diprotodontidae.

As the marsupials evolved in Australia filling the same ecological niches as placental mammals filled elsewhere so, in many cases, they adopted similar morphological solutions to ecological problems. One example of this convergent evolution, the carnivorous Tasmanian wolf or thylacine, is very similar in overall form to wolves and dogs of other continents. The so-called

3

4

5

Marsupial mole is very similar in form to placental moles, likewise burrowing insectivores. The marsupial Sugar glider and the two flying squirrels of North America have developed a similar gliding membrane (patagium).

Today's marsupials are found only in the New World and in the Australian region. In the USA and Canada there are only 2 species in one family, from Panama north, 9 didelphid species. South America has 81 species in three families, mostly small insectivores or omnivores, some of which are arboreal. Australia has about 120 species in 15 families and New Guinea about 53 species in six families. These range from tiny, 5g (0.18oz) or even less, shrew-like insectivorous dasyures to large grass-eating kangaroos (males may reach 90kg/200lb) with, in between, a great variety of medium-sized carnivores, arboreal leaf-eaters and terrestrial omnivores in a range of habitats from desert to rain forest.

Although there are many other anatomical differences, it is their reproduction that sets marsupials apart from other mammals. In its form and early development in the uterus, the marsupial egg is like that of reptiles and birds and quite unlike the egg of placental mammals (eutherians). Whereas eutherian young undergo most of their development and considerable growth inside the female, marsupial young are born very early in development. For example, a female Eastern gray kangaroo of about 30kg (66lb) gives birth after 36 days' gestation to an offspring which weighs about 0.8g (under 0.03oz). This young is then carried in a pouch on the abdomen of its mother where it suckles her milk, develops and grows until after about 300 days it weighs about 5kg (11lb) and is no longer carried in the pouch. After it leaves the pouch the young follows its mother closely and continues to suckle until about 18 months old.

Immediately after birth, the newborn young makes an amazing journing from the opening of the birth canal to the area of the nipples. Forelimbs and head develop far in advance of the rest of the body, and the young is able to move with swimming movements of its forelimbs. Although it is quite blind, the young locates (how is not yet known) a nipple and sucks it into its circular mouth. The end of the nipple enlarges to fit depressions and ridges in the mouth, and the young remains firmly attached to the nipple for 1–2 months until, with further development of its jaws, it is able to open its mouth and let go.

In many marsupials the young is protected by a fold of skin which covers the nipple area, forming the pouch (see diagram).

In marsupials the major emphasis in the nourishment of the young is on lactation, since most development and all growth occurs outside the uterus. For the short time that the embryo is in the uterus, it is nourished by transfer of nutrients from inside the uterus across the wall of the

BODY PLAN OF MARSUPIALS

▼ **Marsupial skulls** generally have a large face area and a small brain-case. There is often a sagittal crest for the attachment of the temporal muscles that close the jaws, and the eye socket and opening for the temporal muscles run together, as in most primitive mammals. There are usually holes in the palate, between the upper molars. The rear part of the lower jaw is usually turned inward, unlike placental mammals.

Many marsupials have more teeth than placental mammals. American opossums for instance have 50. There are usually three premolars and four molars on each side in both upper and lower jaws.

Marsupials with four or more lower incisors are termed polyprotodont. The Eastern quoll (*Dasyurus viverrinus*), has six. Its chiefly insectivorous and partly carnivorous diet is reflected in the relatively small cheek teeth each with two or more sharp cusps and the large canines with a cutting edge. Its dental formula is I4/3, C1/1, P2/2, M4/4 = 42. The largely insectivorous bandicoot (*Perameles*) has small teeth of even size with sharp cusps for crushing the insects which it seeks out with its long pointed snout (I4–5/3, C1/1, P3/3, M4/4 = 46–48).

Diprotodont marsupials have only two lower incisors, usually large and forward-pointing. The broad, flattened skull of the leaf-eating Brush-tailed possum contains reduced incisors, canines and premolars, with simple low-crowned molars (I3/2, C1/0, P2/1, M4/4 = 34). The large wombat has rodent-like teeth and only 24 of them (I1/1, C0/0, P1/1, M4/4), all rootless and ever-growing to compensate for wear in chewing tough fibrous grasses.

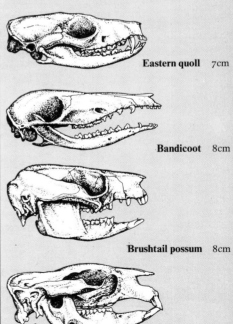

Eastern quoll 7cm

Bandicoot 8cm

Brushtail possum 8cm

Common wombat 18cm

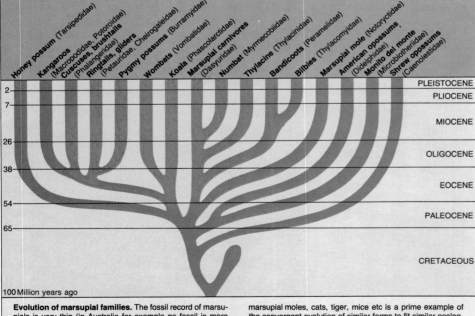

Evolution of marsupial families. The fossil record of marsupials is very thin (in Australia for example no fossil is more than 23 million years old) and the early dates shown here are accordingly approximate. For the same reason much of the left-hand side of this "tree" is speculative. The evolution of the marsupial moles, cats, tiger, mice etc is a prime example of the convergent evolution of similar forms to fit similar ecological "niches" to those exploited on other continents by carnivores, rodents and herbivores.

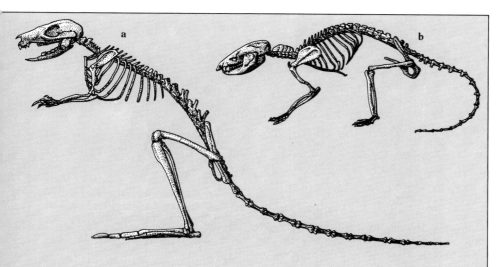

▲ **Skeletons** of (**a**) Tasmanian bettong and (**b**) Virginia opossum. The Virginia opossum is medium-sized, with unspecialized features shared with its marsupial ancestors. These include the presence of all digits in an unreduced state, all with claws. The skull and teeth are those of a "generalist," the long tail is prehensile, acting as "a fifth hand," and there are epipubic or "marsupial bones" that project forward from the pelvis and help support the pouch. The hindlimbs in this quadruped are only slightly longer than the forelimbs. The larger kangaroo has small forelimbs, and larger hindlimbs for leaping. The hindfoot is narrowed and lengthened (hence macro-podid, "large-footed"), and the digits are unequal. Stance is more, or completely, upright, and the tail is long, not prehensile but used as an extra prop of foot.

▼ **Feet of marsupials:** (**a**) opposable first digit in foot of the tree-dwelling Virginia opossum; (**b**) long narrow foot, lacking a first digit, of the kultarr, a species of inland Australia with a bounding gait; both these species have the second and third digits separate (didactylous); in many marsupials (eg kangaroos and bandicoot) these digits are fused (syndactylous), forming a grooming "comb": (**c**) opposable first digit and sharp claws for landing on trees in Feathertail glider; (**d**) first digit much reduced in long foot of terrestrial Short-nosed bandicoot—fourth digit forms axis of foot; (**e**) first digit entirely absent in foot of kangaroo.

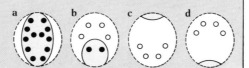

▲ **Pouches** (marsupia) occur in females of most marsupials. Some small terrestrial species have no pouch. Sometimes a rudimentary pouch (**a**) is formed by a fold of skin on either side of the nipple area that helps protect the attached young (eg mouse opossums, antechinuses, quolls). In (**b**) the arrangement is more of a pouch (eg Virginia and Southern opossums, Tasmanian devil, dunnarts). Many of the deepest pouches, completely enclosing the teats, belong to the more active climbers, leapers or diggers. Some, opening forward (**c**), are typical of species with smaller litters of 1–4 (eg possums, kangaroos). Others (**d**) open backward and are typical of digging and burrowing species (eg bandicoots, wombats).

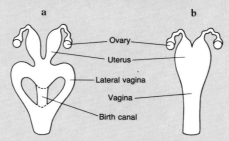

▲ **Anatomy of reproduction**, and its physiology, set marsupials (**a**) apart. In the female, eggs are shed into a separate (lateral) uterus, to be fertilized. The two lateral vaginae are often matched in the male by a two-lobed penis. Implantation of the egg may be delayed, and the true placenta of other mammals is absent. The young are typically born through a third, central, canal; this is formed before each birth in most marsupials, such as American opossums: in the Honey possum and kangaroos the birth canal is permanent after the first birth. The shape of uterus in placental mammals is shown in (**b**).

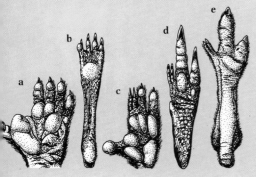

yolksac that makes only loose contact with the uterine wall. In eutherians, nourishment of the young during its prolonged internal gestation occurs by way of the placenta, in which the membranes surrounding the embryo make close contact with the uterine wall, become very vascular and act as the means of transport of material between maternal circulation and embryo.

A further peculiarity of reproduction, similar to the delayed implantation found in some other mammals, occurs in most kangaroos and a few species in other marsupial families. Pregnancy in these marsupials occupies more or less the full length of the estrous cycle but does not affect the cycle, so that at about the time a female gives birth, she also becomes receptive and mates. Embryos produced at this mating develop only as far as a hollow ball of cells (the blastocyst) and then become quiescent, entering a state of suspended animation or "embryonic diapause." The hormonal signal (prolactin) which blocks further development of the blastocyst is produced in response to the sucking stimulus from the young in the pouch. When sucking decreases as the young begins to eat other food and to leave the pouch, or if the young is lost from the pouch, the quiescent blastocyst resumes development, the embryo is born, and the cycle begins again. In some species which do not breed all year round, such as the Tammar wallaby, the period of quiescence of the blastocyst is extended by seasonal variables such as changes in day length. The origin of embryonic diapause may have been to prevent a second young being born while the pouch was already occupied, but it has other advantages, allowing rapid replacement of young which are lost, even in the absence of a male.

In marsupials, the only teeth replaced during the animal's lifetime are the posterior premolars. Relatively unspecialized marsupials such as American opossums, Australian dasyures and bandicoots have many more incisor teeth than placental mammals (10 or eight in the upper jaw and eight in the lower jaw, compared with a maximum of six upper and six lower incisors in the lower jaw). Those marsupials with at least four incisors in the lower jaw are termed polyprotodont, in contrast to the diprotodonts, which have only two incisors, generally large and directed forwards, in the lower jaw. Although the original function of these two teeth may have been for holding and stabbing insect prey, in herbivores they make, with the upper incisors, a wonderfully precise instrument for the selection of individual leaves or, indeed, blades of grass.

The Australian diprotodonts have the second and third toes of the hindfoot joined (syndactylous). So too do the polyprotodont bandicoots, suggesting that they share a syndactylous ancestor.

The most important senses of marsupials are hearing and smell. Most species are nocturnal, so vision is not particularly important. Arboreal species in particular use sound for communication at a distance, and all marsupials seem to live in a world dominated by smells. As well as urine and feces, each species has several odor-producing skin glands which are used to mark important sites, other animals, or themselves. It is known that the Sugar glider can recognize strangers to its group on the basis of scent alone, and it is likely that most marsupials recognize by scent other individuals, places and sexual condition.

The historical accident of their late discovery has led to marsupials being originally classified as a single order of the class Mammalia. Some modern authorities recognize that the antiquity and diversity of the group warrant division into a number of orders within a superorder Marsupialia. However, authorities do not agree on what the orders should be, because the affinities of the marsupial families are still constantly debated and uncertain, owing to the paucity of early fossil history. Here we divide the marsupials into 18 living families grouped into two suborders: the Polyprotodonta and Diprodota (see p824). The numbers of genera and species are only approximate, because new collections and modern work on old specimens are producing many changes in classification.

South American marsupials are mostly small terrestrial or arboreal insectivores or omnivores, but in Australia marsupials have adopted many of the life-styles found in placental mammals in other parts of the world: terrestrial grazing and browsing herbivores (kangaroos and rat kangaroos), arboreal folivores (koala and possums), arboreal omnivores (Sugar glider and pygmy possums), in addition to small terrestrial insectivores/omnivores (dasyurids and bandicoots), and nectarivores (Honey possum).

The "primitive" tag that was applied to marsupials for a long time was taken as sufficient explanation of why marsupial societies do not have the subtleties and complexities of primate and carnivore societies. Now it is clear that there is a range of social systems in marsupials which is the product of an independent evolution in

response to ecological circumstances. However, the starting point of marsupial evolution in this respect was different, because the pattern of reproduction, with the young carried in a pouch or on the abdomen of the mother, was in itself an important factor in social evolution.

The common stereotype of marsupial parental care is a female kangaroo with one large young in her pouch, but there are other very different patterns. For example, in most species of the families Didelphidae (American opossums) and Dasyuridae (Australian carnivores), large litters of eight or more are born and, in contrast to the large deep bag of the kangaroos, the pouch is no more than a raised fold of skin around the nipples and does not enclose the litter. Immediately after birth, young didelphids or dasyurids stay firmly attached to the nipples, and are carried wherever their mother goes. As soon as the young are able to release the teat, the mother leaves the litter in a nest when she goes out to feed. At this stage the young have very little hair, their eyes are not open, and they are unable to stand, but the whole litter may weigh more than their mother, and is just too much for her to carry while she forages.

In the bandicoots and some of the smaller diprotodonts (eg gliders, Mountain pygmy possum, Honey possum), 2–4 young are born and the nipples are completely enclosed by the pouch. The smaller litter is carried until later in development, although it, too, is left in a nest when the young become too heavy for the mother to carry, but before they are able to follow her. Only those species which give birth to a single young carry it in the deep pouch until it is able to keep up with its mother on the ground or ride on her back.

In all mammals, because of the milk produced by the mother, male assistance in feeding the young is less important than in, for example, birds, where both parents may feed the young. In many marsupials, the role of the male is even further reduced because the pouch takes over the functions of carrying and protecting the young and keeping it warm. A female's need for assistance in rearing young does not appear to be an important factor promoting the formation of long-lasting male-female pairs or larger social groups.

The majority of marsupial species mate promiscuously, show few examples of long-term bonds and do not live in groups. That some species do form monogamous pairs or harems suggests that their general absence is due to the lack of external pressures

▶ **Life in the pouch**—the single young of an Australian Common brushtail possum attached to a teat in its mother's deep pouch. At birth the young weigh just 0.2g, or less than a hundredth of an ounce. Unlike other marsupials such as the Virginia opossum, whose much more numerous young are "parked" by the mother while she forages, this one will develop in the pouch for 5–6 months.

▼ **A "North Australian tiger"** depicted in an Aboriginal rock painting in Kakado National Park, Northern Territory. Better known as the Tasmanian wolf or tiger, or thylacine, the species became extinct on the mainland before Europeans arrived, probably as the dingo spread from the northwest. It survived in Tasmania, where there were no dingos, only to be hunted to extinction by Europeans.

For Aboriginal man, the animals of his world were an important part of his view of life, which saw him in a three-sided relationship with mythical beings and nature. Everything in his life was thought to be fitted into a pattern established when the world was made in the dreamtime, enshrined in the lore passed on by the storytellers. Specific reasons for the form and behaviour of every living creature were contained in the dreamtime stories. Every Aboriginal was linked to the mythical beings of the dreamtime by some creature of the present. Membership of a particular totem dictated the whole pattern of life, who should marry whom and which animals could be hunted.

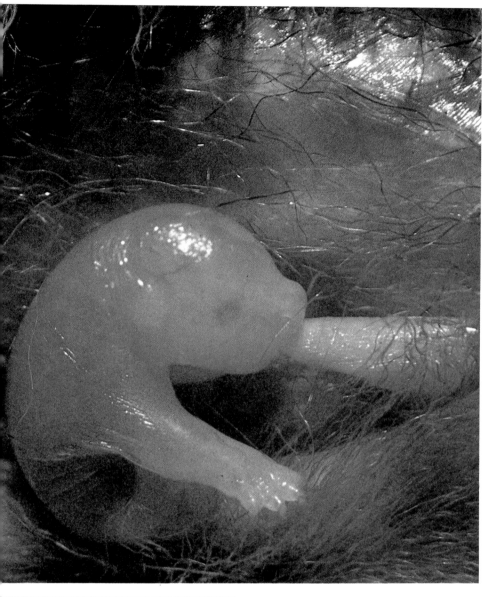

predation and his domestic pets. Hunting (by small family parties with very limited equipment) by itself probably had little effect on marsupial populations. Less certain is the effect of the practice of burning large tracts of land both to drive out game which could be killed and to produce new growth that would attract kangaroos at a later time.

The second period of major change began with the arrival of Europeans with their sheep, cattle, rabbits, foxes, cats, dogs, donkeys and camels, and large-scale modification of habitat for pastoral and agricultural enterprises. Approximately nine species have become extinct and another 15–20 have suffered a gross reduction in range, persisting only in small isolated populations. The species which have suffered most from the changes wrought by European man are small kangaroos, bandicoots and large carnivores such as the thylacine and native cats. Habitat degradation by sheep, cattle and rabbits seems to have had most effect on the species of more arid regions, such as the Greater bilby or Rabbit-eared bandicoot, numbat and various rat kangaroos (eg Brush-tailed and Lesueur's rat kangaroos).

The environmental changes brought about by European man have not all been unfavorable to marsupials. A few of the large grazing herbivores, the kangaroos (eg Red, Eastern and Western gray kangaroos, and wallaroos), have increased in numbers and range with the spread of grasslands and watering points for stock, and in some areas are numerous enough to provide significant competition for sheep or cattle, mainly when food and water are scarce during droughts. These kangaroos, which at times are seen as pests, present the problem of keeping numbers below a level at which competition becomes significant. Harvesting is controlled by the fauna authorities throughout Australia with a quota of kangaroos to be shot each year determined on the basis of population surveys.

Most species of marsupial have little or no importance as pests and their continued existence depends largely on the maintenance of sufficient habitat to support secure populations and also on the control of feral foxes and cats, which have spread over the whole Australian continent and are significant predators of the small marsupials.

Of the marsupials outside Australia, the Virginia opossum in North America appears to coexist happily with man. Despite the widespread destruction of habitat in South America, no species of marsupial is classified as endangered.   EMR

favoring them, and not because marsupials are incapable of their development.

The marsupials lived in Australia without man for more than 45 million years. Since then they have lived first with Aboriginal man and subsequently with European man for less than 100,000 years (see LEFT).

The time since the arrival of man has shown some marked changes in the marsupial fauna of Australia. The first major change was in the late Pleistocene, with the extinction of whole families of some large terrestrial marsupials, most notably *Diprotodon* and the large browsing kangaroos. It is probable that the climatic fluctuations, increasing aridity and reduction of favorable habitat at this time placed many forms under increasing stress, and early man placed the final nails in the coffin with fires,

AMERICAN OPOSSUMS

Family: Didelphidae
Seventy-five species in 11 genera.
Distribution: throughout most of S and
C America, north through eastern N America to
Ontario, Canada. Virginia opossum introduced
into the Pacific coast.

Habitat: wide-ranging, including temperate
deciduous forests, tropical forests, grasslands,
and regions, mountains, and human settlements. Terrestrial, arboreal and semi-aquatic.

Size: ranges from the small Formosan mouse
opossum with head-body length 6.8cm (2.7in),
tail length 5.5cm (2.2in), and Kuns' short-
tailed opossum with head-body length 7.1cm
(2.8in), tail length 4.2cm (1.7in), to the
Virginia opossum with head-body length
33–55cm (13–19.7in), tail length 25–54cm
(9.8–21.3in) and weight 2–5.5kg (4.4–12.1lb).

Gestation: 12–14 days.
Longevity: 1–3 years (to about 8 in captivity).

Coat: either short, dense and fine, or woolly, or
a combination of short underfur with longer
guard hairs. Dark to light grays and browns,
golden; some species with facial masks or
stripes.

Subfamily Didelphinae
Seventy species in 8 genera, distribution as family.
Virginia or **Common opossum** (*Didelphis
viriginiana*), **Southern opossum** (*D. marsupialis*),
and **White-eared opossum** (*D. albiventris*);
mouse opossums, 47 species of *Marmosa*,
including the **Common mouse opossum**
(*M. murina*), **Ashy** (*M. cinerea*), **Elegant**
(*M. elegans*), **Formosan** (*M. formosa*) and **Pale-
bellied** (*M. robinsoni*) **mouse opossums**; the
yapok or **Water opossum** (*Chironectes
minimus*); **Lutrine** or **Little water** or **Thick-
tailed opossum** (*Lutreolina crassicaudata*);
short-tailed opossums, 14 species of
Monodelphis, including the **Gray short-tailed
opossum** (*M. domestica*) and **Kuns' short-tailed
opossum** (*M. kunsi*); **Patagonian opossum**
(*Lestodelphys halli*); **Brown four-eyed opossum**
(*Metachirus nudicaudatus*); **Gray four-eyed
opossum** (*Philander opossum*) and **Mcilhenny's
four-eyed opossum** (*P. mcilhennyi*).

Subfamily Caluromyinae
Five species in 3 genera, from S Mexico through
C America and most of northern S America.
Woolly opossums (*Caluromys philander*,
C. derbianus, *C. lanatus*); **Black-shouldered
opossum** (*Caluromysiops irrupta*); **Bushy-tailed
opossum** (*Glironia venusta*).

When marsupials were first introduced to Europeans in 1500, it was a female Southern opossum from Brazil that the explorer Pinzón presented to the royal court of Spain's Ferdinand and Isabella. The monarchs examined this female with young in her pouch and dubbed her an "incredible mother." Despite this royal introduction, the popular image of the opossum has never been as lofty; they are often viewed as rather slow-witted animals with a dreadful smell. In fact, although not as diverse as the well-known Australian marsupials, the American opossums are a successful group with a variety of different species, ranging from the highly specialized tree-dwelling woolly opossums to generalists such as the Southern and Virginia opossums.

American opossums range in size from that of a mouse to that of a cat. The nose is long and pointed with long tactile hairs (vibrissae). Eyesight is generally well-developed and, in many species, the eyes are round and somewhat protruding. When an opossum is aroused it will often threaten the intruder, with mouth open and lips curled back revealing its 50 sharp teeth. Hearing is acute and the naked ears are often in constant motion as an animal tracks different sounds. Most opossums are proficient climbers, with hands and feet well adapted for grasping. Each foot has five digits and the big toe on the hind foot is opposable. The round tail is generally furred at the base with the remainder either naked or sparsely haired. Most opposums have prehensile tails which are used as grasping organs as animals climb or feed in trees. Unlike the Southern opossum which was introduced to

Spanish royalty, not all female opossums have a well-developed pouch. In some species there are simply two lateral folds of skin on the abdomen, whereas in others the pouch is absent altogether. In males the penis is forked and the pendant scrotum often distinctly colored.

The Virginia or Common opossum of North and Central America, the Southern opossum of Central and South America, and the White-eared opossum of higher elevations in South America are generalized species, occurring in a variety of habitats from grasslands to forests. They have cat-sized bodies, but are heavier with shorter legs. Although primarily terrestrial, these opossums are capable climbers. In tropical grasslands, the Southern opossum becomes highly arboreal during the rainy season when the ground is flooded. Opportunistic feeders, these opossums vary their diets depending upon what is seasonally or locally abundant. Their diet includes fruit, insects, small vertebrates, carrion and garbage. In tropical forests of southeastern Peru the Southern opossum climbs to heights of 25m (80ft) to feed on flowers and nectar during the dry season.

The four-eyed opossums from the forests of Central and northern South America are also rather generalized species. They are smaller than the Virginia opossum with more slender bodies. They have distinct white spots above each eye, from which their common name is derived. These opossums are adept climbers, but the degree to which they climb seems to vary between habitats. The four-eyed opossums are also opportunistic feeders; earthworms, fruit,

▲ **Possum up a tree.** The Virginia opossum is at home on the ground but may inhabit woodlands and is well able to climb trees. Its diet ranges from insects and fruit to carrion. Opposable first digits on hind feet are found in all American opossums.

◄ **A Woolly opossum** (*Caluromys lanatus*) peers through the leaves in the Amazon rain forest. The large forward-facing eyes are those of a specialized tree-dweller. Woolly opossums eat fruit and nectar.

insects and small vertebrates are all eaten.

The yapok, or Water opossum, is the only marsupial highly adapted to the aquatic habit. The hindfeet of this striking opossum are webbed, making the big toe less opposable than in other didelphids. When swimming, the hindfeet alternate strokes while the forefeet are extended in front, allowing them to either feel for prey or carry food items. Yapoks are primarily carnivorous, feeding on crustaceans, fish, frogs and insects. Although yapoks can climb, they do so rarely and the long, round tail is not very prehensile. Both male and female yapoks possess a pouch, which opens to the rear. During a dive, the female's pouch becomes a watertight chamber; fatty secretions and

long hairs lining the lips of the pouch form a seal and strong sphincter muscles close the pouch. In males, the scrotum can be pulled into the pouch when the animal is swimming or moving swiftly.

The Lutrine or Little water opossum is also a good swimmer, although it lacks the specializations of the yapok. Unlike the yapok, which is found primarily in forests, Lutrine opossums often inhabit open grasslands. Known as the ''comadreja'' (weasel) in South America, this opossum has a long, low body with short, stout legs. The tail is very thick at the base and densely furred. Lutrine opossums are able predators, being excellent swimmers and climbers and also agile on the ground. They feed on a variety of prey including small mammals, birds, reptiles, frogs and insects.

The mouse, or murine, opossums are a diverse group, with individual species varying greatly in size, climbing ability and habitat. All are rather opportunistic feeders. The largest species, the Ashy mouse opossum, is one of the most arboreal, whereas others are more terrestrial (eg *Marmosa fuscata*). The tail in most species is long and slender and very prehensile, but in some species (eg the Elegant mouse opossum) can become swollen at the base for fat storage. The large thin ears can become crinkled when the animal is aroused. The females lack a pouch and the number and arrangement of mammae vary between species. Mouse opossums inhabit most habitats from Mexico through South America; they are absent only from the high Andean páramo and puna zones, the Chilean desert and Patagonia. In Patagonia, mouse opossums are replaced by another small species, the Patagonian opossum, which has the most southerly distribution of any didelphid. This opossum broadly resembles the mouse opossums. The muzzle is shorter, which allows for greater biting power. For this, and because insects and fruit are rare in their habitat, Patagonian opossums are believed to be more carnivorous than mouse opossums. The feet, which are stronger and possess longer claws than in mouse opossums, suggest burrowing (fossorial) habits. As in some of the mouse opossums, the tail of the Patagonian opossum can become swollen with fat.

The short-tailed opossums are small didelphids inhabiting forest and grasslands from eastern Panama through most of northern South America east of the Andes. The tail of these shrew-like opossums is short and naked and their eyes are proportionately smaller than in most didelphids and not as protruding. As these anatomical features suggest, short-tailed opossums are primarily terrestrial, but they can climb. Like mouse opossums, females lack a pouch and the number of mammae varies between species. Short-tailed opossums are omnivorous,

The Monito del Monte—a Hibernator

In the forests of south-central Chile is found a small marsupial known as the "monito del monte" or "colocolo." Once thought to belong to the same family as the widespread American opossums (Didelphidae), the monito del monte is now believed to be the only living member of an otherwise fossil family, the Microbiotheriidae. About 20 million years ago at least six other species of this family inhabited southern South America. Now the single remaining species, *Dromiciops australis*, has a limited distribution extending south from the city of Concepción to the island of Chiloe and east from the coast of Chile to the high Andes.

Monitos (little monkeys) have small bodies with short muzzles, small round ears, and thick tails. Their head-body length is 8–13cm (3.2–5.1in), tail length 9–13cm (3.5–5.1in) and they weigh just 16–31g (0.6–1.1oz). Their fur is short, dense and silky. They have a light gray face with black eye-rings. Their upper body is gray-brown with cinnamon on the crown and neck and several light gray patches on the shoulders and hips. This alternation of colors results in a slightly marbled appearance which may help camouflage the animal. The underparts are pale buffy. The tail is covered with dense body fur at the base and brown hair for the remainder.

These arboreal marsupials are found in cool, humid forests, especially in bamboo thickets. Environmental conditions are often harsh in this region, and monitos del monte exhibit various adaptations to the cold. The dense body fur and small, well-furred ears help prevent heat loss. During winter months when temperatures drop and food (primarily insects and other small invertebrates) is scarce, monitos del monte hibernate. Prior to hibernation, the base of the tail becomes swollen with fat deposits. The nests of these marsupials also protect them from the cold. They construct spherical nests out of water-repellent bamboo leaves and line them with moss or grass. The nests are placed in well-protected areas such as tree holes, fallen logs, or under tree roots.

Female monitos del monte have small but well-developed pouches with four teats. Breeding takes place in the spring and the number of young produced ranges from two to four. Initially the young monitos remain in the pouch, then the female leaves them in the nest. After emerging from the nest the young will ride on the female's back and later, mother and young forage together. Both males and females become sexually mature during their second year.

There are various local superstitions about these harmless animals. One is that the bite of a monito del monte is venomous and produces convulsions in humans. Another is that it is bad luck to see a monito del monte; some people have even been reported to burn their house to the ground after seeing a monito del monte in their home.

▷ **A portrait of alertness** OVERLEAF This Southern opossum has the sensitive whiskers on a long snout, naked ears that move as they track sounds, and hands adapted for grasping that are typical of American opossums.

▼ **Representative species of American opossums** (family Didelphidae), plus single species from the families Caenolestidae and Microbiotheriidae. (**1**) Red-sided short-tailed opossum (*Monodelphis brevicaudata*) eating centipede. (**2**) Brown four-eyed opossum (*Metachirus nudicaudatus*) grooming. (**3**) White-eared opossum (*Didelphis albiventris*) showing strength of prehensile tail. (**4**) Ashy mouse opossum (*Marmosa cinerea*) climbing tree. (**5**) Yapok (*Chironectes minimus*) with fish. (**6**) Lutrine or Little water opossum (*Lutreolina crassicaudata*) in aggressive stance. (**7**) Woolly opossum (*Caluromys lanatus*). (**8**) Patagonian opossum (*Lestodelphys halli*) hunting spider. (**9**) Shrew opossum (*Lestoros inca*—family Caenolestidae). (**10**) Gray four-eyed opossum (*Philander opossum*) foraging for fruit. (**11**) Monito del monte or colocolos (*Dromiciops australis*—family Microbiotheriidae) in nest. (**12**) Black-shouldered opossum (*Caluromysiops irrupta*) eating nut. (**13**) Bushy-tailed opossum (*Glironia venusta*).

feeding on insects, earthworms, carrion, fruit etc. Often they will inhabit human dwellings, where they are a welcome predator on insects and small rodents.

The three species of woolly opossum, the Black-shouldered and Bushy-tailed opossum are placed in a separate subfamily (Caluromyinae) from other didelphids on the basis of differences in blood proteins, anatomy of the females' urogenital system, and males' spermatozoa. The woolly opossums and the Black-shouldered opossum are among the most specialized of the didelphids. Highly arboreal, they have large protruding eyes which are directed somewhat forward, making their faces reminiscent of that of a primate. Inhabitants of humid tropical forests, these opossums climb through the upper canopy of trees in search of fruit. During the dry season, they also feed on the nectar of flowering trees and serve as pollinators for the trees they visit. While feeding they can hang by their long prehensile tails to reach fruit or flowers.

Although the Bushy-tailed opossum resembles mouse opossums in general appearance and proportions, dental characteristics, such as the size and shape of the molars, indicate that it is actually more closely related to the woolly and Black-shouldered opossums. This species is known from only a few museum specimens, all of which were taken from humid tropical forests.

One popular misconception was that opossums copulated through the nose and that young were later blown through the nose into the pouch! The male's bifurcated penis, the tendency for females to lick the pouch area before birth, and the small size (1cm/0.13g) of the young at birth all probably contributed to this notion. Reproduction in didelphids is typical of marsupials: gestation is short and does not interrupt the estrous cycle. Young are poorly developed at birth, and most of the development of the young takes place during lactation.

Most opossums appear to have seasonal reproduction. Breeding is timed so that first young leave the pouch when resources are most abundant. For example, the Virginia opossum in North America breeds during the winter and the young leave the pouch in the spring. Opossums in the seasonal tropics breed during the dry season and the first young leave the pouch with the commencement of the rainy season. Up to three litters can be produced in one season, but the last litter is often produced at the beginning of the period of food scarcity and often these young die in the pouch. Opossums in

aseasonal tropical forests may reproduce throughout the year. Year-round breeding may also occur in the White-eared opossum in the arid region of northeast Brazil.

There are no elaborate courtship displays nor long-term pair-bonds. The male typically initiates contact, approaching the female while making a clicking vocalization. A non-receptive female will avoid contact or be aggressive, but a female in estrus will allow the male to mount. In some species courtship behavior involves active pursuit of the female. Copulation can be very prolonged, up to six hours in Pale-bellied mouse opossums.

Many of the newborn young die, as many as never attach onto a teat. A female will often produce more young than she has mammae. For example, most female Virginia opossums have 13 mammae, some of which may not even be functional, but they may produce as many as 56 young. The number of young in an opossum litter which do attach ranges from one to 15, but varies both within and between species. Older females tend to have fewer young and litters born late in the season are often smaller. Litter sizes in the Virginia and Southern opossums seem to increase with increasing latitude. The number of mammae provides an indication of maximum possible litter size. In general, some species (Virginia or Common opossum, short-tailed opossums, Pale-bellied mouse opossum) have comparatively large litters (about seven young), whereas others (Gray four-eyed opossum, Woolly opossum) have 3–5 young. Females of some species (eg Virginia opossum) cannot usually raise a single offspring because there is insufficient stimulus to maintain lactation.

The rearing cycle in species which have been studied ranges from about 70 to 125 days. For example, the Gray four-eyed opossum and Woolly opossum are similar in size (usually about 400g/14oz), but the time from birth to weaning is 68–75 days in the former and 110–125 days in the latter. Initially young remain attached to the teats. Later, the young begin to crawl about the female and/or are left in a nest while the female forages. Toward the end of lactation, young begin to follow the female. Female Pale-bellied mouse opossums will retrieve detached young within a few days of birth; in contrast, female Virginia opossums do not respond to distress calls of detached young until after the young have left the pouch (at about 70 days). During the nesting phase young opossums become more responsive to clicking vocalizations of the female.

Although individual vocalizations and odors allow for some mother-infant recognition, maternal care in opossums does not appear to be restricted solely to a female's own offspring. Female Pale-bellied mouse opossums will retrieve young other than their own, and Virginia opossums and woolly opossums have been observed carrying other females' young in their pouches. Toward the end of lactation, females cease any maternal care and dispersal is rapid. Sexual maturity is attained within six to 10 months. Age at sexual maturity is not related directly to body size. Considering the Gray four-eyed opossum and the woolly opossums again, the former can breed at six months, the latter not until 10 months.

In general, opossums are not long-lived. Few Virginia opossums live beyond two years in the wild and the smaller mouse opossums may not survive much beyond one reproductive season. Woolly opossums and the Black-shouldered opossum may live longer. Although animals kept in captivity may survive longer, females are generally not able to reproduce after two years. Thus, among many of these didelphids there is a trend towards the production of a few large litters during a limited reproductive life. Indeed, a female Pale-bellied mouse opossum may typically reproduce only once in a lifetime.

The American opossums appear to be locally nomadic animals, without defended territories. Radio-tracking studies reveal that individual animals occupy home ranges, but do not exclude others of the same species (conspecifics). How long a home range is occupied varies both between and within species. In the forests of French Guiana, for example, some woolly opossums have been observed to remain up to a year in the same home range whereas others shifted home range repeatedly. Gray four-eyed opossums were more likely to shift home range. In contrast to some other mammals, didelphids do not appear to explore their entire home range on a regular basis. Movements primarily involve feeding and travel to and from a nest site and are highly variable depending upon food resources and/or reproductive condition. Thus home range estimates for Virginia opossums in the central United States vary from 12.5 to 38.8 hectares (31–96 acres). An individual woolly opossum's home range may vary from 0.3 to 1ha (0.75–2.5 acres) from one day to the next. In general, the more carnivorous species have greater movements than similar-sized species which feed more on fruit. During the breeding season

Shrew Opossums of the High Andes

Seven small, shrew-like marsupial species are found in the Andean region of western South America from southern Venezuela to southern Chile. Known sometimes as shrew (or rat) opossums, they are unique among American marsupials in having a reduced number of incisors, the lower middle two of which are large and project forward.

The South American group represents a distinct line of evolution which diverged from ancestral stock before the Australian forms did, and its members are placed in a separate family, the Caenolestidae. Fossil evidence indicates that about 20 million years ago seven genera of caenolestids occured in South America. Today the family is represented by only three genera and seven species. There are five species of *Caenolestes*. They are: the Gray-bellied caenolestid (*C. caniventer*), Blackish caenolestid (*C. convelatus*), Ecuadorian caenolestid (*C. fuliginosus*), Colombian caenolestid (*C. obscurus*) and Tate's caenolestid (*C. tatei*). Placed in separate genera are the Peruvian caenolestid (*Lestoros inca*) and the Chilean caenolestid (*Rhyncholestes raphanurus*). Known head-body lengths of these small marsupials are in the range of 9–14cm (3.5–5.5in), tail lengths mostly 10–14cm (3.9–5.5in), and weights 14–41g (0.5–1.4oz).

The elongated snouts are equipped with numerous tactile whiskers. The eyes are small and vision is poor. The well-developed ears project above the fur. The rat-like tails are about the same length as the body (rather less in the Chilean caenolestid) and are covered with stiff, short hairs. The fur on the body is soft and thick and is uniformly dark brown in most species. Females lack a pouch and most species have four teats (five in the Chilean caenolestid). Caenolestids are active during the early evening and/or night, when they forage for insects, earthworms and other small invertebrates, and small vertebrates. They are able predators, using their large incisors to kill prey. Caenolestids travel about on well-marked ground trails or runways. More than one individual will use a particular trail or runway. When moving slowly, they have a typically symmetrical gait, but when moving faster the Colombian and Peruvian caenolestids, and possibly other species, will bound, allowing the animal to clear obstacles. During locomotion, the tail is used as a counter-balance. Although primarily terrestrial, caenolestids can climb.

The Blackish, Colombian, and Ecuadorian caenolestids are distributed in the high-elevation wet, cold cloud forests, intermontane forests, and páramos in the Andes of western Venezuela, Colombia and Ecuador. In these habitats they are most common on moss-covered slopes and ledges that are protected from the cold winds and rain. The Gray-bellied and Tate's caenolestids of southern Ecuador occur at lower elevations. The Peruvian caenolestid is found at high elevations in the Peruvian Andes, but in drier habitats than that of the other species. Peruvian caenolestids have been trapped in areas with low trees, bushes and grasses. The Chilean caenolestid inhabits the cool humid forests of southern Chile. Prior to the winter months the tail of this species becomes swollen with fat deposits.

Very little is known about the biology of these Marsupials. Shrew opossums inhabit inaccessible and (for humans) rather inhospitable areas, which makes them difficult to study. They have always been considered rare, but recent collecting trips suggest that at least some species may be more common.

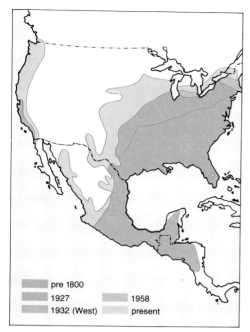

▲ Recolonizing North America where didelphids were once widespread, the Virginia opossum has extended its range well over 2 million sq km (800,000 sq mi) during the past 50 years.

Map legend:
- pre 1800
- 1927
- 1932 (West)
- 1958
- present

◀ One of 47 mouse opossum species, BELOW, this Alston's mouse opossum (*Marmosa alstoni*) of Monteverde, Costa Rica, will eat a variety of foods. Here the victim is a grasshopper.

▼ Harassed mother—Virginia opossum mother and young. Only six offspring are visible here, but the average litter is 10 or more. Newborn may number many more but the female has only 13 teats (some of these may not produce milk), so many newborn American opossums die as they cannot attach to a teat.

male didelphids become more active, whereas reproductive females generally become more sedentary.

Most opossums use several to many nest sites within their range. Nests are often used alternately with conspecifics in the area. Virginia and Southern opossums use a variety of nest sites, both terrestrial and arboreal, but hollow trees are a common location. Four-eyed opossums also nest in trees (in holes and open limbs) and on the ground (rock crevices, tree roots and under fallen palm fronds). Mouse opossums nest either on the ground (under logs and tree roots) or in trees (holes, abandoned birds' nests), depending upon arboreal tendencies and local conditions. Occasionally mouse opossums make nests in banana stalks and more than once animals have been shipped to grocery stores in the United States and Europe! In the open grasslands, the Lutrine opossum constructs globular nests of leaves or uses abandoned armadillo burrows. In more forested areas, these opossums may use tree holes. Unlike other didelphids, yapoks construct more permanent nests. Underground nesting chambers are located near the waterline and are reached through holes dug into stream banks.

Opossums are solitary animals. Although many opossums may congregate at common food sources during periods of food scarcity, there is no interaction unless animals get too close. Typically, when two animals do meet, they threaten each other with open-mouth threats and hissing and then continue on their way. If aggression does persist (usually between males) the hissing changes into a growl and then to a

screech. Communication by smell is very important. Many species have well-developed scent glands on the chest. In addition, male Virginia opossums, Gray four-eyed opossums, and Gray short-tailed opossums have been observed marking objects with saliva. Marking behavior is carried out primarily by males and is thought to advertise their presence in an area.

In tropical forests up to seven species of didelphids may be found at the same locality. Competition between these species is avoided through differences in body size and tendency to climb. For example, woolly opossums and the Ashy mouse opossum may be found in the tree canopy, the Common mouse opossum in lower branches, Southern and Gray four-eyed opossums on the ground and in the lower branches, and short-tailed opossums on the ground. Some species appear to vary their tendency to climb depending upon the presence of similar-sized opossums. For example, in a Brazilian forest where both Gray and Brown four-eyed opossums were found together, the former was more arboreal than the latter. However, in a forest in French Guiana where only the Gray four-eyed opossum was present, it was primarily terrestrial, and elsewhere the Brown four-eyed opossum is primarily arboreal.

North American fossil deposits from 70–80 million years ago are rich in didelphid remains. From North America didelphids probably entered South America and Europe (see p824). By 10–20 million years ago didelphids were extinct in Europe and North America. When about 2–5 million years ago South America again became joined to Central America, northern placental mammals entered South America and many South American marsupials became extinct, but didelphids persisted and even moved northward into Central and North America. During historical times, the early European settlers of North America found no marsupials north of what is now Virginia and Ohio. Since then the Virginia opossum has extended its range to New Zealand and southern Canada. After introductions on the West Coast in 1890, this opossum has expanded from southern California to southern Canada, a change most likely related to man's impact on the environment. Despite being hunted for food and pelts, the Virginia opossum thrives on both farms and in towns and even cities. Elsewhere in the Americas, man's impact on the environment is detrimental to didelphids. Destruction of humid tropical forests results in loss of habitat for the more specialized species. MAO'C

MARSUPIAL CARNIVORES

Family: Dasyuridae

Fifty-one or more species in 18 genera.
Distribution: Australia, New Guinea.

Habitat: diverse, stony desert to forest and alpine heath.

Size: ranges from head-body length 4.6–5.7cm (1.8–2.2in), tail length 5.9–7.9cm (2.3–3.1in) and weight 2–9.4g (0.07–0.33oz) in the Pilbara ningaui, to head-body length 57–65cm (22.5–25.6in), tail length 24.5–26cm (9.6–10in), and weight 6–8g (13.2–17.6lb) in the Tasmanian devil. Males slightly, or in larger species often considerably, heavier than females.

Gestation: Fat-tailed dunnart 13 days; Brown antechinus 27 days; Eastern quoll 16 days.
Longevity: Brown antechinus, males 11.5 months, females 3 years; Tasmanian devil 7 years.

Species include:
10 or more species of *Antechinus*, including the **Brown antechinus** (*A. stuartii*); and **Dusky antechinus** (*A. swainsonii*); 2 species of *Parantechinus*, the **Sandstone antechinus** (*P. bilarni*) and **dibbler** 🔲 (*P. appicalis*); 2 or more species of *Ningaui*, including the **Pilbara ningaui** (*N. timealeyi*) and **Inland** or **Wongai ningaui** (*N. ridei*); the **quolls, native cats** or **tiger cats**, 3 species of *Dasyurus* including *D. viverrinus*, the **Eastern quoll**; **Tasmanian devil** (*Sarcophilus harrisii*); **mulgara** (*Dasycercus cristicauda*); the **dunnarts**, 13 species of *Sminthopsis* including the **Fat-tailed dunnart** (*S. crassicaudata*) and **White-footed dunnart** (*S. leucopus*); **kowari** (*Dasyuroides byrnei*); 2 species of *Phascogale*, including *P. calura*, the **Red-tailed phascogale** or **wambenger** 🔲; and the **planigales**, 5 species of *Planigale*.

The other 8 genera are:
Antechinomys: 1 species, *A. laniger*, the **kultarr**.
Dasykatula: 1 species, *D. rosamondae*, the **Little red antechinus**.
Pseudantechinus: 1 species, *P. macdonnellensis*, the **Fat-tailed** or **Red-eared antechinus**.
Satanellus: 2 species, *S. hallucatus*, the **Northern quoll** or **satanellus**, and *S. albopunctatus*, **New Guinea marsupial cat**.
Murexia: 2 species of **marsupial mouse**.
Myoictis: 1 species, *M. melas*, the **Three-striped marsupial mouse**.
Neophascogale: 1 species, *N. lorentzii*, the **Long-clawed marsupial mouse**.
Phascolosorex: 2 species of **marsupial mouse**.

🔲 Threatened, but status indeterminate.

A VARIETY of marsupials feed upon the flesh of animals, including American opossums, bandicoots, the thylacine (see p841), the numbat (see p844) and the dasyurids. In Australia and New Guinea most of the marsupial carnivores are dasyurids and of the marsupial families inhabiting those lands they are among the most successful. Dasyurids are found in all major terrestrial habitats. A higher proportion of dasyurid species occurs in the deserts of Australia than of any other family of marsupials. They are also found in tropical rain forest, temperate eucalypt forest and woodland, flood plain, alpine and coastal heaths and even in the vegetation of coastal dunes. Yet, despite their success, they are "conservative" in appearance and body form. A large part of their success may, instead, be due to the surprising array of life-cycles they display.

Dasyurids are mostly small and mouse-like. Over 50 percent of the species weigh less than 100g (3.5oz) as adults and they include some of the smallest mammals. The ningauis of central and northwestern Australia may weigh as little as 2g (0.07oz) as adults. Many of these small dasyurids resemble shrews in appearance and habits. They have long conical snouts which presumably enable them to remove their insect prey from crevices.

In the 60–300g (2.1–10.6oz) size range are the kowari and the mulgara of central Australia and, also inland, the two species of *Phascogale* which resemble the tree shrews of Asia (order Scandentia). They too have a pointed snout, but also have conspicuous, large eyes, large ears, and bushy tails which may have a signaling function.

The largest dasyurids, the quolls (Eastern quoll male 0.9–2kg/2–4.4lb) and the much larger Tasmanian devil, bear some resemblance to placental carnivores of similar size, both in appearance and in their coat pattern of dark brown or black with conspicuous white markings.

Dasyurids are easily distinguished from the other major groups of Australian marsupials by a simple combination of characters: they possess three pairs of incisors in the lower jaw, and have front feet with five toes, hind feet with never less than four toes, and all toes are separately developed. These are considered to be primitive marsupial features from which the more specialized dentititon and feet of bandicoots, possums and kangaroos have been derived. Because of this, dasyurids are often considered closest to the stock from which other Australian marsupial groups have arisen. Until recently, the largest of the marsupial carnivores, the thylacine, was also classified as a dasyurid, but although closely related to this group, it is now placed in a separate family, the Thylacinidae.

Most dasyurids are insectivorous, but include items such as earthworms, spiders,

▼ **Representative species of marsupial carnivores** (dasyurids). (**1**) Kultarr (*Antechinomys laniger*. (**2**) Pilbara ningaui (*Ningaui timealeyi*) eating beetle. (**3**) Three-striped marsupial mouse (*Myoictis melas*). (**4**) New Guinea marsupial cat (*Satanellus albopunctatus*). (**5**) Fat-tailed or Red-eared antechinus (*Pseudantechinus macdonnellensis*). (**6**) Marsupial mouse (*Phascolosorex dorsalis*. (**7**) Red-tailed phascogale (*Phascogale calura*). (**8**) Little red antechinus (*Dasykatula rosamondae*) (**9**) Long-tailed marsupial mouse (*Murexia longicaudata*). (**10**) Fat-tailed dunnart (*Sminthopis crassicaudata*). (**11**) Common planigale (*Planigale maculata*) eating caterpillar.

small lizards, flowers and fruit in their diets. Even the larger quolls feed extensively upon beetle larvae. One survey showed that insect remains occurred in 97 percent of the feces of the Eastern quoll, while vertebrate remains occurred in only 17 percent. The largest dasyurid, the Tasmanian devil, is remarkably inefficient in capturing and killing active vertebrates and is thought to subsist on vertebrate carrion, principally from wombats, on Bennet's wallaby and sheep. When feeding upon large carcasses they detach and bolt large portions, crushing bones with their powerful jaws. One description tells how they use their forepaws to cram lengths of intestine into their mouths like eating spaghetti. Only the largest bones remain after they have fed upon a carcass.

The mulgara kills both invertebrate and vertebrate prey with a series of bites, each accompanied by a vigorous shake of the prey. The prey is dropped between bites. With large insects and small lizards the first bites are randomly directed, but the final bite is always directed at the head. There are no preliminary bites with snakes and mice. Mulgara eat mice head first and do not skin the prey. Although there have been few studies of prey capture and feeding by dasyurids, the overall impression is that they capture prey by stealth and not by chase.

Little is known about the social organization of dasyurids but they appear for the most part to be solitary. The Fat-tailed dunnart is found in drier habitats over much of the southern two-thirds of Australia. It nests in groups, but the relationships between individuals in these groups is not known. Out of the breeding season, groups are most commonly of two or three individuals. Nesting in groups is most common in April and May, just before the start of the breeding season, and is lowest in August and November–February, when females are frequently with litters. During the breeding season females, except when in heat (estrus) usually exclude males from the nest.

Recent observations of the Brown antechinus of eastern Australia have shown that males disperse from the maternal nest at weaning and nest with other unrelated females, some of which remain at the maternal nest. As the breeding season approaches, males and females visit a number of nests and increase the number of individuals with which they associate. Males copulate with a number of females.

Many species use mate-attracting calls. Males tend to call at night throughout the

breeding season, but females may confine calling to periods of receptivity. It has been suggested that these calls have arisen as a consequence of the solitary nature of dasyurids and of their occurrence in many open habitats where communication by other means would endanger their lives.

Most small placental mammals ovulate more than once during a breeding season and usually produce more than one litter. A number of dasyurids also show this pattern. For example, the Fat-tailed dunnart breeds between July, in the middle of the Australian winter, and January or February, in summer. Females usually produce two litters during this period, one before and one after October. Gestation lasts about 12–13 days,

▲ **Tools of a carnivore**—powerful jaws and sharp teeth of the Tasmanian devil. This largest of Australian marsupial carnivores will take living prey, including lambs and poultry, but it prefers carrion, and can chew and swallow all parts of a sheep carcass, including the bones.

▶ **Four-month old Eastern quolls** OPPOSITE ABOVE in their grass-lined den. Usually six young of this species attach to teats in the pouch. In mid-August, about 10 weeks after birth, the mother deposits them in the nest.

▶ **"Tiger cat" of Tasmania** BELOW. The Spotted-tailed quoll is nearly as large as the Tasmanian devil and a much more active hunter. It kills its prey—gliders, small wallabies, reptiles—with a bite to the back of the head.

which is short even for a marsupial. The young are at first suckled continuously within the pouch and later within a grass-lined nest located within a hollow log or under a large stone. Lactation occupies 60–70 days. Young Fat-tailed dunnarts mature rapidly for a marsupial and are capable of breeding at six months. Even so, they do not breed in the breeding season of their birth, but in the following season. Few if any individuals live beyond 18 months of age and so do not breed in a second season.

In tropical Australia and New Guinea there are some small dasyurids which breed year-round but little else is known of their life-cycles. It is thought that many dasyurids like the Fat-tailed dunnart coincide breeding with spring and summer because this is the time of greatest insect, and therefore food, abundance. If this is so, then it is likely that year-round reproduction is possible in some tropical species because of the year-round abundance of insects.

The quolls and Tasmanian devil may also be able to ovulate more than once a year but only produce a single litter. This is because it takes these large dasyurids longer to raise a litter. Mating occurs in March in the Tasmanian devil and most births occur in April. The young are suckled for 8–9 months and are weaned in November or December. The young may take two years to reach sexual maturity, rather than the one year typical of most dasyurids, but they may live to at least six years of age.

The Thylacine or Tasmanian Wolf

The thylacine, Tasmanian wolf or Tasmanian tiger was the largest of the recent marsupial carnivores. Fossil thylacines are widely scattered in Australia and New Guinea, but the living animal was confined in historical times to Tasmania, where it now appears to be extinct.

Superficially the thylacine resembled a dog. It stood about 60cm (24in) at the shoulders, head-body length averaged 80cm (31.5in), and weight 15–35kg (33–77lb). The head was dog-like, the neck was short and the body sloped away from the shoulders. The legs were short, as in large dasyurids. The features which clearly distinguish it from dogs are the long (50cm/20in) stiff tail which is thick at the base, and the coat pattern of black or brown stripes across the back on a sandy yellow ground. The thylacine (*Thylacinus cynocephalus*) is placed in its own family, the Thylacinidae.

Most of the information we have on the behavior of the thylacine is either anecdotal or has been obtained from old film. It ran with diagonally opposing limbs moving alternately. It could sit upright on the hindlimbs and tail rather like a kangaroo, and leap 2–3m with great agility. Thylacines appear to have hunted alone or in pairs, and before Europeans settled in Tasmania probably fed upon wallabies, possums, bandicoots, rodents and birds. It is suggested that they caught prey by stealth rather than by chase.

At the time of European settlement the thylacine appears to have been widespread in Tasmania and was particularly common where settled areas adjoined dense forest. It was thought to rest during the day in dense forest on hilly terrain and emerge to feed at night in grassland and woodland. Its extinction on mainland Australia some time in the last 3,000 years may have been a consequence of competition with the dingo.

From the early days of European settlement of Tasmania the thylacine developed a reputation for killing sheep. As early as 1830 bounties were offered for killing thylacines and their destruction led to fears for the species as early as 1850. Even so, the Tasmanian Government introduced its own bounty scheme in 1888 and in the next 21 years, before the last bounty was paid, 2,268 animals were officially accounted for. The number of bounties paid had declined sharply by the end of this period and it is thought that an epidemic disease as well as hunting led to the thylacine's final disappearance.

The last thylacine to be captured was obtained in western Tasmania in 1933 and died in the Hobart zoo in 1936. Since then there have been a number of very thorough searches of Tasmania and despite alleged sightings of this animal, even to this day, the most recent survey concluded that there has been no positive evidence of the survival of thylacines since 1936.   AKL

Roughly one-third of the dasyurid species are unusual among small mammals in that females only ovulate once a year and so are only able to produce a single litter annually. Some of these species, such as the Sandstone antechinus in the north of the Northern Territory, live up to 30–36 months and breed in two breeding seasons. Once again, mating usually occurs in winter and the young are weaned in late spring to early summer when insects are most abundant.

The most unusual life-cycle is found among seven species of *Antechinus* and two of *Phascogale*. These species also ovulate once a year and produce a single litter, but males only live for 11.5 months and mate in one two-week mating period. In the Brown antechinus, females may live to three years of age but they rarely if ever produce more than two litters.

Births in the Brown antechinus usually occur in September or October, that is, in very early spring, and usually within a one- to two-day period within a given population. The young are then attached firmly to teats in a saucer-like pouch for about a month. They grow substantially during this period and hang beneath the body as the female moves about foraging. Subsequently the young are placed in a nest in a tree hollow and suckled for a further 2–3 months. These young are weaned by December or January and mature and mate in the following winter. Mating is intense. Males tend to disperse widely during the mating period, and may mount and copulate with females for prolonged periods. Males caged with a female have been observed to remain mounted for up to 12 hours and may repeat this for up to 13 nights.

At the end of the mating period in July or August, all males die, and the females are free to raise young in their absence. Like the onset of mating, births and weaning, this male die-off (see below) occurs at precisely the same time each year within a population and usually occupies 5–10 days.

The size of the litters varies considerably among dasyurids and is usually greatest in the small species. *Planigale* and *Antechinus* may have litters of 12 young whereas the much larger Tasmanian devil has litters of 2–4 young. However, not all the small dasyurids have large litters; the two New Guinea species of *Antechinus*, which breed year-round, have litters of three or four young.

Because the young attach firmly to a teat for a period after birth, the number of young raised is limited by the number of teats available. Dasyurids produce more young

The Marsupial Mole

The Marsupial mole is the only Australian mammal that has become specialized for a burrowing (fossorial) life. Others, including small native rodents, have failed to adapt to the use of this niche, apart from a few species nesting in burrows.

Because of its extensive and distinct modifications the Marsupial mole (*Notoryctes typhlops*) is placed in a separate family, the Notoryctidae. Its limbs are short stubs. The hands are modified as excavating instruments, with rudimentary digits and greatly enlarged flat claws on the third and fourth digits. Excavated soil is pushed back behind the animal with the hindlimbs, which also give forward thrust to the body and, like the hands, are flattened with reduced digits and three small flat claws on the second, third and fourth digits. The naked skin (rhinarium) on the tip of the snout has been extended into a horny shield over the front of the head, apparently for thrusting through the soil. The coat is pale yellow and silky. The nostrils are small slits, there are no functional eyes or external ears, and the ear openings are concealed by fur. The neck vertebrae are fused together, presumably to provide rigidity for thrusting motions. Females have a rear-opening pouch with two teats. The tail, reduced to a stub, is said to be used sometimes as a prop when burrowing. Head-body length is approximately 13–14.5cm (5.1–5.7in), tail length 2–2.5cm (1in) and weight 40g (1.4oz). Dentition is I4/3, C1/1, P2/3, M4/4 = 44.

Little is known of these moles in the wild. They occur in the central deserts, using sandy soils in river-flat country and sandy spinifex grasslands. They are thought to prefer to feed on insects, particularly burrowing larvae of beetles of the family Scarabaeidae and moths of the family Cossidae. In captivity Marsupial moles will seek out insect larvae buried in the soil and consume them underground. They also feed on the surface. They are not known to make permanent burrows, the soil collapsing behind them as they move forward, and in this respect they are most unusual among fossorial mammals, which usually construct permanent burrows. Animals in captivity have been observed to sleep in a small cavity which collapses after they leave.

Compared to other burrowing animals (eg true moles and golden moles), the Marsupial mole shows differences of detail in the adaptive route it has followed. The head shield is much more extensive than in many others, the eyes are more rudimentary than in most, and the rigid head/neck region with fused vertebrae appears to be specific to the Marsupial moles.   GG

than they have teats and, in most species, all or all-but-one of the teats are occupied by a young. The small species tend to have more teats than the larger species.

How do these life-cycles relate to the habitats used by dasyurids? Most of the species from the desert habitats, such as the Fat-tailed dunnart and the kowari, have 6–8-month breeding seasons in which the females produce two litters, each containing 5–8 young. These habitats are harsh and it is difficult to predict when conditions will favor successful reproduction. By reproducing twice during a breeding season these species increase the chance of successfully rearing at least one litter. They are presumably restricted to breeding in spring and summer by the seasonal distribution of food. Insects would be reduced in abundance during dry and cold periods of fall and winter.

The larger dasyurids are restricted to a single litter annually, probably because it

◄ ▲ ► **Competition in small dasyures.** The Brown antechinus RIGHT is abundant in forests and heathlands of southeastern Australia, where it does however face competition for food. The larger and more terrestrial Dusky antechinus ABOVE will force its Brown relative into more open country, or up into the trees. Consequently Brown antechinus populations may be 70 percent lower where the two overlap. The Brown antechinus competes more successfully with other marsupials such as the smaller Common dunnart (*Sminthopsis murina*) LEFT or the White-footed dunnart, and expels them from favorable habitat.

Brown antechinus females carry more female than male young in the pouch (58 percent of the total). Since females only breed once they can produce more daughters without having to "worry" about future competition from them.

In the Dusky antechinus these percentages are reversed, and more males are carried, probably because most females breed twice in their life-time and the mother can thus reduce competition from stay-at-home daughters. (Males on the other hand disperse from their natal homes before they breed.)

takes longer for large mammals to raise young and this is especially true of marsupials. Here the ability to ovulate more than once during a breeding season provides an opportunity to replace a litter if one is lost. These large dasyurids are found in arid as well as wet forest habitats and their success in a variety of habitats, some of which are harsh, may be related to their size. Generally large mammals live longer than small mammals and are able to spread their breeding over a number of years.

The small dasyurids which can only produce one litter a year fall into two groups. Those species where both males and females reproduce in two years tend to occur where there is some risk of losing a litter, as in habitats marginal to deserts. Others, such as the White-footed dunnart of southern Victoria and Tasmania, use vegetation which is regenerating after a fire and suitable for only a few years. They tend to produce large litters of 10 young, and may have opted to produce one large litter a year rather than two smaller litters.

The second group of such species are those typified by a male die-off after the first breeding season. They are restricted to the forests and heathlands of Australia, where their chances of successfully raising a litter are good. In these habitats the abundance of insects is highly seasonal, reaching peaks in

The Numbat
—Termite-eater

The numbat is highly specialized to feed upon termites and, perhaps because of the diet, it is the only fully day-active Australian marsupial. Because of its distinctive coat markings and delicate appearance, it is also one of the most attractive marsupials.

The numbat (*Myrmecobius fasciatus*) is the sole member of the family Myrmecobiidae. Head-body length averages 24.5cm (9.7in), tail length 17.7cm (7in) and males weigh 0.5kg (1.1lb), females 0.4kg (0.9lb). The black-and-white bars across the rump fade into reddish-brown on the upper back and shoulder. A prominent white-bordered dark bar passes from the base of each ear through the eye to the snout. The long tail is bushy.

The numbat spends most of its active hours searching for food. It walks, stopping and starting, sniffing at the ground and turning over small pieces of wood in its search for shallow underground termite galleries. On locating a gallery, the numbat squats on its hind feet and digs rapidly with its strong clawed forefeet. Termites are extracted with the extremely long, narrow tongue which darts in and out of the gallery. Some ants are eaten but (despite alternative names of Banded or Marsupial anteater) it seems that the numbat usually takes these in accidentally, while picking up the termites. It does not chew its food, and also swallows grit and soil acquired while feeding.

Numbats are solitary for most of the year, each individual occupying a territory of up to 150ha (370 acres). During the cooler months a male and female may share the same territory, but they are still rarely seen together. Hollow logs are used for shelter and refuge throughout the year, although numbats also dig burrows and often spend the nights in them during the cooler months. The burrows and some logs contain nests of leaves, grass and sometimes bark. In summer numbats sunbathe on logs.

Four young are born between January and May, and attach themselves to the nipples of

the female, which lacks a pouch. In July or August the mother deposits them in a burrow, suckling them at night. By October, the young numbats are half grown and are feeding on termites themselves while remaining in their parents' area. They disperse in early summer (December).

Numbats once occurred across the southern and central parts of Australia, from the west coast to the semi-arid areas of western New South Wales. They are now found only in a few areas of eucalypt forest and woodland in the southwest of Western Australia. Destruction of their habitat for agriculture and predation by foxes have probably been the major contributors to this decline. While most of their habitat is now secure from further clearing, remaining numbat populations are so small that the species is considered rare and endangered. Efforts are being made to set up a breeding colony from which natural populations may be reestablished.   AKL

► **Mulgara** eating a locust. This widespread species of inland Australia digs burrows in the sand. A black crest on the short, fat tail identifies the mulgara.

▼ **The kowari** is another small burrowing carnivore of the inland deserts, but restricted to southwestern Queensland. The black brush tips a longer tail than the mulgara's.

spring and early summer and lows in the dry fall and cold winters. Because of the long period required by these small marsupials to raise litters it is probably not feasible for them to raise two litters in a year without a high chance of losing at least one of the litters. Instead they have concentrated their breeding into producing a single annual litter. Males faced with the probability of dying before a second breeding season go all out to father as many young as possible during their first short breeding season and die abruptly as a result of this effort.

Considerable attention has been focussed on the die-off of males which occurs at the conclusion of the breeding season of these dasyurids. Once again, we have most information for the Brown antechinus, in which the death of males seems to result from a variety of diseases, including stomach and intestinal ulcers. These ulcers, often called "stress ulcers," and other evidence suggest that the males are severely stressed during the mating period. At first it was thought that their stress resulted from their behavior during that period, when they may spend much of their time mating, chasing females and fighting with other males. But recently it has become obvious that the stress response in males is to their advantage as breeding animals, allowing them to give up feeding and concentrate upon finding and mating with females. Stressed animals convert body protein into energy which may replace the energy usually obtained from food. But while being stressed may be beneficial in allowing the males to mate with more females and father more offspring, it also makes the males more susceptible to disease.

The ranges of most dasyurids have almost certainly shrunk since the advent of European settlement in Australia. In most instances this has probably resulted from the destruction of habitat by clearing for agriculture or from frequent burning which is used to create pasture for sheep and cattle. Restrictions in the ranges of the larger species such as the quolls have been most obvious and some of these, such as the Eastern quoll, are endangered over parts of their range. Whether or not this reduction in numbers is due to competition from feral cats is unknown.

Five dasyurids are considered threatened with extinction. They are the dibbler, which is restricted to Cheyne Beach, Jerdacuttup, Western Australia, the Red-tailed phascogale of southwestern Australia, and three *Sminthopsis* species. The dibbler may already be extinct.   AKL

BANDICOOTS AND BILBIES

Families: Peramelidae and Thylacomyidae
Nineteen species in 8 genera.
Distribution: Australia and New Guinea.

Habitat: all major habitats in Australia and New Guinea from desert to rain forest and semi-urban areas.

Size: ranges from head-body length 15–17.5cm (6–7in), tail length 11cm (4.3in) in the Mouse bandicoot to head-body length 39–56cm (15–22in), tail length 12–34 (4.7–13.4in) and weight up to 4.7kg (10.4lb) in the Giant bandicoot. Males of larger species 60–80 percent heavier than females.

Gestation: 12.5 days in Long-nosed and Brindled bandicoots; others probably similar.

Longevity: not known, but 3 or more years in Brindled bandicoot.

Bandicoots (family Peramelidae)
Australian long-nosed bandicoots, 4 species of *Perameles*: the **Long-nosed bandicoot** (*P. nasuta*), **Eastern barred bandicoot** (*P. gunnii*); **Desert** or **Orange bandicoot** EX (*P. eremiana*) and **Western barred bandicoot** or **marl** R (*P. bougainville*); **New Guinea long-nosed bandicoots**, 4 species of *Peroryctes* including the **Giant bandicoot** (*P. broadbenti*) and **Striped bandicoot** (*P. longicauda*); **Mouse bandicoot** (*Microperoryctes murina*); **spiny bandicoots**, 3 species of *Echymipera* including the **Rufescent** or **Rufous spiny bandicoot** (*E. rufescens*) and **Spiny bandicoot** (*E. kalubu*); the **Ceram Island bandicoot** (*Rhynchomeles prattorum*); **short-nosed bandicoots**, 3 species of *Isoodon*: the **Brindled** or **Northern brown bandicoot** (*I. macrourus*), **Golden bandicoot** (*I. auratus*) and **Southern brown** or **short-nosed bandicoot** or **quenda** (*I. obesulus*); and the **Pig-footed bandicoot** EX (*Chaeropus ecaudatus*).

Rabbit-eared bandicoots or **bilbies** (family Thylacomyidae)
Two species of *Macrotis*, the **Greater bilby**, **Greater rabbit-eared bandicoot**, or **dalgyte** R (*M. lagotis*) and **Lesser bilby** EX (*M. leucura*).

EX Extinct. R Rare.

Bandicoots have the highest reproductive rate of all marsupials. In this respect they are the marsupials showing greatest similarity with placental groups such as rodents whose life-cycle centers on producing many young with little maternal care. Otherwise these small insectivores and omnivores fit into an ecological niche somewhat similar to that of many members of the order Insectivora (shrews and hedgehogs).

Most bandicoots are rabbit-sized or smaller, have short limbs, a long pointed muzzle, and a thickset body with a short neck. The ears are normally short, the forefeet have three prominent toes with strong flattish claws, and the pouch opens backward. The furthest from this pattern is the recently extinct Pig-footed bandicoot which developed longer limbs and small footpads as adaptations to a more cursorial (running) life on open plains. Ear length is only prominent in the Pig-footed bandicoot and Rabbit-eared bandicoots. Teeth are small, relatively even-sized and with pointed cusps, as in typical insectivore teeth. These animals are basically designed for digging out invertebrate food in the ground, poking into crevices and rooting around with their powerful foreclaws.

Bandicoots are distinguished from all other marsupials by the combined presence of fused (syndactylous) toes on the hindfeet, forming a comb for grooming, and polyprotodont dentition (see p826).

The rear-opening pouch normally has eight teats. It extends forward along the abdomen as the young enlarge, eventually occupying most of the mother's underside and contracting again after departure of the young.

The sense of smell is well developed but eyesight seems variable, as eye diameter varies noticeably between species. Loud calls are either absent or uncommon. Only the Long-nosed bandicoot is known to have a loud vocalization; it produces a sharp squeaky alarm call when disturbed at night. A very low sibilant "huffing" with bared teeth is uttered as a threat by the Brindled bandicoot and probably others.

The dental formula is I5/3, C1/1, P3/3, M4/4 = 48, except in spiny bandicoots, which have four pairs of upper incisors (I4/3). Sexual dimorphism occurs in the larger species, males being about 60–80 percent heavier and 15 percent longer, with larger canines.

In recent (Pleistocene) times bandicoots have to a large extent evolved separately in mainland Australia and New Guinea, as a result of the intermittent separation of the two land masses and the marked habitat differences between the tropical rain-forested island and the drier continent. All genera of the New Guinea region are exclusive to these northern islands, or nearly so. Only one species, the Rufescent bandicoot, extends into northern Australia. The remaining genera are essentially Australian although one species, the Brindled bandicoot, intrudes into the grassy woodlands in lowland southern New Guinea, as the Rufescent bandicoot does in the opposite direction where it uses the small rainforest patches of Cape York Peninsula. These two intrusions indicate that an important—perhaps even the dominant—influence on the different bandicoot fauna of the two land masses is habitat, and not just the water barrier. Where suitable habitat exists, animals have managed to colonize the other land mass in spite of its geographic isolation.

Within New Guinea, different species occur at different altitudes. The Spiny bandicoots range from sea level to about 1,600m (5,250ft). On the other hand the Mouse bandicoot and two of the four New Guinea long-nosed bandicoots are restricted to altitudes above 1,200m (4,000ft).

Within Australia, there are pronounced

▼ ▲ **The short-nosed Brindled bandicoot**
ABOVE inhabits low ground cover in coastal areas of eastern and northeastern Australia and south New Guinea. BELOW One of the large bandicoots, it is active at night and frequently sniffs the air (1) to detect any danger. The usual gait is on all fours, but the larger hindlimbs are used in an aggressive hop (2) characteristic of males. The Brindled bandicoot digs out fungi, tubers, insects, worms and other invertebrate prey with its strong foreclaws (3). After the shortest gestation of perhaps any mammal, the newborn young crawl into their mother's rear-opening pouch (4), where they are carried for seven weeks, by which time the mother's pouch is bulging (5).

climatic influences on distribution of species, which tend to fall into two groups. A number of species are restricted to the dry semi-arid and arid areas; all of these have suffered big population declines in recent times. They include the Desert bandicoot, Western barred bandicoot, the Golden bandicoot, Pig-footed bandicoot and the rabbit-eared bandicoots. The remaining Australian bandicoots occur in coastal or sub-coastal zones. This pattern is an effect, whether direct or indirect, of rainfall. The Brindled bandicoot, a coastal species of eastern and northern Australia, is widely distributed as far inland as about the 75cm (29.5in) isohyet (rainfall line). Beyond this it tends to be largely confined to watercourses (riverine) and extends much farther inland in

this manner, almost to the 60cm (23.6in) isohyet. Other distribution patterns reflect apparent latitudinal influences on distribution in both the above groups.

The classification of bandicoots is not fully resolved, although all genera and the two families are clearly defined. Groups in need of revision include the Southern short-nosed bandicoot and Western barred bandicoot and possibly several others. Characteristic adaptive changes from the primitive condition include lengthening of the rostrum of the skull permitting animals to probe further into holes and crevices; longer ears; larger auditory bullae and longer limbs in species from more open habitats, particularly arid areas; shortening of the tail; and enlarged molars (development of a fourth cusp).

The short-nosed bandicoots are stocky, short-eared, plain-colored animals. They favor areas of close ground cover, tall grass or low shrubbery, although actual habitat may vary from this. They occur throughout Australia. Important differences between species or populations are: an increase in the size of auditory bullae in more arid areas which may assist earlier detection of predators; overall size differences including dwarfed forms, particularly on islands, and in more open areas, perhaps an adaptation to scarcer food; variation in the angle of the ascending rear portion of the lower jaw; and the presence of an extra cusp on the last upper molar. There is little overlap between different species' distributions.

The long-nosed bandicoots are more lightly built, with a relatively longer skull, longer ears, barred body markings, and a preference for areas of open ground cover such as bare forest floors or short grassland, although again habitat use is extremely flexible. They have exploited most of the continent with some noticeable gaps—there has been a failure to fully utilize tropical humid areas other than rain forest. Important variables between species are: increased ear length and bullae size in more arid areas; different degrees of barring, which is absent in the forest species, the Long-nosed bandicoot, and conspicuous in grassland species either as disruptive camouflage in grass cover or as species recognition marks in more open areas.

The Pig-footed bandicoot is highly specialized, with adaptations for a running (cursorial), plains-dwelling life. In the forefeet only the second and third digits remain functional, forming a paired foot pad like an even-toed ungulate (artiodactyl) foot. In the hindfeet, the fourth digit forms a single functional pad. The syndactylous "comb" also remains. The limbs are correspondingly lengthened for better running. This bandicoot also has longer ears.

The several New Guinea genera are poorly known. They tend to be little modified, short-eared forest bandicoots. However, lengthening of the rostrum is marked in some, particularly spiny bandicoots and the Ceram Island bandicoot. Spiny bandicoots also have shorter tails.

Rabbit-eared bandicoots have lengthened ears, rostra and limbs, highly developed auditory bullae with twin chambers, and a long crested tail. They are the only burrowing bandicoots and are an offshoot from the main bandicoot stock that has become highly specialized for arid areas. The main differences between species and populations are: size; tail coloration; coat coloration; and bulla size (larger in Lesser bilby).

Although bandicoots are specialized for feeding on soil invertebrates, the few dietary studies have shown that feeding is opportunistic and omnivorous. Their diets include insects, other invertebrates, fruits, seeds of non-woody plants, subterranean fungi and occasional plant fiber. Diet can also include a high proportion of surface food and it is likely that bandicoots switch to other food

when insects are unavailable. Food in the ground is dug out with the strong foreclaws, usually leaving characteristically conical holes. The elongation of the muzzle is presumably associated with probing down into holes and other crevices and under logs etc for food. The Brindled bandicoot has a characteristic foraging pattern, moving slowly over its whole 1–5ha (2.5–12 acre) range, and spending little time in any one spot. This is an adaptation for finding food that occurs as small isolated and scattered items rather than concentrated in a few areas.

The biology of bandicoots has been poorly studied, although there are indications that the life-cycle of the Brindled bandicoot is typical of the group. Its gestation of 12.5 days is less than half that of most other marsupials and perhaps the shortest of any mammal. Development of the embryo is aided by a form of chorioallantoic placentation that is unique to bandicoots among marsupials. It resembles the placenta of

► **The Long-nosed bandicoot** of eastern Australia, in addition to living up to its common name, has longer ears than the short-nosed bandicoots. It is well known for the conical holes it leaves in the ground after night-time foraging for insects.

► **Distinctive white markings** BELOW RIGHT on the rump give the Eastern barred bandicoot its name. Grasslands are its preferred habitat, in Tasmania and south Victoria.

▼ **Big ears**—the Greater bilby or Rabbit-eared bandicoot is a desert species now much reduced in distribution. Apart from the long ears, the long furry tail and burrowing habits set this rare animal apart from other surviving bandicoots.

eutherian mammals. Other marsupials form only a yolk-sac placenta, whereas bandicoots and eutherians have independently evolved both types of placentation. Young at birth are about 1cm (0.4in) long, and about 0.2g (0.007oz) in weight, with well-developed forelimbs. The allantoic stalk anchors the young to the mother whilst the newborn crawls to the pouch, where it attaches to a nipple. Litter size ranges from 1 to 7 (commonly 2–4). Young leave the pouch at about 49–50 days and are weaned about 10 days later. In good conditions, sexual maturity may occur at about 90 days of age, although it is normally attained much later. Females are polyestrous and breed throughout the year in suitable climates, elsewhere breeding seasonally. Mating can occur when the previous litter is near the end of its pouch life. Since the gestation is 12.5 days, the new litter is born at about the time of weaning the earlier litter.

The reproductive cycle is one of the most distinctive characteristics of bandicoots, setting them well apart from all other marsupials. They have become uniquely specialized for a high reproductive rate and reduced parental care. In the Brindled bandicoot, and possibly in all bandicoots, this is achieved by accelerated gestation, rapid development of young in the pouch, early sexual maturity and a rapid succession of litters in the polyestrous females. In one Brindled bandicoot population with 6–8 month breeding seasons, females produced an average 6.4 young in one season, and 9.6 in the next. Litter size, however, while higher than in many marsupial groups, is not exceptional, being smaller than in others, such as dasyurids.

Bandicoot society is poorly studied in most species, but again is likely to follow a common pattern throughout the group. The Brindled bandicoot is solitary, animals come together only to mate, and there appears to be no lasting attachment between mother and young, contact being lost at weaning or

soon after. Males are larger than females and socially dominant, dominance correlating approximately with body size. Dominance between closely matched males may be established by chases or, rarely, by fights, in which the males approach each other standing on their hind legs. Either the two combatants lock jaws and wrestle onto the ground, or one may jump high above the other and rake out with its hindfoot in an endeavor to wound it. Male home ranges are larger, 1.7–5.2ha (4.2–12.8 acres) in one study, compared to 0.9–2.1ha (2.2–5.2 acres) for females. Characteristically there is a "core area," apparently where most time is spent foraging. The ranges of both sexes overlap, although core areas do not. Males do a rapid tour around most of the home range each night, perhaps as a patrolling action to detect other males or receptive (estrous) females. Caged animals show intense interest in nest sites, and dominant males may commonly evict others from nests. Nests may therefore be a significant focus of social interactions in the wild. Nests consist of heaps of raked-up ground-litter with an internal chamber. A scent gland is present behind the ear of many species in both sexes. The Brindled bandicoot uses it to mark the ground or vegetation during aggressive encounters between males. The ground cover of Brindled bandicoots is subject to frequent destruction by fire or drought. Their high reproductive rate and mobility enable them to colonize quickly as suitable habitat becomes available.

Australian bandicoots have suffered one of the greatest declines of all marsupial groups. All species of the semi-arid and arid zones have either become extinct or suffered massive declines, being reduced now to a few remnant populations that are still endangered. An important feature of most extinctions appears to be grazing by cattle, sheep or rabbits and the consequent changes in the nature of ground cover. Some authorities alternatively blame introduced predators, including foxes and cats. Removal of sheep and cattle is an important conservation measure in these areas. Control of rabbits and introduced predators is desirable but extremely difficult. Most bandicoots of higher rainfall areas have been little affected by European settlement, or are thriving, and are not yet in need of specific conservation measures. An exception is the Eastern barred bandicoot, which has been rendered almost extinct in Victoria by cultivation and grazing on the grassy plains to which it is restricted, but remains common in Tasmania.   GG

CUSCUSES AND BRUSHTAIL POSSUMS

Family: Phalangeridae
Fourteen species in 3 genera.
Distribution: Australia, New Guinea (including Irian Jaya) and adjacent islands west to Sulawesi, east to Solomon Islands. Common brushtail possum introduced to New Zealand, Gray cuscus possibly also to some Solomon Islands.

Habitat: rain forest, moss forest, eucalypt forest; temperate, arid and alpine woodland.

Size: ranges from the "Lesser" Sulawesi cuscus *Phalanger ursinus* with head-body length 34cm (13.4in), tail length 30cm (11.8in) and weight unknown, to the Black-spotted cuscus with head-body length to 70cm (27.6in), tail length 50cm (19.7in) and weight about 5kg (11lb).

Coat: short, dense, gray (Scaly-tailed possum); long, woolly, gray-black (brushtail possums); long, dense, white-black or reddish brown, some species with spots or dorsal stripes (cuscuses).

Gestation: 16–17 days (brushtail possums).
Longevity: to 13 years (at least 17 in captivity).

Cuscuses or phalangers, 10 species of *Phalanger*: **Spotted cuscus** ◉ (*P. maculatus*); **Gray cuscus** ◉ (*P. orientalis*); **Woodlark Island cuscus** ℝ (*P. lullulae*); **Sulawesi cuscuses** (*P. ursinus* and *P. celebensis*); **Ground cuscus** (*P. gymnotis*); **Stein's cuscus** ℝ (*P. interpositus*); **Mountain cuscus** (*P. carmelitae*); **Silky cuscus** (*P. vestitus*), **Black-spotted cuscus** ℝ (*P. rufoniger*).

Brushtail possums, 3 species of *Trichosurus*: **Common brushtail possum** (*T. vulpecula*); **Mountain brushtail possum** or bobuck (*T. caninus*); **Northern brushtail possum** (*T. arnhemensis*).

Scaly-tailed possum, 1 species of *Wyulda*, *W. squamicaudata*.

◉ CITES listed.   ℝ Rare.

▶ **Plain, spotted species?** ABOVE This Spotted cuscus female represents an unspotted color phase.

▶ **The rare Scaly-tailed possum** BELOW was only discovered early this century in the remote Kimberley region of north Western Australia.

▷ **The Spotted cuscus** OVERLEAF is a tree-dwelling species common in New Guinea, rare in Cape York, Queensland, Australia.

DWELLING in the remote outback as well as in the suburbs of most Australian cities, the Common brushtail possum is perhaps the most frequently encountered of all Australian mammals. Yet this species is only one of 14 in the phalanger family, which includes the rare Scaly-tailed possum of the Kimberley region, and the Woodlark Island cuscus of which only eight specimens have been reported.

The phalangers are nocturnal, usually arboreal, and they possess well-developed forward-opening pouches; they are distinguished from other tree-dwelling marsupials by their relatively large size, simple, low-crowned molar teeth, lack of a gliding membrane and variable amount of bare skin on the tail. All species have curved and sharply pointed foreclaws for climbing, and clawless but opposable first hind toes which aid in grasping branches. Most species are leaf-eaters and have a long cecum in the gut, but their relatively unspecialized dentition allows them to eat a wide variety of plant products (leaves, fruits, bark) and occasionally eggs and invertebrates.

Seven subspecies of the Common brushtail possum have been named (the last as recently as 1963), but only three are currently accepted. The most widespread form is found in wooded habitats in all Australian states and varies considerably in size (2–3.5kg/4.4–7.7lb) and color (light gray to black). The other subspecies are slightly heavier (up to 4.1kg/9lb) and form geographically isolated populations in Tasmania and in northeastern Queensland. Population density of the Common brushtail varies with habitat, from 0.4 animals per hectare (1 per acre) in open forest and woodland, to 1.4 per ha (3.5/acre) in suburban gardens and 2.1 per ha (5.2/acre) in grazed open forest.

Like their widespread congenor, the Mountain and Northern brushtail possums are sharp-faced, with medium-large upright ears, and a tail that is fully furred above with the tip naked below. But they are geographically much more restricted and are not split into subspecies. The solidly built Mountain brushtail occupies dense wet forests in southeastern Australia and may attain population densities of 0.4–1.8 per ha (1–4.5/acre), whereas the little-known Northern brushtail occurs in woodland from the top end of the Northern Territory to Barrow Island, Western Australia.

The very wide distribution of the Common brushtail possum is probably due to its considerable flexibility of feeding and nesting behavior and high reproductive potent-

ial. Where it occurs with the larger and more terrestrial Mountain possum, it eats mostly mature eucalypt leaves and obtains only 20 percent of its food from shrubs in the forest understory. However, in the absence of the larger species, the Common brushtail may spend most of its time on the ground eating a wide range of plant leaves, even grass and clover. In suburban gardens, it has developed an unwelcome taste for rose buds.

The Common brushtail is no less flexible in its nesting behavior. Although, like the other two species, it prefers to nest above ground in tree cavities (dens), the Common brushtail also uses hollow logs and holes in creek banks, while in suburbia it hides under house roofs. In the dense forests of New Zealand it is even known to roost koala-like, exposed on tree forks.

Common brushtail females begin to breed at one year and produce 1–2 young annually. In most populations 90 percent of females breed in the fall (March–May), but up to 50 percent may also breed in spring (September–November). Only one young is born at a time, and the annual reproductive rate of females averages 1.4. In the Mountain brushtail, by contrast, females begin to breed at 2–3 years, produce at most only one young each year, in the fall, and reproduce at an annual rate as low as 0.73. The growth rates of the two species also differ markedly. Both give birth to pink, naked young weighing only 0.22g (0.0080z), but the young of the Common brushtail are weaned first at the age of six months (eight months for the Mountain brushtail) and disperse first at 7–18 months (18–36 months for the Mountain brushtail).

Common brushtails are solitary, except when they are breeding and rearing young. By the end of their third or fourth year, individuals establish small exclusive areas—den trees—within their home ranges, which they defend against individuals of the same sex and social status. Individuals of the opposite sex or lower social status are tolerated within the exclusive areas but, even though the home ranges of males (3–8ha/7.5–20 acres) sometimes completely overlap the ranges of females (1–5ha/2.5–12.4 acres), individuals almost always nest alone and overt interactions are rare.

Despite the ability of the Common brushtail to use a wide variety of dens, defense of den trees suggests that preferred nest sites are in short supply. Because few young die before weaning (15 percent), relatively large numbers of independent young enter the population each year. These young use small, poor-quality dens, and up to 80 percent of males and 50 percent of females die or disperse within their first year. In contrast, in the Mountain brushtail, many young die before weaning (56 percent), so the numbers entering the population—and hence competition for scarce dens—are relatively small. About 80 percent of Mountain brushtail young survive each year after becoming independent, and males and females, far from being solitary, appear to form long-term pair-bonds.

The dispersion of Common brushtails, and in particular the defense of den trees, appears to be maintained through scent marking and to a lesser extent by means of calls, and direct aggression. At least nine scent-producing glands have been recorded in the Common brushtail—more than in any other species of marsupial. In males, secretions from mouth and chest glands are wiped on the branches and twigs of trees, especially den trees, and these are thought to advertise both the presence and the status of the marker to potential rivals. Females also advertise themselves, but they distribute scent more passively, in urine and feces. Auditory signals probably play a smaller role in maintaining the dispersion of Common brushtails, but a very wide repertoire of screeches, hisses, grunts, growls and chatters is nevertheless known. Many calls are loud—audible to humans at up to 300m (980ft)—and may be given in face-to-face encounters.

The brushtail possums are of considerable commercial importance. The Mountain brushtail causes damage in exotic pine plantations in Victoria and New South Wales, while in Queensland it frequently raids banana and pecan crops. The Common brushtail also damages pines, and in Tasmania it is believed to damage regenerating eucalypt forest. A potentially much more serious problem is that the Common brushtail may become infected with bovine tuberculosis. This discovery, made in New Zealand in 1970, led to fears that brushtails may reinfect cattle. A widespread and costly poisoning program was set up, but infected brushtails remain firmly established at a couple of dozen sites on the North and South islands of New Zealand.

On the other side of the economic coin, the Common brushtail has long been valued for its fur. The rich, dense fur of the Tasmanian form has found special favor and between 1923 and 1959 over 1 million pelts were exported. Exports from New Zealand have also grown rapidly (see box). However, in eastern Australia the last open season on possums was in 1963, and the future for all three species of brushtail seems to be quite secure.

The Scaly-tailed possum was discovered only in 1917. It is known from seven localities in the Kimberley region of Western Australia, and is distinguished by its naked, prehensile, rasp-like tail, very large eyes, small ears, sharp face and short dense gray fur. The Scaly-tailed possum is strictly nocturnal, solitary and feeds on the flowers and leaves of *Eucalyptus* trees; unlike other phalangers it probably nests among rocks. One female has been found with a single, naked young in her pouch in June, but no further details of reproductive behavior are available. The Scaly-tailed possum occupies one of the remotest corners of Australia and, although quite rare, it is considered to be

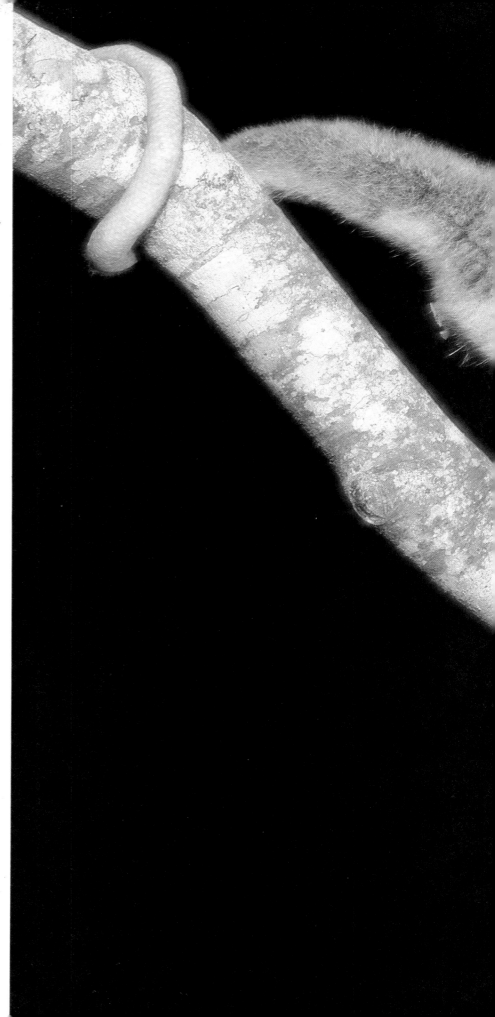

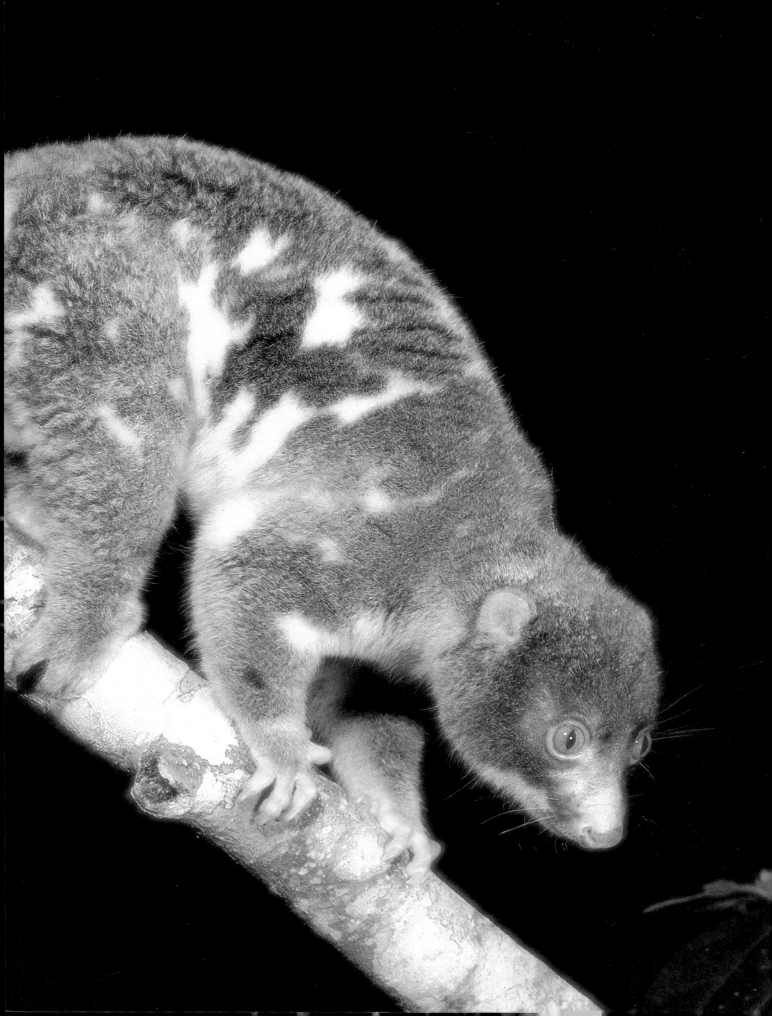

currently safe from the depredations of man.

In terms of numbers of species, the curious, round-headed cuscuses are the most successful representatives of the family. They are small-eared, and the tail is prehensile, and naked for the outer half or two-thirds of its length. Eight species occur in the rain forests of New Guinea (where they evolved) and two further species occur as far west as Sulawesi and neighboring islands. The Sulawesi cuscuses were probably derived from the ancestors of Spotted and Gray cuscuses, which rafted the 750km (470mi) across the Banda Sea long before the end of the last Ice Age about 11,000 years ago. In a later invasion, the Spotted and Gray cuscuses also crossed Torres Strait to establish populations on Cape York Peninsula, northern Queensland.

On mainland New Guinea, the geographical ranges of some species of cuscus overlap, whereas others fall within more or less exclusive (allopatric) altitudinal zones. These allopatric species are very similar in body size and habits and are apparently unable to coexist. The rarity of Stein's cuscus, for example, which is to be found only at altitudes of 1,200–1,500m (about 4,000–5,000ft), has been attributed to competition from the Gray cuscus, which occurs abundantly below 1,200m, and the Mountain cuscus, which occurs at altitudes of 1,500–2,300m (5,000–7,500ft). A fourth species, the Silky cuscus, which differs from

▼ ▲ **Most familiar of Australia's possums**, the Common brushtail is one of three species with large ears, pointed muzzle and furry tail that contrast with the mainly New Guinean cuscuses. They inhabit most areas with trees, through which they move by reaching from one branch to another. The Common brushtail exists in several color phases, including the widespread silver-grey ABOVE and the striking copper-red color of the Queensland form BELOW.

▶ **The Gray cuscus of Cape York, Queensland,** on New Guinea and adjacent islands, is a more lightly built member of the same genus as the Spotted cuscus.

A Marsupial Invader: the Common Brushtail in New Zealand

When the first Australian Common brushtail possums were imported to New Zealand around 1840, it was hoped that they would form the basis of a lucrative fur industry. The venture was manifestly successful. Aided by further importations until 1924 and by freeing of captive-bred animals, populations increased prodigiously, so that sales of pelts are now an important source of revenue: in 1976, 1.5 million pelts worth NZ$4.5m were sold. However, the blessings of this marsupial invader are mixed. As well as carrying bovine tuberculosis (see p852), the possum has recently been shown to have subtle but potentially damaging effects on the indigenous vegetation.

New Zealand forest trees evolved in the absence of leaf-eating mammals and, unlike the Australian eucalypts that produce poisonous oils and phenols, the leaves of most species are palatable and lack defenses against predators. When first introduced to particular New Zealand forests, the possums rapidly exploited the new food source and increased in population density up to 50 animals per ha

(120/acre)—some 25 times more than in Australia. By the time numbers had stabilized at 6–10 per ha (15–25/acre), trees such as ratas (*Metrosideros* species) and konini (*Fuchsia excorticata*) had all but disappeared from many areas, and possums were turning their attention to less favored species.

Possums hasten tree death by congregating on individual trees and almost completely defoliating them. These normally solitary creatures evidently abandon their social inhibitions when food and other resources are abundant and, in contrast to their Australian kin, the New Zealand possums occupy small (1–2ha/2.5–5 acres) and extensively overlapping home ranges.

The final verdict on possum damage is unclear. Young individual ratas and other exploited tree species are appearing in many localities, but they now appear to be distasteful to possums. Presumably possums are conferring a selective advantage on "unpalatable" trees, and so continue to subtly but surely alter the structure of the forest.

CRD

the three lower-altitude species principally in its possession of long dense fur, occupies altitudes above 2,300m. Where two species do overlap, they usually differ in size and habits. Often the larger of the pair is strictly herbivorous and spends most of its life in the forest canopy, whereas the smaller species dwells in the forest understory and feeds to a greater extent on fruits and insects. The Ground cuscus differs markedly from all other cuscuses in being partly carnivorous and semi-terrestrial (on Kobroor Island in the Aru group it reputedly lives in caves); it is accordingly able to coexist with several other cuscuses and its altitudinal range (0–2,700m/8,900ft) is the greatest of any species.

Anecdotal accounts suggest that individuals of all cuscus species usually feed and nest alone. Interactions between individuals are often aggressive; captive Ground cuscuses threaten each other by snarling and hissing, gaping and standing upright, and they attempt to subdue opponents by biting, kicking and cuffing. This species also distributes saliva on the branches and twigs of trees (as well as on itself) and has been observed to dribble urine and drag its cloaca on the ground surface. As in the Common brushtail possum, these activities provide olfactory information for other individuals and presumably mediate social interactions.

Courtship between the male and female cuscus is rather circumspect, and is usually conducted with great deliberation on the limbs of trees. In the Ground cuscus, the male follows the female prior to mating and attempts to sniff her head, flanks and cloaca; it may also utter soft, short clicks. Behavior during copulation has not been described, but the two sexes evidently go separate ways shortly afterwards. In the wild, female Gray and Spotted cuscuses (and probably other species) seldom suckle more than two young, although they possess four teats, and breeding is continuous throughout the year.

Although cuscuses are harmless to man and fortunately of no economic importance, the larger species have long been valued by traditional hunters for their coats and meat. Three species with restricted ranges—the Black-spotted cuscus, Stein's cuscus and the Woodlark Island cuscus—appear to be particularly susceptible to overhunting, and also to the more recent threat of habitat destruction. New Guinea has no national parks, but unless protected areas can be established within the next few years, the continued survival of the susceptible mainland species and the numerous island forms will be gravely threatened. CRD

RINGTAILS, PYGMY POSSUMS, GLIDERS

Families: Pseudocheiridae, Petauridae, Burramyidae
Thirty species in 9 genera.
Distribution: SE, E, N and SW Australia, Tasmania, New Guinea.

Habitat: forests, woodland, shrublands and heathland.

Size: ranges from head-body length 6.4cm (2.5in), tail length 7.1cm (2.8in) and weight 7g (0.25oz) in the Little pygmy possum to head-body length 33–38cm (13–15in), tail length 20–27cm (7.9–10.6in) and weight 1.3–2kg (2.9–4.4lb) in the Rock ringtail possum.

Coat: grays or browns with paler underside; often darker eye patches or forehead or back stripes; tail long, well-furred (most gliders), prehensile and part naked, or feather-like.

Gestation: 12–50 days (all young are less than 1g/0.035oz at birth).

Longevity: 4–15 years (shorter in pygmy possums with large litters and longer in gliders and ringtails with single young).

Ringtail possums
(family Pseudocheiridae)
Sixteen species in 2 genera: 15 species of *Pseudocheirus* (SE, E, N, SW Australia, Tasmania, New Guinea and W Irian). including **Common ringtail possum** (*P. peregrinus*) and **Rock ringtail possum** (*P. dahli*); **Greater glider** (*Petauroides volans*), E Australia.

Gliders (family Petauridae)
Seven species in 3 genera: 4 species of *Petaurus* (Tasmania, SE, E, N, NW Australia, New Guinea), including **Yellow-bellied** or **Fluffy glider** (*P. australis*) and **Sugar glider** (*P. breviceps*); **Leadbeater's possum** [E] (*Gymnobelideus leadbeateri*), Victoria; **striped possums**, 2 species of *Dactylopsila*, NE coastal Queensland, New Guinea.

Pygmy possums
(family Burramyidae)
Seven species in 4 genera: **pygmy possums** (4 species of *Cercartetus*, including the **Eastern pygmy possum**, *C. nanus*), Tasmania, Kangaroo Island, SE, E, NE, SW Australia, New Guinea; **Feathertail glider** or **Pygmy glider** or **Flying mouse** (*Acrobates pygmaeus*), SE to NE Australia; **Feathertail possum** (*Distoechurus pennatus*), New Guinea; **Mountain pygmy possum** [*] (*Burramys parvus*), SE Australia.

[E] Endangered. [*] CITES listed.

THE ringtail possums, gliders and pygmy possums inhabit the unique forested environments of Australia and New Guinea. When the Australian continent was invaded some 40–60 million years ago by primitive possum-like marsupials, it was blanketed in a wet, misty and humid rain forest. Opening of these forests in the mid to late Tertiary (32–5 million years ago) and their gradual replacement by the marginal eucalypt and acacia forests of today, forced this early fauna to seek refuge in the high altitude regions of northern Queensland and Papua-New Guinea, where the ringtail possums radiated to form a diverse family of leaf- and fruit-eating specialists. At the same time the new nectar-, gum- and insect-rich Australian eucalypt and wattle (*Acacia*) forests provided many abundant niches for the predominantly nectarivorous pygmy possums and the sap- and gumivorous petaurid gliders. This diversification has led to some remarkable convergences of form, function and behavior with the arboreal lemurs, bush babies, monkeys and squirrels of other continents.

The ringtail possums, pygmy possums and gliders are arboreal marsupials with hand-like feet, an enlarged opposable big toe on the hindfoot and a range of adaptations suited to moving about in the forest. The modern possum families of Australasia, although superficially similar to one another, differ in external form, internal anatomy and physiology, and the biochemistry of blood proteins as much from one another as from kangaroos and the koala. Previously included with the brushtail possums and cuscuses (the family Phalangeridae), the gliders, pygmy and ringtail possums are now separated into three additional families. In non-gliding species the tail is prehensile and may be used for grasping branches and transporting nest material; a naked undersurface effectively increases friction. In gliders (but not the Feathertail glider) the tail is heavily furred and straight or tapering; it may be used for controlling the direction of flight. Gliding species are specialized for rapid movement in the open forests and are thought to have evolved independently in three families during the mid-late Tertiary. In the nine glider species that survive today, gliding is achieved by use of a thin, furred membrane (patagium) that stretches from fore- to hindlimbs (wrist to ankle in the Sugar glider, wrist to knee in the Feathertail glider), increasing surface area in flight to form a large rectangle. It is retracted when not in use and may be visible

◄ ▼ **Denizens of inland forests** of eastern Australia, these female and male Squirrel gliders (*Petaurus norfolcensis*) LEFT show the long, bushy, soft-furred tail that gives them their name. The larger Yellow-bellied glider BELOW lives in more coastal forest areas with higher rainfall. Both species are uncommon and both have a diet chiefly of gum and sap from *Eucalyptus* trees. Long, sharp claws grip the bark when the animal lands on a trunk at the end of a glide.

as a wavy line along the side of the body. The effective gliding surface area has been increased by lengthening of arm and leg bones and some species are able to cover distances of over 100m (330ft) in a single glide, from the top of one tree to the butt or trunk of another. The heavier Greater glider, with a reduced (elbow-to-ankle) gliding membrane, descends steeply with limited control, but the smaller gliders are accomplished acrobats that weave and maneuver gracefully between trees, landing with precision by swooping upwards. What appears to be a gentle landing to the human eye is in fact shown by slow motion photography to be a high speed collision. The animals bounce backward after impact and must fasten their long claws into the tree trunk to avoid tumbling to the ground. The fourth and fifth digits of the hand are elongated and have greatly enlarged claws which assist clinging after the landing impact.

All these possums are nocturnal and have large, protruding eyes. Most are also quiet, secretive and hence rarely seen. The only audible sign of their presence may be the gentle "plop" of gliders landing on tree trunks, the yapping alarm call of the Sugar glider or the screeching or gurgling flight call of the Yellow-bellied glider. Ringtail possums are generally quiet but occasionally emit soft twittering calls. Most possum species make loud screaming and screeching calls when attacked or handled.

Four major dietary groups can be recognized: folivores, sapivores and gumivores, insectivores, and nectarivores. Ringtail possums and the Greater glider together form a highly specialized group (family Pseudocheiridae) of arboreal leaf-eaters (folivores), characterized by enlargement of the cecum to form a region for microbial fermentation of the cellulose in their highly fibrous diet. Fine grinding of food particles in a battery of well-developed molars with crescent-shaped ridges on the crowns (selenodont molars) enhances digestion. Rates of food intake in these groups are slowed by the time required for cellulose fermentation, and nitrogen and energy is often conserved by slow motion, relatively small litter sizes (1–1.5), coprophagy (reingestion of feces) and adoption of medium to large body size (0.2–2kg/0.4–4.4lb). The preferred diet of the Greater glider of eastern Australia is eucalypt leaves. The greatest diversity of species in this widespread group is in the high-altitude dripping rain forests and cloud forests of northern Queensland and New Guinea.

The four species of petaurid glider and Leadbeater's possum (all of the family

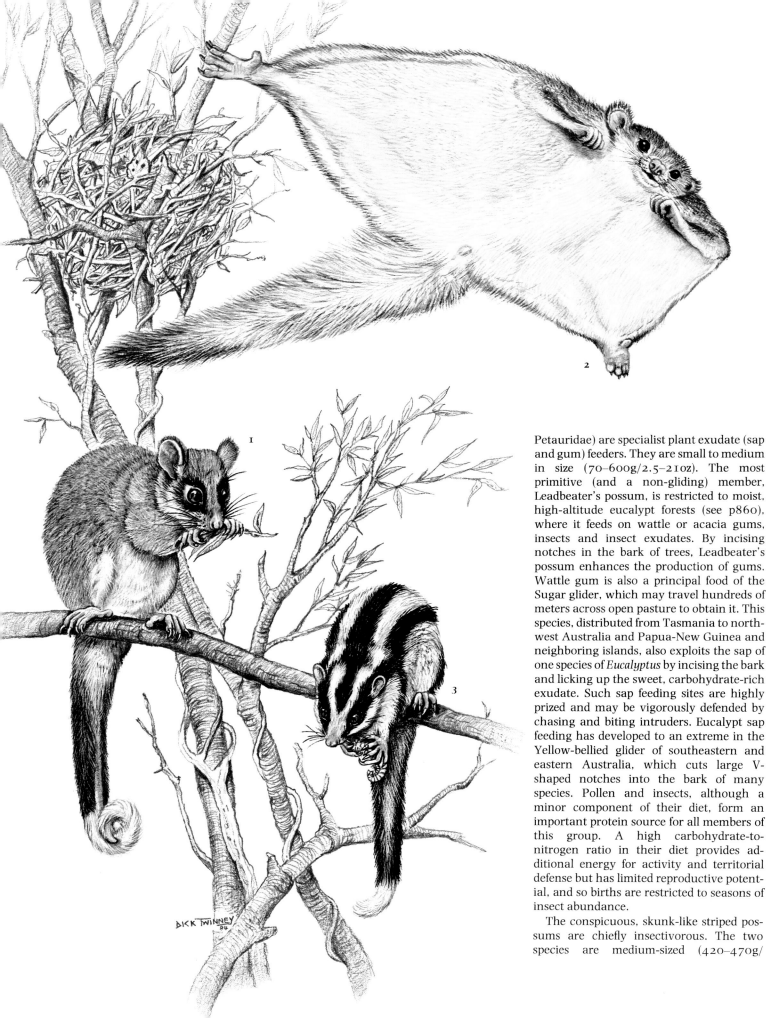

Petauridae) are specialist plant exudate (sap and gum) feeders. They are small to medium in size (70–600g/2.5–21oz). The most primitive (and a non-gliding) member, Leadbeater's possum, is restricted to moist, high-altitude eucalypt forests (see p860), where it feeds on wattle or acacia gums, insects and insect exudates. By incising notches in the bark of trees, Leadbeater's possum enhances the production of gums. Wattle gum is also a principal food of the Sugar glider, which may travel hundreds of meters across open pasture to obtain it. This species, distributed from Tasmania to north-west Australia and Papua-New Guinea and neighboring islands, also exploits the sap of one species of *Eucalyptus* by incising the bark and licking up the sweet, carbohydrate-rich exudate. Such sap feeding sites are highly prized and may be vigorously defended by chasing and biting intruders. Eucalypt sap feeding has developed to an extreme in the Yellow-bellied glider of southeastern and eastern Australia, which cuts large V-shaped notches into the bark of many species. Pollen and insects, although a minor component of their diet, form an important protein source for all members of this group. A high carbohydrate-to-nitrogen ratio in their diet provides additional energy for activity and territorial defense but has limited reproductive potential, and so births are restricted to seasons of insect abundance.

The conspicuous, skunk-like striped possums are chiefly insectivorous. The two species are medium-sized (420–470g/

14.8–16.6oz) and are specialized for exploitation of social insects, ants, bees, termites
and other wood-boring insects in tropical
lowland rain forests of northern Queensland
and New Guinea. A suite of adaptations,
including an extremely elongated fourth
finger (like that of the aye-aye of Madagascar), elongated tongues and enlarged and
forward-pointing upper as well as lower
incisors, aid in the noisy extraction of insects
from deep within wood crevices. Feeding
activity may produce a shower of wood-
chips.

Pygmy possums in the genus *Cercartetus*
and the Feathertail or Pygmy glider form a
fourth group (family Burramyidae) that
has diversified in the nectar-rich sclero-
phyllous Australian heathlands, shrublands
and eucalypt forests. The brush-tipped ton-

gue of the Feathertail glider is used for
sipping nectar from flower capsules, and the
small size (under 35g/1.2oz) and extreme
mobility of all five species increase nectar-
harvesting rates. In poor seasons aggregations of many individuals may be found
on isolated flowering trees and shrubs. Most
species are also thought to take the abundant pollen available at flowers, and occasionally insects, to provide protein. The
small size and abundant dietary nitrogen
permit unusually large litter sizes (4–6) and
rapid growth and development rates, similar
to those of carnivorous marsupials (family
Dasyuridae).

The rare and poorly studied Mountain
pygmy possum is superficially similar. It
spends up to six months of the year active

▲ **Possums and gliders—diet and movement**.
(1) Common ringtail possum (*Pseudocheirus
peregrinus*) eating leaves, with its spherical nest
made of grass and bark in the background.
(2) Greater glider (*Petauroides volans*) gliding.
(3) Striped possum (*Dactylopsila trivirgata*)
eating an insect. (4) Sugar glider (*Petaurus
breviceps*) gliding. (5) The omnivorous
Feathertail possum (*Distoechurus pennatus*)
eating an insect. (6) Yellow-bellied glider
(*P. australis*) feeding on sap of *Eucalyptus* by
biting into bark. (7) Mountain pygmy possum
(*Burramys parvus*), a chiefly nectar- and pollen-
eating species.

▶ **Nectar and pollen** are important in the diet of the Eastern pygmy possum, which uses its brush-tipped tongue to collect nectar from the flowers of *Eucalyptus* (as here): fruits and insects may also be taken.

◀ **Feather-like tail** adds maneuverability and distance to the flight of the Feathertail glider of eastern Australia. Another nectar-eating species, it is the only feather-tailed marsupial in Australasia—the non-gliding Feathertail possum inhabits only New Guinea.

beneath a blanket of snow in the high-altitude heaths of the Snowy Mountains. This scansorial (ground- and tree-foraging) species has a unique diet of seeds, fleshy fruit, some plant foliage, insects and other invertebrates. The remarkable sectorial pre-molar tooth is adapted for husking and cracking seeds. Excess seeds may be cached for use during periods of winter shortage.

The other member of the pygmy possum family, the Feathertail possum of Papua-New Guinea, has a tail like that of the Feathertail glider but is larger (50–55g/1.8–1.9oz) and has no gliding membrane. Its diet includes insects, fruit and possibly plant exudates.

Patterns of social organization and mating behavior in possums and gliders are remarkably diverse, but to some extent predictable from species' body size, diet and habitat characteristics. The larger folivorous ringtail possums and Greater glider are primarily solitary; they sleep singly or occasionally in pairs, in tree hollows or vegetation clumps, by day and emerge to feed on foliage in home ranges of up to 3ha (7.4 acres) at night. Male home ranges are generally exclusive but may partially overlap those of one or two females. The occupation of exclusive home ranges by males and overlapping home ranges by females is associated with greater mortality of males among juveniles and a consequent female-biased sex ratio.

The tendency towards gregariousness increases with decreasing body size, the Common ringtail of eastern Australia forming nesting groups of up to 3 individuals, the Yellow-bellied glider groups of up to 5, the Sugar glider up to 12, and the Feathertail glider up to 25. Most nesting groups consist of mated pairs with offspring, but the

Rediscovery of Leadbeater's Possum

An hour after nightfall one evening in 1961, at a tourist spot in the wet misty mountains just 110km from Melbourne, the attention of a fauna survey group from the National Museum of Victoria was caught by a small, bright-eyed and alert, gray possum leaping nimbly through the forest undergrowth. It's size at first suggested a Sugar glider but the absence of a gliding membrane and the narrow bushy tail led to the exciting conclusion that here, alive, and only the sixth specimen known to science, was the long-lost Leadbeater's possum.

This rare little possum, now the State of Victoria's faunal emblem, was first discovered in 1867 in the Bass River Valley. Only five specimens were collected, all prior to 1909, and in 1921 it was concluded that the destruction of the scrub and forest in the area had resulted in complete extermination.

Surveys following the rediscovery led to its detection at some 40 separate sites within a 25×40km (15.5×25mi) area. Its prefered habitat was the zone of Victoria's Central Highland forests dominated by the majestic Mountain ash (*Eucalyptus regnans*), the world's tallest hardwood and Australia's most valued timber producing tree. Standing beneath such forest giants provides the most reliable method of catching a glimpse of Leadbeater's possum as the animals emerge at dusk from their family retreat in a hollow branch to feed in the surrounding forest. Now, only 22 years after its discovery, the possum is again threatened with extinction by forest clearance, inappropriate forest management, and natural disappearance of the large dead trees that provide nest sites in regrowth forests (in 1939 a fire destroyed two thirds of Victoria's Mountain ash forests). The species' survival, along with that of the only other nationally endangered possum, the Mountain pygmy possum (threatened by ski-run and general tourist development at alpine resorts), depends upon effective government action which is yet to be forthcoming.

petaurids may form truly mixed groups with up to four or more unrelated adults of both sexes (Sugar glider), one male and one or several females (Yellow-bellied glider), or with one female and up to three males (Leadbeater's possum). The chief reason for nesting in groups is thought to be improved energy conservation through huddling during winter. Larger nesting groups of one species, the Sugar glider, disband into small groups during summer. The aggregation of females during winter enables dominant males to monopolize access to up to three females in the petaurid gliders, and a harem-defense mating system prevails.

An entirely different mating system occurs in the closely related Leadbeater's possum. Individual females occupy large nests in hollow trees and actively defend a surrounding 1–1.5ha (2.5–3.7 acres) exclusive territory from other females. Mating is strictly monogamous and male partners assist females in defense of territories. Additional adult males may be tolerated in family groups by the breeding pairs but adult females are not, and an associated higher female mortality results in a male-biased sex ratio. This pattern appears to be associated with the construction of well-insulated nests, avoiding the necessity for females to huddle during winter, and with the occupation of dense, highly productive habitats in which food resources are readily defensible and surplus energy is available to meet the cost of territorial defense.

Selective pressures exerted during competition for mating partners (sexual selection) have led to the prolific development of scent-marking glands in the petaurid group, for use in marking other members of the social group. Leadbeater's possum, the most primitive and only monogamous member, shows least development of special scent glands, and scent-marking between partners involves the mutual transfer of saliva to the tail base with its adjacent anal glands. Males of the promiscuous Sugar glider, in contrast, possess forehead, chest and anal glands. Males use their head glands to spread scent on the chest of females, and females in turn spread scent on their heads by rubbing the chest gland of dominant males. Male Yellow-bellied gliders have similar glands but scent transfer is achieved quite differently, by rubbing the head gland against the female's anal gland. Females in turn rub their heads on the dominant male's anal gland. Such behavior probably facilitates group cohesion by communicating an individual's social status, sex, group membership and reproductive position. AS

Families: Macropodidae and Potoroidae
About 60 species in 16 genera.
Distribution: Australia, New Guinea.
Habitat: wide ranging, from inland plains to tropical rain forests.

Size: ranges from head-body length 28.4cm (11.2in), tail length 14.2cm (5.6in) and weight 0.5kg (1.2lb) in the Musky rat kangaroo to head-body length 165cm (65in), tail length 107cm (42in) and weight 90kg (198lb) in the male Red kangaroo.

Gestation: newborn attach to a maternal teat within a pouch and there further develop for 6–11 months.

Longevity: variable according to species and climatic conditions. Larger species may attain 12–18 years (28 years in captivity).

SOME 224 modern mammals have been described from Australia since the first European settlement in 1788, but the popular image of an Australian mammal both in that country and abroad is still perhaps that of a hopping female kangaroo, with an attractive offspring protruding from its abdominal pouch. Among the approximately 120 species of marsupials in Australia itself, some 45 are recognized as belonging to the families Potoroidae and Macropodidae. There are 10 further kangaroo species in New Guinea and nearby islands, in addition to two also present in Australia.

Most of the kangaroo species had been collected and described by the mid-19th century. Although argument about the taxonomy and nomenclature of some kangaroos is not yet settled, their recorded history is frequently first associated with early explorers. Thus when in 1770 James Cook's vessel struck the Barrier Reef and repairs were undertaken at Endeavour River, near the present site of Cooktown, the party's naturalists, Banks and Solander, together with the artist Parkinson, collected specimens throughout the seven-week delay, including three kangaroos. Descriptions of these specimens aroused great interest in Europe at the time, but nearly 200 years later the descriptions were revealed to be

Families of Kangaroos and Wallabies

Rat kangaroos
(Family Potoroidae)
Genus *Hypsiprymnodon*. One species, the **Musky rat kangaroo** (*H. moschatus*).

Genus *Potorous*. Three species, including the **Long-nosed potoroo** or **rat kangaroo** (*P. tridactylus*) and the **Long-footed potoroo** [I] (*P. longipes*).

Genus *Caloprymnus*. One species, the **Plains** or **Desert rat kangaroo** [I] (*C. campestris*).

Genus *Bettongia* [*] Four species, the **bettongs** or **short-nosed rat kangaroos**: the **Brush-tailed rat kangaroo** or **bettong**, or **woylie** [E] (*B. penicillata*); **Lesueur's** or **Burrowing rat kangaroo**, or **boodie** [R] (*B. lesueur*); **Northern rat kangaroo** or **bettong** (*B. tropica*) and **Gaimard's rat kangaroo** or **Tasmanian bettong** (*B. gaimardi*).

Genus *Aepyprymnus*. One species, the **Rufous rat kangaroo** or **bettong** (*A. rufescens*).

Kangaroos
(Family Macropodidae)
Genus *Dendrolagus*. Seven species, the **tree kangaroos**.

Genus *Lagostrophus*. One species, the **Banded hare wallaby** [R] (*L. fasciatus*).

Genus *Lagorchestes*. Four species, including the **Spectacled hare wallaby** (*L. conspicillatus*); and **Western** or **Rufous hare wallaby** [R] (*L. hirsutus*).

Genus *Onychogalea*. Three species, including the **Bridled nailtail wallaby** [E] (*O. fraenata*) and the **Crescent nailtail wallaby** [EX] (*O. lunata*).

Genus *Petrogale*. Ten species, including the **Yellow-footed** or **Ring-tailed rock wallaby** (*P. xanthopus*).

Genus *Thylogale*. Four species, the **pademelons** or **scrub wallabies**.

Genus *Setonix*. One species, the **quokka** (*S. brachyurus*).

Genus *Wallabia*. One species, the **Swamp** or **Black wallaby** (*W. bicolor*).

Genus *Dorcopsis*. Three species, the **greater forest** or **New Guinea wallabies**.

Genus *Dorcopsulus*. Two species, the **lesser forest** or **mountain wallabies**.

Genus *Macropus*. Fourteen species, including the **Red kangaroo** (*M. rufus*), **Eastern gray kangaroo** or **forester** (*M. giganteus*), **Western gray** or **Mallee** or **Blackfaced kangaroo** (*M. fuliginosus*), the **wallaroo** or **euro**, or **Hill kangaroo** (*M. robustus*), the **Tammar** or **Dama wallaby** (*M. eugenii*), **Whiptail** or **Prettyface wallaby**, or **flier** (*M. parryi*) and **Parma wallaby** (*M. parma*).

[*] CITES listed. [E] Endangered.
[EX] Extinct. [I] Threatened, but status indeterminate.

▲ **Largest and most typical of marsupials—** the Red kangaroo. A family group, with the head, tail and one foot of a joey protruding after it has jumped into its mother's pouch. The single young does not finally leave the pouch until about eight months old. In most kangaroos the male's coat is russet to brick red; female "blue fliers" have blue-grey fur.

◄ **Sheltering from the sun**, a Spectacled hare wallaby hides beneath hummocks of grass on Barrow Island, off Western Australia. In this harsh location, the hare wallaby feeds on tips of spinifex grass leaves and does not drink even when water is present.

composites of all three animals, and it became necessary to identify what were the species collected in order to ensure the application to the correct species of the first used scientific name. The problem was finally settled in 1966 by determination No. 760 of the International Commision on Zoological Nomenclature, which assigned the name *Macropus giganteus* to the Eastern gray kangaroo, the first specimen of a kangaroo collected on the Australian mainland by Europeans for scientific study.

The rat kangaroos are often regarded as ancestral to other kangaroos. They are placed in a separate family, the Potoroidae, on the basis of their dentition (see below), which is adapted to a more generalized diet. Other ancestral features include the pre-

sence of the first toe in the Musky rat kangaroo, and possum-like morphology of the brain. Among the Macropodidae, the use of the names kangaroo and wallaby to indicate large and small species is now largely a matter of tradition. The original discrimination between the two on the basis of length of hindfoot and basal length of the skull has long proved unsatisfactory and the term kangaroo is used here to cover both.

Kangaroos and rat kangaroos have adapted to a great range of habitats, including open plains, woodlands and forests, rocky outcrops, slopes and cliffs, with a few species becoming arboreal. Such adaptations have led to a wide variation in size and form. Common features which distinguish the group from other marsupials include the characteristic body shape and structure of their jaw and teeth. With the exception of the Musky rat kangaroo, which has a simple stomach, all have a large sacculated stomach akin to that found in ruminants. The distinctive shape includes short forelimbs as opposed to elongated hindlimbs adapted to hopping, and a large and often heavy tail used as a balance in this form of locomotion or as an additional prop to support the animal's weight during slow forward movement, particularly when grazing. On horizontal surfaces tree kangaroos can move in a similar hopping manner to the ground-dwelling forms, from which they differ in that their forelimbs are stouter and more muscular, while their hindlimbs are relatively shorter, bearing squat broad feet. Tree kangaroos climb by gripping branches with their stout foreclaws and walking backwards or forwards with alternate movements of their hindfeet. This independent movement of the hindlimbs is a feature not present in other kangaroos, except for swimming when hindlimbs move independently. The forelimbs of all kangaroos have five clawed digits while the long foot bears two major toes, of similar shape and bearing prominent claws, the larger corresponding to the fourth and the smaller to the fifth digit. A first digit equivalent to a "big toe" is absent in all species except the Musky rat kangaroo, while the second and third digits are very small and, except for the claws, are bound in a common sheath. Despite their small size these bound toes provide an extremely flexible grooming organ.

Both males and females have a prominent cloacal protuberance that encloses the rec-tal opening and the uro-genital passage in the female or the retracted penis in the male. In front of the cloacal protruberance a pendulous scrotum is obvious in mature males, while females possess a pouch bearing on the abdominal wall four independent mammary glands each with a teat, to one of which the newborn young attaches after climbing up the outer body wall and over the lip of the pouch.

Apart from two forward-projecting (procumbent) incisors which move laterally against six upper incisors, and a wide gap (diastema) between the incisors and cheek teeth in both upper and lower jaw, the two families possess characteristic grinding (molariform) teeth. In young animals the first two teeth in the cheek row are a shearing (sectorial) premolar and the only "milk" tooth, a molar-like premolar. Both these are eventually shed and replaced by a single sectorial premolar. Four robust molariform teeth follow and erupt in sequence over a relatively long period of the animal's life span, and with advancing age gradually move toward the front of the jaw with an associated loss in turn of those teeth further forward. In rat kangaroos the first upper incisor is much longer than the other two, whereas in kangaroos and wallabies the three teeth tend to be much the same size. The molars of the rat kangaroos bear four cusps and decrease in size towards the rear, but those of kangarooos and wallabies bear two transverse ridges with a prominent longitudinal connecting link between them. The size of successive molars may increase slightly or remain much the same.

Marked sexual dimorphism in size and, in

▶ **The Rufous bettong** or Rufous rat kangaroo spends the day in a nest in the grass, as do other rat kangaroos and potoroos. At night it feeds on grasses, herbs, roots and tubers on the floor of the open forests where it prefers to live.

▼ **Representative species of larger kangaroos and wallabies**, shown in a hopping sequence. (1) Bridled nailtail wallaby (*Onychogalea fraenata*). (2) Wallaroo (*Macropus robustus*). (3) Quokka (*Setonix brachyurus*). (4) Red-legged pademelon (*Thylogale stigmatica*). (5) Yellow-footed rock wallaby (*Petrogale xanthopus*). (6) Gray forest wallaby (*Dorcopsis veterum*).

some cases, coat color, is readily observed among kangaroos. Male and female young develop at much the same rate within the parental pouch but on emergence the rate of male growth increases so that fully developed adult males are larger than females and in some species, such as the Red and both gray kangaroos, may exceed twice the size of females of comparable age. Examples of differences in color are well known; thus the soft dense and woolly coat of Red kangaroos, for example, may be pale to dark russet-red in the male and in the female a distinct blue-gray, while the coarse and shaggy-haired male wallaroos are black and their females silver-gray. However, differences in coat color between the sexes of gray kangaroos are not marked. The eastern species is predominantly gray-colored and the western brown. Both possess soft fur which may vary from short and sparse in the tropics to dense and woolly in colder regions.

Members of this group are primarily associated with habitats containing herbs, grasses and shrubs. In Australia these are found in relatively open communities that merge into forests on the one hand and through woodlands into grasslands on the other. The woodland fauna includes the large kangaroos, large and small wallabies and several of the rat kangaroos. Many of these species are widespread but some have even more specialized requirements. The somewhat sedentary wallaroos or hill kangaroos are usually to be found associated with stony rises, rock hills and escarpments; the colonial rock wallabies, with rock piles, steep rocky hills and cliffs; thus these species'

potential distribution tends to be discontinuous though widespread throughout the woodland zone.

Unfortunately, since the late 18th century human settlement has brought considerable and often dramatic changes to the original habitat, with consequent drastic effects on the distribution and numbers of kangaroos, so that some species are now restricted to a portion of their former ranges, often including offshore islands. Removal or thinning of shrubs and degradation of grasslands by domestic stock and by the introduced rabbit have rendered such areas unsuitable for many of the smaller grass-dwelling and grass-nesting species, as well as exposing them to predation. The total removal of cover for the establishment of improved pastures, crops or monocultures of exotic conifers has had equally disastrous effects on many marsupials. However, in some cases it would appear that the reduction of the original tall coarse dry grasses to a closely cropped sward by introduced stock has allowed the larger kangaroos to thrive, and in other cases some species of forest-dwelling wallabies appear to have prospered when improved pastures have been developed adjacent to timber from which they emerge to graze.

Until quite recently little was known of selection and utilization of native plants by domestic stock, let alone by kangaroos. The possible effects of competition by feral animals including rabbits, goats and an assemblage of other grazing animals was usually overlooked.

Food preferences have been studied in only a few native and domestic species, habitats and widely separate locations. Regrettably the findings of these studies have been extrapolated frequently and uncritically to many other parts of the continent. In general kangaroos prefer to feed from dusk to dawn in habitats with open undergrowth which permits freedom of movement by the animal and promotion of the grasses. These conditions tend to occur on the fringes of forests, in open woodlands or near natural or improved pastures. Investigations to date indicate that domestic and native grazing animals are selective in their choice of food. Kangaroos eat no more than sheep of equivalent size but they do select plants of a lower nitrogen content. At any one time of the year food preferences of sheep differ from those of kangaroos, while different species and sexes of kangaroos within a common habitat may select different varieties or quantities of particular plants.

In one study of the diets of sheep, Red kangaroos and Eastern gray kangaroos within the mulga-box and spinifex plant associations of southwestern Queensland, it

was found that although each group of animals concentrated on particular species of plants, the Eastern gray kangaroos concentrated mainly on grasses (64 percent grass, 36 percent dicotyledons), while the diet of Red kangaroos and sheep had more in common (46 percent grass, 54 percent dicotyledons) through their preference for forbs and browse. With changed seasonal conditions accompanied by a drop in temperature, dicotyledons became less abundant and all these species turned their attention to grasses as the food sources became depleted.

In a further example, involving a different group of species, Yellow-footed rock wallabies near Broken Hill, New South Wales (where they are of restricted distribution and relatively rare) were compared with the wallaroo and two feral competitors, the goat and rabbit. During good seasons considerable overlap was reported in the species eaten by all the herbivores present and this overlap increased with deterioration of the

▷ **Gregarious grazers** OVERLEAF, Eastern gray kangaroos move out from the cover of trees to feed on the grasses that are their preferred food. Larger size and safety in numbers may permit such daytime feeding, while most kangaroos feed exclusively in the safety of dusk or darkness.

▼ **Representative small- and medium-sized** kangaroos and wallabies. (**1**) Papuan or Lesser forest wallaby (*Dorcopsulus macleayi*). (**2**) Musky rat kangaroo (*Hypsiprymnodon moschatus*). (**3**) Lumholtz's tree kangaroo (*Dendrolagus lumholtzi*). (**4**) Desert rat kangaroo (*Caloprymnus campestris*). (**5**) Boodie (*Bettongia lesueur*). (**6**) Rufous hare wallaby (*Lagorchestes hirsutus*). (**7**) Long-nosed potoroo (*Potorous tridactylus*). (**8**) Banded hare wallaby (*Lagostrophus fasciatus*). (**9**) Rufous bettong (*Aepyprymnus rufescens*).

vegetation. In good seasons the largest component of the wallaby's diet was forbs (42–52 percent), chiefly small herbaceous ephemeral species. During drought, browse became the most important dietary component (44 percent), with a marked overlap of dietary components (75 percent) for all categories of plants in wallabies and goats. Competition from the wallaroos appeared least, with limited overlap of dietary components and little variation from the high proportion of grasses ingested throughout the study. The results indicate that small surviving "island" colonies of rock wallabies are subject to considerable competition, particularly with goats capable of removing browse to levels beyond the reach of wallabies, and to a lesser extent with rabbits.

Patterns of reproduction vary greatly. Some species such as the Red kangaroo are opportunistic breeders, mating and producing young when seasonal conditions favor successful rearing of the offspring; others such as the gray kangaroos are capable of breeding throughout the year but tend to be primarily seasonal breeders with most young born during the summer months, the young then leaving the pouch at a most favorable time, the spring of the following year. Still other species have a very restricted breeding season, for example the Tammar wallaby, in which the greatest number of young are born in late January. Onset of sexual maturity may be less than a year in the female Tammar, but exceed 15 months and even attain 2–3 years depending on seasonal conditions in the larger species of kangaroos. Courtship may be restricted to a few hours or may extend over 2–3 days, with the male closely associated with a female coming into heat (estrus). Males generally follow such females, frequently sniffing the opening of the pouch and urogenital area, while pawing at the female's tail. During this interaction male wallabies exhibit a characteristic sideways sinuous swishing of their tails, an activity less noticeable in the larger kangaroos. Attention by other males at this time results in chasing or fighting between the competitors. Mating may be brief or, as in the case of gray kangaroos on occasions, exceed one hour.

The gestation period is about one month (28 days in Tammars; in Eastern gray kangaroos it is one of the longest at 36 days). Young at birth are small and undeveloped and weigh from 0.3g (0.01oz) in smaller wallabies to under 1g (0.035oz) in the larger kangaroos. One young is the usual number born to each female, but twins are

known. The Musky rat kangaroo is an exception in that two young are usually born. Other than in the Swamp wallaby, where the estrous cycle (32 days) is shorter than the gestation period (35 days), the estrous cycle is longer than gestation. In a number of species matings occur when the female is receptive after giving birth (post-partum estrus), in which case a quiescent blastocyst may result, later to develop on vacation of the pouch by the young produced at the preceding mating. Immediate post-partum mating is unknown in the gray kangaroos, where loss of young is followed by return to estrus about one week later. Newborn young attach to one of four teats in the pouch where the young remain for several months until they eventually leave the pouch for short periods and then, depending on species, subsequently vacate the pouch completely some 5–11 months after birth. Young of most species then continue to suckle from the same teat they occupied during pouch life for a further 2–6 months. Young tend to associate with their mothers until they attain sexual maturity.

Like the herbivorous hoofed mammals, kangaroos have social systems which range from solitary to group living according to factors such as habitat, diet, body size and mobility. The principal evolutionary trends within the family have been away from the early forms, believed to have been small omnivorous forest dwellers, probably nocturnal and solitary by nature, towards larger size and grazing habits, with some development of daytime activity and group living. The rat kangaroos are generally solitary. In the Long-nosed potoroo male home ranges (about 19ha/47 acres) overlap those of several females (about 5ha/12.5 acres), but there is some indication that the home ranges of males do not overlap. The Brush-tailed bettong has feeding areas which overlap, but an area of 1–2ha (2.5–5 acres) surrounding several daytime nests in

current use appears to be almost exclusive to individuals, any overlap being between males and adjacent females. Total home ranges of males (27ha/67 acres) are larger than those of females (20ha/49 acres).

The biology of many of the smaller wallabies is little known. Some are solitary, eg in the 4–6kg (9–13lb) Red-necked pademelon the home ranges of about 14ha (35 acres) overlap extensively, but there are no persistent associations. One of the smallest wallabies, the 2–5kg (4.4–11lb) quokka, has individual home ranges of 1.6–9.7ha (4–24 acres) which overlap in areas of suitable habitat. There is some interchange between areas, but the population is in effect divided into subunits. Individuals within an area are not gregarious but remain tolerant of each other except to monopolize particular shelter sites in hot weather.

There is an obvious trend towards increasing sociability in the kangaroos, in which increased size, greater mobility, less completely nocturnal activity and a diet based on the grasses of more open habitat

▲ **The Burrowing bettong or boodie** is the only kangaroo that regularly inhabits burrows. Once common on mainland Australia, it is now extinct there and confined to islands off the coast of Western Australia. It eats tubers, roots, seeds, fruit, fungi and termites.

▼ **Red kangaroo fight**. Before a fight two males may engage in a "stiff-legged" walk (1) in the face of the opponent, and in scratching and grooming (2, 3), standing upright on extended rear legs. The fight is initiated by locking forearms (4) and attempting to push the opponent backward to the ground (5). Fights may occur when one male's monopoly of access to an individual or group of females is challenged. There appears to be no defense of territory for its own sake.

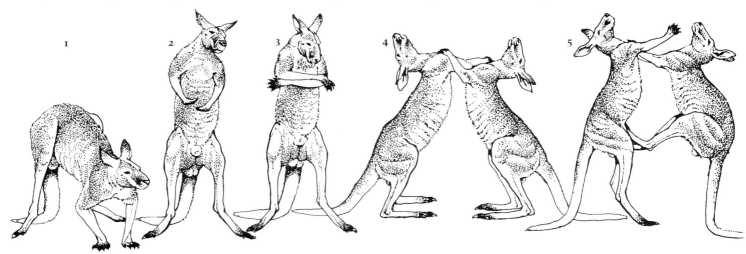

▲ **Climbing kangaroo.** Lumholtz's tree kangaroo (*Dendrolagus lumholtzi*).

▼ **Reproduction in kangaroos** varies according to whether the fertilized egg develops continuously from fertilization to birth or whether it enters a period of dormancy (diapause) before finally developing. Also kangaroos may give birth at any time of the year (aseasonal) or at fixed times (seasonal). Three types are indicated here. (**a**) The Western gray kangaroo is a seasonal breeder with a continuous gestation of 30 days: the young stay in the pouch for 320 days. When one joey has just been born (white bar) another is still at foot. (**b**) The Tammar wallaby is a seasonal breeder and exhibits diapause. When one joey has just been born the mother has another fertilized egg in her uterus, which will not be born until the following season. (**c**) The Red Kangaroo is an aseasonal breeder with diapause. Thus when one joey has just been born, there is another at foot, suckling, while the mother has a fertilized egg in her uterus which will be born soon after the joey leaves the pouch.

are correlated with an increased tendency for the individual to be one of a group. A similar pattern is known in several species, eg Eastern gray kangaroo, Western gray kangaroo, Whiptail wallaby, Red kangaroo and Antilopine wallaroo, which are usually seen in small groups of 2–10, solitary animals being rare. These may be subunits of a group which shares a common home range, or of a large unstable aggregation the members of which may be locally nomadic (as in the Red kangaroo). In the Whiptail wallaby, whose males attain 25kg (55lb) and females 15kg (33lb), up to 50 individuals share a group home range of about 100ha (250 acres) but are generally found in small, continually changing subgroups of 7–10 animals. Individual home ranges for males are about 75ha (185 acres) and for females about 56ha (138 acres).

In all of these species, the main association between individuals is between mother and offspring, which may stay together after the young is weaned. In most kangaroos mating is promiscuous, with males competing for access to females. The largest, most dominant males are able to monopolize a female in heat, and in a group one male may father most of the offspring. It is presumably selection for increased body size in competing males which has led to the marked sexual dimorphism in the larger kangaroos.

At the time of European settlement nomadic Aboriginal man utilized kangaroos as a source of meat and hides but hunting at this level probably had little effect on kangaroo populations. Early European settlers also valued kangaroos as a source of meat and hides but increasing settlement gradually brought changes in the environment. Habitat destruction, coupled with the introduction of predators such as the European Red fox, domestic dog and cat, and of competitors such as domestic stock and the rabbit, all contributed to the depletion in numbers and in some cases the extinction of a few species of kangaroos. However, for some other species and in particular the larger kangaroos there is circumstantial evidence of an increase in numbers following limitation of predation by dingoes, coupled with changes in pasture conditions and composition, and the increasing provision of watering points with the extension of agricultural and pastoral zones. As some species of kangaroos increased in numbers and competed with stock, settlers gradually came to regard them as pests. Initially numbers were reduced on a local scale by organized drives, shooting and occasionally by poisoning, but in recent years the culling of these species of kangaroos has been under State or Territorial control.

Most species are relatively secure. The greatest threats are associated with destruction of habitat or, in the case of smaller species, predation by foxes. Ten species are regarded as endangered. The Parma wallaby is now reduced to restricted, but well-established, populations in wet sclerophyll forest, rain forest and dry sclerophyll forest in northeastern New South Wales. Of the two nailtail wallabies, and four rat kangaroos of the genus *Bettongia*, some are restricted to inland or relict mainland populations and others have not been sighted for many years. The Banded and Western hare wallabies, both once plentiful in the interior, are now restricted to islands in Shark Bay.

WEP

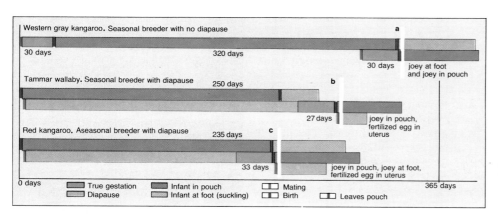

Western gray kangaroo. Seasonal breeder with no diapause   **a**
30 days   320 days   30 days   joey at foot and joey in pouch

Tammar wallaby. Seasonal breeder with diapause   250 days   **b**
27 days   joey in pouch, fertilized egg in uterus

Red kangaroo. Aseasonal breeder with diapause   235 days   **c**
33 days   joey in pouch, joey at foot, fertilized egg in uterus

0 days   365 days

■ True gestation   ■ Infant in pouch   ▫ Mating
■ Diapause   ■ Infant at foot (suckling)   ▫ Birth   ▫ Leaves pouch

KOALA

Phascolarctos cinereus
Sole member of the family Phascolarctidae.
Distribution: mainland E Australia.

Habitat: eucalypt forest below 600m (2,000ft);
highly specific in feeding preference for a few
eucalypt species.

Size: animals from S of range
significantly larger. Head-body
length in S averages 78cm
(30.7in) (male), 72cm (28.3in)
(female); average weight in S
11.8kg (26lb) (male), 7.9kg
(17.4lb) (female); in N 6.5kg
(14.3lb) (male), 5.1kg (11.2lb)
(female).

Coat: gray to tawny; white on chin, chest and
inner side of forelimbs; ears fringed with long
white hairs; rump dappled with white patches;
coat shorter and lighter in N of range.

Gestation: 34–36 days.
Longevity: up to 13 years (18 in captivity).

Subspecies: 3; *P. c. victor* (Victoria), *P. c. cinereus*
(New South Wales), *P. c. adustus* (Queensland).
May be arbitrarily divided, as there is a gradual
south-to-north decrease in body size, hair
length and darkness of coat.

▶ **Mother and young** OPPOSITE – a young koala
of 7–10 months rides on its mother's back. At
birth the single young weighs one fiftieth of an
ounce. At five months, before leaving its
mother's pouch, the young koala feeds on part-
digested leaves provided by the mother. Tree-
fork sites such as the one pictured are daytime
sleeping sites.

▶ **Protected koala** — this animal lives in the
Currumbin Sanctuary, Queensland. At such
sites populations can increase to dangerous
levels that put food trees at risk.

▷ **Koala feeding** OVERLEAF on *Eucalyptus*
leaves, chief food of this specialized leaf-eater.

THE koala was not always the popular and
loved animal that it is today. The early
white settlers in Australia killed millions for
their pelts. This hunting, together with land
clearance and an increased frequency and
scale of forest fires between 1850 and 1900,
so decimated koala populations that by the
early 1930s they were thought to be inexor-
ably headed for extinction. Bans on hunting
in the various states between 1898 and
1927 and intensive management, particularly from 1944 in the southern populations, has reversed this decline and koalas
are now relatively common in their favored
habitat, with densities approaching three
animals per ha (1.2/acre) in some colonies.

Koalas are principally nocturnal and extremely specialized for a life spent almost
exclusively in trees. Their stout body is
covered with dense fur, the tail is reduced to
a stump, the paws are large and the digits
strongly clawed. The first and second digits
of the forepaw are opposable to the other
three and this enables the animal to grip the
smaller branches as it climbs. Koalas ascend
large trees by clasping the bole with the
sharp claws of the forepaws and bringing
the hindfeet up together in a bounding
movement. They are less agile on the
ground but travel using a similar bounding
action or a slower quadrupedal walk.

Males are up to 50 percent heavier than
females, have a broader face, comparatively
smaller ears, and a large chest gland.
Females lack this gland. They have a pouch
which opens to the rear.

Female koalas are sexually mature at two
years of age. Males are fertile at two years
old but usually do not mate until they are
four, because they require longer to become
large enough to compete for females.

The summer breeding season (October–
February) is characterized by a great deal of
aggression between males and their bellowing is heard throughout the night. These
calls, which consist of a series of harsh
inhalations each followed by a resonant,
growling expiration, advertise an individual's presence and warn other males
away. The call of one male usually elicits a
response from all the adult males in the area.
In contrast, the only vocalization commonly
heard from females and subadult males is a
harsh wailing distress call, given usually
when harrassed by adult males.

Koalas are polygynous (males mate with
several females) and relatively sedentary.
Adults occupy fixed home ranges, the males
usually 1.5–3ha (3.7–7.4 acres), females
0.5–1ha (0.1–2.5 acres). The home range of
a breeding male overlaps those of females as
well as of subadult and nonbreeding males.
In the breeding season the adult males are
very active at night and constantly move
through their range, both ejecting male
rivals and mating with any receptive (estrous) females. Copulation is brief, usually
lasting less than two minutes, and occurs in
the tree. The male covers the female and
grasps on to the back of her neck with his
teeth while mating.

The females give birth to a single young
each year with the majority of births occurring in mid-summer (December–January).
The newborn weigh less than 0.5g (0.02oz)
and attach to one of the two nipples in the
pouch. Weaning commences after five
months and is initiated by the young feeding
on partially digested leaf material produced
from the female's anus. This pap is thought
to come from the cecum of the mother and
to innoculate the gut of the young with the
microbes it needs to digest eucalypt leaf.
Growth is rapid once the young begins
feeding on leaves. The young departs the
pouch for good after seven months and
travels around clinging to the mother's back
before becoming independent by eleven
months of age. It may continue to live close
to the mother for a few more months.

Outside the breeding season there is little
obvious social behavior. While neighboring
animals are doubtless aware of each other's
presence, there are few interactions and no

apparent social groupings. While koalas will feed on a large number of eucalypt and non-eucalypt species, the leaves of only a few eucalypt species make up the bulk of their diet. In the south, *Eucalyptus viminalis* and *E. ovata* are the preferred species, while the northern populations feed predominantly on *E. punctata*, *E. camaldulensis* and *E. tereticornis*. An adult koala eats about 500g (1.1lb) daily and its diet of low-protein, highly fibrous eucalypt leaf contains high concentrations of phenolics and volatile oils. The koala has adapted in numerous ways to cope with this diet. The cheek teeth are reduced to a single premolar and four broad, high-cusped molars on each jaw which finely grind the leaves for easier digestion. Some toxic plant compounds appear to be detoxified in the liver through the action of glucuronic acid, and are excreted. Microbial fermentation occurs in the cecum, which is up to four times the body length of the koala and the largest of any mammal in proportion to size. Because of the low quality of the diet, koalas conserve energy by their behavior. They are slow-moving and sleep up to 18 hours out of 24. This has given rise to the popular myth that koalas are drugged by the eucalypt compounds they ingest. Koalas feed from dusk onwards and the animal moves from its favored resting fork to the tree crown to feed. Except in the hottest weather, they obtain all of their water requirements from the leaves.

Koala populations can build up to extremely high densities wherever their favored food species occur. This is illustrated by the fate of a koala population introduced onto a small island off the coast of southeastern Australia. Between 1923 and 1933 a total of 165 koalas were transferred to this island from another colony which was overpopulated. In 1944, when it was apparent that these koalas had multiplied to such numbers that they were killing their food trees and many had already died of starvation, 1,349 koalas were removed. Populations on offshore islands are now managed much more intensively but, because of large-scale clearing of native forest, many of the areas of habitat suitable for koalas now occur in small isolated patches which have similar management problems to island habitats. The future management of these populations is complicated by the shortage of suitable forest areas where surplus animals can be released. Alternative procedures, such as the release of animals into forests with a lower density of preferred *Eucalyptus* species, are now being investigated.   RM

WOMBATS

Family: Vombatidae
Three species in 2 genera.
Distribution: Australia.

Common wombat

Vombatus ursinus
Common, Naked-nosed, Coarse-haired or Forest wombat.
Distribution: SE Australia including Flinders Island and Tasmania.
Habitat: temperate forests, heaths, mountains.

Size: head-body length 90–115cm (35.4–45in); tail length about 2.5cm (1in); height about 36cm (14.2in); weight 22–39kg (48.5–86lb).
Coat: coarse, black or brown to gray; bare muzzle, short rounded ears.
Gestation: unknown.
Longevity: over 5 years (up to 26 years in captivity).
Subspecies: 3.

Southern hairy-nosed wombat

Lasiorhinus latifrons
Southern hairy-nosed or soft-furred wombat or Plains wombat.
Distribution: central southern Australia.
Habitat: savanna woodlands, grasslands, shrub steppes.

Size: head-body length 87–99cm (34.3–39in); weight 19–32kg (42–70lb); tail and height similar to Common wombat.
Coat: fine, gray to brown, with lighter patches; hairy muzzle, longer pointed ears.
Gestation: 20–22 days.
Longevity: unknown (to 18 years in captivity).

Northern hairy-nosed wombat [E]

Lasiorhinus krefftii
Northern or Queensland hairy-nosed or soft-furred wombat.
Distribution: single colony in mid-eastern Queensland.
Habitat: semi-arid woodland.

Size: similar (possibly slightly larger) size, weight and appearance to Southern hairy-nosed wombat, but broader muzzle.
Gestation and longevity: unknown.

[E] Endangered.

▶ **The Common wombat** ABOVE uses its strong foreclaws to excavate burrows—one reason it is still regarded as "vermin" in eastern Australia, where it may damage rabbit-proof fences. Erosion BELOW reveals the complexity of a wombat warren in South Australia.

SHIPWRECKED sailors on Preservation Island in Bass Strait were the first Europeans to discover wombats. They nicknamed them "badgers" because the animals were mainly nocturnal and lived in burrows. When the sailors were rescued in 1798 they brought one wombat back with them to Sydney. This was eventually described as *Vombatus ursinus*, the "bear-like wombat," after the aboriginal name "wombach" for the slightly larger species found around Sydney.

At first sight wombats really do resemble a small bear, or, even more closely, the marmots of the Northern Hemisphere, with their thick, heavy bodies, small eyes, massive, flattened heads and similar teeth. Unlike their closest relative, the smaller arboreal koala (see p872), wombats are completely terrestrial and well equipped with short, powerful legs and long, strong claws (absent on the first toe of the hind foot) for digging their large, often complex burrows. Their dentition (I1/1, C0/0, P1/1, M4/4 = 24), particularly the single pair of upper and lower incisors which like the other teeth are rootless and grow continuously, is unique among marsupials but remarkably similar to that of the rodents. Both sexes are similar in size. Wombats have poor eyesight but keen senses of smell and hearing.

During the Pleistocene (2 million to 10,000 years ago) another two, much larger, types of wombat occurred in Australia. Today the Southern hairy-nosed wombat is abundant in arid to semi-arid saltbush, acacia and mallee shrublands of southern South Australia from the Murray River in the east to a few isolated colonies in the southeast of Western Australia. The rarer Northern hairy-nosed wombat formerly occurred at three widely scattered localities in the semi-arid interior of eastern Australia but the remaining colony of perhaps only 20 individuals is now restricted to the most northerly of these, 130km (80mi) northwest of Clermont, mid-eastern Queensland. In both cases the areas are characterized by high summer temperatures, low irregular rainfalls, frequent droughts, limited freestanding water and highly fibrous grass foods containing little water and protein. To survive in these areas hairy-nosed wombats have adopted similar strategies to the much smaller desert mammals. They live in burrows and mainly emerge to feed at night. The Southern hairy-nosed wombat also has a very low metabolic rate and low rate of water turnover, achieved by concentrating its urine, reducing fecal water loss to very low levels and restricting respiratory loss of water by limiting its activity above ground to periods of more suitable temperature and humidity. Both species also have a low nitrogen requirement which, together with a relatively variable body temperature, low heat loss, basking behavior and failure to ovulate during droughts, further economizes on energy.

The Common wombat mainly inhabits the wetter, subhumid, eucalypt forests from southeastern Queensland along the Great Dividing Range to eastern Victoria. It is rare now in southwestern Victoria but scattered populations still persist in the coastal grasslands and some remnant forest areas in southeastern South Australia. Two smaller subspecies occur on Flinders Island in Bass Strait and in Tasmania. The Common wombat is now extinct on Preservation and Clarke Islands and Cape Barren. Despite their less dry habitat, Common wombats exhibit similar physiological and behavioral adaptations to those of hairy-nosed wombats, although they are less efficient at limiting water loss. In summer they avoid environmental temperatures over 25°C (77°F), above which they begin to lose their ability to regulate their body temperatures, by remaining in their burrows until after sunset. The burrows are up to 50cm (20in) wide and 30m (100ft) long, often with several entrances, side tunnels and resting chambers. In winter, burrow air temperatures rarely fall below 4°C (39°F), and basking and daytime activity become more common. The Common wombat also generates increased body heat in its nocturnal activities, during which it may travel up to 3km (2mi), and decreases its respiration and heartbeat rates when resting in its burrow.

Wombats' teeth are highly adapted to breaking up their tough, highly fibrous food, mainly grasses such as snow tussocks (Common wombats) and spear grass (Southern hairy-nosed wombat). Relative to body weight they eat less than most other marsupials, and a very slow rate of food passage and microbial fermentation of fiber in their colons also help them survive on poor quality food.

The wombats' social behavior and reproduction are also adapted to saving energy. Both southern species have highly stable social relationships involving minimal close-quarter interaction. When a fight does occur the aggressor attempts to bite the other on the ear or flanks while the defender presents its large exceptionally thick-skinned rump to the attacker and kicks out with its hind feet.

The home ranges of Southern hairy

nosed wombats are centered around the warrens, which are often situated around the edges of large claypans, where the thick hard surface has broken away, exposing soft limestone sediments underneath. Home ranges are of 2.5–4.2ha (6.2–10.4 acres), depending on the amount of food present. Common wombat home ranges vary from 5 to 23ha (12.4–57 acres), depending upon the distribution of feeding areas in relation to the burrows, most of which are usually located along slopes above creeks and gullies. Population densities of Southern hairy-nosed wombats can reach 0.2 per ha (0.08/acre), while those of Common wombats may attain 0.5 per ha (0.2/acre), particularly where native forest adjoins open grassy areas.

Southern hairy-nosed wombats are seasonal breeders, giving birth usually to a single young in spring (October–January) in good seasons but none during droughts. Common wombat young appear to be born at any time of the year. Young wombats first leave the pouch at about 6–7 months but may still return to it occasionally over the next three months. Weaning may not occur until they are 15 months old. Southern hairy-nosed wombats are sexually mature at 18 months when they are 60–80cm (23.6–31.5in) in length and weigh 15–20kg (33–44lb), compared with about 23 months for Common wombats weighing some 22kg/48.5lb. Major causes of death include starvation during droughts, outbreaks of mange, predation by dingoes and collisions with road vehicles.

Each species is protected to some degree in the different States, but the Common wombat is still classed as vermin in eastern Victoria, mainly because of its damage to rabbit-proof fences.   JMcI

HONEY POSSUM

Tarsipes rostratus
Sole member of the family Tarsipedidae.
Distribution: SW Australia.

Habitat: heathland, shrubland and open low
woodlands with heath understory.

Size: adult male head-body
length 6.5–8.5cm
(2.6–3.3in), tail length
7–10cm (2.8–3.9in), weight
7–11g (about 0.3oz); adult
female head-body length
7–9cm (2.8–3.5in), tail length
7.5–10.5cm (3–4.1in), weight
8–16g (0.3–0.6oz); females
(average weight 12g (0.4oz)
one-third heavier than males.

Coat: grizzled grayish-brown above, reddish
tinge on flanks and shoulder, next to cream
undersurface. Three back stripes: a distinct
dark brown stripe from back of head to base of
tail, with less distinct, lighter brown stripe on
each side.

Gestation: uncertain, about 28 days.
Longevity: 1–2 years in wild.

▶ **Nothing but nectar and pollen** feature in the
diet of the Honey possum, here feeding on
Banksia. The reduced home range of nursing
Honey possum mothers may center on a single
such shrub. The many large, compound nectar-
rich inflorescences are produced over a long
period and may provide all the food
requirements for the mother and her young.

The pointed snout and brush-tipped tongue
are specializations for this diet, probing deep
into the individual flowers for nectar. With its
grasping hands, feet and tail, and its small size,
the Honey possum can feed on small terminal
flowers on all but the most slender branches.
While nectar is an easily digested source of
energy, pollen grains, the chief source of
protein in the diet, are not. The Honey
possum's complex stomach has two chambers,
but whether these assist in digestion of pollens
is not certain.

THE Honey possum does not eat honey
and is only very distantly related to
possums. It is an animal all on its own: no
fossil history is known earlier than 35,000
years ago and it appears to be the sole surviving representative of a line of marsupials
that diverged very early from the possum-
kangaroo stem (see p826). Because it is
especially adapted to feed on nectar and
pollen, it probably evolved at a time when
heathlands with a great diversity of flowering plants were widespread, about 20 million years ago. Heathlands exist today in
patches around the edge of Australia's arid
center; those in southwestern Australia are
still very varied and with more than 3,600
species of flowering plant there are always
some species in flower to provide enough
food for this animal totally dependent on
nectar.

These tiny shrew-like mammals have a
long pointed snout, and a prehensile tail
longer than head and body together. The
first digit of the hind foot is opposable to the
others and there is a considerable span for
gripping branches: all digits have rough
pads on the tips.

Honey possums move through vegetation
chiefly by fast running; the tail is used for
extra support and stability in climbing, and
when feeding its use frees the forelimbs to
manipulate blossoms. With its grasping
hands, feet and tail and small size the Honey
possum is able to feed on small terminal
flowers of all but the slenderest branches.
Teeth are few in number and very small and
weak. The dental formula is I2/1, C1/0,
P1/0, M3/3 = 22, but the molars are merely
tiny cones.

Honey possums communicate with only a
small repertoire of visual signal postures and
a few high-pitched squeaks, a reflection of
their mainly nocturnal activity. The sense of
smell on the other hand appears to be very
important in social behavior and feeding.

Both sexes mature at about six months.
Births may occur throughout the year but
numbers of births are at a very low level in
mid-summer (December), when few plants
are in flower, before reaching a highly
synchronized peak of births during
January–February. There are two further,
less synchronized peaks at about three-
month intervals, the minimum time required for the rearing of a litter. A second
litter may be born very soon after the first
leaves the pouch or is weaned, since the
Honey possum exhibits embryonic diapause
(see p871), so far the only marsupial outside
the kangaroos and wallabies (families Macropodidae and Potoroidae) known to do so.

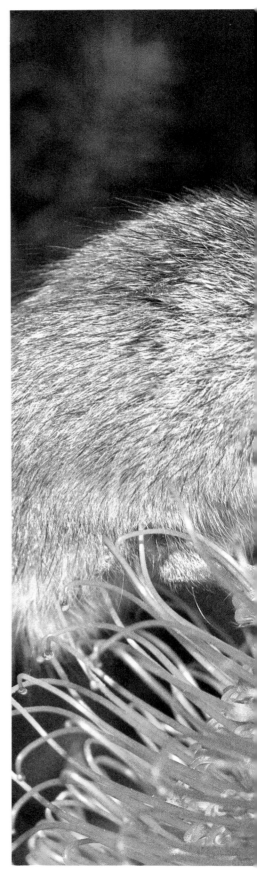

This timing of births is such that the young are ready to leave the pouch and begin to fend for themselves when food is abundant, in autumn, spring and early summer.

Courtship is minimal; the male follows a female that is becoming receptive (approaching estrus) and attempts to mount her, but only when actually in estrus does she stay still long enough for mating to occur. The young at birth are tiny—about 0.005g (0.0002oz)—and their stage of development is typical of young marsupials. The pouch has four teats, and although litters of four may occur, two and three are most common. The young are carried in the mother's deep pouch for about eight weeks, by which time each weighs about 2.5g (about 0.09oz) and has a good covering of fur, including the back stripes; their eyes are open, but they are very shaky on their feet. A litter of four such young weighs nearly as much as their mother. As soon as they venture out of the pouch, their mother leaves them in a nest (an old bird's nest or a hollow branch) while she forages, returning from time to time to nurse them. After a few days, the young are able to ride on their mother's back but she appears to avoid this if she can—with the whole litter on her back she can hardly move. About one week after leaving the pouch, the young follow their mother while she feeds. They cease to suckle at about 11 weeks, and probably disperse soon after.

Most Honey possums live in overlapping home ranges of about 1ha (2.5 acres), but females with large pouch young have a smaller, more or less exclusive, home range of about 0.01ha (120sq yd). In captivity, females are dominant to males and juveniles, and very aggressive to strangers, especially males, suggesting that in the wild the exclusive area of females with large young may be a temporary feeding and nesting territory. Such females are greatly hampered by their young as these grow, and after they leave the pouch their mother needs to return frequently to the nest to nurse them. Honey possums frequently huddle together, an energy-saving behavior which is found in many small mammals. In cold weather when food is short, Honey possums may become torpid.

The Honey possum is at present not endangered, but in the long term a species with such restricted distribution may need special attention. Most of its habitat is in the wetter areas of a very dry continent, where unreserved habitat is still being cleared for agriculture. In reserves, feral cats and, perhaps, "controlled" burning, also pose dangers.   EMR

APPENDIX

The following lists all species, genera and families of the orders Rodentia, Lagomorpha, Macroscelidea, Insectivora, Edentata and Marsupialia, together with their distribution and common names. Orders and families are in the sequence in which they appear in the book, and all genera and species within genera in alphabetical order.

ORDER RODENTIA

SUBORDER SCIUROMORPHA

FAMILY CASTORIDAE

Castor

C. canadensis **North American or Canadian beaver.** N America from Alaska E to Labrador, S to N Florida and Temaulipas (Mexico). Introduced to Europe and Asia

C. fiber **European or Asiatic beaver.** NW and NC Eurasia, in isolated populations from France E to Lake Baikal and Mongolia

FAMILY APLODONTIDAE

Aplodontia

A. rufa **Mountain beaver (beaver, boomer).** Pacific coast, USA and Canada

FAMILY SCIURIDAE

Aeretes

A. melanopterus **Groove-toothed flying or North Chinese flying squirrel.** NE China

Aeromys

A. tephromelas **Black flying squirrel.** S Thailand, Malaya, Sumatra, Borneo

A. thomasi **Thomas' flying squirrel.** Borneo

Ammospermophilus

A. harrisii **Antelope ground squirrel.** Arizona to SW New Mexico and adjoining Mexico

A. insularis **Island antelope squirrel.** Espiritu Santo Is, Mexico

A. interpres **Texas antelope squirrel.** New Mexico and W Texas to Mexico

A. leucurus **White-tailed antelope squirrel.** USA south to Mexico

A. nelsoni **Nelson's antelope squirrel.** S California

Atlantoxerus

A. getulus **Barbary ground squirrel.** Morocco, Algeria

Belomys

B. pearsoni **Hairy-footed flying squirrel.** Sikkim to Indochina, Burma, Thailand, S China, Taiwan

Callosciurus

C. adamsi **Earspot squirrel.** N Borneo

C. albescens **Kloss squirrel.** N Sumatra

C. baluensis **Kinabalu squirrel.** Borneo

C. caniceps **Golden-backed or Grey-bellied squirrel.** Eastern Himalayas, Thailand, Burma, Thailand to Malaya, Formosa

C. erythraeus **Pallas squirrel.** Burma to SE China, Formosa

C. finlaysoni **Finlayson's or Variable squirrel.** Burma, Thailand, Indochina

C. flavimanus **Belly-banded or Mountain red-bellied squirrel.** S China, Malaya, Thailand, Burma

C. inornatus Vietnam, Laos

C. melanogaster **Loga squirrel.** Mentawai Is (Sumatra)

C. nigrovittatus **Black-banded squirrel.** Thailand, Malaya, Sumatra, Java, Borneo

C. notatus **Plantain squirrel.** Thailand, Malaya, Sumatra, Java, Borneo

C. phayrei South Burma

C. prevosti **Prevost's squirrel.** S Thailand, Malaya to Sumatra, Borneo

C. pygerythrus **Irrawaddy squirrel.** Nepal, Assam, Burma to Indochina, SE China

C. quinquestriatus **Anderson's squirrel.** Yunnan (China), Burma

Cynomys

C. gunnisoni **Gunnison's prairie dog.** Colorado to Arizona

C. leucurus **White-tailed prairie dog.** Wyoming, Montana, Utah, NW Colorado

C. ludovicianus **Black-tailed prairie dog.** C USA

C. mexicanus **Mexican prairie dog.** C Mexico

C. parvidens **Utah prairie dog.** Utah

Dremomys

D. everetti **Bornean mountain ground squirrel.** N Borneo

D. lokriah **Orange-bellied Himalayan squirrel.** Nepal to N Burma

D. pernyi **Perny's long-nosed squirrel.** S China to Burma, Taiwan

D. rufigenis **Red-cheeked squirrel.** S China, Malaya, Thailand, Indochina

Epixerus

E. ebii **Temminck's giant squirrel.** Ghana, Sierra Leone

E. wilsoni **African palm squirrel.** Gabon, Cameroun

Eupetaurus

E. cinereus **Woolly flying squirrel.** Kashmir

Exilisciurus

E. concinnus **Pygmy squirrel.** Basilan (Philippines)

E. exilis **Plain pygmy squirrel.** Borneo

E. luncefordi **Pygmy squirrel.** Mindanao (Philippines)

E. samaricus **Samar pygmy squirrel.** Samar, Philippines

E. surrutilus **Mindanao pygmy squirrel.** Mindanao, Philippines

E. whiteheadi **Whitehead's pygmy squirrel.** N Borneo

Funambulus

F. layardi **Layard's striped squirrel.** Sri Lanka, S India

F. palmarum **Indian palm squirrel.** Sri Lanka and peninsular India

F. pennanti **Northern palm squirrel.** India, Nepal, Baluchistan

F. tristatus **Jungle striped squirrel.** Peninsular India

F. sublineatus **Dusky striped squirrel.** Sri Lanka, S India

Funisciurus

F. anerythrus **Thomas tree or Redless squirrel.** Senegal, Nigeria to Uganda, Zaire

F. auriculatus **Matschies squirrel.** Nigeria to Congo

F. bayoni **Bayon's or Bocage's tree squirrel.** N Angola, SW Zaire

F. congicus **Western striped or Kuhl's tree squirrel.** R Zaire–SW Africa

F. isabella **Gray's four striped or Lady Burton's squirrel.** Cameroun–R. Zaire

F. lemniscatus **Leconte's four-striped squirrel.** Cameroun to Gabon

F. leucogenys **Orange-headed squirrel.** Ghana, C African Republic, Rio Muni

F. leucostigma **White spotted squirrel.** Ivory Coast to Cameroun

F. mandingo **Mandingo squirrel.** Gambia to Nigeria

F. mystax **Rope squirrel.** Congo–Zaire

F. pyrrhopus **Red-footed or Cuvier's tree squirrel.** Gambia, Uganda, Angola

F. substriatus **De Winton's tree squirrel.** Ivory Coast to Nigeria

Glaucomys

G. sabrinus **Northern flying squirrel.** Canada, W USA

G. volans **Southern flying squirrel.** E USA to Honduras

Glyphotes

G. canalvus **Gray-bellied sculptor squirrel.** Sarawak.

G. simus **Red-bellied sculptor squirrel.** Mt Kinabalu (Borneo)

Heliosciurus

H. gambianus **Gambian sun squirrel.** Senegal to Ethiopia to Zambia

H. rufobrachium **Red-legged sun squirrel.** Senegal to Kenya to Zimbabwe

H. ruwenzorii **Ruwenzori sun squirrel.** E Zaire, Rwanda, Burundi, Uganda

Hylopetes

H. alboniger **Particolored flying squirrel.** Nepal to Indochina

H. fimbriatus **Kashmir pygmy flying squirrel.** Afghanistan to Kashmir

H. lepidus **Gray-cheeked flying squirrel.** Thailand, S Burma, Malaya, Java, Sumatra, Borneo

H. mindanensis **Mindanao pygmy flying squirrel.** Mindanao, Philippines

H. nigripes **Palawan pygmy flying squirrel.** Palawan, Philippines

H. phayrei **Phayre's flying squirrel.** Burma, Thailand, Laos

H. spadiceus **Red-cheeked flying squirrel.** S Burma, Thailand, Indochina, Malaya, Sumatra, Java, Borneo

Hyosciurus

H. heinrichi **Celebes long-nosed squirrel.** Sulawesi

Iomys

I. horsfieldi **Horsfield's flying squirrel.** Malaya, Sumatra, Java, Borneo

Lariscus

L. hosei **Four-striped ground squirrel.** N Borneo

L. insignis **Three-striped ground squirrel.** Malaya, Sumatra, Java, Borneo, S Thailand

L. niobe **Striped ground squirrel.** Sumatra

L. obscurus **Mentawai ground squirrel.** Mentawai Is (Sumatra)

Marmota

M. bobak **Bobak marmot.** S Russia to Manchuria, Himalayas

M. broweri **Alaska marmot.** N Alaska

M. caligata **Hoary marmot.** Alaska to Idaho

M. camtschatica **Black-capped marmot.** NE Siberia

M. caudata **Long-tailed marmot.** Tien Shan to Kashmir

M. flaviventris **Yellow-bellied marmot.** W USA

M. himalayana **Himalayan marmot.** Nepal

M. marmota **Alpine marmot.** Mountains of C Europe from French Alps eastwards

M. menzbieri **Menzbier's or Tien Shan marmot.** W Tien Shan

M. monax **Woodchuck.** Alaska to Labrador

M. olympus **Olympic marmot.** Olympic Mountains, W Washington

M. sibirica **Siberian or Mongolian marmot.** Siberia, USSR, Mongolia

M. vancouverensis **Vancouver marmot.** Vancouver Is, Canada

M. himalayana **Himalayan marmot.** Nepal

M. marmota **Alpine marmot.** Mountains of C Europe from French Alps east

Menetes

M. berdmorei **Berdmore's or Indochinese ground squirrel.** Burma to Indochina, Thailand

Microsciurus

M. alfari **Alfaro's pygmy squirrel.** S Nicaragua to Panama

M. boquetensis **Boquete pygmy squirrel.** W Panama

M. flaviventer **Yellow-bellied pygmy squirrel.** N Brazil, Peru

M. mimulus **Pygmy squirrel.** Panama to Ecuador

M. santanderensis **Santander pygmy squirrel.** Colombia

Myosciurus

M. pumilio **African pygmy squirrel.** Cameroun, Gabon

Nannosciurus

N. melanotis **Black-eared pygmy squirrel.** Sumatra, Java, Borneo

Paraxerus

P. alexandri **Alexander's bush squirrel.** NE Zaire, Uganda

P. antoniae **African bush squirrel.** Congo

P. boehmi **Boehm's bush squirrel.** E and C Africa

P. cepapi **Smith's bush squirrel.** S Angola, S Tanzania to Transvaal

P. cooperi **Cooper's green squirrel.** Cameroun

P. flavivittis **Striped bush or Eastern striped squirrel.** Mozambique to Kenya

P. lucifer **Black and red bush squirrel.** N Malawi, Tanzania, Zambia

P. ochraceus **Huet's bush squirrel.** Tanzania, S Sudan

P. palliatus **Red bush squirrel.** Natal to Somalia

P. poensis **Small green squirrel.** Sierra Leone to R Zaire

P. vexillarius **Swynnerton's bush squirrel.** Tanzania

P. vincenti **Vincent's bush squirrel.** N. Mozambique

Petaurillus

P. emiliae **Lesser pygmy flying squirrel.** Sarawak (Borneo)

P. hosei **Hose's pygmy flying squirrel.** Sarawak (Borneo)

P. kinlochi **Selangor pygmy flying squirrel.** Selangor (Malaya)

Petaurista

P. alborufus **Red and white giant flying squirrel.** S China to Thailand, Taiwan

P. elegans **Spotted giant flying squirrel.** Himalayas, Burma, S China, Thailand, Malaya, Sumatra, Java

P. leucogenys **Japanese giant flying squirrel.** Japan except Hokkaido; Kansu to Yunnan (China)

P. magnificus **Hodgson's flying squirrel.** Nepal, Sikkim

P. petaurista **Red giant flying squirrel.** Ceylon and peninsular India north to Kashmir, east through Burma and southern China to Taiwan, south to Java, Sumatra, Borneo

Petinomys

P. bartelsi **Bartel's flying squirrel.** Mt Pangrango (Java)

P. crinitus Basilan Is, Philippines

P. electilis **Pygmy flying or Hainan flying squirrel.** Hainan (China)

P. fuscocapillus **Small flying or Travancore flying squirrel.** S India, Sri Lanka

P. genibarbis **Whiskered flying squirrel.** Malaya, Sumatra, Java, Borneo

P. hageni **Hagen's flying squirrel.** Borneo, Sumatra, Mentawai Is

P. sagitta **Indo-Malaysian flying or Arrow-tailed flying squirrel.** Java

P. setosus **White-bellied flying or Temminck's flying squirrel.** Burma, Thailand, Malaya, Sumatra, Borneo

P. vordermanni **Vordermann's flying squirrel.** Malaya, Belitung (Sumatra), Borneo

Prosciurillus

P. abstrusus **Celebes dwarf squirrel.** C and SE Sulawesi

P. elbertae **Elberta squirrel.** S Sulawesi

P. leucomus **Celebes dwarf squirrel.** Sulawesi and adjacent islands

P. murinus **Celebes dwarf squirrel.** Sulawesi

P. weberi **Weber's Celebes dwarf squirrel.** C Sulawesi

Protoxerus

P. aubinnii **Giant forest squirrel.** Africa, Liberia and Ghana

P. stangeri **Slender-tailed giant or Stanger's squirrel.** Africa, Ghana to Kenya and Angola

Pteromys

P. momonga **Small Japanese flying squirrel.** Honshu and Kyushu (Japan)

P. volans **Russian or Siberian flying squirrel.** Finland to Korea, Hokkaido

Pteromyscus

P. pulverulentus **Smoky flying squirrel.** S Thailand, Malaya, Sumatra, Borneo

Ratufa

R. affinis **Common giant or Cream-colored giant squirrel.** S Thailand, Malaya, Sumatra to Borneo

R. bicolor **Black giant squirrel.** E Himalayas, Burma and S China and Indochina to Sumatra, Java and Bali

R. indica **Indian giant or Malabar squirrel.** Peninsular India

R. macroura **Grizzled Indian (giant) squirrel.** Sri Lanka, S India

Rheithrosciurus

R. macrotis **Tufted ground squirrel.** Borneo

Rhinosciurus

R. laticaudatus **Shrew-faced ground or Long-nosed squirrel.** Thailand, Malaya, Sumatra, Borneo

Rubrisciurus

R. rubriventer **Red-bellied squirrel.** Sulawesi

Sciurillus

S. pusillus **Neotropical pygmy or South American pygmy squirrel.** Guianas, NE Brazil, the Amazon basin of Peru

Sciurotamias

S. davidianus **Pere David's rock squirrel.** Northern China

S. forresti **Forrest's rock squirrel.** Yunnan, China

Sciurus

S. aberti **Abert or Tassel-eared squirrel.** W USA to New Mexico

S. aestuans **Brazilian squirrel.** Venezuela to N Argentina

S. alleni **Allen's squirrel.** Mexico

S. anomalus **Persian squirrel.** Turkey, Soviet Transcaucasia, Iran, Syria, Israel

S. apache **Apache fox squirrel.** S Arizona to New Mexico

S. arizonensis **Arizona gray squirrel.** Arizona, New Mexico, Mexico

S. aureogaster **Red-bellied or Guatemalan gray squirrel.** Guatemala to Mexico

S. carolinensis **Eastern gray or Gray squirrel.** E Texas to SE Canada. Introduced to Britain, S Africa

S. chiricahuae **Chiricahuae squirrel.** Arizona

S. colliaei **Collie's squirrel.** Mexico

S. deppei **Deppes squirrel.** Mexico to Costa Rica

S. flammifer Venezuela, Colombia

S. gilvigularis N Brazil

S. granatensis Ecuador, Venezuela to Costa Rica

S. griseus **Western gray squirrel.** Washington to California

S. ignitus Andes of Bolivia and Peru, Brazil, NW Argentina

S. kaibabensis **Kaibab squirrel.** Arizona

S. igniventris **Red-bellied squirrel.** NW Brazil, Colombia, S Venezuela

S. lis **Japanese squirrel.** Honshu, Shikoku, Kyushu (Japan)

S. nayaritensis **Nayarit squirrel.** Mexico to SE Arizona (USA)

S. niger **Fox squirrel.** Texas, N Mexico to Manitoba, E to Atlantic coast

S. oculatus **Peters' squirrel.** Mexico

S. pucheranii Andes of Colombia

S. pyrrhinus Andes of Peru

S. richmondi **Richmond's squirrel.** Nicaragua

S. sanborni **Sanborn's squirrel.** Peru

S. spadiceus Colombia, Ecuador, Peru, Brazil, Bolivia

S. stramineus NE Peru, SE Ecuador

S. variegatoides **Variegated squirrel.** Mexico through Central America to Panama

S. vulgaris **Red or Eurasian red squirrel.** Forested regions of Palearctic from Iberia and Britain east to Kamchatka Peninsula (USSR), south to Mediterranean and Black Sea, N Mongolia, NE China

S. yucatanensis **Yucatan squirrel.** Mexico, Belize, Guatemala

Spermophilopsis

S. leptodactylus **Long-clawed ground squirrel.** Afghanistan, Russian Turkestan, N Iran

Spermophilus

S. adocetus **Tropical ground squirrel.** Mexico

S. alashanicus **Alashan souslik.** Mongolia, N China

S. annulatus **Ring-tailed ground squirrel.** Mexico

S. armatus **Uinta ground squirrel.** NW USA
S. atricapillus **Rock or Baja California rock squirrel.** Mexico
S. beecheyi **Californian ground or Californian rock squirrel.** W Washington to Mexico
S. beldingi **Belding's ground squirrel.** Oregon, Idaho, California, Nevada, Utah
S. brunneus **Idaho ground squirrel.** Idaho
S. citellus **European souslik.** SE Germany and SW Poland to Turkey, Rumania and Ukraine
S. columbianus **Columbian ground squirrel.** British Columbia and Alberta to Oregon, Idaho, Montana
S. dauricus **Daurian ground squirrel.** USSR, Mongolia, N China
S. elegans Nevada, Oregon, Idaho, Montana to Colorado and Nebraska
S. erythrogenys Kazakhstan and Siberia (USSR), Mongolia, Singkiang (China)
S. franklinii **Franklin's ground squirrel.** Great Plains of Canada south to Kansas, Indiana, Illinois
S. fulvus **Large-toothed souslik.** Kazakhstan (USSR) south to NE Iran, Afghanistan, Singkiang (China)
S. lateralis **Golden-mantled ground squirrel.** Montane NW America south to New Mexico, California and Nevada
S. madrensis **Sierra Madre mantled ground squirrel.** Mexico
S. major **Red-cheeked or Russet souslik.** Steppe between Volga and Irtysh rivers (USSR)
S. mexicanus **Mexican ground squirrel.** Mexico to USA
S. mohavensis **Mohave ground squirrel.** S California
S. musicus N Caucasus mountains, USSR
S. parryii **Arctic souslik or Arctic ground squirrel.** NW Canada, Alaska, NE USSR
S. perotensis **Perote ground squirrel.** Mexico
S. pygmaeus **Little souslik.** Kazakhstan and S Ural to Crimea, USSR
S. relictus **Tien Shan souslik.** Tien Shan mtns, USSR
S. richardsonii **Richardson's ground squirrel.** N Great Plains in Canada and N USA
S. saturatus **Cascade golden-mantled ground squirrel.** Cascade mountains of USA and Canada
S. spilosoma **Spotted ground squirrel.** C Mexico to S USA
S. suslicus **Spotted souslik.** Steppes of C and S Europe including Poland, E Rumania, Ukraine, north to Oka river and east to Volga (USSR)
S. tereticaudus **Round-tailed ground squirrel.** Deserts of SE California, Nevada, Arizona and NW Mexico
S. townsendii **Townsend's ground squirrel.** Great Basin of W USA
S. tridecemlineatus **Thirteen-lined ground squirrel.** Great Plains from Texas to Utah, Ohio and SC Canada
S. undulatus **Long-tailed Siberian souslik or Arctic ground squirrel.** E Kazakhstan, Siberia, Transbaikalia (USSR), N Mongolia, W China
S. variegatus **Rock squirrel.** S Nevada, Texas, Utah (USA) to C Mexico
S. washingtoni **Washington ground squirrel.** SE Washington, NE Oregon (USA)
S. xanthoprymnus Soviet Transcaucasia, Turkey, Syria, Israel
Sundasciurus
S. brookei **Brooke's squirrel.** Borneo
S. hippurus **Horse-tailed squirrel.** S Thailand, Malaya, Sumatra, Borneo
S. hoogstraali Basuanga, Philippines
S. jentinki **Jentink's squirrel.** Borneo
S. juvencus Palawan, Philippines
S. lowii **Low's squirrel.** S Thailand, Malaya, Sumatra, Borneo
S. mindanensis **Mindanao squirrel.** Mindanao, Philippines
S. mollendorffi Calamian Is, Philippines
S. philippinensis **Philippine squirrel.** Basilan Mindanao, Philippines
S. rabori Palawan mountains
S. samarensis **Samar squirrel.** Samar, Philippines
S. steeri Palawan, Philippines
S. tenuis **Slender squirrel.** S Thailand, Malaya, Sumatra, Borneo
Syntheosciurus
S. brochus **Panama mountain squirrel.** Panama
S. poasensis **Poas mountain squirrel.** Panama, Costa Rica
Tamias
T. alpinus **Alpine chipmunk.** Alpine zone in Sierra Nevada
T. amoenus **Yellow pine chipmunk.** British Columbia south to California, east to Wyoming and Montana (USA)
T. bulleri **Buller's chipmunk.** Mexico
T. canipes **Gray-footed chipmunk.** Mtns of New Mexico and W Texas
T. cinereicollis **Gray-collared chipmunk.** Mts of Arizona and New Mexico
T. dorsalis **Cliff chipmunk.** W USA through New Mexico and Arizona to Mexico
T. durangae **Durango or Buller's chipmunk.** Mexico
T. merriami **Merriam's chipmunk.** California to Mexico
T. minimus **Least chipmunk.** Canada, USA
T. obscurus **Dusky chipmunk.** California to Mexico
T. ochrogenys California
T. palmeri **Palmer's chipmunk.** Charleston Mt (Nevada)
T. panamintinus **Panamint chipmunk.** Mts of S California and Nevada
T. quadrimaculatus **Long-eared chipmunk.** Sierra Nevada mts
T. quadrivittatus **Colorado chipmunk.** Mts of Colorado and Utah south to New Mexico (USA)
T. ruficaudus **Red-tailed chipmunk.** NW USA and British Columbia
T. senex California, Nevada, Oregon
T. sibiricus **Asiatic or Siberian chipmunk.** N European USSR and Siberia to China and Korea, Hokkaido (Japan)
T. siskiyou **Siskiyou chipmunk.** N California to Oregon
T. sonomae **Sonoma chipmunk.** California
T. speciosus **San Bernardo or Lodgepole chipmunk.** California, W Nevada
T. striatus **Eastern American or Eastern chipmunk.** USA north to Manitoba and Nova Scotia
T. townsendii **Townsend's chipmunk.** British Columbia, NW USA
T. umbrinus **Uinta chipmunk.** California, Arizona, Wyoming, Montana
Tamiasciurus
T. douglasii **Douglas squirrel.** British Columbia to California
T. hudsonicus **American red squirrel.** Alaska and Quebec south, in Rocky Mountains to New Mexico, in Appalachians to S Carolina
Tamiops
T. maritimus China, S Vietnam, Laos, Taiwan
T. macclellandi **Himalayan striped squirrel.** Nepal, Burma, Thailand, Malaya to S China
T. rodolphei **Cambodian striped squirrel.** Thailand, Cambodia, Laos, S Vietnam
T. swinhoei **Swinhoe's striped squirrel.** NE China to Burma and N Vietnam
Trogopterus
T. xanthipes **Complex-toothed flying squirrel.** China
Xerus
X. erythropus **Geoffroy's ground or Western ground squirrel.** Morocco, Senegal to Kenya
X. inauris **Cape ground squirrel.** Africa south of Zambesi
X. princeps **Kaokoveld ground squirrel.** SW Africa, Angola
X. rutilus **Unstriped or Spiny ground squirrel.** Ethiopia to N Tanzania

FAMILY GEOMYIDAE
Geomys
G. arenarius **Desert pocket gopher.** S New Mexico and Texas
G. bursarius **Plains pocket gopher.** C N America from S Canada to Texas
G. personatus **Texas pocket gopher.** S Texas and to Mexico
G. pinetus **Southeastern pocket gopher.** Coastal plains SE USA
G. tropicalis **Tropical pocket gopher.** NE Mexico
Orthogeomys
O. cavator **Chiriqui pocket gopher.** Costa Rica, Panama
O. cherriei **Cherrie's pocket gopher.** Costa Rica
O. cuniculus **Oaxacan pocket gopher.** Oaxaca (Mexico)
O. dariensis **Darien pocket gopher.** Panama
O. grandis **Large pocket gopher.** Pacific coast mexico, Guatemala, El Salvador, Honduras
O. heterodus **Variable pocket gopher.** Costa Rica
O. hispidus **Hispid pocket gopher.** S Mexico S to Honduras
O. lanius **Big pocket gopher.** Veracruz (Mexico)
O. matagalpae **Nicaraguan pocket gopher.** Nicaragua
O. underwoodi **Underwood's pocket gopher.** Costa Rica
Pappogeomys
P. alcorni **Alcorn's pocket gopher.** Jalisco (Mexico)
P. bulleri **Buller's pocket gopher.** SW Mexico
P. castanops **Yellow-faced pocket gopher.** S Great Plains in USA and Mexican Plateau.
P. fumosus **Smoky pocket gopher.** Colima (Mexico)
P. gymnurus **Llano pocket gopher.** Jalisco, Michoacan (Mexico)
P. merriami **Merriam's pocket gopher.** Veracruz, Puebla, Hidalgo (Mexico)
P. neglectus **Queretaro pocket gopher.** Queretaro (Mexico)
P. tylorhinus **Naked-nosed pocket gopher.** Michoacan, Jalisco, Hidalgo, Mexico, and Guanajuato (Mexico)
P. zinseri **Zinser's pocket gopher.** Jalisco (Mexico)
Thomomys
T. bottae **Valley pocket gopher.** Oregon to N Mexico
T. bulbivorus **Camas pocket gopher.** Willamette Valley (Oregon)
T. clusius **Wyoming pocket gopher.** Wyoming
T. idahoensis **Idaho pocket gopher.** N Rocky Mts
T. mazama **Mazama pocket gopher.** NW USA
T. monticola **Mountain pocket gopher.** N Sierra Nevada (California)
T. talpoides **Northern pocket gopher.** Rocky Mts and NW USA and NW Canada
T. townsendii **Townsend pocket gopher.** NW USA
T. umbrinus **Mexican pocket gopher.** Arizona and New Mexico S to C Mexico
Zygogeomys
Z. trichopus **Michoacan pocket gopher.** Michoacan (Mexico)

FAMILY ANOMALURIDAE
Anomalurus
A. beecrofti **Beecroft's scaly-tailed flying squirrel.** Senegal to Uganda, Zaire
A. derbianus **Lord Derby's scaly-tailed flying squirrel.** Sierra Leon to Angola, E to Kenya, S to Mozambique and Zambia
A. peli Sierra Leon to Ghana
A. pusillus S. Cameroun, Gabon, Zaire
Idiurus
I. macrotis **Pygmy scaly-tailed flying squirrel.** Sierra Leon to E Zaire
I. zenkeri S Cameroun to Uganda
Zenkerella
Z. insignis **Flightless scaly-tailed squirrel.** SW Cameroun, Rio Muni, Gabon, Central African Republic

FAMILY HETEROMYIDAE
Dipodomys
D. agilis **Agile kangaroo rat.** SW California, N Baja, California
D. antiquarius **Huey's kangaroo rat.** N Baja California
D. deserti **Desert kangaroo rat.** SW USA
D. elator **Texas kangaroo rat.** Oklahoma, Texas
D. elephantinus **Big-eared kangaroo rat.** SW California
D. gravipes **San Quintin kangaroo rat.** NW Baja California
D. heermanni **Heerman's kangaroo rat.** California
D. ingens **Giant kangaroo rat.** S California
D. insularis **San José Island kangaroo rat.** San José Island, Gulf of California
D. margaritae **Margarita Island kangaroo rat.** Santa Margarita Is, Baja California
D. merriami **Merriam's kangaroo rat.** SW USA to C Mexico
D. microps **Chisel-toothed kangaroo rat.** SW USA
D. nelsoni **Nelson's kangaroo rat,** NC Mexico
D. nitratoides **Fresno kangaroo rat.** C California
D. ordii **Ord's kangaroo rat.** W USA to C Mexico
D. panamintinus **Panamint kangaroo rat.** California
D. paralius **Santa Catarina kangaroo rat.** N Baja California
D. peninsularis **Baja California kangaroo rat.** C Baja California
D. phillipsii **Phillip's kangaroo rat.** C. Mexico
D. spectabilis **Banner-tailed kangaroo rat.** SW USA, N Mexico
D. stephensi **Stephen's kangaroo rat.** S California
D. venustus **Narrow-faced kangaroo rat.** WC California
Heteromys
H. anomalous **South American spiny pocket mouse.** N S America
H. australis **Southern spiny pocket mouse.** S Panama and N S America
H. desmarestianus **Desmarest's spiny pocket mouse.** C Mexico to N S America
H. gaumeri **Gaumer's spiny pocket mouse.** Yucatan peninsula (Mexico)
H. goldmani **Goldman's spiny pocket mouse.** SW Mexico
H. lepturus **Santo Domingo spiny pocket mouse.** N C Mexico
H. longicaudatus **Long-tailed spiny pocket mouse.** NE Mexico
H. nelsoni **Nelson's spiny pocket mouse.** Chiapas (Mexico)
H. nigricaudatus **Goodwin's spiny pocket mouse.** C Mexico
H. oresterus **Mountain spring pocket mouse.** Costa Rica
H. temporalis **Motzorongo spiny pocket mouse.** N C Mexico
Liomys
L. adspersus **Panama spiny pocket mouse.** C Panama
L. irroratus **Mexican spiny pocket mouse.** Mexico
L. pictus **Painted spiny pocket mouse.** Mexico
L. salvini **Salvin's spiny pocket mouse.** S Mexico to Costa Rica
L. spectabilis **Jaliscan spiny pocket mouse.** Jalisco, Mexico
Microdipodops
M. megacephalus **Dark kangaroo mouse.** Nevada
M. pallidus **Pale kangaroo mouse.** Nevada
Perognathus
P. alticola **White-eared pocket mouse.** California
P. amplus **Arizona pocket mouse.** Arizona and NW Mexico
P. anthonyi **Anthony's pocket mouse.** Cerros Is, Baja California
P. arenarius **Little desert pocket mouse.** Baja California
P. artus **Narrow-skulled pocket mouse.** W Mexico
P. baileyi **Bailey's pocket mouse.**SW USA
P. californius **California pocket mouse.** S California to N Baja, California
P. dalquesti **Dalquest's pocket mouse.** S Baja California
P. fallax **San Diego pocket mouse.** S California, N Baja, California
P. fasciatus **Olive-backed pocket mouse.** NC USA and S Canada
P. flavescens **Plains pocket mouse.** C and SW USA
P. flavus **Silky pocket mouse.** C USA to C Mexico
P. formosus **Long-tailed pocket mouse.** SW USA
P. goldmani **Goldman's pocket mouse.** W Mexico
P. hispidus **Hispid pocket mouse.** C USA to C Mexico
P. inornatus **San Joaquin pocket mouse.** C California
P. intermedius **Rock pocket mouse.** SW USA, N Mexico
P. lineatus **Lined pocket mouse.** NE Mexico
P. longimembris **Little pocket mouse,** SW USA
P. nelsoni **Nelson's pocket mouse.** Texas to C Mexico
P. parvus **Great Basin pocket mouse.** W USA
P. penicillatus **Desert pocket mouse.** SW USA to C Mexico
P. pernix **Sinaloan pocket mouse.** W Mexico
P. spinatus **Spiny pocket mouse.** S California, Baja, California
P. xanthonotus **Yellow-eared pocket mouse.** California

FAMILY PEDETIDAE
Pedetes
P. capensis **Springhare or Springhass.** Kenya, Tanzania, Angola, Zimbabwe, Botswana, Namibia, S Africa

SUBORDER MYOMORPHA
FAMILY MURIDAE
SUBFAMILY HESPEROMYINAE
Abrawayaomys
A. ruschii **Brazilian spiny rat.** Brazil, Espirito Santo, Forno Grande, Castelo
Aepeomys
A. fuscatus W, C Andes Colombia
A. lugens Andes from W Venezuela to Ecuador
Akodon **South American Field Mice**
A. aerosus Ecuador, Peru, Bolivia
A. affinis Andes of W Colombia
A. albiventer SE Peru, S to N Chile
A. andinus Mts from Peru, S to N Argentina and Chile
A. azarae S Brazil to NE Argentina
A. boliviensis S Peru through Bolivia to NW Argentina
A. budini Mts of NW Argentina
A. caenosus NW Argentina to S Bolivia
A. cursor SE Brazil, S to N Argentina
A. dolores C Argentina
A. illuteus NW Argentina
A. iniscatus C and S Argentina
A. jelskii S Peru, S to Argentina
A. kempi Islands of the Rio Parana estuary; E Argentina, S Uruguay
A. lanosus S Argentina and Chile
A. llanoi Isla de los Estados, Argentina
A. longipilis Chile and WC Argentina
A. mansoensis Andes of Rio Negro Province, Argentina
A. markhami Isla Wellington, Chile
A. molinae EC Argentina
A. mollis Ecuador, through Peru to Bolivia
A. nigrita E Brazil, through Paraguay to Argentina
A. olivaceus Chile to SW Argentina
A. orophilus Amazonian Peru
A. pacificus Andes of W Bolivia
A. puer W Bolivia to SC Peru
A. reinhardti E Brazil
A. sanborni S Chile and Argentina
A. serrensis SE Brazil to N Argentina

A. surdus Andes of SE Peru
A. urichi Trinidad, Tobago, N Venezuela and Colombia
A. varius Paraguay, Bolivia, W Argentina
A. xanthorhinus S Argentina and Chile
Andalgalomys
A. olrogi Catamarca Province, Argentina
A. pearsoni Paraguay
Andinomys
A. edax **Andean Mouse.** Peru, Bolivia, NW Argentina and N Chile
Anotomys **Fish-eating rats**
A. leander N Ecuador
A. trichotis Mts of Colombia and W Venezuela
Auliscomys
A. boliviensis Mts of W Bolivia, N Chile and S Peru
A. micropus S Argentina, Chile
A. pictus Andes of C Peru to Bolivia
A. sublimis W Bolivia, S Peru to N Argentina, Chile
Baiomys
B. musculus **Southern Pygmy Mouse.** NW Nicaragua to S Mexico
B. taylori **Northern Pygmy Mouse.** Arizona, New Mexico, Texas, S to Veracruz, Mexico
Bibimys **Red-nosed rats**
B. chacoensis N and C Argentina
B. labiosus Minas Gerais, Brazil
B. torresi Delta of Rio Parana, Argentina
Blarinomys
B. breviceps **Brazilian shrew mouse.** CE Brazil
Bolomys
B. amoenus Andes of SE Peru
B. lactens NW Argentina
B. lasiurus E Brazil and Paraguay
B. lenguarum Bolivia, SW Brazil to Argentina
B. obscurus S Uruguay to NC Argentina
B. temchuki Missiones, Argentina
Calomys **Vesper mice**
C. callosus SW Brazil, Bolivia, Paraguay to N Argentina
C. fecundus Bolivia
C. hummelincki Curacao, Aruba, Venezuela
C. laucha SE Brazil, Bolivia, Paraguay, Uruguay to N Argentina
C. lepidus S Peru to NE Chile
C. muriculus E slope of Andes in Bolivia
C. musculinus N Argentina
C. sorellus Peru
Chilomys
C. instans **Colombian forest mouse.** Mts of W Venezuela, Andes of Colombia, S to Ecuador
Chinchilla
C. sahamae **Chinchilla mouse.** S Peru and W Bolivia, S to N Chile and Argentina
Daptomys **Fish-eating Rats**
D. oyapocki French Guiana
D. peruviensis Peru
D. venezuelae Venezuela
Eligmodontia
E. typus **Highland desert mouse.** S Peru to N Chile and Argentina
Euneomys **Patagonian chinchilla mice**
E. chinchilloides S Chile and Argentina
E. fosor NW Argentina
E. mordax WC Argentina to Chile
E. noei Chile
Galenomys
G. garleppi N Chile and adjacent Peru and Bolivia
Graomys
G. domorum E Andes of Bolivia to N Argentina
G. edithae Argentina
G. griseoflavus Argentina, Bolivia, Paraguay
Habromys
H. chinanteco Oaxaca, Mexico
H. lepturus Oaxaca, Mexico
H. lophurus **Crested tailed mouse.** Chiapas, Mexico to El Salvador
H. simulatus **Jico deer mouse.** Veracruz, Mexico
Hodomys
H. alleni **Allen's woodrat.** Sinaloa to Oaxaca, Mexico
Holochilus **Marsh rats**
H. brasiliensis S Brazil through Uruguay to N Argentina
H. chacarius Paraguay to NE Argentina
H. magnus Uruguay to SE Brazil
H. sciureus Colombia, Peru, Venezuela and Guianas, S to Minas Gerais, Brazil
Ichthyomys **Fish-eating Rats**
I. hydrobates Merida, Venezuela
I. pittieri Aragua, Venezuela
I. stolzmanni E Ecuador and Andean Peru
Irenomys
I. tarsalis **Chilean rat.** Argentina and Chile
Isthmomys
I. flavidus W Panama
I. pirrensis E Panama
Juscelimomys
J. candango C Brazil
Kunsia **Giant South American water rats**
K. fronto Argentina to Minas Gerais, Brazil
K. tomentosus E Brazil to Bolivia
Lenoxus
L. apicalis SE Peru to W Bolivia
Megadontomys
M. thomasi C Guerrero to Veracruz, Mexico
Megalomys **Giant rice rats**
M. desmarestii Martinique. Extinct
M. luciae Santa Lucia. Extinct
Microxus
M. bogotensis Colombia and Venezuela
M. latebricola Andes of Ecuador
M. mimus SE Andes of Peru
Neacomys **Bristly mice**
N. guianae Guianas to S Venezuela and N Brazil
N. spinosus SW Brazil to E Ecuador, Colombia and Peru
N. tenuipes E Panama across Colombia to N Venezuela
Nectomys
N. parvipes **South American water rat.** Comte River, French Guiana
N. squamipes Lowland S America to NE Argentina
Nelsonia
N. neotomodon **Diminutive woodrat.** Mts of C Mexico
Neotoma
N. albigula SW USA to C Mexico
N. angustapalata Tamaulipas, Mexico
N. anthonyi Baja California, Mexico
N. bryanti Cedros Island, Baja California
N. bunkeri Coronados Isle, Baja California
N. chrysomelas Matagalpa, Nicaragua
N. cinerea **Bushy-tailed woodrat.** NW Canada to New Mexico
N. floridana **Eastern woodrat.** Texas to Florida, N to Connecticut
N. fuscipes **Dusky-footed woodrat.** Oregon to Baja California
N. goldmani Chihuahua to San Luis Potosi, Mexico
N. lepida **Desert woodrat.** Baja California to SE Idaho
N. martinensis San Martin Isle, Baja California
N. mexicana El Salvador to NC Colorado
N. micropus SW Kansas to N Veracruz
N. nelsoni Perote, Veracruz
N. palatina NC Jalisco, Mexico
N. phenax S Sonora to N Sinaloa, Mexico
N. stephensi NW New Mexico to C Arizona and N to S Utah
N. veria Turner Is, Sonora, Mexico
Neotomodon
N. alstoni **Volcano mouse.** Michoacan to Puebla, Mexico
Neotomys
N. ebriosus **Andean swamp rat.** Peru, S to NW Argentina
Nesoryzomys **Galapagos rice rats**
N. darwini Santa Cruz Is, Galapagos
N. fernandinae Fernandina Is, Galapagos
N. indefessus Santa Cruz Is, Galapagos
N. narboroughi Fernandina Is, the Galapagos
N. swarthi James Is, Galapagos
Neusticomys
N. monticolus **Ecuadorian fish-eating rat.** S Andes of Colombia to N Ecuador
Notiomys
N. angustus W Argentina
N. delfini S Chile, Argentina
N. edwardsii S Argentina
N. marconyx S and W Argentina, E and S Chile
N. megalonyx C Chile
N. valdivianus C Chile and Chiloe Is, W Argentina
Nyctomys
N. sumichrasti **Vesper rat.** Jalisco, S Veracruz to C Panama
Ochrotomys
O. nuttalli **Golden mouse.** E Texas to S Illinois, E to C Virginia and N Florida
Onychomys
O. arenicola New Mexico and W Texas, S to Chihuahua, Mexico
O. leucogaster **Northern grasshopper mouse.** N Tamaulipas, Mexico, W to California and N to Alberta and Saskatchwan
O. torridus **Southern grasshopper mouse.** C California to W Texas, S to Baja California and San Luis Potosi, Mexico
Oryzomys
O. albigularis Costa Rica, S to Peru and E to W Venezuela
O. alfari E Honduras, S to N Ecuador and E to W Venezuela
O. alfaroi S Tamaulipas, Mexico, S to Ecuador
O. altissimus Andes of Ecuador and Peru
O. andinus N Peru, W of the Andes
O. aphrastus Costa Rica
O. arenalis NE Peru
O. argentatus Cudjoe Key, Florida
O. auriventer Peru and Ecuador
O. balneator E and S Ecuador
O. bauri Santa Fe Is, Galapagos
O. bicolor Panama and N S America, S to C Brazil and Bolivia
O. bombycinus E Nicaragua to N Ecuador
O. buccinatus Paraguay to NE Argentina
O. caliginosus **Black mouse.** E Honduras, S to Ecuador
O. capito E Costa Rica, S to NW Argentina
O. caudatus NC Oaxaca, Mexico
O. chacoensis The chaco of Paraguay, Bolivia, N Argentina
O. chaparensis C Bolivia
O. concolor Costa Rica, S to N Peru, N Brazil
O. couesi Hidalgo, Mexico to C Panama
O. delicatus Trinidad and Tobago
O. delticola Uruguay, NE Argentina
O. dimidiatus Nicaragua
O. flavescens S Brazil to N Argentina
O. fornesi N Argentina, Paraguay to S Brazil
O. fulgens Vally of Mexico
O. fulvescens Tamaulipas, Mexico, S to W Venezuela
O. galapagoensis San Cristobal Is, Galapagos
O. gorgasi NW Colombia
O. hammondi Andean Ecuador
O. intectus C Colombia
O. kelloggi SE Brazil
O. lamia SE Brazil
O. longicaudatus Chile, parts of Argentina
O. macconnelli Surinam, Guyana, S Venezuela, S Colombia, E Ecuador, E Peru
O. melanostoma E Peru
O. melanotis Oaxaca, Mexico, S to Honduras
O. minutus N S America
O. munchiquensis W Colombia
O. nelsoni Maria Madre Is, Nayarit, Mexico
O. nigripes E Brazil through Paraguay to Argentina
O. nitidus Ecuador, S to NW Argentina, E to S Brazil
O. palustris **Marsh rice rat.** S New Jersey, S to the gulf Coast of Texas, N to SE Kansas
O. peninsulae S Baja California, Mexico
O. polius N Peru
O. ratticeps S Brazil and Paraguay
O. rivularis NW Ecuador
O. robustulus E. Ecuador
O. spodiurus W Ecuador
O. subflavus E Brazil, N through the Guyanas
O. utiaritensis C Brazil
O. victus St Vincent Isle, the Antilles
O. villosus N Colombia
O. xantheolus W Peru
O. yunganus Bolivia, Peru
O. zunigae W C Peru
Osgoodomys
O. bandaranus Nayarit to Guerrero, Mexico
Ototonyctomys
O. hatti **Yucatan vesper rat.** Yucatan Peninsula, Mexico, S to N Guatemala
Ototylomys
O. phyllotis **Big-eared climbing rat.** Yucatan Peninsula to Costa Rica
Oxymycterus **Burrowing mice**
O. akodontius NW Argentina
O. angularis E Brazil
O. elator Paraguay
O. hispidus NE Argentina to E Brazil
O. iheringi C Brazil to N Argentina
O. inca W Bolivia to SE Peru
O. paramensis NW Argentina to Bolivia and E Peru
O. roberti E Brazil
O. rutilans S Brazil, Paraguay, Uruguay, Argentina
Peromyscus **Deer mice**
P. atwateri SE Kansas to C Texas
P. aztecus C Mexico to Honduras
P. boylii California to W Oklahoma, S to Honduras
P. bullatus Veracruz, Mexico
P. californicus **California mouse.** C California, S to Baja California
P. caniceps S Baja California
P. crinitus **Canyon mouse.** C Oregon to W Colorado, S to Baja California, Sonora, Mexico
P. dickeyi S Baja California, Tortuga Is
P. difficilus Colorado to Oaxaca, Mexico
P. eremicus **Cactus mouse.** S Nevada, E to W Texas, S to San Luis Potosi, Mexico
P. eva Carmen Is, Baja California
P. furvus C Veracruz to NW Oaxaca, Mexico
P. gossypinus S Illinois, E to the Atlantic, S to the Gulf of Mexico and S Florida USA
P. grandis Guatemala
P. guardia Guarda Is, Baja, California
P. guatemalensis Chiapas Mexico to SW Guatemala
P. gymnotis Pacific coast of S Guatemala, Chiapas
P. hooperi Coahuila, Mexico
P. interparietalis San Lorenzo Is, Salsipuedes Is, Gulf of California, Mexico
P. leucopus **White-footed mouse.** Alberta, Nova Scotia, S to Virginia in the E, Arizona in the W
P. madrensis Tres Marias Is, Mexico
P. maniculatus **Deer mouse.** Alaska and N Canada, S to Oaxaca, Mexico
P. mayansis Guatemala
P. megalops Guerrero and Oaxaca, Mexico
P. mekisturus Southern Puebla, Mexico
P. melanocarpus N C Oaxaca, Mexico
P. melanophrys S Durango, S Coahuila, S to Chiapas
P. melanotis SE Arizona, S to Veracruz, Mexico
P. melanurus Oaxaca, Mexico
P. merriami S Arizona to N Sinoloa, Mexico
P. mexicanus San Luis Potosi, S to W Panama
P. ochraventer S Tamaulipas, San Luis Potosi
P. pectoralis S New Mexico, C Texas S to Jalisco, Mexico
P. pembertoni Sonora, Mexico
P. perfulvus C Mexico
P. polionotus SE USA
P. polius Chihuahua, Mexico
P. pseudocrinitus Coronados Is, Baja California
P. sejugis Santa Cruz Is, Gulf of California
P. simulus CW Mexico
P. sitkensis Queen Charlotte Is, Canada
P. slevini Santa Catalina Is, Baja California
P. spicilegus WC Mexico
P. stephani San Estaban Is, Sonora, Mexico
P. stirtoni S Guatemala to Honduras
P. truei **Pinyon mouse.** SW Oregon, S to NC Texas and Oaxaca Mexico
P. winkelmanni Michoacan, Mexico
P. yucatanicus **Yucatan deer mouse.** Yucatan Peninsula, Mexico
P. zarhynchus C Chiapas, Mexico
Phaenomys
P. ferrugineus Eastern Brazil
Phyllotis **Leaf-eared mice**
P. amicus NW Peru
P. andium S Ecuador to N Peru
P. bonaeriensis Buenos Aires Province, Argentina
P. caprinus S Bolivia to N Argentina
P. darwini S Peru to C Chile, W Argentina
P. definitus Peru
P. gerbillus NW Peru
P. haggardi Ecuador
P. magister S Peru to N Chile
P. osgoodi C Chile
P. osilae SE Peru to Bolivia and N Argentina
P. wolffsohni E slopes of Andes in C Bolivia

Podomys
P. floridanus Florida
Podoxymys
P. roraimae Guyana, Venezuela, NC Brazil
***Pseudoryzomys* False rice rats**
P. simplex E Brazil
P. wavrini Bolivia, E to Paraguay, N Argentina
Punomys
P. lemminus **Puna mouse.** S Peru
Reithrodon
R. physodes **Rabbit rat.** Argentina, Chile, Uruguay
***Reithrodontomys* Harvest mice**
R. brevirostris **Short-nosed harvest mouse.** Nicaragua to C Costa Rica
R. burti **Burt's harvest mouse.** C Sinoloa to C Sonora, Mexico
R. chrysopsis **Volcano harvest mouse.** Jalisco to C Veracruz
R. creper **Chiriqui harvest mouse.** Costa Rica to W Panama
R. darienensis **Darien harvest mouse.** Panama to NW Colombia
R. fulvescens **Fulvous harvest mouse.** From Arizona and Missouri S to W Nicaragua
R. gracilis **Slender harvest mouse.** Chiapas and Yucatan, S to N Costa Rica
R. hirsutus **Hairy harvest mouse.** Nayarit and Jalisco, Mexico
R. humulis **Eastern harvest mouse.** SE USA
R. megalotis **Western harvest mouse.** W USA and Canada, S to NW Mexico
R. mexicanus **Mexican harvest mouse.** C Mexico, S to N Ecuador
R. microdon **Small-toothed harvest mouse.** S Mexico to Guatemala
R. montanus **Plains harvest mouse.** C S Dakota, S to N Mexico
R. paradoxus Nicaragua to Costa Rica
R. raviventris **Salt marsh harvest mouse.** Shores of San Francisco Bay, California
R. rodriguezi Costa Rica
R. spectabilis Cozumel Is, Quintana Roo, Mexico
R. sumichrasti Gulf coast of Mexico, S to W Panama
R. tenuirostris **Narrow-nosed harvest mouse.** S Guatemala
Rhagomys
R. rufescens E Brazil
***Rheomys* Water mice**
R. hartmanni Costa Rica to W Panama
R. mexicanus Oaxaca, Mexico
R. raptor Panama
R. thomasi S Mexico to El Salvador
R. underwoodi Costa Rica to W Panama
***Rhipidomys* South American climbing rats**
R. latimanus Venezuela, Colombia, Ecuador
R. leucodactylus S Venezuela, S to NW Argentina
R. macconnelli SE Venezuela through Amazonian Brazil
R. maculipes Bahia, Brazil
R. mastacalis Venezuela and the Guyanas, S to C Brazil
R. scandens E Panama
R. sclateri S Venezuela to Guyana, Brazil
Scapteromys
S. tumidus **Argentinean water rat.** SE Brazil through Paraguay to E Argentina
Scolomys
S. melanops **Ecuadorean spiny mouse.** Ecuador
***Scotinomys* Brown mice**
S. teguina Oaxaca Mexico to W Panama
S. xerampelinus C Costa Rica to W Panama
***Sigmodon* Cotton rats**
S. alleni **Brown cotton rat.** C Mexico
S. alstoni **Groove-toothed cotton mouse.** NE Venezuela, the Guyanas to NE Brazil
S. arizoni Nevada S to Nayarit, Mexico
S. fulviventer New Mexico S to Michoacan, Mexico
S. hispidus **Hispid cotton rat.** S USA, S to Peru and N Venezuela
S. leucotis **White-eared cotton rat.** C Mexico
S. mascotensis Jalisco to Oaxaca, Mexico
S. ochrognathus **Yellow-nosed cotton rat.** Arizona to W Texas, S to Durango, Mexico
***Thomasomys* Paramo rats**
T. aureus Andes of Venezuela, W Colombia, S to Peru
T. baeops W Ecuador
T. bombycinus W Colombia
T. cinerieventer Colombia and Ecuador
T. cinereus SW Ecuador to NW Peru
T. dorsalis SE Brazil to NE Argentina
T. daphne SW Peru to C Bolivia
T. gracilis Ecuador to SE Peru
T. hylophilus N Colombia to W Venezuela
T. incanus Andean Peru
T. ischyurus NW Peru to W Ecuador
T. kalinowskii C Peru
T. ladewi NW Bolivia
T. laniger W Venezuela to C Colombia
T. monochromos NE Colombia
T. notatus SE Peru
T. oenax S Brazil to Uruguay
T. oreas Andean Bolivia
T. paramorum Ecuador
T. pictipes NE Argentina
T. pyrrhonotus S Ecuador to NW Peru
T. rhoadsi Ecuador
T. rosalinda NW Peru
T. taczanowskii NW Peru
T. vestitus W Venezuela
***Tylomys* Central American climbing rats**
T. bullaris Chiapas, Mexico
T. fulviventer E Panama
T. mirae C Colombia to N Ecuador
T. nudicaudus Nicaragua to Veracruz, Mexico
T. panamensis E Panama
T. tumbalensis Chiapas, Mexico
T. watsoni Costa Rica to Panama
Wiedomys
W. pyrrhorhinos **Red-nosed mouse.** E Brazil
Xenomys
X. nelsoni **Magdalena rat.** C W Mexico
***Zygodontomys* Cane mice**
Z. borreroi NC Colombia
Z. brevicauda Costa Rica, N S America
Z. reigi Guyana

SUBFAMILY MICROTINAE
Alticola
A. macrotis **Large-eared or High mountain vole.** Altai, Sayan Mts
A. roylei **Royle's mountain vole.** W Himalayas to Altai
A. stoliczkanus **Stoliczka's mountain vole.** Himalayas to Altai
A. strelzowi **Flat-headed vole.** Altai, E Kazakhstan
Arvicola
A. richardsoni **American water vole.** NW USA, SW Canada
A. sapidus **Southwestern water vole.** Iberia, SW France
A. terrestris **European water or Ground vole.** Europe to E Siberia
Clethrionomys
C. andersoni **Japanese red-backed vole.** Honshu, Japan
C. gapperi **Gapper's or Southern red-backed vole.** Canada (except NW), N USA, Rocky Mts, Appalachians
C. glareolus **Bank vole.** W and C Europe, S to N Spain, N Italy, W to Lake Baikal
C. occidentalis **Western red-backed vole.** Washington to N California
C. rex Hokkaido
C. rufocanus **Gray red-backed vole.** Scandinavia, Siberia, NE China, Hokkaido
C. rutilus **Northern or Ruddy red-backed vole.** N Eurasia, Arctic America
Dicrostonyx
D. hudsonius **Labrador collared lemming.** Labrador, Quebec
D. torquatus **Collared or Arctic lemming.** Tundra of Siberia, Alaska, Greenland, N America
Dinaromys
D. bogdanovi **Martino's snow vole.** Yugoslavia
Ellobius
E. fuscocapillus **Southern or Afghan mole-vole.** SW Turkestan to Baluchistan
E. talpinus **Northern mole-vole.** Ukraine to Sinkiang
Eothenomys
E. chinensis S China
E. custos S China
E. eva W China
E. inez C China
E. lemminus NE Siberia
E. melanogaster **Père David's vole.** S China, Taiwan
E. olitor Yunnan, S China
E. proditor S China
E. regulus Korea, E Manchuria
E. shanseius Shansi, China
E. smithi Japan, except Hokkaido
Hyperacrius
H. fertilis **True's vole.** Kashmir, N Punjab
H. wynnei **Murree vole.** N Pakistan
Lagurus
L. curtatus **Sagebrush vole.** W USA, SW Canada
L. lagurus **Steppe lemming.** Ukraine to Mongolia, Sinkiang
L. luteus **Yellow steppe lemming.** S Mongolia, N Sinkiang
Lemmus
L. amurensis **Amur lemming.** Verkhoyansk Mts to Upper Amur
L. lemmus **Norway lemming.** Mts of Scandinavia and tundra from Lappland to White Sea
L. nigripes **Black-footed lemming.** Pribilof Is
L. sibiricus **Siberian or Brown lemming.** Holoarctic except Scandinavia
Microtus
M. abbreviatus **Insular vole.** St Matthew Is, Hall Is, Bering Sea
M. agrestis **Field or Short-tailed vole.** Europe to R Lena, E Siberia, NW Iberia
M. arvalis **Common vole.** Europe, Orkneys, Guernsey, USSR to upper Yenesi and S to Caucasus
M. bedfordi **Duke of Bedford's vole.** Kansu, China
M. brandti **Brandt's vole.** Mongolia, Transbaikalia
M. cabrerae **Cabrera's vole.** Iberia
M. californicus **California vole.** California and SW Oregon
M. chrotorrhinus **Rock or Yellow nose vole.** E Canada, NE USA
M. clarkei **Clarke's vole.** Yunnan, N Burma
M. coronarius **Coronation Island vole.** Coronation Is, Alaska
M. fortis **Reed vole.** China, SE Siberia
M. fulviventer **Oaxacan vole.** Oaxaca, Mexico
M. gregalis **Narrow-headed or Narrow-skulled vole.** Siberia, C Asia
M. guatemalensis **Guatemalan vole.** Guatemala
M. gud NE Asia Minor, Caucasus
M. guentheri **Gunther's vole.** SE Europe to Israel, Libya
M. irani **Persian vole.** Iran
M. kikuchii **Taiwan vole.** Taiwan
M. kirgisorum **Tien Shan vole.** Tien Shan Mts
M. longicaudus **Long-tailed vole.** SW USA to Alaska
M. ludovicianus **Louisiana vole.** Louisiana, E Texas
M. mandarinus **Mandarin vole.** C China to SE Siberia
M. maximowiczi Upper Amur, E Siberia
M. mexicanus **Mexican vole.** Mexico, SW USA
M. middendorffi **Middendorff's vole.** N Siberia
M. millicens **Szechwan vole.** Szechwan, China
M. miurus **Singing or Alaskan vole.** Alaska, Yukon
M. mongolicus **Mongolian vole.** NE Mongolia
M. montanus **Montane vole.** W USA, British Columbia
M. montebelli **Japanese grass vole.** Japan
M. mujanensis Vitim Basin, E Siberia
M. nivalis **Snow vole.** SW Europe to Iran
M. ochrogaster **Prairie vole.** C USA
M. oeconomus **Tundra or Root vole.** N Eurasia, Alaska, NW Canada
M. oregoni **Creeping or Oregon vole.** NW coast USA
M. pennsylvanicus **Meadow vole or Field mouse.** Canada, Alaska, C and E USA
M. roberti **Robert's vole.** NE Asia Minor, W Caucasus
M. sachalinensis **Sakhalin vole.** Sakhalin Is
M. socialis **Social vole.** Kazakhstan to Palestine
M. subarvalis E Europe, Russia
M. townsendii **Townsend's vole.** W coast USA, Vancouver Is
M. transcaspicus **Transcaspian vole.** SW Turkestan
M. umbrosus **Zempoaltepec or Tarabundi vole.** Oaxaca, Mexico
M. xanthognathus **Taiga or Yellow-cheeked vole.** NW Canada and E and C Alaska
Myopus
M. schisticolor **Wood lemming.** Scandinavia, Siberia, W Mongolia
Neofiber
N. alleni **Florida water rat (Round-tailed muskrat).** Florida
Ondatra
O. zibethicus **Muskrat.** USA except S, Canada except Arctic, introduced to C and N Europe, USSR
Phenacomys
P. albipes **White-footed vole.** W Oregon, NW California
P. intermedius **Heather vole.** Canada, W USA
P. longicaudus **Red-tree vole.** Coastal Oregon, NW California
P. silvicola **Dusky tree vole.** W Oregon
Pitymys
P. afghanus **Afghan vole.** Afghanistan, S Turkestan
P. bavaricus **Bavarian pine vole.** S Germany
P. duodecimcostatus **Mediterranean pine vole.** S, E Iberia to SE France
P. juldaschi **Juniper vole.** Tien Shan, Pamirs
P. leucurus **Blyth's vole.** Tibetan Plateau, Himalayas
P. liechtensteini **Liechtenstein's pine vole.** NW Yugoslavia
P. lusitanicus **Lusitanian pine vole.** NW Iberia, SW France
P. majori **Major's or Asia Minor pine vole.** Caucasus, Asia Minor
P. multiplex **Alpine pine vole.** European Alps
P. pinetorum **American pine or Woodland vole.** S, E USA
P. quasiater **Jalapan pine vole.** EC Mexico
P. savii **Savi's pine vole.** S Europe
P. schelkovnikovi **Schelkovnikov's pine vole.** Elburz, Talysh Mts, S of Caspian Sea
P. sikimensis **Sikkim vole.** Himalayas to W China
P. subterraneus **European pine vole.** France to C Russia
P. tatricus **Tatra pine vole.** Tatra Mts
P. thomasi **Thomas' pine vole.** S Balkans
Prometheomys
P. schaposchnikowi **Long-clawed mole-vole.** Caucasus, NE Asia Minor
Synaptomys
S. borealis **Northern bog lemming.** Canada, Alaska
S. cooperi **Southern bog lemming.** NE USA, SE Canada

SUBFAMILY MURINAE
***Acomys* Spiny mice**
A. cahirinus **Cairo spiny mouse.** Circum-Sahara, Ethiopia, Kenya, Israel to Pakistan
A. cilicicus Asia Minor
A. minous Crete
A. russatus **Golden spiny mouse.** NE Egypt to Jordan, E Arabia
A. spinosissimus Zaire, Tanzania to Mozambique and Botswana
A. subspinosus Sudan to South Africa
A. wilsoni Sudan, Ethiopia, Somalia, Kenya, Uganda
Aethomys
A. bocagei Angola, Zaire
A. chrysophilus **Red veld rat.** SE Kenya, Angola to S Africa
A. hindei **Hind's bush rat.** N Zaire, Uganda, Kenya, Sudan, N Tanzania, N Nigeria
A. kaiseri **Kaiser's rat.** E Zaire to Kenya to Malawi, Angola
A. namaquensis **Namaqua rock rat.** S Africa to Zambia, Mozambique and Angola
A. nyikae **Nyika rat.** NE Zambia, Malawi, Angola, Zaire, Zimbabwe
A. selindensis **Selinda rat.** Mt Selinda, Zimbabwe
A. thomasi W C Angola
Anisomys
A. imitator **New Guinea giant rat.** New Guinea
Anonymomys
A. mindorensis Philippines
Apodemus
A. agrarius **Striped field mouse.** C Europe to China
A. argenteus **Small Japanese field mouse.** Japan
A. draco S China to Burma and Assam
A. flavicollis **Yellow-necked mouse.** Europe to the Urals, Asia Minor
A. gurka **Himalayan field mouse.** Nepal
A. krkensis **Krk mouse.** Krk Is, Yugoslavia
A. latronum **Big-eared wood mouse.** SW China, N Burma
A. microps **Pygmy field mouse.** E Europe, Asia Minor
A. mystacinus **Broad-toothed mouse.** SE Europe, Asia Minor, Iraq, Iran
A. peninsulae **Korean field mouse.** Manchuria, Korea, Kansu, Shansi, Hokkaido
A. semotus **Formosan field mouse.** Taiwan
A. speciosus **Large Japanese field mouse.** Japan, Kunashir Is (USSR)
A. sylvaticus **Wood mouse.** Europe except N Scandinavia, E to the Altai and Himalayas and S into Arabia and N Africa
Apomys
A. abrae Luzon, Philippines
A. datae Philippines
A. hylocoetes Philippines
A. insignis Philippines
A. littoralis Mindanao, Philippines
A. microdon Philippines
A. musculus Luzon, Philippines
A. petraeus Mindanao, Philippines
***Arvicanthis* African grass rats**
A. abyssinicus W Africa to Ethiopia to Zambia
A. blicki Ethiopia

A. niloticus Egypt
A. somalicus Somalia, Ethiopia, Kenya
A. testicularis W Africa to Ethiopia
Bandicota
B. bengalensis **Lesser bandicoot rat.** Sri Lanka, S India to Burma, Sumatra, Java
B. indica **Greater bandicoot rat.** India to S China to Java, Sri Lanka, Taiwan
B. savilei Burma to Vietnam
Batomys
B. dentatus **Luzon forest rat.** Luxon, Philippines
B. granti **Luzon forest rat.** Luzon, Philippines
B. salomonseni **Mindanao rat.** Mindanao, Philippines
Berylmys
B. berdmorei **Small white-toothed rat.** S Burma, Thailand, Vietnam
B. bowersi **Bower's rat.** NW India to S China to Malaya
B. mackenziei **Kenneth's white-toothed rat.** Assam to Thailand
B. manipulus **Manipur rat.** Assam to Burma, Malaya
Bullimus
B. bagopus Mindanao, Philippines
B. luzonicus Luzon, Philippines
B. rabori Mindanao, Philippines
Bunomys
B. andrewsi Sulawesi
B. chrysocomus Sulawesi
B. fratrorum Sulawesi
B. penitus Sulawesi
Carpomys
C. melanurus **Luzon rat.** Luzon, Philippines
C. phaeurus **Luzon rat.** Luzon, Philippines
Chiromyscus
C. chiropus **Fea's tree rat.** Vietnam, Laos, E Burma, Thailand
Chiropodomys
C. calamianensis Palawan, Philippines
C. gliroides **Pencil-tailed tree mouse.** NE India, Burma and S China, S to Sumatra, Java, Bali, Borneo
C. karlkoopmani N Pagai Is, Indonesia
C. major Borneo
C. muroides Borneo
Chiruromys
C. forbesi **Greater tree mouse.** SE Papua, NE New Guinea and adjacent Is
C. lamia **Broad-skulled tree mouse.** SE New Guinea
C. vates **Lesser tree mouse.** SE New Guinea
Conilurus
C. albipes **White-footed tree rat.** E S Australia
C. penicillatus **Brush-tailed tree rat.** N NW Australia, New Guinea
Crateromys
C. paulus Mindoro, Philippines
C. schadenbergi **Bushy-tailed cloud rat.** N Luzon, Philippines
Cremnomys
C. blanfordi Sri Lanka, India
C. cutchicus **Cutch rat.** India
C. elvira SE India
Crunomys
C. fallax Luzon, Philippines
C. melanius Mindanao, Philippines
Dacnomys
D. millardi **Millard's rat.** Nepal, Bengal, Assam, Laos
Dasymys
D. incomtus **African marsh rat.** Sierra Leone to Sudan and Kenya, S to South Africa and Namibia
Diomys
D. crump **Crump's mouse.** NE India
Diplothrix
D. legatus Ryukyu Is, Japan
Echiothrix
E. leucura **Sulawesi spiny rat.** Sulawesi
Eropeplus
E. canus **Sulawesi soft-furred rat.** Sulawesi
Golunda
G. ellioti **Indian bush rat.** India, Sri Lanka, Nepal, Bhutan, Pakistan
Hadromys
H. humei **Manipur bush rat.** Assam, India
Haeromys
H. margarettae **Ranee mouse.** Borneo
H. minahassae N Sulawesi
H. pusillus **Lesser ranee mouse.** Borneo
Hapalomys
H. delacouri **Marmoset rat.** Laos, Vietnam, Hainan, Kwangsi (China)
H. longicaudatus **Marmoset rat.** Thailand, Burma, Malaya
Hybomys
H. trivirgatus **Temmincks striped mouse.** Guinea to W Nigeria
H. univittatus **Peter's striped mouse.** Guinea to Uganda
Hyomys
H. goliath **Rough-tailed giant rat.** New Guinea
Komodomys
K. rintjanus Lesser Sunda Is
Leggadina
L. forresti **Forrest's mouse.** Arid zones of Australia
L. lakedownensis E Queensland
Lemniscomys
L. barbarus **Striped grass mouse.** Senegal to Tanzania, Morocco, Algeria, Tunisia, Sudan
L. bellieri Ivory Coast
L. griselda **Single-striped mouse.** Angola
L. linulus Ivory Coast, Senegal
L. macculus Ethiopia, S Sudan, Zaire to Tanzania
L. rosalia Transvaal to Namibia, Zimbabwe to Kenya
L. roseveri Zambia
L. striatus **Punctated grass mouse.** Sierra Leone to Tanzania and N Angola
Lenomys
L. meyeri **Sulawesi giant rat.** Sulawesi
Lenothrix
L. canus **Gray tree rat.** Malaya, Borneo
Leopoldamys
L. edwardsi **Edwards rat.** W Himalayas to S China, S to Malaya and Sumatra
L. neilli **Neill's rat.** Thailand
L. sabanus **Long-tailed giant rat.** S China to Java, Borneo, Sulawesi
L. siporanus Mentawi Is, Indonesia
Leporillus
L. apicalis **White-tipped stick-nest rat.** C Australia
L. conditor **Stick-nest rat.** Franklin Is, S Australia
Limnomys
L. sibuanus Mindanao, Philippines
Lophuromys **Harsh-furred rats**
L. cinereus E Zaire
L. flavopunctatus **Speckled harsh-furred rat.** Ethiopia to S Zaire, Angola, Zambia, Malawi and Mozambique
L. luteogaster NE Zaire
L. medicaudatus EC Zaire
L. melanonyx Ethiopia
L. nudicaudus Cameroun, Equatorial Guinea
L. rahmi E Zaire
L. sikapusi **Rusty-bellied rat.** Sierra Leone to S Sudan, S to Zaire, N Angola and N Tanzania
L. woosnami E Zaire, W Uganda
Lorentzimys
L. nouhuysi **New Guinea jumping mouse.** New Guinea
Macruromys
M. elegans **Western small-toothed rat.** W New Guinea
M. major **Eastern small-toothed rat.** Mts of E New Guinea
Malacomys **African swamp rats**
M. edwardsi **Edward's long-footed rat.** Sierra Leone to Nigeria
M. longipes **Long-footed rat.** Ivory Coast to Uganda, Zambia
M. verschureni **Verschuren's long-footed rat.** E Zaire
Mallomys
M. rothschildi **Smooth-tailed giant rat.** New Guinea
Margaretamys
M. beccarii Sulawesi
M. elegans Sulawesi
M. parvus Sulawesi
Mastacomys
M. fuscus **Broad-toothed rat.** SE Australia, Tasmania
Maxomys
M. alticola **Mountain spiny rat.** Borneo
M. baeodon **Small spiny rat.** Borneo
M. bartelsi Java
M. dollmani Sulawesi
M. hellwaldi Sulawesi
M. hylomyoides Sumatra
M. inas **Malayan mountain spiny rat.** Mts of Malaya
M. panglima Palawan
M. rajah **Brown spiny rat.** Thailand, Vietnam, Malaya to Java, Borneo
M. surifer **Red spiny rat.** S Burma, Vietnam to Java, Borneo
M. inflatus Sumatra
M. moi Vietnam
M. musschenbroeki Sulawesi
M. orchraceiventer **Chestnut-bellied spiny rat.** Borneo, Sulawesi
M. pagensis Mentawai Is, Indonesia
M. whiteheadi **Whitehead's rat.** Thailand to Sumatra and Borneo
Melasmothrix
M. naso **Lesser shrew-rat.** Sulawesi
Melomys
M. aerosus Ceram
M. albidens **White-toothed Melomys.** W C New Guinea
M. arcium **Rossel Island Melomys.** Rossel Is, New Guinea
M. burtoni S New Guinea
M. capensis NE Queensland
M. cervinipes **Fawn-footed melomys.** E NE Australia
M. fellowsi **Red-bellied melomys.** NE New Guinea
M. fraterculus Ceram
M. fulgens Ceram
M. leucogaster **White-bellied melomys.** New Guinea
M. levipes **Long-nosed melomys.** New Guinea
M. lorentzi **Long-footed melomys.** New Guinea
M. lutillus **Little melomys.** SE New Guinea, N Queensland
M. moncktoni **Southern melomys.** New Guinea
M. obiensis Obi Is, Halmahera
M. platyops **Lowland melomys.** New Guinea
M. porculus Guadalcanal, Solomon Is
M. rubex **Highland melomys.** New Guinea
M. rubicola **Bramble Cay melomys.** Bramble Cay, E New Guinea
M. rufescens **Rufescent melomys.** New Guinea, Solomon Is
Mesembriomys
M. gouldi **Black-footed tree rat.** N Australia
M. macrurus **Golden-backed tree rat.** NW N Australia
Micromys
M. minutus **Harvest mouse.** Palearctic region from Britain to Japan
Millardia
M. gleadowi **Sand-colored rat.** Pakistan, NW India, Afghanistan
M. kathleenae Burma
M. kondana India
M. meltada **Soft-furred field rat.** India, Sri Lanka, Pakistan, Nepal
Muriculus
M. imberbis Ethiopia
Mus
M. abbottii Macedonia, Turkey, N Iran
M. baoulei Ivory Coast
M. booduga **Indian field mouse.** India, Sri Lanka, Burma
M. bufo E Zaire, Uganda
M. callewaerti **Callewaert's mouse.** S Zaire, Angola
M. caroli **Ryukyu mouse.** Ryukyu Is, Taiwan, China, Vietnam to Java, Flores
M. castaneus S and E Asia (cities)
M. cookii **Cook's mouse.** Assam to Vietnam
M. crociduroides Sumatra
M. domesticus Europe to Himalayas
M. dunni India
M. famulus S India
M. fernandoni Sri Lanka
M. fulvidiventris Sri Lanka
M. goundae N Central African Republic
M. gratus W Uganda
M. haussa Senegal to Nigeria
M. hortulanus Austria, Yugoslavia, trans-Caucasia, USSR
M. industus S Africa, Botswana
M. lepidoides Burma
M. mahomet Ethiopia, Somalia
M. mattheyi Senegal to Ghana
M. mayori Sri Lanka
M. minutoides **Pygmy mouse.** Africa S of the Sahara
M. musculus **House mouse.** Worldwide
M. oubanguii Central African Republic
M. pahari **Gairdner's shrew-mouse.** Sikkim to Vietnam
M. phillipsi India
M. platythrix **Flat-haired jungle mouse.** India
M. poschiavinus Switzerland
M. procondon Ethiopia, Somalia, NE Zaire
M. saxicola India, Pakistan, Nepal
M. setulosus Guinea to Gabon, Central African Republic, Ethiopia
M. setzeri Namibia, Botswana, Zambia
M. shortridgei **Shortridge's mouse.** Burma to Vietnam
M. sorella Tanzania, Zambia, Uganda, Kenya, Malawi
M. spretus SW Europe, NW Africa
M. tenellus Sudan, Somalia, Tanzania, Ethiopia
M. terricolor Nepal to Pakistan
M. triton **Larger pygmy mouse.** Zaire to Kenya, Malawi, Mozambique
M. vulcani Java
Mylomys
M. dybowskii **Three-toed grass rat.** Ivory coast to Kenya, Zaire
Nesokia
N. indica **Short-tailed bandicoot-rat.** Egypt to NW India, Sinkiang, S USSR
Niviventer
N. andersoni SW China
N. brahma Assam, N Burma
N. bukit **Malayan white-bellied rat.** S China to Bali
N. confucianus **Chinese white-bellied rat.** S Manchuria to Vietnam, Taiwan
N. coxingi Taiwan
N. cremoriventer **Dark-tailed tree rat.** Malaya to Bali, Borneo
N. eha **Smoke-bellied rat.** Nepal to SW China
N. excelsior SW China
N. fulvescens **Chestnut rat.** Nepal, Burma, Fukien, Kansu, Yunnan, S to Malaya
N. hinpoon Thailand
N. langbianus Assam to Vietnam
N. lepturus Java
N. niviventer **White-bellied rat.** Nepal
N. rapit **Long-tailed mountain rat.** Malaya, Sumatra, Borneo
N. tenaster Assam, Burma, Vietnam
Notomys
N. alexis **Spinifex hopping mouse.** C Australia
N. amplus **Short-tailed hopping mouse.** C Australia
N. aquilo **Northern hopping mouse.** N Queensland
N. cervinus **Fawn-colored hopping mouse.** Queensland, Northern territory, S Australia
N. fuscus **Dusky hopping mouse.** C Australia
N. longicaudatus **Long-tailed hopping mouse.** SW C Australia
N. macrotis **Big-eared hopping mouse.** SW Australia
N. mitchelli **Mitchell's hopping mouse.** S Australia
Oenomys
O. hypoxanthus **Rufous-nosed rat.** Sierra Leone to Ghana, S Nigeria to Sudan and Kenya, Ethiopia, Zaire, Angola
Papagomys
P. armandvillei **Flores giant rat.** Flores, Lesser Sunda Is
Paruromys
P. dominator Sulawesi
Pelomys
P. campanae **Huet's groove-toothed swamp rat.** Zaire, W Angola
P. fallax **Groove-toothed swamp rat.** Kenya–Botswana, Angola
P. harringtoni **Harrington's groove-toothed swamp rat.** C. Ethiopia
P. hopkinsi **Hopkin's groove-toothed swamp rat.** C Ethiopia
P. isseli **African creek rat.** Is of Kome, Bugala, Bunyama, Lake Victoria
P. minor **Lesser creek rat.** Zaire, Angola, Zambia
Phloeomys
P. cumingi **Cuming's slender-tailed cloud rat.** NW Luzon, Philippines
P. pallidus Luzon, Philippines
Pithecheir
P. melanurus Sumatra, Java
P. parvus **Malayan tree rat.** Malaysia
Pogonomelomys
P. bruijni **Lowland brush mouse.** W S New Guinea
P. mayeri **Shaw Mayer's brush mouse.** W New Guinea
P. ruemmleri **Rümmler's brush mouse.** W New Guinea
P. sevia **Highland brush mouse.** NE New Guinea
Pogonomys
P. loriae New Guinea
P. macrourus **Long-tailed tree mouse.** New Guinea
P. sylvestris **Gray-bellied tree mouse.** New Guinea
Praomys **African soft-furred rats**
P. albipes Ethiopia
P. alleni **Climbing wood-mouse.** Guinea to E Zaire, Uganda and Kenya
P. angolensis **Angola rat.** Angola, Zaire
P. baeri **Baer's wood mouse.** Ivory Coast
P. carillus C Africa

P. daltoni **Dalton's mouse.** Senegal to Cameroun, Sudan
P. delectorum **Mlanje rat.** Kenya to Malawi
P. denniae C E Africa
P. derooi Ghana to W Nigeria
P. erythroleucus Morocco, Senegal to Sudan
P. fumatus **African meadow rat.** E Africa, SW Arabia
P. fumosus C Africa
P. hartwigi Cameroun
P. huberti Senegal to Zaire
P. jacksoni **African forest rat.** Cameroun E to S Sudan, Uganda, Kenya and Tanzania, Zambia
P. morio **Cameroun soft-furred rat.** Cameroun, Central African Republic
P. natalensis **Multimammate rat.** Africa S of the Sahara
P. parvus Gabon, Zaire
P. pernanus S Kenya, N Tanzania
P. shortridgei **Shortridge's mouse.** Namibia, Botswana
P. tullbergi **Tullberg's soft-furred rat.** Guinea to Gabon
P. verreauxi **Verreaux's rat.** Cape Province
Pseudomys
P. albocinereus **Ash-gray mouse.** SW W Australia
P. apodemoides S Australia
P. australis **Eastern mouse.** E S Australia
P. chapmani W Australia
P. delicatulus **Little native mouse.** NE N NW Australia
P. desertor **Brown desert mouse.** WC Australia
P. fieldi **Alice Springs mouse.** Alice Springs, C Australia
P. fumeus **Smokey mouse.** Victoria
P. glaucus Queensland
P. gouldi **Gould's mouse.** SW S Australia
P. gracilicaudatus **Eastern chestnut mouse.** C S Queensland, New South Wales
P. hermannsburgensis **Pebble mound mouse.** Australia
P. higginsi **Tasmanian mouse.** Tasmania
P. nanus **Western chestnut mouse.** W N Australia
P. novaehollandiae **New Holland mouse.** E New South Wales, Victoria, Tasmania
P. occidentalis **Western mouse.** SW Australia
P. oralis **Hastings river mouse.** New South Wales, Queensland
P. pilligaensis New South Wales
P. praeconis **Shark Bay mouse.** Shark Bay, W Australia
P. shortridgei **Shortridge's native mouse.** Victoria, SW Australia
Rattus
R. annandalei **Annandale's rat.** S Thailand, Malaya, Sumatra
R. argentiventer **Ricefield rat.** Vietnam to Java, Borneo, Philippines
R. atchinus Sumatra
R. baluensis **Summit rat.** Borneo, Sumatra
R. blangorum Sumatra
R. bontanus Sulawesi
R. burrus Nicobar Is
R. callitrichus Sulawesi
R. ceramicus Ceram
R. culionensis Culion, Philippines
R. dammermani Sulawesi
R. doboensis Aru Is
R. elephinus Sulawesi
R. enganus Sumatra
R. everetti Philippines
R. exulans **Polynesian rat.** Burma S to Sumatra and Java, Philippines, New Guinea and Pacific Is
R. feliceus Ceram
R. faramineus Sulawesi
R. fuscipes **Southern bush rat.** E S SW Australia
R. hamatus Sulawesi
R. hoffmanni Sulawesi
R. hoogerwerfi Sumatra
R. hoxaensis Vietnam
R. infraluteus **Mountain giant rat.** Borneo, Sumatra, Java
R. latidens Luzon, Philippines
R. leucopus **Mottle-tailed rat.** New Guinea, N Queensland
R. losea **Lesser ricefield rat.** SE China to Malaya, Taiwan
R. lutreolus **Australian swamp rat.** E SE Australia, Tasmania
R. macleari Christmas Is, Indian Ocean
R. marmosurus Sulawesi
R. maxi Java
R. montanus Sri Lanka
R. morataiensis Moluccas
R. muelleri **Müller's rat.** Malaya, Sumatra, Borneo, Palawan
R. mindorensis Philippines
R. nativitatis Christmas Is, Indian Ocean
R. niobe **Moss-forest rat.** New Guinea
R. nitidus **Himalayan rat.** Himalayas to Vietnam, Philippines
R. norvegicus **Norway or Common or Brown rat.** Worldwide
R. omichlodes New Guinea
R. owiensis Owi Is, Papua
R. palmarum Nicobar Is
R. praetor New Guinea, Solomon Is
R. pulliventer Nicobar Is
R. punicans India
R. ranjiniae India
R. rattus **Roof rat.** Originated in SE Asia, now worldwide
R. remotus Thailand
R. rennelli Solomon Is
R. richardsoni **Richardson's rat.** New Guinea
R. rogersi S Andaman Is
R. salocco Sulawesi
R. sikkimensis Nepal, Burma, Vietnam, S China
R. simalurensis Mentawi Is, Indonesia
R. sordidus **Australian dusky field rat.** N NE Australia, New Guinea
R. stoicus Andaman Is
R. taerae Sulawesi
R. tiomanicus **Malaysian field rat.** Malaya to Borneo
R. tunneyi **Tunney's rat.** W N NE C Australia
R. turkestanicus Russian Turkestan to SW China
R. tyrannus Philippines

R. verecundus **Slender rat.** New Guinea
R. xanthurus Sulawesi
Rhabdomys
R. pumilio **Four-striped grass mouse.** S Africa to C Angola and Malawi; discontinuously distributed in Mozambique, Tanzania, Kenya, Uganda
Srilankamys
S. ohiensis **Ohiya rat.** Sri Lanka
Solomys
S. ponceleti Bougainville Is, Solomons
S. salebrosus Bougainville Is, Solomons
S. sapientis Santa Ysabel and Choiseul Is, Solomons
Stenocephalemys
S. albocaudata **Ethiopian narrow-headed rat.** Ethiopia
S. griselcauda Ethiopia
Stochomys
S. defua **Defua rat.** Guinea to Ghana
S. longicaudatus **Target rat.** Zaire to Nigeria
Taeromys
T. arcuatus Sulawesi
T. celebensis Sulawesi
Tarsomys
T. apoensis Mindanao, Philippines
Tateomys
T. rhinogradoides **Tate's rat.** Sulawesi
Thallomys
T. paedulcus **Acacia rat.** S Africa to Somalia
Thamnomys
T. cometes Mozambique–Kenya
T. dolichurus **Tree rat.** Guinea and Mali to S Sudan and S to Angola, Zimbabwe and South Africa
T. rutilans **Peter's arboreal forest rat.** Guinea to Uganda, Angola
T. venustus **African mountain thicket rat.** E Zaire, Uganda, Rwanda
Tokudaia
T. osimensis **Ryukyu spiny rat.** Ryukyu Is, Amamioshima Is, Sumiyo-mura (Japan)
Tryphomys
T. adustus **Mearn's Luzon rat.** Philippines
Uranomys
U. ruddi Senegal to Kenya to Mozambique
Uromys
U. anak **Black-tailed tree rat.** C NE New Guinea
U. caudimaculatus **Giant white-tailed rat.** New Guinea, NE Queensland
U. imperator Guadalcanal Is, Solomons
U. neobritannicus New Britain Is
U. rex Guadalcanal Is, Solomons
U. salamonis Florida Is, Solomons
Vandeleuria
V. nolthenii **Long-tailed climbing mouse.** Sri Lanka
V. oleracea **Palm mouse.** Sri Lanka, S India, Kumaon, Nepal, Assam, Burma, Vietnam, Thailand
Vernaya
V. fulva **Verney's climbing mouse.** Yunnan (China), N Burma
Xenuromys
X. barbatus **Mimic tree rat.** New Guinea
Zelotomys
Z. hildegardeae **Broad-headed mouse.** Angola to Malawi to Uganda to Central African Republic
Z. woosnami **Woosnam's desert rat.** NW Botswana, E C Namibia
Zyzomys
Z. argurus **Common Australian rock rat.** W N Australia
Z. pedunculatus **Macdonnell range rock rat.** N Australia
Z. woodwardi **Woodward's rock rat.** NW N Australia

SUBFAMILY SPALACINAE
Nannospalax
N. ehrenbergi **Ehrenberg's mole-rat.** Syria and Israel to coastal region of Libya
N. leucodon **White-toothed mole-rat.** Yugoslavia and Greece to SW Ukraine
N. nehringi **Nehring's mole-rat.** Asia Minor
Spalax
S. arenarius **Sandy mole-rat.** S Ukraine (Kherson)
S. giganteus **Giant mole-rat.** NW of Caspian Sea, S Urals
S. graecus **Bukovin mole-rat.** N Rumania, Ukraine
S. microphthalmus **Common Russian mole-rat.** S Russia from E Ukraine to the R. Volga
S. polonicus **Podolsk mole-rat.** W Ukraine

SUBFAMILY CRICETOMYINAE
Beamys
B. hindei **Lesser pouched rat.** S Kenya, Malawi and NE Zambia
Cricetomys
C. emini **Forest giant pouched rat.** Sierra Leone to Zaire
C. gambianus **Savanna giant pouched rat.** South Africa through East Africa to Senegal
Saccostomys
S. campestris **Southern short-tailed pouched rat.** Zambia to South Africa
S. mearnsi **Northern short-tailed pouched rat.** Tanzania to Ethiopia

SUBFAMILY OTOMYINAE
Otomys
O. anchietae Angola and S Tanzania
O. angoniensis **Angoni vlei rat.** South Africa to Angola and Kenya
O. denti NE Zambia to Uganda
O. irroratus **Vlei rat.** South Africa to Zimbabwe
O. laminatus **Laminate vlei rat.** South Africa
O. maximus **Large vlei rat.** SW Zambia
O. saundersae **Saunder's vlei rat.** South Africa, mainly Cape Province
O. sloggetti **Sloggett's vlei rat.** Mts of South Africa
O. tropicalis Kenya to NE Zaire, Mt Cameroon
O. typus **Common East African swamp rat.** Zambia through East Africa to Ethiopia

O. unisulcatus **Bush karroo rat.** Cape Province, South Africa
Parotomys
P. brantsii **Brant's whistling rat.** South Africa, in Cape Province, Botswana, Namibia
P. littledalei **Littledale's whistling rat.** Cape Province and Namibia

SUBFAMILY LOPHIOMYINAE
Lophiomys
L. imhausii **Crested rat.** Mts of East Africa in Kenya, Somalia, Ethiopia and E Sudan

SUBFAMILY DENDROMURINAE
Delanymys
D. brooksi **Delany's swamp mouse.** SW Uganda and adjacent part of Zaire
Dendromus
D. kahuziensis **Kahuzi climbing mouse.** Mts of E Zaire
D. lovati **Pygmy Ethiopian climbing mouse.** Mts of Ethiopia
D. melanotis **Gray pygmy climbing mouse.** South Africa through Ethiopia to Guinea
D. mesomelas **Brant's climbing mouse.** South Africa through East Africa to Cameroun
D. mystacalis **Chestnut climbing mouse.** South Africa through East Africa to Nigeria
D. nyikae **Nyika climbing mouse.** E Transvaal, Zimbabwe, Malawi, Zambia, Angola
Dendroprionomys
D. rousseloti Congo Republic
Deomys
D. ferrugineus **Link rat.** Cameroun through Zaire to Uganda
Leimacomys
L. buettneri Togo
Malacothrix
M. typica **Large-eared mouse.** South Africa to S Angola
Megadendromus
M. nikolausi **Large Ethiopian climbing mouse.** Mts of E Ethiopia
Petromyscus
P. collinus **Pygmy rock mouse.** Cape Province to S Angola
P. monticularis **Brukkaros rock mouse.** Namibia
Prionomys
P. batesi **Dollman's tree mouse.** Cameroun and Central African Republic
Steatomys
S. cuppedius Senegal to Niger
S. eaurinus Senegal to Nigeria
S. jacksoni Ghana and W Nigeria
S. krebsi South Africa to Angola and Zambia
S. parvus **Tiny fat mouse.** Namibia to Somalia and Senegal
S. pratensis **Common fat mouse.** Natal through East Africa to Senegal

SUBFAMILY RHIZOMYINAE
Cannomys
C. badius **Bay bamboo rat.** E Nepal through Burma to Thailand
Rhizomys
R. pruinosus **Hoary bamboo rat.** S China to Malaya
R. sinensis **Chinese bamboo rat.** S China to Burma
R. sumatrensis **Large bamboo rat.** Burma and N Vietnam to Malaya and Sumatra
Tachyoryctes
T. splendens **East African root rat.** Kenya, N Tanzania, Uganda, E Zaire
T. macrocephalus **Giant Ethiopian root rat.** Highlands of Ethiopia

SUBFAMILY NESOMYINAE
Brachytarsomys
B. albicauda E Madagascar
Brachyuromys
B. betsileoensis Madagascar
B. ramirohitra **Ramirohitra.** E Madagascar
Eliurus
E. minor E Madagascar
E. myoxinus Madagascar
Gymnuromys
G. roberti E Madagascar
Hypogeomys
H. antimena **Votsotsa.** W Madagascar
Macrotarsomys
M. bastardi W Madagascar
M. ingens NW Madagascar
Mystromys
M. albicaudatus **White-tailed rat.** South Africa
Nesomys
N. rufus Madagascar

SUBFAMILY PLATACANTHOMYINAE
Platacanthomys
P. lasiurus **Spiny dormouse.** S India
Typhlomys
T. cinereus **Chinese dormouse.** S China and N Vietnam

SUBFAMILY MYOSPALACINAE
Myospalax
M. aspalax **East Siberian zokor.** SE Siberia and N Mongolia
M. fontanierii **Common Chinese zokor.** C China from Sichuan to near Peking
M. myospalax **West Siberian zokor.** Altai, Mts and parts of W Siberia
M. psilurus **Manchurian zokor.** Manchuria to Shandong, China
M. rothschildi **Rothschild's zokor.** Gansu and Hubei, C China
M. smithii **Smith's zokor.** Gansu, China

SUBFAMILY HYDROMYINAE
Celaenomys
C. siliceus **Luzon shrew-rat.** Mts of Luzon, Philippines
Chrotomys
C. mindorensis **Mindoro striped rat.** Is of Mindoro and Luzon, Philippines

C. whiteheadi **Luzon striped rat.** Mts of Luzon, Philippines
Crossomys
C. moncktoni **Earless water rat.** E New Guinea
Hydromys
H. chrysogaster **Australian water rat.** Australia except for arid interior
H. habbema **Mountain water rat.** W New Guinea
H. hussoni **Husson's water rat.** W New Guinea
H. neobritannicus **New Britain water rat.** New Britain
Leptomys
L. elegans **Long-footed hydromyine.** New Guinea
Mayermys
M. ellermani **One-toothed shrew-mouse.** NE New Guinea
Microhydromys
M. richardsoni **Groove-toothed shrew-mouse.** W New Guinea
Neohydromys
N. fuscus **Short-tailed shrew-mouse.** Mts of NE New Guinea
Parahydromys
P. asper **Coarse-haired water rat.** Mts of New Guinea
Paraleptomys
P. rufilatus **Red-sided hydromyine.** N New Guinea
P. wilhelmina **Short-footed hydromyine.** W New Guinea
Pseudohydromys
P. murinus **Eastern shrew-mouse.** NE New Guinea
P. occidentalis **Western shrew-mouse.** W New Guinea
Rhynchomys **Shrew-rats**
R. isarogensis **Isarog shrew-rats.** Mt Isarog, Luzon, Philippines
R. soricoides **Mt Data shrew-rat.** Mt Data, Luzon, Philippines
Xeromys
X. myoides **False swamp rat.** SW coast of Queensland and N coast of N Australia

SUBFAMILY CRICETINAE
Calomyscus
C. bailwardi **Mouse-like hamster.** Iran, Transcaucasia, S Turkmenia, Afghanistan, N Pakistan
C. baluchi Baluchistan, E Afghanistan
C. hotsoni Baluchistan (Bangladesh)
C. mystax S Turkmenia, NW Iran, Armenia
C. urartensis Azerbaidzhan, NW Iran, probably Armenia
Cricetulus
C. alticola Kashmir, Ladak
C. barabensis S Siberia, from Irtysh River to Ussuri region S to NW Mongolia, NE China and Korea
C. curtatus S Mongolia, N of Altai Mts to Honan and Ningsia
C. eversmanni Kazakstan, steppes from Volga River to Tarbagatai
C. griseus NW China, Inner Mongolia
C. kamensis Tibet, Tsinghai, Kansu
C. longicaudatus Altai and Tuva, W and S Mongolia, Singkiang to Hopei and Szechuan
C. migratorius SE Europe through Asia Minor, Transcaucasia, Kazakhstan to Mongolia and Ningsia, S to Israel, Iraq, Iran, Pakistan
C. obscurus Kansu to Shansi, S and E Mongolia
C. pseudogriseus Transbaikalia, N Mongolia, N Inner Mongolia
C. triton Korean gray rat. Kansu and Kiangsu to NE China, Korea and upper Ussuri region
Cricetus
C. cricetus **Common or Black-bellied hamster.** Belgium, C Europe, W Siberia, N Kazakhstan to Upper Yenesei and Altai and Sinkiang
Mesocricetus
M. auratus **Golden hamster.** Asia Minor to Syria, Lebanon, Israel
M. brandti N Transcaucasia, Kurdistan, Lebanon and Israel
M. newtoni E Bulgaria, E Rumania
M. raddei N Caucasus to Don River and Sea of Azor
Mystromys
M. albicaudatus **White-tailed rat.** Cape Province, Transvaal
Phodopus
P. campbelli Transbaikalia, Mongolia, Heilungkiang, Inner Mongolia, Hopei and Singkiang
P. roborovskii Tuva and E Kazakhstan, Mongolia, N China
P. sungorus **Dzungarian hamster.** E Kazakhstan and SW Siberia

SUBFAMILY GERBILLINAE
Ammodillus
A. imbellis **Somali gerbil.** Somalia, S Ethiopia
Brachiones
B. przewalskii **Przewalski's gerbil.** Sinkiang – Kansu
Desmodillus
D. auricularis **Cape short-eared gerbil.** S Africa
Desmodilliscus
D. braueri **Pouched gerbil.** Sudan to Senegal
Dipodillus
D. maghrebi **Greater short-tailed gerbil.** N Morocco
D. simoni **Lesser short-tailed gerbil.** Algeria, Tunisia
D. zakariai Tunisia
Gerbillurus
G. paeba **South African pygmy gerbil.** SW Angola to Cape Province
G. setzeri Namib Desert
G. tytonis S Namib, SW Africa
G. vallinus **Tassel-tailed pygmy gerbil.** SW Angola to Orange R
Gerbillus
G. agag C Mali to E Sudan
G. aquilus S Afghanistan
G. amoensus **Charming gerbil.** N Africa
G. andersoni **Anderson's gerbil.** Tunisia to Israel
G. aureus NW Libya
G. bottai Sudan, Kenya
G. campestris **Large North African gerbil.** Morocco to Somalia, N of Sahara
G. cheesmani **Cheesman's gerbil.** Arabia to Pakistan
G. dasyurus **Wagner's gerbil.** Arabia, Syria, Sinai
G. dunni Ethiopia, Somalia
G. famulus **Black-tufted gerbil.** SW Arabia
G. gerbillus Sahara, S Israel
G. gleadowi **Indian hairy-footed gerbil.** Pakistan, NW India
G. henleyi **Pygmy gerbil.** Algeria to W Arabia
G. hesperinus Morocco
G. hoogstrali Morocco
G. jamesi Tunisia
G. latastei **Hairy-footed gerbil.** E Egypt, S Israel, Jordan, Saudi Arabia
G. mackilligini **Mackelligin's gerbil.** N Africa
G. mauritaniae Mauritania
G. mesopotamiae **Mesopotamian gerbil.** Lower Tigris/Euphrates
G. muriculus Sudan
G. nancillus Sudan
G. nanus **Baluchistan gerbil.** Pakistan to Algeria
G. occiduus Morocco
G. perpallidus NW Egypt
G. poecilops **Large Aden gerbil.** SW, W Arabia
G. pulvinatus Ethiopia
G. pusillus Kenya
G. pyramidium **Greater Egyptian gerbil.** Senegal to S Israel
G. riggenbachi NW Sahara
G. ruberrimus N Kenya, Somalia, E Ethiopia
G. syrticus Libya
G. watersi E Sudan, Somalia
Meriones
M. chengi N China and Mongolia
M. crassus Sahara–Afghanistan
M. hurrianae **Indian desert gerbil.** NW India to SE Iran
M. libycus **Libyan jird.** Sahara to Sinkiang
M. meridianus **Midday gerbil.** Russian Turkestan to China
M. persicus **Persian jird.** Asia Minor to Pakistan
M. rex **King jird.** SW Arabia
M. sacramenti **Buxton's jird.** S Israel
M. shawi **Shaw's jird.** Morocco to Egypt
M. tamariscinus **Tamarisk gerbil.** Lower Volga to W Kansu
M. tristrami **Tristram's jird.** Sinai to Asia Minor, NW Iran
M. unguiculatus **Mongolian gerbil (Clawed jird).** Mongolia, Sinkiang to Manchuria
M. vinogradovi **Vinogradov's jird.** E Asia Minor to N Iran
M. zarudnyi NE Iran etc
Microdillus
M. peeli **Somali pygmy gerbil.** Somalia
Pachyuromys
P. duprasi **Fat-tailed gerbil.** N Sahara
Psammomys
P. obesus **Fat sand rat.** Algeria to Arabia
P. vexillaris **Pale fat sand rat.** Algeria, Tunisia
Rhombomys
R. opimus **Great gerbil.** Caspian Sea to Sinkiang, Pakistan
Sekeetamys
S. calurus **Bushy-tailed jird.** E Egypt, S Israel to C Arabia
Tatera
T. afra **Cape gerbil.** SW Cape Province
T. boehmi **Boehm's gerbil.** S Kenya to Zambezi, Angola
T. brandtsii **Highveld gerbil.** S Africa
T. inclusa **Gorongoza gerbil.** Mozambique, E Tanzania
T. indica **Indian gerbil.** W India to Asia Minor, Sri Lanka
T. leucogaster **Bushveld gerbil.** S, C Africa
T. nigricauda **Black-tailed gerbil.** E Africa
T. robusta **Fringe-tailed gerbil.** Guinea to Somalia, Tanzania
T. valida **Savanna gerbil.** Senegal to Sudan to Angola
Taterillus
T. arenarius Mauritania, Niger, Mali
T. congicus Cameroun to Sudan
T. emini Chad to Kenya
T. gracilis W Africa
T. harringtoni Sudan to NE Tanzania
T. lacustris L Chad etc
T. pygargus Senegal to Mali

FAMILY GLIRIDAE
Dryomys
D. nitedula **Tree dormouse.** C and E parts of Europe and parts of Asia Minor
Eliomys
E. quercinus **Garden or Orchard dormouse.** Most parts of Europe from S Spain to W USSR and Asia Minor, North Africa
E. melanurus **Asiatic garden dormouse.** Asia Minor, North Africa
Glirulus
G. japonicus **Japanese dormouse.** Japan (Kyushu, Shikoku, Honshu)
Glis
G. glis **Edible or Fat or Squirrel-tailed dormouse.** From N Spain to W S and C USSR
Graphiurus
G. murinus **African or Black and white dormouse.** Africa from the Southern Sahara area to South Africa
G. nanus **Dwarf dormouse.** Africa from S Sahara to South Africa
G. ocularis Africa from S Sahara to South Africa
Muscardinus
M. avellanarius **Common dormouse or Hazel mouse.** Europe from England to W USSR and Asia Minor
Myomimus
M. personatus **Mouse-like or Asiatic dormouse.** C Asia and Bulgaria

FAMILY SELEVINIIDAE
Selevinia
S. betpakdalensis **Desert dormouse.** E Kazakh (USSR)

FAMILY ZAPODIDAE
Eozapus
E. setchuanus **Szechuan jumping mouse.** W, SW China
Napaeozapus
N. insignis **Woodland jumping mouse.** S E Canada, NE USA
Sicista
S. betulina **Northern birch mouse.** N, C Europe to E Siberia
S. caucasica W Caucasus, Armenia (USSR)
S. caudata **Chinese birch mouse.** Ussuri, Sakhalin Is (USSR), NE China
S. concolor **Chinese birch mouse.** W China, Kashmir
S. kluchorica USSR
S. napaea **Altai birch mouse.** N Altai Mts
S. pseudonapaea S Altai Mts
S. subtilis **Southern birch mouse.** E Europe to L Baikal
S. tianshanica Tien Shan Mts (USSR, China)
Zapus
Z. hudsonius **Meadow jumping mouse.** Canada, N, C USA
Z. princeps **Western jumping mouse.** W N America
Z. trinotatus **Pacific jumping mouse.** W Coast USA

FAMILY DIPODIDAE
Alactagulus
A. pumilio R Don – Inner Mongolia, N Iran
Allactaga
A. bobrinskii **Bobrinski's jerboa.** Kizil-kum, Kara-kum Deserts, Turkestan
A. bullata S, W Mongolia
A. elater **Small five-toed jerboa.** E Asia Minor to Baluchistan
A. euphratica **Euphrates jerboa.** Jordan to Afghanistan
A. firouzi Isfahan, Iran
A. hotsoni **Hotson's jerboa.** Persian Baluchistan
A. major **Great jerboa.** Ukraine to Tien Shan
A. nataliae S Mongolia
A. severtzovi **Severtzov's jerboa.** Turkestan
A. sibirica **Mongolian five-toed jerboa.** R Ural to Manchuria
A. tetradactyla **Four-toed jerboa.** Libya, Egypt
Cardiocranius
C. paradoxus **Five-toed pygmy jerboa.** Mongolia
Dipus
D. sagitta **Northern three-toed jerboa.** S European Russia to N Iran – China
Euchoreutes
E. naso **Long-eared jerboa.** W Sinkiang to Inner Mongolia
Jaculus
J. blanfordi **Blanford's jerboa.** E, S Iran, W Pakistan
J. jaculus **Lesser Egyptian jerboa.** Sahara, Arabia
J. lichtensteini **Lichtenstein's jerboa.** Caspian Sea to L Balkhash
J. orientalis **Greater Egyptian jerboa.** Morocco to Israel
J. turcmenicus **Turkmen jerboa.** Ukraine to Mongolia
Paradipus
P. ctenodactylus **Comb-toed jerboa.** SW Turkestan
Pygeretmus
P. platyurus **Lesser fat-tailed jerboa.** Kazakhstan
P. shitkovi **Greater fat-tailed jerboa.** E Kazakhstan
P. vinogradovi SE Kazakhstan
Salpingotulus
S. michaelis **Baluchistan pygmy jerboa.** NW Baluchistan
Salpingotus
S. crassicauda **Thick-tailed pygmy jerboa.** S Mongolia, E Turkestan
S. heptneri **Heptner's pygmy jerboa.** Uzbekistan, S of Aral Sea
S. kozlovi **Koslov's pygmy jerboa.** S Mongolia
S. thomasi Afghanistan, NE India, S Tibet
Stylodipus
S. telum **Thick-tailed three-toed jerboa.** Ukraine – Mongolia

SUBORDER CAVIOMORPHA
FAMILY ERETHIZONTIDAE
Coendou
C. bicolor **South American tree porcupine.** S Panama, W Colombia, E Ecuador, E Peru, C Bolivia
C. prehensilis **Prehensile-tailed porcupine.** E Venezuela, the Guyanas, Surinam, N Brazil
Echinoprocta
E. rufescens **Upper Amazonian porcupine.** C Colombia
Erethizon
E. dorsatum **North American porcupine.** Alaska to Mexico
Sphiggurus
S. insidiosus **Black-tailed tree porcupine.** Surinam, N Brazil
S. mexicanus **Mexican tree porcupine.** States of San Luis, Potosi and Yucatan (Mexico) to W Panama
S. pallidus **West Indian tree porcupine.** W Indies
S. spinosus **South American tree porcupine.** SE Brazil, N Uruguay, E Paraguay NE Argentina
S. vestitus **Colombia tree porcupine.** Colombia, W Venezuela
S. villosus **Brown tree porcupine.** SE Brazil

FAMILY CAVIIDAE
Cavia
C. aperea **Cavy or cobaia.** Colombia, Venezuela to N Argentina, Uruguay
C. fulgida **Cobaia or prea.** E Brazil
C. nana W Bolivia
C. porcellus **Domestic guinea pig.** Domestic state only
C. tschudii S Peru, S Bolivia, N Argentina, N Chile
Dolichotis
D. patagonum **Patagonian hare or Mara.** C and S Argentina
D. salinicolum **Patagonian hare or Mara.** C and S Argentina
Galea
G. flavidens **Preá.** E Brazil
G. musteloides **Cuis.** S Peru, S Bolivia, N Chile, N Argentina
G. spixii **Preá.** NE Brazil to E Bolivia
Kerodon
K. rupestris **Mocó or rock cavy.** NE Brazil
Microcavia
M. australis **Cuis chico.** NW Argentina, S to Santa Cruz
M. niata **Cuis.** Andes of Bolivia, N Chile, SW Peru
M. shiptoni **Cuis or coi.** NW Argentina

FAMILY HYDROCHOERIDAE
Hydrochoerus
H. hydrochaeris **Capybara.** S America, E of the Andes from Panama to NE Argentina

FAMILY MYOCASTORIDAE
Myocastor
M. coypus **Coypu.** S Brazil, Paraguay, Uruguay, Bolivia, Argentina, Chile; introduced to N America, N Asia, E Africa, Europe

FAMILY CAPROMYIDAE–HUTIAS
Capromys
C. angelcabrerae Cuba
C. arboricolus Cuba
C. auritus Cuba
C. garridoi Cuba
C. melanurus Cuba
C. nanus Cuba
C. piloroides Cuba, Isle of Pines
C. prehensilis Cuba, Isle of Pines
C. sanfelipensis Cuba
Geocapromys
G. browni Jamaica
G. ingrahami Bahamas, East Plana Key
Isolobodon
I. portoricensis Puerto Rico, Hispaniola
Plagiodontia
P. aedium Hispaniola

FAMILY DINOMYIDAE
Dinomys
D. branicki **Pacarana or False paca.** Colombia, Ecuador, Peru, Brazil, Bolivia

FAMILY AGOUTIDAE
Agouti
A. paca **Paca.** Mexico to Paraguay and S Brazil
A. taczanowski **Mountain paca.** Andes of Peru, Ecuador, Colombia, Venezuela

FAMILY DASYPROCTIDAE
Dasyprocta
D. azarae **Azara's agouti.** E, C and S Brazil, Paraguay, N Argentina
D. coibae Goiba Is, Panama
D. cristata **Crested agouti.** Guianas
D. fuliginosa Colombia, S Venezuela, Surinam, N Brazil
D. guamara Orinoco Delta, Venezuela
D. kalinowski SE Peru
D. leporina Lesser Antilles, Venezuela, Guianas, N and E Brazil; introduced to Virgin Is
D. mexicana Mexico; introduced to Cuba
D. prymnolopha NE Brazil
D. punctata S Mexico to S Bolivia, N Argentina, SW Brazil; introduced to Cuba, Cayman Is
D. ruatanica Roatan Is, Honduras
Myoprocta
M. acouchi Guianas, N Brazil, Ecuador, N Peru, S Venezuela, Colombia
M. exilis Brazil, Guianas, S Venezuela and Colombia, N Peru, E Ecuador

FAMILY ABROCOMIDAE
Abrocoma
A. bennetti **Chinchilla rat.** S Chile
A. cinerea **Chinchilla rat.** SE Peru, W Bolivia, N Chile, NW Argentina

FAMILY ECHIMYIDAE – Spiny Rats
Carterodon
C. sulcidens E and C Brazil
Clyomys
C. laticeps E and C Brazil
Dactylomys
D. boliviensis **Coro-coro.** C Bolivia, SE Peru
D. dactylinus Upper Amazon of Brazil, Peru, Ecuador, Colombia
D. peruanus SE Peru
Diplomys
D. caniceps **Soft-furred spiny rat.** W Colombia, N Ecuador
D. labilis Panama to E Colombia
D. rufodorsalis NE Colombia
Echimys
E. blainvillei SE Brazil
E. braziliensis E Brazil
E. chrysurus Guianas, N Brazil
E. dasythrix SE Brazil
E. grandis Amazonian Brazil, NE Peru
E. macrurus Brazil, S of Amazon
E. nigrispinus E Brazil
E. saturnus Ecuador, E of Andes
E. semivillosus N Colombia, Venezuela
E. unicolor Brazil
Euryzygomatomys
E. spinosus **Guira.** S, C and E Brazil, N Argentina, Paraguay
Hoplomys
H. gymnurus **Thickspined rat.** C Honduras to N Ecuador
Isothrix
I. bistriatus **Toros.** W Brazil, S Venezuela, Colombia
I. pictus Bahia, Brazil
I. villosus Peru
Kannabateomys
K. amblyonyx Sao Paulo, Brazil
Lonchothrix
L. emiliae Brazil, S of Amazon
Makalata
M. armata N Ecuador, Colombia, Venezuela, Guianas, Tobago, Trinidad
Mesomys
M. didelphoides Brazil
M. hispidus Amazonian Brazil
M. obscurus Brazil
M. stimulax N Brazil, Surinam
Proechimys
P. albispinus **Guiara.** Bahia, Brazil

P. amphicoricus S Venezuela, adjacent Brazil
P. brevicauda E Peru, NW Brazil
P. canicollis Colombia
P. cuvieri Guyana, Surinam, French Guiana
P. dimidiatus Rio de Janeiro, Brazil
P. goeldii Para, Brazil
P. guiarae Venezuela
P. guyannensis Colombia, Guianas, Peru, Bolivia, Brazil
P. hendeei NW Peru to S Colombia
P. hoplomyoides Venezuela
P. iheringi E Brazil
P. longicaudatus C and E Peru, Bolivia, Paraguay, Brazil
P. myosurus Bahia, Brazil
P. oris C Brazil
P. poliopus NW Venezuela, Colombia
P. quadruplicatus E Ecuador, Peru
P. semispinosus Honduras to Peru and N Brazil
P. setosus E Brazil
P. trinitatus Trinidad
P. urichi N Venezuela
P. warreni Guyana, Surinam
Thrichomys
T. apereoides E Brazil, Paraguay
Thrinacodus
T. albicauda NE and C Colombia
T. edax W Venezuela, N Colombia

FAMILY CHINCHILLIDAE – Chinchillas
Chinchilla
C. brevicaudata Andes of S Bolivia, S Peru, NW Argentina, Chile
C. lanigera N Chile
Lagidium
L. peruanum **Mountain viscacha.** S and C Peru
L. viscacia W Argentina, S and N Bolivia, N Chile, S Peru
L. wolffsohni SW Argentina, adjacent Chile
Lagostomus
L. maximus **Plains viscacha.** S Paraguay, N Argentina

FAMILY OCTONDONTIDAE
Aconaemys
A. fuscus **Rock rat.** Andes of Chile and Argentina
Octodon
O. bridgesi **Bridge's degu.** Andes of Chile
O. degus **Degu.** W Andes of Peru
O. lunatus Coastal Mts of Peru
Octodontomys
O. gliroides **Choz choz.** Andes of N Chile, SW Bolivia, NW Argentina
Octomys
O. barrerae **Viscacha rat.** Argentina, Mendoza Province
O. mimax Foothills of Andes in W Argentina
Spalacopus
S. cyanus **Coruros.** Chile, W of Andes

FAMILY CTENOMYIDAE – Tuco-tucos
Ctenomys
C. australis E Argentina
C. azarae La Pampa province, Argentina
C. boliviensis Santa Cruz Dept, Bolivia
C. brasiliensis Minas Gerais, Brazil
C. colburni Santa Cruz Province, Argentina
C. conoveri Paraguayan chaco
C. dorsalis W Paraguay
C. emilianus Nequen Province, Argentina
C. frater Jujuy Province, Argentina
C. fulvus NW Argentina, N Chile
C. knighti N and W Argentina
C. latro NW Argentina
C. leucodon W Bolivia, E Peru
C. lewisi S Bolivia
C. magellanicus S Chile, S Argentina
C. maulinus SC Chile
C. mendocinus E Argentina
C. minutus SW Brazil, Uruguay, NW Argentina
C. nattereri Mato Grosso, Brazil
C. occultus N Argentina
C. opimus NW Argentina, SW Bolivia, S Peru, N Chile
C. perrensis NE Argentina
C. peruanus High Andes of S Peru
C. pontifex W Argentina
C. porteousi E Argentina
C. saltarius N Argentina
C. sericeus SW Argentina
C. steinbachi E Bolivia
C. talarum E Argentina
C. torquatus S Brazil, Uruguay, NE Argentina
C. tucomax NW Argentina
C. tucomanus NW Argentina
C. validus Mendoza Province, Argentina

FAMILY THRYONOMYIDAE – Cane rats
Thryonomys
T. gregorianus Cameroun, Central African Republic, Zaire, Sudan, Ethiopia, Kenya, Uganda, Tanzania, Malawi, Zambia, Zimbabwe, Mozambique
T. swinderianus Africa S of Sahara

FAMILY PETROMYIDAE
Petromus
P. typicus **African Rock Rat.** W S Africa, SW Angola

FAMILY HYSTRICIDAE
Atherurus
A. africanus **African brush-tailed porcupine.** Gambia to Zaire, W Uganda, Sudan, W Kenya
A. macrourus **Asiatic brush-tailed porcupine.** Sumatra, Malay States and adjacent Is, Indo-China, Haiman, Szechuan (China), Tenasserium, Assam

Hystrix
H. africaeaustralis **Cape porcupine.** S, C, E Africa
H. brachyura **Malayan porcupine.** Malay states, Borneo, Sumatra
H. cristata **African porcupine.** N Africa to Mediterranean coast including Zaire, Uganda, Kenya, N and C Tanzania, Somalia, Ethiopia
H. hodgsoni **Crestless Himalayan porcupine.** Nepal, Assam, Burma, Tenasserim, Siam, Indo-China, Yunnan, SE China; recently introduced to Britain
H. indica **Indian porcupine.** Sri Lanka, India, Iran, Iraq, Palestine, Syria, Israel
Thecurus
T. crassispinis **Bornean porcupine.** N Borneo
T. pumilis **Philippine porcupine.** Palawan and Busuanga Is
T. sumatrae **Sumatran porcupine.** Sumatra
Trichys
T. lipura **Bornean long-tailed porcupine.** Borneo, Sumatra, Malaya

FAMILY BATHYERGIDAE
Bathyergus
B. janetta **Namaqua dune mole-rat.** Namaqualand of S Africa
B. suillus **Cape Dune mole-rat.** Sand SW Cape of S Africa
Crytomys
C. hottentotus **Common mole-rat.** S Africa
C. mechowi **Giant Angolan mole-rat.** NW Zambia to C Angola and S Zaire
C. ochraeocinereous **Ocher mole-rat.** N Uganda, Sudan to Ghana
Georychus
G. capensis **Cape mole-rat.** SE to SW Cape of S Africa
Heliophobius
H. argenteocinereous **Silvery mole-rat.** Mozambique, E Africa, Zaire
H. spalax **Thomas' silvery mole-rat.** Taveta (Kenya)
Heterocephalus
H. glaber **Naked mole-rat (sand puppy).** C Somalia, C and E Kenya, Ethiopia

FAMILY CTENODACTYLIDAE
Ctenodactylus
C. gundi **North African gundi.** SE Morocco, N Algeria, Tunisia, Libya
C. vali **Desert or Sahara gundi.** SE Morocco, NW Algeria, Libya
Felovia
F. vae **Felou gundi.** SW Mali, Mauritania
Massoutiera
M. mzabi **Mzab or Lataste's gundi.** Algeria, Niger, Chad
Pectinator
P. spekei **Speke's or East African gundi.** Ethiopia, Somalia, N Kenya

ORDER LAGOMORPHA
FAMILY LEPORIDAE
(see entry Rabbits and hares)
FAMILY OCHOTONIDAE
Ochotona
O. alpina **Altai pika (Alpine pika).** USSR (Altai Mts, Sayan Mts, E Transbaikalia, NW Mongolia)
O. collaris **Collared pika.** Alaska, NW Canada
O. daurica **Daurian pika.** USSR (SE Altai, S Transbaikalia), Mongolia, N China
O. hyperborea **Northern pika.** USSR (Urals to Siberia as far N as 72°, Kamchatka Peninsula, Sakhalin Is), Mongolia, NE China, N Korea, N Japan
O. koslowi **Koslow's pika.** China (N Tibet)
O. ladacensis **Ladak pika.** India (Kashmir), China (Tibet, Turkestan)
O. macrotis **Large-eared pika.** USSR Ala-Tau Mts, Tien-Shan Mts, E Pamiers, Nepal, China, Tibet
O. pallasi **Mongolian, Pallas's or Price's pika.** USSR (Kazakhstan, Altai) Mongolia, NW China (Sinkiang)
O. princeps **North American pika.** USA (Rocky Mts, Cascade Mts, Sierra Nevada Mts), W Canada (Rocky Mts)
O. pusilla **Steppe or Small pika.** USSR (Upper R Volga, S Urals E through N Kazakhstan)
O. roylei **Royle's pika.** N Pakistan, N India (Kashmir), Nepal, China (Tibet, Yunnan, Szechuan), N Burma
O. rufescens **Afghan pika.** USSR (Turkmenia), Iran, Afghanistan, Pakistan (Baluchistan)
O. rutila **Red pika.** USSR (W Pamirs, Ala-Tau Mts), W China (Tien-han Mts, Tibet, Tsinghai)
O. thibetana **Moupin pika.** India (Sikkim), China (Yunnan, Szechuan, Shensi, Shansi, Kansu, Hupeh)

ORDER MACROSCELIDIA
FAMILY MACROSCELIDIDAE
Elephantulus
E. brachyrhynchus **Short-snouted elephant-shrew.** C, E and S Africa
E. edwardi **Cape elephant-shrew.** S Africa
E. fuscipes **Dusky-footed elephant-shrew.** S Sudan, NE Zaire, Uganda
E. fuscus **Dusky elephant-shrew.** SW Zambia, S Malawi, lower Zambezi Valley in Mozambique
E. intufi **Bushveld or Pale elephant-shrew.** SW Africa, Botswana
E. myurus **Rock or Eastern rock elephant-shrew.** SE Africa
E. revoili **Somali elephant-shrew.** N Somalia
E. rozeti **North African elephant-shrew.** Morocco E – W Libya
E. rufescens **Rufous or Spectacled elephant-shrew.** E Africa
E. rupestris **Rock or Western rock elephant-shrew.** SW Africa
Macroscelides
M. proboscideus **Short-eared or Black-eared elephant-shrew.** SW Africa
Petrodromus
P. tetradactylus **Four-toed or Knob-bristled elephant-shrew.** E Africa

Rhynchocyon
R. *chrysopygus* **Golden-rumped or Yellow-rumped elephant-shrew.** Kenya
R. *cirnei* **Chequered elephant-shrew.** E Africa
R. *petersi* **Black and rufous or Black and red elephant-shrew.** E Africa

ORDER INSECTIVORA

FAMILY ERINACEIDAE
Echinosorex
E. *gymnurus* **Greater moonrat.** Thailand, Malaysia, Sumatra, Borneo
Erinaceus
E. *albiventris* **Four-toed hedgehog.** Senegal to Zambesi
E. *algirus* **Algerian hedgehog.** NW Africa, SW Europe
E. *amurensis* **Manchurian hedgehog.** NE China
E. *concolor* **Eastern European hedgehog** SE Europe to Syria
E. *europaeus* **Western European hedgehog.** W, N Europe, introduced to New Zealand
E. *frontalis* **South African hedgehog.** S Africa to Zambesi
E. *sclateri* **Somali hedgehog.** N Somalia
Hemiechinus
H. *auritus* **Long-eared hedgehog.** Libya to Mongolia, NW India
H. *dauuricus* **Daurian hedgehog.** E of Gobi desert
Hylomys
H. *suillus* **Lesser moonrat.** Burma, Indochina, Thailand, Malaysia, Sumatra, Java, Borneo
Neotetracus
N. *sinensis* **Shrew-hedgehog.** S China and Indochina
Neohylomys
N. *hainanensis* **Hainan moonrat.** Hainan Is, China
Paraechinus
P. *aethiopicus* **Desert hedgehog.** Sahara, Arabian deserts
P. *hypomelas* **Brandt's hedgehog.** Iran to Pakistan
P. *micropus* **Indian hedgehog.** W India
Podogymnura
P. *truei* **Mindanao moonrat.** Mindanao Is

FAMILY TALPIDAE
Condylura
C. *cristatus* **Star-nosed mole.** Manitoba to Labrador in Canada, S to Minnesota, Wisconsin, Indiana, N Carolina, Georgia
Desmana
D. *moschata* **Russian desman.** European Russia: rivers Don, Volga, lower Ural; introduced to Dnepr
Galemys
G. *pyrenaicus* **Pyrenean desman.** Pyrenees, NW Spain, N Portugal
Neurotrichus
N. *gibbsi* **American shrew-mole.** W coast N America from S British Columbia to C California
Parascalops
P. *breweri* **Hairy tailed mole.** N America: S Quebec and Ontario to C Ohio, S to N Carolina
Scalopus
S. *aquaticus* **Eastern American mole.** C, E USA
S. *inflatus* **Tamaulipan mole.** NE Mexico
S. *montanus* **Coahuilan mole.** NE Mexico
Scapanulus
S. *oweni* **Kansu mole**
Scapanus
S. *latimanus* **California or Broad-footed mole.** California
S. *orarius* **Pacific or Coast mole.** W USA
S. *townsendi* **Townsend mole.** W coast USA
Scaptonyx
S. *fusicaudus* **Long-tailed mole.** S China, N Burma
Talpa
T. *altaica* **Siberian mole.** W, C Siberia
T. *caeca* **Mediterranean mole.** Parts of Iberia, S alps, Balkans, Asia Minor, Caucasus
T. *caucasica* **Caucasian mole.** Caucasus
T. *europaea* **European mole.** Britain, S Sweden, Europe, E through Russia to rivers Ob and Irtish
T. *latouchei* SE China, Hainan
T. *leucura* Assam
T. *micrura* Nepal, S China, Malaya
T. *mizura* **Japanese mountain mole.** Honshu, Japan
T. *moschata* **Short faced mole.** E China
T. *robusta* **Large Japanese mole.** Japan, Korea
T. *romana* **Roman mole.** Italy, Balkans
T. *streeti* **Persian mole.** NW Iran
T. *wogura* **Japanese mole.** Japan
Uropsilus
U. *soricipes* **Chinese or Asiatic shrew-mole.** S China, N Burma
Urotrichus
U. *pilirostris* **Lesser Japanese shrew-mole**
U. *talpoides* **Greater Japanese shrew-mole.** Japan

FAMILY TENRECIDAE
Dasogale
D. *fontoynonti* E Madagascar
Echinops
E. *telfairi* **Lesser hedgehog tenrec.** SW Madagascar
Geogale
G. *aurita* **Large-eared tenrec.** SW and NE Madagascar
Hemicentetes
H. *semispinosus* **Streaked tenrec.** E Madagascar
H. *nigriceps* **Streaked tenrec.** C, E Madagascar
Limnogale
L. *mergulus* **Aquatic tenrec.** Madagascar
Microgale **Long-tailed tenrecs**
M. *brevicaudata* NE Madagascar
M. *cowani* E Madagascar
M. *crassipes* S Madagascar
M. *decaryi* S Madagascar
M. *dobsoni* E, N Madagascar
M. *drouhardi* N Madagascar

M. *gracilis* E Madagascar
M. *longicaudata* E Madagascar
M. *occidentalis* W Madagascar
M. *parvula* N Madagascar
M. *principula* SE Madagascar
M. *prolixicaudata* E Madagascar
M. *pusilla* E Madagascar
M. *sorella* E Madagascar
M. *taiva* E Madagascar
M. *talazaci* E Madagascar
M. *thomasi* E Madagascar
Micropotamogale
M. *lamottei* **Mount Nimba least otter shrew.** Mt Nimba (Ivory coast, Liberia)
M. *ruwenzori* **Ruwenzori least otter shrew.** Ruwenzori Mts, Zaire, Uganda
Oryzoryctes
O. *talpoides* **Rice tenrec.** NE, NW Madagascar
O. *tetradactylus* **Rice tenrec.** NC Madagascar
O. *hova* **Rice tenrec.** NC Madagascar
Potamogale
P. *velox* **Giant otter shrew.** W, C Africa
Setifer
S. *setosus* **Greater hedgehog tenrec.** NW, E Madagascar
Tenrec
T. *ecaudatus* **Common tenrec.** Madagascar, introduced to Comoros, Mascarenes, and Seychelles

FAMILY SOLENODONTIDAE
Solenodon
S. *cubanus* **Cuban solenodon.** Cuba
S. *paradoxurus* **Hispaniola solenodon.** Hispaniola

FAMILY CHRYSOCHLORIDAE
Amblysomus
A. *gunningi* **Gunning's golden mole.** Transvaal, S Africa
A. *hottentotus* **Hottentot golden mole.** S, E Cape Province, Natal, Swaziland, N, E Orange Free State, S E Transvaal
A. *iris* **Zulu golden mole.** S Cape Province, Natal, SE Transvaal
A. *julianae* E of Pretoria, Numbi gate, Kruger National Park
Calcochloris
C. *obtusirostris* **Yellow golden mole.** Zululand, NE Kruger National Park, N to Save river, Mozambique, SE Zimbabwe
Chlorotalpa
C. *arendsi* **Arend's golden mole.** E Zimbabwe
C. *duthiae* **Duthie's golden mole.** Knysna – Port Elizabeth, S Cape Province, S Africa
C. *sclateri* **Sclater's golden mole.** W, C, E Central Cape Province, E Orange Free State, SE Transvaal, Pretoria, Lesotho
C. *leucorhina* **Congo golden mole.** N Angola, S Congo, Cameroun
C. *tytonis* Giohar, Somalia
Chrysochloris
C. *asiatica* **Cape golden mole.** Robben Is, W Cape Province, Damaraland
C. *visagiei* Gouna, Calvinia, SW Cape
C. *stuhlmanni* C, E Africa, Cameroun
Chrysospalax
C. *trevelyani* **Giant golden mole.** E Cape Province
C. *villosus* **Rough-haired golden mole.** Transvaal, Natal, E Cape Province
Cryptochloris
C. *wintoni* **De Winton's golden mole.** Port Nolloth, Little Namaqualand, S Africa
C. *zyli* **Van Zyl's golden mole.** Near Lambert's Bay, S W Cape Province
Eremitalpa
E. *granti* **Grant's desert golden mole.** SW Cape Province, Little Namaqualand, Namib desert

FAMILY SORICIDAE
Anourosorex
A. *squamipes* **Mole-shrew.** S China to N Thailand, Taiwan
Blarina
B. *brevicauda* **American short-tailed shrew.** E USA, SE Canada
B. *carolinensis* **Southern short-tailed shrew.** SE USA
B. *telmalestes* **Swamp short-tailed shrew.** Virginia
Blarinella
B. *quadraticauda* **Chinese short-tailed shrew.** China
Chimarrogale **Oriental water shrews**
C. *himalayica* Himalayas, Malaya, S China
C. *phaeura* Sumatra, Borneo
C. *styani* Szechuan, China
Crocidura
C. *aequicauda* Sumatra, Malaya
C. *allex* Kenya, N Tanzania
C. *andamanensis* S Andaman Is, Indian Ocean
C. *arethusa* N Nigeria
C. *attenuata* Himalayas to S China, Taiwan
C. *baileyi* Ethiopia
C. *beatus* Mindanao, Philippines
C. *beccarii* Sumatra
C. *bicolor* Sudan to S Africa
C. *bloyeti* C Tanzania
C. *bottegi* W Africa, Ethiopia, N Kenya
C. *bovei* Zaire
C. *butieri* Sudan, N Kenya, S Somali
C. *buettikoferi* W Africa
C. *caligirea* NE Zaire
C. *cinderella* Gambia, Mali
C. *congobelgica* NE Zaire
C. *crenata* Zaire
C. *crossei* Nigeria to Ivory Coast
C. *cyanea* Ethiopia to S Africa
C. *denti* C Africa
C. *dolichura* W, C Africa
C. *douceti* Ivory Coast, Guinea

C. *dzinezumi* Japan, Taiwan
C. *edwardsiana* Jolo Is, Philippines
C. *eisentrauti* Mt Cameroun
C. *elgonius* Kenya, N Tanzania
C. *elongata* Celebes
C. *erica* W Angola
C. *fischeri* S Ethiopia to Zaire
C. *flavescens* Africa
C. *floweri* **Flower's shrew.** Egypt
C. *foucauldi* Morocco
C. *foxi* Nigeria to Ghana, Guinea
C. *fuliginosa* S China to Malaya, Sumatra, Java, Borneo, Celebes
C. *fumosa* E Africa
C. *glassi* Harar, Ethiopia
C. *gracilipes* Africa
C. *grandis* Mindanao, Philippines
C. *grassei* C Africa
C. *grayi* Luzon, Philippines
C. *greenwoodi* S Somalia
C. *gueldenstaedti* Crete, W Asia
C. *halconis* Mindoro, Philippines
C. *hirta* S Africa to S Somalia
C. *hispida* **Andaman spiny shrew.** Middle Andaman Is, Indian Ocean
C. *horsfeldi* Sri Lanka, Indochina, Hainan, Taiwan, Ryuku Is
C. *jacksoni* E Africa
C. *kivuana* Kivu, E Zaire
C. *lamottei* Ivory Coast
C. *lanosa* E Zaire, Rwanda
C. *lasia* Asia Minor
C. *lasiura* Korea to Ussuri
C. *latona* E Zaire
C. *lea* Celebes
C. *lepidura* Sumatra
C. *leucodon* **Bicolored white-toothed shrew.** CS Europe to Israel
C. *levicula* Celebes
C. *luluae* S Zaire
C. *luna* Angola to Ethiopia
C. *lusitania* W Sahara
C. *macarthuri* Kenya
C. *maconi* Mt Nyiro, N Kenya
C. *maquassiensis* Transvaal, Zimbabwe
C. *mariquensis* S Africa to Zambia
C. *maurisca* Uganda, Kenya
C. *mindorus* Mindoro, Negros, Philippines
C. *miya* **Ceylon long-tailed shrew.** Sri Lanka
C. *monax* Kenya, Tanzania
C. *monticola* Borneo, Java to Flores
C. *mutesae* Uganda
C. *nana* Somalia, Ethopia, Egypt
C. *nanilla* W, E Africa
C. *neglecta* Sumatra
C. *nicobarica* Gt Nicobar Is, Indian Ocean
C. *nigricans* Angola
C. *nigripes* Celebes
C. *nigrofusca* Zaire
C. *niobe* Ruwenzori, Uganda
C. *nimbae* Ivory Coast, Liberia
C. *odorata* **African forest shrew.** W Africa
C. *palawanensis* Palawan, Philippines
C. *paradoxura* Sumatra
C. *parvacauda* Mindanao, Philippines
C. *pasha* Sudan to Mali
C. *pergrisea* Asia Minor to Tein Shan
C. *phaeura* S Ethiopia
C. *picea* Cameroun
C. *pitmani* Zambia
C. *planiceps* N Uganda, S Sudan
C. *poensis* W, C Africa
C. *religiosa* **Egyptian pygmy shrew.** Egypt
C. *rhoditis* Celebes
C. *roosevelti* Uganda to Angola
C. *russula* **White-toothed shrew.** CS Europe, N Africa
C. *sansibarica* Zanzibar, Penbar Is
C. *sericea* Kenya to S Morocco
C. *sibirica* C Asia
C. *sicula* Sicily
C. *somalica* Somalia
C. *suahelae* Kenya, Tanzania
C. *suaveolens* **Lesser white-toothed shrew.** SW Europe, N Africa to Korea, China, Taiwan
C. *susiana* SW Iran
C. *tenuis* Timor
C. *tephra* W Africa
C. *theresae* W Africa
C. *turba* Zambia, N Angola
C. *vosmaeri* Bangka Is, Sumatra
C. *vulcani* Mt Cameroun
C. *weberi* Sumatra
C. *whitakeri* NW Africa
C. *wimmeri* W Africa
C. *xantippe* Kenya, Tanzania
C. *zaodon* Zambia, S Sudan, W Africa
C. *zaphiri* Kenya, S Ethiopia
C. *zarudnyi* Afghanistan, Baluchistan
C. *zimmeri* S Zaire
Cryptotis
C. *avia* Colombia
C. *endersi* **Ender's small-eared shrew.** W Panama
C. *goldmani* S Mexico
C. *goodwini* **Goodwin's small-eared shrew.** S Guatemala
C. *gracilis* **Talamanean small-eared shrew.** Honduras to Panama
C. *magna* **Big small-eared shrew.** Oaxaca, S Mexico
C. *mexicana* **Mexican small-eared shrew.** SE Mexico
C. *montivaga* S Ecuador
C. *nigrescens* **Blackish small-eared shrew.** S Mexico to Panama

C. parva **Lesser short-tailed shrew.** C, E USA to Panama
C. squamipes W Colombia
C. surinamensis Surinam
C. thomasi Ecuador to Venezuela
Diplomesodon
D. pulchellum **Piebald shrew.** Russian Turkestan
Feroculus
F. feroculus **Kelaart's long-clawed shrew.** Sri Lanka
Microsorex
M. hoyi **Hoy's pygmy shrew.** Canada
M. thompsoni **Thompson's pygmy shrew.** SE Canada, NE USA
Myosorex **Mouse shrews**
M. blarina C Africa
M. cafer S Africa
M. eisentrautii Cameroun, Fernando Poo
M. geata Tanzania
M. longicaudatus Knysna, S Africa
M. narae Aberdare Mts, Kenya
M. polulus Mt Kenya
M. preussi Mt Cameroun
M. pulli Kasai, Zaire
M. varius S Africa
Nectogale
N. elegans **Elephant water shrew.** Sikkim to Shenshi
Neomys
N. anomalus **Miller's water shrew.** SE Europe
N. fodiens **Eurasian water shrew.** Europe, N Asia
N. schelkovnikovi Armenia, Georgia
Notiosorex
N. crawfordii **Desert or gray shrew.** SW USA, Mexico
N. phillipsi **Oaxaca desert shrew.** Oaxaca, Mexico
Paracrocidura
P. schoutedeni Cameroun to Ruwenzori
Podihik
P. kura **Sri Lanka shrew.** Sri Lanka
Scutisorex
S. somereni **Armored shrew.** C Africa
Solisorex
S. pearsoni **Pearson's long-clawed shrew.** Sri Lanka
Sorex
S. alaskanus **Glacier Bay water shrew.** S Alaska
S. alpinus **Alpine shrew.** Europe
S. araneus **European common shrew.** Europe to river Yenesei
S. arcticus **Arctic shrew.** Siberia, N America
S. arizonae **Arizona shrew.** Arizona
S. asper **Tien Shan shrew.** C Asia
S. bedfordiae **Lesser striped shrew.** Kansu to Himalayas
S. bendirii **Pacific water shrew.** W coast of USA
S. buchariensis **Pamir shrew.** Tibet
S. caecutiens **Laxmann's shrew.** E Europe to Japan
S. caucasicus **Caucasian shrew.** Caucasus
S. cinereus **Masked shrew.** N America, NE Siberia
S. coronatus France
S. cylindricauda **Greater striped shrew.** Szechuan, China
S. daphaenodon **Large-toothed Siberian shrew.** Siberia, NE China
S. dispar **Long-tailed shrew.** NE USA
S. fumeus **Smoky shrew.** SE Canada, NE USA
S. gaspensis **Gaspé shrew.** Gaspé Peninsula, Quebec
S. gracillimus **Slender shrew.** NE Asia, N Japan
S. granarius Spain
S. hosonoi **Azumi shrew.** C Honshu, Japan
S. jacksoni **St Lawrence Island shrew.** St Lawrence Is, Alaska
S. juncensis **Tule shrew.** Mexico, Baja California
S. longirostris **Southeastern shrew.** SE USA
S. lyelli **Mount Lyell shrew.** California
S. macrodon **Large-toothed shrew.** Veracruz, Mexico
S. merriami **Merriam's shrew.** W USA
S. milleri **Carmen mountain shrew.** NE Mexico
S. minutissimus **Least shrew.** Scandinavia, Siberia, S China, Japan
S. minutus **European pygmy shrew.** Europe to Himalayas
S. mirabilis **Giant shrew.** E Siberia, N Korea
S. monticolus **Dusky shrew.** N America, NW Mexico
S. nanus **Dwarf shrew.** Wyoming to Mexico
S. oreopolis **Mexican long-tailed shrew.** Mexico
S. ornatus **Ornate shrew.** Mexico, California
S. pacificus **Pacific shrew.** W USA
S. palustris **American water shrew.** N America
S. planiceps **Kashmir shrew.** W Himalayas
S. preblei **Preble's shrew.** E Oregon
S. pribilofensis **Pribilof shrew.** St Paul Is, Alaska
S. raddei **Radde's shrew.** Caucasus
S. saussurei **Saussure's shrew.** E Mexico to Guatemala
S. sclateri **Sclater's shrew.** Chiapas, Mexico
S. sinalis N Europe to N China
S. sinuosus **Suisun shrew.** Grizzly Is, California
S. stizodon **San Cristobal shrew.** Chiapas, Mexico
S. tenellus **Inyo shrew.** Nevada, California
S. trowbridgii **Trowbridge's shrew.** W USA
S. unguiculatus **Long-clawed shrew.** E Siberia, N Japan
S. vagrans **Vagrant shrew.** W USA, S Mexico
S. veraepacis **Verapaz shrew.** SE Mexico
S. vir **Flat-skulled shrew.** E Siberia
Soriculus
S. caudatus **Hodgson's brown-toothed shrew.** Himalayas, Indochina
S. fumidus Taiwan
S. gruberi **Gruber's shrew.** Nepal
S. hypsibius **de Winton's shrew.** China
S. leucops **Indian long-tailed shrew.** S China, Himalayas
S. lowei **Lowe's shrew.** Indochina
S. nigrescens **Himalayan shrew.** Himalayas to N Burma
S. salenskii **Salenski's shrew.** S China
S. smithi **Smith's shrew.** S China
Sylvisorex
S. granti Cameroun to E Africa
S. johnstoni Cameroun, Gabon, Fernando Poo
S. lunaris C Africa
S. megalura W, C, E Africa
S. morio Mt Cameroun, Fernando Poo
S. ollula Cameroun to Zaire
S. suncoides C Africa
Suncus
S. ater **Black shrew.** Mt Kinabalu, Borneo
S. dayi S India
S. etruscus **Pygmy white-toothed shrew (Savi's pygmy shrew, Etruscan shrew).** Mediterranean to India, Sri Lanka,
S. hosei Sarawak, Borneo
S. infinitessimus S Africa to Kenya, W Africa
S. lixus Kenya to Angola, Transvaal
S. luzoniensis Luzon, Philippines
S. malayanus Malaya, Borneo
S. mertensi Flores
S. murinus **House shrew.** S Asia, E Africa
S. occultidens S Philippines
S. palawensis Palawan, Philippines
S. remyi Gabon
S. stoliczkanus India
S. varilla S Africa to Tanzania, Zaire

ORDER EDENTATA

FAMILY DASYPODIDAE
Cabassous
C. centralis **Northern naked-tailed armadillo.** Honduras to Colombia, W Venezuela
C. chacoensis **Chacoan naked-tailed armadillo.** NW Argentina, W Paraguay, S Bolivia
C. tatouay **Greater naked-tailed armadillo.** Mato Grosso (C and S Brazil) to Uruguay and N Argentina
C. unicinctus **Southern naked-tailed armadillo.** N America, E of Andes to S Brazil
Chaetophractus
C. nationi **Andean hairy armadillo.** W Bolivia
C. vellerosus **Screaming or Hairy armadillo.** S Bolivia, Paraguay, to Argentina
C. villosus **Larger hairy armadillo.** N Paraguay to C Argentina, E Chile
Chlamyphorus
C. retusus **Greater fairy armadillo.** SE Bolivia, W Paraguay, N Argentina
C. truncatus **Lesser fairy armadillo.** C Argentina
Daspypus
D. hybridus **Southern lesser long-nosed armadillo.** N Argentina, Paraguay, Uruguay, S Brazil
D. kappleri **Greater long-nosed armadillo.** Colombia E of Andes to the Guianas, S to Amazon Basin of Brazil and N Bolivia
D. novemcinctus **Common long-nosed or Nine-banded armadillo.** S USA, Mexico, C America, S America E of Andes to Uruguay and Argentina, Trinidad, Tobago, Grenada, Margarita
D. pilosus **Hairy long-nosed armadillo.** E slopes of Peruvian Andes
D. sabanicola **Northern lesser long-nosed armadillo.** Venezuela, Colombia
D. septemcinctus **Brazilian lesser long-nosed armadillo.** Amazon delta to W Brazil, S to N Argentina
Euphractus
E. sexcinctus **Yellow or Six-banded armadillo.** C Brazil to SE Bolivia, Paraguay, N Argentina, Uruguay, possibly S Surinam
Priodontes
P. maximus **Giant armadillo.** America E of Andes to N Argentina and Paraguay
Tolypeutes
T. matacus **Southern three-banded armadillo.** E Bolivia, Paraguay, C Brazil to S Argentina
T. tricinctus **Brazilian three-banded armadillo.** NE Brazil
Zaedyus
Z. pichiy **Pichi.** C and S Argentina to Strait of Magellan, S Chile

FAMILY MYRMECOPHAGIDAE
Cyclopes
C. didactylus **Silky anteater.** S Mexico to Amazon Basin of S America, Bolivia, N W Peru
Myrmecophaga
M. tridactyla **Giant anteater.** C America E from Belize and Guatemala, N S America to Uruguay and N Argentina
Tamandua
T. mexicana **Northern tamandua.** S Mexico through C America to NW S America W of Andes
T. tetradactyla **Southern tamandua.** S America E of Andes, Brazil, Trinidad

FAMILY MEGALONICHYDAE
Bradypus
B. torquatus **Maned sloth.** Coastal S Brazil
B. tridactylus **Pale-throated three-toed sloth.** C Venezuela, S and E to Amazon Basin
B. variegatus **Brown-throated three-toed sloth.** C America E from Honduras, and N S America to N Argentina
Choloepus
C. didactylus **Linné's two-toed sloth.** Colombia to the Guianas, S to N Peru and Brazil
C. hoffmanni **Hoffmann's two-toed sloth.** Nicaragua to NW S America

ORDER CHIROPTERA

SUBORDER MEGACHIROPTERA

FAMILY PTEROPODIDAE
Acerodon
A. celebensis **Celebes flying fox.** Celebes
A. humilis **Talaud flying fox.** Talaud Is, N Moluccas
A. jubatus Philippines
A. lucifer Panay Is, C Philippines
A. mackloti **Sunda flying fox.** Lesser Sunda Is
Aethalops
A. alecto **Pygmy fruit bat.** Malaya, Sumatra, Borneo
Alionycteris
A. paucidentata Mindanao, Philippines
Aproteles
A. bulmerae E New Guinea
Balionycteris
B. maculata **Spotted-winged fruit bat.** S Thailand, Malaya, Borneo
Boneia
B. bidens N Celebes
Casinycteris
C. argynnis **Short-palate fruit bat.** Cameroun to E, NE Zaire
Chironax
C. melanocephalus **Black-capped fruit bat.** S Thailand to Java, Celebes
Cynopterus
C. archipelagus Polillo Is
C. brachyotis **Lesser dog-faced fruit bat.** S China to Java, Borneo, Philippines, Celebes, Sri Lanka, Andamans, Nicobars
C. horsfieldi **Horsfield's fruit bat.** S Thailand to Java, Borneo
C. minor Celebes
C. sphinx **Short-nosed fruit bat.** India to S China to Java, Timor
Dobsonia
D. anderseni S Bismarck Archipelago
D. beauforti Waigeo Is
D. crenulata N Molucca Is
D. exoleta **Celebes naked-backed bat.** Celebes
D. inermis **Solomons naked-backed bat.** Solomons
D. minor **Lesser naked-backed bat.** W New Guinea
D. moluccensis **Greater naked-backed bat.** Moluccas, New Guinea, N Queensland
D. pannietensis Trobriand, D'Entrecasteaux, Louisiade Is
D. peroni **Western naked-backed bat.** Lesser Sunda Is
D. praedatrix **New Britain naked-backed bat.** Bismarck Archipelago
D. remota **Trobriand naked-backed bat.** Trobriand Is, New Guinea
D. viridis **Greenish naked-backed bat.** Negros Is, S Moluccas
Dyacopterus
D. brooksi Sumatra
D. spadiceus **Dayak fruit bat.** Malaya to Borneo, Luzon
Eidolon
E. helvum **Straw-colored flying fox.** Africa S of Sahara, Ethiopia, SW Arabia, Madagascar
Eonycteris
E. major Borneo
E. robusta Philippines
E. rosenbergi **Celebes dawn bat.** N Celebes
E. spelaea **Dawn bat (Cave fruit bat).** Burma to Java, Sumba, Borneo, Philippines
Epomophorus
E. angolensis **Angolan epauletted fruit bat.** S Angola, NW Namibia
E. anurus **Eastern epauletted fruit bat.** Nigeria, Uganda to Tanzania
E. crypturus **Peter's epauletted fruit bat.** Angola to S Tanzania to S Africa
E. gambianus **Gambian epauletted fruit bat.** Senegal to S Ethiopia, S Zaire, Zambia, Zimbabwe
E. labiatus **Little epauletted fruit bat.** N Ethiopia, S Sudan
E. minor **Lesser epauletted fruit bat.** S Ethiopia to Zambia, Malawi
E. pousarguesi **Pousargue's epauletted fruit bat.** Central African Republic
E. reii **Garua epauletted fruit bat.** Cameroun
E. wahlbergi **Wahlberg's epauletted fruit bat.** Somalia to S Africa, Angola, Zaire, Cameroun
Epomops
E. buettikoferi **Büttikofer's fruit bat.** Guinea to Ghana
E. dobsoni **Dobson's fruit bat.** W, C Angola to S Zaire, NE Botswana, Zambia
E. franqueti **Franquet's flying fox (Singing fruit bat).** Ivory Coast to S Sudan to Angola
Haplonycteris
H. fischeri Philippines
Harpyionycteris
H. celebensis Celebes
H. whiteheadi **Harpy fruit bat.** Philippines
Hypsignathus
H. monstrosus **Hammer-headed bat.** Gambia to SW Sudan to Zaire, NE Angola
Latidens
L. salimalii S India
Lissonycteris
L. angolensis **Bocage's fruit bat (Angola fruit bat).** Guinea to Kenya, Angola, Zimbabwe, Zambia, Mozambique
Macroglossus
M. fructivorus Mindanao, Philippines
M. minimus **Common long-tailed fruit bat (Lesser long-tongued fruit bat).** Indochina, Philippines, Solomons, NW, N Australia
M. sobrinus **Hill long-tongued fruit bat (Greater long-tongued fruit bat).** NE India to Java, Bali
Megaerops
M. ecaudatus **Tail-less fruit bat.** Thailand to Sumatra, Borneo
M. kusnotoi Java
M. wetmorei Mindanao, Philippines
Megaloglossus
M. woermanni **African long-tongued fruit bat.** Guinea to Uganda to N Angola
Melonycteris
M. melanops **Black-bellied fruit bat.** E New Guinea, Bismarck Archipelago
Micropteropus
M. grandis **Sanborn's epauletted fruit bat.** NE Angola, Congo Republic

M. intermedius **Hayman's epauletted fruit bat.** NE Angola, S Zaire
M. pusillus **Dwarf epauletted fruit bat.** Senegal to Ethiopia, Angola to Zambia
Myonycteris
M. brachycephala **Sào Tomé collared fruit bat.** Sào Tomé Is, Gulf of Guinea
M. torquata **Little collared fruit bat.** Sierra Leone to Angola, Zambia
Nanonycteris
N. veldkampi **Veldkamp's dwarf fruit bat.** Guinea to Cameroun
Neopteryx
N. frosti **Small-toothed fruit bat.** W Celebes
Nesonycteris
N. aurantius **Orange fruit bat.** Choiseul, Florida Is, Solomons
N. woodfordi **Woodford's fruit bat.** Solomons
Notopteris
N. macdonaldi **Long-tailed fruit bat.** New Hebrides, New Caledonia, Fiji
Nyctimene
N. aello **Broad-striped tube-nosed bat.** Halmahera, New Guinea
N. albiventer **Common tube-nosed bat.** N Moluccas, New Guinea, Admiralty Is to Solomon Is, NE Australia
N. cephalotes **Pallas' tube-nosed bat.** Celebes, Timor to NW New Guinea, Admiralty Is
N. cyclotis **Round-eared tube-nosed bat.** New Guinea
N. major **Greater tube-nosed bat.** E New Guinea, Bismarck Archipelago
N. malaitensis **Malaita Island tube-nosed bat.** Malaita Is, E Solomon Is
N. minutus **Lesser tube-nosed bat.** Celebes, Buru
N. robinsoni NE Australia
N. sanctacrucis Santa Cruz Is
Otopteropus
O. cartilagonodus Luzon, Philippines
Paranyctimene
P. raptor **Lesser tube-nosed bat.** New Guinea
Penthetor
P. lucasi **Lucas' short-nosed fruit bat.** Malaya, Borneo
Plerotes
P. anchietai **Anchieta's fruit bat.** Angola, S Zaire, Zambia
Ptenochirus
P. jagori Philippines
P. minor Mindanao, Palawan
Pteralopex
P. acrodonta Taveuni Is, Fiji Is
P. anceps Bougainville, Choiseul Is, Solomons
P. atrata **Cusp-toothed flying fox.** Ysabel, Guadalcanal Is, Solomons
Pteropus
P. admiralitatum **Admiralty flying fox.** Admiralty Is, Solomons
P. alecto **Central flying fox (Black flying fox).** Celebes to S New Guinea, NW, N, NE Australia
P. anetianus New Hebrides, SW Pacific
P. argentatus **Silvery flying fox.** Celebes, Amboina Is
P. arquatus **Dubious flying fox.** C Celebes
P. balutus Balut Is, Sarangani Is, S Philippines
P. brunneus Percy Is
P. caniceps **Ashy-headed flying fox.** Celebes, N Moluccas, etc
P. chrysoproctus **Amboina flying fox.** Sanghir Is, S Moluccas
P. cognatus Solomons
P. conspicillatus **Spectacled flying fox.** N Moluccas, New Guinea to NE Queensland
P. dasymallus **Ryukyu flying fox.** S Japan, Ryukyu Is, Taiwan
P. faunulus Car Nicobar Is
P. fundatus Banks Is, New Hebrides
P. giganteus **Indian flying fox.** India to Burma, W China, Sri Lanka, Maldive Is
P. gilliardi **Gilliard's flying fox.** New Britain, Bismarck Archipelago
P. griseus **Gray flying fox.** Timor to Celebes, Luzon
P. howensis Ontong Java Atoll, Solomons
P. hypomelanus **Small flying fox.** S Burma to Solomons, Philippines, C Maldive Is
P. insularis Ruck Is, Caroline Is
P. intermedius S Burma
P. leucopterus Luzon, Philippines
P. leucotis Busuanga Is, Palawan, S Philippines
P. livingstonei **Comoro black flying fox.** Johanna Is, Comoro Is
P. lombocensis **Lombok flying fox.** Lesser Sunda Is
P. loochoensis Okinawa, Ryukyu Is
P. lylei **Lyle's flying fox.** Thailand, Indochina
P. macrotis **Big-eared flying fox.** S New Guinea, Aru Is
P. mahaganus **Lesser flying fox.** Ysabel Is, Bougainville Is, Solomons
P. mariannus **Guam flying fox.** Marianne Is, W Pacific
P. mearnsi Mindanao, Basilan, S Philippines
P. melanopogon **Black-bearded flying fox.** Sanghir Is, S Moluccas, New Guinea
P. melanotus Andaman Is, Nicobar Is to Christmas Is, Indian Ocean
P. molossinus E Caroline Is
P. neohibernicus **Bismarck flying fox.** New Guinea, Bismarck Archipelago
P. niger **Greater Mascarene flying fox.** Reunion, Mauritius
P. nitendiensis Santa Cruz Is, SW Pacific
P. ocularis **Ceram flying fox.** Buru, Ceram Is, S Moluccas
P. ornatus New Caledonia, Loyalty Is
P. pelewensis Palau Is
P. personatus **Masked flying fox.** Celebes, N Moluccas
P. phaeocephalus Mortlock Is, E Caroline Is
P. pilosus Palau Is
P. pohlei **Geevink Bay flying fox.** W New Guinea
P. poliocephalus **Gray-headed flying fox.** E Australia, Tasmania
P. pselaphon Bonin Is, Volcano Is
P. pumilus Palmas Is, S Philippines
P. rayneri **Solomon flying fox.** Solomon Is
P. rodricensis **Rodriguez flying fox.** Rodriguez Is
P. rufus **Madagascar flying fox.** Madagascar
P. samoensis **Somoan flying fox.** Fiji, Samoa
P. scapulatus W, N, E Australia
P. seychellensis **Seychelles flying fox.** Comoro Is, Aldabra Is, Seychelles, Mafia Is
P. speciosus Philippines
P. subniger Reunion, Mauritius
P. tablasi **Taylor's flying fox.** Tablas Is, C Philippines
P. temmincki **Temminck's flying fox.** S Moluccas, Bismarck Archipelago
P. tokudae Guam Is
P. tonganus **Insular flying fox.** Karkar Is to Samoa, Cook Is
P. tuberculatus Vanikoro Is, Santa Cruz Is
P. ualanus Ualan Is, E Caroline Is
P. vampyrus **Large flying fox (Common flying fox).** S Burma to Java, Philippines, Borneo, Timor
P. vanikorensis Vanikoro Is, Santa Cruz Is
P. vetulus New Caledonia
P. voeltzkowi **Pemba flying fox.** Pemba Is
P. woodfordi **Least flying fox.** Solomons
P. yapensis Yap, Mackenzie Is, W Caroline Is
Rousettus
R. aegyptiacus **Egyptian rousette.** S Africa to Senegal, Ethiopia, Egypt to Lebanon to Pakistan, Cyprus
R. amplexicaudatus **Geoffroy's rousette.** S Burma to Solomons, Philippines
R. celebensis **Celebes rousette.** Celebes, Sanghir Is
R. lanosus **Ruwenzori long-haired rousette.** S Ethiopia to Tanzania
R. leschenaulti **Leschenault's rousette.** Pakistan to Thailand to S China, Sri Lanka
R. madagascariensis **Madagascar rousette.** Madagascar
R. obliviosus Comoro Is
R. stresemanni **Streseman's rousette.** New Guinea
Scotonycteris
S. ophiodon **Pohle's fruit bat.** Liberia to Congo Republic
S. zenkeri **Zenker's fruit bat.** Liberia to E Zaire
Sphaerias
S. blanfordi **Blanford's fruit bat.** NE India to NW Thailand
Styloctenium
S. wallacei **Stripe-faced fruit bat.** Celebes
Syconycteris
S. australis **Southern blossom bat.** E New Guinea, NE Australia
S. crassa **Common blossom bat.** S Moluccas to New Guinea, Bismarck Archipelago, D'Entrecasteaux Is
S. naias Woodlark Is, Trobriand Is
Thoopterus
T. nigrescens **Swift fruit Bat.** N Celebes, Morotai Is, N Moluccas

SUBORDER MICROCHIROPTERA

FAMILY RHINOPOMATIDAE

Rhinopoma
R. hardwickei **Lesser mouse-tailed bat.** Morocco, Mauretania to Thailand
R. microphyllum **Greater mouse-tailed bat.** Senegal to India, Sumatra
R. muscatellum Oman to S Afghanistan

FAMILY CRASEONYCTERIDAE

Craseonycteris
C. thonglongyai **Kitti's hog-nosed bat.** S Thailand

FAMILY EMBALLONURIDAE

Balantiopteryx
B. infusca Ecuador
B. io **Thomas' sac-winged bat.** S Mexico to Guatemala
B. plicata **Peter's bat.** N Mexico to Costa Rica
Centronycteris
C. maximiliani **Shaggy-haired bat.** S Mexico to Ecuador, E Peru, Brazil
Coleura
C. afra **African sheath-tailed bat.** Africa, Aden
C. seychellensis **Seychelles sheath-tailed bat.** Seychelles
Cormura
C. brevirostris **Wagner's sac-winged bat.** Nicaragua to Peru, Brazil
Cyttarops
C. alecto Costa Rica, Nicaragua, Guyana, Brazil
Depanycteris
D. isabella Venezuela, Brazil
Diclidurus
D. albus E Peru, Venezuela, Surinam, Brazil, Trinidad
D. ingens Colombia, Venezuela
D. scutatus Venezuela to Surinam, Peru, Brazil
D. virgo **White bat.** W Mexico to Colombia, Venezuela
Emballonura
E. alecto **Philippine sheath-tailed bat.** Philippines, Borneo to S Moluccas
E. atrata **Peter's sheath-tailed bat.** Madagascar
E. beccarii **Beccari's sheath-tailed bat.** New Guinea
E. dianae **Rennell Island sheath-tailed bat.** Malaita, Rennell Is, Solomons
E. furax **Greater sheath-tailed bat.** New Guinea
E. monticola **Lesser sheath-tailed bat.** S Burma to Celebes
E. nigrescens Celebes to Solomons
E. raffrayana **Raffray's sheath-tailed bat.** Ceram, NW New Guinea, Solomons, Santa Cruz Is
E. semicaudata New Hebrides to Palau, Samoa, Fiji
E. sulcata Caroline Is
Peronymus
P. leucopterus Venezuela, Surinam, E Peru, Brazil
Peropteryx
P. kappleri **Greater sac-winged bat.** S Mexico to E Peru, Surinam
Rhynchonycteris
R. naso **Proboscis bat (Tufted bat).** S Mexico to Peru, Brazil
Saccopteryx
S. bilineata **Sac-winged bat (Greater white-lined bat).** W, E Mexico to Peru, Brazil
S. canescens Colombia to Peru, Brazil
S. gymnura Brazil
S. leptura **Lesser white-lined bat.** S Mexico to Peru, Brazil
Taphozous
T. australis **Gould's pouched bat.** New Guinea, N Queensland
T. capito Catanduanes Is, N Philippines
T. flaviventris **Yellow-bellied pouched bat.** Australia
T. georgianus **Sharp-nosed pouched bat.** W, N Australia
T. hamiltoni **Hamilton's tomb bat.** Sudan, NW Kenya
T. hildergardeae **Hildegards's tomb bat.** Kenya
T. longimanus **Long-winged tomb bat.** India to Java, Flores
T. mauritianus **Mauritian tomb bat.** Africa S of Sahara, Madagascar, Aldabra, Mauritius, Reunion, Assumption
T. melanopogon **Black-bearded tomb bat.** India to Java, Lesser Sundas, Borneo, Philippines
T. mixtus **Troughton's pouched bat.** S, E New Guinea, N Queensland
T. nudicluniatus **Naked-rumped pouched bat.** S, E New Guinea, Solomons, NE Queensland
T. nudiventris **Naked-rumped tomb bat.** Senegal to Somalia, Tanzania, Israel, Arabia, E Afghanistan to Malaya
T. peli **Pel's pouched bat.** Liberia to NE Angola, Kenya
T. perforatus **Egyptian tomb bat (Perforated sheath-tailed bat).** Senegal to Somalia, Mozambique, Zimbabwe to India
T. pluto Mindanao, Philippines
T. saccolaimus **Blyth's tomb bat.** India to Java, Borneo
T. solifer China
T. theobaldi **Theobald's tomb bat.** India to Vietnam

FAMILY NYCTERIDAE

Nycteris
N. arge **Bate's slit-faced bat.** Sierra Leone to SW Sudan, W Kenya, NW Angola, Fernando Poo
N. gambiensis **Gambian slit-faced bat.** Senegal to Ghana, Togo
N. grandis **Large slit-faced bat.** Guinea to Tanzania, Mozambique, Zimbabwe
N. hispida **Hairy slit-faced bat.** Senegal to Ethiopia to S Africa
N. javanica **Javan slit-faced bat.** S Burma to Java, Borneo, Timor
N. macrotis **Dobson's slit-faced bat.** Gambia to Somalia to Zimbabwe, Mozambique
N. major **Ja slit-faced bat.** Cameroun, E, S Zaire
N. nana **Dwarf slit-faced bat.** Ghana to W Kenya to NE Angola
N. parisii **Parisi's slit-faced bat.** Cameroun, Ethiopia, Somalia
N. thebaica **Egyptian slit-faced bat.** Africa S of Sahara, Morocco, Egypt, Israel, Arabia
N. woodi **Wood's slit-faced bat.** Tanzania, Zambia, Zimbabwe

FAMILY MEGADERMATIDAE

Cardioderma
C. cor **Heart-nosed bat.** Ethiopia to N Tanzania, Zanzibar
Lavia
L. frons **Yellow-winged bat.** Gambia to Somalia to Zambia
Macroderma
M. gigas **Australian false vampire (Ghost bat).** W, N Australia
Megaderma
M. lyra **Greater false vampire.** E Afghanistan to S China, Malaya, Sri Lanka
M. spasma **Lesser false vampire.** India to Java, Celebes, Philippines to N Moluccas, Sri Lanka

FAMILY RHINOLOPHIDAE

Rhinolophus
R. acuminatus **Acuminate horseshoe bat.** Thailand to Java to Lombok, Borneo, Palawan
R. adami Congo Republic
R. affinis **Intermediate horseshoe bat.** N India to S China to Java, Lesser Sundas
R. alcyone **Halcyon horseshoe bat.** Senegal to NE Zaire, Gabon
R. anderseni Palawan, Luzon, Philippines
R. arcuatus **Arcuate horseshoe bat.** Philippines, Borneo, Sumatra
R. blasii **Blasius' horseshoe bat.** N Africa, Italy to Afghanistan, Ethiopia to Transvaal, Mozambique
R. borneensis **Bornean horseshoe bat.** Cambodia, Borneo
R. capensis **Cape horseshoe bat.** Zambia, Zimbabwe, S Africa
R. celebensis **Celebes horseshoe bat.** S Celebes
R. clivosus **Geoffroy's bat.** Afghanistan to Algeria, Ethiopia to Zambia, Mozambique
R. coelophyllus **Peter's horseshoe bat (Croslet horseshoe bat)** Burma to Malaya
R. cognatus Andaman Is
R. cornutus **Little Japanese horseshoe bat.** Japan, Ryukyu Is
R. creaghi **Creagh's horseshoe bat.** Borneo, Java, Madura Is
R. darlingi **Darling's horseshoe bat.** Tanzania to Namibia, Transvaal, Mozambique
R. denti **Dent's horseshoe bat.** Guinea, Botswana, Namibia, S Africa, S Sudan to S Somalia to N Tanzania
R. eloquens S Sudan to Somalia to N Tanzania
R. euryale **Mediterranean horseshoe bat.** Portugal, Morocco to Iran
R. euryotis **Broad-eared horseshoe bat.** Moluccas, Kei Is, Aru Is
R. feae Burma
R. ferrumequinum **Greater horseshoe bat.** Britain to Morocco to N India, Japan
R. fumigatus **Rüppell's horseshoe bat.** Senegal to N Ethiopia to S Africa
R. gracilis SE India
R. hildebrandti **Hildebrandt's horseshoe bat.** Ethiopia, Somalia to Botswana
R. hipposideros **Lesser horseshoe bat.** Britain, Ireland to N India, NE Africa
R. hirsutus Guimaras Is, Philippines
R. importunus Java
R. inops Mindanao, Philippines
R. javanicus **Javan horseshoe bat.** Java
R. keyensis **Insular horseshoe bat.** Moluccas, Wetter Is, Kei Is
R. landeri **Lander's horseshoe bat.** Gambia to Somalia to Angola, Transvaal, Mozambique

R. lepidus **Blyth's horseshoe bat.** Afghanistan to S China, Thailand
R. luctus **Woolly horseshoe bat.** N India to S China to Java, Borneo
R. maclaudi **Maclaud's horseshoe bat.** Guinea, E Zaire, W Uganda
R. macrotis **Big-eared horseshoe bat.** Nepal to Malaya, Sumatra, Philippines
R. madurensis Madura Is
R. malayanus **North Malayan horseshoe bat.** Thailand to Vietnam, Malaya
R. marshalli **Marshall's horseshoe bat.** Thailand
R. megaphyllus **Southern horseshoe bat.** SE New Guinea, E Australia
R. mehelyi **Mehely's horseshoe bat.** Spain, Morocco to Iran
R. minutillus Anamba Is
R. mitratus C India
R. monoceros Taiwan
R. nereis Anamba Is, Natuna Is
R. osgoodi Yunnan, China
R. paradoxolophus **Bourret's horseshoe bat.** Thailand, Vietnam
R. pearsoni **Pearson's horseshoe bat.** NE India to Indochina
R. petersi India
R. philippinensis **Philippine horseshoe bat.** Philippines, Borneo, Celebes, NE Queensland
R. pusillus **Least horseshoe bat.** NE India, S China to Java
R. refulgens **Glossy horseshoe bat.** S Thailand to Sumatra
R. rex S China
R. robinsoni **Peninsular horseshoe bat (Robinson's horseshoe bat).** S Thailand, Malaya
R. rouxi India, S China, Sri Lanka
R. rufus Philippines
R. shamili **Shamel's horseshoe bat.** Burma, Indochina, Malaya
R. sedulus **Lesser woolly horseshoe bat.** Malaya, Borneo
R. simplex **Lombok horseshoe bat.** Lesser Sunda Is
R. simulator **Bushveld horseshoe bat.** Nigeria to Ethiopia to Transvaal
R. stheno **Lesser brown horseshoe bat.** Thailand to Java
R. subbadius N India to Vietnam
R. subrufus Philippines
R. swinnyi **Swinny's horseshoe bat.** S Zaire, Tanzania to S Africa
R. thomasi **Thomas' horseshoe bat.** Burma, Yunnan, Indochina
R. toxopeusi **Buru horseshoe bat.** Buru Is, S Moluccas
R. trifoliatus **Trefoil horseshoe bat.** N India to Java, Borneo
R. virgo S Philippines
R. yunanensis **Dobson's horseshoe bat.** NE India to Yunnan, Thailand

FAMILY HIPPOSIDERIDAE

Anthops
A. ornatus **Flower-faced bat.** Solomon Is
Asellia
A. patrizii **Patrizi's trident bat.** Ethiopia
A. tridens **Trident bat.** Morocco, Senegal to Pakistan
Aselliscus
A. stoliczkanus **Stoliczka's trident bat.** Burma, S China to Indochina, Penang Is
A. tricuspidatus **Dobson's trident bat.** Moluccus to New Hebrides
Cloeotis
C. percivali **Short-eared trident bat (Percival's trident bat).** Kenya to SE Botswana to Mozambique
Coelops
C. frithi **Tail-less leaf-nosed bat.** India to Java, Bali, Taiwan
C. robinsoni **Malayan taI-less leaf-nosed bat.** S Thailand, Malaya, Mindoro, Philippines
Hipposideros
H. abae **Aba leaf-nosed bat.** Guinea to S Sudan, Uganda
H. armiger **Himalayan leaf-nosed bat (Great leaf-nosed bat).** N India to Malaya, Taiwan
H. ater **Dusky leaf-nosed bat.** India to Java to NW, N Australia, Philippines
H. beatus **Dwarf leaf-nosed bat.** Liberia to SW Sudan, Gabon
H. bicolor **Bicolored leaf-nosed bat.** India to Java, Celebes, Philippines, Sri Lanka
H. breviceps N Pagi Is, Mentawei Is
H. caffer **Sundevall's leaf-nosed bat.** Africa, Arabia
H. camerunensis **Greater cyclops bat.** Cameroun, E Zaire
H. calcaratus **Spurred leaf-nosed bat.** New Guinea, Bismarck Archipelago to Solomon Is
H. cineraceus **Least leaf-nosed bat.** N India to Malaya, Borneo
H. commersoni **Commerson's leaf-nosed bat (Giant leaf-nosed bat).** Gambia to Somalia to Angola, Mozambique
H. coronatus Mindanao, Philippines
H. coxi **Cox's leaf-nosed bat.** Sarawak, Borneo
H. crumeniferus **Timor leaf-nosed bat.** Timor Is
H. cupidus **Eaga leaf-nosed bat.** New Guinea to Solomon Is
H. curtus **Short-tailed leaf-nosed bat.** Nigeria, Cameroun, Fernando Poo
H. cyclops **Cyclops leaf-nosed bat.** Guinea to Gabon to SW Kenya
H. diadema **Diadem leaf-nosed bat.** Burma to Java, Philippines, Borneo, New Guinea, Solomons, NE Queensland
H. dinops **Fierce leaf-nosed bat.** Celebes, Peling Is, Solomons
H. doriae Sarawak, Borneo
H. dyacorum **Dyak leaf-nosed bat.** Borneo
H. fuliginosus **Sooty leaf-nosed bat.** Ghana to Cameroun to Ethiopia
H. fulvus **Fulvous leaf-nosed bat.** Afghanistan to Vietnam
H. galeritus **Cantor's leaf-nosed bat (Fawn leaf-nosed bat).** Sri Lanka, India to New Hebrides, N Queensland
H. inexpectatus **Crested leaf-nosed bat.** N Celebes
H. jonesi **Jones' leaf-nosed bat.** Guinea to Nigeria
H. lankadiva **India, Sri Lanka**
H. larvatus **Horsfield's leaf-nosed bat.** Burma to Java, Sumba, Borneo
H. lekaguli **Lekagul's leaf-nosed bat.** S Thailand
H. lylei **Shield-faced leaf-nosed bat.** Burma to Malaya
H. marisae **Aellen's leaf-nosed bat.** Ivory Coast, Guinea
H. megalotis **Big-eared leaf-nosed bat.** Ethiopia, Kenya
H. muscinus **Fly River leaf-nosed bat.** Papua New Guinea
H. nequam **Malay leaf-nosed bat.** Malaya
H. obscurus Philippines
H. papua **Geelvink Bay leaf-nosed bat.** W New Guinea
H. pratti **Pratt's leaf-nosed bat.** SW China, Vietnam
H. pygmaeus Philippines
H. ridleyi **Ridley's leaf-nosed bat.** Malaya
H. ruber **Noack's leaf-nosed bat.** Senegal to Ethiopia to Angola
H. sabanus **Sabah leaf-nosed bat.** Malaya, Sumatra, Borneo
H. schistaceus India
H. semoni **Semon's leaf-nosed bat.** E New Guinea, N Queensland
H. speoris **Schneider's leaf-nosed bat.** India, Sri Lanka
H. stenotis **Narrow-eared leaf-nosed bat.** NW, N Australia
H. turpis **Lesser leaf-nosed bat.** S Thailand, Ryukyu Is
H. wollastoni **Wollaston's leaf-nosed bat.** S New Guinea
Paracoelops
P. megalotis Vietnam
Rhinonycteris
R. aurantius **Orange leaf-nosed bat.** NW, N Australia
Triaenops
T. furculus **Trouessart's trident bat.** Madagascar, Aldabra, Cosmoledo Is
T. humbloti **Humblot's trident bat.** Madagascar
T. persicus **Persian trident bat.** Congo Republic, Mozambique to S Arabia to Iran
T. rufus **Madagascan red trident bat.** Madagascar

FAMILY MORMOOPDAE

Mormoops
M. blainvillei **Antillean ghost-faced bat (Blainville's leaf chinned bat).** Greater Antilles, C Bahamas
M. megalophylla **Peter's ghost-faced or leaf chinned bat.** S USA to Ecuador, Venezuela, Trinidad etc
Pteronotus
P. davyi **Davy's naked-backed bat.** Mexico to Peru; Brazil, Lesser Antilles, Trinidad
P. fuliginosus **Sooty moustached bat.** Greater Antilles
P. gymnonotus **Big naked-backed bat.** S Mexico to E Peru, Brazil
P. macleayi **Macleay's moustached bat.** Cuba, Jamaica
P. parnelli **Parnell's moustached bat.** N Mexico to E Peru, Brazil, Trinidad, Greater Antilles
P. personatus **Wagner's moustached bat.** N Mexico to Brazil, Trinidad

FAMILY NOCTILIONIDAE

Noctilio
N. albiventris **Lesser bulldog bat (Southern bulldog bat).** Honduras to N Argentina
N. leporinus **Mexican bulldog bat (Fisherman bat).** W, S Mexico to N Argentina, Antilles, Trinidad

FAMILY MYSTACINIDAE

Mystacina
M. robusta **Greater short-tailed bat.** Stewart Is, New Zealand
M. tuberculata **Lesser short-tailed bat.** New Zealand

FAMILY PHYLLOSTOMATIDAE

Ametrida
A. centurio Venezuela, Guianas, Brazil, Trinidad
Anoura
A. brevirostrum Colombia, E Peru
A. caudifer Colombia, Venezuela to E Peru, E Brazil
A. cultrata Costa Rica, Panama, Venezuela
A. geoffroyi **Geoffroy's long-nosed bat.** N Mexico to NW Argentina, Trinidad, Grenada
A. werckleae Costa Rica
Ardops
A. nichollsi **Tree bat.** Lesser Antilles
Ariteus
A. flavescens **Jamaican fig-eating bat.** Jamaica
Artibeus
A. anderseni E Peru, W Brazil
A. aztecus C, NE Mexico to W Panama
A. cinereus **Gervais' fruit-eating bat.** Colombia, Venezuela to Bolivia, E Peru, Brazil, Trinidad, Tobago, Grenada
A. concolor Colombia, Venezuela to E Peru, Brazil, Surinam
A. fraterculus S Ecuador, N Peru
A. fuliginosus Peru to Colombia to Guianas
A. glaucus NW Colombia to C Peru
A. hirsutus **Hairy fruit-eating bat.** W Mexico
A. inopinatus El Salvador, Honduras, Nicaragua
A. jamaicensis **Jamaican fruit-eating bat.** N Mexico to Colombia, Venezuela, Bahamas, Antilles, Trinidad
A. lituratus **Great fruit-eating bat.** NE Mexico to N Argentina, Lesser Antilles, Trinidad, Tobago
A. phaeotis **Dwarf fruit bat.** W Mexico to E Peru
A. planirostris Peru, Bolivia to Guianas
A. toltecus NE Mexico to N Ecuador
A. watsoni S Mexico to Colombia
Barticonycteris
B. daviesi Costa Rica, E Peru, Guyana
Brachyphylla
B. cavernarum **St Vincent fruit-eating bat.** Puerto Rico, Lesser Antilles
B. nana **Cuban fruit-eating bat.** Cuba, Grand Cayman, S Bahamas, Hispaniola, Jamaica
Carollia
C. brevicauda E Mexico to E Peru, Bolivia, NE Brazil
C. castanea **Allen's short-tailed bat.** Honduras to E Peru, Bolivia
C. perspicillata **Seba's short-tailed bat (Short-tailed bat).** S Mexico to S Brazil, Paraguay, Trinidad, Tobago, Grenada, Jamaica
C. subrufa W Mexico to Nicaragua
Centurio
C. senex **Wrinkle-faced bat.** NE Mexico to Panama, Venezuela, Trinidad
Chiroderma
C. doriae E Brazil
C. improvisum Guadeloupe, Lesser Antilles
C. slavini **Salvin's white-lined bat.** W, C Mexico to Ecuador
C. trinitatum Panama, E Peru, Bolivia, Brazil, Trinidad
C. villosum S Mexico to E Peru, Bolivia, Brazil
Choeroniscus
C. godmani **Godman's bat.** W Mexico to Colombia, Venezuela
C. intermedius Trinidad, E Peru
C. minor E Peru to Colombia to Guianas, Brazil
C. periosus W Colombia (coast)
Choeronycteris
C. mexicana **Mexican long-nosed bat.** S California, S Arizona, S New Mexico to Honduras, NW Venezuela
Chrotopterus
C. auritus **Peter's woolly false vampire bat.** S Mexico to Paraguay, N Argentina
Ectophylla
E. alba **Honduran white bat.** Nicaragua to W Panama
E. macconnelli Costa Rica to E Peru, Bolivia, Brazil, Trinidad
Enchisthenes
E. harti **Little fruit-eating bat.** NE Mexico to E Peru, S Arizona
Erophylla
E. sezekorni **Buffy flower bat.** Bahamas, Cuba, Cayman Is, Jamaica, Hispaniola, Puerto Rico
Glossophaga
G. alticola C Mexico to Costa Rica
G. commissarisi W Mexico to Panama
G. longirostris **Miller's long-tongued bat.** Colombia, Venezuela, Lesser Antilles, Trinidad, etc
G. soricina **Pallas' long-tongued bat.** N Mexico to N Argentina, Trinidad, Jamaica, Bahamas
Hylonycteris
H. underwoodi **Underwood's long-tongued bat.** W Mexico to W Panama
Leptonycteris
L. curasoae Colombia, Venezuela, Curaçao Is
L. nivalis **Mexican long-nosed bat.** S Texas to Guatemala
L. sanborni **Sandborn's long-tongued bat.** S Arizona, New Mexico to El Salvador
Lichonycteris
L. degener NE Brazil
L. obscura Guatemala to Peru, Surinam
Lionycteris
L. spurrelli Panama, E Peru, N Brazil
Lonchophylla
L. bokermanni S Brazil
L. concava **Goldman's long-tongued bat.** Costa Rica, Panama, Peru
L. hesperia Peru
L. mordax Ecuador, Bolivia, Brazil
L. robusta **Panama long-tongued bat.** Nicaragua to Peru
L. thomasi **Thomas' long-tongued bat.** Panama, Peru, Bolivia, Brazil, Surinam
Lonchorhina
L. aurita **Tomes' long-eared bat.** S Mexico to E Peru, Bolivia, Brazil, Trinidad, Bahamas
L. marinkellei Colombia
L. orinocoensis C Venezuela
Macrophyllum
M. macrophyllum **Long-legged bat.** S Mexico to N Argentina
Macrotus
M. californicus **California leaf-nosed bat.** S California, S Nevada, Arizona, NW Mexico
M. waterhousei **Waterhouse's leaf-nosed bat.** N Mexico to Guatemala, Bahamas, Greater Antilles
Micronycteris
M. belni E Peru, C Brazil
M. brachyotis S Mexico to C Brazil
M. hirsuta **Hairy big-eared bat.** Honduras to E Peru, Guyana, Trinidad
M. megalotis **Brazilian big-eared bat.** NE Mexico to E Venezuela, Peru
M. minuta Nicaragua to E Peru, Brazil, Trinidad, Grenada
M. nicefori Nicaragua to NE Peru, N Brazil, Trinidad
M. pusilla E Colombia, N Brazil
M. schmidtorum **Schmidt's big-eared bat.** S Mexico to Colombia
M. sylvestris **Brown big-eared bat.** W, S Mexico to E Peru, NE Brazil, Trinidad
Mimon
M. bennetti Guyana, Surinam, NE Brazil
M. cozumelae **Cozumel spear-nosed bat.** S Mexico to N Colombia
M. crenulatum S Mexico to Peru, Bolivia, Brazil
Monophyllus
M. plethodon **Barbados long-tongued bat.** Lesser Antilles, Puerto Rico
M. redmani **Jamaican long-tongued bat.** Greater Antilles, S Bahamas
Musonycteris
M. harrisoni **Banana bat.** W Mexico
Phylloderma
P. stenops **Peters' spear-nosed bat.** S Mexico to Peru, NE Brazil
Phyllonycteris
P. aphylla **Jamaican flower bat.** Jamaica
P. major **Puerto Rican flower bat.** Puerto Rico
P. obtusa **Hispaniola flower bat.** Hispaniola
P. poeyi **Cuban flower bat,** Cuba
Phyllops
P. falcatus **Cuban fig-eating bat.** Cuba
P. haitiensis **Dominican fig-eating bat.** Hispaniola
Phyllostomus
P. discolor **Pale spear-nosed bat.** S Mexico to N Argentina
P. elongatus Colombia, Venezuela to E Peru, Bolivia, SE Brazil
P. hastatus **Greater spear-nosed bat.** Honduras to Peru, Bolivia, SE Brazil
P. latifolius SE Colombia, Guyana

Platalina
P. genovensium Peru
Pygoderma
P. bilabiatum **Ipanema bat.** Surinam to Paraguay, N Argentina
Rhinophylla
R. alethina W Colombia
R. fischerae Colombia to E Peru, NW Brazil
R. pumilio E Peru to Colombia to Surinam
Scleronycteris
S. ega Venezuela, Brazil
Sphaeronycteris
S. toxophyllum Colombia, E Peru to Venezuela, Bolivia
Stenoderma
S. rufum **Red fruit bat.** Puerto Rico, Virgin Is
Sturnira
S. aratathomasi SW Colombia, W Ecuador
S. bidens Colombia, E Ecuador, E Peru
S. bogotensis Colombia, Venezuela
S. erythromos E Peru, Venezuela
S. lilium **Yellow-shouldered bat.** N Mexico to N Argentina, Uruguay, Chile, Lesser Antilles, Jamaica
S. ludovici **Anthony's bat.** NE Mexico to E Peru, Venezuela
S. magna Amazonian Colombia to E Peru
S. mordax **Hairy-footed bat.** Costa Rica
S. nana C Peru
S. thomasi **Thomas' epauletted bat.** Guadeloupe, Lesser Antilles
S. tildae E Peru to Brazil, Trinidad
Tonatia
T. bidens **Spix's round-eared bat.** Guatemala to E Peru, E Brazil, Trinidad, Jamaica
T. brasiliense E Peru, C, E Brazil
T. carrikeri Venezuela, E Peru, Bolivia, Surinam
T. evotis S Mexico to Honduras
T. minuta S Mexico to E Peru, Trinidad
T. sylvicola **D'Orbigny's round-eared bat.** S Mexico to N Argentina
T. venezuelae Venezuela
Trachops
T. cirrhosus **Fringe-lipped bat.** S Mexico to E Peru, Bolivia, S Brazil
Uroderma
U. bilobatum **Tent-making bat.** S Mexico to S Peru, Bolivia, SE Brazil, Trinidad
U. magnirostrum S Mexico to E Peru, N Bolivia
Vampyressa
V. bidens Colombia to E Peru, N Brazil, Guyana
V. brocki Colombia, Guyana
V. melissa E Peru
V. nymphaea **Big yellow-eared bat.** Nicaragua to W Colombia
V. pusilla S Mexico to E Peru, SE Brazil
Vampyrodes
V. caraccioloi **Great stripe-faced bat.** S Mexico to E Peru, N Brazil
Vampyrops
V. aurarius E Venezuela
V. brachycephalus Colombia to E Peru, Guyana
V. dorsalis Costa Rica to E Peru, Venezuela
V. helleri **Heller's bat.** S. Mexico to Peru, Bolivia, Brazil, Trinidad
V. infuscus Colombia to E Peru, Brazil
V. lineatus Colombia, E Peru, C, E Brazil to N Argentina, Uruguay
V. recifinus E Brazil
V. umbratus Colombia, Venezuela
V. vittatus **Greater white-lined bat.** Costa Rica to E Peru, Venezuela
Vampyrum
V. spectrum **False vampire bat (Linnaeus' false vampire bat).** S. Mexico to E Peru, C Brazil, Trinidad, Jamaica

FAMILY DESMODONTIDAE

Desmodus
D. rotundus **Common vampire.** N Mexico to C Chile, N Argentina, Uruguay, Trinidad
Diaemus
D. youngi **White-winged vampire.** NE Mexico to E Peru, Bolivia, Brazil, Trinidad
Diphylla
D. ecaudata **Hairy-legged vampire bat.** S Texas to E Peru, S Brazil

FAMILY MOLOSSIDAE

Cheiromeles
C. parvidens C Celebes, Philippines
C. torquatus **Naked or Hairless bat.** Malaya to Java, Borneo, Philippines
Eumops
E. auripendulus **Slouch-eared bat.** S Mexico to N Argentina, Trinidad, Jamaica
E. bonariensis **Peter's mastiff bat.** S Mexico to C Argentina
E. dabbenei N Venezuela, N Colombia, N Argentina, Paraguay
E. glaucinus **Wagner's mastiff bat.** S Florida, C Mexico to Paraguay, SE Brazil, Cuba, Jamaica
E. hansae Costa Rica to Brazil, Guyana
E. maurus **Guianan mastiff bat. Guyana, Surinam**
E. perotis **Greater mastiff bat.** S USA to C Mexico, Venezuela to N Argentina
E. underwoodi **Underwood's mastiff bat.** S Arizona to Honduras
Molossops
M. abrasus Venezuela
M. aequatorianus W Ecuador
M. brachymeles Peru to N Argentina, Guianas
M. greenhalli W Mexico to Venezuela, Trinidad
M. milleri Peru
M. paranum C Mexico, Venezuela, Brazil
M. planirostris **Dog-faced bat.** Panama to Guyana, Brazil
M. temmincki Colombia to N Argentina, Uruguay
Molossus
M. ater **Red mastiff-bat.** N Mexico to Guyana to Peru, N Argentina, Trinidad
M. barnese French Guiana, Brazil
M. bondae **Bonda mastiff bat.** Honduras to N Colombia, NW Venezuela
M. molossus **Pallas' mastiff bat.** N Mexico to N Argentina, Trinidad, Antilles
M. pretiosus **Miller's mastiff bat.** Nicaragua to Venezuela
M. sinaloae **Allen's mastiff bat.** W. Mexico to N Venezuela
M. trinitatis **Trinidad mastiff bat.** Trinidad
Myopterus
M. albatus **Banded free-tailed bat.** Ivory Coast, NE Zaire
M. daubentoni **Daubenton's free-tailed bat.** Senegal
M. whitleyi **Bini free-tailed bat.** Ghana to Zaire, Uganda
Neoplatymops
N. mattogrossensis C Venezuela, S Guyana, C Brazil
Otomops
O. martiensseni **Giant mastiff bat (Martienssen's free-tailed bat).** Ethiopia to Angola, Natal, Madagascar
O. papuensis **Big-eared mastiff bat.** C New Guinea
O. secundus **Mantled mastiff bat.** NC New Guinea
O. wroughtoni **Wroughton's free-tailed bat.** S India
Platymops
P. setiger **Peters' flat-headed bat.** SE Sudan, S Ethiopia, Kenya
Promops
P. centralis **Thomas' mastiff bat.** W Mexico to Peru, Paraguay
P. nasutus N Argentina to E Brazil, Trinidad
P. pamana C Brazil
Sauromys
S. petrophilus **Robert's flat-headed bat.** Namibia to Zimbabwe, Mozambique, S Africa
Tadarida
T. acetabulosus **Natal wrinkle-lipped bat.** Ethiopia, Natal, Madagascar, Mauritius, Reunion
T. aegyptiaca **Egyptian free-tailed bat.** Africa, Arabia, Iran to India, Sri Lanka
T. africana **Giant African free-tailed bat.** Sudan, Ethiopia to Mozambique, Transvaal
T. aloysiisabaudiae **Duke of Abruzzi's free-tailed bat.** Ghana, Gabon, N Zaire, Uganda
T. ansorgei **Ansorge's free-tailed bat.** Cameroun, Ethiopia to Angola
T. aurispinosa Mexico, Colombia, Peru, E Brazil
T. australis **Southern mastiff bat.** E, SW, S Australia
T. beccarii **Beccari's mastiff bat.** Amboina, New Guinea, Queensland
T. bemmeleni **Gland-tailed free-tailed bat.** Liberia to S Sudan to N Tanzania
T. bivittatta **Spotted free-tailed bat.** Ethiopia to Zambia, Mozambique
T. brachypterus Mozambique
T. brasiliensis **Brazilian free-tailed (Mexican free-tailed bat).** W, S USA to C Chile, Argentina, Bahamas, Antilles
T. chapini **Chapin's free-tailed bat.** W, NE Zaire, Uganda to Namibia, Ethiopia
T. condylura **Angola free-tailed bat.** Gambia to Somalia to Angola, Mozambique, Madagascar
T. congica **Medje free-tailed bat.** Nigeria, Cameroun, NE Zaire
T. demonstrator **Mongalla free-tailed bat.** Upper Volta, Sudan, NE Zaire, Uganda
T. doriae Sumatra
T. europs Venezuela, Brazil, Trinidad
T. femorosacca **Pocketed free-tailed bat.** S USA to S Mexico
T. fulminans **Madagascar large free-tailed bat.** E Zaire, Kenya to Zimbabwe, Madagascar
T. gallagheri C Zaire
T. gracilis Venezuela, Brazil
T. jobensis **Northern mastiff bat.** New Guinea, N Australia, Solomons, Fiji
T. johorensis **Dato Meldrum's bat.** Malaya, Sumatra
T. jugularis **Peter's wrinkle-lipped bat.** Madagascar
T. kalinowskii Peru, N Chile
T. kuboriensis **Small-eared mastiff bat.** SE New Guinea
T. lanei Mindanao, Philippines
T. laticaudata **Broad-tailed bat.** NE Mexico to Venezuela to Paraguay, Cuba
T. leonis **Sierra Leone free-tailed bat.** Sierra Leone to E Zaire
T. lobata **Big-eared Kenya free-tailed bat.** Kenya, Zimbabwe
T. macrotis **Big free-tailed bat.** C USA to Brazil, Paraguay, Greater Antilles
T. major **Lappet-eared free-tailed bat.** Ghana, Mali to S Sudan to Tanzania
T. midas **Midas bat.** Senegal to Ethiopia to Zimbabwe, SW Arabia, Madagascar
T. minuta Cuba
T. mops **Malayan free-tailed bat.** Malaya, Sumatra, Borneo, Java
T. nanula **Dwarf free-tailed bat.** Sierra Leone to Ethiopia, Kenya
T. nigeriae **Nigerian free-tailed bat.** Nigeria to Ethiopia to Namibia, Zambia
T. niveiventer **White-bellied free-tailed bat.** Zaire to Angola, N Botswana, Zambia, Madagascar
T. norfolkensis **Norfolk Island scurrying bat,** SE Queensland, Norfolk Is
T. phrudus Peru
T. planiceps **Little flat bat.** N, W, S Australia, SE New Guinea
T. plicata **Wrinkle-lipped bat.** Sri Lanka, India to S China to Java, Borneo, Cocos-Keeling Is, Philippines, Hainan
T. pumila **Little free-tailed bat.** Senegal to Somalia to Angola, Natal, Madagascar, SW Arabia
T. pusillus Aldabra Is
T. russata **Russet free-tailed bat.** Ghana, Cameroun, NE Zaire
T. sarasinorum **Celebes mastiff bat.** C Celebes
T. teniotis **European free-tailed bat.** S Europe, N Africa to N India, China, Korea, Japan, Taiwan
T. thersites **Railer bat.** Sierra Leone to SE Zaire, Zanzibar
T. trevori S Sudan, NE Zaire, Uganda
T. ventralis **Giant African free-tailed bat.** S Sudan, Ethiopia to Malawi, Transvaal
T. yucatanica **Yucatan free-tailed bat.** S Mexico to Guatemala
Xiphonycteris
X. spurrelli **Spurrell's free-tailed bat.** Ghana, Togo, Rio Muni, Fernando Poo, Zaire

FAMILY NATALIDAE

Natalus
N. brevimanus Old Providence Is
N. lepidus **Gervais' long-legged bat.** Bahamas, Cuba, Isle of Pines
N. macer Cuba
N. major Cuba, Jamaica, Hispaniola
N. micropus **Jamaican long-legged bat.** Jamaica
N. stramineus **Mexican funnel-eared bat.** Mexico to Brazil, Guianas, Lesser Antilles
N. tumidifrons Bahamas
N. tumidirostris Colombia, Venezuela, Trinidad, Curacao Is

FAMILY THYROPTERIDAE

Thyroptera
T. discifera **Honduran disk-winged bat.** Nicaragua to E Peru, French Guiana
T. tricolor **Spix's disk-winged bat.** S Mexico to E Peru, Guianas, S, E Brazil, Trinidad

FAMILY FURIPTERIDAE

Amorphochilus
A. schnabli W Ecuador to N Chile
Furipterus
F. horrens Costa Rica to E Peru, Guianas, SE Brazil, Trinidad

FAMILY VESPERTILIONIDAE

Antrozous
A. dubiaquercus Tres Marias Is, Veracruz, Mexico, Honduras
A. koopmani Cuba
A. pallidus **Pallid bat.** British Colombia, W, C USA to W Mexico
Baeodon
B. alleni **Allen's baeodon.** W, C Mexico
Barbastella
B. barbastellus **Western barbastelle.** England, France, Morocco to Caucasus
B. leucomelas **Eastern barbastelle.** Caucasus to N India, W China, Japan, NE Africa
Chalinolobus
C. dwyeri **Large-eared pied bat.** S Queensland to C New South Wales
C. gouldi **Gould's wattled bat.** Australia, New Caledonia
C. morio **Chocolate bat.** S Australia, Tasmania
C. nigrogriseus **Hoary bat (Frosted bat).** SE New Guinea, Fergusson Is, N Australia
C. picatus **Little pied bat.** S Queensland to New South Wales
C. tuberculatus **Long-tailed bat.** New Zealand
Dasypterus
D. ega **Southern yellow bat.** SW USA to Argentina
D. egregius Panama, Brazil
D. intermedius **Northern yellow bat (Eastern yellow bat).** SE Virginia to Honduras, Cuba
Eptesicus
E. bobrinskoi Kazakhstan, NW Iran
E. bottae **Botta's serotine.** NE Egypt to Arabia to Turkestan
E. brasiliensis **Brazilian brown bat.** C Mexico to Peru, C, SE Argentina
E. brunneus **Dark brown serotine.** Ivory Coast to C Zaire
E. capensis **Cape serotine.** Africa S of Sahara, Ethiopia, Madagascar
E. chiriquinus **Chiriqui brown bat.** Panama
E. demissus **Surat serotine.** S Thailand
E. diminutus E, SE Brazil to N Argentina, Uruguay
E. douglasi W Australia
E. flavescens **Yellow serotine.** Angola
E. floweri **Horn-skinned bat.** Mali, S Sudan
E. furinalis Mexico to N Argentina
E. fuscus **Big brown bat.** Alaska, S Canada to Colombia, Venezuela, Bahamas, Cuba, Hispaniola, Puerto Rico
E. guadeloupensis Guadeloupe, Lesser Antilles
E. guineensis **Tiny serotine.** Senegal to Sudan, NE Zaire
E. hottentotus **Long-tailed house bat.** Namibia to Mozambique, S Africa
E. innoxius W Ecuador, W Peru, Panama
E. loveni **Loven's serotine.** W Kenya
E. lynni **Lynn's brown bat.** Jamaica
E. melckorum **Melck's house bat.** Zambia, Mozambique, SW Cape Province
E. nasutus **Sind bat.** S Arabia to Pakistan
E. nilssoni **Northern bat.** France, Norway to E Siberia, Japan, Iraq, Tibet
E. pachyotis **Thick-eared bat.** Assam to N Thailand
E. platyops **Lagos serotine.** Senegal, Nigeria
E. pumilus **Little bat.** N, C, E Australia
E. regulus SW, SE Australia
E. rendalli **Rendall's serotine.** Gambia to Somalia to Mozambique, Botswana
E. sagittula SE Australia, Lord Howe Is
E. serotinus **Serotine.** Morocco, W Europe to Thailand, China, Korea
E. somalicus **Somali serotine,** Somalia to Upper Volta, Togo, NE Zaire, Kenya, Cameroun, Namibia
E. tenuipinnis **White-winged serotine.** Guinea to Kenya to Angola
E. vulturnus SE Australia, Tasmania
E. walli **Wall's serotine.** W Iraq, SW Iran
E. zuluensis **Aloe bat.** Namibia, Zambia to S Africa
Euderma
E. maculatum **Spotted bat (Pinto bat).** W, S USA to N, C Mexico
Eudiscopus
E. denticulus **Disc-footed bat.** Burma, Laos
Glauconycteris
G. alboguttatus **Allen's striped bat.** E Zaire
G. argentata **Silvered bat.** Cameroun to Kenya to NE Angola, Tanzania
G. beatrix **Beatrix bat.** Cameroun to Uganda
G. egeria **Bibundi bat.** Cameroun
G. gleni Cameroun, Uganda
G. machadoi **Machado's butterfly bat.** E Angola

G. poensis **Abo bat.** Ghana to C, E Zaire, Fernando Poo
G. superba **Pied bat.** Ghana, NE Zaire, Uganda
G. variegata **Butterfly bat.** Ghana to Somalia to Namibia, Mozambique
Glischropus
G. javanus Java
G. tylopus **Thick-thumbed pipistrelle.** Burma to Sumatra, Borneo, Palawan
Harpiocephalus
H. harpia **Hairy-winged bat.** India to Vietnam, Sumatra, Java, S Moluccas
Hesperoptenus
H. blanfordi **Blanford's bat.** S Burma to Malaya
H. doriae **False serotine bat.** Malaya, Borneo
H. tickelli **Tickell's bat.** India to Thailand, Andaman Is, Sri Lanka, S China
H. tomesi Malaya, Borneo
Histiotus
H. alienus Brazil, Uruguay
H. laephotis Bolivia
H. macrotus Chile, Peru
H. montanus Colombia to Chile, Argentina
H. velatus Brazil
Ia
I. io **Great evening bat.** Assam to S China, Indochina
Idionycteris
I. phyllotis **Allen's big-eared bat.** Arizona, W New Mexico to C Mexico
Kerivoula
K. africana **Tanzanian woolly bat.** Tanzania
K. agnella **Louisiade trumpet-eared bat.** Sudest Is, SE New Guinea, St Aignan's Is, Louisiade Archipelago
K. argentata **Damara woolly bat.** S Kenya to Namibia to Natal
K. cuprosa **Copper woolly bat.** S Cameroun, Kenya, Zaire
K. eriophora Ethiopia
K. hardwickei **Hardwicke's forest bat.** Sri Lanka, India to S China to Java, Lesser Sundas, Philippines, Celebes
K. lanosa **Lesser woolly bat.** Liberia to Ethiopia to S Africa
K. minuta **Least forest bat.** S Thailand, Malaya
K. muscina **Fly River trumpet-eared bat.** SE New Guinea
K. myrella **Bismarck trumpet-eared bat.** Admiralty Is, Bismarck Archipelago
K. papillosa **Papillose bat.** NE India to Java, Borneo
K. pellucida **Clear-winged bat.** Philippines, Malaya, Borneo, Sumatra, Java
K. phalaena **Spurrell's woolly bat.** Liberia, Ghana, Cameroun, Zaire
K. picta **Painted bat.** Sri Lanka, S India to S China to Java, Lesser Sundas, Borneo, Ternate Is
K. smithi **Smith's woolly bat.** Nigeria to E Zaire, Kenya
K. whiteheadi S Thailand, Borneo, Philippines
Laephotis
L. angolensis Angola, S Zaire
L. botswanae S Zaire, Zambia, NW Botswana
L. namibensis Namibia
L. wintoni **De Winton's long-eared bat.** Ethiopia, Kenya
Lamingtona
L. lophorhina **Lamington free-eared bat.** SE New Guinea
Lasionycteris
L. noctivagans **Silver-haired bat.** Alaska, S Canada to NE Mexico
Lasiurus
L. borealis Red bat. S Canada to C Chile, Argentina, Bahamas, Greater Antilles, Puerto Rico
L. brachyotis Galapagos Is
L. castaneus Panama
L. cinereus **Hoary bat.** S Canada to C Chile, N Argentina, Hawaii
L. seminolus **Seminole bat.** E USA
Mimetillus
M. moloneyi **Moloney's flat-headed bat.** Sierra Leone to W Kenya to Angola
Miniopterus
M. australis **Little long-fingered bat (Lesser bent-winged bat).** Thailand, Philippines to E Australia, New Caledonia, Loyalty Is
M. fraterculus **Lesser long-fingered bat.** Malawi to S Africa, Zambia
M. inflatus **Greater long-fingered bat.** Cameroun to Somalia, Zambia, Mozambique
M. medius **SE Asian long-fingered or bent-winged bat.** Thailand to Java, Philippines, New Caledonia, Loyalty Is
M. minor **Least long-fingered bat.** Congo Republic to Tanzania, Madagascar
M. robustior Loyalty Is
M. schreibersi **Schreiber's bent-winged bat or long-fingered bat.** Africa, Madagascar, SW Europe to China, Japan, Philippines, Solomons, NW, N, E Australia
M. tristis Philippines, New Guinea, Solomons, New Hebrides
Murina
M. aenea **Bronze tube-nosed bat.** Malaya
M. aurata **Little tube-nosed bat.** Nepal to China, Korea, Sakhalin, Japan
M. balstoni Java
M. cyclotis **Round-eared tube-nosed bat.** N India to S China to Malaya, Hainan, Sri Lanka, Philippines, Borneo
M. florium **Flores tube-nosed bat.** Lesser Sunda Is, S Moluccas
M. grisea **Peters' tube-nosed bat.** NW India
M. huttoni **Hutton's tube-nosed bat.** Himalayas to S China, Malaya
M. leucogaster **Greater tube-nosed bat.** NE India, S China to E Siberia, Japan
M. suilla **Brown tube-nosed bat.** Malaya to Java, Borneo
M. tenebrosa Tsushima Is, Japan
M. tubinaris Kashmir to Vietnam
Myotis
M. abei Sakhalin Is
M. adversus **Large-footed bat.** Borneo, Java to Solomon Is, New Hebrides, N, E Australia
M. aelleni Argentina
M. albescens **Paraguay myotis.** S Mexico to Paraguay, Uruguay
M. altarium S China
M. annectans **Hairy-faced bat.** NE India to NE Thailand
M. argentatus **Silver-haired myotis.** Veracruz, S Mexico
M. atacamensis S Peru, N Chile
M. auriculus **Mexican long-eared bat.** SW New Mexico, SE Arizona to SC Mexico
M. australis **Small-footed myotis.** E Australia
M. austroriparius **South-eastern myotis.** N Carolina, Kentucky to Louisiana, Florida
M. bartelsi Java
M. bechsteini **Bechstein's bat.** Spain, England to W Russia, Caucasus
M. blythi **Lesser mouse-eared bat.** Spain, Morocco to Afghanistan to S China
M. bocagei **Rufous mouse-eared bat.** Liberia to Kenya to Angola, Mozambique
M. brandti **Brandt's bat.** Spain, Britain to Urals
M. browni Mindanao, Philippines
M. californicus **California myotis.** Alaska to S Mexico
M. capaccinii **Long-fingered bat.** N Africa, Spain to Iran
M. chiloensis Chile, Costa Rica, Panama
M. chinensis **Large myotis.** S China, N Thailand
M. cubanensis **Cuban myotis.** Guatemala
M. dasyeneme **Pond bat.** Netherlands to Manchuria
M. daubentoni **Daubenton's bat.** Spain, Britain to E Siberia, Manchuria, Sakhalin, Hokkaido
M. dominicensis Dominica, Lesser Antilles
M. dryas S Andaman Is, Indian Ocean
M. elegans C Mexico to Costa Rica
M. emarginatus **Geoffroy's bat.** SW Europe to Russian Turkestan, E Iran, Morocco
M. evotis **Long-eared myotis.** SW Canada to NW Mexico
M. fimbriatus Fukien, SE China
M. findleyi **Findley's myotis.** Tres Marias Is, W Mexico
M. formosus **Hodgson's bat.** E Afghanistan to Korea, S China, Taiwan
M. fortidens **Cinnamon myotis.** W, S Mexico
M. frater Turkestan to E Siberia, SE China, Japan
M. goudoti **Malagasy mouse-eared bat.** Madagascar, Anjouan Is, Comoro Is
M. grisescens **Gray bat.** Oklahoma to Kentucky to Georgia
M. hasselti **Lesser large-footed bat.** Sri Lanka, Thailand to Malaya, Java, Borneo
M. hermani NW Sumatra
M. herrei Luzon, Philippines
M. horsfieldi **Deignan's bat.** S China to Java, Bali, Borneo, Celebes
M. hosonoi N, C Honshu, Japan
M. ikonnikovi E Siberia, N Korea, Sakhalin, Hokkaido
M. jeannea Mindanao, Philippines
M. keaysi NE Mexico to Venezuela to Peru, Trinidad
M. keeni **Keen's myotis.** Alaska to Washington, Manitoba to Newfoundland to Florida
M. larensis NW Venezuela
M. leibi **Least brown bat (Small-footed myotis).** SW Canada, USA (except SE), N Mexico
M. levis S Brazil to Paraguay, Uruguay, Argentina
M. longipes Afghanistan, Kashmir, Vietnam
M. lucifugus **Little brown myotis or bat.** Alaska, S Canada to C Mexico
M. macrodactyius E Siberia, S Kurile Is, Japan
M. macrotarsus Philippines, Borneo
M. martiniquensis Martinique, Lesser Antilles
M. milleri **Miller's myotis.** Baja California, Mexico
M. montivagus **Burmese whiskered bat.** S India, Burma, S China, Malaya
M. morrisi Ethiopia
M. muricolo N India to Java to New Guinea
M. myotis **Large mouse-eared bat.** SW Europe to Syria
M. mystacinus **Whiskered bat.** Ireland to Japan, N Iran, Tibet, Morocco
M. nathalinae Spain, France, Switzerland
M. nattereri **Natterer's bat.** Morocco, W Europe to SE Siberia, Japan
M. nigricans **Black myotis.** W, NE Mexico to N Argentina, Trinidad, Tobago, Grenada
M. occultus **Arizona myotis.** SW USA
M. oreias **Singapore whiskered bat.** Malaya
M. oxyotus Costa Rica to Peru, N Bolivia
M. ozensis C Honshu, Japan
M. patriciae Mindanao, Philippines
M. peninsularis Baja California
M. pequinius Hopei, Shantung, China
M. peshwa India
M. planiceps **Flat-headed myotis.** N Mexico
M. pruinosus N Honshu, Japan
M. ricketti **Rickett's big-footed bat.** E China
M. ridleyi **Ridley's bat.** Malaya, Sumatra
M. riparius Honduras to Peru to Uruguay
M. rosseti **Thick-thumbed myotis.** Thailand, Cambodia
M. ruber SE Brazil, Paraguay
M. rufopictus Philippines
M. scotti **Scott's mouse-eared bat.** Ethiopia
M. seabrai **Angola wing-gland bat.** Angola to Cape Province
M. sicarius Nepal, Sikkim
M. siligorensis **Himalayan whiskered bat.** N India to S China to Malaya
M. simus Panama to E Peru, Brazil
M. sodalis **Indiana bat.** C, E USA
M. stalkeri **Kei myotis.** Kei Is, New Guinea
M. surinamensis Surinam
M. thysonodes **Fringed myotis.** SW Canada to S Mexico
M. tricolor **Cape hairy bat.** Ethiopia to Zaire, S Africa
M. velifer **Cave myotis.** S USA to Honduras
M. volans **Long-legged bat (Hairy-winged bat).** Alaska to S Mexico
M. weberi **Orange-winged myotis.** S Celebes
M. welwitschi **Welwitsch's bat.** Ethiopia to Zaire, Mozambique, S Africa
M. yumanensis **Yuma myotis.** British Columbia to C USA to C Mexico
Nyctalus
N. aviator Korea, Japan
N. lasiopterus **Giant noctule.** SW Europe to Iran
N. leisleri **Leisler's or Lesser noctule.** Madeira, Azores, W Europe to N India
N. montanus E Afghanistan to N India
N. noctula **Noctule.** Britain, Morocco, W Europe to China, Japan, Taiwan, Malaya
Nycticeius
N. balstoni **Balston's broad-nosed bat.** W, N, E, SC Australia, New Guinea
N. greyi **Little broad-nosed bat (Grey's bat).** W, N Australia
N. humeralis **Evening bat (Twilight bat).** C, SE USA to E Mexico, Cuba
N. influatus **Hughenden broad-nosed bat.** C Queensland
N. rueppelli **Rüppell's broad-nosed bat.** E Queensland, E New South Wales
N. schlieffeni **Schlieffen's bat.** Mauretania to Egypt to Namibia, Mozambique, SW Arabia
Nyctophilus
N. arnhemensis **Arnhem Land long-eared bat.** N Northern Territories
N. bifax **Northern long-eared bat.** N Australia
N. geoffroyi **Lesser long-eared bat.** C, S Australia, Tasmania
N. micordon **Small-toothed long-eared bat.** SE New Guinea
N. microtis **Papuan long-eared bat.** SE New Guinea
N. timoriensis **Greater long-eared bat.** Australia, Tasmania, Timor
N. walkeri N Northern Territories
Otonycteris
O. hemprichi **Hemprich's long-eared bat.** Algeria to Egypt to Afghanistan
Pharotis
P. imogene **Big-eared bat.** SE New Guinea
Philetor
P. brachypterus **New Guinea brown bat.** Malaya to New Guinea, Borneo, Java
Phoniscus
P. aerosa **Dubious trumpet-eared bat.** SE Asia
P. atrox **Groove-toothed bat.** S Thailand, Malaya, Sumatra
P. jagori **Peter's trumpet-eared bat.** Samar Is, Philippines, Java, Celebes
P. papuensis **Papuan trumpet-eared bat.** E New Guinea, NE Queensland
Pipistrellus
P. aero NW Kenya, Ethiopia
P. affinis **Chocolate bat.** NE Burma, Yunnan
P. anchietai **Anchieta's pipistrelle.** Angola, S Zaire, Zambia
P. angulatus **Greater New Guinea pipistrelle.** New Guinea, Solomons, S Moluccas, N, W Australia, N Celebes
P. anthonyi N Burma
P. ariel **Desert pipistrelle.** Egypt, Sudan
P. babu Pakistan, N, C India
P. bodenheimeri **Bodenheim's pipistrelle.** Israel, SW Arabia, Socotra Is
P. cadornae **Thomas' pipistrelle.** NE India to Thailand
P. ceylonicus **Kelaart's pipistrelle.** Pakistan to S China, Borneo, Sri Lanka
P. circumdatus **Gilded black pipistrelle.** N Burma to Java
P. coromandra **Indian pipistrelle.** E Afghanistan to S China, Vietnam, Sri Lanka
P. crassulus **Broad-headed pipistrelle.** Cameroun, C Zaire
P. deserti **Desert pipistrelle.** Algeria to Egypt, N Sudan, Upper Volta
P. eisentrauti **Eisentratu's pipistrelle.** Cameroun
P. endoi Honshu, Japan
P. hesperus **Western pipistrelle (Canyon bat).** Washington to C Mexico
P. imbricatus **Brown pipistrelle.** Malaya to Java, Celebes, Philippines
P. inexspectatus **Aellen's pipistrelle.** Cameroun, Sudan, Zaire
P. javanicus **Javan pipistrelle.** Japan, E Siberia to Java, Borneo, Celebes, Philippines, N Australia
P. joffrie Burma
P. kitcheneri Borneo
P. kuhli **Kuhl's pipistrelle.** Africa, SW Europe to Kashmir
P. lophurus S Thailand, S Burma
P. macrotis Sumatra
P. maderensis **Madeira pipistrelle.** Madeira, Canary Is
P. mimus **Indian pygmy pipistrelle.** Pakistan, India to N Vietnam, Sri Lanka
P. minahassae **Minahassa pipistrelle.** N Celebes
P. mordax Java, Sri Lanka, India
P. murrayi Christmas Is, Cocos-Keeling Is
P. musciculus **Least pipistrelle.** Cameroun, C Zaire, Gabon
P. nanulus **Tiny pipistrelle.** Nigeria to NE Zaire
P. nanus **Banana bat.** Sierra Leone to Somalia to S Africa, Madagascar
P. nathusii **Nathusius' pipistrelle.** Spain to Urals, Caucasus
P. peguensis Pegu, S Burma
P. permixtus **Dar-es-Salaam pipistrelle.** Tanzania
P. petersi **Peters' pipistrelle.** N Celebes, Buru, S Moluccas
P. pipistrellus **Common pipistrelle.** W Europe, Morocco to Kashmir, Korea
P. pulveratus **Chinese pipistrelle.** S China, Thailand
P. rueppelli **Rüppell's bat.** Senegal to Tanzania to Botswana, Egypt, Iraq
P. rusticus **Rusty bat.** Ghana to Ethiopia to Namibia, Transvaal
P. savii **Savi's pipistrelle.** Canary, Cape Verde Is, Iberia, Morocco to Korea, Japan, Burma
P. socientatis Malaya
P. stenopterus Malaya, Sumatra, Borneo, Philippines

P. subflavus **Eastern pipistrelle.** SE Canada to Honduras
P. tasmaniensis **Tasmanian pipistrelle.** E Australia, Tasmania
P. tenuis **Least pipistrelle.** S Thailand to Java, Borneo, Philippines, Celebes, Timor
Pizonyx
P. vivesi **Fish-eating bat (Mexican fishing bat).** NW Mexico, coasts and islands
Plecotus
P. auritus **Brown or Common long-eared bat.** Britain, France to NE China, Korea, Japan, N India
P. austriacus **Gray long-eared bat.** Spain, S England to W China, Cape Verde Is, N Africa
P. mexicanus NW, NE, C Mexico
P. rafinesquei **Raffinesques big-eared bat (Eastern lump-nosed bat).** SE USA
P. townsendi **Townsend's big-eared bat (Lump-nosed bat).** SW Canada, W USA to Mexico
Rhogeessa
R. gracilis **Slender yellow bat.** W Mexico
R. minutilla N Venezuela, NE Colombia
R. mira Michoacan, C Mexico
R. parvula **Little yellow bat.** W Mexico
R. tumida E Mexico to Ecuador, S Brazil, Bolivia
Scotoecus
S. albofuscus **Light-winged lesser house bat.** Senegal to Tanzania, S Malawi, Mozambique, Zambia
S. hindei Nigeria to Somalia to Tanzania to Angola
S. hirundo Senegal to Ethiopia
S. pallidus Pakistan, N India
Scotomanes
S. emarginatus India
S. ornatus **Harlequin bat.** N India to S China, Vietnam
Scotophilus
S. borbonicus Reunion, Madagascar
S. celebensis **Celebes yellow bat.** N Celebes
S. dinganii **African yellow house bat.** Senegal to Ethiopia to S Africa, Madagascar
S. heathi **Asiatic greater yellow house bat.** Afghanistan to S China, Vietnam, Sri Lanka
S. kuhli **Asiatic lesser yellow house bat.** Pakistan to Hainan to Timor, Borneo, Philippines, Taiwan
S. leucogaster Senegal to Somalia, Aden
S. nigrita **Greater brown bat.** Senegal to S Sudan to Mozambique
S. nigritellus Niger, Mali to Ivory Coast, Ghana, Togo
S. viridis Tanzania to Angola, S Africa
Scotozous
S. dormeri India
Tomopeas
T. ravus NW Peru
Tylonycteris
T. pachypus **Bamboo bat (Lesser club-footed bat).** India, S China to Java, Lesser Sundas, Borneo, Philippines
T. robustula **Greater club-footed bat (Flat-headed bat).** SW China to Java, Borneo, Celebes, Timor
Vespertilio
V. murinus **Particolored bat.** Scandinavia, Siberia to Iran, Afghanistan
V. orientalis E China, Honshu, Taiwan
V. superans E Siberia, E China, Japan

FAMILY MYZOPODIDAE
Myzopoda
M. aurita **Sucker-footed bat.** Madagascar

ORDER MARSUPIALIA

FAMILY DIDELPHIDAE
Caluromys
C. derbianus **Derby's woolly opossum.** Veracruz, Mexico to W Colombia to N Ecuador
C. lanatus **Ecuadorian woolly opossum.** E Colombia, W Venezuela, Brazil, Bolivia, Paraguay, E Ecuador, E Peru
C. philander **Bare-tailed woolly opossum.** N S America E of Andes: Brazil, Venezuela, French Guiana, Guyana, Surinam, Trinidad
Caluromysiops
C. irrupta **Black-shouldered opossum.** SE Peru
Chironectes
C. minimus **Yapok (Water opossum).** S Mexico to N half of S America inc W Colombia, Venezuela, Guianas, Surinam, E Peru, NE and SW Brazil, NE Argentina
Didelphis
D. albiventris **White-eared opossum.** Andes of W Venezuela, Colombia, Ecuador, Peru; E Brazil, N and C Argentina, W Bolivia, Uruguay, Paraguay
D. marsupialis **Southern opossum.** E Mexico through C America to Peru, Bolivia, E Paraguay, NE Argentina; absent from high Andes
D. virginiana **Virginia or Common opossum.** N America from S Canada through C and E USA to N Costa Rica; introduced in W coast of USA
Glironia
G. venusta **Bushy-tailed opossum.** Amazon regions of N Bolivia, E Ecuador, Peru
Lestodelphys
L. halli **Patagonian opossum.** S Argentina
Lutreolina
L. crassicaudata **Lutrine or Little water opossum.** Guianas, Bolivia, Paraguay, S Brazil, NE Argentina
Marmosa
M. aceramarcae **Mouse opossum.** C Bolivia
M. agilis **Agile mouse opossum.** Brazil, Paraguay, N Argentina, Uruguay; W Bolivia, E Peru
M. agricolai **Mouse opossum.** C Brazil
M. alstoni **Alston's mouse opossum.** Belize, Honduras to W Colombia
M. andersoni **Anderson's mouse opossum.** Nr Cuzco, Peru
M. canescens **Grayish mouse opossum.** SW and S Mexico
M. cinerea **Ashy mouse opossum.** E Colombia, Venezuela, French Guiana, Guyana, Surinam; E and S Brazil, Paraguay, NE Argentina
M. constantia **Mouse opossum.** C Brazil, W Bolivia, NW Argentina
M. cracens **Narrow-headed mouse opossum.** Falcon, Venezuela
M. domina **Mouse opossum.** Amazonian Brazil
M. dryas **Mouse opossum.** W Venezuela, E Colombia
M. elegans **Elegant mouse opossum.** N, C Chile, S, SW Boliva, NW Argentina, S Peru
M. emiliae **Paran mouse opossum.** NE Brazil, Surinam
M. formosa **Formosan mouse opossum.** N Argentina
M. fuscata **Mouse opossum.** Venezuela, N, C Colombia, Trinidad
M. germana **Mouse opossum.** E Ecuador; E Peru
M. grisea **Long-tailed mouse opossum.** Paraguay
M. handleyi **Handley's mouse opossum.** Colombia
M. impavida **Pale mouse opossum.** Mts of Panama to Venezuela, W Bolivia, S Peru
M. incana **Mouse opossum.** E Brazil
M. invicta **Panama mouse opossum.** Mts of Panama
M. juninensis **Mouse opossum.** C Peru
M. karimii **Mouse opossum.** NE, C Brazil
M. lepida **Radiant mouse opossum.** Surinam to Bolivia, E Peru, Ecuador, Colombia
M. leucastra **Mouse opossum.** N Peru
M. mapiriensis **Mouse opossum.** Mts of NW Bolivia, Peru
M. marica **Venezuelan mountain opossum.** Andes of W Venezuela
M. mexicana **Mexican mouse opossum.** Coastal E, S Mexico to W Panama
M. microtarsus **Small-footed mouse opossum.** E Brazil
M. murina **Common mouse opossum.** French Guiana, Guyana, Surinam, Amazonian and NE Brazil, Trinidad, Venezuela, Colombia, Ecuador, N Peru
M. noctivaga **Mouse opossum.** Amazonian Brazil to E Ecuador, W Bolivia, C, E Peru
M. ocellata **Spectacled mouse opossum.** C Bolivia
M. parvidens **Mouse opossum.** Guyana, E Venezuela
M. phaea **Mouse opossum.** W slopes of Andes in Ecuador, Colombia
M. pusilla **Dwarf mouse opossum.** N Argentina, SW Bolivia, Paraguay
M. quichua **Quichuan mouse opossum.** E Peru
M. rapposa **Vulpin's mouse opossum.** NW Bolivia, SE Peru
M. regina **Mouse opossum.** C Colombia
M. robinsoni **Pale-bellied mouse opossum.** Honduras, Belize; Panama to N Venezuela, Trinidad, N and W Colombia, W Ecuador, NW Peru
M. rubra **Red mouse opossum.** E Ecuador, NE Peru
M. scapulata **Mouse opossum.** E C Brazil
M. tatei **Tate's mouse opossum.** W slopes of Andes, C Peru
M. tyleriana **Mouse opossum.** SE, S Venezuela
M. unduaviensis **Mouse opossum.** W Bolivia
M. velutina **Velvety mouse opossum.** E Brazil
M. xerophila **Orange mouse opossum.** Coastal NE Colombia, NW Venezuela
M. yungasensis **Mouse opossum.** NW Brazil to Ecuador, W Bolivia
Metachirus
M. nudicaudatus **Brown four-eyed opossum.** Nicaragua to Peru, E Bolivia, Paraguay, NE Argentina, S Brazil; Guyana, French Guiana, Surinam
Monodelphis
M. adusta **Cloudy short-tailed opossum.** Panama to C Colombia, Ecuador, E Peru, W Bolivia
M. americana **Three-striped short-tailed opossum.** French Guiana, Guyana, Surinam to SE and C Brazil
M. brevicaudata **Red-sided short-tailed opossum.** Amazon Basin in Brazil, French Guiana, Guyana, Surinam, Venezuela, E Colombia
M. dimidiata **Eastern short-tailed opossum.** C Brazil, Uruguay, Argentina pampas
M. domestica **Gray short-tailed opossum.** E and C Brazil, Bolivia, Paraguay
M. henseli **Hensel's short-tailed opossum.** S Brazil, SE Paraguay, NE Argentina
M. iheringi **Short-tailed opossum.** S Brazil
M. kunsi **Kuns' short-tailed opossum.** Bolivia
M. maraxina **Short-tailed opossum.** Marajo Is, mouth of Amazon, Brazil
M. orinoci **Orinoco short-tailed opossum.** Orinoco Basin in S Venezuela, W Guyana, N Brazil
M. scalops **Red-headed short-tailed opossum.** SE Brazil
M. touan **Short-tailed opossum.** Brazil, Paraguay, N Argentina, French Guiana, Guyana, Surinam
M. umbristriata **Short-tailed opossum.** E Brazil
M. unistriata **One-striped short-tailed opossum.** SE Brazil (Sao Paulo)
Philander
P. mcilhennyi **Mcilhenny's four-eyed opossum.** E Peru near border with Brazil
P. opossum **Gray four-eyed opossum.** Mexico to E Peru, W Bolivia, Paraguay, NE Argentina, Brazil

FAMILY DASYURIDAE
Antechinomys
A. laniger **Kultarr.** Inland Australia
Antechinus
A. bellus **Fawn antechinus.** N of Northern Territory
A. flavipes **Yellow-footed antechinus.** Inland E and SW Australia
A. godmani **Atherton antechinus.** NE Queensland
A. leo **Cinnamon antechinus.** Cape York
A. melanurus **Black-tailed marsupial mouse.** New Guinea
A. minimus **Swamp antechinus.** S Victoria, Tasmania
A. naso **Long-nosed marsupial mouse.** New Guinea
A. stuartii **Brown antechinus.** E Australia
A. swainsonii **Dusky antechinus.** E Australia, Tasmania
A. wilhelmina **Lesser marsupial mouse.** Central range, New Guinea
Dasycercus
D. cristicauda **Mulgara.** Inland Australia
Dasykatula
D. rosamondae **Little red antechinus.** NW Australia
Dasyuroides
D. byrnei **Kowari.** C Australia
Dasyurus
D. geoffroii **Western quoll (or Native or Tiger cat).** Inland and Western Australia
D. maculatus **Tiger quoll.** E Australia, Tasmania
D. viverrinus **Eastern quoll.** SE Australia, Tasmania
Murexia
M. longicaudata **Short-haired marsupial mouse.** New Guinea
M. rothschildi **Broad-striped marsupial mouse.** SW New Guinea
Myoictis
M. melas **Three-striped marsupial mouse.** New Guinea
Neophascogale
N. lorentzii **Long-clawed marsupial mouse.** Central range, New Guinea
Ningaui
N. ridei **Inland (or Wongai) ningaui.** C Australia
N. timealeyi **Pilbara ningaui.** NW Australia
Parantechinus
P. apicalis **Dibbler.** Cheyne Beach, Jerdacuttup, Western Australia
P. bilarni **Sandstone antechinus.** N of Northern Territory
Phascogale
P. calura **Red-tailed phascogale or wambenger.** SW Australia
P. tapoatafa **Brush-tailed phascogale.** Inland Australia
Phascolosorex
P. doriae **Red-bellied marsupial mouse.** W New Guinea
P. dorsalis **Narrow-striped marsupial mouse.** Central range, New Guinea
Planigale
P. gilesi **Paucident planigale.** NE South Australia and NW New South Wales
P. ingrami **Long-tailed planigale.** Northern Territory, N Queensland
P. maculata **Common planigale.** Northern Territory, Queensland
P. novaeguineae **Papuan marsupial mouse.** C and E of S New Guinea lowlands
P. tenuirostris **Narrow-nosed planigale.** Inland E Australia
Pseudantechinus
P. macdonnellensis **Fat-tailed or Red-eared antechinus.** C Australia, Kimberley, Western Australia
Satanellus
S. albopunctatus **New Guinea marsupial cat.** New Guinea
S. hallucatus **Northern quoll.** N Australia
Sarcophilus
S. harrisii **Tasmanian devil.** Tasmania
Sminthopsis
S. butleri **Carpentarian dunnart.** Kimberley, Western Australia
S. crassicaudata **Fat-tailed dunnart.** S Australia
S. granulipes **White-tailed dunnart.** Inland SW Australia
S. hirtipes **Hairy-footed dunnart.** C Australia, inland Western Australia
S. leucopus **White-footed dunnart.** S Victoria, Tasmania
S. longicaudata **Long-tailed dunnart.** C, NW Australia
S. macroura **Stripe-faced dunnart.** N and C Australia
S. murina **Common dunnart.** E and SW Australia
S. ooldea **Ooldea dunnart.** C Australia
S. psammophila **Sandhill dunnart.** C Australia, Eyre Peninsula
S. douglasi **Julia Creek dunnart.** NW Queensland
S. virginiae **Red-cheeked dunnart.** N and NE Australia
S. youngsoni **Lesser hairy-footed dunnart.** C Australia, inland W Australia

FAMILY MICROBIOTHERIIDAE
Dromiciops
D. australis **Monito del monte (colocolos).** W coast to high Andes, Chile, S of Concepcion to Chiloe Is

FAMILY CAENOLESTIDAE
Caenolestes
C. caniventer **Gray-bellied caenolestid ("shrew" or "rat" opossum).** Lower Andes, S Ecuador
C. convelatus **Blackish caenolestid.** High Andes, W Venezuela
C. fuliginosus **Ecuadorian caenolestid.** High Andes, Ecuador
C. obscurus **Colombian caenolestid.** High Andes, Colombia
C. tatei **Tate's caenolestid.** Over Andes, S Ecuador
Lestoros
L. inca **Peruvian caenolestid.** High Andes of Peru
Rhyncholestes
R. raphanurus **Chilean caenolestid.** S Chile

FAMILY THYLACINIDAE
Thylacinus
T. cynocephalus **Thylacine (Tasmanian wolf or tiger).** Tasmania

FAMILY MYRMECOBIIDAE
Myrmecobius
M. fasciatus **Numbat (Banded anteater, walpurti).** SW Western Australia

FAMILY NOTORYCTIDAE
Notoryctes
N. typhlops **Marsupial mole.** C deserts, Australia

FAMILY PERAMELIDAE
Chaeropus
C. ecaudatus **Pig-footed bandicoot.** S Australia
Echymipera
E. clara **White-lipped bandicoot.** N half of New Guinea
E. kalubu **Spiny bandicoot.** New Guinea, Bismarck Archipelago
E. rufescens **Rufescent bandicoot.** New Guinea, Cape York, Kei and Aru Is
Isoodon
I. auratus **Golden bandicoot.** C, NW Australia

I. macrourus   **Brindled or Northern brown bandicoot.** E New South Wales, Queensland, N Northern Territory, Kimberleys in NW Australia

I. obesulus   **Southern or Brown short-nosed bandicoot or quenda.** S Australia from Sydney to SW of Western Australia, Tasmania, Cape York, Barrow Is

Microperoryctes
M. murina   **Mouse bandicoot.** W Irian
Perameles
P. bougainville   **Bougainville's or Western barred bandicoot or marl.** S Australia from E New South Wales to far W of Western Australia
P. eremiana   **Desert or Orange bandicoot.** C Australia, W Northern Territory, and South Australia, E Western Australia
P. gunnii   **Barred or Eastern barred bandicoot.** Tasmania, SW Victoria
P. nasuta   **Long-nosed bandicoot.** E, SE Australia
Peroryctes
P. broadbenti   **Giant bandicoot.** New Guinea
P. longicauda   **Striped bandicoot.** New Guinea
P. papuensis   **Papuan bandicoot.** SE New Guinea
P. raffrayanus   **Raffray's bandicoot.** New Guinea
Rhyncomeles
R. prattorum   **Ceram Island bandicoot.** Ceram Is

FAMILY THYLACOMYIDAE
Macrotis
M. lagotis   **Greater bilby (Greater rabbit-eared bandicoot, dalgyte).** Northern Territory, Western Australia, W Queensland
M. leucura   **Lesser bilby (Lesser rabbit-eared bandicoot).** C Australia

FAMILY PHALANGERIDAE
Phalanger
P. carmelitae   **Mountain cuscus.** New Guinea
P. celebensis   **"Sulawesi cuscus."** Sulawesi, Peleng Is, Halmahera, Obi, Sanghir Is
P. gymnotis   **Ground cuscus.** New Guinea, Aru, Biak, Japen Is, possibly Wetar and Timor
P. interpositus   **Stein's cuscus.** New Guinea, Japen Is
P. lullulae   **Woodlark Island cuscus.** Woodlark Is, New Guinea, Province of Milne Bay
P. maculatus   **Spotted cuscus.** NE Queensland, New Guinea, and adjacent Islands
P. orientalis   **Gray common cuscus.** NE Queensland, New Guinea and adjacent islands
P. rufoniger   **Black-spotted cuscus.** New Guinea
P. ursinus   **"Sulawesi cuscus."** Sulawesi incl. Peleng and Togian Is, Talaud Is
P. vestitus   **Silky cuscus.** New Guinea
Trichosurus
T. arnhemensis   **Northern brushtail possum.** N Northern Territory, Kimberley, across to Barrow Is
T. caninus   **Mountain brushtail possum (bobuck).** SE Australia, C Victoria to SE Queensland
T. vulpecula   **Common brushtail or Silver gray possum.** Australia except Kimberley, Barrow Is and parts of C Australia; introduced to New Zealand
Wyulda
W. squamicaudata   **Scaly-tailed possum.** Kimberley

FAMILY PSEUDOCHEIRIDAE
Petauroides
P. volans   **Greater glider.** E Australia
Pseudocheirus
P. albertisi   **D'Albertis' ringtail possum.** New Guinea
P. archeri   **Green ringtail possum.** Queensland
P. canescens   **Lowland ringtail possum.** Papua, W Irian
P. caroli   **Weyland ringtail possum.** New Guinea
P. convolutor   **Tasmanian ringtail possum.** Tasmania
P. corinnae   **Eastern ringtail possum.** E New Guinea
P. cupreus   **Coppery ringtail possum.** E New Guinea
P. dahli   **Rock ringtail possum.** N Northern Territory, N Western Australia
P. forbesi   **Moss forest ringtail possum.** E New Guinea
P. herbertensis   **Herbert river ringtail possum.** Queensland
P. lemuroides   **Lemuroid ringtail possum.** Queensland
P. mayeri   **Pygmy ringtail possum.** W New Guinea
P. occidentalis   **Western ringtail possum.** SW Western Australia
P. peregrinus   **Common ringtail possum.** E Australia
P. schlegeli   **Arfak ringtail possum.** NW New Guinea

FAMILY PETAURIDAE
Dactylopsila
D. palpator   **Long-fingered possum.** New Guinea
D. trivirgata   **Common striped possum.** Queensland, New Guinea
Gymnobelideus
G. leadbeateri   **Leadbeater's possum.** Victorian C Highlands
Petaurus
P. abidi   **Northern glider.** NW coastal New Guinea

P. australis   **Yellow-bellied glider.** SE, E Australia
P. breviceps   **Sugar glider.** Tasmania, SE South Australia, Victoria, E New South Wales and Queensland, N Northern Territory, NE Western Australia, New Guinea and surrounding islands
P. norfolcensis   **Squirrel glider.** SE, E Australia

FAMILY BURRAMYIDAE
Acrobates
A. pygmaeus   **Feathertail or Pygmy glider (flying mouse).** SE South Australia to N Queensland
Burramys
B. parvus   **Mountain pygmy possum.** Victoria, New South Wales
Cercartetus
C. caudatus   **Long-tailed pygmy possum.** S C New Guinea
C. concinnus   **Western pygmy possum.** SW Western Australia, S South Australia, Kangaroo Is, W Victoria
C. lepidus   **Little pygmy possum.** SE South Australia, W Victoria, Tasmania
C. nanus   **Eastern pygmy possum.** SE and E Australia, Tasmania
Distoechurus
D. pennatus   **Feathertail possum.** New Guinea

FAMILY POTOROIDAE
Aepyprymnus
A. rufescens   **Rufous rat kangaroo (or bettong).** Newcastle (New South Wales) to Cairns (Queensland)
Bettongia
B. gaimardi   **Gaimard's or Tasmanian rat kangaroo.** E coast from SE Queensland to Victoria (not recorded last 60 years), Tasmania
B. lesueur   **Lesueur's rat or Burrowing kangaroo (boodie or bettong).** Barrow, Boodie, Bernier and Dorri Is off Western Australia
B. penicillata   **Brush-tailed rat kangaroo (bettong or woylie).** SW Western Australia
B. tropica   **Northern rat kangaroo (or bettong).** N Queensland
Caloprymnus
C. campestris   **Plains kangaroo (Desert rat kangaroo).** E Lake Eyre, South Australia, SW Queensland
Hypsiprymnodon
H. moschatus   **Musky rat kangaroo.** N coastal Queensland from Ingham N to S of Cooktown
Potorous
P. longipes   **Long-footed potoroo or rat kangaroo.** E Gippsland, Victoria
P. platyops   **Broad-faced potoroo.** King George Sound, Western Australia
P. tridactylus   **Long-nosed potoroo.** SE Queensland, N New South Wales, Victoria and SE South Australia, Bass Strait Is, Tasmania

FAMILY MACROPODIDAE
Dendrolagus
D. bennettianus   **Bennett's tree kangaroo.** N Queensland, near Cooktown
D. dorianus   **Dusky or Unicolored tree kangaroo.** NE New Guinea
D. goodfellowi   **Ornate tree kangaroo.** NE New Guinea
D. inustus   **Grizzled tree kangaroo.** NW New Guinea
D. lumholtzi   **Lumholtz's tree kangaroo.** N Queensland
D. matschiei   **Matschie's tree kangaroo.** NE New Guinea, Huon Peninsula
D. ursinus   **Black tree kangaroo.** NW New Guinea, Vogelkop
Dorcopsis
D. atrata   **Black forest or Black dorcopsis wallaby.** Goodenough Is
D. hageni   **Greater forest or North New Guinea wallaby.** N New Guinea
D. veterum   **Gray forest wallaby.** S, NW New Guinea, Mysol and Salawatti Is
Dorcopsulus
D. macleayi   **Lesser forest wallaby (Papuan or Macleay's dorcopsis).** New Guinea
D. vanheurni   **Lesser forest wallaby.** New Guinea
Lagorchestes
L. asomatus   **Desert (or Central) hare wallaby.** Northern Territory
L. conspicillatus   **Spectacled hare wallaby.** Barrow Is, Western Australia; C W to N Queensland, parts of Northern Territory
L. hirsutus   **Western (or Rufous) hare wallaby.** Bernier, Dorre Is, Western Australia; Tanami desert, Northern Territory
L. leporides   **Brown (Eastern) hare wallaby.** E South Australia, W New South Wales
Lagostrophus
L. fasciatus   **Banded hare wallaby.** Bernier and Dorre Is, Dirk Hartog Is, Western Australia
Macropus
M. agilis   **Agile wallaby.** New Guinea, N Australia from N Western Australia to Queensland, Stradbroke and Peel Is

M. antilopinus   **Antilopine wallaroo.** Kimberleys (Western Australia), Arnhem Land (Northern Territory), Cape York (Queensland)
M. bernardus   **Black wallaroo.** Arnhem Land escarpment, Northern Territory
M. dorsalis   **Black-striped wallaby.** NE New South Wales, E Queensland N to Rockhampton
M. eugenii   **Tammar or Dama or Scrub wallaby.** SW Western Australia; Eyre Peninsula, S Australia; Houtman's, Abrolhos, Garden Is, Recherche Archipelago in Western Australia; Kangaroo, Greenly Is, South Australia
M. fuliginosus   **Western gray or Mallee or Blackfaced kangaroo.** Kangaroo Is, S Western and South Australia, W Victoria, New South Wales
M. giganteus   **Eastern gray or Great gray or Scrub kangaroo.** E states of Australia, Tasmania, SE South Australia, W of Broken Hill, New South Wales
M. greyi   **Toolache wallaby.** SE South Australia
M. irma   **Western bush wallaby.** SW Western Australia
M. parma   **Parma or White-fronted wallaby.** E New South Wales
M. parryi   **Whiptail or Pretty-face or Parry's wallaby.** Cape York to NE New South Wales
M. robustus   **Wallaroo, euro or Hill kangaroo (or biggada).** Australia-wide, except most of Victoria, and Tasmania
M. rufogriseus   **Bennett's or Red-necked wallaby.** King Is, Tasmania, E and SE coastal Australia
M. rufus   **Red kangaroo.** Inland Australia
Onychogalea
O. fraenata   **Bridled nailtail wallaby.** C Queensland
O. lunata   **Crescent nailtail wallaby.** SW, S and C Australia
O. unguifera   **Northern nailtail wallaby.** N Western Australia, N Northern Territory, N Queensland
Petrogale
P. brachyotis   **Short-eared rock wallaby.** N Kimberley, N Western Australia, to N Northern Territory
P. burbidgei   **Burbidge's rock wallaby (warabi).** W Kimberley, N Western Australia, incl. Mitchell Plateau and Bigge Is
P. concinna   **Little rock wallaby (nabarlek).** N Kimberley, N Western Australia; N Northern Territory
P. godmani   **Godman's rock wallaby.** Cape York Peninsula, Queensland
P. inornata   **Unadorned rock wallaby.** SE Cape York Peninsula; Great Palm and Magnetic Is
P. lateralis   **Black-flanked or Black-footed rock wallaby.** Western and South Australia
P. penicillata   **Brush-tailed rock wallaby.** SE Queensland, S New South Wales, E Victoria, Grampian Mts
P. persephone   **Proserpine rock wallaby.** Proserpine, C Queensland
P. rothschildi   **Rothschild's rock wallaby.** W Western Australia; Enderby, Rosemary Is
P. xanthopus   **Yellow-footed or Ring-tailed rock wallaby.** Gawler, Andamooka, Flinders ranges, N of Olary in South Australia; NE Broken Hill, New South Wales; near Adavale, Queensland
Setonix
S. brachyurus   **Quokka.** SW Western Australia, Rottnest and Bald Is
Thylogale
T. billardierii   **Red-bellied pademelon or scrub wallaby.** Tasmania, Bass Strait Is
T. brunii   **Dusky pademelon.** Aru Is; S and N New Guinea; New Britain
T. stigmatica   **Red-legged pademelon.** S New Guinea; E Australia from Cape York to S Newcastle
T. thetis   **Red-necked pademelon.** SE Queensland to S Sydney, New South Wales
Wallabia
W. bicolor   **Swamp or Black or Black-tailed wallaby.** E Australia, Cape York, Queensland, E coast through Victoria to SE South Australia

FAMILY PHASCOLARCTIDAE
Phascolarctos
P. cinereus   **Koala.** E Australia

FAMILY VOMBATIDAE
Lasiorhinus
L. krefftii   **Northern or Queensland hairy-nosed wombat.** Mid E Queensland
L. latifrons   **Southern hairy-nosed or Plains wombat.** S Australia
Vombatus
V. ursinus   **Common or Forest or Naked-nosed wombat.** SE Australia, incl. Flinders Is and Tasmania

FAMILY TARSIPEDIDAE
Tarsipes
T. rostratus   **Honey possum.** SW Western Australia

Bibliography

The following list of titles indicates key reference works used in the preparation of this volume and those recommended for further reading. The list is divided into five categories: general mammology and those titles relevant to each of the main sections hoofed mammals, rodents and lagomorphs, insectivores, bats, and monotremes and marsupials.

General

Boyle, C. L. (ed) (1981) *The RSPCA Book of British Mammals*, Collins, London.
Corbet, G. B. and Hill, J. E. (1980) *A World List of Mammalian Species*, British Museum and Cornell University Press, London and Ithaca, N.Y.
Dorst, J. and Dandelot, P. (1972) *Larger Mammals of Africa*, Collins, London.
Grzimek, B. (ed) (1972) *Grzimek's Animal Life Encyclopedia*, Vols 10, 11 and 12, Van Nostrand Reinhold, New York.
Hall, E. R. and Kelson, K. R. (1959) *The Mammals of North America*, Ronald Press, New York.
Harrison Matthews, L. (1969) *The Life of Mammals*, vols 1 and 2, Weidenfeld & Nicolson, London.
Honacki, J. H., Kinman, K. E. and Koeppl, J. W. (eds) (1982) *Mammal Species of the World*, Allen Press and Association of Systematics Collections, Lawrence, Kansas.
Kingdon, J. (1971–82) *East African Mammals*, vols I–III, Academic Press, New York.
Morris, D. (1965) *The Mammals*, Hodder & Stoughton, London.
Nowak, R. M. and Paradiso, J. L. (eds) (1983) *Walker's Mammals of the World* (4th edn), 2 vols, Johns Hopkins University Press, Baltimore and London.
Vaughan, T. L. (1972) *Mammalogy*, W. B. Saunders, London and Philadelphia.
Young, J. Z. (1975) *The Life of Mammals: their Anatomy and Physiology*, Oxford University Press, Oxford.

Hoofed Mammals

Chaplin, R. E. (1977) *Deer*, Blandford, Poole, Dorset, England.
Chapman, D. and Chapman, N. (1975) *Fallow Deer – Their History, Distribution and Biology*, Terrence Dalton, Lavenham, Suffolk, England.
Clutton-Brock, T. H., Guinness, F. E. and Albon, S. D. (1982) *Red Deer – Behaviour and Ecology of Two Sexes*, Edinburgh University Press, Edinburgh.
Dagg, A. I. and Foster, J. B. (1976) *The Giraffe – its Biology, Behavior and Ecology*, Van Nostrand Reinhold, New York.
Eltringham, S. K. (1982) *Elephants*, Blandford, Poole, Dorset, England.
Gauthier-Pilters, H. and Dagg, A. I. (1981) *The Camel – its Evolution, Ecology, Behavior and Relationship to Man*, University of Chicago Press, Chicago.
Geist, V. (1971) *Mountain Sheep – a Study in Behavior and Evolution*, University of Chicago Press, Chicago.
Groves, C. P. (1974) *Horses, Asses and Zebras in the Wild*, David and Charles, Newton Abbot, England.
Haltenorth, T. and Diller, H. (1980) *A Field Guide to the Mammals of Africa Including Madagascar*, Collins, London.

Hoofed Mammals continued

Kingdon, J. (1979) *East African Mammals*, vol III, parts B, C, D, Academic Press, London and New York.
Laws, R. M., Parker, I. S.C. and Johnstone, R. C. B. (1975) *Elephants and Their Habitats – the Ecology of Elephants in North Bunyoro, Uganda*, Clarendon Press, Oxford.
Leuthold, W. (1977) *African Ungulates – a Comparative Review of their Ethology and Behavioral Ecology*, Springer-Verlag, Berlin.
Mloszewski, M. J. (1983) *The Behaviour and Ecology of the African Buffalo*, Cambridge University Press, Cambridge.
Moss, C. (1976) *Portraits in the Wild – Animal Behaviour in East Africa*, Hamish Hamilton, London.
Nievergelt, B. (1981) *Ibexes in an African Environment – Ecology and Social System of the Walia Ibex in the Simen Mountains, Ethiopia*, Springer-Verlag, Berlin.
Schaller, G. B. (1967) *The Deer and the Tiger – a Study of Wildlife in India*, University of Chicago Press, Chicago.
Schaller, G. B. (1977) *Mountain Monarchs – Wild Sheep and Goats of the Himalaya*, University of Chicago Press, Chicago.
Sinclair, A. R. E. (1977) *The African Buffalo*, University of Chicago Press, Chicago.
Spinage, C. A. (1982) *A Territorial Antelope – The Uganda Waterbuck*, Academic Press, London.
Walther, F. R., Mungall, E. C. and Grau, G. A. (1983) *Gazelles and their Relatives – A Study in Territorial Behavior*, Noyes Publications, Park Ridge, New Jersey.
Whitehead, G. K. (1972) *Deer of the World*, Constable, London.

Rodents and Lagomorphs

Barnett, S. A. (1975) *The Rat: A Study in Behavior*, University of Chicago Press, Chicago and London.
Berry, R. J. (ed) (1981) *Biology of the House Mouse*, Academic Press, London.
Calhoun, J. B. (1962) *The Ecology and Welfare of the Norway Rat*, US Public Health Service, Baltimore.
Curry-Lindahl, K. (1980) *Der Berglemming*, A. Ziemsen Verlag, Wittenberg.
Delany, M. J. (1975) *The Rodents of Uganda*, British Museum (Natural History), London.
Eisenberg, J. F. (1963) *The Behavior of Heteromyid Rodents*, University of California Publications in Zoology, vol 69, pp1–100.
Ellerman, J. R. (1940) *The Families and Genera of Living Rodents*, British Museum (Natural History), London.
Ellerman, J. R. (1961) *The Fauna of India: Mammalia*, vol 3, Delhi.
Elton, C. (1942) *Voles, Mice and Lemmings*, Oxford University Press, Oxford.
Errington, P. L. (1963) *Muskrat Populations*, University of Iowa Press, Iowa City.

Rodents and Lagomorphs continued

de Graff, G. (1981) *The Rodents of Southern Africa*, Butterworth, Durban.
Harrison, D. L. (1972) *The Mammals of Arabia*, vol 3, Ernest Benn, London.
King, J. (ed) (1968) *The Biology of Peromyscus*, Special Publication no. 2, American Society of Mammalogists, Oswego, N.Y.
Kingdon, J. (1971) *East African Mammals: An Atlas of Evolution*, vol 2B, Academic Press, London and New York.
Laidler, K. (1980) *Squirrels in Britain*, David and Charles, Newton Abbot and North Pomfret, Vermont.
Linsdale, J. M. (1946) *The California Ground Squirrel*, University of California Press, Berkeley.
Linsdale, J. M. and Tevis, L. P. (1951) *The Dusky-footed Woodrat*, University of California Press, Berkeley.
Lockley, R. M. (1976) *The Private Life of the Rabbit* (2nd edn), André Deutsch, London.
Menzies, J. I. and Dennis, E. (1979) *Handbook of New Guinea Rodents*, Handbook no. 6, Wau Ecology Institute, Wau, New Guinea.
Morgan, L. H. (1868) *The American Beaver and his Works*, Burt Franklin, New York.
Niethammer, J. and Krapp, F. (1978, 1982) *Handbuch der Säugetiere Europas*, vols 1 and 2, Akademische Verlagsgesellschaft, Wiesbaden.
Ognev, S. I. (1963) *Mammals of the USSR and Adjacent Countries*, vols 4 and 6, Israel Program for Scientific Translation, Jerusalem.
Orr, R. T. (1977) *The Little-known Pika*, Collier Macmillan, New York.
Prakash, I. and Ghosh, P. K. (eds) (1975) *Rodents in Desert Environments*, Monographae Biologicae, W. Junk, The Hague.
Rosevear, D. R. (1969) *The Rodents of West Africa*, British Museum (Natural History), London.
Rowlands, I. W. and Weir, B. (1974) *The Biology of Hystricomorph Rodents*, Zoological Society Symposia, no. 34, Academic Press, London and New York.
Watts, C. H. S. and Aslin, H. J. (1981) *The Rodents of Australia*, Angus and Robertson, Sydney and London.

Monotremes and Marsupials

Archer, M. (ed) (1982) *Carnivorous Marsupials*, Royal Zoological Society of New South Wales, Sydney.
Augee, M. L. (ed) (1978) *Monotreme Biology*, Royal Zoological Society of New South Wales, Sydney.
Fleay, D. M. (1980) *The Paradoxical Platypus*, Jacaranda Press, Brisbane.
Frith, H. J. and Calaby, J. H. (1969) *Kangaroos*, F. W. Cheshire, Melbourne.
Grant, T. R. (1983) *The Platypus*, University of New South Wales Press, Kensington.

Monotremes and Marsupials continued

Griffiths, M. E. (1978), *The Biology of Monotremes*, Academic Press, New York.
Hunsaker II, D. (ed) (1977), *The Biology of Marsupials*, Academic Press, New York.
Mares, M. A. and Genoways, H. H. (eds) *Mammalian Biology in South America*, University of Pittsburgh, Pennsylvania.
Marlow, B. J. (1965) *Marsupials of Australia*, Jacaranda Press, Brisbane.
Ride, W. D. L. (1970) *The Native Mammals of Australia*, Oxford University Press, Melbourne.
Smith, A. and Hume, I. (in press 1984) *Possums and Gliders*, Royal Zoological Society of New South Wales, Sydney.
Stonehouse, B. (ed) (1977) *The Biology of Marsupials*, Macmillan, London.
Strahan, R. (ed) (1983), *The Complete Book of Australian Mammals*, Angus and Robertson, Sydney.
Troughton, E. le G. (1941), *Furred Mammals of Australia*, Angus and Robertson, Sydney.
Tyler, M. J. (ed) (1978) *The Status of Endangered Australian Wildlife*, Royal Zoological Society of New South Wales, Sydney.
Tyler, M. J. (ed) (1979) *The Status of Endangered Australasian Wildlife*, Royal Zoological Society of South Australia, Adelaide.
Tyndale-Biscoe, C. H. (1973) *Life of Marsupials*, Edward Arnold, London.
Wood-Jones, F. (1923–25) *The Mammals of South Australia*, 3 vols, Government Printer, Adelaide.

Insectivores, Edentates and Allies

Battastini, R. and Richard-Vindard G. (eds) (1972) *Biogreography and Ecology in Madagascar*, Dr. W. Junk B.V., The Hague.
Crowcroft, P. (1957) *The Life of the Shrew*, Max Reinhart, London.
Eisenberg, J. F. (1970) *The Tenrecs: a Study in Mammalian Behavior and Evolution*, Smithsonian Institution, Washington DC.
Godfrey, G. K. and Crowcroft, P. (1960) *The Life of the Mole*, Museum Press, London.
Meester, J. and Setzer, H. W. (1971) *The Mammals of Africa. An Identification Manual*, Smithsonian Institution, Washington DC.
Mellanby, K. (1976) *Talpa: Story of a Mole*, Collins, London.
Montgomery, G. G. (ed) (1978) *The Ecology of Arboreal Folivores*, Smithsonian Institution, Washington DC.
Montgomery, G. G. (ed) (in press) *The Evolution and Ecology of Sloths, Anteaters and Armadillos*, Smithsonian Institution, Washington DC.

Bats

Allen, G. M. (1939) *Bats*, Harvard University Press, Cambridge.

Barbour, R. W. and Davis, W. H. (1969) *Bats of America*, University of Kentucky Press, Lexington, Kentucky.

Corbet, G. B. and Southern, H. N. (1977) *Handbook of British Mammals*, 2nd edn, Blackwell Scientific Publications, Oxford.

Fenton, M. B. (1983) *Just Bats*, University of Toronto Press, Toronto.

Griffin, D. R. (1958) *Listening in the Dark*, Yale University Press, New Haven.

Kunz, T. H. (ed) (1982) *Ecology of Bats*, Plenum Publishing Corp., New York.

Leen, N. and Norvic, A. (1969) *The World of Bats*, Holt Rinehart and Winston, New York.

Rosevear, J. R. (1965) *The Bats of West Africa*, British Museum (Natural History), London.

Turner, D. E. (1975) *The Vampire Bat, a Field Study in Behaviour and Ecology*, Johns Hopkins University Press, Baltimore.

Wimsatt, W. A. (ed) (1970) *Biology of Bats*, Vols 1 & 2, Academic Press, New York.

Wimsatt, W. A. (ed) (1977) *Biology of Bats*, Vol 3, Academic Press, New York.

Yalden, D. W. and Morris, P. A. (1975) *The Lives of Bats*. David and Charles, Newton Abbot.

Picture Acknowledgements

Key: *t* top. *b* bottom. *c* centre. *l* left. *r* right.
Abbreviations: A Ardea. AH Andrew Henley. AN Nature, Agence Photographique. ANT Australasian Nature Transparencies. BC Bruce Coleman Ltd GF George Frame. J Jacana. FL Frank Lane Agency. OSF Oxford Scientific Films. PEP Planet Earth Pictures. FS Fiona Sunquist. SA Survival Anglia.

451 H. Hoeck. 453 WWF/M. Boulton. 454 W. Ervin, Natural Imagery. 455*t* NHPA. 456–457 A. 458–459, 458, 459 P. D. Moehlman. 460*t* SA. 460*b* BC 461 WWF/M. Boulton. 462–463, 464, 465 H. Hoeck. 467 SA. 468–469 BC. 476–477 Musée du Périgord, Périgeux. 478–479 Alan Hutchison. 479*b* G. Herrmann. 482 M. P. Kahl. 483 Woodfall Wildlife Pictures. 486*t* SA. 486–487 PEP. 487*t* J. 490 BC. 491*t* M. P. Kahl. 491*b* Woodfall Wildlife Pictures. 494–495 Nature Photographers. 495*b* WWF. 496 J. 497 GF. 500*t* Nature Photographers. 500*b* SA. 501 BC. 502, 504*b* Leonard Lee Rue III. 504–505 AH. 506*b* M. 506–507 SA. 507*b* R. M. Laws. 508–509 PEP. 510 Nature Photographers. 511 R. M. Laws. 512 AN. 513 AH. 514–515, 515*b*, 516 AN. 517 BC. 520 AN. 521 Leonard Lee Rue III. 522 J. MacKinnon. 523 P. Newton. 525 BC. 527 W. Ervin, Natural Imagery. 530*b* BC. 530–531 J. 531*t*, 532 FL. 533 GF. 534 J. 535 Leonard Lee Rue III. 536–537 M. P. Kahl. 538 J. 539 R. Pellew. 540–541 M. P. Kahl. 541*t* J. 542, 543 D. Kitchen. 544*b* NHPA. 544*t*, 545 BC. 546–547 PEP. 548, 549 AH. 550–551*t* PEP. 550*c* AH. 550–551*b* PEP. 554–555 D. Lott. 558–559 A. Bannister. 560 PEP. 561 P. Wirtz. 564–565 A. Bannister. 565*b* Leonard Lee Rue III. 566–567, 568 PEP. 568–569 Eric and David Hosking. 569 SA. 572, 573 M. Stanley Price. 574–575 A. Bannister. 575*b* W. Ervin, Natural Imagery. 578–579 PEP. 578*b* J. 579*l* W. Ervin, Natural Imagery. 579*r* GF. 582 M. P. Kahl. 583 W. Ervin, Natural Imagery. 586–587 NHPA. 587*b* Leonard Lee Rue III. 590–591 BC. 591*b* G. di Nunzio. 592 AN. 594 E. Beaton. 595 AH. 598–599 A. 600 AN. 601 ICI Plant Protection Division. 602 Mansell Collection. 602–603 ICI Plant Protection Division. 603 AH. 604–605 A. Bannister. 606–607 Aquila. 607*b*, 608, 609*b* BC. 609*t* J. Kaufmann. 611 BC. 612–613, 613, 614, 614–615 A. 615*b* K. Sugowara, Orion Press. 618 Aquila. 619*b* M. Fogden. 619*t*, 620–621, 623*t* W. Ervin, Natural Imagery. 623*b* NHPA. 624, 625 Bio-Tec Images. 626–627 OSF. 627 SA. 630*b* A. 630–631 BC. 633 A. 634 NHPA. 635 J. 637 AH. 638–639 AN. 640 R. W. Barbour. 641*t* M. Fogden. 641*b* BC. 642 M. Fogden. 643, 648*b* BC. 648*t*, 649 M. Fogden. 650–651 NHPA. 651*b* BC. 655 NHPA. 656 Bodleian Library. 657*t* J. 657*b* Naturfotografernas Bildbyrå. 658–659 BC. 662 AH. 663 NHPA. 664*t*

Scala. 664–665 A. 665*t* GF. 668–669 BC. 672 NHPA. 673 BC. 674–675, 678 A. 679 BC. 680–681 AN. 682 R. W. Barbour. 683 AN. 684 A. Bannister. 685 BC. 686–687 Tony Morrison. 687*b* BC. 688 GF. 689 BC. 691*t* T. Owen-Edmunds. 691*b* Tony Morrison. 692–693 A. 694–695 BC. 694*b* D. Macdonald. 695*c* BC. 695*b* NHPA. 696, 697, 698, 699*t* D. Macdonald. 699*b*, 700–701 A. 701*b* BC. 705*t* A. Bannister. 705*b* A. 706, 707 W. George. 708–709 A. Bannister. 709*t* P. van den Elzen. 711 J. U. M. Jarvis. 713 PEP. 716 AN. 717*t* M. Fogden. 717*b* A. 718–719 BC. 722 A. 723*t* BC. 723*b* NHPA. 724–725, 725*b*, 726, 727 BC. 728*l* W. Ervin, Natural Imagery. 728–729 BC. 728*r* W. Ervin, Natural Imagery. 733 BC. 734–735 P. Morris. 736 A. Bannister. 738–739 AN. 739*t* OSF. 742 Bodleian Library. 743*t* A. 743*b* OSF. 746*b* NHPA. 746–747 BC. 748 J. 749 A. 750–751 A. Bannister. 751*b* J. Payne. 753 OSF. 754–755 BC. 756*t* NHPA. 756*b* BC. 757 A. 758 BC. 759 OSF. 760–761 Eric and David Hosking. 761*b* BC. 762–763, 763*b* D. R. Kuhn. 765 A. Bannister. 768*b* D. R. Kuhn. 768–769 A. 769*b* D. R. Kuhn. 772–773 A. 773*b* BC. 774*t* M. Fogden. 774*b* BC. 775 OSF. 776, 777, 778–779 M. Fogden. 780–781 J. 781*b* BC. 782*t* OSF. 782*b* PEP. 783 J. 784–785 BC. 785*t* A. 786, 787 ANT. 792 AH. 793 NHPA. 794*b* Frithfoto. 794–795 ANT. 795*b* J. 796 BC. 797 G. F. McCraken. 798 A. 799*t* ANT. 799*b* Aquila. 800 AH. 801 NHPA. 802 WWF/S. Yorath. 803*t* ANT. 803*b* A. 806 M. Fogden. 808 BC. 809 R. Stebbings. 810–811 M. Tuttle. 812–813 OSF. 812*b* J. 813*b*, 814, 815 OSF. 816 A. 816–817 J. 818–819 ANT. 819 BC. 820 AH. 821*t* Frithfoto. 821*b* AH. 822*b* E. Beaton. 822–823 G. Mazza. 828–829 ANT. 829*t* A. 830 Tony Morrison. 831 BC. 834–835, 836, 837 OSF. 840, 841*t* AH. 841*c* WWF, Switzerland. 841*b* ANT. 842. 843*b* AH. 843*t*, 844 A. 844–845 AN. 845*t* J. 847 A. 848–849, 849*t* ANT. 849*b* AH. 850–851 E. Beaton. 851*t* ANT. 852–853 A. 854*t* E. Beaton. 854*b* AH. 855 J. 856–857, 860*t* AH. 860*b* A. Smith. 861 AH. 862*b* E. Beaton. 862–863 NHPA. 865, 868–869 AN. 870 E. Beaton. 871 ANT. 872 AH. 873 AN. 874–875, 877*t* ANT. 877*b* BC. 878–879 A. Smith. 880 BC.

Artwork

All artwork © Priscilla Barrett unless stated otherwise below.
Abbreviations: SD Simon Driver. ML Michael Long. AEM Anne-Elise Martin. DO Denys Ovenden. CW Carol Wells. DT Dick Twinney.

450 ML. 452 Malcolm McGregor. 454 SD. 470, 471 SD. 472*t* ML. 472 SD. 473 ML. 474, 475, 476 ML. 481*t* AEM. 481 ML. 499*t* AEM. 499 ML, SD. 510, 520, 533 SD. 597*c*, 604 SD. 605 AEM. 608, 610, 622, 624, 636 SD. 637 AEM. 654, 655, 669, 684 SD. 685 AEM. 711, 712, 713, 722, 724 SD. 740*b*, 741*b* ML. 740*t*, 740*c* SD. 744 DO. 753, 756, 764 SD. 767 DO. 788*c*, 789*l* CW. 789*b* SD. 789*r* ML. 790 Graham Allen. 793 CW. 808 SD. 824, 825, 826*r* ML. 826*l*, 827 SD. 832 DT. 837 CW. 839, 842, 858, 859, 864, 866 DT. 871 CW. Maps and scale drawings SD.

GLOSSARY

Abomasum   the final chamber of the four sections of the RUMINANT ARTIODACTYL stomach (following the RUMEN, RETICULUM and OMASUM). The abomasum alone corresponds to the stomach "proper" of other mammals and the other three are elaborations of its proximal part.

Adaptive radiation   the pattern in which different species develop from a common ancestor (as distinct from CONVERGENT EVOLUTION, a process whereby species from different origins became similar in response to the same SELECTIVE PRESSURES).

Adult   a fully developed and mature individual, capable of breeding, but not necessarily doing so until social and/or ecological conditions allow.

Aestivate   (noun: aestivation) to enter a state of dormancy or torpor in seasonal hot, dry weather, when food is scarce.

Allantoic stalk   a sac-like outgrowth of the hinder part of the gut of the mammalian fetus, containing a rich network of blood vessels. It connects fetal circulation with the PLACENTA, facilitating nutrition of the young, respiration and excretion. (See CHORIOALLANTOIC PLACENTATION.)

Allopatry   condition in which populations of different species are geographically separated (cf SYMPATRY).

Alpine   of the Alps or any lofty mountains; usually pertaining to altitudes above 1,500m (4,900ft).

Altricial   young that are born at a rudimentary stage of development and require an extended period of nursing by parent(s). See also PRECOCIAL.

Amphibious   able to live on both land and in water.

Amynodont   a member of the family Amynodontidae, large rhinoceros-like mammals (order Perissodactyla), which became extinct in the Tertiary.

Anal gland or sac   a gland opening by a short duct either just inside the anus or on either side of it.

Ancestral stock   a group of animals, usually showing primitive characteristics, which is believed to have given rise to later, more specialized forms.

Anthracothere   a member of the family Anthracotheriidae (order Artiodactyla), which became extinct in the late Tertiary.

Antigen   a substance, whether organic or inorganic, that stimulates the production of antibodies when introduced into the body.

Antrum   a cavity in the body, especially one in the upper jaw bone.

Aquatic   living chiefly in water.

Arboreal   living in trees.

Arthropod   the largest phylum in the animal kingdom in number of species, including insects, spiders, crabs etc. Arthropods have hard, jointed exoskeletons and paired, jointed legs.

Artiodactyl   a member of the order Artiodactyla, the even-toed ungulates.

Astragalus   a bone in the ungulate tarsus (ankle) which (due to reorganization of ankle bones following reduction in the number of digits) bears most of the body weight (a task shared by the CALCANEUM bone in most other mammals).

Baculum (os penis or penis bone)   an elongate bone present in the penis of

Baculum contd.
certain mammals.

Bifid (of the penis)   the head divided into two parts by a deep cleft.

Biotic community   a naturally occurring group of plants and animals in the same environment.

Blastocyst   see IMPLANTATION.

Boreal region   a zone geographically situated south of the Arctic and north of latitude 50°N; dominated by coniferous forest.

Bovid   a member of the cow-like artiodactyl family, Bovidae.

Brachydont   a type of short-crowned teeth whose growth ceases when full-grown, whereupon the pulp cavity in the root closes. Typical of most mammals, but contrast the HYPSODONT teeth of many herbivores.

Brindled   having inconspicuous dark streaks or flecks on a gray or tawny background.

Brontothere   a member of the family Brontotheriidae (order Perissodactyla), which became extinct in the early Tertiary.

Browser   a herbivore which feeds on shoots and leaves of trees, shrubs etc, as distinct from grasses (cf GRAZER).

Bullae (auditory)   globular, bony capsules housing the middle and inner ear structures, situated on the underside of the skull.

Bunodont   molar teeth whose cusps form separate, rounded hillocks which crush and grind.

Câche   a hidden store of food; also (verb) to hide food for future use.

Calcaneum   one of the tarsal (ankle) bones which forms the heel and in many mammalian orders bears the body weight together with the ASTRAGALUS.

Camelid   a member of the camel family, Camelidae, of the Artiodactyla.

Cameloid   one of the South American camels.

Cannon bone   a bone formed by the fusion of METATARSAL bones in the feet of some families.

Caprid   a member of the tribe Caprini of the Artiodactyla.

Carnivore   any meat-eating organism (alternatively, a member of the order Carnivora, many of whose members are carnivores).

Carpals   wrist bones which articulate between the forelimb bones (radius and ulna) and the METACARPALS.

Caudal gland   an enlarged skin gland associated with the root of the tail. Subcaudal: placed below the root; supracaudal: above the root.

Cecum   a blind sac in the digestive tract, opening out from the junction between the small and large intestines. In herbivorous mammals it is often very large; it is the site of bacterial action on cellulose. The end of the cecum is the appendix; in species with reduced ceca the appendix may retain an antibacterial function.

Cellulose   the fundamental constituent of the cell walls of all green plants, and some algae and fungi. It is very tough and fibrous, and can be digested only by the intestinal flora in mammalian guts.

Cementum   hard material which coats

Cementum contd.
the roots of mammalian teeth. In some species, cementum is laid down in annual layers which, under a microscope, can be counted to estimate the age of individuals.

Cervid   a member of the deer family (Cervidae), of the Artiodactyla.

Cervix   the neck of the womb (cervical—pertaining to neck).

Chalicothere   a member of the family Chalicotheriidae (order Perissodactyla), which became extinct in the Pleistocene.

Cheek-teeth   teeth lying behind the canines in mammals, comprising premolars and molars.

Chorioallantoic placentation   a system whereby fetal mammals are nourished by the blood supply of the mother. The chorion is a superficial layer enclosing all the embryonic structures of the fetus, and is in close contact with the maternal blood supply at the placenta. The union of the chorion (with its vascularized ALLANTOIC STALK and YOLK SAC) with the placenta facilitates the exchange of food substances and gases, and hence the nutrition of the growing fetus.

Class   taxonomic category subordinate to a phylum and superior to an order.

Clavicle   the collar bone.

Cloaca   terminal part of the gut into which the reproductive and urinary ducts open. There is one opening to the body, the cloacal aperture, instead of a separate anus and urinogenital opening.

Cloud forest   moist, high-altitude forest characterised by dense UNDERSTORY growth, and abundance of ferns, mosses, orchids and other plants on the trunks and branches of the trees.

Colon   the large intestine of vertebrates, excluding the terminal rectum. It is concerned with the absorption of water from feces.

Colonial   living together in colonies. In bats, more usually applied to the communal sleeping habit, in which tens of thousands of individuals may participate.

Concentrate selector   a herbivore which feeds on those plant parts (such as shoots and fruits) which are rich in nutrients.

Congenor   a member of the same species (or genus).

Conspecific   member of the same species.

Convergent evolution   the independent acquisition of similar characters in evolution, as opposed to possession of similarities by virtue of descent from a common ancestor.

Crepuscular   active in twilight.

Crustaceans   members of a class within the phylum Arthropoda typified by five pairs of legs, two pairs of antennae, head and thorax joined, and calcareous deposits in the exoskeleton, eg crayfish, crabs, shrimps.

Cricetine   adjective and noun used to refer to (a) the primitive rodents from which the New World rats and mice, voles and lemmings, hamsters and gerbils are descended, (b) these modern rodents. In some taxonomic classification systems these subfamilies of the family Muridae are classified as members of a separate family called Cricetidae, with members of the Old World rats and mice alone constituting the Muridae.

Crypsis   an aspect of the appearance of an organism which camouflages it from the view of others, such as predators or competitors.

Cryptic (coloration or locomotion)   protecting through concealment.

Cue   a signal, or stimulus (eg olfactory) produced by an individual which elicits a response in other individuals.

Cursorial   being adapted for running.

Cusp   a prominence on a cheek-tooth (premolars or molar).

Delayed implantation   see IMPLANTATION.

Dental formula   a convention for summarizing the dental arrangement whereby the numbers of each type of tooth in each half of the upper and lower jaw are given; the numbers are always presented in the order: incisor (I), canine (C), premolar (P), molar (M). The final figure is the total number of teeth to be found in the skull. A typical example for Carnivora would be I3/3, C1/1, P4/4, M3/3 = 44.

Dentition   the arrangement of teeth characteristic of a particular species.

Dermis   the layer of skin lying beneath the outer epidermis.

Desert   areas of low rainfall, typically with sparce scrub or grassland vegetation or lacking vegetation altogether.

Dicerathere   a member of the family Diceratheriidae (order Perissodactyla), which became extinct in the Miocene.

Dicotyledon   one of the two classes of flowering plants (the other class comprises monocotyledons), characterized by the presence of two seed leaves in the young plant, and by net-veined, often broad leaves, in mature plants. Includes deciduous trees, roses etc.

Digit   a finger or toe.

Digital glands   glands that occur between or on the toes.

Digitgrade   method of walking on the toes without the heel touching the ground (cf PLANTIGRADE).

Diprotodont   having the incisors of the lower jaw reduced to one functional pair, as in possums and kangaroos (small, non-functional incisors may also be present). (cf POLYPROTODONT.)

Disjunct or **discontinuous distribution**   geographical distribution of a species that is marked by gaps. Commonly brought about by fragmentation of suitable habitat, especially as a result of human intervention.

Dispersal   the movements of animals, often as they reach maturity, away from their previous home range (equivalent to emigration). Distinct from dispersion, that is, the pattern in which things (perhaps animals, food supplies, nest sites) are distributed or scattered.

Display   any relatively conspicuous pattern of behavior that conveys specific information to others, usually to members of the same species; can involve visual and or vocal elements, as in threat, courtship or "greeting" displays.

Distal   far from the point of attachment or origin (eg tip of tail).

Diurnal   active in daytime.

Dormancy   a period of inactivity; many bears, for example, are dormant for a period in winter; this is not true HIBERNATION, as pulse rate and body

Dormancy contd.
temperature do not drop markedly.

Dorsal on the upper or top side or surface (eg dorsal stripe).

Echolocation the process of perception, often direction finding, based upon reaction to the pattern of reflected sound waves (echoes).

Ecology the study of plants and animals in relation to their natural environmental setting. Each species may be said to occupy a distinctive ecological NICHE.

Ecosystem a unit of the environment within which living and nonliving elements interact.

Ecotype a genetic variety within a single species, adapted for local ecological conditions.

Edentate a member of an order comprising living and extinct anteaters, sloths, armadillos (XENARTHRANS), and extinct paleanodonts.

Elongate relatively long (eg of canine teeth, longer than those of an ancestor, a related animal, or than adjacent teeth).

Embryonic diapause the temporary cessation of development of an embryo (eg in some bats and kangaroos).

Emigration departure of animal(s), usually at or about the time of reaching adulthood, from the group or place of birth.

Entelodont a member of the family Entelodontidae, Oligocene artiodactyls which represent an early branch of the pig family, Suidae.

Epidermis the outer layer of mammalian skin (and in plants the outer tissue of young stem, leaf or root).

Erectile capable of being raised to an erect position (erectile mane).

Estrus the period in the estrous cycle of female mammals at which they are often attractive to males and receptive to mating. The period coincides with the maturation of eggs and ovulation (the release of mature eggs from the ovaries). Animals in estrus are often said to be "on heat" or "in heat." In primates, if the egg is not fertilized the subsequent degeneration of uterine walls (endometrium) leads to menstrual bleeding. In some species ovulation is triggered by copulation and this is called **induced ovulation**, as distinct from spontaneous ovulation.

Eucalpt forest Australian forest, dominated by trees of the genus *Eucalyptus*.

Eutherian a mammal of the subclass Eutheria, the dominant group of mammals. The embryonic young are nourished by an ALLANTOIC PLACENTA.

Exudate natural plant exudates include gums and resins; damage to plants (eg by marmosets) can lead to loss of sap as well.

Facultative optional (cf OBLIGATE).

Family a taxonomic division subordinate to an order and superior to a genus.

Feces excrement from the bowels; colloquially known as droppings or scats.

Feral living in the wild (of domesticated animals, eg cat, dog).

Fermentation the decomposition of organic substances by microorganisms. In some mammals, parts of the digestive tract (eg the cecum) may be inhabited by

Fermentation contd.
bacteria that break down cellulose and release nutrients.

Fetal development rate the rate of development, or growth, of unborn young.

Flehmen German word describing a facial expression in which the lips are pulled back, head often lifted, teeth sometimes clapped rapidly together and nose wrinkled. Often associated with animals (especially males) sniffing scent marks or socially important odors (eg scent of estrous female). Possibly involved in transmission of odor to JACOBSON'S ORGAN.

Folivore an animal eating mainly leaves.

Follicle a small sac, therefore (a) a mass of ovarian cells that produces an ovum, (b) an indentation in the skin from which hair grows.

Forbs a general term applied to ephemeral or weedy plant species (not grasses). In arid and semi-arid regions they grow abundantly and profusely after rains.

Fossorial burrowing (of life-style or behavior).

Frugivore an animal eating mainly fruits.

Gallery forest luxuriant forest lining the banks of watercourses.

Gamete a male or female reproductive cell (ovum or spermatozoon).

Generalist an animal whose life-style does not involve highly specialized strategems (cf SPECIALIST); for example, feeding on a variety of foods which may require different foraging techniques.

Genotype the genetic constitution of an organism, determining all aspects of its appearance, structure and function.

Genus (plural genera) a taxonomic division superior to species and subordinate to family.

Gestation the period of development within the uterus; the process of **delayed implantation** can result in the period of pregnancy being longer than the period during which the embryo is actually developing (See also IMPLANTATION.)

Glands (marking) specialized glandular areas of the skin, used in depositing SCENT MARKS.

Graviportal animals in which the weight is carried by the limbs acting as rigid, extensible struts, powered by extrinsic muscles; eg elephants and rhinos.

Grazer a herbivore which feeds upon grasses (cf BROWSER).

Grizzled sprinkled or streaked with gray.

Gumivorous feeding on gums (plant exudates).

Harem group a social group consisting of a single adult male, at least two adult females and immature animals: a common pattern of social organization among mammals.

Heath low-growing shrubs with woody stems and narrow leaves (eg heather), which often predominate on acidic or upland soils.

Herbivore an animal eating mainly plants or parts of plants.

Heterothermy a condition in which the internal temperature of the body follows the temperature of the outside environment.

Hibernation a period of winter inactivity during which the normal physiological process is greatly reduced and thus during which the energy requirements of the animal are lowered.

Hindgut fermenter herbivores among which the bacterial breakdown of plant tissue occurs in the CECUM, rather than in the RUMEN or foregut.

Holarctic realm a region of the world including North America, Greenland, Europe, and Asia apart from the southwest, southeast and India.

Home range the area in which an animal normally lives (generally excluding rare excursions or migrations), irrespective of whether or not the area is defended from other animals (cf TERRITORY).

Hybrid the offspring of parents of different species.

Hypothermy a condition in which internal body temperature is below normal.

Hypsodont high-crowned teeth, which continue to grow when full-sized and whose pulp cavity remains open; typical of herbivorous mammals (cf BRACHYDONT).

Hyracodont a member of the family Hyracodontidae (order Perissodactyla) which became extinct in the Oligocene.

Implantation the process whereby the free-floating blastocyst (early embryo) becomes attached to the uterine wall in mammals. At the point of implantation a complex network of blood vessels develops to link mother and embryo (the placenta). In **delayed implantation** the blastocyst remains dormant in the uterus for periods varying, between species, from 12 days to 11 months. Delayed implantation may be obligatory or facultative and is known for some members of the Carnivora and Pinnipedia and others.

Induced ovulation see ESTRUS.

Inguinal pertaining to the groin.

Insectivore an animal eating mainly arthropods (insects, spiders).

Interdigital pertaining to between the digits.

Interfemoral a membrane stretching between the femora, or thigh bones in bats.

Intestinal flora simple plants (eg bacteria) which live in the intestines, especially the CECUM, of mammals. They produce enzymes which break down the cellulose in the leaves and stems of green plants and convert it to digestible sugars.

Introduced of a species which has been brought, by man, from lands where it occurs naturally to lands where it has not previously occurred. Some introductions are accidental (eg rats which have traveled unseen on ships), but some are made on purpose for biological control, farming or other economic reasons (eg the Common brush-tail possum, which was introduced to New Zealand from Australia to establish a fur industry).

Invertebrate an animal which lacks a backbone (eg insects, spiders, crustaceans).

Ischial pertaining to the hip.

Jacobson's organ a structure in a foramen (small opening) in the palate of many vertebrates which appears to be involved in olfactory communication. Molecules of scent may be sampled in these organs.

Juvenile no longer possessing the characteristics of an infant, but not yet fully adult.

Kopje a rocky outcrop, typically on otherwise flat plains of African grasslands.

Labile (body temperature) an internal body temperature which may be lowered or raised from an average body temperature.

Lactation (verb: lactate) the secretion of milk, from MAMMARY GLANDS.

Lamoid Llama-like; one of the South American cameloids.

Larynx dilated region of upper part of windpipe, containing vocal chords. Vibration of cords produces vocal sounds.

Latrine a place where feces are regularly left (often together with other SCENT MARKS); associated with olfactory communication.

Lek a display ground at which individuals of one sex maintain miniature territories into which they seek to attract potential mates.

Lipotyphlan an early insectivore classification; menotyphlan insectivores possess a cecum, lipotyphlans do not. Only lipotyphlans are now classified as Insectivora.

Llano South American semi-arid savanna country, eg of Venezuela.

Loph a transverse ridge on the crown of molar teeth.

Lophiodont a member of the family Lophiodontidae (order Perissodactyla) which became extinct in the early Tertiary.

Lophodont molar teeth whose cusps form ridges or LOPHS.

Mallee a grassy, open woodland habitat characteristic of many semi-arid parts of Australia. "Mallee" also describes the multi-stemmed habit of eucalypt trees which dominate this habitat.

Mamma (pl. mammae) **mammary glands** the milk-secreting organ of female mammals, probably evolved from sweat glands.

Mammal a member of the CLASS of VERTEBRATE animals having MAMMARY GLANDS which produce milk with which they nurse their young (properly: Mammalia).

Mammilla (pl. mammillae) nipple, or teat, on the MAMMA of female mammals; the conduit through which milk is passed from the mother to the young.

Mandible the lower jaw.

Mangrove forest tropical forest developed on sheltered muddy shores of deltas and estuaries exposed to tide. Vegetation is almost entirely woody.

Marine living in the sea.

Masseter a powerful muscles, subdivided into parts, joining the MANDIBLE to the upper jaw. Used to bring jaws together when chewing.

Melanism darkness of color due to presence of the black pigment melanin.

Menotyphlan see Lipotyphlan.

Metabolic rate the rate at which the chemical processes of the body occur.

Metabolism the chemical processes occurring within an organism, including the production of PROTEIN from amino acids, the exchange of gasses in respiration, the liberation of energy from

Metabolism contd.

foods and innumerable other chemical reactions.

Metacarpal bones of the hand, between the CARPALS of the wrist and the phalanges of the digits.

Metapodial the proximal element of a digit (contained within the palm or sole). The metapodial bones are METACARPALS in the hand (manus) and METATARSALS in the foot (pes).

Metatarsal bones of the foot articulating between the tarsals of the ankle and the phalanges of the digits.

Microhabitat the particular parts of the habitat that are encountered by an individual in the course of its activities.

Midden a dunghill, or site for the regular deposition of feces by mammals.

Migration movement, usually seasonal, from one region or climate to another for purposes of feeding or breeding.

Monogamy a mating system in which individuals have only one mate per breeding season.

Monotreme a mammal of the subclass Monotremata, which comprises the platypus and echidnas. The only egg-laying mammals.

Monotypic a genus comprising a single species.

Montane pertaining to mountainous country.

Montane forest forest occurring at middle altitudes on the slopes of mountains, below the alpine zone but above the lowland forests.

Morphology (morphological) the structure and shape of an organism.

Moss forest moist forest occuring on higher mountain slopes—eg 1,500–3,200m (4,900–10,500ft) in New Guinea—characterized by rich growths of mosses and other plants on the trunks and branches of the trees.

Murine adjective and noun used to refer to members of the subfamily (of the family Muridae) Murinae, which consists of the Old World rats and mice. In some taxonomic classification systems this subfamily is given the status of a family, Muridae, and the members then are sometimes referred to as murids. See also CRICETINE.

Mutation a structural change in a gene which can thus give rise to a new heritable characteristic.

Natal range the home range into which an individual was born (natal = of or from one's birth).

Nectivore (nectivorous) an animal that feeds principally on nectar.

Niche the role of a species within the community, defined in terms of all aspects of its life-style (eg food, competitors, predators, and other resource requirements).

Nicker a vocalization of horses, also called neighing.

Nocturnal active at nighttime.

Nose-leaf characteristically shaped flaps of skin surrounding the nasal passages of horseshoe, or nose-leaf bats (Family Rhinolophidae). Ultrasonic cries are uttered through the nostrils, with the nose leaves serving to direct the echo-locating pulses forwards.

Obligate required, binding (cf FACULTATIVE).

Occipital pertaining to the occiput at back of head.

Olfaction, olfactory the olfactory sense is the sense of smell, depending on receptors located in the epithelium (surface membrane) lining the nasal cavity.

Omasum third of the four chambers in the RUMINANT ARTIODACTYL stomach.

Omnivore an animal eating a varied diet including both animal and plant tissue.

Opposable (of first digit) of the thumb and forefinger in some mammals, which may be brought together in a grasping action, thus enabling objects to be picked up and held.

Opportunist (of feeding) flexible behavior of exploiting circumstances to take a wide range of food items; characteristic of many species. See GENERALIST; SPECIALIST.

Order a taxonomic division subordinate to class and superior to family.

Oreodont a member of the family Oreodontidae (order Artiodactyla), which became extinct in the late Tertiary.

Ovulation (verb ovulate) the shedding of mature ova (eggs) from the ovaries where they are produced (SEE ESTRUS).

Pair-bond an association between a male and female, which lasts from courtship at least until mating is completed, and in some species, until the death of one partner.

Palearctic a geographical region encompassing Europe and Asia north of the Himalayas, and Africa north of the Sahara.

Paleothere a member of the family Paleotheriidae (order Perissodactyla), which became extinct in the early Tertiary.

Palmate palm shaped.

Pampas Argentinian steppe grasslands.

Papilla (plural: papillae) a small nipple-like projection.

Páramo alpine meadow of northern and western South American uplands.

Parturition the process of giving birth (hence *post partum*)—after birth).

Patagium a gliding membrane typically stretching down the sides of the body between the fore- and hindlimbs and perhaps including part of the tail. Found in colugos, flying squirrels, bats etc.

Pecoran a ruminant of the intra-order Pecora, which is characterized by the presence of horns on the forehead.

Perissodactyl a member of the Perissodactyla (the odd-toed ungulates).

Perineal glands glandular tissue occuring between the anus and genitalia.

Pheromone secretions whose odors act as chemical messengers in animal communication, and which prompt a specific response on behalf of the animal receiving the message (see SCENT MARKING).

Phylogeny a classification or relationship based on the closeness of evolutionary descent.

Phylogenetic (of classification or relationship) based on the closeness of evolutionary descent.

Phylum a taxonomic division comprising a number of classes.

Physiology study of the processes which go on in living organisms.

Pinna (plural: pinnae) the projecting cartilaginous portion of the external ear.

Placenta, placental mammals a structure that connects the fetus and the mother's womb to ensure a supply of nutrients to the fetus and removal of its waste products. Only placental mammals have a well-developed placenta; marsupials have a rudimentary placenta or none and monotremes lay eggs.

Plantigrade way of walking on the soles of the feet, including the heels (cf DIGITIGRADE).

Polyandrous see POLYGYNOUS.

Polyestrous having two or more ESTRUS cycles in one breeding season.

Polygamous a mating system wherein an individual has more than one mate per breeding season.

Polygynous a mating system in which a male mates with several females during one breeding season (as opposed to polyandrous, where one female mates with several males).

Polymorphism occurrence of more than one MORPHOLOGICAL form of individual in a population. (See SEXUAL DIMORPHISM.)

Polyprodont having more than two well-developed lower incisor teeth (as in bandicoots and carnivorous marsupials). (cf DIPROTODONT.)

Population a more or less separate (discrete) group of animals of the same species within a given BIOTIC COMMUNITY.

Post-orbital bar a bony strut behind the eye-socket (orbit) in the skull.

Post-partum estrus ovulation and an increase in the sexual receptivity of female mammals, hours or days after the birth of a litter (see ESTRUS).

Pouch (marsupial) a flap of skin on the underbelly of female marsupials, which covers the MAMMILLAE. The pouch may be a simple, open structure as in most carnivorous marsupials, or a more enclosed pocket-like structure as in phalangers and kangaroos.

Prairie North American steppe grassland between 30°N and 55°N.

Predator an animal which forages for live prey; hence "anti-predator behavior" describes the evasive actions of the prey.

Precocial of young born at a relatively advanced stage of development, requiring a short period of nursing by parents (see ALTRICIAL).

Prehensile capable of grasping.

Pre-orbital in front of the eye socket.

Preputial pertaining to the prepuce or loose skin covering the penis.

Proboscidean a member of the order of primitive ungulates, Proboscidea.

Proboscis a long flexible snout.

Process (anatomical) an outgrowth or protuberance.

Procumbent (incisors) projecting forward more or less horizontally.

Promiscuous a mating system wherein an individual mates more or less indiscriminately.

Pronking movement where an animal leaps vertically, on the spot, with all four feet off the ground. Also called stotting. Typical of antelopes, especially when alarmed.

Protoceratid a member of the family Protoceratidae (order Artiodactyla), which became extinct in the late Tertiary.

Protein a complex organic compound made of amino acids. Many different kinds of proteins are present in the muscles and tissues of all mammals.

Proximal near to the point of attachment or origin, (eg the base of the tail).

Pseudoallantoic placentation a kind of placenta shown only by the marsupial bandicoots. Compared with the true eutherian kind of plancentation, transfer of food and gas across the chorioallantoic/placental interface is inefficient, as contact between the fetal and maternal membranes is never close.

Puberty the attainment of sexual maturity. In addition to maturation of the primary sex organs (ovaries, testes), primates may exhibit "secondary sexual characteristics" at puberty. Among higher primates it is usual to find a growth spurt at the time of puberty in males and females.

Puna a treeless tableland or basin of the high Andes.

Pylorus the region of the stomach at its intestinal end, which is closed by the pyloric sphincter.

Quadrate bone at rear of skull which serves as a point of articulation for lower jaw.

Quadrupedal walking on all fours, as opposed to walking on two legs (BIPEDAL) or moving suspended beneath branches in trees (SUSPENSORY MOVEMENT).

Race a taxonomic division subordinate to subspecies but linking populations with similar distinct characteristics.

Radiation see ADAPTIVE RADIATION.

Rain forest tropical and subtropical forest with abundant and year-round rainfall. Typically species rich and diverse.

Range (geographical) area over which an organism is distributed.

Receptive state of a female mammal ready to mate or in ESTRUS.

Reduced (anatomical) of relatively small dimension (of certain bones, by comparison with those of an ancestor or related animals).

Refection process in which food is excreted and then reingested a second time to ensure complete digestion, as in the Common shrew.

Reingestion process in which food is digested twice, to ensure that the maximum amount of energy is extracted from it. Food may be brought up from the stomach to the mouth for further chewing before reingestion, or an individual may eat its own feces (see REFECTION).

Reproductive rate the rate of production of offspring; the net productive rate may be defined as the average number of female offspring produced by each female during her entire lifetime.

Resident a mammal which normally inhabits a defined area, whether this is a HOME RANGE or a TERRITORY.

Reticulum second chamber of the RUMINANT ARTIODACTYL four-chambered stomach (SEE RUMEN, OMASUM, ABOMASUM). The criss-crossed (reticulated) walls give rise to honeycomb tripe.

Retractile (of claws) able to be withdrawn into protective sheaths.

Rodent a member of the order Rodentia, the largest mammalian order, which includes rats and mice, squirrels, porcupines, capybara etc.

Rostrum a forward-directed process at the front of the skull of some whales and dolphins, forming a beak.

Rumen first chamber of the RUMINANT ARTIODACTYL four-chambered stomach. In the rumen the food is liquefied, kneaded by muscular walls and subjected to fermentation by bacteria. The product, cud, is regurgitated for further chewing; when it is swallowed again it bypasses the RUMEN and RETICULUM and enters the OMASUM.

Ruminant a mammal with a specialized digestive system typified by the behavior of chewing the cud. Their stomach is modified so that vegetation is stored, regurgitated for further maceration, then broken down by symbiotic bacteria. The process of rumination is an adaptation to digesting the cellulose walls of plant cells.

Rupicaprid a member of the tribe Rupicaprini, the chamois etc of the Artiodactyla.

Rut a period of sexual excitement; the mating season.

Satellite male an animal excluded from the core of the social system but loosely associated with the periphery, in the sense of being a "hanger-on" or part of the retinue of more dominant individuals.

Scent gland an organ secreting odorous material with communicative properties; see SCENT MARK.

Scent mark a site where the secretions of scent glands, or urine or FECES, are deposited and which has communicative significance. Often left regularly at traditional sites which are also visually conspicuous. Also the "chemical message" left by this means; and (verb) to leave such a deposit.

Sclerophyll forest a general term for the hard-leafed eucalypt forest that covers much of Australia.

Scute a bony plate, overlaid by horn, which is derived from the outer layers of the skin. In armadillos, bony scute plates provide armor for all the upper, outer surfaces of the body.

Secondary sexual character a characteristic of animals which differs between the two sexes, but excluding the sexual organs and associated structures.

Sedentary pertaining to mammals which occupy relatively small home ranges, and exhibiting weak dispersal or migratory tendencies.

Selective pressure a factor affecting the reproductive success of individuals (whose success will depend on their fitness, ie the extent to which they are adapted to thrive under that selective pressure).

Selenodont molar teeth with crescent-shaped cusps.

Sella one of the nasal processes of leaf-nose bats; an upstanding central projection which may form a fluted ridge running backwards from between the nostrils.

Septum a partition separating two parts of an organism. The nasal septum consists of a fleshy part separating the nostrils and a vertical, bony plate dividing the nasal cavity.

Serum blood from which corpuscles and clotting agents have been removed; a clear, almost colourless fluid.

Sexual dimorphism a condition in which males and females of a species differ consistently in form, eg size, shape. (See POLYMORPHISM.)

Serology the study of blood sera; investigates ANTIGEN-antibody reactions to elucidate responses to disease organisms and also PHYLOGENETIC relationships between species.

Siblings individuals who share one or both parents. An individual's siblings are its brothers and sisters, regardless of their sex.

Sinus a cavity in bone or tissue.

Sirenia an order of herbivorous aquatic mammals, comprising the manatees and dugong.

Sivathere a member of a giraffe family which became extinct during the last Ice Age.

Solitary living on its own, as opposed to social or group-living in life-style.

Sonar sound used in connection with Navigation (SOund NAvigation Ranging).

Sounder the collective term for a group of pigs.

Specialist an animal whose life-style involves highly specialized stratagems; eg feeding with one technique on a particular food.

Species a taxonomic division subordinate to genus and superior to subspecies. In general a species is a group of animals similar in structure and which are able to breed and produce viable offspring.

Speciation the process by which new species arise in evolution. It is widely accepted that it occurs when a single-species population is divided by some geographical barrier.

Sphincter a ring of smooth muscle around a pouch, rectum or other hollow organ, which can be contracted to narrow or close the entrance to the organ.

Spinifex a grass which grows in large, distinctive clumps or hummocks in the driest areas of central and Western Australia.

Stotting see PRONKING.

Stridulation production of sound by rubbing together modified surfaces of the body.

Subadult no longer an infant or juvenile but not yet fully adult physically and/or socially.

Subfamily a division of a FAMILY.

Suborder a subdivision of an order.

Subspecies a recognizable subpopulation of a single species, typically with a distinct geographical distribution.

Successional habitat a stage in the progressive change in composition of a community of plants, from the original colonization of a bare area towards a largely stable climax.

Suid a member of the family of pigs, Suidae, of the Artiodactyla.

Supra-orbital pertaining to above the eye (eye-socket or orbit).

Sympatry a condition in which the geographical ranges of two or more different species overlap (cf ALLOPATRY).

Syndactylous pertaining to the second and third toes of some mammals, which are joined together so that they appear to be a single toe with a split nail (as opposed to didactylous). In kangaroos, these

Syndactylous contd.
syndactyl toes are used as a fur comb.

Taiga northernmost coniferous forest, with open boggy rocky areas in between,

Tarsal pertaining to the tarsus bones in the ankle, articulating between the tibia and fibia of the leg and the metatarsals of the foot (pes).

Terrestrial living on land.

Territory an area defended from intruders by an individual or group. Originally the term was used where ranges were exclusive and obviously defended at their borders. A more general definition of territoriality allows some overlap between neighbors by defining territoriality as a system of spacing wherein home ranges do not overlap randomly—that is, the location of one individual's, or group's home range influences those of others.

Testosterone a male hormone synthesized in the testes and responsible for the expression of many male characteristics (contrast the female hormone estrogen produced in the ovaries).

Thermoneutral range the range in outside environmental temperature in which a mammal uses the minimum amount of energy to maintain a constant internal body temperature. The limits to the thermoneutral range are the lower and upper critical temperatures, at which points the mammals must use increasing amounts of energy to maintain a constant body temperature. (cf HETEROTHERMY)

Thermoregulation the regulation and maintenance of a constant internal body temperature in mammals.

Tooth-comb a dental modification in which the incisor teeth form a comb-like structure.

Thoracic pertaining to the thorax or chest.

Torpor a temporary physiological state in some mammals, akin to short-term hibernation, in which the body temperature drops and the rate of METABOLISM is reduced. Torpor is an adaptation for reducing energy expenditure in periods of extreme cold or food shortage.

Tragus a flap, sometimes moveable, situated in front of the opening of the outer ear in bats.

Trypanosome a group of protozoa causing sleeping sickness.

Tylopod a member of the suborder Tylopoda (order Artiodactyla) which includes camels and llamas.

Umbilicus navel.

Underfur the thick soft undercoat fur lying beneath the longer and coarser hair (guard hairs).

Understory the layer of shrubs, herbs and small trees beneath the forest canopy.

Ungulate a member of the orders Artiodactyla (even-toed ungulates), Perissodactyla (odd-toed ungulates), Proboscidea (elephants), Hyracoidea (hyraxes) and Tubulidentata (aardvark), all of which have their feet modified as hooves of various types (hence the alternative name, hoofed mammals). Most are large and totally herbivorous, eg deer, cattle, gazelles, horses.

Unguligrade locomotion on the tips of the "fingers" and "toes," the most distal phalanges. A condition associated with reduction in the number of digits to one or two in the perissodactyls and artiodactyls

Unguligrade contd.
(cf DIGITIGRADE and PLANTIGRADE).

Vascular of, or with vessels which conduct blood and other body fluids.

Vector an individual or species which transmits a disease.

Velvet furry skin covering a growing antler.

Ventral on the lower or bottom side or surface; thus ventral or abdominal glands occur on the underside of the abdomen.

Vertebrate an animal with a backbone; a division of the phylum Chordata which includes animals with notochords (as distinct from invertebrates).

Vestigial a characteristic with little or no contemporary use, but derived from one which was useful and well developed in an ancestral form.

Vibrissae stiff, coarse hairs richly supplied with nerves, found especially around the snout, and with a sensory (tactile) function.

Vocalization calls or sounds produced by the vocal cords of a mammal, and uttered through the mouth. Vocalizations differ with the age and sex of mammals but are usually similar within a species.

Withers ridge between shoulder blades, especially of horses.

Xenarthrales bony elements between the lumbar vertebrae of XENARTHRAN mammals, which provide additional support to the pelvic region for digging, climbing etc.

Xenarthran a member of the suborder Xenarthra, which comprises the living armadillos, sloths and anteaters (see EDENTATE).

Yolk sac a sac, usually containing yolk, which hangs from the ventral surface of the vertebrate fetus. In mammals, the yolk sac contains no yolk, but helps to nourish the embryonic young via a network of blood vessels.

INDEX

A **bold number** indicates a major section of the main text, following a heading: a **bold italic** number indicates a fact box on a single species or group of species; a single number in (parentheses) indicates that the animal name or subjects are to be found in a boxed feature and a double number in (parentheses) indicates that the animal name or subjects are to be found in a spread special feature. *Italic* numbers refer to illustrations.